Coevolution of Parasitic Arthropods and Mammals

Coevolution of Parasitic Arthropods and Mammals

Edited by

KE CHUNG KIM
Professor of Entomology
Curator, The Frost Entomological Museum
Department of Entomology
The Pennsylvania State University
University Park, Pennsylvania

A Wiley-Interscience Publication
JOHN WILEY & SONS
New York • Chichester • Brisbane • Toronto • Singapore

Library of Congress Cataloging in Publication Data:
Main entry under title:

Coevolution of parasitic arthropods and mammals.

   "A Wiley-Interscience publication."
   Includes index.
   1. Mammals—Evolution.   2. Mammals—Parasites—
Evolution.   3. Arthropoda—Evolution.   4. Coevolution.
I. Kim, Ke Chung.
QL708.5.C64 1985   599'.05249   85-641
ISBN 0-471-08546-4

Printed in the United States of America

10 9 8 7 6 5 4 3 2 1

To

G. H. E. HOPKINS

*for his contributions to
the study of mammalian ectoparasites*

Contributors

Peter H. Adler, *Department of Entomology, Clemson University, Clemson, South Carolina*

Barbara L. Clauson, *Department of Mammals, Chicago Zoological Society, Brookfield Zoo, Brookfield, Illinois*

K. C. Emerson, *Department of Entomology, National Museum of Natural History, Smithsonian Institution, Washington, D.C.*

Alex Fain, *Institut de Medicine Tropicale, Antwerp, Belgium*

Harry Hoogstraal, *Department of Medical Zoology, U.S. Naval Medical Research Unit No. 3, Fleet Post Office, New York 09527*

Kerwin E. Hyland, Jr., *Department of Zoology, University of Rhode Island, Kingston, Rhode Island*

Daniel H. Janzen, *Department of Biology, University of Pennsylvania, Philadelphia, Pennsylvania*

Ke Chung Kim, *Department of Entomology, The Frost Entomological Museum, The Pennsylvania State University, University Park, Pennsylvania*

William B. Nutting, *Department of Zoology, University of Massachusetts, Amherst, Massachusetts*

Roger D. Price, *Department of Entomology, University of Minnesota, St. Paul, Minnesota*

Frank J. Radovsky, *Department of Entomology, Bernice P. Bishop Museum, Honolulu, Hawaii*

Robert M. Timm, *Division of Mammals, Field Museum of Natural History, Chicago, Illinois*

Robert Traub, *Department of Microbiology, University of Maryland, School of Medicine, Baltimore, Maryland*

Preface

Associations of parasitic arthropods with humans and animals have been known perhaps since prehistoric times. Since the late nineteenth century, the important roles played by parasitic arthropods in the transmission of numerous diseases of humans and animals have been recognized; thus an emphasis has been placed on the study of the relationships between parasitic arthropods and mammals in medical and veterinary entomology. During the last 30 years, particularly, many significant advancements have been made by medical entomologists in the epidemiology of arthropod-borne diseases, zoonoses, and host specificity, and mammalian ectoparasites in tropical Africa, America, and Southeast Asia have been studied extensively by many scientific groups, for example, as described in *Ectoparasites of Panama* by R. L. Wenzel and V. J. Tipton (1966). However, the data accumulated from these studies on taxonomy, host associations, behavior, ecology, and bionomics have been published mainly in diverse specialized journals and publications of medical and veterinary entomology. Furthermore, parasitic arthropods are treated superficially or excluded completely from most parasitology books. As a result, the data base for parasitic arthropods has not been available to most biological and medical scientists beyond the bounds of specialists in entomology.

Recently, two major works have brought together information on parasitic arthropods: *The Ecology of Ectoparasitic Insects* by A. G. Marshall (1981) and *Parasite–Host Relationships of Arthropods with Terrestrial Vertebrates* by Yu. S. Balashov (1982) (in Russian). In 1971 R. R. Askew included general information on parasitic insects of mammals in his book *Parasitic Insects*. However, no comprehensive book has been devoted to the evolution and coevolution of parasitic arthropods and mammals. The First Symposium on Host Specificity of Vertebrate Parasites, held at Neuchâtel in 1957, provided the first sound basis for conceptualizing the parasite–host relationships in an evolutionary context. The Second Symposium on the same subject, held 13–17 April 1981 at the Musée National d'Histoire Naturelle in Paris, summarized the up-to-date information on the host specificity of vertebrate parasites.

My original idea for the *Coevolution of Parasitic Arthropods and Mammals* was conceived for a symposium at the Sixteenth International Congress of Entomology, 3–9 August 1980, Kyoto, Japan, and my plan was to have its proceedings published. Unfortunately, only four speakers were able to participate in the symposium. Although the presentations were limited to only a few groups of parasitic arthropods, I felt that the information presented was significant and should be available to the scientific community. Accordingly, after the Congress I modified the scope of the original project and continued to press for its completion. The leading scientists who agreed to participate in the project continued their writing efforts, and this book is the culmination of their dedication and those efforts.

Coevolution is a process and also a manifestation of the evolution between two interactive species. Many of the intimate associations between parasitic arthropods and mammals are undoubtedly the result of their evolutionary interactions and perhaps coevolution. Thus coevolutionary studies entail experiments and analyses of parasite–host associations and adaptations. However, coevolutionary processes are most difficult to observe directly in parasitic arthropod–mammal systems. Therefore in this book we analyze the parasitic arthropod–mammal relationships as the first step toward developing coevolutionary paradigms for the associations of parasitic arthropods with mammals. The evolution of these symbiotic relationships is inferred from the study of their host associations, adaptations, ecology, geographical distributions, and the paleontological data for mammals. Accordingly, studies of the relationships between parasitic arthropods and mammals, which are still foreign to many nonparasitologists, may provide excellent models for the dynamics and evolution of animal communities and populations and for coevolution, and they may also shed some light on many outstanding questions in ecology and evolutionary biology.

This book brings together information on the diversity, distribution, and adaptations of parasitic insects and acarines on mammals and thus provides the basis for studies on the evolution of biological relationships (perhaps coevolutionary) between parasites (consumers) and hosts (suppliers). Primary focus is placed on obligate, permanent parasites. They include Anoplura, Mallophaga, Siphonaptera, and the minor taxa, Dermaptera, Hemiptera, Diptera, and Coleoptera (Insecta), and numerous Mesostigmata, Metastigmata, Prostigmata, and Astigmata (Acari). Parasitic insects of minor orders are discussed briefly together. Because the level of knowledge currently available varies greatly by different taxonomic groups, many different approaches must be used to study evolutionary relationships between these parasites and their hosts. Thus the approach used in each chapter differs considerably from others. The term *coevolution* is used somewhat differently by different authors, but it primarily refers to the concepts of coaccommodation, coadaptation, and cospeciation. This represents a parasitological concept of coevolution, differing from the restrictive definition used for reciprocal evolution of plant–insect interactions. My use of the term *coevolution* in this book is deliberate in that it will promote interest and research among parasitologists and evolutionary biologists in the evolution of parasite–host associations, although many phenomena of parasite–host interactions may not be coevolutionary. Such studies will provide the basis for a better understanding of the evolution of mammalian parasites and a synthesis of evolutionary relationships between parasitic arthropods and mammals. This book will, I hope, eventually lead to an articulation of the concept of coevolution in parasite–host systems. Moreover, my intent is to facilitate the understanding of such evolutionary paradigms, not only by parasitologists but also by evolutionary biologists, ecologists, geneticists, and general biologists.

The book is divided into four parts. In the first part biological and evolutionary relationships between parasitic arthropods and mammals are analyzed and various factors are considered. As a point of reference and as a background for better understanding of the evolution of parasite–host relationships, the phylogeny of mammals relative to the evolution of parasitic arthropods is described. In the second and third parts the host associations and adaptations of parasitic insects, mites, and ticks are documented and analyzed to elucidate the evolution and perhaps coevolution of parasitic arthropods and mammals. The last part provides an overview by discussing the conceptual problems of coevolution related to parasite–host relationships and coevolutionary paradigms and presenting theoretical alternatives for the evolution of the associations between arthropod parasites and mammalian hosts. Finally, the appendixes provide lists of parasite–host associations for reference.

I am much indebted to the contributing authors for their enthusiasm and dedication, which made the completion of this book possible. My special thanks are due to Harry Hoogstraal for his generous invitation to and hospitality in Cairo, Egypt, to prepare the manuscript for Chapter 10. I acknowledge the following publishers and individuals for permission to use the figures indicated: Academic Press Inc. (London) (3.1; 4.3); Bernice P. Bishop Museum (6.7; 8.115–8.117; 9.1, 9.4, 9.5); Brigham Young University Press (6.1–6.3, 6.5, 6.6); Cambridge University Press (3.2–3.7; 6.8); Centre National de la Recherche Scientifique (12.5–12.6, 12.7); CRC Press, Inc. (11.2, 11.4); Entomological Society of America (5.6); Florida Entomological Society (6.4); Harper & Row Publishers, Inc. (1.4); Koniklijke Maatschappij voor Dierkunde van Antwerpen (12.3, 12.4); Royal Institute of Natural Sciences of Belgium (12.1, 12.2).

I thank Ms. Kelly L. Morris for her excellent assistance in editing and Dr. Charles W. Pitts, Jr. for his active support and encouragement, without which this book could not have been completed. I acknowledge the patience and industrious assistance of Jackie Wolfe, Judi Hicks, and Thelma Brodzina in the Department of Entomology, The Pennsylvania State University for typing and completing the manuscript. I also thank Mary M. Conway and the Production Staff of John Wiley and Sons, Inc. for their excellent efforts in all aspects of the making of this book.

KE CHUNG KIM

State College, Pennsylvania
May 1985

Contents

Coevolution of Parasitic Arthropods and Mammals

PART ONE

EVOLUTION OF THE PARASITE–HOST SYSTEM

Introduction

Chapter 1

Evolutionary Relationships of Parasitic Arthropods and Mammals

Ke Chung Kim

Authorized for publication as paper number 7035 in the Journal Series of the Pennsylvania Agricultural Experiment Station, University Park, PA 16802, U.S.A. A contribution from the Frost Entomological Museum, Department of Entomology, The Pennsylvania State University, University Park, PA (AES Proj. No. 2594).

INTRODUCTION

Diverse groups of Arthropoda are associated with mammals, forming dynamic interspecific interactions. The Insecta (insects) and the Acari (mites and ticks) account for the majority of the arthropod associations with mammals, although some crustaceans, like copepods (Penellidae) and amphipods (Cyamidae), are also parasitic on the skin of whales and dolphins (order Cetacea) (Rohde 1982).

The associations of parasitic arthropods and mammals have definite organizations and specific patterns. These patterns have been molded by long evolutionary processes, involving a succession of interactive and perhaps reciprocal responses between parasitic arthropods and mammals ever since their initial associations were established. Certain parasitic arthropods are closely associated with particular groups of mammals, and particular mammals harbor sets of specific arthropods. Furthermore, certain arthropods are more likely to coexist with a particular set of parasites on a specific group of mammals (Kim and Adler, Chapter 4). The mammals, which as hosts provide habitats and food for parasitic arthropods, have responded to the parasite load over time. Some of these responses were undoubtedly specific responses to particular parasite species, whereas others were general adaptations to the multispecies load of parasites. These adaptations might have included hosts' counteradaptations to parasites. They have directly influenced the structure-function, behavior, ecology, and development of parasitic arthropods, which are biological responses to selection pressures generated by the host and host-related parameters. However, it is difficult to assess and measure such interactive responses in mammalian hosts, and as a result very little information is available in this area.

The traditional approach to the study of parasite–host relationships has been primarily to describe the structure-functions and life cycles of individual species and list host–parasite associations. It often lacks any reference to other symbiotes or parasites and their interactions, even when they are on the same host. The synthesis of these empirical data should provide a basis for new approaches to the study of parasite–host relationships. However, the use of such data requires added care because published records often contain stragglers, contaminants, and misidentifications. Parasite–host associations are often used in taxonomic inference in classification and phylogenetic analysis of parasites or hosts. This also requires special care because phylogenetic reasoning could easily become circular when host associations are used to infer parasite phylogeny and then the relationships of parasites are employed to make quasi-phylogenetic inferences about mammalian evolution (Kim and Ludwig 1978a).

Coevolution is well documented for plant–pathogen (Day 1974), plant–insect (Gilbert and Raven 1975), and insect–insect (Price 1975) systems. However, little direct evidence exists to support reciprocal coevolution in parasitic arthropod–mammal systems (Waage 1979), although it has been suggested that coevolution is the rule for parasites and symbiotes (Price 1977, 1980; Thompson 1982; Futuyma and Slatkin 1983). This situation is aptly described by May and Anderson (1983): "Both the data and the formal mathematical models tell us that a great variety of different patterns are possible, but neither empiricists nor theorists have yet produced a crisp and ineluctable codification of these patterns." This chapter describes ecological and evolutionary relationships between parasitic arthropods and mammals to provide the basis for coevolutionary synthesis.

PARASITE–HOST RELATIONSHIPS

Organism Associations

Most parasitic arthropods are intimately associated with specific mammals. These associations are dynamic biological relationships, involving at least two sets of organisms living together; such association is called *symbiosis* (deBary 1879). In this context, the term *symbiote* refers to the small, dependent partner (e.g., arthropods), and the large, independent partner (e.g., mammals) is called the *symbiont* (Nutting, Chapter 11). Symbiosis ranges from casual contact for feeding to permanent dwelling for food and shelter. This may differ by the level of symbiote dependence (*facultative* or *obligate*), the duration of contact (*temporary* or *permanent*), and the number of species involved (*oligophilic* or *polyphilic*) (Starr 1975; Boucher et al. 1982). The wide range of these relationships makes the definition and categorization of the associations between arthropods and mammals difficult.

Diverse organism associations have been identified by different terms: phoresy, mutualism, commensalism, inquilinism, and parasitism. Unfortunately, these terms have been applied carelessly to different phenomena; this terminological confusion has been reviewed and discussed by Hertig, Taliaferro, and Schwarts (1937), Starr (1975), Smyth (1976), and Boucher et al. (1982).

Within symbiotic relationships, the terms phoresy and mutualism represent relatively loose or casual associations, whereas commensalism, inquilinism, and parasitism indicate intimate associations. *Phoresy* refers to the casual transport of the symbiote by another organism. For example, chewing lice (Mallophaga) attach to hippoboscid flies for transport from one host to another (Keirans 1975; Marshall 1981a), and the human bot fly [*Dermatobia hominis* (Linnaeus, Jr.)] glues eggs on carriers (many species of flies and one tick) to be transported to suitable vertebrate hosts (Guimaraes and Papavero 1966). *Mutualism* refers to an association in which both partners benefit, as when the symbiote feeds on other parasitic arthropods or the symbiont's wastes while living on it, and the symbiont in turn is benefitted by the reduction of its parasite load and its wastes (Nutting, Chapter 11). The term *commensalism* describes interspecific interactions where the symbiote gains nutrients and shelter without causing obvious harm to the symbiont; this may include *inquilinism* in which the symbiote gains shelter (Nutting, Chapter 11).

Parasite and Host

Of all the organismal associations, *parasitism* is very difficult to define precisely because it represents a wide range of metabolic and physiological dependence (Smyth 1976), pathological effects and mortality (Crofton 1971a, b), and ecological relationships (Kennedy 1975; Price 1980). A *parasite* is defined here as *a symbiote that lives, throughout a part or the entire period of its life, in or on the host from which its food and other biological necessities are derived* (Caullery 1952; Rogers 1962; Dogiel 1964; Smyth 1976; Thompson 1982). To be considered a parasite, this definition requires that an arthropod must live in and gain metabolic needs from a host for a certain period of its life span (Figs. 1.1 and 1.2). Thus bed bugs ["nest-burrow blood suckers" of Balashov (1982)] and most biting flies, like mosquitoes, black flies, and deer flies ["free-living blood suckers" of Balashov (1982, 1984)], are not considered here as parasites.

To search for general principles in interactions between organisms, the term parasite has been applied broadly to include the interactions between phytophagous insects and plants by Price (1975, 1977, 1980) and others (e.g., Thompson 1982). The broad usage of the term ignores the impor-

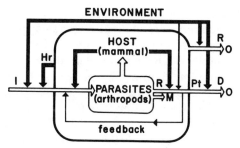

Figure 1.1 A model of the parasite–host system for obligate parasitic arthropods, showing the parasite flow paths (thick hollow lines) and control factors (thick solid lines). D, dispersal; Hr, host response; I, input; M, mortality; O, output; P, population; Po; initial parasite population; Pt, parasite population leaving the host after feeding; R, reproduction.

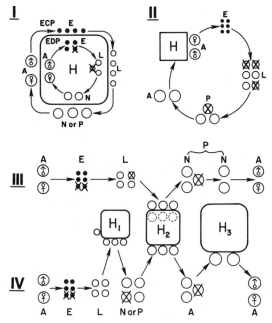

Figure 1.2 Models of four types of the parasite–host associations and life cycles in parasitic arthropods. I, permanent parasites, showing endoparasites (EDP) within the host and ectoparasites (ECP) on the host. II, temporary parasites, associated with a host in adult stage. III, temporary parasites, with a host (H₂) in larval stage. IV, temporary parasites, requiring three hosts (H₁, H₂, H₃) during their life cycles. E, eggs; L, larvae; N, nymphs; P, pupae; A, adults; X, mortality.

tance of distinctive differences between plant and animal hosts, although it may be useful in searching for general principles in symbiosis. Janzen points out these differences in Chapter 2.

Historically, parasitologists have focused on the aspect of tissue damage and pathological effects on the host by parasites, their primary interest being to control or eradicate parasites from man and animals. Recently, however, parasitism has been defined in terms of energy loss, lowered survival, or reduced reproductive potential in the host (Anderson 1978; Anderson and May 1978a, b, 1979; Price 1980; Thompson 1982). These definitions imply that a parasite must have a detrimental effect on the survival and intrinsic reproductive rate of its host population. This concept has become a central premise in ecological modeling of host–parasite interactions by theoretical ecologists (Anderson and May 1978a, b, 1979, 1982; May 1983a; May and Anderson 1983). In the extreme, Crofton (1971a, b) defines a parasite in terms of its potential to kill the host. Although there are a few examples that seem to support such a conceptualization of the host–parasite relationship, most parasites do not inflict severe damage or have a detrimental effect on their host. In stable parasite–host systems parasites do not seem to lower the fitness or reproductive potential of the host, and there are few data available to support these theoretical premises and hypotheses.

On the other hand, a host is often considered, in the context of parasite–host relationships, as a passive black box that simply provides a habitat, food, and other biological necessities for parasites (May 1983a). The mammalian host is not merely a passive receiver of the impacts by parasites, but an actively interacting partner of the dynamic parasite–host system (Odening 1976). Within the ecological time scale, the direct effect of parasites may be the depression of the host population dynamics, but within the evolutionary time scale the host population would invariably respond genetically to the effect of parasites and perhaps develop a reciprocal counter-response against parasites.

In a parasite–host system, parasites exploit the resources of the host, and, conversely, the host reacts to counter this exploitation. In this process parasites attempt to obtain food, shelter, and other biological requirements at minimal cost, whereas the host invests its behavioral, physiological, and energy resources to minimize its losses (Fig. 1.1). Contemporary relationships between parasitic arthropods and mammals are the manifestation of such interactions over long evolutionary time.

The state of parasite–host relationships is determined by a degree of tension between regulatory and destabilizing processes that involves over-dispersion of parasites (density) per host, density-dependent constraints on parasite population growth in or on individual hosts, and host mortality induced by parasite burden per host (Anderson 1978; Anderson and May 1978a). Parasite–host associations involving parasitic arthropods usually maintain an ecological stability, with stable parasite population density and little host mortality (Marshall 1981a). Evolutionary processes in parasitic arthropod–mammal associations tend to favor stable parasite–host relationships with a strong regulatory force counteracting a destabilizing influence (Kennedy 1975, 1976; Holms et al. 1977; Anderson 1978).

Populations, Community, and Diversity

Parasites in a black box of the parasite–host model (Fig. 1.1) represent a multitude of dynamic components involving different levels of parasite assemblages within the host. They include all individuals of different parasite species and their infra- and interspecific interactions. As documented by many ectoparasite surveys (e.g., Elton et al. 1931; Holdenried et al. 1951; Murray 1957; Timm 1975), most individual hosts harbor only a part of the total parasite assemblage in the host territory. For example, in Bagley Wood, Oxford, Elton et al. (1931) recorded 41 species of parasites from small rodents, *Apodemus, Clethrionomys* (= *Evotomys*), and *Microtus*, of which 24 species represented ectoparasitic arthropods associated with *Apodemus sylvaticus*. They included 9 species of fleas, 12 species of mites, and one species each of sucking lice, beetles, and ticks. Thus individual hosts of *A. sylvaticus* may harbor a set of ectoparasitic arthropods in any combination from a pool of 24 species.

The term *population* is defined here, in referring to a parasite species in a local host population, as an assemblage of all parasites including all developmental stages in or on the hosts at a particular time (Marshall 1981a); for example, all individuals of *Polyplax serrata* Burmeister on *A. sylvaticus* in Bagley Wood. Thus an *infrapopulation* (parasitome of Pavlovsky 1966; body population of Marshall 1981a) is an assemblage of all individuals of a parasite species in or on a host individual, such as all *P. serrata* found on a *A. sylvaticus* host. All of the parasite populations in one host species may be called a *suprapopulation*; for example, all *P. serrata* found on *A. sylvaticus* throughout its range (Esch et al. 1975; Holmes et al. 1977).

The assemblage of all multispecific parasites in a host individual is considered here as a *community* (parasitocoenose of Pavlovsky 1966; alpha-diversity of Whittaker 1972; faunule). Thus the term *supracommunity* may be used to describe the total assemblage of all parasites on a local host population (gamma-diversity of Whittaker 1972); for example, 24 species of ectoparasites associated with *A. sylvaticus* population in Bagley Wood. The assemblage of all parasites found in a specific habitat or site in or on the host may be called an *infracommunity*; for example, fleas, lice, and mites found in the fur of the host belly. The term *fauna* may be used to describe the total assemblage of component parasites found in a host community; for example, 41 species of parasites in Bagley Wood; a *faunule* (= community) represents species richness in a host individual. Likewise, the total assemblage of all component parasite species associated with a host species throughout the range of host distribution is called the parasite *diversity* in the sense of species richness (Whittaker 1972, 1977).

Host Associations and Specificity

Most parasitic arthropods are closely associated with specific mammalian hosts. Such host associations are remarkably specific and consistent throughout the range of their distribution (see Chapters 4, 5, and 7). The present pattern of host associations and specificity, which is a culmination of long evolutionary interactions between parasitic arthropods and mammals, is established only by accurate taxonomy of both parasites and hosts.

Many parasitic arthropods are specific to certain microhabitat within a host individual, as observed in other parasites (Rohde 1979), and this specificity is consistent among individuals of the same species and species of the same higher taxon. For example, the pinniped-infesting sucking lice, *Antarctophthirus*, are specific to their hosts' naked or thinly furred skin (Kim 1975). The mesostigmatid Halarachnidae are confined to the respiratory organs of their mammal hosts: the species of *Pneumonyssus* are found in the lungs and maxillary sinuses of monkeys (Kim 1977), whereas *Halarachne* and *Orthohalarachne* are found in the respiratory organs of Pinnipedia (Kim et al. 1980; Furman and Daily 1980).

Some obligate parasites frequent only one host species (called *monoxeny*), as commonly observed among permanent parasites like sucking lice and chewing lice, whereas others infest two or more host species (called *polyxeny*). Polyxeny may involve two or more congeneric host species (*oligoxeny*), for example, *Solenopotes burmeisteri* (Fahrenholz) on several species of red deer *Cervus*, or two or more heterogeneric host species in the same family (*pleioxeny*), for example, *Orthohalarachne attenuata* Banks, found in the respiratory organs of many otariid seals including *Callorhinus*, *Eumetopias*, and *Zalophus*. Parasitic arthropods, particularly permanent parasites, are usually monoxenous or oligoxenous, but pleioxeny may occur: see Kim and Adler (Chapter 4) for the parasitic dermapterans, Kim (Chapter 5) for Anoplura, Emerson and Price (Chapter 6) for Mallophaga. Except for the Tungidae and some unusual fleas, the species of Siphonaptera are more commonly pleioxenous and frequently oligoxenous (Chapter 8). The host associations of parasitic acarines are usually oligoxenous or pleioxenous, but those highly specialized acarines like Demodicidae, Myobiidae, and Dermanyssidae are monoxenous (see Chapters 9, 11, 12).

Most parasitic arthropods are consistently associated with a specific host taxon at various taxonomic levels (Ludwig 1982). A parasite species is specific to a single host at species level; for example, *Pthirus pubis* (Linn.) is found only on man, and *P. fluctus* (Ferris) is a specific parasite of *Callorhinus ursinus*. At this level, host specificity for parasitic arthropods may be measured by the number of parasites per individual host examined (infestation rate, parasite index, percentage occurrence, or specificity index) (Miles et al. 1957; Stark and Kinney 1969; Marshall 1981a). Likewise, a parasite genus-taxon may exclusively associate with a host family-group or higher taxon; *Proechinophthirus* is found on the otariid Arctocephalinae (Kim et al. 1975), *Enderleinellus* on Sciuridae, and *Prolinognathus* on the order Hyracoidea. Similarly, a parasite family-taxon may also be exclusively associated with a host family or order-taxon, for example, Enderleinellidae in Sciuridae and Echinophthiriidae in Pinnipedia (Kim and Ludwig 1978a).

ASSOCIATING ARTHROPODS OF MAMMALS

Most symbiotic arthropods associated with mammals are members of the class Insecta and the arachnid subclass Acari and particularly of the orders Hemiptera, Dermaptera, Coleoptera, Lepidoptera, Diptera, Anoplura, Mallophaga, and Siphonaptera (Insecta): Parasitiformes and Acariformes (Acarina) (for detail, see the list of parasitic arthropods and their hosts given in the Appendix). Waage (1979) and Marshall (1981a) also reviewed parasitic insects of birds and mammals.

Parasitism in Arthropods

The relationships between parasitic arthropods and mammals represent a wide spectrum of parasitism, from occasional associations for parasite feeding (*facultative parasitism*) (e.g., ticks and fleas) to a complete dependence of the parasite on the host for its survival and development in the host for its survival and development (*obligate parasitism*) (e.g., sucking lice and follicle mites). Obligate parasites cannot survive long off the host, whereas facultative parasites can survive without the host for a long period.

Certain obligate parasites, called *temporary parasites* ("nest ectoparasites" of Nelson et al. 1975), are associated with the host for a short period in the life cycle (Fig. 1.2, II, III, IV). On the other hand, *permanent parasites* ("host ectoparasites" of Nelson et al. 1975) spend their entire lives in or on the host, are usually highly host specific, and usually require specific microhabitats (Fig. 1.2, I).

All of the Anoplura and Mallophaga are permanent parasites, as are many other insects and mites; for example, the Polyctenidae (Hemiptera), Platypsyllidae (Coleoptera), Streblidae and Nycterebiidae (Diptera); Spinturnicidae, Spelaeorhynchidae, Halarachnidae (Parasitiformes), Myobiidae, Psorergatidae, Demodicidae, and Gastronyssidae (Acariformes) (Waage 1979; Marshall 1981a; Krantz 1978).

Many parasitic arthropods, like sucking lice, ticks, and rat mites (*Ornithonyssus*), inhabit the body surface of a host, and are called *ectoparasites*. *Endoparasites* are those parasites invading the skin tissues, the digestive

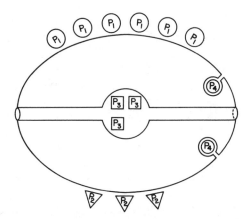

Figure 1.3 A model of a mammalian host, showing distribution and microhabitats of four species, P_1, P_2, P_3, P_4, of parasitic arthropods. P, parasite species; P_1 and P_2, ectoparasites on the body surface; P_3, endoparasites within a host organ system; P_4, endoparasites in the epidermis.

tracts, respiratory passages, and other tissues and internal organs of a host, such as bot flies (*Gasterophilus*) in the equid digestive tracts, nasal mites (*Orthohalarachne*) in the respiratory tract of seals, and lung mites (*Pneumonyssus*) in the lungs of monkeys and baboons (Fig. 1.2, I; Fig. 1.3). Some parasitic arthropods, like mange mites (*Sarcoptes*), may even be ectoparasitic in one life stage and endoparasitic in another.

Haematophagous and parasitic arthropods have been differently classified by various workers based on different criteria: Tatchell (1969b) used the speed of feeding, Marshall (1976, 1981) used the criteria of mobility and the intimacy of the host relationship, Nelson et al. (1975) classified them on the basis of their mode of life or ecological niche, and Beklemishev (1945, 1951, 1954) and Balashov (1982, 1984) on the basis of ecological and behavioral criteria. The Beklemishev–Balashov system identifies the following five groups of parasitic arthropods: (1) free-living bloodsuckers (e.g., biting flies, including mosquitoes, black flies, tsetse); (2) nest-burrow bloodsuckers (e.g., Cimicidae, most Argasidae, many gamasine mites, some fleas, and Reduviidae); (3) temporary ectoparasites with rapid feeding (e.g., pupiparous Diptera, some fleas, and gamasine mites); (4) temporary ectoparasites with slow feeding (e.g., Ixodidae, some chiggers, some Macronyssidae, and some fleas); and (5) permanent ectoparasites (e.g., Anoplura, Mallophaga, Polyctenidae, some Hippoboscidae, many acarines).

Commensals and Bloodsuckers

The lepidopterans, mostly members of the Geometridae, Noctuidae, Notodontidae, and Pyralidae, have primarily phoretic, commensal, or opportunistic associations with mammals (Bänziger and Buttiker 1969; Bänziger 1972, 1975; Waage and Montgomery 1976; Waage 1979). Parasitic dermapterans, Hemimeridae (suborder Hemimerina) associated with rats that feed on dried skin and exudates from skin, eyes, ears, mouth, and anus, and the Arixeniidae (suborder Arixeniina) closely associated with bats (*Cheiromeles torquatus*), apparently do not harm the host (Ashford 1970; Marshall 1977). Their associations perhaps better represent mutualistic or commensal relationships than parasitism (Kim and Adler, Chapter 4).

The beetles associated with mammals are represented by Languriidae with Nearctic rodents, Leiodidae with Palearctic lagomorphs, Scarabaeidae with sloths, Leptinidae with rodents and insectivores, Platypsyllidae with beavers, and the amblyopine Staphylinidae with rodents and marsupials. The associations of Languriidae, Leiodidae, Scarabaeidae, and Staphylinidae (*Myotyphlus*) with Tasmanian murid are commensal (Marshall 1981a). However, Leptinidae and Platypsyllidae on beavers, mice, and water moles, and Staphylinidae Amblyopini are definitely ectoparasites (Kim and Adler, Chapter 4).

The true bugs of four families, all Cimicidae (Usinger 1966), all Polyctenidae (Usinger 1946; Maa 1964; Ryckman and Casdin 1977), a few Reduviidae, and certain Lygaeidae are associated with mammals. The Cimicidae, Polyctenidae, and a few Reduviidae are bloodsuckers (Marshall 1981a), and certain Lygaeidae may be parasitic on mammals (Miller 1971). The Cimicidae (beg bugs) are night prowling and bloodsucking, feeding on birds and bats, occasionally man (Usinger 1966). The triatomine Reduviidae, *Triatoma* and *Panstrongylus*, are not true ectoparasites as defined here, whereas Polyctenidae are true ectoparasites of bats (Kim and Adler, Chapter 4).

Bloodsuckers are found widely throughout the order Diptera. In the Nematocera, many species of Simuliidae, Psychodidae, Ceratopogonidae, and Culicidae are bloodsucking, and they are important vectors of numerous pathogens including viruses, protozoans, and nematodes. Many Tabanidae of the Brachycera are avid bloodsuckers also involved in transmission of viral, protozoan, filarioid, and bacterial parasites in man and animals. In the Cyclorrhapha there are a number of families that include bloodsucking species. The family Chloropidae include *Hippelates* and *Siphunculina* which may play a role in pathogen transmission (e.g., anaplasmosis of cattle), and the Muscidae include *Musca* (face fly), *Glossina* (tsetse), *Stomoxys* (stable fly), and *Haematobia* (horn fly). Some species of the Calliphoridae and Sarcophagidae like *Cochliomyia* and *Wohlfahrtia* are also sanguinivorous and pose economic importance (Harwood and James 1979). *Mystacinobia zelandica* Holloway, the sole species of Mystacinobiidae associated with a bat in New Zealand has a commensal association (Holloway 1976).

The acarine associations of mammals range from phoretic and commensal relationships to obligate parasitism. Many parasitiform acarines are nest-inhabiting scavengers which include Laelapidae (e.g., *Haemolaelaps, Hirstionyssus,* and *Haemogamasus*) and Dermanyssidae (e.g., *Dermanyssus*). The nest-inhabiting acarines of the Acariformes include Cheyletidae (e.g., *Acaropsis, Cheyletiella,* and *Cheyletia*), Acaridae (e.g., *Acarus* and *Tyrophagus*), Glycyphagidae (e.g., *Glycyphagus*), and Pyroglyphidae (e.g., *Dermatophagoides*) (Evans et al. 1961; Krantz 1978; Radovsky, Chapter 9; Nutting, Chapter 11; Fain and Hyland, Chapter 12).

Endoparasites and Ectoparasites

Obligate parasites are found in a limited number of insect orders, namely, Hemiptera (Polyctenidae), Coleoptera (Platypsyllidae, Leptinidae, and Staphylinidae), the calyptrate Diptera, Anoplura, Mallophaga, and many Siphonaptera. Parasitic acarines are found in the suborders Mesostigmata and Metastigmata (Ixodoidea) (Order Parasitiformes), and Prostigmata and Astigmata (Order Acariformes).

Parasitic Diptera

Parasitic Diptera associated with mammals are found in the calyptrate families, Hippoboscidae (Bequaert 1953, 1954–1957; Maa 1963), Nycteribiidae, Streblidae, Calliphoridae, Sarcophagidae, Cuterebridae, Gasterophilidae, Oestridae, and Hypodermatidae (Kim and Adler, Chapter 4).

Of the parasitic Diptera, the larvae of many Calliphoridae, Sarcophagidae, Cuterebridae, Gasterophilidae, Oestridae, and Hypodermatidae are obligate parasites (Fig. 1.2, III) which cause myiasis in man and animals (Zumpt 1965). These larvae may occur subdermally as do Cuterebridae and Hypodermatidae or within the internal organs like digestive system, urogenital tract, auricular, and nasopharynx where they feed on living tissues and mucus, as do the Gasterophilidae and Oestridae (Zumpt 1965; Papavero 1977; Harwood and James 1979; Kim and Adler, Chapter 4).

Anoplura and Mallophaga

The Anoplura (sucking lice) and the Mallophaga (chewing lice), often collectively called parasitic Psocodea or Phthiraptera (lice), are obligate, permanent ectoparasites (Fig. 1.2, I) (Kim and Ludwig 1982). All the Anoplura are mammalian ectoparasites (Kim and Ludwig 1978; Kim, Chapters 5 and 7), whereas the Mallophaga are primarily ornithophilic ectoparasites, and only a small number of species (about 18% of the total diversity) are parasitic on mammals (Emerson and Price 1981, and Chapter 6 of this book).

Anoplura are avid bloodsuckers, parasitic on all major groups of living eutherian mammals except the Edentata, Chiroptera, Pholidota, Proboscidea, Cetacea, and Sirenia (Kim and Ludwig 1978; Kim, Chapters 5 and 7).

The estimated diversity of Anoplura is about 1000 species, but only about one-half of the diversity (500 species) is known so far. They are classified into 15 families and 45 genera.

Mallophaga consist of three suborders, Amblycera, Ischnocera, and Rhynchophthirina for 205 genera and about 2590 known species (Marshall 1981a). Of the seven amblyceran families recognized, the Abrocomophagidae, Boopiidae, Gyropidae, and Trimenoponidae are mammalian ectoparasites, including 23 genera and 107 species (the 75 genera and 836 species known for Amblycera). Similarly, whereas the Philopteridae with 90 genera and 1460 species are ornithophilic ischnocerans, the Trichotectidae and Trichophilopteridae are parasites of mammals, representing 17% (292 species) of the known Ischnocera (about 1752 species) (Marshall 1981a). The Haematomyzidae are a monotypic family for the Rhychophthirina with two species (Emerson and Price 1981 and Chapter 6).

Siphonaptera

The adults of Siphonaptera (fleas) are largely obligate, facultative ectoparasites of rodents, exclusively bloodsucking, but their larvae are free living (Fig. 1.2, II), feeding on a variety of organic materials in the areas associated with the host. Of the extant diversity estimated as 3000 species (Marshall 1981a), 15 families, 212 genera, and 2018 species and subspecies are known for the order Siphonaptera (Lewis 1972–1975). About 74% of known fleas are recorded from rodents. It is followed by 8% from Insectivora, 5% each in Marsupialia and Chiroptera, 3% each in Carnivora and Lagomorpha, and <1% each in Monotremata, Edentata, Pholidota, Hyracoidea, and Artiodactyla. Only 6% of the total diversity is ornithophilic fleas (Marshall 1981a).

The associations of fleas and mammals range widely from nest dwellers (e.g., *Conorphinopsylla*) and stationary ectoparasites (e.g., *Echidnophaga* on dogs) to permanent intracutaneous parasites (e.g., *Tunga*). However, most fleas are good jumpers, easily moving from one host to another (Traub Chapter 8).

Parasitiform Acari

Of the Parasitiformes, the Tetrastigmata are predators, but many of the Mesostigmata and all Metastigmata (ticks) are vertebrate parasites. The Acariformes include diverse mites, suborders Prostigmata, Astigmata, and Cryptostigmata (Whittaker and Wilson 1974; Krantz 1978), of which the Cryptostigmata are not parasitic. The mites of the mesostigmate Dermanyssoidea and the prostigmate Cheyletoidea are vertebrate symbiotes. Most parasitic mites are usually specific to particular mammal groups.

Of the mesostigmate mites the Spinturnicidae and Spelaeorhynchidae are exclusively parasitic on bats inhabiting skin, and the Halarachnidae are endoparasites of respiratory organs of Pinnipedia and Primates (Krantz 1978; Furman 1979). Many dermanyssoids which are broadly parasitic on diverse vertebrates are also found in specific mammals: Macronyssidae are bat parasites, Dermanyssidae and Laelapidae are parasites of rodents, Hystrichonyssidae are parasitic on porcupines. The Raillietidae are commonly found in ear cavity of artiodactyls, and the Dasyponyssidae are parasites of edentates (Radovsky, Chapter 9).

The Metastigmata are large mites, feeding intermittently on the blood of reptiles, birds, and mammals for all postembryonic stages (Fig. 1.2, IV) (Hoogstraal and Kim, Chapter 10). Two major families, Ixodidae (hard ticks) and Argasidae (soft ticks), are both found on mammals.

Acariform Acari

The suborders Prostigmata and Astigmata include mites associated with vertebrates. The Prostigmata are cosmopolitan mites, found in a broad range of habitats. The Astigmata are a homogenous assemblage of fungivorous, saprophytic, predatory, gramnivorous, and parasitic mites (Krantz 1978).

Of the symbiotic Prostigmata, mammalian parasites are found mostly in the Cheyletoidea and Trombidioidea (Nutting, Chapter 11). Some members of the tydeoid Ereynetidae are found in the respiratory passages of amphibians, birds, and mammals. Some of the ereynetid Speleognathinae are parasitic on respiratory passages of rodents and insectivores (Krantz 1978). The Cheyletoidea are parasitic mites of arthropods, reptiles, birds, and mammals, except for the Cheyletidae which includes nonparasitic species. Of the parasitic Cheyletidae many species of *Cheyletiella*, *Chelacaropsis*, and *Hemicheyletus* are found in the fur of rodents, rabbits, and carnivores. The Myobiidae are ectoparasites of rodents, bats, and insectivores. The Psorergatidae are ectoparasites of sheep, murid rodents, porcupines, and monkeys (Fain et al. 1966). The Demodicidae, which are parasitic on many different mammals, burrow into hair follicles and feed on sebum and other subcutaneous secretions (Nutting, Chapter 11). The Trombidioidea include mites parasitic on insects and vertebrates as larvae and predaceous on arthropods as nymphs and adults (Krantz 1978). Of the trombidioids, larvae of the Trombiculidae, commonly known as *chiggers*, attack most of vertebrates; of these *Trombicula, Eutrombicula, Acomatacarus, Schöngastia, Euschöngastia,* and *Apolonia* cause dermatitis on mammals, including man (Baker et al. 1956). Chiggers of *Doloisia* are also found in the nasal passages of bats (Yunker and Brennan 1962). For feeding, chiggers attach to the skin and form a characteristic feeding tube or *stylostome* at the attachment site (Cross 1964).

Many astigmatic mites, primarily Listrophoroidea, Psoroptoidea, and Sarcoptoidea, are associated with mammals as ectoparasites or as scavengers. A few species of the Astigmata inhabit the respiratory passages of different mammals (Krantz 1978; O'Conner 1982; Fain and Hyland, Chapter 12). The Lemurnyssidae are parasites of the respiratory passage of Primates, and Pneumocoptidae and Yunkeracaridae live in the respiratory passages of rodents. The Myocoptidae and some Pyroglyphidae are exclusively associated with rodents; the Chirorhynchobiidae, Gastronyssidae, Rosensteiniidae, and Teinocoptidae are parasites of bats. The Lobalgidae and Psoralgidae are associated with edentates. The skin parasites of Primates are found in the families Audicoptidae, Psoroptidae, and Rhyncoptidae (Krantz 1978).

The Listrophoroidea include four families—the Listrophoridae, Myocoptidae, Rhyncoptidae, and Chirorhychobiidae. The Listrophoridae are found widely in a variety of mammals, including rodents, bats, and carnivores.

The Psoroptoidea, with the six families, Psoroptidae, Lobalgidae, Yunkeracaridae, Lemurnyssidae, Audycoptidae, and Pyroglyphidae, are primarily parasites of mammals. The Pyroglyphidae, which are found in a variety of habitats, are primarily nest inhabitants or associated with stored products (Fain 1967a, b). The Psoroptidae are serious pests of domestic and wild animals. The species of Audycoptidae are found in the hair follicles of lip tissues of Primates (Lavoipierre 1964; Fain 1968a, b; O'Conner 1982).

The Sarcoptoidea, including four families, Sarcoptidae, Knemidocoptidae, Teinocoptidae, and Evansacaridae, are skin parasites of both birds and mammals (Krantz 1978; O'Conner 1982). The species of Sarcoptidae are serious skin parasites of mammals, including man, and the Teinocoptidae are parasitic on bats, whereas the Knemidocoptidae and Evansacaridae are bird ectoparasites.

The Pneumocoptidae and Gastronyssidae are members of the Cytoditoidea, which also include two other families found on birds, Cytoditidae and Laminosioptidae. The species of Pneumocoptidae inhabit the respiratory passages of rodents, whereas the Gastronyssidae are found in the stomach, intestines, nasal cavities, or eyes of bats (Fain 1956). *Rodentopus* (Labidophoridae), described by Fain (1968b) as Hypoderidae from an African rodent, is a member of the Acarioidea (Krantz 1978).

THE PARASITE'S HABITATS AND ENVIRONMENTS

The body and various organs of mammals constitute the habitats and provide niches for parasitic arthropods. They include the skin and fur, hair follicles, respiratory and digestive tracts, and the bladder and urinary passages. The physical environment in these habitats is relatively constant, although physiochemical conditions may fluctuate because of recurrent changes in host behavior, ecology, and physiology. These environments are influenced directly by host physiological changes and indirectly by environmental changes surrounding the host animals. Parasitic arthropods are well adapted to the structure and conditions of these habitats and utilize them effectively as the basis for their niches (Marshall 1981a; Timm 1983; Timm and Clauson, Chapter 3; Traub, Chapter 8; and Nutting, Chapter 11).

The Host's Integument

The Skin

In mammals the skin and its associated integumentary structures protect the body from external environmental changes, prevent water and heat loss, and function as a receptor of external sensory signals and as an excretory organ. The structure of the skin is highly adapted to different environments and thus varies considerably among species (Champion et al. 1970; Sokolov 1982). Morphological and behavioral adaptations of parasitic arthropods are highly correlated with the variations of mammalian skin (Timm and Clauson, Chapter 3; Traub, Chapter 8). The mammalian skin provides numerous microhabitats for different parasitic arthropods to reside and obtain food.

The skin consists of the stratified outer layer, the *epidermis*, and the thick inner *dermis*, with other accessory structures (Fig. 1.4). The epidermis is composed of surface layers of flat, dead cornified cells, which are continually replaced, and the inner layers where new cells are germinated (Sims 1970; Tortora and Anagnostakos 1981). Ectoparasites like sucking lice inhabit the skin surface and fur, whereas tissue-dwelling mites like Sarcoptidae and Teinocoptidae (Mellanby 1944, 1972; Lavoipierre et al. 1967) and the skin "burrowers" like Tungidae (Jordan 1962; Barnes and Radovsky 1969) invade the epidermis. The dermis includes hair roots, follicles, glands and associated muscles, connective tissues, nerves, and blood vessels, and beneath it is a subcutaneous layer of fat tissues (Tortora and Anagnostakos 1981). Dermis thickness varies considerably with different species and in different parts of the body (Gillman 1970). A network of blood vessels and sweat glands in the dermis is involved in mammalian thermoregulation; hair and subcutaneous fat tissues provide insulation. The dermis provides the habitat for parasitic mites like *Demodex*, whereas fly larvae like warble flies (*Hypoderma*) and bot flies (*Cuterebra*) occur beneath the skin (dermis) (James 1948; Zumpt 1965).

The skin glands in mammals are the *sebaceous* or *oil gland* and *sudoriferous* or *sweat gland*. The size and shape of these glands may vary by different parts of the body and also differ considerably between species, whereas some mammals like rodents and elephants have no sweat glands (Ebling 1970; Champion 1970). These glands are not favored habitats for parasitic arthropods.

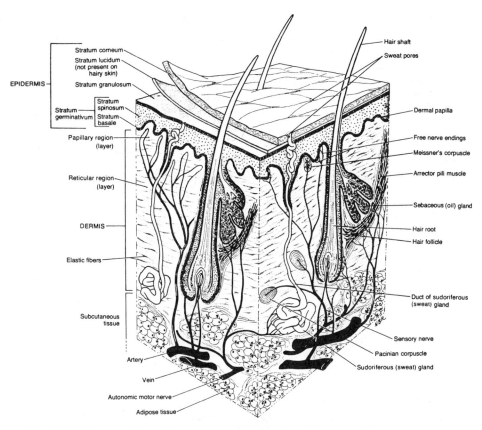

Figure 1.4 Diagram of mammalian skin and its associated organs. (From F. J. Tortora and N. P. Anagnostakos. 1981. *Principles of Anatomy and Physiology*, 3rd ed., Harper & Row, New York, p. 106, by permission)

Hair

Hair is an eidermal derivative exclusively found in mammals. Although its basic morphology is alike in all mammals, the hair varies greatly among different taxa and ecological types in anatomical details, type, length, thickness, arrangement, distribution, and direction (Toldt 1935; Noback 1950; Carter 1965; Sokolov 1982). The study of ectoparasitic arthropods requires a close analysis of these variations in relation to adaptations of the parasites.

The hair is made of a *shaft* and a *root* (Fig. 1.4). The shaft consists of three principal parts, the middle *cortex*, surrounded by the *medulla*, and the outermost *cuticular scales*. The hair root is the part below the skin surface that penetrates into the dermis and is surrounded by the hair follicle (Rook 1970; Tortora and Anagnostakos 1981). The active follicles contain mucopolysaccharides, sudanophilic lipids, glycogen, and various enzymes of which the amount and activity differ by different species (Sokolov 1982). The Demodicidae invade these hair follicles (Nutting 1975 and Chapter 11).

The hair may be spiniform, rough or soft, long or short, thick or thin, and colored or colorless. The shaft may be straight or bent, wavy or curly, and spinal or crumpled. Hairs are usually clustered by groups of one guard or pile hair with some fur hairs on each side, although the distribution of these hair types is quite uneven (Sokolov 1982). Correlations have been established between the average diameter of the host hair and the comb spacing in fleas and parasitic bugs (Polyctenidae) (Humphries 1967a, b; Amin et al. 1974; Amin and Sewell 1977; Kim and Adler, Chapter 4).

The pelage is made of the guard, pile, intermediate, and fur hairs. The guard hairs are longest and thick, but only few. The intermediate hairs are shorter and thinner than the pile hairs but longer and thicker than fur hairs. The fur hairs are short, thin, and usually more numerous, forming the underfur. Their morphology, density, and distribution in mammals vary considerably by different species, locations, and seasons. The pelage is periodically replaced, although hair growth and molting are usually completed before sexual maturity. The characteristics and periodicity of the pelage is usually species specific (Sokolov 1982).

Although pelage color is determined by the amount of melanine granules present in the skin and hairs, no single hypothesis can explain why mammals have different colors and what selection pressure is responsible for such coloration (Burtt 1981). Solar heat gain in mammals is independent of pelage color, and dark pelage color may either increase or decrease heat gain depending upon anatomical characteristics of the skin and hairs and their environmental properties (Walsberg et al. 1978; Finch et al. 1980; Walsberg 1983).

Adaptations in the Skin

All skin variation in mammals reflects their successful adaptations to a wide variety of environmental demands, such as habitats (terrestrial, arboreal, subternaean, and aquatic) or modes of life (flying, climbing, burrowing, and swimming) (Table 1.1). Variation in the skin and its associated organs occurs not only among individuals and species but also within individual animals between seasons (Sokolov 1982).

Rodents have a relatively uniform pattern of hairs covering the entire body. Hair length is relatively constant, with the hairs on the back longer than on the belly and short on the head, ears, and extremities. The body hairs slope backward from the head, and the limb hairs are directed distally (Parnell 1950). This hair pattern is considered primitive, and different mammals have developed great variations from this pattern (Toldt 1935; Sokolov 1982).

Considerable variations exist in the density and skin structure among

Table 1.1 Adaptations of Mammal Skin to Environment

Skin Characteristics	Terrestrial Mammals	Aquatic Mammals		Burrowing Mammals
		Densely Furred	Sparsely Furred	
Thickness of skin	Thin	Thick	Thick	Thick
Development of subcutaneous fat tissues	Poorly	Highly	Thick, variable	Variable
Back skin	Thickest	Thin	Thick	Thick
Belly skin	Thin	Thick	Thin	Thick
Epidermis	Thin	Thin	Thick	Thick
Dermis	Variable	Thick	Thick	Loose, variable

Source: Basic data from Sokolov (1982).

species and taxa. In terrestrial mammals, heavily furred artiodactyls have 233–6916 hairs per square centimeter with the epidermis 17–112 μ in thickness, whereas the sparsely furred wild boar has 874–1190 hairs per square centimeter with the epidermis 209–396 μ thick and with highly developed alveoli. Although sweat glands are a common feature in terrestrial mammals, Marsupialia, Insectivora, Carnivora, Perissodactyla, and Artiodactyla, numerous taxa like rodents and elephants lack them (Sokolov 1982). Dense hairs seem to be characteristics of mammals living in areas of relatively high humidity and greater wind velocity; high hair density is necessary for protection against cooling, as shown in arctic mammals, which have about nine times more insulation than do tropical animals.

Among aquatic mammals, skin evolution has taken two lines, one to dense pelage and the other to sparse pelage. The aquatic mammals with dense pelage are more common, including amphibian mammals of Monotremata, Marsupialia, Insectivora, Rodentia, Carnivora, and many Pinnipedia. The amphibian mammals with sparse fur include *Thalarctos maritimus* and sparsely furred Pinnipedia like walruses. In the densely furred amphibious mammals, the skin is relatively thick, and the skin on the belly is thicker than on the back (e.g., *Lutra lutra*), with fur hairs usually much thicker. The pelage of these mammals grows in tufts and groups, and naturally has more hairs than that of most terrestrial mammals of similar size: for example, *Ondatra zibethicus* has 11,000–12,000 hairs per square centimeter, whereas the otter has 35,000–51,000. All of these features permit aquatic mammals to carry an air layer in the pelage while in water. In the sparsely furred Pinnipedia, the hair reduction varies with different species. The skin is thick, with compact collagen plexus in the dermis for protection, and thick subcutaneous fat tissues providing the insulating function for these animals. Skin glands usually are well developed in these mammals. However, aquatic mammals like whales (Cetacea) and manatees (Sirenia) have smooth skins, minimal pelage, and no skin glands (Sokolov 1982).

Microclimate in the Skin Habitat

Mammalian skin and hairs provide a relatively constant microclimate and reasonable protection from the influence of external environment for parasitic arthropods. Nevertheless, the microclimate in the skin habitat is naturally influenced by the ambient climate immediately surrounding the host and the host's thermoregulatory activity, and the degree of such influence is largely determined by the property of the skin and pelage.

Mammals maintain normal body temperature (e.g., 37–39°C for eutherian mammals) by regulating heat production and loss (Mount 1979). Body heat is generated primarily by metabolic activities in cells and tissues, and heat loss is made through the skin by radiation in the form of infrared heat, conduction, and convection. The body also loses heat by water evaporation from the skin and through breathing and panting (Carlson and Hsieh 1974; Robinson and Wiegman 1974; Hole 1978).

The body core temperature can be influenced by environmental temperature which varies with geographical locations and in the same location with seasonal and daily changes. The instability of body temperature changes the rate and qualitative characters of biochemical reactions in cells and tissues. Mammals have developed thermoregulatory mechanisms to cope with acute and prolonged exposure to temperature extremes, resulting in specific life history patterns. These behavioral and life history patterns mold the microhabitat environment of the parasitic arthropods. Skin temperature is directly influenced by peripheral blood circulation due to thermoregulatory activity of the host, as in Weddel seals and Southern elephant seals.

Skin temperature varies from site to site on the same body surface (Murray 1957a, b; Jensen and Roberts 1966; Kim 1972). For example, northern fur seal pups had a skin temperature of 33.6°C in naked areas and 31.6°C in furred area with a rectal temperature of 39.7°C at the ambient temperature of 20°C. Temperatures of the naked area, such as nose and eyelids, fluctuate considerably as the air and rectal temperatures change (Kim 1975). A temperature gradient also exists vertically within the skin and pelage layer at each site (Kim 1975).

Microclimate in the fur environment may vary by host activity and changes in its surrounding environment. For example, both temperature and relative humidity in nests and burrows, which provide habitats for many temporary parasites, are relatively more stable than they are outside, and the presence of breeding adults produces a higher temperature (Cotton and Griffiths 1967) and relative humidity which in turn influences fur temperature.

The Host's Internal Organs

Respiratory System

The respiratory system in mammals consists of the nose, pharynx, larynx, trachea, bronchi, and lungs. The nose connects the pharynx with the outside through two nostrils, internal nares, and nasal cavities (four paranasal sinuses and the nasolacrimal ducts). Two main nasal cavities are divided by the nasal septum. Each nasal cavity is largely filled with thin, complex scrolls of bone called the cochae (turbinates) which are attached to the surrounding walls of the nasal cavity. They are a small *dorsal concha (nasoturbinates)*, a *ventral concha (maxilloturbinates)*, and the middle, large, complicated folds, *middle concha (ethnoturbinates)*. Between the turbinates and the hard palate is the *ventral nasal meatus*, through which air passes. Similarly, the *dorsal nasal meatus* and *common meatus* lies between the conchae and the nasal bones. Mucous membrane lines the cavity and its shelves, and mucus secreted by the goblet cells moistens the passage. Here reside the larvae of nasal mites (*Orthohalarachne*) in otariid seals (Fig. 1.5).

The *internal naris* is connected to the pharynx (Fig. 1.6). The anteriormost portion of the pharynx is the *nasopharynx* and the middle portion the *oropharynx*. The posterior pharynx is the *laryngopharynx* which empties into the posterior esophagus and into the larynx anteriorly. Parasitic mites

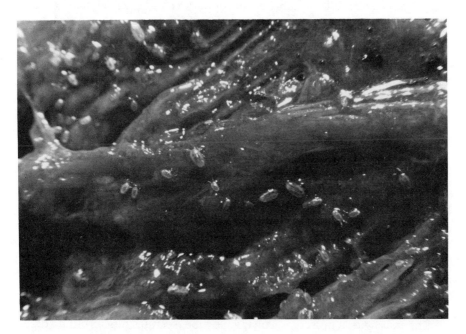

Figure 1.5   Middle conchae and meati of the northern fur seal (subadult), showing larvae of *Orthohalarachne attenuata* (Banks) (large mites) and *O. diminuata* (Doetschmann) (small mites).

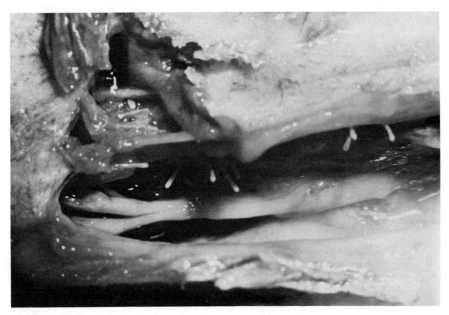

Figure 1.6 Nasopharynx of the northern fur seal (subadult), showing adult opisthosomae of *Orthohalarachne attenuata* (Banks).

like *O. attenuata* adults inhabit the nasopharynx and internal naris, embedding themselves into mucous tissues (Fig. 1.5).

The *trachea* is a tubular organ for air passage, lined with ciliated epithelium of columnar cells, goblet cells, and basal cells. It is divided into two primary bronchi which connect to the lungs. The *lungs* are lined by a ciliated columnar epithelium on incomplete rings of cartilage. Upon entering the lungs, the primary bronchi further divide to form the secondary and tertiary bronchi, which in turn divide into bronchioles and terminal bronchioles. In the lungs, parasitic mites like *O. diminuata* adults inhabit secondary bronchi and bronchioles (Kim et al. 1980).

Alimentary Canal

The *alimentary canal* or gastrointestinal tract is a continuous tube composed of the mouth, pharynx, esophagus, stomach, small intestine, large intestine, and anus. Food is broken down in the alimentary canal physically by muscular contractions and chemically by digestive enzymes. The walls of the alimentary canal, from the esophagus to the anus, are made of four tissue layers or tunics: the mucosa, submucosa, muscularis, and serosa. The inner lining of the tract is a mucous membrane.

The *esophagus*, a muscular, collapsible tube, secretes mucus but no digestive enzymes, and involuntarily pushes food through to the stomach by peristalsis. No parasitic arthropods inhabit this organ. The *stomach* is a

Figure 1.7 *Gasterophilus* larvae and scars on lining of the horse stomach. (Photograph from Merck Co.)

bulbous enlargement of the alimentary canal connecting the esophagus and the duodenum. It has large folds, the *rugae,* which are covered with mucosa. The mucosa consists of a layer of simple columnar epithelium lined with zymogenic, parietal, and mucous cells which produce mucus, pepsinogen, hydrochloric acid, and other gastric juices. The secretion of gastric juices is regulated by sensory, nervous, and hormonal mechanisms (Smyth 1976). Bot flies (*Gasterophilus*) reside in these habitats (Fig. 1.7). The *small intestine,* a long tube connected to the stomach and opening into the large intestine, is where the major portion of digestion and absorption occurs. The three organs of chemical digestion, pancreas, liver, and gall-bladder, are connected to it. The *large intestine* completes nutrient absorption, manufactures vitamins, and forms the feces.

Environment in the Host's Internal Organs

The respiratory tract provides stable habitats for many parasitic arthropods, such as the mesostigmate Halarachnidae (Kim et al. 1980). The complicated scrolls of the conchae in the mammal nasal cavities and bronchioles in the lung, where gas exchanges are made, abound with mucus and capillaries providing abundant space and food for parasitic arthropods.

On the other hand, the alimentary canal in mammals provides one of the most volatile and hazardous habitats for parasitic arthropods (Befus and Podesta 1976). Because of regular physiological changes due to feeding, the environmental conditions of these habitats are highly variable, for example, the level of various enzymes that break down carbohydrates, protein, and fats gyrates and the pH ranges from 1.5 to 8.4. It is almost oxygen free; and it moves constantly and frequently with violent contractions, although it provides a diversity of habitats (Code 1968; Smyth 1967; Morton 1967). Each of these factors presents hazards to the parasites living in it, and these parasites, such as *Gasterophilus*, have developed adaptations to counter this environmental adversity (Smyth 1976). To understand the parasite–host relationships for these parasites the properties of the habitat and its environment must be related to the adaptations of the parasites.

BIOLOGY OF PARASITIC ARTHROPODS

Our knowledge on the biology of parasitic arthropods still is fragmentary, and the major part of the published information is limited to a relatively small number of species of medical and veterinary importance. Recently, Nelson et al. (1975, 1977), Marshall (1981a), and Balashov (1982) have made major effects in synthesizing the available data on the biology and ecology of parasitic arthropods on vertebrates. This section briefly considers the biological parameters essential in discussing evolutionary relationships of parasitic arthropods and mammals.

Reproduction and Life Cycles

Parasitic arthropods usually follow the classic patterns of bisexual reproduction and life cycle. Parthenogenesis is known in several ischnoceran Mallophaga (Matthysse 1944, 1946; Hopkins 1949; Murray 1963; Andrews 1972; Mock 1974; Price and Timm 1979) and occurs facultatively in various orders of Acari (Krantz 1978). Limited *arrhenotoky*, the production of males from unfertilized eggs, has been documented for the Mesostigmata and Prostigmata, and *thelytoky*, production of females from unfertilized eggs, has been observed in the Prostigmata, some ixodids (Metastigmata), and some of the lower Cryptostigmata (Evans et al. 1961). Both phenomena were also observed in the astigmatid Anoetoidea (Hughes and Jackson 1958).

In the majority of parasitic arthropods typical oviposition is most common; most species deposit their eggs in the microhabitat of the host, usually in the nest or runways (Krantz 1978; Marshall 1981a). Some, like lice, polyctenids, and some dermapterans lay their eggs on the host, usually attached on hairs (Weber 1969; Marshall 1981a). The number of eggs laid per day and the total fecundity varies by species, season, and temperature. Sucking lice like *Pediculus humanus* L. may deposit 4.6 eggs per day or 81.5 eggs per female for a life span of 17.6 days (Gooding 1963), whereas ticks like *Dermacenter variabilis* (Say) may lay 4000–6000 eggs on the ground in 4–10 days (Harwood and James 1979).

Many parasitic arthropods exhibit definite specialization in the mode of reproduction. Ovoviviparity and viviparity occur in the Polyctenidae, parasitic Dermaptera, pupiparous Diptera (Marshall 1981a; Kim and Adler, Chapter 4), and Mesostigmata like Laelapidae and Prostigmata (Strandtmann and Wharton 1958; Filipponi and Fracaviglia 1963; Filipponi 1965; Cross 1965).

All arthropods go through a series of postembryonic changes after emerging from the egg to reach the reproductive stage of adult. Exopterygote insects, such as Dermaptera, Anoplura, Mallophaga, and Hemiptera, undergo gradual postembryonic development, with increase in size and structural complexity after each molt, through three distinct stages: *egg*, *nymph* (a series of instars), and *imago* (Fig. 1.8A, I). Endopterygote insects, like Diptera, Siphonaptera, and Coleoptera, after emerging from the *egg* go through two distinct immature stages, the active *larva* (a series of stages) and the resting *pupa*, to attain the sexually mature condition of the adult (Fig. 1.8A, II). Similarly, the developmental cycle of the Acari includes one or more immature stages between the egg and the adult. The larva hatch-

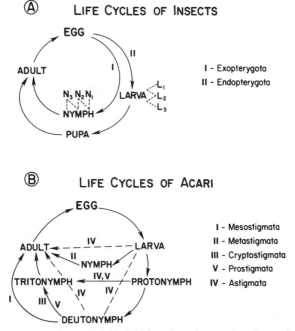

Figure 1.8 Diagrammatic representation of life cycles of parasitic arthropods. (*A*) Insects: I, Exopterygota; II, Endopterygota; L_1, L_2, L_3, larval instars 1, 2, 3; N_1, N_2, N_3, nymphal instars 1, 2, 3. (*B*) Acari: I, Mesostigmata; II, Metastigmata; III, Cryptostigmata; IV, Prostigmata; V, Astigmata.

ing from the egg is an active hexapod, which is followed by as many as three active octopod stages before the acarine reaches sexual maturity, namely, protonymph, deutonymph, and tritonymph (Fig. 1.8B).

Mesostigmatid mites like Laelapidae and Halarachnidae have four active immature stages in the life cycle, larva, protonymph, deutonymph, and adult (Fig. 1.8B, I). In certain parasitic mesostigmatids like *Orthohalarachne*, the octopod protonymph and deutonymphs, which are inactive, nonfeeding, have a very short duration (Furman and Smith 1973; Kim et al. 1980), or the protonymph directly transforms into the adult without going through the deutonymphal stage (Evans et al. 1961). The Metastigmata or ticks have a much simpler life cycle, which includes a larva and one nymphal stage before reaching sexual maturity, except for some Argasidae whose nymph may go through a series of molts (Fig. 1.8B, II).

Four distinct types of life cycles are recognized for the Prostigmata based on the number of active immature stages (Fig. 1.8B, IV): (1) egg–larva–adult (e.g., Tarsonemina); (2) egg–larva–nymph–adult (e.g., Parasitengona–Trombidiidae); (3) egg–larva–protonymph–tritonymph–adult (e.g., Tetranychoidea); (4) egg–larva–protonymph–deutonymph–tritonymph–adult (e.g., Bdellidae and Tydeidae). Astigmatid mites have two nymphal stages after the larvae, protonymph and tritonymph, and in many groups they have an additional hypopial stage between two nymphal stages (Fig. 1.8B, V). A hypopus is nonfeeding and may be active or inactive. The Cryptostigmata have four active immature stages, larva–protonymph–deutonymph–tritonymph (Evans et al. 1961) (Fig. 1.8B, III).

Some taxa of parasitic arthropods are parasitic on mammals only as larva or adult, whereas others inhabit the host throughout the postembryonic life as parasites (Fig. 1.2). Thus the annual life cycle of permanent ectoparasites is often synchronized with annual migration and reproductive patterns of the host, as observed in sucking lice on the northern fur seal (Kim 1972, 1975). Many parasitic acarines like obligate parasitic insects reside and feed upon the host throughout most of their life cycle: some Mesostigmata, such as Halarachnidae; a few Metastigmata like *Boophilus*; some astigmatids like Sarcoptidae and Listrophoridae; and the Prostigmata like Demodicidae and Myobiidae (Fig. 1.2, I). The Siphonaptera are parasitic on the host as adults (Fig. 1.2, II), whereas the parasitic stage is the larva for some Prostigmata (e.g., Trombiculidae) and for myiasis-producing Diptera (Fig. 1.2, III).

Among parasitic arthropods life cycles tend to be abbreviated by ovoviviparity or viviparity, reduction in the number of molts, and shortening of the time for certain immature stages. In the pupiparous Diptera (Hippoboscidae, Streblidae, and Nycteribiidae), the eggs hatch and the first three larval stages develop within the female. The female deposits a third-stage larva which soon pupates (Evans 1950; Kim and Adler, Chapter 4). Parasitic arthropods undergo three to four instars in nymphal or larval stages (Marshall 1981a; Kim and Adler, Chapter 4; Kim, Chapter 5). In parasitic acarines the active nymphal stages may be replaced by inactive resting stages, such as the nymphochrysalis and imagochrysalis in the prostigmatid Trombiculidae (Sasa 1961). Certain astigmatid mites like Listrophoridae lack an active deutonymph or have a reduction in the number of life stages (Evans et al. 1961), and some mesostigmatids like *Orthohalarachne* go through protonymphal and deutonymphal stages very quickly (Furman and Smith 1973).

The Ixodidae usually require a number of hosts for completing the life cycle (Fig. 1.2, IV). Most species of *Ixodes*, for example, are three-host ticks, each instar requiring a different host, and each molt takes place on the ground. Other ticks, such as *Rhipicephalus*, require two hosts (two-host ticks); on the first host the larva feeds and molts into the nymph which after engorgement drops off to the ground to molt, and then the adult seeks a new host. The highly specialized ticks like *Boophilus* are one-host ticks; the larvae, nymphs, and adults all engorge and molt on the same host (Hoogstraal and Aeschlimann 1982; Hoogstraal and Kim, Chapter 10).

Host and Microhabitat Location

Parasitic arthropods must first locate the host and then find a suitable microhabitat on or in it. Permanent parasites, such as lice and polyctenids, do not have much problem in locating a host because they are transferred to newborn hosts during parturition and nursing (Kim 1975) or may disperse between host individuals (Marshall 1981a; Durden 1983). Many temporary parasites, however, must locate a host as adults or immatures (Fig. 1.2, II, III, IV).

Dispersal

Parasitic arthropods usually disperse by walking, but most fleas reach a host by jumping, and most pupiparous Diptera get to it by flying. Most parasites have well-developed legs with large claws which are used primarily for attachment but may also be modified for use in copulation and feeding (Marshall 1981a). Tarsal claws are better developed in permanent than in temporary parasites, and the shape of claws is closely related to their use; for example, the acuminate claws found in *Antarctophthirus* are sunk into the skin whereas the blunt claws in *Proechinophthirus* are used for grasping hair (Kim 1975). Mammal-infesting species usually have a single tarsal claw as in Anoplura and Ischnocera, whereas avian species have double claws. In Mallophaga the forelegs are often modified for feeding; the mid- and hindlegs are for locomotion and attachment to the host (Stenram 1964; Richards and Davies 1977).

Fleas disperse by walking or jumping. Their jumping, which involves the wing-hinge ligaments, the tergo-trochanteral depressor, the subalar and the basalar muscles, is a modified form of the flight mechanism used by other pterygote insects (Bennet-Clark and Lucey 1967; Rothschild et al. 1972, 1975; Rothschild 1973).

Most of the Streblidae and Hippoboscidae are fully winged, but many lose their wings by abrasion after reaching the host. Adults of myiasis producers such as *Gasterophilus* and *Cochliomyia* are usually strong fliers, and individual females lay a large number of eggs on hairs, wounds, or sores in or near the haunts of the host.

Sensory Organs

The responses of parasitic arthropods to external stimuli are highly variable, and published information on the subject is fragmented. As Marshall (1981a) comprehensively summarized for ectoparasitic insects, parasitic arthropods are responsive to the host-related stimuli such as warmth, odor, CO_2, light, air movement, and sound (Dethier 1957; Hocking 1971). They are equipped with a diversity of complex sense organs to receive such stimuli, including sensory setae, sensillae, eyes, and specialized organs such as the Haller's organ in ticks and chordotonal organs in sucking lice, although most of these organs are of unknown function (Dethier 1957, 1961; Hocking 1971; Richards and Davies 1977).

Light is received by dermal photoreceptors, ocelli, or compound eyes, some of which may be lacking in parasitic arthropods (Dethier 1957). Dermal photoreceptors are perhaps distributed over the body, although Camin (1953) found photoreceptors on the puvilli of the forelegs of snake mite *Ophionyssus*.

Mechanoreceptors include sensory hairs found on all segments of the antennae, around the mouthparts, and on the legs in parasitic insects (Wigglesworth 1941; Richards and Davies 1977), the trichobothria and eupathedia in parasitic mites (Evans et al. 1961), the sensilla campaniformes on the trochanter, the basal segments of legs and even on the wings and specialized chordotonal organs in the femur, tibia, and tarsus of legs in insects (Wigglesworth 1941; Richards and Davies 1977).

Auditory organs identified in insects include the tympanal organ, auditory hairs, and Johnston's organ; they have not been reported for other parasitic arthropods except *Ornithodoros concanensis* Cooley and Kohls, which possesses sound receptors on the forelegs (Webb et al. 1977).

The receptors of chemical and olfactory stimuli are different sensillae located on insect antennae (Al-Abbasi 1981) and mouthparts (Dethier 1957, 1963), and the solenidia in acarines and palpal organ in ticks (Evans et al. 1961). Some sensillae such as tuft organs on the antennae in *Pediculus* (Wigglesworth 1941) and the anterior eupathidia of the Haller's organ in ticks (Lees 1948) are sensitive to humidity and temperature.

Host Finding

Permanent parasites are transmitted primarily from mother to young as observed in the Echinophthiriidae (Kim 1975) and Demodicidae (Gaafar 1967). Movement to the pup is probably stimulated by their higher body temperature and sparse hairs (Murray and Nicholls 1965; Murray et al. 1965; Kim 1972, 1975). Transfer between host individuals also occurs through body contact (Beresford-Jones 1967; Durden 1983).

Temporary parasites, on the other hand, require host-related stimuli to locate a host. Initiation of movement, searching, and settling on the host involve environmental cues such as low wind velocity and turbulence, falling barometric pressure, changing light intensity, and the shape and movement of the host (Hocking 1971; Arthur 1976). Responses to CO_2 are variable among parasitic arthropods; low concentrations activate and/or orient fleas (Benton and Lee 1965), mites (Sasa and Wakasugi 1957), and ticks (Burgdorfer 1969). Thermal stimuli are important in many parasitic arthropods (Hocking 1971; Marshall 1981a) including Anoplura (Weber 1929; Wigglesworth 1941); parasitic Dermaptera (Ashford 1970; Marshall 1982); Siphonaptera (Sgonina 1935; Benton and Lee 1965; Askew 1971; Greenwood and Holdich 1979; Marshall 1981a); ticks (Lees 1948; Lees and Milne 1951); and mites (Cross and Wharton 1964a, b). Fleas also respond to other stimuli such as air movement or vibration from a host, or urine, or other host-related odors (Sgonina 1935; Benton et al. 1959; Iqbal 1974; Vaughan and Mead-Briggs 1970; Greenwood and Holdich 1979).

Microhabitat Selection

Parasitic arthropods select and aggregate in certain sites or microhabitats on or in the host. For example, sucking lice on mice are usually restricted to the head (Murray 1961). On man, *Pediculus humanus capitis* DeGeer is restricted to the head hair, whereas *P. h. humanus* Linnaeus is distributed on clothings and other parts of the body (Buxton 1946).

Predation or dislodgement by the host and other host activities are the most important factors controlling the choice of microhabitat by ectoparasitic arthropods (Murray 1961; Bell and Clifford 1964; Bell et al. 1966). Other important factors include hair structure, coat thickness (Murray 1957a, b), skin structure (Wharton and Fuller 1952; Nutting 1965), and host acquired resistance (Nelson and Bainborough 1963; Bell et al. 1966). Temperature of the host's body surface, which changes by season, also plays an important role in microhabitat selection in Anoplura and Mallophaga (Murray 1957a, b, 1961, 1963; Murray and Nicholls 1965; Murray et al. 1965; Kim 1972, 1975). This may also involve kairomones (Whittaker and Feeny 1971; Hocking 1971).

Food and Feeding

Parasitic arthropods feed on various body fluids, tissues, and blood (Good-ing 1972; Marshall 1981a). Those with piercing-sucking mouthparts, such as Anoplura, Polyctenidae, Hippoboscidae, Nycterebiidae, Streblidae, and adult Siphonaptera, feed primarily upon blood. Tungid fleas usually feed on fluid exudates first (Lavoipierre et al. 1979). Also feeding on blood are those insects with highly specialized chewing mouthparts like the Rhyn-chophthirina and some Amblycera (Nelson et al. 1975). Other parasitic insects, such as Amblycera, Ischnocera, Arixeniidae and Hemimeridae, and Platypsyllidae, feed largely upon hair, skin, debris, or exudates from the skin or secretions from exocrine glands (Arthur 1976; Marshall 1981a). The larvae of myiasis-producing Diptera feed on mucus, tissues, and even blood (Zumpt 1965).

The parasitic Mesostigmata, such as Laelapidae, Dermanyssidae, Mac-ronyssidae, and Spinturnicidae, are basically polyphagous. For example, *Haemogamasus ambulans* (Thorell) and *Brevisterna utahensis* (Ewing) utilize both fluid and dried blood of vertebrates, flea feces, and living or dead arthropods, but rarely penetrate the skin. *Echinolaelaps echidninus* (Berlese) and *Haemolaelaps glasgowi* (Ewing) are also general feeders, but usually penetrate the skin (Furman 1959). The Metastigmata feed on the blood and lymph fluids. The endoparasitic mesostigmatid Halarachnidae are primar-ily mucus tissue feeders.

Ticks pierce the host skin with the dentate chelicerae, and this action allows the insertion of the barbed hypostome as the palps sway out. The tick then sucks the blood in along the hypostomal groove with the muscu-lar pharyngeal action (Arthur 1946, 1951). The ixodid ticks are slow feeders with well-developed hypostome and chelicerae, whereas the soft ticks (Ar-gasidae) are rapid feeders with poorly developed hypostomal teeth and large chelicerae (Bennington and Kemp 1980).

Among the parasitic prostigmatid Cheyletoidea, the Demodicidae living in hair follicles feed on subcutaneous secretions, particularly sebum (Smith 1961), and the Myobiidae usually feed on sebaceous secretions at the base of hairs (Ewing 1938) and interstitial fluid (Wharton 1954). The parasitic Trombidioidea, such as Trombiculidae, are parasitic as larvae and free living in the nymphal and adult stages. The larvae feed on serous tissue substances including blood cells. The larva attaches itself to the thin skin of the favorite area and inserts its scimitarlike chelicerae into the epidermis of the skin. When firmly attached, the larva injects salivary fluids that cause histolysis of cellular components (Jones 1950). This leads to the formation of a feeding tube, called *stylostome* or *cytostome,* by the hardening of the skin in which the mouthparts remain until the mite drops to the ground. The stylostome may penetrate the dermis. The trombiculid larvae feed by sucking in lymph and the contents of disintegrated cells through the sty-lostome (Wharton and Fuller 1952).

The parasitic Astigmata, such as Sarcoptidae, Psoroptidae, and Listrophoridae, feed on epidermal detritus and secretions of dermal glands, and some may feed on blood (Evans et al. 1961). *Sarcoptes* burrows into the cornified epidermis by attaching itself to the substratum with ambulacral suckers and cutting a channel into the skin with its chelicerae and edges of the foretibiae (Mellanby 1944). The Listrophoridae feed on dermal detritus and sebaceous secretions as they secure themselves to single hairs (Hughes 1954).

Bloodsucking arthropods have two basic types of blood-feeding apparatus (Hocking 1971). In the first type a single channel is used alternately for outward salivary secretion and inward bloodsucking, as in mites and ticks (Gregson 1960, 1967). The second type found in bloodsucking insects has two channels: dorsal, for blood ingestion, and ventral, usually within the hypopharynx, for salivary secretion (Hocking 1971). Bloodsucking arthropods feed either directly from vessels, such as venules, small veins, or lymphatic vessels, or from a blood pool resulting from the laceration of blood vessels by the probing of mouthparts (Lavoipierre 1965; Nelson et al. 1975). Vessel feeders or *solenophages* include Anoplura (Lavoipierre 1967), Cimicidae, many fleas (Lavoipierre and Hamachi 1961), and at least one hippoboscid *Melophagus ovinus* (Nelson and Petrunia 1969). Pool feeders or *telmophages* include most pupiparous Diptera (Nelson and Petrunia 1969), a few fleas, ticks (Gregson 1960), and parasitic mesostigmatid Macronyssidae (Lavoipierre and Beck 1967). Ticks may be modified vessel or pool feeders, which feed from a pool of blood resulting from extensive laceration of blood vessels (Lavoipierre and Riek 1955). To aid the blood feeding, some parasitic arthropods such as Anoplura and ixodid ticks have histamine, histaminelike compounds, and various enzymes, with proteolytic, anticoagulant, cytolytic, and spreading action, secreted by salivary glands (Gaafar 1966; Gooding 1972; Wakelin 1976; Bennington and Kemp 1980). They are actively produced during feeding. In many haematophagous insects a temperature gradient or differential is considered the most important signal inciting probing, and the presence of nucleotides, particularly ATP, signals the location of a blood source (Friend and Smith 1977).

POPULATION PATTERNS OF PARASITIC ARTHROPODS

For parasitic arthropods a population is defined as an assemblage of all conspecific parasites including all developmental stages in or on the host of a local host population at a particular time. Thus population studies of parasitic arthropods are usually made by analyzing parasite infrapopulations (body population of Marshall 1981a) from certain numbers of the hosts in a particular location (host territory).

Implicit in such studies is that the estimated pattern of infrapopulations reflects that of the population and suprapopulation of the parasite species.

However, unless the collecting technique is reliable, accurate, and consistently applied, such estimation may be quite misleading (Cook and Beer 1955; Janzen 1963; Kim 1972). Population analysis must take account of inherent variation in biological parameters in different species, particularly for temporary parasites; for example, the amount of time spent on the host differs greatly among different species of fleas. Within a species, flea attraction to the host's body may differ by age and sex and may also change with space and time (Marshall 1981a).

It is exceedingly difficult to identify the critical factors involved in population regulation of parasitic arthropods because the parasite–host relationship is complex and closely intertwined. Different mechanisms may be operating on different stages in the life cycle or at the infrapopulation, population, and suprapopulation levels, although regulation at the infrapopulation level can influence regulation at the population level (Holmes et al. 1977; Esch et al. 1977). At the same time, there are very few comprehensive and reliable works on population ecology of parasitic arthropods.

Populations of parasitic arthropods are usually defined by the following measures:

1. Infestation rate (incidence rate of Marshall 1981a)—the percentage of host individuals infested (total number of hosts examined/total number of hosts infested × 100) (Table 1.2).

2. Population rate (infestation rate of Marshall 1981a, or "parasite index" or "flea index")—the mean number of parasites per host examined (Tables 1.2, 1.3).

3. Dispersion rate (frequency distribution)—the percentage of hosts infested with a specific number of parasites (Table 1.3).

Infestation and Dispersion

Infestation and population rates provide a general description of the parasite population in a local host population. However, the infestation rate by itself does not provide much information on parasite populations, particularly when it is very high, and the population rate does not reveal an accurate picture of parasite population density and distribution for clumped populations without the dispersion rate (Marshall 1981a).

Infestation and population rates are usually underestimated because of inherent bias and errors in collecting techniques and parasite departure in response to behavioral and physiological changes in captured hosts. Marked differences have been observed between estimates of infestation and population rates by visual examination and by Cook's dissolving technique (Cook 1954a, b; Mitchell 1964; Kim 1972; Miller et al. 1973). With adult seals, visual examination revealed no louse infestation, whereas the

Table 1.2 Infestation Rates (%) and Population Rate (Mean Number of Parasites Per Host Examined) of Nine Parasitic Arthropods[a]

Parasites	Hosts	Infestation rate (%)					Population rate				
		♂	♀	A	J	T	♂	♀	A	J	T
1. Es (P:H)	Ms (CHI)	0.91	0.81	0.85	—	0.85	11.7	15.1	13.7	—	13.7
2. Hh (H:A)	Pm (ROD)	0.34	0.27	0.31	—	0.31	5.4	5.4	5.4	—	5.4
3. Ac (E:A)	Cu (PIN)	0.33	1.00	0.81	1.00	0.91	0	39.0	28.8	33.5	32.9
4. Pf (E:A)	Cu (PIN)	0.67	1.00	0.83	0.85	0.84	1.5	25.8	18.9	17.8	17.9
5. Bh (N:D)	Tp (CHI)	0.45	0.27	0.35	—	0.35	0.8	0.3	0.5	—	0.5
6a. Cf (P:S)	Ha (CAN)	0.67	0.43	0.55	—	0.58	3.9	1.5	2.7	—	3.4
6b. Cf (P:S)	Lc (LAG)	0.63	0.74	0.69	0.45	0.57	4.2	3.5	3.9	1.4	2.3
7. So (M:M)	Ef (CHI)	1.00	1.00	1.00	0.95	0.97	4.8	34.4	32.2	29.6	30.9
8. Od (H:M)	Cu (PIN)	1.00	1.00	1.00	0	1.00	1468	322	831	0	831
9. Oa (H:M)	Cu (PIN)	1.00	1.00	1.00	0	1.00	918	1066	1000	0	1000

[a]A, adult; J, juvenile; T, total. 1. Es (P:H), *Eoctenes spasmae* (Polyctenidae: Hemiptera); Ms (CHI), *Megaderma spasma* (Chiroptera) (Marshall 1982). 2. Hh (H:A), *Hoplopleura hesperomydis* (Hoplopleuridae: Anoplura); Pm (ROD), *Peromyscus maniculatus* (Rodentia) (Beer and Cook 1968). 3. Ac (E:A), *Antarctophthirus callorhini* (Echinophthiriidae: Anoplura); Cu (PIN), *Callorhinus ursinus* (Pinnipedia). 4. Pf (E:A), *Proechinophthirus fluctus* (Echinophthiriidae: Anoplura); Cu (PIN) (Kim 1972, 1975). 5. Bh (N:D), *Basilia hispida* (Nycteribiidae: Diptera); Tp (CHI), *Tylonycteris pachypus* (Chiroptera) (Marshall 1971). 6a. Cf (P:S), *Ctenocephalides felis* (Pulicidae: Siphonaptera); Ha (CAN), *Herpestes auropunctatus* (Canivora) (Haas 1966). 6b. Cf (P:S), Lc (LAG), *Lepus capensis* (Lagomorpha) (Flux 1972). 7. So (M:M), *Steatonyssus occidentalis* (Macronyssidae: Mesostigmata); Ef (CHI), *Eptesicus fuscus* (Chiroptera) (Miller et al. 1973). 8. Od (H:M), *Orthohalarachne diminuata* (Halarachnidae: Mesostigmata); Cu, *Callorhinus ursinus* (Pinnipedia). 9. Oa (H:M), *Orthohalarachne attenuata* (Halarachnidae: Mesostigmata); Cu, *Callorhinus ursinus* (Pinnipedia) (Kim 1983).

dissolving technique revealed an infestation rate of 71.4% (Kim 1972). Ectoparasites like lice and mites are often active for many hours after host death and temporary parasites like fleas may leave the host almost at once (Cook 1954a, b; Olson 1969). Many fleas also leave a struggling or agitated host leading to the underestimation of infestation and population rates (Stark and Kinney 1962; Olson 1969; Gross and Bonnet 1949). These sampling errors also influence the dispersion rate (frequency distribution).

Infestation and population rates vary greatly among parasitic arthropods by species (Tables 1.2, 1.3), location, season, and host age and sex (Marshall 1981a, b). In equatorial Malaysia, the polyctenid *Eoctenes spasmae* (Waterhouse) shows an 85.2% infestation rate on the bat *Megaderma spasma* with 13.9 bugs per bat and no significant difference between male and female or between seasons. However, males of the Malaysian bats *Tylonycteris pachypus* and *T. robustula* had consistently higher infestation rates of the nycteribiid *Basilia hispida* Theodor compared

Table 1.3 Infestation Rates, Population Rates, and Frequency Distributions of Eight Parasitic Arthropods[a]

Parasite; order: family	A:H	D:N	D:N	D:S	D:S	S:P	S:H	P:M
Parasite species	Hh	Bh	Pb	Ma	Tc	Xc	Cn	Rs
Host order	ROD	CHIR	CHIR	CHIR	CHIR	ROD	ROD	ROD
Host species	Pl	Tp	Ma	Aj	Pt	Rr	As	Pl
Infestation rate	0.41	0.35	0.47	0.53	0.77	0.27	0.36	0.90
Number parasites collected	339	370	47	347	809	>203	605	>666
Population rate	3.3	0.5	0.2	0.8	2.6	>1.0	0.8	7.2
Maximum number parasites/host	57	8	9	6	12	>10	20	70
Dispersion rates (%); hosts infested with specific number of parasites 0	0.58	0.65	0.53	0.46	0.23	0.73	0.64	0.10
1	0.04	0.23	0.27	0.32	0.17	0.12	0.19	0.22
2	0.04	0.08	0.13	0.17	0.15	0.03	0.06	0.09
3	0.04	0.02	0.03	0.04	0.16	0.02	0.04	0.08
4	0.04	0.01	0.03	0.01	0.10	0.01	0.03	0.07
5	0.03	0.01	0.01	0	0.05	0.01	0.01	0.05
6	0.05	—	—	—	0.05	0.01	0.01	0.09
7	0.05	—	—	—	0.04	0.01	—	0.03
8	0.01	—	—	—	0.03	0.01	—	0.03
9	0.02	—	—	—	0.01	0.01	—	0.02
>10	11	—	—	—	0.01	0.04	—	0.33

[a]Infestation rate = percent hosts infested; population rate = mean number of parasites per host examined. A:H, Anoplura:Hoplopleuridae; Hh, *Hoplopleura hesperomydis*; ROD, Rodentia; Pl, *Peromyscus leucopus* (Combs and Kim 1977). D:N, Diptera: Nycteribiidae; Bh, *Basilia hispida*; CHIR, Chiroptera; Tp, *Tylonycteris pachypus* (Marshall 1971). D:S, Diptera:Streblidae; Ma, *Megistopa aranea*; CHIR, Chiroptera; Aj, *Artibeus jamaicensis* (Overal 1980). D:S, Diptera: Streblidae; Tc, *Trichobius corynorhini*; CHIR, Chiroptera; Pt, *Plecotus townsendii* (Kunz 1976). S:P, Siphonaptera:Pulicidae; Xc, *Xenopsylla cheopis*; ROD, Rodentia; Rr, *Rattus rattus* (Olson 1969). S:H, Siphonaptera:Hystrichopsyllidae; Cn, *Ctenophthalmus nobilis*; ROD, Rodentia; As, *Apodemus sylvaticus* (Evans and Freeman 1950). P:M, Acariformes:Prostigmata:Myobiidae; Rs, *Radfordia subliger*; ROD, Rodentia, Pl, *Peromyscus leucopus* (Combs and Kim 1977).

with female bats, and *T. pachypus* was more heavily infested than *T. robustula* (Marshall 1971). Infestation rates for streblids and other nycteribiids on New Hebridean bats were also quite variable, from 9% to 100% (Maa and Marshall 1981). In the temperate zone, for the sucking louse *Hoplopleura hesperomydis* (Osborn), on *Peromyscus maniculatus*, the infestation rate averaged 31% with the population rate of 5.4 and varied greatly year by year, although no marked difference was observed between male and female hosts (Beer and Cook 1968). *Hoplopleura acanthopus* (Burmeister) on *Microtus pennsylvanicus* showed a significant seasonal variation in infestation rate with an April peak, while little difference was observed between male and female hosts (Cook and Beer 1958).

The cosmopolitan flea *Ctenocephalides felis* was studied on the mongoose *Herpestes auropunctatus* in Hawaii (Haas 1966) and on the hare *Lepus capensis* in Kenya (Flux 1972) (Table 1.2). The infestation rates were essentially similar in both studies, with the population rate higher on mongooses than on hares and the infestation rate was higher on male mongooses than females and higher on adult than juveniles. Both studies showed considerable variations in flea populations by locality and time.

For mesostigmatid mites, infestation rates by *Laelaps nuttalli* Hirst on *Rattus exulans* and *R. rattus* were 100% and 97% respectively, whereas infestation rates by *L. echidninus* on the same hosts were 68% and 56%. The population rates were significantly different between host species: 43.4 *L. nuttalli* from 25 *R. exulans* and 11.3 mites from 64 *R. rattus*; 6.0 *L. echidninus* from *R. exulans* and 2.6 mites from *R. rattus*, respectively (Mitchell 1964).

On aquatic carnivore Pinnipedia the infestation rates of sucking lice (Echinophthiriidae) were similar between sexes and age groups, but the population rates were generally higher on pups and juveniles than on adult seals and on adult females than on males (Murray and Nicholls 1965; Murray et al. 1965; Kim 1972, 1975). In the northern fur seal *Callorhinus ursinus* both species of respiratory mites (*Orthohalarachne*) showed infestation rates 100% and no significant difference in the population rate between adult males and females, while practically no pups were infested with mites (Kim et al. 1980). This shows a temporal progression of the mite population increase, as the host gets older. Although ectoparasitic arthropods usually have low population rate (mean population density), the population rate for Anoplura and Mallophaga is relatively high (Cook and Beer 1958; Kim 1972; Rust 1974). Marine carnivores such as northern fur seal are heavily infested with mesostigmatid *Orthohalarachne* in their respiratory organs: the population rate of 831 mites for *O. diminuata* (Doetschmann) with a maximum of 7400 mites, and 1000 for *O. attenuata* (Banks) with a maximum of 5642; a 6-year-old male had the largest number of 10,628 mites of both species (Kim et al. 1980; Kim 1983).

The population rate does not provide an accurate picture of the population of parasitic arthropods because of their overdispersed distribution and distortion of the pattern by one or a few heavily infested hosts. There are many examples of unusually high numbers of parasites on one or a few hosts for different parasitic arthropods: for large mammals, over a million *Bovicola ovis* (Schrank) (Ischnocera; Mallophaga) were recorded from a domestic sheep (Murray 1965) and 20,000 *Trichodectes canis* (DeGeer) from an injured dog; and for small mammals, 932 *Archaeopsylla erinacei* Bouche from a hedgehog *Erinaceus europaeus* (Smit 1958), and 769 *Spilopsyllus cuniculi* (Dale) from a rabbit *Oryctolagus cuniculus* (Shepherd and Edmunds 1976).

The levels of parasite infestation within a host population or even a single host age group differ in different individuals, and thus the distributions of parasitic arthropods within the host population are rarely random and usually overdispersed (Table 1.3, Fig. 1.9), as in other kinds of parasites (Williams 1964; Crofton 1971a, b; Anderson 1978; Anderson and Gordon 1982). Many host individuals may have no parasitic arthropods; for example, 58% of *P. leucopus* had no *H. hesperomydis*, 65% of *T. pachypus* was without *B. hispida*, 73% of *R. rattus* was without *X. cheopis* (Table 1.3), whereas some carry heavy parasite loads (Fig. 1.9).

For the distribution patterns of parasites with the host populations, three distinct levels are recognized in a dispersion spectrum (Anderson and Gordon 1982): underdispersed, random, overdispersed. The Poisson probability distribution describes the random distribution of parasites, whereas binomial distribution fits the dispersed population patterns, the positive binomial for underdispersed distribution and the negative binomial for overdispersed. The degree of parasite dispersion is commonly measured by the formula $S^2/\bar{x}$, where S^2 is the sample variance and $\bar{x}$ is the sample mean of parasite numbers per host.

Parasitic arthropods are usually randomly distributed only for a short period when they invade new hosts. It is also probable that at the beginning of primordial host associations parasitic arthropods were randomly distributed in the populations of primitive mammals.

The distribution patterns of parasite population constantly change in time and space owing to opposing mechanisms that increase or decrease overdispersion. These mechanisms include two types of factors, the *demographic stochasticity* involving population growth (e.g., natality, mortality, dispersal rates) and the *environmental stochasticity*, which includes variable environmental factors such as climate, microclimate, habitat morphology, host susceptibility, or behavior (Anderson and Gordon 1982).

Overdispersed distributions of parasitic arthropods are essentially caused by heterogeneity in the probability of parasite infestation of a host. Thus, for obligate parasites, particularly permanent parasites, differences in host behavior, morphology, susceptibility, and immunity are most important, whereas for temporary parasites heterogeneity in climate and demographic parameters is critical (Nelson et al. 1975, 1977; Kennedy 1975; Marshall 1981a).

Overdispersion is important for maintaining a stable parasite–host system. The high degree of overdispersion minimizes the deleterious effects

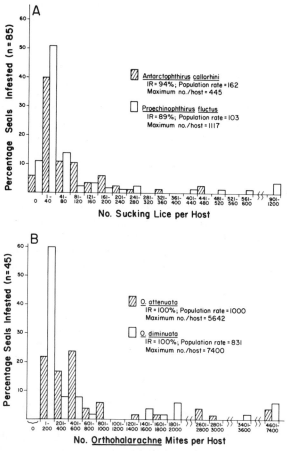

Figure 1.9 Frequency distribution of sucking lice and mites (*Orthohalarachne*) of northern fur seal (*Callorhinus ursinus*) on St. Paul Island, Alaska (Kim 1983). (*A*) Sucking lice (Echinophthiriidae: Anoplura) extracted from seal skins of both sexes and all age groups (2-day to 18-year-old) collected between 1969 and 1978. (*B*) Mesostigmatid mites (Halarachnidae) in the respiratory organs of adult seals of both sexes and 4–18-year-old age groups collected during the summer 1978. IR, infestation rate, which equals the total number of hosts examined divided by the total number of hosts infested.

of the parasites on the host population and selects susceptible or diseased individuals with little or no effect on the host population. It removes a larger proportion of parasites with little effect on the parasite population by the death of heavily infested host individuals (Crofton 1971a; Anderson and May 1978a, b; Kennedy 1982; Anderson and Gordon 1982). Thus overdispersion in relation to the selection and density of both parasites and hosts may play a significant role in population regulation and coevolution of both associating organisms.

Population Structure

Parasite population structure is usually expressed by a percentage breakdown of life stages of parasite infrapopulations such as eggs, various immature instars, and adults. For a given species, population structure varies
with time under the influence of both demographic and environmental
mechanisms and thus usefully describes the temporal stability of that
population. However, much of the data on population structure of
parasitic arthropods so far documented do not illustrate the dynamic nature of such information but rather merely show an average pattern for the
species (Marshall 1981a).

The parasite infrapopulation on a given host is a dynamic system, begun
with a founder population on the newborn host and supplemented by
immigrants and generations of offspring (Figs. 1.1, 1.10). Theoretically,
with normal mortality in successive life stages, there are more eggs than
the first immature instars, and in turn more first instars than the second
instars (Evans and Smith 1952). Likewise, there are fewer adults than any

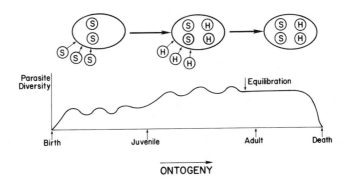

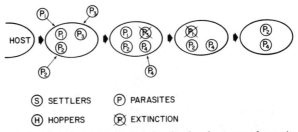

Figure 1.10 Ecological and evolutionary models for the development of parasite community,
host association, and diversity.

other life stage (Fig. 1.2). Thus the population structure of parasitic arthropods forms a "pyramid of numbers of individuals" for a given generation. However, the population structures for various parasitic arthropods documented often do not show the pyramid because of overlapping generations where life stages of one generation are mixed with those of a succeeding generation or because of temporal changes in demographic and environmental forces.

Of 148 individuals of *Hesperoctenes fumarius* (Westwood) on bats 34% were nymphs and 66% adults (30% females and 36% males) (Ueshima 1972). Marshall (1981a) observed that in viviparous arixeniids and polyctenids about two-thirds of an active population are adults, whereas in the oviparous cimicids adults account for only one quarter.

In the oviparous Anoplura and Mallophaga nymphal instars commonly outnumber adults (about 67% for 30 species), although the percentage of nymphs (18–88%) and different nymphal instars is highly variable (Marshall 1981a). For *Pediculus humanus* Linnaeus a stable population structure was calculated as 67.9% eggs, 15.2% first-stage nymphs, 7.0% second-stage nymphs, 4.3% third-stage nymphs, and 5.7% adults (82% nymphs: 18% adults) (Evans and Smith 1952). Population structure among sucking lice may vary seasonally, as in two species on northern fur seal, *Proechinophthirus fluctus* and *Antarctophthirus callorhini* (Kim 1972). The percentage of adults was distinctly different among three seasons, March–April, July, and October; for *P. fluctus* this was 48%, 69%, and 2% and for *A. callorhini* 40%, 37%, and 0%, respectively.

For the oviparous mesostigmatids the population structure of *Laelaps nuttalli* on *R. rattus* was 90% adult females, 4% adult males, and 6% nymphs, and on *R. exulans* 82%, 12%, and 6% respectively. Similarly, for *L. echidninus* the percentage abundance of adult females, adult males, and nymphs was 95%, 2%, and 3% on *R. rattus*, and 86%, 7%, and 7% on *R. exulans* respectively (Mitchell 1964). On the big brown bat *Eptesicus fuscus* the populations of the macronyssid *Steatonyssus occidentalis* (Ewing) were 59.4% adults and 40.6% nymphs on adult bats for April, and 41.3% adults and 58.7% nymphs on adult bats in July when immature bats were infested by more nymphs (82.7%) than adults (17.3%) (Miller et al. 1973). For the three-host tick *Dermacenter andersoni* Stiles, the ratio of larvae to nymphs on the rodents *Neotoma cinerea* and *Spermophilus lateralis* was 40:30.6% and 35.1:60.3% respectively (Sonenshine et al. 1976).

For the ovoviparous mesostigmatid mites of marine Carnivora, the population structures of the halarachnid *Orthohalarachne* species in the respiratory system of *Callorhinus ursinus* showed that the *O. attenuata* (Banks) population was 95% larvae and 5% adults in subadult males, 69% larvae and 31% adults with only few nymphs in adult females, and 76% larvae and 24% adults in silver pups, whereas the ratio of larvae to adults of the *O. diminuata* (Doetschmann) population was 99:1% in subadult males, 84:16% in adult females respectively, and 100% larvae in silver pupa.

Small infestations of both mites in 3 to 4-month-old pups represented initial founder populations and early stage of progressive population buildup (Kim et al. 1980).

Sex Ratio

Parasitic arthropods usually emerge as adults in equal numbers of male and female (Maynard Smith 1978; Marshall 1981a, b; Wharton and Fuller 1952). In natural populations of both temporary and permanent parasites, however, sex ratio is generally imbalanced and females predominate. This phenomenon is expected on theoretical and genetic grounds (Hamilton 1967). Of 359 collections of 245 species of ectoparasitic insects surveyed, 30% had nearly equal sex ratio, 63% had significantly more females, and only 7% significantly more males (Marshall 1981b). Likewise, in natural populations of parasitic acarines females usually predominate. Females constituted 90% of the populations of the mesostigmatid L. nuttalli on R. rattus, 82% of the same species on R. exulans, 95% of L. echidninus on R. rattus, and 86% of the same species on R. exulans (Mitchell 1964). Among adult mites of Orthohalarachne in the respiratory organs of northern fur seals, 92% of O. attenuata and 88% of O. diminuata were females (Kim et al. 1980).

Skewed sex ratios are caused by many different factors, such as collecting technique, season, climate, host, species, behavior, population density, and nutritional status (Marshall 1981a, b). Some examples of the imbalance are related to the collecting method used: (1) small sample size (Marshall 1981b); (2) body size, females usually being larger and easier to detect; (3) escape behavior, males being faster and more likely to drop off the host. Differential longevity and mortality may be a factor: in many parasitic arthropods the female lives longer, but the male is more active than the female and thus likely to suffer from predation by the host and less likely to withstand climatic and nutritional disturbances (Marshall 1981b).

Collections with significantly more males include the Hippoboscidae, Streblidae, and Cimicidae (Marshall 1981b). The imbalance in the Streblidae collections made by Wenzel (1976) was caused by the way these flies were collected and does not represent a true sexual imbalance. Males were more readily collected from the host bodies than females. A skewed ratio in Cimicidae is relatively common and perhaps caused by the longer lifespan of males and female mortality related to repeated copulation (Usinger 1966). The sex ratio may also change by season in Anoplura, Mallophaga, Hippoboscidae, and Siphonaptera. In the ischnoceran louse, Tricholipeurus parallelus (Osborn) on Odocoileus virginianus, males were more predominant during the winter and at minimum during the spring (Samuel and Trainer 1971). A similar pattern is observed in the Streblidae

(Overal 1980), Nycteribiidae (Funakoshi 1977), and Hippoboscidae (Corbet 1956).

The complete absence or rare presence of males has been recorded among the trichodectid Mallophaga, *Bovicola, Damalinia,* and *Geomydoecus* (Hopkins 1949, 1957; Matthyssee 1944, 1946; Murray 1963; Andrews 1972; Mock 1974, Westrom et al. 1976; Price and Timm 1979), and the mesostigmatid mites, *Laelaps* and *Haemolaelaps* (Furman 1966; Mitchell 1968). This imbalance is attributed to arrhenokotous parthenogenesis. Arrhenotoky commonly occurs in the Mesostigmata, although thelytoky is found in two species of machrochelids (Filiponni 1965). In *Laelaps myonyssognathus* Grochovskaya and Nguyen, progeny from virgin females were males (Mitchell 1968). Arrhenotoky is generally considered biologically advantageous for parasites like trichodectids with considerable inbreeding (Hamilton 1967).

POPULATION DYNAMICS OF PARASITIC ARTHROPODS

The relationships between parasitic arthropods and mammal hosts, like other parasite–host systems, are generally stable, each host usually having a small parasite population (Bradley 1974), although some infrapopulations can be large (Tables 1.2, 1.3, 1.4). The infrapopulation distribution within a host individual is often clumped (Kim 1972, 1975, 1983; Kim et al. 1980), and the frequency distribution within a host population is usually overdispersed (Table 1.3, Fig. 1.9). Many hosts harbor no specific parasites. In most instances, the load of parasitic arthropods seldom affects the health of individual hosts or the host population, perhaps with some exceptions for domestic animals (Nelson et al. 1977; Marshall 1981a).

A Systems Model

The parasitic arthropod population on a mammalian host may be considered as a dynamic control system with an input Po and output Pt (Fig. 1.1). This system involves two sets of associating organisms, the parasites (parasitic arthropods) and the host (mammal), that interact under the regulating influence of the microhabitat characteristics and its associated environmental factors and the environmental conditions surrounding the host. The *input* is the number of parasitic arthropods reaching a host, $I = Po$, which includes initial parasite population that invades a host (e.g., newborn litter) and subsequent transfers. The *output* has two components: the number of eggs produced by female adults, R, for permanent parasites and the number of parasites leaving the host, Pt.

Interactions may occur between (1) conspecific parasites, (2) parasites of different synhospitalic species, (3) the parasite and the host, (4) the para-

Table 1.4   Species Composition and Mean Density per Host of the Ectoparasite Collections from Live *Rattus*[a] in Three Localities

Ectoparasites	Honolulu, HI (1934)	Savannah, GA (1932)	Dothan, AL (1933–1934)
Number live *Rattus:*	6382	5242	4832
Fleas			
Xenopsylla cheopis	2.53	12.25	5.08
Echidnophaga gallinacea	3.87	3.45	3.03
Ctenocephalides felis	0.09	0.13	0.07
Leptopsylla segnis	—	3.31	2.26
Nosopsyllus fasciatus	—	0.66	0.04
Others	0.00	0.02	0.09
Lice			
Polyplax spinulosa	0.31	7.24	1.72
Hoplopleura spp.	0.06	0.01	0.12
Mites			
Laelaps nuttalli	0.94	6.25	0.00
Laelaps echidninus	0.08	0.94	2.88
Ornithonyssus bacoti	0.00	6.39	0.07
Others	0.00	0.37	0.20

Source: Data from Cole and Koepke (1947).
[a] Primarily *R. norvegicus.*

site population and the host, or (5) the total parasite load (a parasite community) and the host. These interactions are influenced by some or all of the major biological parameters of associating organisms as follows:

1. Parasite
    a. Habitat: microhabitat, microclimate
    b. Structure and functions: adaptations
    c. Host locating and feeding behavior
    d. Reproduction: type, oviposition site and behavior, rate
    e. Dispersal and transfer
    f. Life cycle: duration, stages, molting, host requirement
    g. Population dynamics: density, structure, seasonality
    h. Interspecific relationships: synhospitalic associating organisms
2. Host
    a. Habitat, territory, and range
    b. Age and sex
    c. Life history
    d. Physiology

e. Behavior
f. Immune system
g. Population dynamics
h. Interspecific associations

The system assumes that the mammalian host provides habitats for parasitic arthropods which as an individual or a population cost the host a certain loss of energy and induce a host response *Hr* (Fig. 1.1). Host responses and environmental changes are control factors that usually do not operate in a feedback manner. However, host responses, including host behavior and immune reactions, may reduce the number of subsequent transfers, prevent the continual increase of the parasite population, and affect parasite dispersal *D*. Here they function as a negative feedback control on the parasite population in the output flow path (Holmes et al. 1977). Parasite fecundity *R* and mortality *M* as part of the output may also prevent the continual buildup of the parasite population and reduce the number of transfers as the negative feedback control.

Life History Strategies

The life history strategies of parasitic arthropods do not fit neatly within the concept of *r*- and *K*-selection, but instead lie along the *r–K* continuum (Pianka 1970; Gadgil and Solbrig 1972; Esch et al. 1977). Permanent parasites such as Anoplura and Halarachnidae possess attributes of *r*-strategists including rapid development, high resource thresholds, rapid reproduction, short longevity, small body size, semelparity, and, also low fecundity, a characteristic of *K*-strategists. Unlike other *r*-strategists, permanent parasites have relatively constant, predictable environment and directed mortality. Temporary parasites have the characteristics of *K*-strategists but often have high fecundity of *r*-strategists in such taxa as ticks. They also are exposed to variable, unpredictable environments away from the host and catastrophic, often nondirected, mortality by density-independent factors.

Population Regulation

Populations of parasitic arthropods, as in other animals, are regulated by the interplay of three differential processes: (1) fecundity to increase population size, (2) mortality to check excessive population increase, and (3) transmission and immigration. These processes vary with space and time (Holmes et al. 1977; Marshall 1981a). Bradley (1972, 1974) outlined three possible ways to regulate parasite populations: (1) by transmission with no negative feedback mechanisms (Type I), (2) by regulatory mechanisms such as immune response or overdispersion at the level of the host population (Type II), and (3) by regulatory mechanisms such as partial immune response at the level of the host individual (Type III).

To persist, a parasite population must be transmitted from one host to another (Type I regulation). Transmission is a successional process through which newborn hosts are first invaded by a founder population of permanent ectoparasites such as sucking lice during parturition and nursing. As the founder parasites become established, a second wave of temporary ectoparasites such as ticks and chiggers and endoparasites like nasal mites may invade the host (Figs. 1.1; 1.10, I). The secondary invaders usually leave after a transient residence and feeding, whereas permanent parasites once established, remain with the host throughout its life. Transmission and dispersal are greatly influenced by (1) behavioral, physiological, and immune activities of the host and (2) environmental factors associated with the host habitat (Marshall 1981a).

Type II regulation involves responses of the host population to the exponential increase of permanent parasites. These responses include "sterile" immune response and overdispersion-mortality mechanisms which are highly dependent on the size and spatial structure of the host population (Holmes et al. 1977). The result of both processes, which Bradley considers precarious in the real world, is the survival of individual hosts with protective immunity or a small number of parasitic arthropods (Crofton 1971a; Bradley 1974). Type III regulation of parasite populations involves density-dependent factors at the level of the host individual; here, the parasitic arthropod population is regulated by behavioral, physiological, and immune responses of host individuals and perhaps by infra- and interspecific competition of parasites (Holmes et al. 1977).

Food and space generally are not major limiting factors for parasitic arthropods. Transmission and dispersal usually have little significance in population fluctuation at the population and suprapopulation levels. Parasitic arthropod populations are primarily host limited and differ between host individuals, subspecies, age groups, and sexes. Population differences between host individuals and subspecies may be due to the differences in immunological capacity, anatomical and physiological parameters, and health condition which affects grooming behavior (Nelson et al. 1975, 1977; Marshall 1981a). Young mammal hosts are usually more heavily infested with ectoparasites than are adults, because of behavioral and physiological differences, although sometimes no age difference exists. This is particularly true in bloodsucking taxa, such as Hippoboscidae (Graham and Taylor 1941; Bequaert 1953), Anoplura (Murray and Nicholls 1965; Murray et al. 1965; Kim 1972), fleas (Combs 1977), and some mesostigmatid mites (Combs 1977). Juvenile mammals are not as efficient at grooming as adults, and resistance to infestation may increase with age (Nelson 1962a, b; Scharff 1962; Nelson et al. 1970, 1975). On the other hand, for some fleas, sucking lice, and mites, adult mice harbored more ectoparasites than did juveniles (Combs 1977; Combs and Kim 1977). In some non-blood-feeding ectoparasites, such as chewing lice and platypsyllids, host age does not seem to affect populations consistently; it may either decrease or increase ectoparasite population size (Samuel and Trainer 1971; Janzen 1963).

In some ectoparasites, such as sucking lice and fleas, host body size seems to influence population size. Smaller species of rodents harbor fewer ectoparasites than do closely related larger species, and smaller hosts of the same species support smaller populations (Holdenried et al. 1951; Linsdale and Davis 1956; Morlan 1952; Stark and Miles 1962; Mohr and Stumpf 1964). However, the real causes for these differences are not yet clear; they may be due to the size of host nests and home ranges, host age, or competition for space.

Effects of host sex on the population rate of parasitic arthropods may vary with (1) morphological factors such as relative body size and differences in the skin and fur; (2) physiological factors such as differences in blood hormonal levels due to reproductive or stress conditions; and (3) behavioral factors such as differences in grooming, nesting, and mobility (Marshall 1981a). Frequently, in some fleas, parasitic flies, sucking lice, and mites, male hosts harbor more ectoparasites than do females, and this pattern is also changed by season (Holdenried et al. 1951; Mohr and Stumpf 1964; Cowx 1967; Tipton and Mendez 1968; Ulmanen and Myllymäki 1971; Marshall 1971; Combs 1977). With some fleas, some sucking lice, and parasitic beetles no significant difference in population rate exists between host sexes (Cook and Beer 1958; Beer and Cook 1958; Beer et al. 1959; Mohr and Stumpf 1964; Hilton and Mahrt 1971; Combs 1977). However, in certain species of Anoplura and some nycteribiids and streblids, the female hosts harbor greater populations (Hurka 1964; Murray and Nicholls 1965; Murray et al. 1965; Kim 1972, 1975; Kunz 1976).

Populations of temporary parasites, such as fleas, parasitic flies, many ticks and chiggers, and nest-associated ectoparasites, are subjected to the influence of characteristics and climate of the host habitat which also change by season (Lees and Milne 1951; Janzen 1963; Furman 1968; Stark and Kinney 1969; Jackson and Defoliart 1975a, b, c; Sonenshine et al. 1976).

INTERACTIONS AMONG PARASITIC ARTHROPODS

The Community of Parasitic Arthropods

A parasite community in a host individual (a mobile habitat) is a natural ecological unit in which infrapopulations of component species interact among themselves and with the host (Fig. 1.1). The structure of parasite communities may vary with different host individuals and species and also with space and time, whereas the species diversity of parasites for a host species is usually constant. Unlike other biological communities, however, the parasite community does not exhibit any definable trophic structure, consisting primarily of scroungers. Furthermore, in any parasite community there are dominant species which, by their size or abundance, affect the structure and dynamics of the whole community (McNaughton and Wolf 1970). The dominants may differ by host species and locality (Elton et al. 1931; Beer et al. 1959; Combs 1977). Within the range of host distribution, the dominants, which appear when the community is in equilibration, are usually the same species. The present structure and function of a parasite community is a historical manifestation through the evolution of parasite–host associations (Fig. 1.10).

The community structure of parasitic arthropods may differ by individual host, area, and season because the ectoparasite fauna of a mammalian community consists of different types of ectoparasites. Three primary categories of ectoparasites were recognized by Timm (1975): (1) taxon-specific parasites, including those specific to a host species (e.g., many Anoplura) and those exclusively associated with a specific higher taxon [e.g., *Orchopeas caedens* (Jordan), a flea found on the Sciuridae]; (2) habitat-specific parasites which are associated with a number of unrelated hosts that occupy similar habitats (e.g., *Ctenophthalmus pseudagyrtes* Baker on *Blarina brecauda, Tamias striatus, Eutamias minimus, Peromyscus maniculatus,* and *Clethrionomys gapperi* in the aspen-birch, maple-aspen-birch, and the fir-birch community-type habitats); and (3) cosmopolitan parasites which are found on diverse host taxa and in various habitats [e.g., *Androlaelaps fahrenholzi* (Berlese) on rodents, insectivores, rabbits, carnivores, and others in many different habitats].

Many other organisms are also members of a community in a host individual in addition to parasitic arthropods; they include microorganisms, fungi, Protozoa, Cestoda, Nematoda, and other arthropods. Many parasitic arthropods are infected with virus, rickettsia, spirochate, and bacteria (Marshall 1981a). Some such as rickettsiae (e.g., *Rickettsia*) and spirochaetes (e.g., *Borrelia*) in Anoplura (e.g., *Pediculus humanus*) (Weyer 1960) and bacteria (e.g., *Yersinia pestis*) in fleas (Jenkins 1964) are parasitic, but others such as protozoans (e.g., *Leptomonas*) in fleas (Steinhaus 1963; Brooks 1964; Wallace 1966; Brooks 1974) are symbiotic. Various tapeworms use the ischnoceran Mallophaga and fleas as intermediate hosts, and some nematodes are found in species of Mallophaga, Hippoboscidae, and fleas (Marshall 1981a). Parasitic mites such as Laelapidae and Saproglyphidae are recorded on arixeniids and hemimerids (Nakata and Maa 1974), and others are phoretic (Fain and Beaucournu 1976; Marshall 1981a). There are few records of parasitism on parasitic arthropods by Hymenoptera, for example, Chalcidoidea on immatures of Hippoboscidae, Nycteribiidae, and fleas (Marshall 1981a). Predation upon parasitic arthropods is primarily by the host, although predations by arachnids and insects are documented, for example, *Cheyletiella parasitivorax* (Megnin) feeding on Listrophoridae and other ectoparasitic mites (Evans et al. 1961) and *Haemogamasus ambulans* (Thorell) on Anoplura and flea larvae (Furman 1959).

The diversity of parasitic arthropods on a host species is usually well defined and consistent throughout the range of its host distribution. Even for highly mobile fleas no significant differences exist among the faunules on rodents in different habitats (Murray 1957). For example, the ectoparasite diversity of *Rattus norvegicus* consists of seven species of fleas, two species of sucking lice, and seven species of mites (Cole and Koepke 1947). The community structure of ectoparasites on *R. norvegicus*, except for local habitat-specific temporary parasites and cosmopolitan species, are identical in three distant localities (Table 1.4). Approximately 97 species of arthropods have been recorded from *Peromyscus leucopus* (Whittaker 1968), including commensals and stragglers as well as facultative and obligate parasites. Although community structure and diversity differ by locality, 27 species of parasitic insects and acarines in Wallingford, Connecticut (Combs 1977), 11 species in Massachusetts (Parsons 1962), and 26 species from Indiana (Basolo and Funk 1974), the following species were consistently found on *P. leucopus*: two flea species (temporary parasites), *Orchopeas leucopus* (Baker) and *Epitedia wenmanni* (Rothschild); one sucking louse, *Hoplopleura hesperomydis* (Osborn) (permanent parasite); one bot fly (temporary parasite), *Cuterebra fontinella* Clark; two mites (permanent parasites), *Radfordia subuliger* Ewing and *Androlaelaps fahrenholzi* (Berlese) (cosmopolitan species). Similarly, free-ranging black bears (*Ursus americanus*) in Upper Michigan during 1968 and in northeastern Minnesota from 1969 to 1974 consistently harbored *Dermacenter variabilis* (Say), *D. albipictus* (Packard), *Trichodectus pinguis euarctidos* Hopkins, and numerous fleas (Rogers 1975).

Analysis of parasitic arthropod diversity reveals consistent associations or exclusions between particular positive taxa (association specificity) (see Chapter 4). For example, parasitic beetles and chewing lice never occur with other ectoparasites, as sucking lice are not associated with them (Wenzel and Tipton 1966; Kim and Adler, Chapter 4; and Kim, Chapter 7). Exclusion of parasitic arthropods from a host species may be due to competitive exclusion or perhaps phylogenetic and ecological isolation (Hopkins 1949, 1957; Wenzel and Tipton 1966).

Interactions and Coexistence

A mature parasitic arthropod community in a host individual has a definite organization that includes certain species richness, population density, and microhabitat selection. Such community organization results from successive invasions and colonizations in which the parasitic arthropods continually interact among themselves and with the host (Fig. 1.10, I).

A host individual provides general habitats as well as numerous microhabitats for specific infrapopulations. For example, *Antidorcas marsupialis* (South African springbock) provides different habitats for six Anoplura species, five *Linognathus*, and one *Solenopotes* (Weisser 1975). Habitat structure, resources (food and space), and microclimate between host individuals are relatively constant, although different hosts may vary in physiological characteristics, immune capacity, and behavior. Species richness and population density of parasitic arthropods are usually small in healthy hosts, and competition does not seem to occur there. High diversity is confined to a few hosts, especially those unhealthy or injured. However, when the level of populations is very high, as observed in *Orthohalarachne* mites in the respiratory organs of northern fur seals (Kim et al. 1980), parasitic arthropods are expected to interact closely. Such interactions, including infraspecific crowding and interspecific competition may result in competitive exclusion (Wenzel and Tipton 1966) or change of microhabitat (Nelson 1972; Lewis et al. 1967) of one or more species and further induce host reactions that are more harmful to the species evoking them (Rohde 1979). For example, the number of good sites for lice, such as *Polyplax serrata* (Burm.) and *Geomydoecus oregonus* Price and Emerson on small mammals, are restricted by host grooming, hair size and density, molting frequency, and regional body surface temperature (Murray 1961; Bell et al. 1962; Bell and Clifford 1964; Rust 1974; Nelson et al. 1975). Parasite species may avoid competition by resource partitioning or niche restriction in the habitat; the existence of microhabitat restriction does not necessarily indicate that competition occurs, however, because microhabitat restriction frequently occurs without the presence of competing species (Rohde 1979; Roughgarden 1976; Schoener 1971, 1974). Microhabitat or niche restriction is a necessity for survival of the parasite species to increase infraspecific contact and to improve the chance of mating because population densities of most parasitic arthropods are low (Tables 1.2, 1.3) and parasite populations are overdispersed (Rohde 1979). Furthermore, association specificity and competitive exclusion may be part of the chemical interplay that leads to (1) a close chemical fit between parasite and host, (2) evolution of chemical defenses by hosts against parasites, and (3) evolution of the host and parasite pair toward tolerance of one another's chemical characteristics, hence toward a stable relationship (Whittaker and Feeny 1971).

The relationship between parasitic arthropods and northern fur seal illustrates resource partition and microhabitat restriction already described. The sucking lice *Antarctophthirus callorhini* (Osborn) and *Proechinophthirus fluctus* (Ferris) are transmitted primarily from cows to pups during parturition and nursing, whereas the respiratory mites *Orthohalarachne attenuata* (Banks) and *O. diminuata* (Doetschmann) are slowly established in the respiratory organs, reaching infestation rates of 100% and more than 150 mites per host by the postmolting period (Kim et al. 1980).

The infestation rate of sucking lice is 100% for pups and about 80% for adult seals. Counts of mean populations of the sucking lice were highest on the pups, with 169 lice per host, decreasing with age to 48 lice per adult

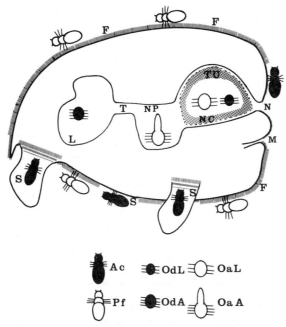

Figure 1.11 Distribution of parasitic arthropods within a northern fur seal adult. Ac, *Antarctophthirus callorhini*; Pf, *Proechinophthirus fluctus*; Od, *Orthohalarachne diminuata*; Oa, *Orthohalarachne attenuata*; L, larva; A, adult. M, mouth; N, nostril; F, fur; s, skin; Tu, nasal turbinates; NC, nasal cavity; NP, nasopharynx; T, tracheae; L, lungs.

seal (Kim 1972). Mean populations of nasal mites, however, were higher as seals got older, with 22 mites per pup and 1808 mites per subadult male (Kim et al. 1980) (Fig. 1.11). Unlike the sucking lice (Kim 1972, 1975), the respiratory mites show no evidence of competition. Although the *O. attenuata* adults inhabit the nasopharynx (Fig. 1.6) and *O. diminuata* adults are found primarily in the lungs, the larvae of both species occupied the mucus-filled turbinates (Fig. 1.5) (Kim et al. 1980).

Although no experimental data exist to support any specific factors responsible for resource partition or microhabitat restriction, there are definite correlations between microhabitat preference and morphological traits. *A. callorhini* has dense scales on the abdomen, a modified spiracular atrium, and pointed short claws on the mid- and hindlegs, which are useful adaptations to naked or thinly furred skin, whereas *P. fluctus* is suited for rapid movement in the fur environment with long, thin abdominal setae without scales, and blunt claws on the mid- and hindlegs for grasping hairs (Kim 1975). Additionally, the larvae of the two species of *Orthohalarachne* occurring together in the turbinates have a distinct size difference. The *attenuata* larvae are about twice as large as the *diminuata* although a considerable variation in idiosomal measurements exists within each species (Kim et al. 1980): the *attenuata* idiosoma length 1.13 mm and

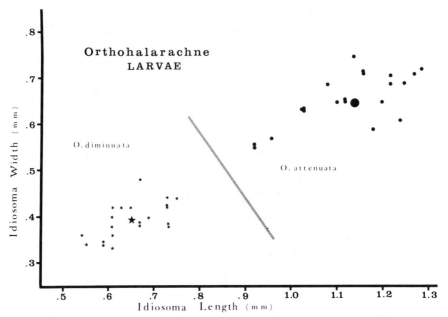

Figure 1.12   Scatter diagram of the larvae of *Orthohalarachne diminuata* (stars) and *O. attenuata* (solid circles). Large star and circles represent the mean.

width 0.65 mm; the *diminuata* idiosoma length 0.67 mm and width 0.38 mm (Fig. 1.12).

Development of Parasite Communities

The process of community development (microsuccession) in parasitic arthropods depends on organismal diversity and environmental conditions of the host territory as well as on historical determinants involving associating organisms and the host. The origin and development of interspecific interactions including coadaptation, cospeciation, and coaccommodation over evolutionary time (Fig. 1.10, II) underlie the ontogeny of a contemporary parasite community (Fig. 1.10, I) (Brooks 1979; Thompson 1982). A parasite community on a host organism is structured by the existing biotic makeup surrounding the host in a given habitat. Specifically, during the life span of a host organism, successive colonizations of parasitic species take place with intermittent extinction (Fig. 1.10, I). For a given host, the rate of colonization and accumulation of parasite species eventually falls to zero when a characteristic community structure or "climax" community is attained (MacArthur and Wilson 1967; Whittaker 1977). In other words, the parasite community of a mature host usually is in a stage of equilibration (but not necessarily saturation).

INTERACTIONS BETWEEN PARASITES AND THEIR HOSTS

In a parasite–host system the host is not only the provider of food and habitat but also a major component regulating resident parasites (Fig. 1.1). The parasites usually induce specific reactions and changes in host behavior, physiology, and immunity as they reside and forage in the habitat (Trail 1980). The host uses these changes to counter the parasite attack. These interactions are triggered by mechanical and chemical stimuli, perhaps involving allelochemicals as Whittaker and Feeny (1971) suggested, and mediated by specific adaptations in both associating organisms. The parasites use behavioral, physiological, and morphological adaptations to survive and reproduce in the habitat, whereas the host employs these and immunological adaptation to counter and regulate the parasite load.

Adaptations and Parasitic Arthropods

Parasitic arthropods of mammals must remain in or on the host to survive and sustain specific parasite–host relationships. To accomplish this they possess morphological, behavioral, and physiological adaptations developed as evolutionary responses to host-related parameters. Some are general adaptations for parasitism, whereas others are specific responses to traits and reactions of the particular host. Adaptations in taxa such as *Tunga* (chigoe fleas) and *Demodex* are site specific, and they are usually correlated with the closeness of the parasite–host association (Marshall 1981a).

There are a number of adaptations common to most parasitic arthropods, despite their remote phylogenetic origin, including size and shape of body and head, eyes, setal type, and other traits (Marshall 1980, 1981a; Kim and Ludwig 1982; Kim and Adler, Chapter 4). Most parasitic insects and acarines are dorsoventrally flattened, with head and other structures modified accordingly, such as the displacement of antennae, mouthparts, and thoracic spiracles. However, fleas and some nycteribiid and streblid flies are laterally compressed to allow them to move about readily in dense fur. Setal adaptations particularly related to combs or ctenidia among parasitic insects have been dealt with extensively by Marshall (1980, 1981a), Traub (1980 and Chapter 8 of this book), Amin and his co-workers (1974, 1977, 1983), and Kim and Adler in Chapter 4, and thus are not discussed here.

The major morphological adaptations of parasitic acarines are the attachment and feeding organs (Fain 1969; Dubinina 1969; Radovsky 1969, 1982; Bennington and Kemp 1980). Development of the attachment organs greatly varies. In the Ixodidae and some Mesostigmata the long, toothed hypostome or the long, barbed corniculi is deeply sunk into the skin, and attachment is aided by the chericeral digits with recurbed teeth and tarsal suckers. Also involved in most ixodid ticks is the cement produced by salivary glands (Tatchell 1969a, b; Bennington and Kemp 1980). In the acariform and some specialized mesostigmatid mites, such as Trombiculidae larvae and Spelaeorhynchidae, the highly modified chelicerae and tarsal suckers are used for anchoring. In the fur mites, Myobiidae (Trombidiformes) and Listrophoridae (Sarcoptiformes), the attachment organs are the gnathosoma and legs, and in Labidophoridae hypopi they are the suckers or claspers on the ventral side of the opisthosoma (Fain 1969; Dubinina 1969; Fain and Hyland, Chapter 12).

In many parasitic Mesostigmata the chelicerae are the major component of feeding organs with modified terminal chelae and slender shafts used to pierce the skin (Radovsky 1969, 1982, and Chapter 9; Theron 1974). The first pair of legs with modified claws are also involved in feeding in certain species of rhinonyssid nasal mites and Spinturnicidae (Radovsky 1982).

Reproduction and life history are also adapted in many parasitic arthropods. Ovovivipary and vivipary, which allow females to invest more energy in fewer offspring (Marshall 1981a), are common in the Polyctenidae, parasitic Dermaptera, and pupiparous Diptera, and in endoparasitic mites and some ectoparasitic mites. In the endoparasitic mites Halarachnidae, Rhinonyssidae, and Entonyssidae, ovoviviparity or viviparity are the rule, and some or all the nymphal stages may disappear (Fig. 1.8). The ectoparasitic Mesostigmata and Sarcoptiformes are primarily oviparous, but specialized taxa such as the Spelaeorhynchidae, Omentolaelapidae, Spinturnicidae, and Teinocoptidae are ovoviviparous or viviparous (Fain 1969). Reproductive cycles of the rabbit fleas, *Spilopsyllus cuniculi* (Dale) and *Cediopsylla simplex* (Baker), are closely aligned with the reproductive cycle of their hosts in which host hormones, such as corticosteroids and estrogens, directly influence the maturation of eggs and oviposition (Rothschild and Ford 1964, 1966, 1972). Annual life cycles of seal lice (Echinophthiriidae) are closely synchronized with the annual migration and reproductive patterns of their hosts (Kim 1972, 1975; Kim and Keyes 1984; Murray and Nicholls 1965; Murray et al. 1965). The postembryonic development of echinophthiriids is delayed during the pelagic migration of the northern fur seal, but when seals reach the breeding ground on St. Paul Island, Alaska, the sucking lice mature to adults or late third-stage instars ready to build up the population (Kim 1975).

Host Reactions and Counteradaptation

The central aspect of parasitic associations is nutrient transfer and energy exchange (Wakelin 1976). These interactions have both trophic and informational significance. As they attach and feed on the host, parasitic arthropods not only take nutrients such as blood and epidermal products from the host but also introduce their own salivary secretions into the host.

The introduction of these substances sends a signal of the parasite's presence to the host and triggers defensive responses. Such host responses in turn provide new informational stimuli for the parasite to make proper countermeasures. This sequence of responses and counterresponses continues throughout the history of their associations, and such processes result in the induction of adaptations in behavior, physiology, development, and even morphology for both parasite and host (Whitfield 1979).

The effects of parasitic arthropods on the host and host responses to parasitic invasion have been reviewed by Feingold et al. (1968), Benjamini and Feingold (1970), Nelson et al. (1977) and Wikel (1982) specifically on arthropods, and Wakelin (1976, 1979) on broader aspects including genetic considerations.

Effects on the Host

Parasitic arthropods may cause annoyance, irritation, tissue damage, blood loss, or even death to their hosts, and their activities may be involved in the transmission of pathogens. Thus in theory these effects may decrease host fitness and regulate host population growth by reducing host reproductive potential (Anderson 1978; Anderson and May 1978a, b; May 1983a), and they may be related to host nutrition (Good 1981). However, the effects of parasitic arthropods and host responses in mammals are not well established, and those documented are limited only to several species (Nelson et al. 1977; Wikel 1982).

Exsanguination and the injection of toxins and other foreign substances into the blood circulation induce immediate and delayed responses in behavior and immunity from the host. These responses help to detach or kill the parasites, limit the infestation, and control the parasite load (Wakelin 1976). Heavy infestations cause irritation which may result in loss of sleep and activity and haematologic changes which may result in anemia (Nelson et al. 1977). These effects may lower production of milk and meat, causing economic loss. The skin and fur of the host may also be damaged by bleeding and wounds caused by parasitic arthropods, and such damage may lead to dermatitis and predisposition to secondary infection (Nelson et al. 1977; Kim 1977; Marshall 1981a). The larvae of warble flies like *Hypoderma* cause irritation as they migrate through the body of the host, and subsequent emergence through the skin leaves an open running wound that may persist for a long time. Some endoparasites such as nasal mites and bot fly maggots may block the nasal passages and retard the growth and deteriorate the general body condition of the host (Harwood and James 1979). Heavy infestation with *Orthohalarachne* mites may result in impairment of respiration and cause expiratory dyspnea owing to the blockage of the nasal passages in fur seals (Kim et al. 1980).

Behavioral Responses

The activity of parasitic arthropods induces the behavioral responses from the host in the form of grooming and scratching (Nelson et al. 1977). Scratching is a response to irritation caused by injection of toxins and antigens, whereas grooming is an innate behavior to clean the hair and skin of dirt, grease, and other foreign materials, including ectoparasites as well as epidermal by-products.

In mammals self-grooming is done by use of the mouth, hands, or feet and varies from a casual shaking of the body to the precise removal of individual arthropods (Snowball 1956; Bell et al. 1962; Bennett 1969; Stebbings 1974; Nelson et al. 1975; Marshall 1981a). In many captive animals, mutual grooming may account for the lower density of parasitic arthropods (Marshall 1981a).

Immune Reactions

Immunity is defined here as the sum of all physiological potential of a host animal that prevents it from parasitic infestation. Thus *resistance* and *susceptibility* depend on the extent of immunity, which varies greatly between individuals (Smyth 1976; Willadsen 1980; Cohen and Warren 1982). Immunity may be an innate mechanism against parasites or may be acquired through an initial parasite infestation ("acquired immunity"). Innate immunity is determined by geographical or ecological isolation, foraging pattern, and anatomical and physiological differences (Sprent 1969). Nutritional immunity may be added to these mechanisms as in, for example, a vertebrate host defense mechanism that withholds growth-essential iron or zinc from the invading parasite while retaining host access to these metals (Weinberg 1979; Good 1981). Immune responses involve complex interactions of immunocompetent cells (e.g., t-lymphocytes), accessory cells (e.g., macrophages), and many other mediators (e.g., complement and basoactive amines). They are distributed throughout the body but directly involve the thymus, lymph nodes, spleen, and bone marrow. All of these cells, particularly the lymphocytes, can be activated by contact with foreign ("nonself") material (Smyth 1976).

The complex interactions of many elements constitute host immune reactions: (1) *phagocytosis*, (2) *inflammation*, and (3) *adaptive immunity* (Wakelin 1976; Smyth 1976; Nelson et al. 1976, 1977). The first host reaction to the foreign material is phagocytosis. This involves macrophages (including monocytes in the bloodstream, histiocytes in the connective tissue, and reticulum cells in the spleen) and microphages (polymorphonuclear leucocytes) which engulf and digest foreign particulate matter. This mechanism can be facilitated by the action of opsonizing antibodies, especially IgG and IgM, and by serum complement (Smyth 1976). These allow the presentation of antigens to immunocompetent cell.

Inflammation is the primary mechanism through which host responses and antiparasitic processes are initiated at the site of parasite invasion. Through this process, which is closely linked to tissue repair and fibroblast activities, the parasites are sequestered and killed, infestation is limited, and the parasite load is reduced (McCall 1971; Wakelin 1976). Inflammation usually appears as swelling due to vasodilation of the area of parasite invasion, followed by abnormal growth of the affected area (Smyth 1976).

Inflammation and phagocytosis are nonspecific responses to tissue damage or the introduction of a foreign substance, whereas adaptive immune reactions are specific responses to parasitic invasion and the most important mechanism to limit the parasite infestation (Wakelin 1976).

Invasion of the host by a parasite or other foreign material provokes serological or humoral reactions involving antigens and antibodies. An antigen-antibody reaction not only provides immunity to the host but also may produce hypersensitivity (or allergy). *Hypersensitivity*, defined as "a heightened response to antigen in an animal which has already been sensitized to that antigen" (Smyth 1976), may be antibody mediated (immediate) or cell mediated (delayed) (Lepow and Ward 1972). This is a complex subject; for details consult texts such as Roitt (1980) and Sell (1980).

Parasitic arthropods elicit a wide range of immune reactions in mammals (Parish 1970a, b; Wakelin 1979; Willadsen 1980; Wikel 1982). Parasite antigens may differ with species (Gaafar 1972) since the specificity of antigen depends upon the configuration of the amino acids or saccharides exposed to its host (Parish 1970a). The same antigen may also induce varying immune responses in different individuals and strains of the same species or at different interfaces of the same host (Wikel 1982). Thus immune reactions at the epidermis–parasite interface (Parish 1970a, b) may be different from those at the mucosa–parasite interface in internal organs such as the respiratory passages and digestive tracts (Befus 1982; Castro 1982). Conversely, phylogenetically unrelated parasites all mimic the same host antigenic determinant-convergent evolution at molecular level (Damian 1964). Furthermore, immune reactions depend largely on the nature of the immunogen, the route of introduction, the site of interactions, the immune response capabilities of the host, and the history of prior exposure to the antigen (Wakelin 1979; Willadsen 1980; Wikel 1982).

Host immune responses to bloodsucking arthropods take two primary forms, desensitization and resistance (Marshall 1981a). Desensitization is derived from the allergic response to repeated bites through a five-stage pattern: (1) no skin reaction, induction of hypersensitivity; (2) delayed reaction; (3) immediate then delayed reaction; (4) immediate reaction only; and (5) no reaction (Feingold et al. 1968). This response is often associated with temporary parasites (Marshall 1981a). Acquired resistance (protective immunity) related to parasitic arthropods often appears as hypersensitivity (Wakelin 1979) and is usually associated with permanent parasites, such as the sheep ked *Melophagus ovinus* (Linn.) and the short-nosed cattle louse *Haematopinus eurysternus* Denny (Nelson 1962; Bell et al. 1966; Nelson et al. 1970, 1977; Marshall 1981a). Host immune reactions to parasitic arthropods are both antibody- and cell-mediated responses in which biologically active mediators include complement components, vasoactive amines, and prostaglandins (Willadsen 1980; Wikel 1982).

Although our knowledge of mammalian immune system–parasitic arthropod interactions still is scanty and superficial, the following papers provide the most up-to-date information: Parish (1970a, b) for comprehensive description of host skin reactions; Tatchell (1969b) and Nelson et al. (1975, 1977) for interactions between ectoparasites and hosts; Wakelin (1979) for genetic control of immune responses including those of arthropods; Wakelin (1976) and Wikel (1982) for the most recent review of immune responses to arthropods. The histopathology of feeding lesions of arthropods is well documented: for example, Larrivee et al. (1964) and Lavoipierre et al. (1979) on fleas; Nelson and Bainborough (1963) on the hippoboscids; Nelson et al. (1970) and Lavoipierre (1967) on sucking lice; Lavoipierre and Beck (1967), Lavoipierre et al. (1967), and Lavoipierre and Rajamanickam (1968) on mites; Hoeppli and Schumacher (1962) on chiggers; Nutting (1975) on demodicosis; Lavoipierre and Riek (1955), Tatchell and Moorhouse (1968), Tatchell (1969a, b), and Arthur (1970, 1973) on ticks.

Counteradaptation

Behavioral and immune responses are used by mammalian host to prevent the establishment of parasitic arthropods. In the face of these adverse selection pressures, parasitic arthropods must have developed counterstrategies against some or all of these host responses. Stable ecological balance, with small parasite load and sustainable host responses, maintained in most relationships between parasitic arthropods and mammals supports the reality of counteradaptations in parasitic arthropods. Such ecological stability is the result of long evolutionary interactions between parasitic arthropods and mammals manifested by selection and adaptation. Thus the present state of unresponsive submission of the host to parasites must have been reached only after a period of active interactions, which Sprent (1962) called "adaptation tolerance."

Adaptation tolerance is defined as "a dual modification of host and parasite, operating through natural selection, comprising (1) a stabilization and modification of parasite antigens toward conformity with related substances in the host, and (2) a selective obliteration of corresponding antigen-combining sites in the host, which thereby becomes the natural host and is rendered tolerant to the parasite" (Sprent 1962, 1969).

The parasites must develop various immune evasion strategies to limit and overcome the multiplicity of immunological effector mechanisms used by the parasitized mammalian host. Three broad mechanisms of immune evasion are identified: (1) reduced or altered parasite antigenicity/immunogenicity, (2) modification of induction or expression of host immune responses, and (3) modification of intramacrophage environment (Mitchell 1982). These are well documented in helminths and protozoans (Wakelin 1979; Damian 1979), but only a few are so far known in parasitic arthropods. In *Rhipicephalus sanguineus sanguineus* (Latreille) a histamine-blocking agent detected in the salivary gland regulates the excessive accumulation of histamine that causes adverse host allergic reaction (Chinery 1981). Wikel (1985) reports that tick feeding induces immune suppression in the host whose immune responses return a few days after ticks are removed. Of these mechanisms the concept of molecular mimicry or host–parasite antigen sharing has greatly influenced immunoparasitology in recent years (Damian 1979). Three hypotheses have been proposed to explain how parasites synthesize hostlike antigens (Smyth 1973): (1) natural selection (Damian 1962, 1964), (2) antigen induction (Capron et al. 1968), and (3) antigen masking (Smithers et al. 1969). Damian's hypothesis suggests that natural selection acts on fortuitous cross-reactivity in potential symbiotic partners to fix the capacity to synthesize hostlike antigens already fixed in the parasite genome and that his cross-reactivity is a preadaptation for parasitism (Damian 1964). The antigen induction also involves genetic fixation, but induced synthesis of antigens is adapted to the particular host species with which the parasite is associated. The antigen-masking hypothesis differs from the other two in that the shared antigens are actually host antigens, acquired in the intimate contact between host and parasite (Damian 1979).

Host–parasite antigen sharing may result in host susceptibility to parasite infestation (Jenkin 1963), development of host antigenic polymorphisms (Damian 1962, 1964), and infraspecific variation in host response to parasite invasion (Wakelin 1979). Wakelin (1979) considered low responsiveness or unresponsiveness to infection among a host population an important factor in bringing about parasite overdispersion.

EVOLUTION OF PARASITIC ARTHROPOD/MAMMAL ASSOCIATIONS

Mammal associations of parasitic arthropods have arisen independently in diverse orders of the Insecta and Acari, and they represent the evolutionary manifestation of interspecific interactions between symbiotes and symbionts. The intimacy of the host associations is shown by the degree of host specificity. All parasitic arthropods have acquired behavioral, physiological, and morphological adaptations geared to promote their parasitic relationships with the host, principally those involved in physical associations and feeding, and they show considerable variability in their life histories and host specificity (Waage 1979). In many groups of arthropods a single taxon contains both parasitic and free-living species with numerous transitional forms exhibiting evolutionary gradients of adaptations and host associations: for example, Cimicoidea (Cimicidae and Polyctenidae); cyclorrhaphous Diptera which include many free-living flies and myiasis producers.

The development and organization of a parasitic arthropod/mammal ecosystem are controlled by contemporary factors such as environmental conditions, biological properties of the host, and species diversity. The diversity of parasitic arthropods, defined here as the total assemblage of all parasite species associated with a host species throughout its range, is the result of evolutionary processes involving interspecific interactions, selection, and speciation (Whittaker 1977; Fowler and MacMahon 1982). Thus critical analyses of contemporary parasite–host associations, host specificity, species diversity, and adaptations and phylogenetic relationships of associating species provide useful clues to the evolution of parasitic arthropod and mammal associations (Ghiselin 1972) as follows:

1. Cladistic analysis of key morphological characters relative to the adaptive zone (Van Valen 1971) of the parasitic taxon; for example, Nutting (1965), Fain (1968a, 1969), Kim and Ludwig (1978a, b), Traub (1980).

2. Comparative study of morphological, ecological, and life history traits of both parasitic and free-living species with various transitional forms involved in their adaptations to different habitats within a higher taxon at familial or ordinal rank (or even generic); for example, Usinger (1934, 1966), Zumpt (1965), Kim and Ludwig (1978a, b, 1982), Schaefer (1979), Waage (1979).

3. Analysis of host specificity and geographical distribution; for example, Jellison (1942), Vanzolini and Guimaraes (1955), Traub (1980 and Chapter 8), Kim (Chapters 5 and 7).

Evolutionary Succession

The great radiation of the Cenozoic mammals during the Paleocene and Eocene epochs must have provided ample new habitats for opportunistic species of various arthropod groups. It is reasonable to assume that early mammals were continually invaded and colonized by many different arthropods, some of which were perhaps parasitic on premammalian hosts and others scavengers, saprophages, or even chance feeders. Those that successfully colonized the mammalian host were likely to have been opportunistic species.

Populations of opportunistic species are usually physiological generalists, with a high intrinsic rate of increase, short generation time (MacArthur 1960, 1972), and preadaptations suited for parasitic life (Pickett 1976). They may rapidly increase in numbers when environmental factors such as space, temperature, and food availability become favorable. However, they are very unstable and may also become extinct very quickly (Levinton 1970). They are not resource limited, and the population size is below the carrying capacity of the habitat. Morphological and physiological preadaptations permit these colonizers to shift their realized niches to avoid competition and to make use of new adaptive opportunities (Pickett 1976).

In the early stage of evolutionary succession, many founder populations of opportunistic species successfully colonized "vacant" habitats in mammals (Fig. 1.10, II). They were subjected to differential selection pressure resulting in selective extinction, each species having distinct genetic characteristics and a different probability of becoming extinct regardless of its evolutionary age (Foin et al. 1975).

As "vacant niches" were occupied, parasite communities became more complex and stable, perhaps reaching a climax (Valentine 1972) in which extinction exceeded immigration and their vulnerability to invasion by other immigrant species decreased (MacArthur and Wilson 1963, 1967; Robinson and Valentine 1979). By this time, most parasite–host associations had become obligatory by coadaptive evolutionary processes through which the ecotopes (niche-habitat hyperspace) of component species are narrowed by selection (Whittaker et al. 1973, Whittaker 1977; Price 1980) and the host specificity is defined by genetic interactions (Damian 1979; Wakelin 1979). While others become extinct, some taxa in the obligate parasite–host associations persist for long periods of evolutionary time. They may become more specialized by selection for efficient physical association with and feeding on the host. Such associations result in high host specificity for parasitic arthropods.

The structure and functioning of a parasite community in a mammalian host may change through time because of the immigration of dominant species or the extinction of some species (Valentine 1972). Qualitative changes may also occur due to both sympatric speciation which involves crowding and competition and allopatric speciation which may involve dispersal and movement or evoked by the speciation of mammalian hosts (Bush 1975a, b).

Host mortality may eliminate parasite communities from the parasite–host system. Populations of opportunistic species would leave the dead host and move to other conspecific hosts. However, populations of highly specialized species remaining with the dead host would face extinction.

Origin of Arthropod/Mammal Associations

All parasitic arthropods have been through a series of transitional stages from free-living to parasitic states. Through such evolutionary processes behavioral, physiological, and morphological adaptations have been developed to promote physical association with the host (type "A" adaptations) and capacity to feed on it (type "F" adaptations) (Waage 1979). The type A adaptations include host-finding behavior, habitat selection, attachment, grasping and dispersal, whereas the type F adaptations involve the mouthparts, salivary secretion, feeding behavior and physiology, digestion, and counterreaction.

The diversity of arthropod/mammal associations suggests that the progenitors of mammal associates had diverse lifestyles; some were predators, parasites, nest and roost associates or dung dwellers, and others free living. Two hypothetical pathways have been proposed for insects associated with vertebrates (Waage 1979):

1.   Pathway I: Type A adaptations precede type F adaptations. Physical association involves a dependence on the host dwelling such as roost, nest, or runway, or on other products such as carrion or excrement, which provides parasitic insects with easier access to the host and nutritious food sources such as host tissues and secretions. Subsequent adaptations of feeding behavior and mouthparts permit ingestion of more nutritious food such as blood. Most obligate parasites that evolved through pathway I show the following characteristics: (1) Their free-living relatives are associated with nests, roosts, dungs, and the like. (2) Their association with the host or its immediate environment is prolonged, perhaps through several life stages. (3) Secondary acquisition of behavior and mouthparts leading to haematophagy is demonstrated.

2.   Pathway II: Type F adaptations precede type A adaptations. Predators and plant feeders with preadapted mouthparts and behavior feed on nutritious host tissues like blood. Such adaptations lead to selection for regular host association. Bloodsucking Diptera such as mosquitoes and black flies, parasitic Hemiptera, and eye-frequenting moths have evolved through this pathway.

Some early arthropods were scavengers feeding on organic debris such as wastes and remains of the host's meal within lairs, burrows, nests, or roosts. The progenitors of the dermapteran Arixeniidae, Anoplura, Mallophaga, parasitic Coleoptera, commensal flies Mystacinobiidae and Mormotomyidae, gamasid Mesostigmata, and Argasidae were such scavengers (James and Harwood 1969; James 1969; Kim and Ludwig 1978b, 1982; Waage 1979). Such association easily leads to a closer physical association with the host. Once their physical association is established, these mammalian associates evolve the type F adaptations which lead to the keratinophagy observed in Mallophaga or mucophagy in the mesostigmatid Halarachnidae. Keratinophages and mucophages may further evolve to the haematophagy as in Anoplura.

Certain arthropods such as cyclorrhaphous Diptera and commensalistic

Pyralidae (*Cryptoses* and *Bradypophila*) were saprophages feeding on carrion and excrement. Such associations then lead to sarcophagy and perhaps ultimately to haematophagy, which may further evolve to a derma-subdermal parasite (Zumpt 1965; James 1969). Terrestrial predators like Cimicoidea and Reduvioidea usually attack other arthropods. In some cases, however, such predatory habits may encounter mammalian tissues. Such encounters further evolve to blood feeding on mammals (James and Harwood 1969; Waage 1979). Nectar feeders like nematoceran Diptera and liquid-sucking insects like eye-frequenting moths (Pyralidae, Noctuidae, and Geometridae) initially may encounter proteinaceous secretions through chance feeding. Such chance encounters, may evolve into obligate feeding on animal tissue fluids or secretions such as lachrymal secretions (Bänziger and Büttiker 1969; Bänziger 1972, 1975).

A few parasitic arthropods such as pupiparous Diptera (Hippoboscidae, Nycteribiidae, and Streblidae) and Siphonaptera are difficult to place in any one of the four proposed models. The progenitors of the Hippobocidae and Siphonaptera which are presently associated with the body and nest of their hosts could have been scavengers, predators, or even parasites. Similarly, the progenitors of the Streblidae and Nycteribiidae, which are parasites of bats and closely associated with their roosts, might have been either scavengers or saprophages.

Evolution of the Parasitic Associations with Mammals

Archaic mammals, such as symmetrodonts, triconodonts, and multituberculates, were widespread in Africa, Asia, and possibly in North America by the late Triassic and early Jurassic periods as the mammallike reptile Synaspida became extinct (Clemens et al. 1979; Kemp 1982). Some of these vertebrates were already invaded by arthropods such as the Ixodoidea. In the late Paleozoic or early Mesozoic era primitive argasids were parasitic on the Reptilia (Hoogstraal 1978). Although their parasitic status is not clear, there were prostigmate Trombidioidea in the Devonian period (Petrunkevitch 1955).

Although both primitive mammals and many of the modern insect orders were present in the late Mesozoic era, no evidence supports any obligate associations between insects and mammals until the early Tertiary period. Parasitic Psocodea date back to the late Jurassic or early Cretaceous periods (Kim and Ludwig 1982b), and stephanocircid and doratopsylline fleas and certain pygiopsyllids were already present in the Cretaceous period (Traub, Chapter 8). The Cretaceous fauna of arthropods included most of the modern orders such as Dermaptera, Psocoptera (Psocodea), Hemiptera, Coleoptera, Lepidoptera, and Diptera (Hennig 1969; Carpenter 1977). Similarly, late in the Cretaceous period primitive mammals that were small and insectivorous began to rapidly radiate and continued through the middle Tertiary. By the late Cretaceous period, marsupials were predominant in North America, and many placentals were present in Asia (Schmidt-Nielsen et al. 1980; Kemp 1982).

The Cenozoic mammals radiated greatly during the Paleocene and Eocene epochs, reaching the highest ordinal and familial diversity in the late Eocene and the early Oligocene epochs when most archaic mammals disappeared. The total number of orders and families of mammals per continent has been remarkably uniform from the early Oligocene to the Recent epoch, although the total diversity of major continents suffered a decline in the last half of the Oligocene and a rebound in the early Miocene (Lillegraven 1972). Parasitic arthropods also rapidly diversified during the early Tertiary period. Rapid radiation took place for the Anoplura during the Paleocene and Eocene epochs (Kim and Ludwig 1982a) and for the Siphonaptera such as Pulicidae and Ceratophyllidae during the Eocene and Oligocene epochs (Traub, Chapter 8). The rhipicephaline ticks evolved in the early Tertiary period (Hoogstraal and Kim, Chapter 10). These correlations and other data suggest that most of the associations between parasitic arthropods and mammals began during the late Eocene and early Oligocene epochs.

SUMMARY

Associations of arthropods with mammals take many forms, ranging from casual contact to permanent residence for food and shelter. Included are phoresy, mutualism, commensalism, inquilinism, and parasitism. In this chapter parasitic arthropods refer to obligate parasites that live in or on and gain metabolic needs from the host for a certain period of life. They are members of the insect orders Hemiptera, Coleoptera, Diptera, Anoplura, Mallophaga, and Siphonaptera, and the acarine orders Parasitiformes (Mesostigmata and Metastigmata) and Acariformes (Prostigmata and Astigmata). Some, such as Anoplura, Mallophaga, and many Mesostigmata, are permanent parasites, whereas others, like most fleas and ticks, which are associated with the host for a short period of time, are temporary parasites.

Parasitic insects and acarines are specific to particular mammalian taxa. Host specificity is distinct for each taxon at various taxonomic levels. Some species are monoxenous, whereas others are specific to host taxa at generic, familial, or even ordinal level.

The integument (the skin and hairs) and the digestive and respiratory tracts of mammals provide the primary habitats for parasitic arthropods. The basic structure, variation, and environment of these habitats are described and discussed with regard to their arthropod inhabitants.

Parasitic arthropods possess morphological, behavioral, and physiological adaptations for reproduction, life cycle, host finding, microhabitat selection, and feeding, while retaining the basic pattern of structure/function and life history. The basic models and modifications of reproduction and life cycles are described for major parasitic insects and acarines. Also discussed are blood-feeding apparatuses and mechanisms.

Infestation and population structure of parasitic arthropods are reviewed, and population patterns are described by infestation rate, population rate, and dispersion rate, each of which is defined. Overdispersion is a common pattern of distribution for parasitic arthropods, and is important for maintaining a stable parasite–host system. Although the population structure theoretically forms a "pyramid of numbers of individuals" for a given generation, the pyramid often does not occur in the actual population structures of parasitic arthropods because of overlapping generations and temporal changes in demographic and environmental forces. The sex ratio is usually unbalanced, with females predominating.

A systems model is proposed to describe the structure and dynamics of parasitic arthropod populations on a mammalian host. Three mechanisms proposed by Bradley (1972, 1974) for population regulation in the parasitic arthropod/mammal system are discussed. Biological parameters including life history strategies and dispersion are considered in the context of population regulation. Parasitic arthropod populations are primarily host limited and differ by host individual, subspecies, age group, and sex.

The structure and functioning of parasite communities are conceptualized in terms of interactions between parasitic arthropods in or on a host individual which provides mobile habitats. Also discussed are the composition of associating organisms and species diversity. Niche and microhabitat restrictions are documented for several parasitic arthropods. A model is proposed for the development of parasite communities.

Interactions between parasites and their mammalian hosts are discussed in terms of parasitic adaptations, host reactions, and counteradaptations. The discussion includes physical and physiological effects of parasite infestation on mammalian hosts and behavioral and immune reactions of the host against parasitic infestation.

Most parasitic arthropods have begun their associations with mammals through a series of transitional states from a free-living to obligate, permanent parasitic state, likely during the late Eocene and early Oligocene epochs. Through the evolutionary processes, they have acquired two general types of adaptations: (1) those involving physical association with the host (type A adaptations) and (2) others promoting the capacity to feed on the host, including behavioral, physiological, and morphological adaptations (type F adaptations). The progenitors of parasitic arthropods were scavengers, saprophages, predators, or chance feeders, and their life history strategies evolved into keratinophagy, mucophagy, saprophagy, and ultimately haematophagy.

ACKNOWLEDGMENT

Much that is presented in this chapter is a culmination of discussions with former students and colleagues for the last 15 years through my seminars and lectures. Here, I dedicate this paper to all of those colleagues who have contributed to my thoughts for this chapter. I am indebted to Peter H. Adler, Allen L. Norrbom, and Steve K. Wikel for their efforts in reading the manuscript. Their editorial suggestions were invaluable and helped to improve the final paper. My special appreciation goes to Judi Hicks for her excellent typing efforts.

REFERENCES

Al-Abbasi, S. H. 1981. *Comparative morphology and phylogenetic significance of antennal organs in parasitic Psocodea (Insecta).* Ph.D. Thesis, The Pennsylvania State University, University Park, PA, 135 pp.

Amin, O. W. and R. G. Sewell. 1977. Comb variation in the squirrel and chipmunk fleas, *Orchopeas h. howardii* (Baker) and *Megabothris acerbus* (Jordan) (Siphonaptera), with notes on the significance of pronotal comb patterns. *Am. Midl. Nat.* **98**:207–212.

Amin, O. W. and M. E. Wagner. 1983. Further notes on the function of pronotal combs in fleas. *Ann. Entomol. Soc. Am.* **76**:232–234.

Amin, O. W., T. R. Wells, and H. L. Gately. 1974. Comb variation in the cat flea, *Ctenocephalides f. felis* (Bouche). *Ann. Entomol. Soc. Am.* **67**:831–834.

Anderson, R. M. 1978. The regulation of host population growth by parasitic species. *Parasitology* **76**:119–157.

Anderson, R. M. and D. M. Gordon. 1982. Processes influencing the distribution of parasite numbers within host populations with special emphasis on parasite-induced mortalities. *Parasitology* **85**:373–398.

Anderson, R. M. and R. M. May. 1978a. Regulation and stability of host–parasite population interactions. I. Regulatory processes. *J. Anim. Ecol.* **47**:219–247.

Anderson, R. M. and R. M. May. 1978b. Regulation and stability of host–parasite population interactions. II. Destabilizing processes. *J. Anim. Ecol.* **47**:249–267.

Anderson, R. M. and R. M. May. 1979. Population biology of infectious diseases: I. *Nature* (London) **280**:361–367.

Anderson, R. M. and R. M. May. 1982. Coevolution of hosts and parasites. *Parasitology* **85**:411–426.

Andrews, J. R. H. 1972. Description of the hitherto unknown males of *Damalinia longicornis* (Nitzsch, 1918) and *Damalinia hemitragi* (Cummings, 1916), Trichodectidae; Mallophaga. *J. Nat. Hist.* **6**:153–157.

Arthur, D. R. 1946. The feeding mechanisms of *Ixodes ricinus* L. *Parasitology* **37**:154–162.

Arthur, D. R. 1951. The capitulum and feeding mechanisms in *Ixodes hexagonus* Leach. *Parasitology* **41**:82–90.

Arthur, D. R. 1970. Tick feeding and its implications. *Adv. Parasitol.* **8**:275–292.

Arthur, D. R. 1973. The histopathology of skin following bites by *Hyalomma rufipes* (Koch, 1884), and a theory on feeding by this tick. *J. Entomol. Soc. S. Afr.* **36**:117–124.

Arthur, D. R. 1976. Interactions between arthropod ectoparasites and warm blooded hosts. Pages 163–183 in C. R. Kennedy (Ed.), *Ecological Aspects of Parasitology.* North-Holland, Amsterdam.

Ashford, R. N. 1970. Observations on the biology of *Hemimerus talpoides* (Insecta: Dermaptera). *J. Zool.* (London) **162**:413–418.

Askew, R. R. 1971. *Parasitic Insects.* Heinemann, London.

Baker, E. W., T. M. Evans, D. J. Gould, W. B. Hull, and H. L. Keegan. 1956. *A Manual of Parasitic Mites of Medical and Economic Importance*. Natl. Pest Control Assoc. Tech. Publ., 170 pp.

Balashov, Yu S. 1982. *Parasite-Host Relationships of Arthropods with Terrestrial Vertebrates. Proc. Zool. Inst. Acad. Sci.*, Leningrad, USSR, 27:1–296 (in Russian).

Balashov, Yu. S. 1984. Interaction between blood-sucking arthropods and their hosts, and its influence on vector potential. *Annu. Rev. Entomol.* 29:137–156.

Bänziger, H. 1972. Biologie der lachriphagen Lepidopteren in Thailand and Malaya. *Rev. Suisse Zool.* 79:1381–1469.

Bänziger, H. 1975. Skin-piercing blood-sucking moths. I: ecological and ethological studies on *Calpe eustrigata* (Lepidoptera, Noctuidae). *Acta Trop.* 32:125–144.

Bänziger, H. and W. Büttiker. 1969. Records of eye-frequenting Lepidoptera from man. *J. Med. Entomol.* 6:53–58.

Barnes, A. M. and F. J. Radovsky. 1969. A new *Tunga* (Siphonaptera) from the Nearctic Region with description of all stages. *J. Med. Entomol.* 6:19–36.

Basolo, F., Jr. and R. C. Funk. 1974. Ectoparasites from *Microtus ochrogaster, Peromyscus leucopus*, and *Cryptotis parva* in Coles County, Illinois. *Trans. Ill. State Acad. Sci.* 67:211–221.

Beer, J. R. and E. F. Cook. 1958. The louse populations on some deer mice from western Oregon. *Pan-Pac. Entomol.* 34:155–158.

Beer, J. R. and E. F. Cook. 1968. A ten-year study of louse populations on deer mice. *J. Med. Entomol.* 5:85–90.

Beer, J. R., E. F. Cook, and R. G. Schwab. 1959. The ectoparasites of some mammals from the Chiricahua Mountains, Arizona. *J. Parasit.* 45:605–613.

Befus, A. D. 1982. Mechanisms of host resistance at the mucosa–parasite interface. Pages 34–36 in D. F. Mettrick and S. S. Desser (Eds.), *Parasites—Their World and Ours*. Elsevier, Amsterdam.

Befus, A. D. and R. B. Podesta. 1976. Chapter 15. Intestine. Pages 303–325 in C. R. Kennedy (Ed.), *Ecological Aspects of Parasitology*. North-Holland, Amsterdam.

Beklemishev, V. N. 1945. The principles of comparative parasitology applied to hematophagous arthropods. *Med. Parazitol. Mosk.* 17:385–400 (in Russian, cited by Balashov 1982).

Beklemishev, V. N. 1951. Parasitism of arthropods on terrestrial vertebrates. I. Trends in its development. *Med. Parazitol. Mosk.* 20:261–288 (in Russian, cited by Balashov 1982).

Beklemishev, V. N. 1954. Parasitism of arthropods on terrestrial vertebrates. II. Basic trends of its development. *Med. Parazitol. Mosk.* 23:3–20 (in Russian, cited by Balashov 1982).

Bell, J. F. and C. Clifford. 1964. Effects of limb disability on lousiness in mice. II. Intersex grooming relationships. *Exp. Parasitol.* 15:340–349.

Bell, J. F., W. L. Jellison, and C. R. Owen. 1962. Effects of limb disability on lousiness. 1. Preliminary studies. *Exp. Parasitol.* 12:176–183.

Bell, J. F., C. M. Clifford, G. J. Moore, and G. Raymond. 1966. Effects of limb disability on lousiness in mice. III. Gross aspects of acquired resistance. *Exp. Parasitol.* 18:49–60.

Benjamini, E. and B. F. Feingold. 1970. Immunity to arthropods. Pages 1061–1134 in G. J. Jackson, R. Herman, and I. Singer (Eds.), *Immunity to Parasitic Animals*, vol. 2. Appleton-Century-Crofts, New York.

Bennet-Clark, H. C. and E. C. A. Lucey. 1967. The jump of the flea: a study of energetics and a model of the mechanism. *J. Exp. Biol.* 47:59–76.

Bennett, G. F. 1969. *Boophilus microplus* (Acarina: Ixodidae): experimental infestations on cattle restrained from grooming. *Exp. Parasitol.* 26:323–328.

Bennington, K. C. and D. H. Kemp. 1980. Role of tick salivary glands in feeding and disease transmission. *Adv. Parasitol.* **18**:315–339.

Benton, A. H., R. Cerwonka, and J. Hill. 1959. Observations on host perception in fleas. *J. Parasit.* **45**:614.

Benton, A. H. and S. Y. Lee. 1965. Host finding reactions of some Siphonaptera. *Proc. Int. Congr. Entomol. 12th, London* **1964**:792.

Bequaert, J. C. 1953. The Hippoboscidae or louse-flies (Diptera) of mammals and birds. Part I. Structure, physiology and natural history. *Entomol. Am.* (N.S.) **32**:1–209; **33**:211–442.

Bequaert, J. C. 1954–1957. The Hippoboscidae or louse-flies (Diptera) of mammals and birds. Part II. Taxonomy, evolution and revision of American genera and species. *Entomol. Am.* (N.S.) **34**(1954):1–232; **35**(1955):233–416; **36**(1957):417–611.

Beresford-Jones, W. P. 1967. Observations on the transmission of the mite *Psorergates simplex* Tyrrell 1883 in laboratory mice. Pages 277–280 in E. J. L. Soulsby (Ed.), *The Reaction of the Host to Parasitism.* Proc. 3rd Int. Conf. World Assoc. Adv. Vet. Parasitol., Lyons, France.

Boucher, D. H., S. James, and K. H. Keeler. 1982. The ecology of mutualism. *Annu. Rev Ecol. Syst.* **13**:315–347.

Bradley, D. J. 1972. Regulation of parasite populations. A general theory of the epidemiology and control of parasitic infections. *Trans. R. Soc. Trop. Med. Hyg.* **66**:697–708.

Bradley, D. J. 1974. Stability in host–parasite systems. Pages 71–81 in M. B. Usher and M. H. Williamson (Eds.), *Ecological Stability.* Chapman and Hall, London.

Brooks, D. R. 1979. Testing the context and extent of host–parasite coevolution. *Syst. Zool.* **28**:299–307.

Brooks, M. A. 1964. Symbiotes and the nutrition of medically important arthropods. *Bull. WHO* **31**:555–559.

Brooks, W. M. 1974. Protozoan infections. Pages 237–300 and G. E. Cantwell (Ed.), *Insect Diseases*, Vol. 1. Marcel Dekker, New York.

Burgdorfer, W. 1969. Ecology of tick vectors of American spotted fever. *Bull. WHO* **40**:375–381.

Burtt, E. H. 1981. The adaptiveness of animal colors. *Bioscience* **31**:723–729.

Bush, G. L. 1975a. Sympatric speciation in phytophagous parasitic insects. Pages 187–206 in P. W. Price (Ed.), *Evolutionary Strategies of Parasitic Insects.* Plenum, London.

Bush, G. L. 1975b. Modes of animal speciation. *Annu. Rev. Ecol. Syst.* **6**:339–364

Buxton, P. A. 1956. *The Louse.* Williams & Wilkins, Baltimore.

Camin, J. H. 1953. Observations on the life history and sensory behavior of the snake mite, *Ophionyssus natricis* (Gervais) (Acarina: Macronyssidae). *Spec. Publ. Chicago Acad. Sci.* **10**:1–75.

Capron, A., J. Biquet, A. Vernes, and D. Afchain. 1968. Structure antigenique des helminthes. Aspectes immunologique des relations hote–parasite. *Pathol. Biol.* **16**:121–138.

Carlson, L. D. and A. C. L. Hsieh. 1974. Chapter 4. Temperature and humidity. Pages 61–118 in N. B. Slonim (Ed.), *Environmental Physiology.* C. V. Mosby, St. Louis.

Carpenter, F. M. 1977. Geological history and evolution of the insects. *Proc. XV Int. Congr. Entomol.*, Washington, D.C., Aug. 19–27, 1976, pp. 63–70.

Carter, H. B. 1965. Chapter 2. Variation in the hair follicle population of the mammalian skin. Pages 25–33 in A. G. Lyne and B. F. Short (Eds.), *Biology of the Skin and Hair Growth.* Angus and Roberston, Sydney.

Castro, G. A. 1982. Immunophysiological interactions in the small intestine. Pages 242–244 in D. F. Mettrick and S. S. Desser (Eds.), *Parasites—Their World and Ours.* Elsevier, Amsterdam.

Caullery, M. 1952. *Parasitism and Symbiosis.* Sidgwick and Jackson, London.

Champion, R. H. 1970. Chapter 12. Sweat glands. Pages 175–183 in R. H. Champion et al. (Eds.), *An Introduction to the Biology of the Skin*. Blackwell, Oxford.

Champion, R. H., T. Gillman, A. J. Rook, and R. T. Sims. 1970. *An Introduction to the Biology of the Skin*, Blackwell, Oxford, UK.

Chinery, W. A. 1981. Observation on the saliva and salivary gland extract of *Haemaphysalis spinigera* and *Rhipicephalus sanguineus sanguineus*. *J. Parasitol.* **67**:15–19.

Clemens, W. A., J. A. Lillegraven, E. H. Lindsay, and G. G. Simpson. 1979. Where, when and what—A survey of known Mesozoic mammal distribution. Pages 7–58 in J. A. Lillegraven, Z. Kielan-Jaworowska, and W. A. Clemens (Eds.), *Mesozoic Mammals: The First Two-Thirds of Mammalian History*. University of California Press, Berkeley.

Code, C. F. (Ed.). 1968. *Alimentary Canal in Handbook of Physiology*, Vol. 5. Am. Physiol. Soc., Washington, D.C.

Cohen, S. and K. S. Warren. 1982. *Immunology of Parasitic Infections*. 2nd ed. Blackwell, Oxford, UK.

Cole, L. C. and J. A. Koepke. 1947. Problems of interpretation of the data of rodent-ectoparasite surveys and studies of rodent ectoparasites in Honolulu, T. H., Savannah, Ga., and Dothan, Ala. *Public Health Rep. Suppl.* **202**:1–71.

Combs, D. D. 1977. *Population ecology of ectoparasites (Arthropoda) on Peromysus leucopus noveboracensis (Fischer) (Cricetidae, Rodentia)*. A D.Ed. Thesis in Biological Science, The Pennsylvania State University, University Park, PA. 121 pp.

Combs, D. D. and K. C. Kim. 1977. Unpublished data.

Cook, E. F. 1954a. A modification of Hopkins' technique for collecting ectoparasites from mammal skins. *Entomol. News* **15**:35–37.

Cook, E. F. 1954b. A technique for preventing post-mortem ectoparasite contamination. *J. Mammal.* **35**:266–267.

Cook, E. F. and J. R. Beer. 1955. Louse populations of some cricetid rodents. *Parasitology* **45**:409–420.

Cook, E. F. and J. R. Beer. 1958. A study of louse populations on the meadow vole and deer mouse. *Ecology* **39**:645–659.

Corbet, G. B. 1956. The life history and host-relations of a hippoboscid fly *Ornithomyia fringillina* Curtis. *J. Anim. Ecol.* **25**:403–420.

Cotton, M. J. and D. A. Griffiths. 1967. Observations on temperature conditions in vole nests. *J. Zool.* (London) **153**:541–544.

Cowx, N. C. 1967. Some aspects of the ecology and biology of some small mammal fleas from Yorkshire. *J. Biol. Educ.* **1**:75–78.

Crofton, H. D. 1971a. A quantitative approach to parasitism. *Parasitology* **62**:179–193.

Crofton, H. D. 1971b. A model of parasite–host relationships. *Parasitology* **63**:343–364.

Cross, E. A. 1965. The generic relationships of the family Pyemotidae (Acarina: Trombidiformes). *Univ. Kans. Sci. Bull.* **45**:29–275.

Cross, H. F. 1964. Observations on the formation of the feeding tube by *Trombicula splendens* larvae. *Acarologia* **61**:255–261.

Cross, H. F. and G. W. Wharton. 1964a. A comparison of the number of tropical rat mites and tropical fowl mites that feed at different temperatures. *J. Econ. Entomol.* **57**:439–443.

Cross, H. F. and G. W. Wharton. 1964a. A comparison of the number of tropical rat mites and tropical fowl mites that fed under varying conditions of humidity. *J. Econ. Entomol.* **57**:443–445.

Day. P. R. 1974. *Genetics of Host–Parasite Interactions*. W. H. Freeman, San Francisco, CA.

Damian, R. T. 1962. A theory of immunoselection for eclipsed antigens of parasites and its implications for the problem of antigenic polymorphism in man. *J. Parasitol.* **48**:16.

Damian, R. T. 1964. Molecular mimicry: Antigen sharing by parasite and host and its consequences. *Am. Nat.* **98**:129–149.

Damian, R. T. 1979. Molecular mimicry in biological adaptation. Pages 103–126 in B. B. Nichol (Ed.), *Host–Parasite Interfaces*. Academic Press, New York.

deBary, A. 1879. *Die Erscheinung der Symbiose*. Karl J. Trubner, Strasburg.

Dethier, V. G. 1957. The sensory physiology of blood-sucking arthropods. *Exp. Parasit.* **6**:68–122.

Dethier, V. G. 1961. *The Physiology of Insect Senses*. John Wiley, New York.

Dogiel, V. A. 1964. *General Parasitology*. Oliver and Boyd, Edinburgh and London.

Dubinina, Kh. V. 1969. Certain adaptations of listrophorid mites (Fam. Listrophoridae) to parasitism in the hair cover of hosts—rodents. *Proc. 2nd Int. Congr. Acarol.* **1967**:299–300.

Durden, L. A. 1983. Sucking louse (*Hoplopleura erratica*: Insecta, Anoplura) exchange between individuals of a wild population of Eastern chipmunk, *Tamias striatus*, in central Tennessee, U.S.A. *J. Zool.* (London) **201**:117–123.

Ebling, F. J. G. 1970. Chapter 13. Sebaceous glands. Pages 184–196 in R. H. Champion et al. (Eds.), *An Introduction to the Biology of the Skin*. Blackwell, Oxford, UK.

Elton, C., E. B. Ford, and J. R. Baker. 1931. 36. The health and parasites of a wild mouse population. *Proc. Zool. Soc. London*, **1931**(3):657–721.

Emerson, K. C. and R. D. Price. 1981. A host–parasite list of the Mallophaga on mammals. *Misc. Publ. Entomol. Soc. Am.* **12**(1):1–72.

Esch, G. W., J. W. Gibbons, and J. E. Bourgue. 1975. An analysis of the relationship between stress and parasitism. *Am. Midl. Nat.* **93**:339–353.

Esch, G. W., T. C. Hazen, and J. M. Aho. 1977. Parasitism and *r*- and *K*-selection. Pages 9–62 in G. W. Esch (Ed.), *Regulation of Parasite Populations*. Academic Press, New York.

Evans, F. C. and R. B. Freeman. 1950. On the relationship of some mammal fleas to their hosts. *Ann. Entomol. Soc. Am.* **43**:320–333.

Evans, F. C. and F. E. Smith. 1952. The intrinsic rate of natural increase for human louse, *Pediculus humanus* L. *Am. Nat.* **86**:299–310.

Evans, G. O. 1950. Studies on the bionomics of the sheep ked, *Melophagus ovinus* L., in West Wales. *Bull. Ent. Res.* **40**:459–478.

Evans, G. O., J. G. Sheals, and D. McFarlane. 1961. *The Terrestrial Acari of the British Isles*. An Introduction to Their Morphology, Biology and Classification. Vol. 1. Introduction and Biology. British Museum (N.H.), London.

Ewing, H. E. 1938. North American mites of the family Myobiinae, new subfamily (Arachnida). *Proc. Entomol. Soc. Wash.* **40**:180–197.

Fain, A. 1956. Une nouvelle famille d'acariens endoparasites des chauves-souris: Gastronyssidae fam. nov. *Ann. Soc. Belg. Med. Trop.* **36**(1):87–98.

Fain, A. 1967a. Diagnoses d'Acariens Sarcoptiformes nouveaux. *Rev. Zool. Bot. Afr.* **75**(3–4):378–382.

Fain, A. 1967b. Le genre *Dermatophagoides* Bogdanov 1864 son importance dans allergies respiratoires et cutanees chez l'homme (Psoroptidae: Sarcoptiformes). *Acarologia* **9**(1):179–225.

Fain, A. 1968a. Notes sur trois Acariens remarguables (Sarcoptiformes). *Acarologia* **10**(2):276–291.

Fain, A. 1968b. Un hypope de la famille Hypoderidae Murray 1877 vivant sous la peau d'un rongeur (Hypoderidae: Sarcoptiformes). *Acarologia* **10**(1):111–115.

Fain, A. 1969. Adaptation to parasitism in mites. *Acarologia* **11**:429–446.

Fain, A. and J. C. Beaucournu. 1976. Trois nouveaux hypopes du genre *Psylloglyphus* Fain, phoretique sur des puces et un *Hemimerus* (Acarina: Saproglyphidae). *Rev. Zool. Afr.* 90:181–187.

Fain, A., F. Lukoschus and P. Hallmann. 1966. Le genre *Psorergates* chez les murides. Description de trois especies nouvelles (Psorergatidae: Trombidiformes). *Acarologia* 8(2):251–274.

Feingold, B. F., E. Benjamini, and D. Michaeli. 1968. The allergic responses to insect bites. *Annu. Rev. Entomol.* 13:137–158.

Filipponi, A. 1965. Facultative vivipary in Macrochelidae (Acari: Mesostigmata). *Proc. 12th Int. Congr. Entomol.*, London, Sect. 5, pp. 309–310.

Filipponi, A. and G. Francaviglia. 1963. Oviparita e larviparita in *Macrocheles peniculatus* Berl. (Acari: Mesostigmata) regolate da fattori ecologici. *Parassitologia* 24:81–104.

Finch, V. A., R. Daniel, R. Boxman, A. Shkolnik, and C. R. Taylor. 1980. Why black goats in host desert? Effects of coat color on heat exchange of wild and domestic goats. *Physiol. Zool.* 53:19–25.

Flux, J. E. C. 1972. Seasonal and regional abundance of fleas on hares in Kenya. *J. E. Afr. Nat. Hist. Soc.* 29:1–8.

Foin, T. C., J. W. Valentine, and J. F. Ayala. 1975. Extinction of taxa and Van Valen's law. *Nature* (London) 257:514–515.

Fowler, C. W. and J. A. MacMahon. 1982. Selective extinction and speciation: Their influence on the structure and functioning of communities and ecosystems. *Am. Nat.* 119:480–498.

Friend, W. G. and J. J. B. Smith. 1977. Factors affecting feeding by bloodsucking insects. *Annu. Rev. Entomol.* 22:309–331.

Funakoshi, K. 1977. Ecological studies on the bat fly, *Penicillidia jenynsii* (Diptera: Nycteribiidae), infesting the Japanese long-fingered bat, with special reference to the adaptability of the life cycle of their hosts (in Japanese). *Jpn. J. Ecol.* 27:125–140.

Furman, D. P. 1959. Feeding habits of symbiotic mesostigmatid mites of mammals in relation to pathogen-vector potentials. *Am. J. Trop. Med. Hyg.* 8:5–12.

Furman, D. P. 1966. Biological studies on *Haemolaelaps centrocarpus* Berlese (Acarina: Laelapidae) with observations on its classification. *J. Med. Entomol.* 2:331–335.

Furman, D. P. 1968. Effects of the microclimate on parasitic nest mites of the dusky footed wood rat, *Neotoma fuscipes* Baird. *J. Med. Entomol.* 2:160–166.

Furman, D. P. 1979. Specificity, adaptation and parallel evolution in the endoparasitic Mesostigmata of mammals. *Recent Adv. Acarol.* 2:329–337.

Furman, D. P. and M. D. Dailey. 1980. The genus *Halarachne* (Acari: Halarachnidae), with the description of a new species from the Hawaiian monk seal. *J. Med. Entomol.* 17:352–359.

Furman, D. P. and A. W. Smith. 1973. In vitro development of two species *Orthohalarachne* (Acarina: Halarachnidae) and adaptations of the life cycles for endoparasitism in mammals. *J. Med. Entomol.* 10:414–416.

Futuyma, D. J. and M. Slatkin (Eds.). 1983. *Coevolution*. Sinauer Associates, Sunderland, MA.

Gaafar, S. M. 1966. Pathogenesis of ectoparasites. Pages 229–236 in E. J. L. Soulsby, *Biology of Parasites*. Academic Press, New York.

Gaafar, S. M. 1967. Pathogenesis of Cannine demodicosis. Pages 59–71 in E. J. L. Soulsby, *The Reaction of the Host to Parasitism*, Proc. 3rd Int. Conf. World Assoc. Adv. Vet. Parasit., Lyons, France.

Gaafar, S. M. 1972. Immune response to arthropods. Pages 273–285 in E. J. L. Soulsby (Ed.), *Immunity to Animal Parasites*. Academic Press, New York.

Gadgil, M. and O. T. Solbrig. 1972. The concept of *r*- and *K*-selection: evidence from wild flowers and some theoretical considerations. *Am. Nat.* 106:14–31.

Ghiselin, M. T. 1972. 7. Models in phylogeny. Pages 130–145 in J. M. Schopf (Ed.), *Models in Paleobiology*. Freeman, Cooper, and Co., San Francisco, CA.

Gilbert, L. E. and R. H. Ravin. (Eds.). 1975. *Coevolution of Animals and Plants.* University of Texas Press, Austin.

Gillman, T. 1970. Chapter 6. The dermis. Pages 76–113 in R. H. Champion et al. (Eds.), *An Introduction to the Biology of the Skin.* Blackwell, Oxford, UK.

Good, R. A. 1981. Nutrition and immunity. *J. Clin. Immunol.* **1:**3–11.

Gooding, R. H. 1963. Studies on the effect of frequency of feeding upon the biology of a rabbit-adapted strain of *Pediculus humanus. J. Parasit.* **49:**516–521.

Gooding, R. H. 1972. Digestive process of haematophagous insects. I. Literature review. *Quaest. Entomol.* **8:**5–60.

Graham, N. P. H. and K. L. Taylor. 1941. Studies on some ectoparasites of sheep and their control. 1. Observations on the bionomics of the sheep ked (*Melophagus ovinus*). *C.S.1.R.O., Aust. Pam.* **108:**9–26.

Greenwood, M. T. and D. M. Holdich. 1979. A structural study of the sensilium of two species of bird flea, *Ceratophyllus* (Insecta: Siphonaptera). *J. Zool.* (London) **187:**21–38.

Gregson, J. D. 1960. Morphology and functioning of the mouthparts of *Dermacenter andersoni* Stiles. *Acta Trop.* **17:**48–49.

Gregson, J. D. 1967. Observations on the movement of fluids in the vicinity of the mouthparts of naturally feeding *Dermacenter andersoni* Stiles. *Parasitology* **57:**1–8.

Gross, B. and D. D. Bonnet. 1949. Snap trap versus cage traps in plague surveillance. *Public Health Rep.* **64:**1214–1216.

Guimaraes, J. H. and N. Papavero. 1966. A tentative annotated bibliography of *Dermatobia hominis* (Linneaus Jr., 1781) (Diptera, Cuterebridae). *Arq. Zool.* (Sao Paulo) **14:**223–294.

Haas, G. E. 1966. Cat flea–mongoose relationships in Hawaii. *J. Med. Entomol.* **2:**361–362.

Hamilton, W. D. 1967. Extraordinary sex ratios. *Science* **156:**477–488.

Harwood, R. F. and M. T. James. 1979. *Entomology in Human and Animal Health,* 7th ed. Macmillan, New York.

Hennig, W. 1969. *Die Stammesgeschichte der Insekten.* Verlag von Waldemar Kramer in Frankfurt an Main.

Hertig, M., W. H. Taliaferro, and B. Schwartz. 1937. The terms *symbiosis, symbiont* and *symbiote.* Report of the Committee on Terminology (American Society of Parasitologists). *J. Parasitol.* **23:**326–329.

Hilton, D. F. J. and J. L. Mahrt. 1971. Ectoparasites from three species of *Spermophilus* (Rodentia: Sciuridae) in Alberta. *Can. J. Zool.* **49:**1501–1504.

Hocking, B. 1971. Blood-sucking behavior of terrestrial arthropods. *Annu. Rev. Entomol.* **16:**1–26.

Hoeppli, R. and H. H. Schumacher. 1962. Histological reactions to trombiculid mites, with special reference to "natural" and "unnatural" hosts. *Z. Tropenmed. Parasitol.* **13:**419–428.

Holdenried, R., F. C. Evans, and D. S. Longanecker. 1951. Host–parasite–disease relationships in a mammalian community in the central coast of California. *Ecol. Monogr.* **21:**1–18.

Hole, J. W., Jr. 1978. *Human Anatomy and Physiology.* Wm. C. Brown, Dubuque, IA.

Holloway, B. A. 1976. A new bat-fly family from New Zealand. (Diptera: Mystacinobiidae). *N. Z. J. Zool.* **3:**279–301.

Holmes, J. C., R. P. Hobbs, and T. S. Leong. 1977. Populations in perspective: community organization and regulation of parasite populations. Pages 209–245 in G. W. Esch (Ed.), *Regulation of Parasite Populations.* Academic Press, New York.

Hoogstraal, H. 1978. Biology of ticks. Pages 3–14 in J. K. H. Wilde (Ed.), *Tick-borne Diseases and Their Vectors. Proc. Int. Conf.* (Edinburgh, September–October 1976).

Hoogstraal, H. and A. Aeschlimann. 1982. Tick–host specificity. *Bull. Soc. Entomol. Suisse* **55:**5–32.

Hopkins, G. H. E. 1949. The host-associations of the lice of mammals. *Proc. Zool. Soc. Lond.* **119**:387–604.

Hopkins, G. H. E. 1957. The distribution of Phthiraptera on mammals. Pages 88–119 in J. G. Baer (Ed.), *First Symposium on Host Specificity Amongst Parasites of Vertebrates*, Inst. Zool., University of Neuchatel, Neuchatel.

Hughes, R. D. and C. G. Jackson. 1958. A review of the Anoetidae (Acari). *Va. J. Sci.* **9**,n.s.:5–198.

Hughes, T. E. 1954. Internal anatomy of the mite *Listrophorus leuckarti* (Pagenstecher, 1861). *Proc. Zool. Soc. Lond.* **124**:239–256.

Humphries, D. A. 1967a. The function of combs in fleas. *Entomol. Mon. Mag.* **102**:232–236.

Humphries, D. A. 1967b. Function of combs in ectoparasites. *Nature* (London) **215**:319.

Hurka, K. 1964. Distribution, bionomy and ecology of European bat flies with special regard to the Czechoslovak fauna (Diptera: Nycteribiidae). *Acta Univ. Carol., Biol.* **1964**:167–234.

Iqbal, Q. J. 1974. Host-finding behaviour of the rat flea *Nosopsyllus fasciatus* (Bosc.). *Biologia* (Lahore) **20**:147–150.

Jackson, J. O. and G. R. Defoliart. 1975a. Some quantitative relationships of certain nidicolous acari and the white-footed mouse, *Peromyscus leucopos*. *J. Med. Entomol.* **12**:323–332.

Jackson, J. O. and G. R. Defoliart. 1975b. Relationships of the white-footed mouse, *Peromyscus leucopus*, and its associated fleas (Siphonaptera) in southeastern Wisconsin. *J. Med. Entomol.* **12**:351–356.

Jackson, J. O. and G. R. Defoliart. 1975c. Relationships of immature *Dermacenter variabilis* (Say) (Acari: Ixodidae) with the white-footed mouse, *Peromyscus leucopus* in southwestern Wisconsin. *J. Med. Entomol.* **12**:409–412.

James, M. T. 1948. Flies that cause myiasis in man. *USDA Misc. Publ.* **631**:1–175.

James, M. T. 1969. A study in the origin of parasitism. *Bull. Entomol. Soc. Am.* **15**:251–253.

James, M. T. and R. F. Harwood. 1969. *Herm's Medical Entomology*. Macmillan, New York.

Janzen, D. H. 1963. Observations on populations of adult beaver beetles *Platypsyllus castoris* (Platypsyllidae: Coleoptera). *Pan-Pac. Entomol.* **39**:215–228.

Jellison, W. L. 1942. Host distribution of lice on native American rodents north of Mexico. *J. Mammal.* **23**:245–250.

Jenkin, C. R. 1963. Heterophile antigens and their significance in the host–parasite relationship. *Adv. Immunol.* **3**:351–376.

Jenkins, D. W. 1964. Pathogens, parasites and predators of medically important arthropods. Annotated list and bibliography. *Bull. WHO* **30**(suppl.):1–150.

Jensen, R. E. and J. E. Roberts. 1966. A model relating microhabitat temperatures to seasonal changes in the little blue louse (*Solenopotes capillatus*) population. *Ga. Agr. Exp. Sta., Univ. Ga., Coll. Agr. Tech. Bull. N.S.* **55**:4–22.

Jones, B. M. 1950. The penetration of the host tissue by the harvest mite, *Trombicula autumnalis* Shaw. *Parasitology* **41**:229–248.

Jordan, K. 1962. Notes on *Tunga caecigena* (Siphonaptera: Tungidae). *Bull. Br. Mus. Nat. Hist.* (Ent.) **12**:353–364.

Keirans, J. E. 1975. A review of the phoretic relationship between Mallophaga (Phthiraptera: Insecta) and Hippoboscidae (Diptera: Insecta). *J. Med. Entomol.* **12**:71–76.

Kemp, T. S. 1982. *Mammal-like Reptiles and the Origin of Mammals*. Academic Press, London.

Kennedy, C. R. 1975. *Ecological Animal Parasitology*. John Wiley, New York.

Kennedy, C. R. (Ed.). 1976. *Ecological Aspects of Parasitology*. North-Holland, Amsterdam.

Kennedy, C. R. 1982. Biotic factors. Pages 293–302 in D. F. Mettrick and S. S. Desser (Eds.), *Parasites—Their World and Ours*. Elsevier, Amsterdam.

Kim, J. C. S. 1977. Pathobiology of pulmonary acariasis in Old World monkeys. *Acarologia* **19**:371–383.

Kim, K. C. 1972. Louse populations of the northern fur seal (*Callorhinus ursinus*). *Am. J. Vet. Res.* **33**:2027–2036.

Kim, K. C. 1975. Ecology and morphological adaptation of the sucking lice (Anoplura: Echinophthiriidae) on the northern fur seal. *Rapp. p.-v. Reun. Cons. Perm. Int. Explor. Mer* **169**:504–515.

Kim, K. C. 1983. Unpublished data.

Kim, K. C. and M. C. Keyes. 1984. Unpublished data.

Kim, K. C. and H. W. Ludwig. 1978a. The family classification of the Anoplura. *Syst. Entomol.* **3**:249–284.

Kim, K. C. and H. W. Ludwig. 1978b. Phylogenetic relationships of parasitic Psocodea and taxonomic position of the Anoplura. *Ann. Entomol. Soc. Am.* **71**:910–922.

Kim, K. C. and H. W. Ludwig. 1982. Parallel evolution, cladistics, and classification of parasitic Psocodea. *Ann. Entomol. Soc. Am.* **75**:537–548.

Kim, K. C., V. L. Haas, and M. C. Keyes. 1980. Populations, microhabitat preference and effects of infestation of two species of *Orthohalarachne* (Halarachnidae: Acarina) in the northern fur seal. *J. Wildl. Dis.* **16**:45–52.

Kim, K. C., C. A. Repenning, and G. V. Moorejohn. 1975. Specific antiquity of the sucking lice and evolution of otariid seals. *Rapp. p.-v. Reun. Cons. Perm. Int. Exp. Mer* **169**:544–549.

Krantz, G. W. 1978. *A Manual of Acarology*, 2nd ed. OSU Book Stores, Inc., Corvallis, OR.

Kunz, T. H. 1976. Observations on the winter ecology of the bat fly *Trichobius corynorhini* Cockerell (Diptera: Streblidae). *J. Med. Entomol.* **12**:631–636.

Larrivee, D. H., E. Benjamini, B. F. Feingold, and M. Shimizu. 1964. Histologic studies of guinea pig skin: different stages of allergic reactivity to flea bites. *Exp. Parasitol.* **15**:491–502.

Lavoipierre, M. M. J. 1964. A new family of acarines belonging to the Suborder Sarcoptiformes parasitic in the hair follicles of primates. *Ann. Natal Mus.* **16**:191–208.

Lavoipierre, M. M. J. 1965. Feeding mechanism of bloodsucking arthropods. *Nature* **208**:302–303.

Lavoipierre, M. M. J. 1967. Feeding mechanism of *Haematopinus suis* on the transilluminated mouse ear. *Exp. Parasitol.* **20**:303–311.

Lavoipierre, M. M. J. and A. J. Beck. 1967. Feeding mechanism of *Chiroptonyssus robustipes* on the transilluminated bat wing. *Exp. Parasitol.* **20**:312–320.

Lavoipierre, M. M. J. and M. Hamachi. 1961. An apparatus for observations on the feeding mechanism of the flea. *Nature* **192**:998–999.

Lavoipierre, M. M. J. and C. Rajamanickam. 1968. The skin reactions of two species of Southeast Asian Chiroptera to notoedrid and teinoeoptid mites. *Parasitology* **58**:515–530.

Lavoipierre, M. M. J. and R. F. Riek. 1955. Observations on the feeding habits of argasid ticks and on the effect of their bites on laboratory animals, together with a note on the production of coxal fluid by several of the species studied. *Ann. Trop. Med. Parasitol.* **49**:96–113.

Lavoipierre, M. M. J., F. J. Radovsky, and P. D. Budwiser. 1979. The feeding process of a tungid flea, *Tunga monositus* (Siphonaptera: Tungidae) and its relationship to the host inflammatory and repair response. *J. Med. Entomol.* **15**:187–217.

Lavoipierre, M. M. J., C. Rajamanickam, and P. Ward. 1967. Host–parasite relationships of acarine parasites and their vertebrate hosts. 1. The lesions produced by *Bakerocoptes cynopteris* in the skin of *Cynocopterus brachyotis*. *Acta Trop.* **24**:1–18.

Lees, A. D. 1948. The sensory physiology of the sheep tick. *Ixodes ricinus* L. *J. Exp. Biol. Camb.* **5**:145–207.

Lees, A. D. and A. Milne. 1951. The seasonal and diurnal activities of individual sheep ticks (*Ixodes ricinus* L.). *Parasitology* **41**:189–208.

Lepow, I. H. and P. A. Ward. (Eds.). 1972. *Inflammation: Mechanisms and Control.* Academic Press, New York.

Levinton, J. S. 1970. The paleoecological significance of opportunistic species. *Lethaia* **3**:69–78.

Lewis, R. E. 1972. Notes on the geographic distribution and host preferences in the order Siphonaptera. Part 1. Pulicidae. *J. Med. Entomol.* **9**:511–520.

Lewis, R. E. 1973–1975. Notes on the geographical distribution and host preferences in the order Siphonaptera. Part 2. Rhopallopsyllidae, Malacopsyllidae, and Vermipsyllidae (1973); Part 3. Hystricopsyllidae (1974a); Part 4. Coptopsyllidae, Pygiopsyllidae, Stephanocircidae, and Xiphiopsyllidae (1974b); Part 5. Ancistropsyllidae, Chimaeropsyllidae, Ischnopsyllidae, Leptopsyllidae, and Macropsyllidae (1974c); Part 6. Ceratophyllidae (1975). *J. Med. Entomol.* **10**(1973):255–260; **11**(1974):147–167, 403–413, 525–540; **11**(1975):658–676.

Lewis, L. F., D. M. Christenson, and G. W. Eddy. 1967. Rearing the long-nosed cattle louse and cattle-biting louse on host animals in Oregon. *J. Econ. Entomol.* **60**:755–757.

Lillegraven, J. A. 1972. Ordinal and familial diversity of Cenozoic mammals. *Taxon* **21**:261–274.

Lindsdale, J. M. and B. S. Davis. 1956. Taxonomic appraisal and occurrence of fleas at the Hastings Reservation in central California. *Univ. Calif. Publ. Zool.* **54**:293–370.

Ludwig, H. W. 1982. Host specificity in Anoplura and Coevolution of Anoplura and Mammalia. *Mem. Mus. Nat. Hist. Nat. (Paris), N.S., Ser. A, Zool.* **123**:145–151.

Maa, T. C. 1963. Genera and species of Hippoboscidae (Diptera): types, synonymy, habitats, and natural groupings. *Pac. Inst. Monogr.* **6**:1–186.

Maa, T. C. 1964. A review of the Old World Polyctenidae (Hemiptera: Cimicoidea). *Pac. Inst.* **6**:494–516.

Maa, T. C. and A. G. Marshall. 1981. Diptera pupipara of the New Hebrides: taxonomy, zoogeography, host and ecology. *Q. J. Taiwan Mus.* **34**:213–232.

MacArthur, R. H. 1960. On the relative abundance of species. *Am. Nat.* **94**:25–36.

MacArthur, R. H. 1972. *Geographical Ecology. Patterns in the Distribution of Species.* Harper & Row, New York.

MacArthur, R. H. and E. O. Wilson. 1963. An equilibrium theory of insular zoogeography. *Evolution* **17**:373–387.

MacArthur, R. H. and E. O. Wilson. 1967. *The Theory of Island Biogeography.* Princeton University Press, Princeton, NJ.

Marshall, A. G. 1971. The ecology of *Basilia hispida* (Diptera: Nycteribiidae) in Malaysia. *J. Anim. Ecol.* **40**:141–154.

Marshall, A. G. 1976. Host-specificity amongst arthropods ectoparasitic upon mammals and birds in the New Hebrides. *Ecol. Entomol.* **1**:189–199.

Marshall, A. G. 1977. Interrelationships between *Arixenia esau* (Dermaptera) and molossid bats and their ectoparasites in Malaysia. *Ecol. Entomol.* **2**:285–291.

Marshall, A. G. 1980. The function of combs in ectoparasitic insects. Pages 79–87 in R. Traub and H. Starcke (Eds.), *Fleas.* Proc. Int. Conf. Fleas, Ashton Wold/Peterborough/UK, 21–25 June 1977, A. A. Balkema, Rotterdam.

Marshall, A. G. 1981a. *The Ecology of Ectoparasitic Insects.* Academic Press, London.

Marshall, A. G. 1981b. The sex ratio in ectoparasitic insects. *Ecol. Entomol.* **6**:155–174.

Marshall, A. G. 1982. The ecology of *Eoctenes spasmae* (Hemiptera: Polyctenidae). *Biotropica* **14**:50–55.

Matthysse, J. G. 1944. Biology of cattle biting louse and notes on cattle sucking lice. *J. Econ. Entomol.* **37**:436–442.

Matthysse, J. G. 1946. Cattle lice, their biology and control. *Bull. Cornell Univ. Agric. Exp. Stn.* **832**:1–67.

May, R. M. 1983a. Parasitic infections as regulators of animal populations. *Am. Sci.* **71**:36–45.

May, R. M. and R. M. Anderson. 1983. Chapter 9. Parasite–host coevolution. Pages 186–206 in D. J. Funtuyma and M. Slatkin (Eds.), *Coevolution.* Sinaur Assoc. Inc., Sunderland, MA.

Maynard Smith, J. 1978. *The Evolution of Sex.* Cambridge University Press, Cambridge, UK.

McCall, C. E. 1971. Host–parasite interaction. Pages 133–161 in M. F. LaVia and R. B. Hill (Eds.), *Principles of Pathobiology.* Oxford University Press, London.

McNaughton, S. J. and L. L. Wolf. 1970. Dominance and the niche in ecological ecosystem. *Science* **167**:131–139.

Mellanby, K. 1944. The development of symptoms, parasitic infection and immunity in human scabies. *Parasitology* **35**:197–206.

Mellanby, K. 1972. *Scabies,* 2nd ed. Classey, Hampton, UK.

Miles, V. I., A. R. Kinney, and H. E. Stark. 1957. Flea–host relationships of associated *Rattus* and native wild rodents in San Francisco Bay Area of California, with special reference to plague. *Am. J. Trop. Med. Hyg.* **6**:752–760.

Miller, J. R., G. E. Jones, and K. C. Kim. 1973. Populations and distribution of *Steatonyssus occidentalis* (Ewing) (Acarina: Macronyssidae) infesting the big brown bat, *Eptesicus fuscus* (Chiroptera: Vesperlitionidae). *J. Med. Entomol.* **10**:609–613.

Miller, N. C. E. 1971. *The Biology of Heteroptera.* Classey, London.

Mitchell, C. J. 1964. Population structure and dynamics of *Laelaps nuttalli* Hirst and *L. echidninus* Berlese (Acarina: Laelapidae) on *Rattus rattus* and *R. exulans* in Hawaii. *J. Med. Entomol.* **1**:151–153.

Mitchell, C. J. 1968. Biological studies of *Laelaps myonyssognathus* G. & N. (Acarina: Laelapidae). *J. Med. Entomol.* **5**:99–107.

Mitchell, G. F. 1982. Effector mechanisms of host-protective immunity to parasites and evasion by parasites. Pages 24–33 in D. F. Mettrick and S. S. Desser (Eds.), *Parasites—Their World and Ours.* Elsevier, Amsterdam.

Mock, D. E. 1974. *The cattle-biting louse, Bovicola bovis (Linn.). I. In vitro culturing, seasoning population fluctuations, and role of the male. II. Immune response of cattle.* Unpublished doctoral dissertation, Cornell University.

Mohr, C. O. and W. A. Stumpf. 1964. Louse and chigger infestations as related to host size and home ranges of small mammals. *Trans. 29th N. Am. Wildl. Nat. Res. Conf.* 1964:181–195.

Morlan, H. B. 1952. Host relationships and seasonal abundance of some southwest Georgia ectoparasites. *Am. Midl. Nat.* **48**:74–93.

Morton, J. 1967. *Guts.* Arnold, London.

Mount, L. E. 1979. *Adaptation to Thermal Environment: Man and His Productive Animals.* Arnold, London.

Murray, K. F. 1957. An ecological appraisal of host–ectoparasite relationships in a zone of epizootic plague in Central California. *Am. J. Trop. Med. Hyg.* **6**:1068–1086.

Murray, M. D. 1957a. The distribution of eggs of mammalian lice on their hosts. III. The distribution of the eggs of *Damalinia ovis* (L.) on the sheep. *Aust. J. Zool.* **5**:173–182.

Murray, M. D. 1957b. The distribution of eggs of mammalian lice on their hosts. IV. The distribution of the eggs of *Damalinia equi* (Denny) and *Haematopinus asini* (L.) on the horse. *Aust. J. Zool.* **5**:183–187.

Murray, M. D. 1961. The ecology of the louse *Polyplax serrata* (Burm.) on the mouse *Mus musculus* L. *Aust. J. Zool.* **9**:1–13.

Murray, M. D. 1963. Influence of temperature on the reproduction of *Damalinia equi* (Denny). *Aust. J. Zool.* **11**:183–189.

Murray, M. D. 1965. The diversity of the ecology of mammalian lice. *Proc. 12th Int. Congr. Entomol.* **1964**:366–367.

Murray, M. D. and D. G. Nicholls. 1965. Studies on the ectoparasites of seals and penguins. I. The ecology of the louse *Lepidophthirus macrorhini* Enderlein on the elephant seal, *Mirounga leonina* (L.) *Aust. J. Zool.* **13**:437–454.

Murray, M. D., M. S. R. Smith, and Z. Soucek. 1965. Studies on the ectoparasites of seals and penguins. II. The ecology of the louse *Antarctophthirus ogmorhini* Enderlein on the Weddell seal, *Leptonychotes weddeli* Lesson. *Aust. J. Zool.* **13**:761–771.

Nakata, S. and T. C. Maa. 1974. A review of the parasitic earwigs (Dermaptera: Arixeniina: Hemimerina). *Pac. Inst.* **16**:307–374.

Nelson, B. C. 1972. A revision of the New World species of *Ricinus* (Mallophaga) occurring on Passeriformes (Aves). *Univ. Calif. Publ. Entomol.* **68**:1–175.

Nelson, W. A. 1962. Development in sheep of resistance to the ked *Melophagus ovinus* (L.). I. Effects of seasonal manipulation of infestations. II. Effects of adrenocorticotrophic hormone and cortisone. *Exp. Parasitol.* **12**:41–44, 45–51.

Nelson, W. A. and A. R. Bainborough. 1963. Development in sheep of resistance of the ked *Melophagus ovinus* (L.). III. Histopathology of sheep skin as a clue to the nature of resistance. *Exp. Parasitol.* **18**:274–280.

Nelson, W. A. and D. M. Petrunia. 1969. *Melophagus ovinus:* feeding mechanisms on transilluminated mouse ear. *Exp. Parasitol.* **26**:308–313.

Nelson, W. A., J. A. Shemanchuk, and W. O. Haufe. 1970. *Haematopinus eurysternus:* blood of cattle infested with the short-nosed cattle louse. *Exp. Parasitol.* **28**:263–271.

Nelson, W. A., J. F. Bell, C. M. Clifford, and J. E. Keirans. 1977. Interaction of ectoparasites and their hosts. *J. Med. Entomol.* **13**:389–428.

Nelson, W. A., J. E. Keirans, J. F. Bell, and C. M. Clifford. 1975. Host–ectoparasite relationships. *J. Med. Entomol.* **12**:143–166.

Noback, C. R. 1950. Morphology and phylogeny of hair. *Ann. N.Y. Acad. Sci.* **53**:476–492.

Nutting, W. B. 1965. Host parasite relations: Demodicidae. *Acarologia* **7**:301–307.

Nutting, W. B. 1975. Pathogenesis associated with hair follicle mites (Acari: Demodicidae). *Acarologia* **17**:493–507.

Odening, K. 1976. Conception and terminology of hosts in parasitology. *Adv. Parasitol.* **14**:1–95.

O'Conner, B. M. 1982. Evolutionary ecology of astigmatid mites. *Annu. Rev. Entomol.* **27**:385–409.

Olson, W. P. 1969. Rat-flea indices, rainfall, and plague outbreak in Vietnam, with emphasis on the Pleiku area. *Am. J. Trop. Med. Hyg.* **18**:621–628.

Overal, W. L. 1980. Host-relations of the batfly *Megistopoda aranea* (Diptera: Streblidae) in Panama. *Univ. Kansas Sci. Bull.* **52**:1–20.

Papavero, N. 1977. *The World Oestridae (Diptera), Mammals and Continental Drift.* Vol. 14, Series Entomologica. E. Schimitschek and K. A. Spencer (Eds.), W. Junk, The Hague.

Parish, W. E. 1970a. 19. Characteristics of the types of immunological reactions. Pages 273–297 in R. H. Champion et al. (Eds.), *An Introduction to the Biology of Skin.* Blackwell, Oxford, UK.

Parish, W. E. 1970b. 20. Fundamental patterns of histological changes in the skin. Pages 298–

317 in R. H. Champion et al. (Eds.), *An Introduction to the Biology of Skin*. Blackwell, Oxford, UK.

Parnell, J. P. 1950. Hair pattern and distribution in mammals. *Ann. N.Y. Acad. Sci.* **53**:493–497.

Parsons, M. A. 1962. *A Survey of the Ectoparasites of the Wild Mammals of New England and New York State*, Masters Thesis, University of Massachusetts, Amherst (Unpublished).

Pavlovsky, E. N. 1966. *Natural Nidality of Transmissable Diseases*. University of Illinois Press, Urbana, IL.

Petrunkevitch, A. 1955. Arachnida. Chelicerata with sections on Pycnogonida and Palaeoisopus. In R. C. Moore (Ed.), *Treatise on Invertebrate Paleontology, Part P, Arthropoda* **2**:42–162.

Pianka, E. R. 1970. On *r*- and *K*-selection. *Am. Nat.* **104**:592–597.

Pickett, S. T. A. 1976. Succession: an evolutionary interpretation. *Am. Nat.* **110**:107–119.

Price, P. W. 1975. Introduction: The parasite way of life and its consequences. Pages 1–13 in P. W. Price (Ed.), *Evolutionary Strategies of Parasitic Insects and Mites*. Plenum Press, New York.

Price, P. W. 1977. General concepts on the evolutionary biology of parasites. *Evolution* **31**:405–420.

Price, P. W. 1980. *Evolutionary Biology of Parasites*. Princeton University Press, Princeton, NJ.

Price, R. D. and R. M. Timm. 1979. Description of the male *Geomydoecus scleritus* (Mallophaga: Trichodectidae) from the south-eastern pocket gopher. *J. Ga. Entomol. Soc.* **14**:162–165.

Radovsky, F. J. 1969. Adaptive radiation in the parasitic Mesostigmata. *Acarologia* **11**:450–478.

Radovsky, F. J. 1982. Adaptive radiation in the parasitic Mesostigmata. Pages 213–215 in D. F. Mettrick and S. S. Desser (Eds.), *Parasites—Their World and Ours*. Elsevier, Amsterdam.

Richards, O. W. and R. G. Davies. 1977. *Imm's General Textbook of Entomology*, 10th ed. Chapman and Hall, London.

Robinson, J. V. and W. D. Valentine. 1979. The concepts of elasticity, invulnerability and invadability. *J. Theor. Biol.* **81**:91–104.

Robinson, S. and D. L. Wiegman. 1974. Chapter 4. Temperature and humidity. Part B: Heat and humidity. Pages 84–118 in N. F. Slonim (Ed.), *Environmental Physiology*. C. V. Mosby, St. Louis, MO.

Rogers, L. L. 1975. Parasites of black bears of the Lake Superior Region. *J. Wildl. Dis.* **11**:189–192.

Rogers, W. P. 1962. *The Nature of Parasitism*. Academic Press, New York and London.

Rohde, K. 1979. A critical evaluation of intrinsic and extrinsic factors responsible for niche restriction in parasites. *Am. Nat.* **114**:648–671.

Rohde, K. 1982. *Ecology of Marine Parasites*. University of Queensland Press, St. Lucia, Queensland.

Roitt, I. M. 1980. *Essential Immunology*. Blackwell, Oxford, UK.

Rook, A. J. 1970. Chapter 11. Hair. Pages 164–174 in R. H. Champion et al. (Eds.), *An Introduction to the Biology of the Skin*. Blackwell, Oxford, UK.

Rothschild, M. 1973. The flying leap of the flea. *Sci. Am.* **229**:92–100.

Rothschild, M. and R. Ford. 1964. Maturation and egg-laying of rabbit flea (*Spilopsyllus cuniculi* Dale) induced by the external application of hydrocortisone. *Nature* (London) **203**:210–211.

Rothschild, M. and R. Ford. 1966. Hormones of the vertebrate host controlling ovarian regression and copulation of the rabbit flea. *Nature* (London) **211**:261–266.

Rothschild, M. and R. Ford. 1972. Breeding cycle of the flea *Cediopsylla simplex* is controlled by breeding cycle of host. *Science* **178**:625–626.

Rothschild, M., Y. Schlein, K. Parker, and S. Sternberg. 1972. Jump of the Oriental rat flea *Xenopsylla cheopis* (Roths.). *Nature* (London) **239**:45–46.

Rothschild, M., J. Schlein, K. Parker, C. Neville, and S. Sternberg. 1975. The jumping mechanism of *Xenopsylla cheopis*. III. Execution of the jump and activity. *Phil. Trans. R. Soc. London* (B) **271**:499–515.

Roughgarden, J. 1976. Resource partitioning among competing species—a coevolutionary approach. *Theor. Pop. Biol.* **9**:388–424.

Rust, R. W. 1974. The population dynamics and host utilization of *Geomydoecus oregonus*, a parasite of *Thomomys bottae*. *Oecologia* **15**:287–304.

Ryckman, R. E. and M. A. Casdin. 1977. The Polyctenidae of the World, a checklist with bibliography. *Calif. Vector Views* **24**:25–31.

Samuel, W. M. and D. O. Trainer. 1971. Seasonal fluctuations of *Tricholipeurus parallelus* (Osborn, 1896) (Mallophaga: Trichodectidae) on white-tailed deer *Odocoileus virginianus* (Zimmerman, 1780) from South Texas. *Am. Midl. Nat.* **85**:507–513.

Sasa, M. 1961. Biology of Chiggers. *Annu. Rev. Entomol.* **6**:221–241.

Sasa, M. and M. Wakasugi. 1957. Studies on the effect of carbon dioxide as the stimulant on the tropical rat mite *Bdellonyssus bacoti* (Huist, 1913). *Jap. J. Exp. Med.* **27**:207–215.

Schaefer, C. W. 1979. Feeding habits and hosts of calyptrate flies (Diptera: Brachycera: Cyclorrhapha). *Entomol. Gen.* **5**:193–200.

Scharff, D. K. 1962. An investigation of the cattle louse problem. *J. Econ. Entomol.* **55**:684–688.

Schmidt-Nielsen, K., L. Bolis, and C. R. Taylor. (Eds.). 1980. *Comparative Physiology: Primitive Mammals*. Cambridge University Press, Cambridge, UK.

Schoener, T. W. 1971. Theory of feeding strategies. *Annu. Rev. Ecol. Syst.* **2**:369–404.

Schoener, T. W. 1974. Resource partitioning in ecological communities. *Science* **185**:27–39.

Sell, S. 1980. *Immunology, Immunopathology and Immunity*, 3rd ed. Harper & Row, Hagerstown, MD.

Sgonina, K. 1935. Die reizphysiologie des igelflohs (*Archaeopsylla erinacei* Banche) und seiner larve. *Z. Parasitenk.* **7**:539–571.

Shepherd, R. C. H. and J. W. Edmunds. 1976. The establishment and spread of *Spilopsyllus cuniculi* (Dale) and its location on the host, *Oryctolagus cuniculus* (L.) in the Mallee Region of Victoria. *Aust. Wildl. Res.* **3**:29–44.

Sims, R. T. 1970. Chapter 5. The epidermis. Pages 61–75 in R. H. Champion et al. (Eds.), *An Introduction to the Biology of the Skin*. Blackwell, Oxford, UK.

Smit, F. G. A. M. 1958. *Fleas, Their Medical and Veterinary Importance*. British Museum of Natural History, Econ. Ser. No. 3A, London.

Smith, H. J. 1961. *Demodicidiosis in Large Domestic Animals*. Health Anim. Dir., Can. Agr. Ottawa, 56 pp.

Smithers, S. R., R. J. Terry, and D. J. Hockley. 1969. Host antigens and schistosomiasis. *Proc. R. Soc.* B. **171**:483–494.

Smyth, D. H. (Ed.) 1967. Intestinal absorption. *Br. Med. Bull.* **23**:205–290.

Smyth, J. D. 1973. Some interface phenomena in parasitic protozoa and platyhelminths. *Can. J. Zool.* **51**:367–377.

Smyth, J. D. 1976. *Introduction to Animal Parasitology*. John Wiley, New York.

Snowball, G. J. 1956. The effect of self-licking by cattle on infestations of the cattle tick *Boophilus microplus* (Canestrini). *Aust. J. Agric. Res.* **7**:227–232.

Sokolov, V. E. 1982. *Mammalian Skin*. University of California Press, Berkeley.

Sonenshine, D. E., C. E. Yunker, C. M. Clifford, G. M. Clark, and J. A. Rudbach. 1976. Contribution to the ecology of Colorado tick fever virus. 2. Population dynamics and host utilization of immature stages of the Rocky Mountain wood tick. *Dermacenter andersoni. J. Med. Entomol.* **12:**651–656.

Sprent, J. F. A. 1962. Parasitism, immunity and evolution. Pages 149–165 in G. S. Leeper (Ed.), *The Evolution of Living Organisms.* Melbourne University Press, Melbourne, Australia.

Sprent, J. F. A. 1969. Evolutionary aspects of immunity in zooparasitic infections. Pages 3–62 in G. J. Jackson (Ed.), *Immunity to Parasitic Animals,* Vol. 1. North-Holland Publ. Co., Amsterdam.

Stark, H. E. and V. I. Miles. 1962. Ecological studies of wild rodent plague in the San Francisco Bay area of California. VI. The relative abundance of certain flea species and their host relationships on coexisting wild and domestic rodents. *Am. J. Trop. Med. Hyg.* **11:**525–534.

Stark, H. E. and A. R. Kinney. 1962. Abandonment of disturbed hosts by their fleas. *Pan-Pac. Entomol.* **28:**249–251.

Stark, H. E. and A. R. Kinney. 1969. Abundance of rodents and fleas as related to plague in Lava Beds National Monument, California. *J. Med. Entomol.* **6:**287–294.

Starr, B. P. 1975. A generalized scheme for classifying organismic associations. Pages 1–20 in D. H. Jennings and D. L. Lee. (Eds.), *Symbiosis. Symp. Soc. Exp. Biol.* No. 29. Cambridge University Press, Cambridge, UK.

Stebbings, J. H., Jr. 1974. Immediate hypersensitivity: A defense against arthropods? *Perspect. Biol. Med.* **17:**233–239.

Steinhaus, E. A. (Ed.). 1963. *Insect Pathology: an Advanced Treatise.* Academic Press, New York.

Stenram, H. 1964. The evolution of the Mallophaga and the phylogeny of their hosts. *Zool. Rev.* **26:**23–32. (in Swedish).

Strandtmann, R. W. and G. W. Wharton. 1958. *A Manual of Mesostigmatid Mites Parasitic on Vertebrates.* Inst. Acarol. Contrib. **4:**1–330.

Tatchell, R. J. 1969a. The significance of host–parasite relationships in the feeding of the cattle tick, *Boophilus microplus* (Canestrini). *Proc. 2nd Int. Congr. Acarol.,* Nottingham **1967:**341–345.

Tatchell, R. J. 1969b. Host–parasite interactions and the feeding of bloodsucking arthropods. *Parasitology* **59:**93–104.

Tatchell, R. J. and D. E. Moorhouse. 1968. The feeding processes of the cattle tick *Boophilus microplus* (Canestrini). *Parasitology* **58:**441–459.

Theron, P. D. 1974. The functional morphology of the gnathosoma of some liquid and solid feeders in the Trombidiformes, Cryptostigmata and Astigmata (Acarina). *Proc. 4th Int. Congr. Acarol.* **1974:**575–579.

Thompson, J. N. 1982. *Interaction and Coevolution.* John Wiley, New York.

Timm, R. M. 1975. Distribution, natural history, and parasites of mammals of Cook County, Minnesota. *Bell Mus. Nat. Hist. Univ. Minn.,* Occas. Pap. **14:**1–56.

Timm, R. M. 1979. The *Geomydoecus* (Mallophaga: Trichodectidae) parasitizing pocket gophers of the *Geomys* complex (Rodentia: Geomyidae). Unpubl. Ph. D. Dissertation, University of Minnesota, St. Paul.

Timm, R. M. 1983. Fahrenholz's Rule and Resource Tracking: A study of host-parasite coevolution. Pages 225–265 in M. H. Nitecki (Ed.), *Coevolution,* University of Chicago Press, Chicago.

Timm, R. M., and E. F. Cook. 1979. The effect of bot fly larvae on reproduction in white-footed mice, *Peromyscus leucopus. Amer. Midl. Nat.* **101:** 211–217

Timm, R. M., and R. E. Lee, Jr. 1981. Do bot flies, *Cuterebra* (Diptera: Cuterebridae), emasculate their hosts? *J. Med. Entomol.* **18:** 333–336.

Timm, R. M., and R. D. Price. 1980. The taxonomy of *Geomydoecus* (Mallophaga: Trichodectidae) from the *Geomys bursarius* complex (Rodentia: Geomyidae). *J. Med. Entomol.* **17**: 126–145.

Tipton, V. J. and E. Mendez. 1968. New species of fleas (Siphonaptera) of Panama. Pages 289–385 in R. L. Wenzel and V. J. Tipton (Eds.), *Ectoparasites of Panama*. Field Museum of Natural History, Chicago, IL.

Toldt, K., Jr. 1935. *Aufbau und naturliche Farbung des Haarkleides der Wildsaugetiere*. Deutsch. Gesellsch. f. Kleintier-und Pelztierzucht. Leipzig. 291 pp.

Tortora, G. J. and N. P. Anagnostakos. 1981. *Principles of Anatomy and Physiology*, 3rd ed. Harper & Row, New York.

Trail, D. R. Smith. 1980. Behavioral interactions between parasites and hosts: host suicide and the evolution of complex life cycles. *Am. Nat.* **116**:77–91.

Traub, R. 1980. Some adaptive modifications in fleas. Pages 33–67 in R. Traub and H. Starcke (Eds.), *Fleas*, Proc. Int. Conf. on Fleas, Ashton Wold/Peterborough/UK, 21–25 June 1977, A. A. Balkema, Rotterdam.

Ueshima, N. 1972. New World Polyctenidae (Hemiptera) with special reference to Venezuelan species. *Brigham Young Univ. Sci. Bull., Biol. Ser.* **17**:13–21.

Ulmanen, I. and A. Myllyamaki. 1971. Species composition and numbers of fleas (Siphonaptera) in a local population of the field vole *Microtus agrestis* (L.). *Ann. Zool. Fenneci* **8**:374–384.

Usinger, R. L. 1934. Bloodsucking among phytophagous Hemiptera. *Can. Entomol.* **66**:97–100.

Usinger, R. L. 1946. Polyctenidae. *Gen. Cat. Hemiptera* **5**:1–18.

Usinger, R. L. 1966. Monograph of Cimicidae (Hemiptera-Heteroptera). *Entomol. Soc. Am. Thomas Say Found.* **7**:1–585.

Valentine, J. W. 1972. 10. Conceptual models of ecosystem evolution. Pages 192–215 in J. M. Schopf (Ed.), *Models in Paleobiology*, Freeman, Cooper, and Company, San Francisco, CA.

Vanderplank, J. E. 1982. *Host–Pathogen Interactions in Plant Diseases*. Academic Press, New York.

Van Valen, L. 1971. Adaptive zones and the orders of mammals. *Evolution* **25**:420–428.

Vanzolini, P. E. and L. R. Guimaraes. 1955. Lice and the history of South American land mammals. *Rev. Brasil. Entomol.* **3**:13–46.

Vaughan, J. A. and A. R. Mead-Briggs. 1970. Host-finding behaviour of the rabbit flea, *Spilopsyllus cuniculi* with special reference to the significance of urine as an attractant. *Parasitology* **61**:397–409.

Waage, J. K. 1979. The evolution of insect/vertebrate associations. *Biol. J. Linn. Soc.* **12**:187–224.

Waage, J. K. and G. G. Montgomery. 1976. *Cryptoses choloepi*: a coprophagous moth that lives on a sloth. *Science* **193**:157–158.

Wakelin, D. 1976. Chapter 6. Host Responses. Pages 115–141 in C. R. Kennedy (Ed.), *Ecological Aspects of Parasitology*, North-Holland, Amsterdam.

Wakelin, D. 1979. Genetic control of susceptibility and resistance to parasitic infection. *Adv. Parasitol.* **17**:219–308.

Wallace, F. G. 1966. The trypanosomatid parasites of insects and arachnids. *Exp. Parasitol.* **18**:124–193.

Walsburg, G. E. 1983. Coat color and solar heat gain in animals. *Bioscience* **33**:88–91.

Walsburg, G. E., G. S. Campbell, and J. R. King. 1978. Animal coat color and radiative heat gain: a re-evaluation. *J. Comp. Physiol.* **126**:211–222.

Webb, J. P., J. E. George, and B. Cook. 1977. Sound as a host-detection cue for the soft tick *Ornithodoros concanensis*. *Nature* (London) **265**:443–444.

Weber, H. 1929. Biologische Untersuchungen an der Schweinelaus (*Haematopinus suis* L.) unter besonder Berucksichtigung der Sinnesphysiologie. *Z. Vgl. Physiol.* **9**:564–612.

Weber, H. 1969. Die Elefantlaus *Haematomyzus elefantis* Piaget 1869. Versuch einer konstruktionsmorphologischen Analyse. *Zoologica* (Stuttg.) **41**:1–154.

Weinberg, E. D. 1979. Metal starvation of pathogens by hosts. *Bioscience* **25**:314–318.

Weisser, C. F. 1975. *A Monograph of the Linognathidae, Anoplura, Insecta (excluding Prolinognathus)*. Doctoral dissertation, University of Heidelberg, Heidelberg.

Wenzel, R. L. 1976. The streblid bat flies of Venezuela (Diptera: Streblidae). *Brigham Young Univ. Sci. Bull.* **20**:1–177.

Wenzel, R. L. and V. J. Tipton. 1966. *Ectoparasites of Panama*. Field Museum of Natural History, Chicago, IL.

Westrom, D. R., B. C. Nelson, and G. E. Connolly. 1976. Transfer of *Bovicola tibialis* (Piaget) (Mallophaga: Trichodectidae) from the introduced fallow deer to the Columbian black-tailed deer in California. *J. Med. Entomol.* **13**:169–173.

Weyer, F. 1960. Biological relationships between lice (Anoplura) and microbial agents. *Annu. Rev. Entomol.* **5**:405–420.

Wharton, G. W. 1954. Life-cycle and feeding habits of *Myobia musculi*. *J. Parasitology* **40**:29.

Wharton, G. W. and H. S. Fuller. 1952. A manual of chiggers. *Mem. Entomol. Soc. Wash.* **41**:1–185.

Whitfield, P. J. 1979. *The Biology of Parasitism: an Introduction to the Study of Associating Organisms*. University Park Press, Baltimore, MD.

Whittaker, J. O. 1968. 7. Parasites. Pages 254–311 in J. A. King (Ed.), *Biology of Peromyscus (Rodentia)*. Spec. Publ. 2, Am. Soc. Mamm.

Whittaker, J. O. and N. Wilson. 1974. Host and distribution lists of mites (Acari), parasitic and phoretic, in the hair of wild mammals of North America, north of Mexico. *Am. Mid. Nat.* **91**:1–67.

Whittaker, R. H. 1972. Evolution and measurement of species diversity. *Taxon* **21**:213–251.

Whittaker, R. H. 1977. Evolution of species diversity in land communities. *Evol. Biol.* **10**:1–66.

Whittaker, R. H. and P. P. Feeny. 1971. Allelochemics: Chemical interactions between species. *Science* **171**:757–770.

Whittaker, R. H. and G. M. Woodwell. 1972. Evolution of natural communities. Pages 137–157 in J. A. Wien (Ed.), *Ecosystem Structure and Function*, Proc. 31st Ann. Biol. Colloq. Oregon State University Press, Corvalis.

Whittaker, R. H., S. A. Levin, and R. B. Root. 1973. Niche, habitat, and ecotype. *Am. Nat.* **107**:321–338.

Wigglesworth, V. B. 1941. The sensory physiology of the human lice *Pediculus humanus corporis* deGeer (Anoplura). *Parasitology* **33**:67–109.

Wikel, S. K. 1982. Immune responses to arthropods and their products. *Annu. Rev. Entomol.* **27**:21–48.

Wikel, S. K. 1985. Effects of tick infestation on the plague-forming cell response to a thymic-dependent antigen. *Ann. Trop. Med. Parasitol.* (in press).

Willadsen, P. 1980. Immunity to ticks. *Adv. Parasitol.* **18**:293–313.

Williams, C. B. 1964. *Patterns in the Balance of Nature*. Academic Press, London.

Yunker, C. E. and J. M. Brennan. 1962. Endoparasitic chiggers: II. Rediscovery of *Doloisia synoti* Oudemans, 1910, with descriptions of a new subgenus and two new species (Acarina: Trombiculidae). *Acarologia* **4**(4):570–576.

Zumpt, F. 1965. *Myiasis in Man and Animals in the Old World*. Butterworths, London.

Chapter 2

Coevolution as a Process

What Parasites of Animals and Plants Do Not Have in Common

Daniel H. Janzen

INTRODUCTION

I detest reading definitions, but we must begin with one. As used here, *coevolution* is the event where the members of one species select for a change in another species successfully and then in turn evolutionarily respond to that change. This is what was meant by Erlich and Raven's (1965) seminal paper on the subject, although their example of butterflies radiating onto host plant families is largely not an example of coevolution

(Janzen 1980, 1981). *Diffuse coevolution* is the same process, except that one or both species in the foregoing definition are replaced by a suite of species. Defined in this manner, the evolution of elaborate tarsi for holding onto the host's hairs is in itself explicitly *not* an example of coevolution (unless one wishes to argue that hair shape is an evolutionary attempt to make life difficult for the parasite). Likewise, a cataloging of the host specificity of Mallophaga is not in itself a coevolutionary study. In addition to requiring such a hardening up of the word *coevolution* to make it operational, my intent is to draw attention to the obvious fact that many of the tightly evolved relationships between parasites and hosts clearly display no evidence of complementary evolutionary change by the host, and furthermore, in many cases there is no theoretical reason to expect such change. The same applies to mutualisms such as seed dispersal and pollination by animals.

The assignment for this chapter was to discuss coevolution as a process with respect to mammalian ectoparasites. To examine the coevolutionary relationships between a pair of groups usefully, one needs to know who interacts with whom, what group 1 does to group 2, and how group 2 responds to group 1. There is already a very competent and respectable review literature telling us where we stand on these three points with mammalian ectoparasites (in addition to the chapters in the present volume, see Barbehenn 1978; Schad 1963; Catts 1982; Valdivieso and Tamsitt 1970; Freeland 1976; Mansour 1979; Kennedy 1975, 1976; Price 1975, 1977, 1980; Holmes and Price 1980; Brooks 1980; Kuris 1974; Wakelin 1979; Wikel and Allen 1976; Randolf 1979; Jackson 1969; Marshall 1981; Rothschild and Clay 1952; Nelson et al. 1975, 1977; Fain 1979; Johnston 1975). Perhaps the most distinctive trait of this literature is the imbalance in favor of knowing who parasitizes whom: more emphasis has been on the effect of parasites than on how hosts have evolutionarily responded to parasites. Likewise, there is a very large body of information on the ectoparasites of plants (e.g., see reviews in Rosenthal and Janzen 1979; Janzen 1973; Levin 1971, 1973, 1976a, b; Edmunds and Alstad 1978; Price 1975; Jermy 1976; Atsatt 1977; Strong and Levin 1979; Strong 1979; Dugdale 1977; Sutherland 1977; Russell 1977; Farrell 1977; Grandison 1977; Friend and Threlfall 1976; Vinson 1976; Clausen 1976; Niemela and Haukioja 1980; Bryant 1981; Gilbert 1977; van Emden and Way 1973; Southwood 1973; Haukioja and Niemela 1979; Powell 1980; Mound and Waloff 1978; Wallace and Mansell 1976; Atsatt and O'Dowd 1976). Price (1980) offers the first general review that is based on these two literatures, which have evolved at least 100 years with virtually no cross-fertilization.

Rather than to attempt a superficial regurgitation of the literature of mammalian ectoparasitology, my goal is here better served by comparing a caricature of the ectoparasites of mammals with the ectoparasites of plants, and in this manner underlining some of the ways parasites and hosts do and do not evolve or coevolve. Table 2.1 lists a few of the more glaring ways in which these two large groups of ectoparasites differ, emphasizing

Table 2.1 General Differences between Mammalian and Plant Ectoparasites

Mammal Ectoparasites	Plant Ectoparasites
1. Aside from host defenses, subject to virtually no predation and parasitization while on the host.	1. Subject to severe predation and parasitization while on the host.
2. Chemical content of food has relatively low variation among all mammal species.	2. Chemical content of food is enormously variable among all plant species.
3. If more than a small fraction of the living tissues of the host is removed, the host dies.	3. Very large fractions of host tissue may be removed without killing host.
4. Feeding on virtually any member of the host population is likely to generate selective pressure favoring traits that deter the parasite.	4. Many members of the population are genetically dead even if they continue to live and be food for parasites.
5. Close body contact with host is commonplace, such that the host can chemically identify the parasite.	5. Body contact with host is minimal, and where it occurs, opportunity to identify parasite is minimal.
6. Hosts contribute strongly to interhost movements by parasites.	6. At best, hosts play only a passive role (unavoidably produce distinctive chemical/odor fingerprints) in interhost movements by parasites.
7. Allospecific and conspecific individuals compete largely through the medium of the resource budget and immunological system.	7. Allospecific and conspecific individuals often compete directly by eating that portion of the plant some other parasite would have.
8. Commonly occur in large numbers on a host.	8. Rarely occur in large numbers on a host.
9. Spend most if not all their life on or very close to the host.	9. Spend major portions of their life at long distances from the host.

those that I suspect have resulted in substantial differences in evolution or coevolution.

PARASITES (PARASITOIDS) AND PREDATORS OF ECTOPARASITES

There appears to be virtually no predation or parasitoidization of mammalian ectoparasites while they are on the host other than that of the host itself as part of its own defense. To be sure there is the odd case of allogrooming by such things as oxpeckers on zebras, but by and large the fleas, for example, on rodents are not sought after by fur-gleaning birds, ants, or monkeys. Ectoparasites of mammals are not generally sought by ichneumonids, tachinids, braconids, chalcids, and so forth. This is in strong contrast to the very severe mortality bestowed on foliage-feeding insects by whole communities of birds, lizards, monkeys, parasitic wasps, ants, parasitic flies, viruses, fungi, bacteria, and other carnivores.

Surely then, one of the generally unappreciated advantages of evolutionarily moving from a scavenger, omnivore, or free-living carnivore diet to an ectoparasitic mode of life is increased freedom from generalist and many kinds of specialist carnivores. The close proximity to the host must often result in increased mortality, however, from a particular carnivore (the host). Such mortality has been responded to through the evolution of a variety of behavioral and morphological traits that are functional in avoiding predation from a specific animal. Since this evolution occurs on a specific animal with a vested interest in being able to prey on a very particular ectoparasite morphotype, there have been a few cases of what appear to be diffuse coevolution: grooming toe morphologies, social grooming, scratching behavior, itching physiology, and so forth. However, each of these traits is functional in other aspects of cleaning behavior as well, and, therefore, the selective pressure is extremely diffuse and certainly generated by more than just selection by the ectoparasites.

It is likewise conspicuous that mammalian ectoparasites are totally lacking in the complex of aposematic and mimicking insects so prominent among plant ectoparasites. There are several suspect causes. The diets of mammalian ectoparasites are not such as to preadapt the insect to an aposematic mode of existence, in contrast to plant tissue which may fill the gut of the most edible insect with the most inedible materials. Second, the ectoparasite is not under selection so much by visually orienting as tactile predators. Third, the predator of the ectoparasite is not so much after a meal as it wants to be lethal; even when a flea or tick is distasteful, it can be spit out, once killed.

It can be argued with respect to plants that one of the selective pressures for the production of odoriferously conspicuous secondary compounds, which are for the most part defensive chemicals, is that it makes the plant more olfactorily conspicuous to the hymenopterous and dipterous parasitoids that will search it for caterpillars. The possibility for this kind of coevolution simply does not appear to exist in the case of mammalian ectoparasites, unless various I-am-ready-to-be-groomed displays are analogous to such plant chemical signals.

The demography of animal populations such as ectoparasites that spend major portions of their lives in an environment with no threat of general predators or any kind of parasites is bound to be very different from that of free-living animals: hence, much tighter coevolutionary interactions between parasite and host should be possible in mammals and their ectoparasites than in plants and their ectoparasites. Mammalian ectoparasites that are developing coevolutionary (as well as evolutionary) interactions do not have a large variety of other carnivores selectively tugging in other directions. The distorted morphologies of mammalian ectoparasites, as well as those of other vertebrate ectoparasites, are undoubtedly in part the result of this relatively monolithic direction of selection. A caterpillar has to deal not only with the traits of its host, but also with avoiding birds, wasps, and ants—each of which calls for quite different escape behaviors and morphologies.

FOOD DIVERSITY FOR ECTOPARASITES

If a flea were to sample the blood of the 30 species of mammals in its habitat, the diversity of potentially dangerous chemicals it encounters in no way would equal the diversity of secondary compounds that would be encountered by a caterpillar sampling a random selection of the leaves of 30 species of plants. To be sure, there are parts of the plant (apical meristems, cambial cells, some phloem and xylem saps, ovules, seeds encased in very hard nuts) that may be relatively bland, and of course some species of plants have similar chemical defense characteristics (especially closely related ones). Likewise, hormone differences, antibody differences, sugartiters, and other blood traits render each species of mammal's blood a different diet. However, as a general rule, a louse, tick, or flea making an evolutionary hop from one species to another, especially if within families or genera of mammals, will have proportionately fewer diet composition problems than will a caterpillar or sucking bug making the analogous hop across plant taxa.

A major aspect of the coevolution of host–parasite relationships in both mammals and plants is what the parasite does when a host mutant appears that drastically lowers the parasite's fitness. The more easily a mutant parasite can hop to another species of host on which it has a fitness even a bit higher than on the mutant host, the less likely the appearance of the initial mutant host is to generate an act of coevolution (once the parasite has made the hop to a new host, it no longer can coevolve with the old one). The more similar are hosts, the easier the hop. The question then becomes whether the mammalian barrier of mild food differences and strong behavioral/morphological differences, preening, itching, fur texture, and immune responses, are on the average equal to strong food chemical differences and weak morphological differences of plants (problems in caterpillar crypticity on the foliage of the new host, etc.). Although there are many cases in mammalian ectoparasites where a particular group seems to have been locked evolutionarily into one taxon, there are also many plant–insect pairs. Further, the widely held view that restriction of a distinctive parasite group to a particular host taxon means a long evolutionary history is logically quite indefensible. Ectoparasites of plants have quite clearly explosively radiated across large sets of genera or species of hosts, and have probably done it over a time period so short that little or no change in the hosts need have taken place; there is no reason why the same could not have occurred with mammals and their hosts. It should be mentioned that such a radiation is certainly not evidence for coevolution, and in fact is unlikely to represent it. If a species of louse makes the evolutionary hop onto one member of a species-rich genus of rodents and then radiates onto all of them, generating a number of louse species along the way, it is precisely the failure of a host response that allows the louse mutant genotype to hop from mouse species to mouse species— presuming that the closely related mice have similar defenses. Were the mice to start coevolving with the lice, one very possible result would be the elimination of the lice or their restriction to only a few species of mice, the very case that is often thought of as not suggesting *coevolution*.

BITE SIZES

Plants wear their stomachs on the outside, have an enormous surface area to volume ratio for living tissue, and have enormous regenerative power when compared with mammals. Associated with these three traits in concert, plant ectoparasites can, and often do, periodically harvest enormous pulses of food from their hosts, followed by a period when the host recovers while being subjected to little or no parasitism. Plants can sustain ectoparasite species that for seasonal, predatory, or other reasons require periodic peaks of density far greater than could be sustained by the host on a full-time basis. When a caterpillar eats all the leaves off its host plant and then disappears for a year, it has been allowed a life-style equivalent to a population of fleas that thoroughly exsanguinated most or all of the rabbits in a habitat at 12-month intervals. Of course, there are occasional events where a mammalian ectoparasite builds up to a level where it kills the host, but even here the amount of tissue that has been removed is quite small compared with what a population of defoliating caterpillars removes. Furthermore, such an event is generally not viewed as the normal form of the parasite–host interaction.

It is tempting to suspect that the amount of tissue grazed off by mammalian ectoparasites has been evolutionarily or coevolutionarily adjusted downward because these ectoparasites are much more dependent on the host for habitat than are plant ectoparasites. However, an examination of that hypothesis is greatly confounded by the much superior repair and regeneration properties of plants. It is also confounded by the fact that a gram of material removed by a mammalian ectoparasite is of considerably greater usable value than is a gram of material removed by a plant ectoparasite, with the obvious exceptions of mallophaga on the one hand, and specialists on cambial tissue and ovaries on the other hand.

It is easy to suspect that the evolution of strong regenerative powers in plants came about as one of a variety of defensive responses to herbivores. It is hard to imagine how herbivores have evolved into taking amounts of food that do not push the plant out of the habitat; in fact, there is considerable suspicion that herbivores do push this or that species of plant out of a habitat, or substantially alter its density. Coevolution does not readily fall out of the interaction. With mammalian ectoparasites, however, it is easy to see how selection could have resulted in moderation of the amount of tissue removed since so much of the life of the ectoparasite is spent on the host. If there are microdemes of parasites on each individual host, there may even have been a form of group selection possible. However, it is much less clear whether mammals have evolved regenerative powers in direct response to the usual damage done by an ectoparasite. When a warble fly larva exits from the usual host, the wound normally closes cleanly, whereas it may be purulent when the larva exits from an artificial host. But is this because the usual host has evolved such a response or because the antibiotics, anesthetics, and other tissue-altering drugs released by the fly larva were simply not evolutionarily fine-tuned to the biochemistry of the artificial host?

STRUCTURED HOST POPULATION

If a mammalian ectoparasite depresses its host's fitness in any way, it is not hard to imagine that there will be selection for traits to eliminate the parasite, but whether the selection will be manifest in a change in the genotype is quite another question. But what about the hosts that would have died, for example, before reaching reproductive status through mortality factors quite unrelated to the presence of the ectoparasite? If a usual level of fleas does not influence a vole's chances of being taken by a fox before the age at which the vole would have been reproductive were it not to have fleas, have the fleas on that vole gotten a selective "free ride" from each fox-eaten vole? In other words, is the vole population, when under heavy predation by foxes, supporting a major flea population that is then not selectively felt by the voles? This weird-looking question derives directly from a consideration of how plant ectoparasites may interact with their hosts.

The preceding question can be answered in the affirmative *if* the ectoparasites distribute themselves on the host population in partial or total response to the eventual fate of the host. The members of a plant population, excluding seeds for the moment, commonly can be partitioned into two groups. There are those seedlings and saplings that are growing in sites where they have some chance of becoming adults. There are also those seedlings and saplings that are growing in sites in which they are guaranteed never to become reproductive; that is to say, they are evolutionarily dead, irrespective of how green they are. The most extreme case of this would be a tree seedling growing in the dim light of a cave where its seed was dropped by a frugivorous bat. A more commonplace case is that of seedlings and saplings growing directly below the crown of a middle-aged parent tree, where the young plants must attain the canopy or die in a period much shorter than the usual remaining life span of the parent. Here, if the ecological herbivore is a species that specializes on such plants, it becomes an evolutionary detritivore.

The key point is whether the ectoparasite is wholly or largely restricted to the plants that are doomed to die. The evolution of such a feeding preference is easy to visualize. A herbivore immigrates into a new habitat. Its traits cause it quite serendipitously to use that portion of the plant population growing in very heavy shade rather than the portion of the plant population growing in tree falls, edges, creek banks, and so forth. Such a herbivore is much less likely to select for an evolved defense response by the plant population than is the herbivore that initially moves onto the members of the plant population with a high chance of surviving to reproductive status. Additionally, the herbivore need not have come from some other habitat but could also have appeared through an evolutionary move from some other host species. The final outcome of such a process should be the accumulation of a higher equilibrium density of species on those portions of the plant population that are doomed than on those that are not. That is to say, it is the *lack* of a coevolutionary response by the plant that may result in this pattern.

The question now becomes one of whether structure such as that described for plants can be recognized in the interactions of mammals with their ectoparasites. The key data are those telling us how ectoparasites are distributed among the members of a mammalian population and the environmental processes that maintain that distribution. Since most mammals can move about, it may well be that *all* members of the population have some chance of survival to reproductive status, and therefore the scenario for plants cannot be reasonably applied to animals. We need studies of how parasite individuals are distributed over the various fate-classes of their hosts.

HOST RECOGNITION OF ECTOPARASITES

Although the array of host-generated facultative immune responses does not seem to be as spectacular with mammalian ectoparasites as with mammal endoparasites, the long periods of intimate body contact and the feeding mode of mammal ectoparasites make it possible quite often for the host to identify the ectoparasite chemically and therefore respond specifically to it. The chances for coevolution are great. Inflammation from chiggers on humans and the lack of inflammation on usual hosts provide a familiar example. How long a mosquito can feed without itching is no accident, although one wonders to what degree the mammal host has evolved, if at all, toward adjusting its sensitivity to mosquito fluids. It is not surprising to find that the chemical defenses of mammals against ectoparasites are often quite parasite specific; the specificity of the immune system is perhaps its greatest structural difference from the chemical defense systems of plants, be they standing or facultative.

Ectoparasites of plants, however, by and large feed in such a manner that there is only a minimal chance of the plant knowing much more than that it had its leaf eaten off. When the ectoparasite is very sedentary (e.g., scale insects, mealybugs), then there is the chance of internal chemical changes that begin to approximate mammalian immune responses. However, such cases are both rare in species and rare in individual cases when compared with the great amount of foliage browsing and sapsucking done by more mobile herbivores. The only way a plant can "know" who is eating its leaves is through the evolutionary form of learning whereby most of the herbivory committed during previous generations was carried out by that set of specialist herbivores, as well as generalists, that have to some degree breached the chemical defenses of the plant. That is to say, whoever is eating you today is likely to be who was eating you yesterday. When one considers that most plant species have very large suites of herbivore species that feed on them, the possibility of species-level recognition is even more distant.

It should be cautioned that plants do have a variety of facultative defenses that are turned on when tissues are eaten and damaged. And in some cases, the level or kind of chemistry may vary depending on whether it is a scissors or a cow or beetle that dripped its saliva on the leaf when eating that leaf. Nevertheless, there appears to be nothing that approximates the very chemical specificity invented by the mammalian immune response. Why has this system not diffusely coevolved in plants? I suspect that a major reason is that caterpillars, as well as many other kinds of herbivores, feed on a plant for a few weeks or months and then leave for a variety of reasons quite independent of the plant–ectoparasite interaction, for example, bad weather, lack of time for another generation, predator/parasite buildup, and so forth. Furthermore, a plant that is heavily damaged at one time is quite likely to be heavily damaged again by a quite different species of insect, and it may be many years before the first species again commits severe herbivory to that species. Finally, there is enormous heterogeneity in a plant population as to which individuals are fed on in a given year, a heterogeneity that is driven by many more factors than just the chemistry of the foliage. In strong contrast, a mite population that is not depressed or slowed by some sort of active immune or other response by its host mammal is likely to increase to lethal or severely debilitating levels and to spread thoroughly through the remainder of the mammalian population as well. The very traits that make a mammal different from a plant, such as endothermy, high edibility, and active microenvironment modifiers, create a high-quality parasite/predator-free microenvironment for the exploding mite population. In short, mammals have the environment much less on their side than do plants when it comes to dealing with ectoparasites, and they make up for it with their immune system and behavior. It is hard to coevolve a fine watch if God keeps pouring sand into the works.

HOSTS DISPERSE ECTOPARASITES OF MAMMALS

Although a mammalian ectoparasite is capable of movements between hosts of a few centimeters to meters, or of sitting in one place until a host passes by, quite analogous to a caterpillar, it generally lacks the movement abilities of the winged phase of plant ectoparasites. Of course, some parasitic Diptera have wings and plant mites lack them, but, in general, mammalian ectoparasites depend on their hosts for geographic displacement, and plant ectoparasites both search actively through the habitat and move readily across a wide variety of nonhabitats. Clearly, plant ectoparasites have a much higher chance of being panmictic within the general habitat they occupy while mammalian ectoparasites have a much higher chance of existing in microdemes at the level of nests, individual animals, family groups, and so forth. On the other hand, if there are detriments to decreased outcrossing, mammalian ectoparasites are more likely to suffer them than are plant ectoparasites. One wonders why gravid ticks are so eager to abandon house and home!

As briefly mentioned earlier, this means that mammalian ectoparasites have a greater chance to proceed with (co)evolutionary changes through not only the usual kind of selection, but also through some sort of group selection, where the monospecific faunulet of a given host is the unit of selection. This would occur in plants only where the ectoparasites are exceptionally sedentary like some mites, scale insects, and mealybugs, and it would be helped along by having many of the plants in the population so well defended that they were totally unacceptable, thereby rendering the acceptable individuals even more insular. I suspect that any process that tightens up the ecological interaction between host and ectoparasite raises the chance of coevolutionary processes.

In quite a different vein, the adult stage of most herbivores is a specialist at locating hosts or host parts that are widely and often cryptically scattered in space and time. This results in two somewhat different lineages of evolved or coevolved interactions. A plant may have one set of chemical interactions with the caterpillars, once there, and quite a different set of chemical interactions with the ovipositing moth. The very chemicals that render a plant inedible to a large suite of caterpillars may be the unavoidably conspicuous olfactory cues that the ovipositing adult can use to find a rare host plant. Such conflicting effects of the same trait can be very disruptive to the evolution of the one-on-one interactions characteristic of coevolved systems. Perhaps the mammalian analogue would be that the very social proximity that leads to social grooming likewise leads to rapid and thorough spread of ectoparasites among the members of a social group. On the other hand, a mother mammal's ectoparasites are given to offspring about as thoroughly as are the mother's genes; plants start their independent life quite clean. One of the prices paid by juvenile mammals for the milk subsidy is a healthy dose of the mother's ectoparasites, yet they may gain from the opportunity for highly evolved or coevolved interactions that come about through the fidelity and omnipresence of the inoculation.

COMPETING THROUGH THE RESOURCE BUDGET

It is quite evident that mammalian parasites compete with each other through the medium of the resource budget of the host and that they induce immune responses that exclude other parasites. Plant ectoparasites do the same, and likewise in both evolutionary and contemporary time scales, though the facultative chemical responses by plants do not begin to have the specificity and cross-immunity traits of mammals. However, plant ectoparasites also compete inter- and intraspecifically by physically removing the food that would have been eaten by some other herbivore. When a leafcutter ant colony strips the leaves off a tree, it does not spare those with moth eggs or caterpillars and does not leave behind enough food for those larvae to complete their development. When a large number of gypsy moth larvae defoliate a tree, the first to get the leaves do not spare leaves for latecomers. A large caterpillar eating a leaf is likely to consume the young immatures of other insects or physically force them off the leaf.

I suspect that this very direct style of competition is generally missing from mammalian ectoparasite interactions. Feeding fleas and mosquitoes do not generally shoulder each other out of the way or suck the well dry, though physical space may be limiting for mammalian ectoparasites that densely fill small spaces. The consequence should be that three-way interactions between two ectoparasites and a mammalian host are less likely to be focused on the details of the two parasites' traits than is the case with evolved or coevolved three-way interactions between two ectoparasites and a plant host. For example, the timing and location of oviposition by a moth may well be determined by having to wait to see where some other species of moth is taking its bite out of the plant population before being able to determine the best place to lay its eggs.

LARGE NUMBERS PER HOST

The standing crop of ectoparasites on a mammal is commonly numerically very high. Another way to put it is that mammalian ectoparasite populations tend to be made of very many small individuals rather than a few large ones; this is a way to evolutionarily sneak a lion into a mouse colony, and it can be done by selection for individuals small enough to escape the search behavior of the host. To put it another way, if you are going to be as large as a flea, then you cannot ride around by the hundreds clinging to mouse hairs.

With the exception of sometimes being more desirable or conspicuous to foliage-gleaning predators, an increase in size per se is generally not as directly dangerous to a plant ectoparasite. This means that more morphological options are open to the plant ectoparasite in evolving or coevolving its interactions with the host and its environment. In some cases, such as 10–20-g moth caterpillars and leaf-eating beetles, it is clear that selection has pushed the system very far in the a-few-large-individuals direction. On the other hand, there are a few cases where plant ectoparasites have moved in the other direction, with aphids being the most omnipresent, at least in extra-tropical habitats, and conspicuous. A beaver ear stuffed with mites bears a decided resemblance to a cherry inflorescence stuffed with aphids. There are two conspicuous differences, however. First, there are no ants tending the mites in the beaver ear. Second, the mites need not have been produced parthenogenetically, while the aphids were. Aphids are specialists at moving onto a temporarily abundant poorly defended food source and strongly subdividing themselves, with the pieces suffering high carnivory but the whole organism having a very high survivorship once the parthenogenetic beast has had a bit of time to get the initial subdivision well under way. The degree to which they are analogous to the large populations of mammalian ectoparasites sustained by a single mammal depends on the degree to which the mammalian ectoparasites are the parthenogenetically produced offspring of a few initial colonizers. Aphids are well known to have evolved many specific chemical and behavioral adaptations that fine-tune them to the biology of particular species of host plants. However, the degree to which their hosts have responded specifically to aphids with complementary traits is largely unknown. Even gall forming by aphids may be nothing more than a purely one-way manipulation of the host plant's biochemistry by aphid-released hormones and other signals, and therefore not coevolve at all. Incidentally, it is striking that mammalian ectoparasites are not gall formers, unless warble fly warbles and delayed itching of mosquito bites may be viewed as a very crude beginning, whereas this trait has evolved numerous times in all the major taxa of arthropodian ectoparasites of plants.

ALL OF LIFE TIED TO HOST

As has been obliquely referred to on several previous occasions, the great part of the life cycle of mammalian ectoparasites is spent on or very near to the host. Ticks are the only major exception, but this exception is confounded by the observation that ticks are so willing to leave hosts, only to climb right back on another, that it is as if there is some extrinsic value to exchanging hosts and the time spent off the host is an unavoidable by-product of this exchange. Plant ectoparasites normally spend only one portion of the life cycle on the host. Full-grown caterpillars so commonly actively vacate the area of the host that the location of the pupa is almost guaranteed *not* to be the host individual or species. Adult herbivorous insects more often feed on different hosts or host parts than do juveniles. Adult holometabolous insects commonly have nothing to do with larval hosts except as an oviposition site. For major seasons of the year the feeding stages of a plant ectoparasite may well be totally missing from the habitat, even when the plant is a growing and apparently resource-rich substrate.

The result is that the density of feeding ectoparasites to arrive at a host may be largely or entirely determined by density-dependent and density-independent mortality processes connected in no direct manner with the biology of the host. I am sure that this fact goes a long way toward disrupting parasite–plant evolving or coevolving interactions as they begin. If the number of leaves you lose to a caterpillar is determined not so much by how your leaves taste as by whether there happened to be the right species of nectar-bearing flowers seven months ago in a different habitat, and next year it is determined by whether some other species of caterpillar suffered a disease epidemic to which your caterpillar is also susceptible, then evolutionary, to say nothing of coevolutionary, processes have a tough time maintaining the linkage they need to persist.

CONCLUSION

What we know of mammalian ectoparasites derives from medical and taxonomic interests. What we know of plant ectoparasites derives from agricultural and taxonomic interests. The medical world has by and large been more fascinated with the biochemistry of the interaction than has the agricultural world, probably because of the relative different net worths to humans of the individual patients. We now need, and see appearing on the horizon, extensive inquiries into the ecology of the interaction itself in the context of the habitats in which it evolved, coupled with detailed observations of what hosts do when ectoparasites are feeding on them. Stating that the host is sick, dead, or debilitated is not a detailed observation. *All* ectoparasitism has a cost, otherwise the ectoparasites are simply detritivores. The question is one of how this cost compares with the budgetary and resource noise in the system and with the cost of eliminating that ectoparasitism, and whether the possible defenses are compatible with other traits of the animal. The identification of the relative role of evolution versus coevolution in such an inquiry is trivial. Of much greater importance is the now well-established understanding that both members of any interaction are potential evolvers, and that neither evolves solely in response to the selective pressures of the other.

SUMMARY

Although ectoparasites of mammals and of plants have a great deal in common, they also differ in many aspects of their ecology, population biology, potential for evolutionary and coevolutionary change, and host interactions. These differences range from the nearly complete lack of predators and parasitoids of mammalian ectoparasites to the possibility that plants may sustain large ectoparasite populations with little or no selective effect owing to the way plant populations are structured.

ACKNOWLEDGMENTS

This study was supported by NSF DEB 80-11558 and was initiated at the request of K. C. Kim. J. C. Holmes, W. J. Freeland, G. A. Shad, T. Richie, K. C. Kim, W. Hallwachs, and K. R. Barbehenn aided in locating references.

REFERENCES

Atsatt, P. R. 1977. The insect herbivore as a predictive model in parasitic seed plant biology. *Am. Nat.* **111**:579–586.

Atsatt, P. R. and D. J. O'Dowd. 1976. Plant defense guilds. *Science* **193**:24–29.

Barbehenn, K. R. 1978. Concluding comments: from the worm's view, "ecopharmacodynamics," and 2000 A.D. pages 231–236 in D. P. Snyder (Ed.), *Populations of*

Small Mammals under Natural Conditions, Vol. 5. Special Publication Series Pymatuning Laboratory of Ecology, University of Pittsburgh.

Brooks, D. R. 1980. Allopatric speciation and non-interactive parasite community structure. *Syst. Zool.* **30**:192–203.

Bryant, J. P. 1981. Phytochemical deterrence of snowshoe hare browsing by adventitious shoots of four Alaskan trees. *Science* **213**:889–890.

Catts, E. P. 1982. Biology of New World bot flies: Cuterebridae. *Annu. Rev. Entomol.* **27**:313–338.

Clausen, C. P. 1976. Phoresy among entomophagous insects. *Annu. Rev. Entomol.* **21**:343–368.

Dugdale, J. S. 1977. Some characteristics of phytophagous insects and their hosts in New Zealand. *N. Z. Entomol.* **6**:213–221.

Edmunds, G. F. and D. N. Alstad. 1978. Coevolution in insect herbivores and conifers. *Science* **199**:941–945.

Ehrlich, P. R. and P. H. Raven. 1965. Butterflies and plants: a study in coevolution. *Evolution* **18**:586–608.

Fain, A. (Ed.). 1979. Specificity and parallel evolution of host–parasite in mites. *Rec. Adv. Acar.* **2**(5):321–384.

Farrell, J. A. K. 1977. Plant resistance to insects and selection of resistant lines. *N. Z. Entomol.* **6**:244–261.

Freeland, W. J. 1976. Pathogens and the evolution of primate sociality. *Biotropica* **8**:12–24.

Friend, J. and D. R. Threlfall (Eds.). 1976. *Biochemical Aspects of Plant–Parasite Relationships.* Academic Press, New York.

Gilbert, L. E. 1977. Development of theory in the analysis of insect–plant interactions. Pages 117–154 in D. J. Horn, G. R. Stairs, and R. D. Mitchell (Eds.), *Analysis of Ecological Systems.* Ohio State University Press, Columbus, Ohio.

Grandison, G. S. 1977. Relationships of plant-parasitic nematodes and their hosts. *N. Z. Entomol.* **6**:262–267.

Haukioja, E. and P. Niemela. 1979. Birch leaves as a resource for herbivores: seasonal occurrence of increased resistance in foliage after mechanical damage of adjacent leaves. *Oecologia* **39**:151–159.

Hewett, E. W. 1977. Some effects of infestation on plants: a physiological viewpoint. *N. Z. Entomol.* **6**:235–243.

Holmes, J. C. and P. W. Price. 1980. Parasite communities: the roles of phylogeny and ecology. *Syst. Zool.* **29**:203–213.

Jackson, G. J. (Ed.). 1969. *Immunity to Parasitic Animals*, Vol. 1. North Holland, Amsterdam.

Janzen, D. H. 1973. Host plants as islands. Competition in evolutionary and contemporary time. *Am. Nat.* **107**:786–790.

Janzen, D. H. 1980. When is it coevolution? *Evolution* **34**:611–612.

Janzen, D. H. 1981. Patterns of herbivory in a tropical deciduous forest. *Biotropica* **13**:271–282.

Jermy, T. 1976. Insect–host–plant relationship—co-evolution or sequential evolution? *Symp. Biol. Hung.* **16**:109–113.

Johnston, D. E. (Ed.). 1975. The evolution of mite–host relationships. *Misc. Publ. Entomol. Soc. Am.* **9**:229–254.

Kennedy, C. R. 1975. *Ecological Animal Parasitology.* John Wiley, New York.

Kennedy, C. R. (Ed.). 1976. *Ecological Aspects of Parasitology.* North Holland, Amsterdam.

Kuris, A. M. 1974. Trophic interactions: similarity of parasitic castrators to parasitoids. *Q. Rev. Biol.* **49**:129–148.

Levin, D. A. 1971. Plant phenolics: an ecological perspective. *Am. Nat.* **105**:157–181.

Levin, D. A. 1973. The role of trichomes in plant defense. *Q. Rev. Biol.* **48**:3–15.

Levin, D. A. 1976a. The chemical defenses of plants to pathogens and herbivores. *Annu. Rev. Ecol. Syst.* **7**:131–159.

Levin, D. A. 1976b. Alkaloid-bearing plants: an ecogeographic perspective. *Am. Nat.* **110**:261–284.

Mansour, T. E. 1979. Chemotherapy of parasitic worms: new biochemical strategies. *Science* **205**:462–469.

Marshall, A. G. 1981. *The Ecology of Ectoparasitic Insects.* Academic Press, London.

Mound, L. A. and N. Waloff (Eds.). 1978. *Diversity of Insect Faunas.* Symposia R. Entomol. Soc. Lond. No. 9.

Nelson, W. A., J. F. Bell, C. M. Clifford, and J. E. Keirans. 1977. Interaction of ectoparasites and their hosts. *J. Med. Entomol.* **13**:389–428.

Nelson, W. A., J. E. Keirans, J. F. Bell, and C. M. Clifford. 1975. Host–ectoparasite relationships. *J. Med. Entomol.* **12**:143–166.

Niemela, P., J. Tuomi, and E. Haukioja. 1980. Age-specific resistance in trees: defoliation of tamaracks *(Larix laricina)* by larch bud moth *(Zeiraphera improbana)* (Lep. Tortricidae). *Rep. Kevo Subarct. Res. Stat.* **16**:49–57.

Powell, J. A. 1980. Evolution of larval food preferences in microlepidoptera. *Annu. Rev. Entomol.* **25**:135–159.

Price, P. W. (Ed.). 1975. *Evolutionary Strategies of Parasitic Insects and Mites.* Plenum Press, New York.

Price, P. W. 1977. General concepts on the evolutionary biology of parasites. *Evolution* **31**:405–420.

Price, P. W. 1980. *Evolutionary Biology of Parasites.* Princeton University Press, Princeton.

Randolf, S. H. 1979. Population regulation in ticks: the role of acquired resistance in natural and unnatural hosts. *Parasitology* **79**:141–156.

Rosenthal, G. A. and D. H. Janzen. 1979. *Herbivores. Their Interaction with Secondary Plant Metabolites.* Academic Press, New York.

Rothschild, M. and T. Clay. 1952. *Fleas, Flukes and Cuckoos: A Study of Bird Parasites.* Collins, London.

Russell, G. B. 1977. Plant chemicals affecting insect development. *N. Z. Entomol.* **6**:229–234.

Schad, G. A. 1963. Immunity, competition, and natural regulation of helminth populations. *Am. Nat.* **100**:359–363.

Southwood, T. R. E. 1973. In H. F. van Emden, *The Insect/Plant Relationship: An Evolutionary Perspective.* (Ed.), John Wiley, New York.

Southwood, T. R. E. 1973. The insect/plant relationship: an evolutionary perspective. In H. F. van Emden (Ed.). *Insect/Plant Relationships.* John Wiley, New York.

Strong, D. R. 1979. Biogeographic dynamics of insect-host plant communities. *Annu. Rev. Entomol.* **24**:89–119.

Strong, D. R. and D. A. Levin. 1979. Species richness of plant parasites and growth form of their hosts. *Am. Nat.* **114**:1–22.

Sutherland, O. R. W. 1977. Plant chemicals influencing insect behaviour. *N. Z. Entomol.* **6**:222–228.

Valdivieso, D. and J. R. Tamsitt. 1970. A review of the immunology of parasitic nematodes. *Rev. Biol. Trop.* **17**:1–25.

Van Emden, H. F. and M. J. Way. 1973. Host plants in the population dynamics of insects. In H. F. van Emden (Ed.), *Insect/Plant Relationships.* John Wiley, New York.

Vinson, S. B. 1976. Host selection by insect parasitoids. *Annu. Rev. Entomol.* **21**:109–134.

Wakelin, D. 1979. Genetic control of susceptibility and resistance to parasitic infection. *Adv. Parasitol.* **17**:219–308.

Wallace, J. W. and R. L. Mansell (Eds.). 1976. Biochemical interaction between plants and insects. *Rec. Adv. Phytochem.* **10**:1–425.

Wenzel, R. L. and V. J. Tipton. 1966. Some relationships between mammal hosts and their ectoparasites. Pages 677–723 in R. L. Wenzel and V. J. Tipton (Eds.). *Ectoparasites of Panama*. Field Museum of Natural History, Chicago.

Wikel, S. K. and J. R. Allen. 1976. Acquired resistance to ticks. 1. Passive transfer of resistance. *Immunology* **30**:311–318.

Table 3.1   The Geological Time Scale and Major Phylogenetic Events in Mammals

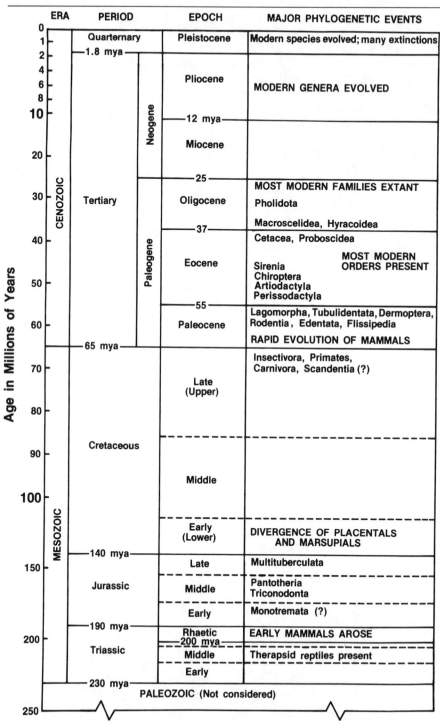

ERA	PERIOD		EPOCH	MAJOR PHYLOGENETIC EVENTS
CENOZOIC	Quarternary		Pleistocene	Modern species evolved; many extinctions
	—1.8 mya—			
	Tertiary	Neogene	Pliocene	MODERN GENERA EVOLVED
			—12 mya—	
			Miocene	
			—25—	MOST MODERN FAMILIES EXTANT
		Paleogene	Oligocene	Pholidota
			—37—	Macroscelidea, Hyracoidea
			Eocene	Cetacea, Proboscidea
				Sirenia MOST MODERN ORDERS PRESENT
				Chiroptera
				Artiodactyla
				Perissodactyla
			—55—	Lagomorpha, Tubulidentata, Dermoptera,
			Paleocene	Rodentia, Edentata, Flissipedia
				RAPID EVOLUTION OF MAMMALS
	—65 mya—			
MESOZOIC	Cretaceous		Late (Upper)	Insectivora, Primates, Carnivora, Scandentia (?)
			Middle	
			Early (Lower)	DIVERGENCE OF PLACENTALS AND MARSUPIALS
	—140 mya—		Late	Multituberculata
	Jurassic		Middle	Pantotheria Triconodonta
			Early	Monotremata (?)
	—190 mya—		Rhaetic —200 mya—	EARLY MAMMALS AROSE
	Triassic		Middle	Therapsid reptiles present
			Early	
	—230 mya—		PALEOZOIC (Not considered)	

Age in Millions of Years

0, 1, 2, 4, 6, 8, 10, 20, 30, 40, 50, 60, 70, 80, 90, 100, 150, 200, 250

CHAPTER **3**

Mammals as Evolutionary Partners

Robert M. Timm and Barbara L. Clauson

INTRODUCTION

Mammals have been available as potential hosts for parasites for approximately 190 million years. Although we know very little about the parasites that might have occurred on early mammals, we do know that both hosts and parasites underwent a tremendous radiation. Mammals have evolved to occupy a wide variety of niches and are found on all land masses and throughout the oceans and freshwaters. Concurrently, arthropods have evolved to occupy a wide array of niches available on the mammalian body. The radiation of mammals was paralleled by a corresponding radiation of the parasitic arthropods, both in morphological and taxonomic diversity, a process referred to as *coevolution*.

Coevolution is a popular term in widespread use in today's biological literature, and has been invoked to describe a variety of observed phenomena. However, in many cases the term has been undefined and used loosely. The prefix "co-," meaning "with" or "together" added to the base word evolution implies that two (or more) organisms are evolving together or with one another. Janzen (1980:611) defined coevolution as "an evolutionary change in a trait of the individuals in one population in response to a trait of the individuals of a second population, followed by an evolutionary response by the second population to the change in the first." Although Janzen's article is aimed primarily at plant–animal interactions, his contention is that the current use of the term coevolution is misleading in many instances. He suggests that we adopt the strict definition of coevolution in which both organisms have evolved responses to each other. Brooks (1979) described *coevolution* in host–parasite systems as encompassing two phenomena, which he termed *co-accommodation* and *co-speciation*. He defined co-accommodation as a "mutual adaptation of a given parasite species and its host(s) through time [which] includes such parameters as pathogenicity, host specificity, and synchrony of life cycles stages," and co-speciation as "cladogenesis of an ancestral parasite species as a result of, or concomitant with, host cladogenesis" (Brooks 1979:300). He viewed co-speciation as an outcome of allopatric speciation and the vicariance biogeography model.

Coevolution between a host and parasite, in its strictest sense, occurs as a result of each exerting a selective force on the other. Such an interaction is difficult to demonstrate, and in many host–parasite systems, coevolution in the strict sense may not occur. Often the host exerts selective pressure on the parasite, but the parasite does not have a corresponding effect on the host. Descriptions of host–parasite systems often use the term coevolution, whereas co-speciation or co-accommodation may more accurately describe the situation. Previously used terms such as "interactions," "symbiosis," "mutualism," and "animal–plant interactions" are not synonymous with coevolution and perhaps better describe many of the interactions observed (Janzen 1980).

An example of co-speciation of host and parasite where the parasite exhibits an evolutionary response to the host but the host exhibits no counterevolutionary response is found in pocket gophers (*Geomys*) and their parasitic lice (*Geomydoecus*) (Timm 1979, 1983; Timm and Price 1980). Population levels of lice, *Geomydoecus* (Mallophaga: Trichodectidae), were found to vary seasonally on pocket gophers of the genus *Geomys* (Rodentia: Geomyidae). In the northern pocket gophers, there was an average of 500 lice per individual gopher during the summer months, with some individual pocket gophers harboring as many as 2000 lice. All stages of the lice are found on the host; the eggs are glued individually to the base of the hair shaft. First, second, and third instars each last for approximately 10 days, and the adults probably overwinter. All stages of lice occur most abundantly on the back of the head and nape of the neck, presumably because these are the most difficult areas for the gopher to reach while grooming (Timm 1983). That the lice have adapted or evolved specializations to their hosts is obvious from their morphology (extreme dorsoventrally flattened bodies, tarsal claws to grasp individual hairs, posteriorly projecting setae), and from their success as parasites measured in terms of both individual populations and taxonomic diversity. Every one of several thousand pocket gophers examined had high numbers of lice, and to date some 102 specific and subspecific taxa of *Geomydoecus* have been recognized. As pocket gophers are solitary, dispersal of lice from one host to another presents additional problems. A young pocket gopher captured just after dispersal from the natal tunnel system (probably within the preceding 24 hours) already had a population of 350 lice. Additionally, the lice appear to be cueing their reproduction to the reproductive cycle of their hosts (Timm 1983). Close, parallel responses of lice and their hosts suggests that the lice have evolved "traits" in response to their hosts.

However, have the pocket gophers undergone an evolutionary change in response to their parasitic lice? Here the answer is probably no. In work with both captive and wild hosts there was no indication that the lice (even as many as 2000 on a single pocket gopher) had any impact upon their hosts. Lice of the genus *Geomydoecus* feed on dead tissue, probably scrapings of skin, and thus incur no direct cost to the host while feeding. They apparently transmit neither diseases nor endoparasites to their hosts.

The one behavior pattern of the host that affects the lice is grooming. But we cannot say that grooming by pocket gophers evolved in response to their harboring Mallophaga. Grooming probably reflects a need to keep the fur clean from dirt, or else may be a direct response to the larger bloodsucking parasites such as fleas or mesostigmatid mites. The fact that grooming by pocket gophers is the main factor controlling louse populations is an incidental by-product of host grooming behavior that is not directed toward the lice.

It can be argued that additional grooming would be a cost to the host in terms of energy expenditure. Increased grooming could make any host more vulnerable to predation, either while grooming or while spending additional time foraging to compensate for the increased energy expenditure. However, it is likely that *Geomys* does not spend more time grooming in response to a high louse population (personal observation). Even if there was an increase in time spent grooming, it probably would have little negative effect on the pocket gophers because they live entirely within an enclosed tunnel system. Both grooming and foraging take place well below the surface. The major predators on an adult *Geomys* are sit-and-wait predators like badgers (*Taxidea taxus*) that open the tunnel and wait for the pocket gopher to repair the damage, or long-tailed weasels (*Mustela frenata*) and snakes that can enter a tunnel system opened by a gopher actively working on it. Weasels and snakes cannot open a closed tunnel system. Pocket gophers on the surface, especially young dispersing animals, are susceptible to predation by raptors, but this is quite independent of louse populations. Further, during the period of highest louse populations (early summer and late fall), most populations of pocket gophers have unlimited supplies of food. Any additional energy expenditure caused by the parasites would be insignificant.

Thus at this time we must conclude that lice of the genus *Geomydoecus*, although they are considered parasitic and their populations may be high, have no effect upon their host. The selective pressure has been exerted on only one member of the system, and therefore coevolution is not an appropriate term. The concordant patterns of speciation between gophers and their lice (Heaney and Timm 1983; Timm 1983) are best thought of in the sense of co-speciation or co-accommodation, which is not synonymous with coevolution.

Documenting coevolution in nature is no trivial task, especially if we use increased or decreased reproductive success by both host and parasite as our measure. Coevolution measured in terms of maximized reproductive output in both host and parasite has been demonstrated for bot flies (Diptera: Cuterebridae) of the genus *Cuterebra* and their rodent hosts. Cuterebrids are subcutaneous parasites commonly found on North American cricetine and sciuromorph rodents.

Bot flies are relatively large in proportion to their hosts; a mature larvae may be 30–35 mm on a white-footed mouse, *Peromyscus*, whose body length is 80–90 mm. It has been demonstrated that excessive numbers of bots artificially inflicted on a host can kill it (Timm and Lee 1981). Smith (1978) demonstrated that deer mice (*Peromyscus maniculatus*) with multiple infestations of bot flies (*Cuterebra approximata*) were more vulnerable to predation by short-tailed weasels (*Mustela erminea*) than were mice with one or no bots. He concluded that "*Peromyscus maniculatus* infected with single *C. approximata* larvae did not appear to be any more vulnerable to shorttail weasel predation than did uninfected control mice" (Smith 1978:47). Additionally, he was able to quantify the cost of bot fly parasitism to the host in terms of calories. During its developmental period, a single bot fly larva consumed 27 kcal, two bots consumed 47 kcal, and three bots consumed 61 kcal of energy (Smith 1975). Larvae from multiple infestations received proportionally less energy than those in single infestations and hence were significantly smaller than those found singly. Immature larvae on a host that either died or was killed by a predator would not survive.

The egg-laying strategy of female bots also suggests an evolved response. Female bots lay 1000–3000 eggs, but scatter them widely so that the infestation rate per individual mouse averages only one or two. Thus there is a definite cost to the host infected with bot flies and a reciprocal cost to the bots if the host is too heavily infected. Bots of larger body size probably have an advantage in overwintering and in successfully mating the following summer.

Timm and Cook (1979) showed that white-footed mice (*Peromyscus leucopus*) have evolved a tolerance for bot fly parasitism. One or two larvae had no effect on the breeding condition of adult male mice, and adult females showed no decrease in number of embryos, corpora lutea, or placental scars (see also Timm and Lee 1981). That this tolerance is evolved is suggested by the fact that Old World hosts, not normally exposed to New World cuterebrids, may be killed if infected with bot fly larvae (Catts 1982). Thus a host–parasite system has "coevolved" adaptations by both the host and parasite to maximize reproductive output.

Other perspectives on coevolution, co-accommodation, co-adaptation, and co-speciation can be found in Price (1980), Thompson (1982), Futuyma and Slatkin (1983), and Nitecki (1983).

MAMMALS AS HABITAT

Hair

The presence of hair is a unique characteristic of mammals, and all mammals have at least some hairs during their development. Hair provides mammals with protection in the form of insulation, antiabrasion, and defense (e.g., quills of porcupines). Specialized hairs are used as tactile organs (vibrissae) and for communication. The hair itself is of epidermal origin, the follicle invaginates the dermis as the hair arises. A typical hair is composed of numerous tightly compacted, keratinized cells forming three distinct layers: an inner medulla consists of the cornified remnants of epithelial cells; the cortex, which is formed of keratinized cells fused into a hyaline mass, comprises the main constituent of the hair shaft and contains most of the pigment; and the cuticle, which may comprise numerous scales that wrap around the hair shaft, forms a thin outer protective layer composed of heavily keratinized cells. The body of most mammals is covered with two main types of hair: stiff guard hairs which provide protection, and softer underfur which primarily serves as insulation. Hairs are typically associated with a variety of sebaceous and sudoriferous glands, and the surface of the hair is often oily. One or more sebaceous glands are found within each hair follicle.

In a few mammalian groups adults are essentially devoid of hair. This is especially true in the Cetacea, but is also true in one species of east African rodent, the naked mole rat (*Heterocephalus glaber*), and in one genus of southeast Asian molossid bats that is represented by two species, both referred to as naked or hairless bats (*Cheiromeles*). Hairlessness must be considered a secondarily derived condition.

For ectoparasitic arthropods, hair provides concealment and in some cases food. However, hair presents its own unique set of problems to the parasite. It is a stiff proteinaceous structure whose rigidity is provided by disulfide bonds. Its composition may be difficult for arthropods both to ingest and digest, and hair is probably not a highly nutritious resource. The stiff, dense nature of hair on the mammalian body creates a forest that can hinder an arthropod's ability to maneuver across the skin. The laterally compressed bodies of fleas and the dorsoventrally compressed bodies of lice are adaptations in response to intense selection pressure. The ability to move rapidly through hair is necessary to avoid dislodgement while the host is grooming. Another problem with hair is that it is an ephemeral resource shed (lost) on a routine basis. This hair loss-replacement pattern is termed molt. In some mammals (for example, carnivores), hair loss and replacement involves individual hairs scattered over the body surface. In most rodents the molting pattern starts as a small patch and sweeps across the entire body within a few weeks. Molt may occur seasonally or annually. Arthropod eggs glued to hairs that are being shed will be lost, as would instars or adults firmly attached to individual hairs. Molt patterns, however, are triggered by environmental cues such as changing day length, and thus are predictable. The normal molt pattern may be delayed if the animal is in poor body condition, or, as in the case of pregnant females, when energy reserves are stressed.

Most arthropods broadly lumped into the category of mammalian ecto-parasites spend at least some portion of their life cycle within the hair; however, few species actually consume the hair itself. Perhaps the only insects to feed on hair are the parasitic beetles *Catopidius* (Leiodidae), *Loberopsyllus* (Languriidae), and *Leptinus* (Leptinidae) (Marshall 1981; Peck 1982). Although it was commonly believed that chewing lice (Mallophaga) fed upon the hair of their hosts, Marshall (1981:129) states unequivocally, "There is no evidence that lice feed upon hair." However, many species of Ischnocera on birds feed partially or exclusively on feathers. Perhaps feathers are a more nutritious food source than hair. Certainly, feathers offer a more structurally diverse environment for ectoparasites than does hair. Another conspicuous difference in the utilization of feathers and hair by arthropods is that several families of mites and lice have colonized the air space or hollow interior of the feather shaft, whereas no arthropods have been able to utilize the hollow hair shafts found in a few species of mammals. The mammalian hair shaft has no umbilicus as does the feather shaft, thus there is no easy entrance point in hair.

Skin

Skin is the protective layer that covers the body surface of mammals. It serves a primary function in thermoregulation, to prevent desiccation, and to provide protection (Sokolov 1982). The glands within the skin function in excretion, and the nerves receive stimuli from the environment. Mammalian skin, like that in other vertebrates, is composed of two discrete layers, the epidermis and the dermis. The surface layer, the epidermis, is composed of epithelial tissue, called stratified epithelium, divided into four layers. The outermost, that exposed to the surface and parasitic arthropods, consists of dead hardened cells that are compressed into a cornified layer termed the stratum corneum. Immediately beneath the stratum corneum is a thin, clear layer, the stratum lucidum, comprising several layers of flattened compacted cells. Beneath that is a layer termed the stratum granulosum, which is actually composed of three to five distinct layers of living epithelial cells ranging from cuboidal to columnar in shape. The innermost layer, the stratum malpighii, is the only layer that contains actively dividing cells undergoing mitosis. Skin (epithelial) cells are formed in the stratum malpighii and are pushed outward by additional actively dividing cells; as they migrate outward, they undergo a complex process called keratinization. All the cell organelles, including the nuclei, disappear, and the cells flatten and fill with granules of keratohyalin. The resulting tissue forms a dense, tough, outer protective layer that wears away, and is continuously replaced. There are no blood vessels found within the epidermis. The dermis, lying beneath the epidermis, is composed of connective tissue, muscles, fat deposits, and vascular and nervous tissue. Beneath the dermis is a layer of loose connective and adipose tissue. Arthropods seeking a blood meal must be able to penetrate through the epidermis to reach the rich supply of blood vessels found within the dermis.

Of all the parasitic arthropods that live on the skin of mammals, relatively few feed directly upon the skin itself. These include insects of the orders Dermaptera (families Arixeniidae and Hemimeridae), Coleoptera (Leiodidae, Leptinidae, Platypsyllidae, Staphylinidae, Languriidae, and Scarabaeidae), Amblycera, and Ischnocera; and mites of the family Psoroptidae.

Glands

A wide variety of glands having diverse functions are found within the skin of mammals, but they fall basically within two types: the sebaceous glands (oil glands) and the sudoriferous glands (sweat glands). Glands serve a variety of functions, including excretion of metabolic wastes, evaporative cooling, communication, and nourishment of young.

Sebaceous glands lie in the dermis and function primarily in lubrication, protection of the skin surface, and reduction of water loss. The secretion from sebaceous glands is produced by the breakdown of the glandular epithelial cells and includes both cell debris and lipids (Montagna 1974). All hair follicles in mammalian skin have sebaceous glands associated with them that open directly into the follicle and whose secretions protect the hair from becoming dry and brittle. Sebaceous glands are also important to mammals in scent marking, both for intra- and interspecific recognition and territoriality. The location of these scent glands varies among the different groups of mammals. For example, shrews (*Blarina, Cryptotis,* and *Sorex*) have midventral and lateral scent glands (Bee et al. 1980); kangaroo rats (*Dipodomys*) have an enlarged scent gland along the midline of the back (Quay 1953); some species of voles (*Microtus*) have lateral scent glands called hip glands (Jannett 1975; Quay 1968; and Tamarin 1981); and many carnivores have an enlarged anal gland for scent marking. Anal glands are prominently developed in the Mustelidae (skunks), Canidae (dogs), and Viverridae (civets). The Meibomian glands lying within the eyelids are modified sebaceous glands that lubricate the eye.

Sudoriferous glands (sweat glands) also lie within the dermis and are found throughout the body surface on many species of mammals. However, they are not known to occur in several groups of mammals, notably the edentates, pinnipeds, and cetaceans. Sudoriferous glands are of two basic types: apocrine glands that are usually located adjacent to hair follicles, and eccrine glands that open directly to the skin surface, independent of hair follicles. The secretions from sudoriferous glands are a product of cellular metabolism, and not a result of cellular decomposition as in sebaceous glands. These secretions often include water, salts, and fatty substances, and their function is generally thermoregulation. Many of the types of glands found in mammals have no counterparts in other vertebrates; for example, mammary glands, highly modified sudoriferous glands, are unique to mammals. Mammary glands are thought to have evolved from sudoriferous glands. The most primitive mammary glands are found in the most primitive mammals, the monotremes, and consist of roughly 100 lobules within the skin that open directly to the ventral surface. Each lobule is associated with a stiff hair and the young lick the milk off the hairs. Bats have perhaps the most diverse array of both sudoriferous and sebaceous glands found within the Mammalia. Glands are found throughout much of the skin surface on the lips, face, throat, chest, and wing membranes (Quay 1970). In recent years, this rich diversity of glands has been found to harbor a diverse fauna of associated mites. A recent review of scent glands in mammals and their functions was provided by Müller-Schwarze (1983).

A few groups of mites, especially demodicids, have been able to successfully invade hair follicles and the sebaceous and sudoriferous glands. *Demodex* punctures the cells of the glands and hair follicles with its stylet chelicerae and feeds upon cell contents. Although demodicids may occur in extremely high numbers, there seldom is any major tissue destruction (Nutting 1965; Nutting and Woolley 1965). Apparently no insects have invaded the glands of mammals.

Respiratory Tract

The respiratory tract of mammals includes the nasal cavity, pharynx, larynx, trachea, and bronchi. These organs are basically tubes for the passage of air from the external environment to the lungs. The respiratory tract of mammals is warm, moist, and has rich supplies of blood and oxygen. The respiratory system can be entered easily through the open external nares, and the first tissue encountered by invading parasites is the soft nasal mucosa.

The nasal chamber of mammals is composed of thin bone (turbinates), cartilage, muscles, connective tissue, and a passageway for air. Sebaceous glands, especially those secreting mucus, are large and well developed. A rich supply of blood vessels is associated with the olfactory epithelial cells. The primary function of this complex arrangement, in addition to olfaction, is the conservation of heat and water during respiratory exchange. There is a distinct temperature gradient within the nose, with the tip approaching ambient temperature. The nasal region is warm and moist, and has a rich supply of oxygen continuously moving across it. The mucosa is easy for an arthropod to burrow through. The turbinal bones provide support as do other parts of the skeletal system, yet because of their extreme thinness do not prohibit arthropod burrowing. Thus the nasal region can provide an ideal environment for development of soft-bodied arthropods. Nasal mites generally are found within the mucous membranes and either feed on blood or the nasal mucosa.

Several families represented by numerous genera of mites and at least three genera of bot flies are known to occur in mammalian nasal passages. Mites parasitic within the nasal mucosa include the following families and genera: Ereynetidae (*Neospeleognathopsis, Paraspeleognathopsis, Speleognathus,* and *Speleorodens*) (Fain 1962, 1963); Gastronyssidae (*Mycteronyssus, Opsonyssus, Rodhainyssus, Sciuracarus,* and *Yunkeracarus*) (Fain 1964a); Halarachnidae (*Halarachne, Orthohalarachne, Pneumonyssoides, Pneumonyssus,* and *Rhinophaga*); Lemurnyssidae (*Lemurnyssus* and *Mortelmansia*) (Fain 1964b); and Trombiculidae (*Alexfainia, Asoschoengastia, Blix, Crotonasis, Doloisia, Euschoengastia, Gahrliepia, Kymocta, Leptotrombidium, Microtrombicula, Myxacarus, Nasicola, Rhinibius, Schoutedenichia, Traubacarus, Vergrandia,* and *Whartonia*).

Nasal mites of the genus *Orthohalarachne* (Halarachnidae) are found in the mucosa of eared seals (Otariidae) and walruses (Odobenidae); the genus *Halarachne* is found in the earless seals (Phocidae) and in sea otters (*Enhydra lutris*) (Fay and Furman 1982). Generally, the impact of parasitic arthropods has been thought to be insignificant to the health of marine mammals. However, in northern fur seals (*Callorhinus ursinus*), Kim et al. (1980:45) reported that all subadult and adult seals examined ($N = 81$) contained high populations of both *Orthohalarachne attenuata* and *O. diminuata* and that "The heavy infestation with these mites appeared to result in impairment of respiration in fur seals, and could also cause lesions in the lungs and secondary alveolar emphysema, predispose to more serious diseases, or even kill the host animal." Additionally, Kenyon et al. (1965:960) reported a captive sea otter harbored "over 3000 *H[alarachne] mioungae*" and that "The nasal passages were crowded with mites and the mucosa showed severe reddening. Also, the turbinates had been destroyed, leaving a nearly unobstructed void from external nares to posterior nasal passages." Other species of nasal mites apparently cause little or no damage to their hosts.

The nasal bot and warble flies found in the nasal fossae, frontal sinuses, and in the nasopharyngeal pouches of mammals include two genera of Oestridae: *Cephenemyia* on cervids and *Oestrus* on bovids.

Several species of mites in the genera *Pneumocoptes* (Pneumocoptidae), *Orthohalarachne* and *Pneumonyssus* (Halarachnidae), and *Lemurnyssus* and *Mortelmansia* (Lemurnyssidae) are known as pulmonary mites and live within the lungs, bronchi, and trachea of the upper respiratory system in mammals. When found, they are generally abundant, but do little damage to the host. Most species feed directly upon blood.

Subcuticular Layer

The subcuticular niche within the mammalian body has not been colonized by very many groups of arthropods. Some conditions necessary for arthropod development are ideal in this environment: warm constant temperature, moisture, and a rich nutrient supply. However, free oxygen levels are low. Penetration of both the epidermis and dermis to form a direct link with the surface is necessary if the parasite's waste products are extensive and a rich oxygen supply is necessary. Bot flies, blow flies, and flesh flies of the families Calliphoridae, Cuterebridae, Oestridae, and Sarcophagidae are the only insects that are found in this niche. Perhaps because of their larger body size, insects have been more successful than other groups of arthropods in penetrating the epidermis and dermis. Most species do not burrow directly through the skin to obtain access to the subcuticular regions, but rather enter through preexisting openings, such as the eyes, nose, mouth, and hair follicles. In the calliphorid flies, especially *Cochliomyia*, the female deposits her eggs on an open wound in the host where the epidermal and dermal layers have already been broken. The human bot (*Dermatobia hominis*) apparently burrows through the dermal layers within a hole originally made by a female mosquito. Mites of the family Cloacaridae (*Epimyodex*) have been found subcutaneously in the fascia in the lower back region of moles and several other species of small mammals (Fain and Orts 1969). It is believed that they enter the host's body through the genitalia and then migrate to this position, rather than penetrate the skin directly.

Other Organs

Mites of several families have proven successful in invading a wide array of mammalian organs. Within the digestive tract, two genera of Demodicidae (*Rhinodex* and *Stomatodex*) and two genera of Sarcoptidae (*Chirnyssus* and *Nycteridocoptes*) are found within the oral cavity of bats and lemurs. In bats *Stomatodex* is often found inside the lips, beneath the tongue, and near the epiglottis. Phillips et al. (1969:1368) reported that the long-nosed bat, *Leptonycteris nivalis* (Phyllostomidae), harbored high populations of mites, *Radfordiella* (Macronyssidae), within the oral mucosa, and that "Osteolysis of hard palate and odontolysis of teeth result from infestations of mites adjacent to the upper premolars and molars." Interestingly, neither mites nor the associated bone damage were found in the closely related and sympatric species, *Leptonycteris sanborni*.

Gastronyssus bakeri (Gastronyssidae) lives within the epithelial lining of the stomach and intestines of megachiropterans (Fain 1955). *Paraspinturnix globosus* (Spinturnicidae) has been described only from within the anus of North American vespertilionid bats of the genus *Myotis* (Rudnick 1960), but may represent an overwintering stage of a more typical spinturnicid.

Demodex canis (Demodicidae) is known from the spleen, kidney, blood, tongue muscles, urine, bladder, liver, intestinal wall, and thyroid glands of dogs, in addition to being found in the skin and associated glands.

A few species of mites live on the surface of the skin and feed on the secretions of the eye; several species of *Demodex* live within the Meibomian glands of the eyelid (Lombert et al. 1983); and *Opsonyssus* (Gastronyssidae) lives within the orbit of the eye itself. Ocular secretions of large mammals are fed on by a few species of tropical and subtropical Lepidoptera of the families Geometridae, Noctuidae, and Pyralidae.

The ear canal has been invaded by a few species of Astigmata and Mesostigmata: *Otodectes* in carnivores (Sweatman 1958a); *Psoroptes* in ungulates (Sweatman 1958b); and *Noteodres* in the dwarf galago, *Galago demidovi*, and a single species of laelapid, *Raillietia hopkins*, in the ears of cattle, sheep, goats, and some African antelope. *Demodex marsupiali* recently was described from the pilocerumen gland complex in the opossum, *Didelphis marsupialis*. Nutting et al. (1980:83) reported that "Pathogenesis is limited to epithelial cell destruction, minor orifice occlusion, and some keratinization . . . mites occasionally penetrate into the dermis, without host cellular response."

EVOLUTION OF MAMMALS

Origins

The first mammals appeared on earth during the late Triassic period (Rhaetic) of the Mesozoic era (see Table 3.1), approximately 190 million years ago. These early mammals evolved from the extinct mammal-like reptiles of the subclass Synapsida, order Therapsida. The exact mammalian ancestor within this group of reptiles is not known. It has been suggested that mammals evolved from several related lines of therapsids (Olson 1959). Such a polyphyletic origin is suggested by the fact that several groups of therapsids possessed mammalian characteristics. A monophyletic origin of the class Mammalia is supported by many authors, based on differing interpretations of the fossil record. Hopson (1967) concluded that mammals probably arose from only one group of advanced late Triassic cynodont therapsids. This view of a monophyletic origin was independently arrived at by Parrington (1967) and Crompton and Jenkins (1973) further supported this view and stated that, although many of the therapsids showed parallel evolution of mammal-like traits, only a single family of cynodonts gave rise to all mammals. Three types of cynodonts possess many mammalian characteristics and may be sister groups to the early mammals: *Probainognathus*, the Tritheledontidae, and the Tritylodontidae (Kemp 1982). Current evidence supports this theory, and a monophyletic origin is now widely accepted (Clemens 1970; Crompton and Jenkins 1973).

The continent on which mammals evolved remains unknown. The oldest known fossils commonly accepted as "good" mammals are from the late Triassic period of western Europe (Clemens et al. 1979). The only earlier fossil that might be mammalian is from the early late Triassic period in southern Brazil. Named *Therioherpeton*, it is either a very reptile-like mammal or a mammal-like reptile (Bonaparte and Barberena 1975). However, given the fragmentary nature of the fossil record, it is not possible to denote a specific location for the origin of mammals. Late Triassic and early Jurassic fossil mammals are also known from Africa, Asia, and North America, indicating that early primitive mammals were widespread and morphologically diverse (Clemens et al. 1979; Jenkins et al. 1983).

Early mammals were small, and most were probably insectivores or carnivores. Although the fossil evidence is sparse, it appears that little radiation occurred, at least in terms of fossilized hard structures, until late in the Mesozoic era. A period of great mammalian radiation began in the late Cretaceous (late Mesozoic) period and continued on into the Paleocene (early Cenozoic) epoch. During this time, many of the modern mammalian orders first appeared. This radiation continued throughout the Cenozoic era, and by the Oligocene epoch the majority of modern families had developed. Many modern genera were present as early as the Pliocene epoch. Modern species or their direct ancestors appeared during the Pleistocene epoch. For excellent and more detailed reviews of this subject, see Crompton and Jenkins (1979) and Kemp (1982).

Mesozoic Era

From the time of their first appearance in the late Triassic period (Rhaetic) until the end of the Mesozoic era, a period of approximately 120 million years, mammals formed only a small fraction of the earth's terrestrial fauna. All Mesozoic mammals so far discovered have been small in size and rather similar in body form. However, during this period several major evolutionary changes occurred. During the Mesozoic era the two major modern mammalian infraclasses, the Eutheria (placentals) and Metatheria (marsupials), originated.

By the end of the Triassic period, mammals had diverged into three major families: the Haramiyidae, the Morganucodontidae, and the Kuehneotheriidae (Clemens 1970; Crompton and Jenkins 1973, 1979; Hopson and Crompton 1969; McKenna 1969).

The Haramiyidae is a poorly known family of uncertain affinities. It appeared in the fossil record earlier than the morganucodontids or kuehneotheriids, having similarities to primitive multituberculates that may indeed be considered ancestral to that group. The Morganucodontidae were ancestral to the non-therian mammals, including the early triconodonts and docodonts and possibly the multituberculates. Although the monotremes do not appear in the fossil record until the Miocene epoch (Clemens 1979), they have similarities with the Morganucodontidae and may have arisen from this group. The Kuehneotheriidae gave rise to therian mammals, including the symmetrodonts, pantotheres, marsupials, and eutherian mammals. Figure 3.1 shows the major groups of Mesozoic mammals; also see the review by Crompton and Jenkins (1979).

Cenozoic Era

The beginning of the Cenozoic era, approximately 65 million years ago, marks the start of the great radiation of mammals. At the end of the Mesozoic era, global climates changed, primarily expressed as major cooling trends (Savin 1977; Lillegraven et al. 1979). During this time the dominant reptilian group, the dinosaurs, became extinct. Such cooling trends may have favored the homeothermic mammals over their reptilian counterparts, and with the extinction of the dinosaurs mammals radiated into niches they had not previously occupied. A tremendous explosion in terms of rapid evolution of new forms occurred during the 65 million years of the Cenozoic era as compared to the previous 125 million years of the Mesozoic after mammals first evolved, even though the Cenozoic covers a time span only half as long. Although they were evolving rapidly, mammals were still not a dominant group in the Paleocene epoch, and all were

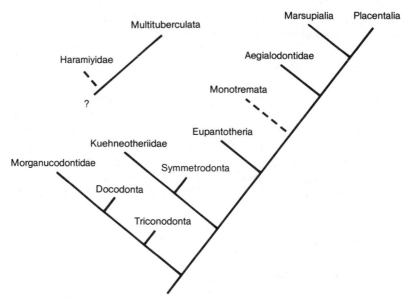

Figure 3.1 A phylogenetic tree for early mammals. (From T. S. Kemp, *Mammal-Like Reptiles and the Origin of Mammals.* Academic Press, London, 1982.)

small in size. By the late Paleocene and early Eocene epochs there was a dramatic increase in overall numbers of orders, and during this time nearly all extant mammalian orders evolved (Kurtén 1972). A period of stability in terms of numbers of orders present at any given time occurred during the Eocene, Oligocene, Miocene, and Pliocene epochs. This was a period of diversification, with new groups evolving as others became extinct. There was a slight decrease in the total number of different types of mammals at the end of the Pleistocene epoch due to the widespread extinction of many large mammals without ecological replacement.

Plate Tectonics and the Geographic Radiation of Mammals

The face of the earth has undergone dramatic changes during the 190 million years since the mammals first arose from reptiles, through the processes of continental drift and plate tectonics. Changes in the locations and configurations of the land masses affected the distribution of the earth's flora and fauna. Populations were split and isolated, local climates changed, and migration and gene flow of populations between continents either prevented or initiated. Thus continental drift had a great effect on the distribution of extant mammals through its effect on their ancestors (Cracraft 1974; Lillegraven et al. 1979).

During the Triassic period the major land masses were joined into one large supercontinent called Pangaea (Fig. 3.2). In the early Jurassic period,

Figure 3.2   Continents at the Rhaetic-late Triassic period, 200 million years ago.

Pangaea began to split into a northern land mass, Laurasia, composed of the North American and Eurasian continental plates, and a southern land mass, Gondwanaland, made up of the South American, African, Indian, Australian, and Antarctic plates (Fig. 3.3). At approximately the same time, Gondwanaland also began to split into an eastern section (Antarctica and Australia) and a western section (South America and Africa). The South American and African plates began to separate from one another by the late Jurassic period and were well separated by the early to mid-Cretaceous period (Fig. 3.4). India had separated from Australia–Antarctica, and Madagascar was free from Africa by the late Cretaceous period. By the end of the Paleocene epoch, eastern North America and western Europe were no longer joined (Fig. 3.5). Australia and Antarctica became separated by the late Paleocene or early Eocene epoch. By the Oligocene or early Miocene epoch, India had moved into its present position against the southern part of the Eurasian plate (Fig. 3.6); the impact of the collision and

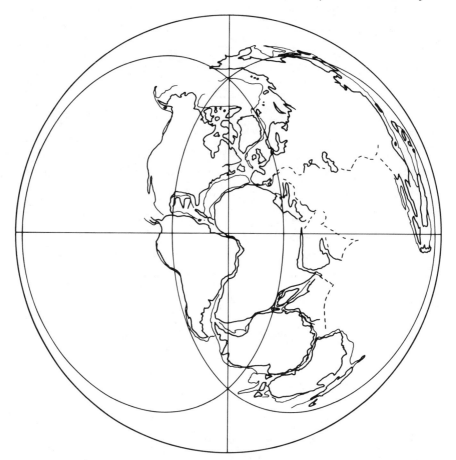

Figure 3.3 The breakup of the world land masses at the mid-Jurassic period, 160 million years ago.

continued northward movement of India created the Himalaya mountains. By the late Miocene epoch the continents had more or less assumed their present positions (Fig. 3.7). There is some disagreement among authors as to exact continental configurations and times of separations, but the general positions and plate movements are widely accepted (see Smith and Briden 1979 for more details).

Another factor affecting the dispersal and distribution of organisms during this long time span was the existence, at various times and in various places, of huge epicontinental seas (Lillegraven et al. 1979). These large bodies of water could effectively prevent the migration of species from one area of a continent to another, and thence on to another continental plate. High mountain ranges would have had a similar effect on dispersal.

Figure 3.4 Continents at the early Cretaceous period, 120 million years ago.

Plio-Pleistocene Epoch

The Pleistocene epoch began approximately 1.8 million years before the present and ended approximately 8000 years ago (Kurtén and Anderson 1980; Savage and Russell 1983). During this epoch most modern species or their direct ancestors appeared. It was during the late Pliocene epoch that North and South America became connected by the Panamanian land bridge at approximately 3.0 million years before the present (Marshall et al. 1982), allowing a major faunal interchange between the two continents. Prior to the formation of this land bridge, between what is now Panama and Colombia, some interchange between the two continents had occurred by waif dispersal. Early immigrants from North America into South America include the caviomorph rodents in the mid-Tertiary and procyonid carnivores in the late Tertiary periods (Kurtén and Anderson 1980;

Figure 3.5   Continents at the Paleocene epoch, 60 million years ago.

Marshall 1980; Simpson 1980). There were also faunal exchanges between Eurasia and North America during the Plio-Pleistocene epoch by way of the Beringian land bridge (Hopkins et al. 1982). These exchanges occurred in several episodes at approximately 3.5, 1.8, and 0.08 million years before the present (Kurtén and Anderson 1980) owing to opening and closing of the land connection by glacially controlled eustatic fluctuations (Hopkins 1967).

The Pleistocene epoch was a time of widespread extinction, resulting in a decrease in overall numbers of species of mammals, due to lack of ecological replacement. These extinctions occurred throughout the Pleistocene and affected both large and small mammals, although the extinction of the megafauna has received more attention. Some entire families that became extinct on one continent survived on other continents. Many causes for these extinctions have been proposed. Early Pleistocene extinctions in

Figure 3.6   Continents at the early Miocene epoch, 20 million years ago.

North America seem largely due to increasing aridity and the subsequent vegetational changes (Dort and Jones 1970). Late Pleistocene extinctions have been attributed both to climatic changes and (for the megafauna) to hunting pressure by humans (Martin 1982; Mosimann and Martin 1975). However, recent evidence seems to indicate that overkill by human hunters was not the major cause of extinctions (Gillespie et al. 1978; MacNeish 1976). Kurtén and Anderson (1980:363) attribute these extinctions to a variety of local conditions, stating that "Extinction did not occur uniformly across the continent. Local conditions affected local populations. No one cause can account for it; rather, a mosaic of adverse conditions prevailed. We believe that changes in vegetation, sudden storms, droughts, loss of habitat, interspecific competition, low reproduction rates, and over-specialization . . . reduced or weakened populations, making them vulnerable to environmental pressures, including man."

Figure 3.7   Present continents.

DIVERSITY AND DISTRIBUTION OF MAMMALS

Subclass Prototheria

Order Monotremata

The prototherians, the duckbilled platypus and the echidnas, are the most primitive of all mammals. Three genera of modern-day monotremes are recognized. The single species of duckbilled platypus, *Ornithorhynchus anatinus*, represents a monotypic family, the Ornithorhynchidae. Three species of Recent echidnas are known, belonging to two genera and representing the family Tachyglossidae. The monotremes inhabit Australia, New Guinea, and Tasmania. The fossil record of the three genera is well represented in Pleistocene deposits of Australia; however, earlier monotreme fossils consist of only two isolated teeth from the middle Miocene epoch of South Australia. The only modern-day monotreme to possess even rudiments of teeth is the duckbilled platypus, which has a few pre-molars and molars that are lost early in development. The Miocene teeth have been assigned to the Ornithorhynchidae, although they shed little light on the evolution of the modern monotremes. Clemens (1979:309) summarized the three groups of hypotheses concerning the origin of monotremes as follows: "The first includes those that recognize a basic evolutionary dichotomy subdividing the Mammalia into nontherian and therian lineages and allocate the monotremes to the nontherians without nominating a known Mesozoic order as their ancestral stock. Another group of hypotheses also recognizes the fundamental dichotomy between therian and nontherian mammals and allocates the monotremes to the nontherian group . . . indicating that among nontherian orders mono-tremes are most . . . closely related to multituberculates. . . . The third group of hypotheses suggests a special phylogenetic relationship between monotremes and marsupials."

There is very little support today for associating monotremes and mar-supials. Thus all we can say about the evolution of the monotremes is that they are an old group of uncertain ancestry. Using a cladistic analysis of a wide array of characters, Marshall (1979:369) demonstrated that "the most basic division of the Mammalia is the dichotomy into the subclasses Pro-totheria (including Monotremata, Multituberculata, Triconodonta, Doco-donta) and Theria (including Metatheria, Eutheria, Pantotheria and Sym-metrodonta). Two major groups exist among living viviparous mammals, the Metatheria and Eutheria; in a cladistic framework these are sister-groups. It is demonstrated that there is no special (sister-group) relation-ship between monotremes and marsupials, and there is no justification for placing them in a group Marsupionta." Excellent summaries of the biology of monotremes may be found in Griffiths (1968, 1978).

Subclass Theria

Infraclass Metatheria

Marsupials. The marsupials are an old group, the marsupial/placental dichotomy having taken place in the early Cretaceous period (Lillegraven 1974). Mid- and late Cretaceous fossils, mostly isolated teeth, are especially abundant in North American faunas until the Paleocene epoch (Clemens 1979). Today the marsupials are found in two areas: the Australian region, including Australia, New Guinea, Tasmania, and nearby islands; and the Neotropics, with one species, *Didelphis virginiana*, now found as far north as the northern United States.

The continent on which marsupials originated continues to be a matter of debate that can only be settled with the discovery of additional fossil material. G. G. Simpson earlier argued that marsupials originated in North America and then migrated to South America, Europe, and Australia; however, recently (Simpson 1980) he conceded that a South American origin is also possible. Most recent authors believe the marsupials evolved in South America or the larger Gondwanaland land mass and then colonized North America, Europe, and finally Australia through Antarctica. Fossil marsupials recently were discovered from the Eocene epoch of Antarctica by Woodburne and Zinsmeister (1982), who suggested that the new fossils support the theory that marsupials originated in South America and migrated across Antarctica to Australia while the three continents were still joined together (prior to 56 million years ago). In North America, marsupial remains are abundant in early Cretaceous formations; the earliest South American fossils are from the late Cretaceous period of southwestern Bolivia.

The classification of the marsupials has been relatively stable at the generic and family level for some time, although the classification at the ordinal level is currently in a state of flux. Historically, all have been classified as a single order, the Marsupialia, and many recent authors still treat them as such (i.e., Honacki et al. 1982). However, in 1959, Cain (1959:214) stated that "Because of their peculiar features [marsupials] are always ranked as a single order of mammals within a separate class, although the briefest inspection is enough to show that there is at least as much difference between a kangaroo and a dasyure (for example) as between an insectivore and a rodent, let alone a rodent and a lagomorph." In studying both fossil and Recent marsupials, Ride (1964) proposed recognizing four orders of Marsupialia: the Marsupicarnivora, the Paucituberculata, the Peramelina, and the Diprotodonta, with the two species of marsupial moles, *Notoryctes* (Notoryctidae), remaining *incertae sedis*. Ride (1964:125) said of the marsupial mole, "*Notoryctes*, the marsupial mole, is unknown as a fossil and it is so highly specialized in dentition, skull structure, and limb structure that arguments as to its affinities which are based upon these features can only produce tentative results." Kirsch (1977a, b) and Marshall (1981) provided reviews of the previous classifications of both fossil and Recent marsupials, with Kirsch recognizing only three orders of marsupials. Recently Szalay (1982) also proposed that four orders of Recent and fossil marsupials be recognized, two New World and two Australian; however, his arrangement of families within those orders differs from that proposed by Ride. Szalay organized the four orders into two newly described cohorts, the Ameridelphia and the Australidelphia. The Ameridelphia are entirely New World in distribution, and the Australidelphia are distributed throughout Australia (and adjacent areas), except for the south temperate *Dromiciops australis*. Kirsch and Archer (1982) however, stated that the serological and dental data align *Dromiciops* with the American didelphids.

The classification of marsupials used here (Table 3.2) includes the four orders recognized by Ride (1964); however, the Notoryctidae are here included as a family within the order Marsupicarnivora. Two orders, Diprotodontia (107 Recent species) and Peramelina (19 Recent species) are restricted to the Australian region. The order Paucituberculata (7 Recent species) is restricted to the Neotropics. The order Marsupicarnivora (130 Recent species) is found both in the Australian and Neotropical regions.

The marsupials are not the primitive form from which eutherians evolved, but rather a separate and equally highly evolved lineage. Of this distinction Kirsch (1977c:900) concluded "In summary, marsupials and placentals probably represent two different solutions to some of the problems of being a mammal; as is usually the case, both solutions have advantages as well as disadvantages. Metatherians and eutherians represent a true bifurcation of the Theria and not respectively sequential adaptive stopping-points on the lineage."

Infraclass Eutheria

Order Edentata (Xenarthra). The edentates are a highly specialized group of five Recent families of some 32 species. All are New World in distribution, most being found in the Neotropical region, although a few species are now found in the southern Nearctic. There is little doubt that the center of evolution of this group was in South America, where they remained isolated from perhaps the late Cretaceous through the Pleistocene period. It is thought that the edentates evolved from the primitive suborder Palaeanodonta in North America and migrated to South America by the late Cretaceous period or early Paleocene epoch, or that they arose in South America during the early Paleocene epoch. The Edentata (*sensu lato*) are thought to have been the first lineage to split from the Eutheria in the Cretaceous period. The Recent edentates include the Central and South American anteaters, armadillos, and tree sloths. Two-toed sloths (*Choloepus*) originally were included with the three-toed forms (*Bradypus*) in the family Bradypodidae. Recently it has been suggested that *Choloepus* is most closely related to the extinct giant ground sloths of the North American Pleistocene epoch, and should be considered a member of the family Megalonychidae (Webb in press). Recent reviews on ecology, systematics, and evolution of edentates were provided by Montgomery (in press) and Wetzel (1982).

Order Insectivora. The Insectivora are a diverse group of small mammals sharing many primitive characters. There has been much debate as to the relationships between families. In the past many families of uncertain affinity were placed in the order Insectivora. Historically, the Tupaiidae (tree shrews) and Macroscelididae (elephant shrews) were classified as

Table 3.2 Diversity and Geographical Distribution of Recent Mammals[a]

Order and Family	Neotropical	Nearctic	Palearctic	Ethiopian	Oriental	Australian	Oceanic	Recent Genera	Recent Species
Order Monotremata						X		3	3
Family Tachyglossidae						X		2	2
Ornithorynchidae						X		1	1
Order Diprotodontia						X		32	107
Family Burramyidae						X		4	7
Macropodidae						X		16	57
Phalangeridae						X		3	15
Phascolarctidae						X		1	1
Petauridae						X		5	23
Tarsipedidae						X		1	1
Vombatidae						X		2	3
Order Peramelina						X		8	19
Family Peramelidae						X		7	17
Thylacomyidae						X		1	2
Order Marsupicarnivora	X	X				X		28	130
Family Dasyuridae						X		13	50
Didelphidae	X	X						11	76
Microbiotheriidae	X							1	1
Myrmecobiidae						X		1	1
Notoryctidae						X		1	1
Thylacinidae						X		1	1
Order Paucituberculata	X							3	7
Family Caenolestidae	X							3	7
Order Edentata (Xenarthra)	X	X						16	32
Family Myrmecophagidae	X							3	4
Bradypodidae	X							1	3
Choloepidae	X							1	2
Megalonychidae (extinct)	X							3	3
Dasypodidae	X	X						8	20
Order Insectivora	X	X	X	X	X		O	66	396
Family Solenodontidae	X							1	2
Nesophontidae	X							1	6
Tenrecidae				X				12	33
Chrysochloridae				X				7	18

124

Table 3.2 (Continued)

Order and Family	Neotropical	Nearctic	Palearctic	Ethiopian	Oriental	Australian	Oceanic	Recent Genera	Recent Species
Erinaceidae			X	X	X			9	18
Soricidae	O	X	X	X	X	O		21	288
Talpidae		X	X		X			15	31
Order Macroscelidea			O	X				4	15
Family Macroscelididae			O	X				4	15
Order Scandentia					X			5	16
Family Tupaiidae					X			5	16
Order Dermoptera					X			1	2
Family Cynocephalidae					X			1	2
Order Chiroptera	X	X	X	X	X	X	X	176	917
Family Pteropodidae			O	X	X	X	O	42	160
Rhinopomatidae			X	X	X			1	3
Emballonuridae	X		O	X	X	X	O	12	48
Craseonycteridae					X			1	1
Nycteridae			O	X	X	O		1	14
Megadermatidae				X	X	X		4	5
Rhinolophidae (incl. Hipposiderinae)			X	X	X	X	O	10	127
Noctilionidae	X							1	2
Mormoopidae	X							2	8
Phyllostomidae (incl. Desmodontinae)	X	O						48	138
Natalidae	X							1	5
Furipteridae	X							2	2
Thyropteridae	X							1	2
Myzopodidae				X				1	1
Vespertilionidae	X	X	X	X	X	X	O	36	315
Mystacinidae						X		1	1
Molossidae	X	X	X	X	X	X	O	12	86
Order Primates	X	X	X	X	X	X	X	54	181
Family Cheirogaleidae				X				4	7
Lemuridae				X				4	16
Indriidae				X				3	4
Daubentoniidae				X				1	1

Table 3.2   (Continued)

Order and Family	Neotropical	Nearctic	Palearctic	Ethiopian	Oriental	Australian	Oceanic	Recent Genera	Recent Species
Lorisidae				X	X			4	5
Galagidae				X				2	8
Tarsiidae					X	O		1	3
Callithricidae	X							4	15
Callimiconidae	X							1	1
Cebidae	X							11	31
Cercopithecidae			O	X	X	O		11	76
Hylobatidae					X			1	9
Pongidae				X	X			3	4
Hominidae	X	X	X	X	X	X	X	1	1
Order Rodentia	X	X	X	X	X	X	O	385	1728
Family Aplodontidae		X						1	1
Sciuridae	X	X	X	X	X	O		49	261
Geomyidae	X	X						5	38
Heteromyidae	X	X						5	63
Castoridae		X	X					1	2
Anomaluridae				X				3	7
Pedetidae				X				1	1
Cricetidae	X	X	X	X	X			105	530
Spalacidae			X					1	3
Rhizomyidae			X	X	X			3	6
Arvicolidae		X	X					20	128
Muridae			X	X	X	X	O	108	437
Gliridae			X	X				7	15
Seleviniidae			X					1	1
Zapodidae		X	X					4	14
Dipodidae			X		O			11	30
Hystricidae			X	X	X	O		4	11
Erethizontidae	X	X						5	12
Caviidae	X							5	14
Hydrochaeridae	X							1	1
Heptaxodontidae (extinct)	X							2	2
Dinomyidae	X							1	1

Table 3.2 (Continued)

Order and Family	Neotropical	Nearctic	Palearctic	Ethiopian	Oriental	Australian	Oceanic	Recent Genera	Recent Species
Agoutidae	X							1	2
Dasyproctidae	X							2	13
Chinchillidae	X							3	6
Capromyidae	X							4	13
Myocastoridae	X							1	1
Octodontidae	X							5	8
Ctenomyidae	X							1	33
Abrocomidae	X							1	2
Echimyidae	X							13	55
Thryonomyidae				X				1	2
Petromyidae				X				1	1
Bathyergidae				X				5	9
Ctenodactylidae				X				4	5
Order Lagomorpha	X	X	X	X	X			12	65
Family Ochotonidae		X	X					2	19
Leporidae	X	X	X	X	X			10	45
Order Carnivora (incl. Pinnipedia)	X	X	X	X	X	X	X	108	271
Family Canidae	X	X	X	X	X	X		11	35
Ursidae (incl. *Ailuropoda*)	X	X	X		X			5	9
Procyonidae (incl. *Ailurus*)	X	X	X					7	19
Mustelidae	X	X	X	X	X		O	23	63
Viverridae				X	X			19	34
Herpestidae			X	X	X			17	36
Protelidae				X				1	1
Hyaenidae			X	X	X			2	3
Felidae	X	X	X	X	X			5	37
Otariidae							X	7	14
Odobenidae							X	1	1
Phocidae			O				X	10	19
Order Cetacea								39	77
Family Platanistidae	X	O			X		O	4	5
Delphinidae					O		X	17	33
Phocoenidae		O					X	3	6

Table 3.2   (Continued)

Order and Family	Neotropical	Nearctic	Palearctic	Ethiopian	Oriental	Australian	Oceanic	Recent Genera	Recent Species
Monodontidae		O	O				X	2	2
Physeteridae							X	2	3
Ziphiidae							X	6	18
Eschrichtidae							X	1	1
Balaenopteridae							X	2	6
Balaenidae							X	2	3
Order Sirenia	X	O	O	X	O	O	X	3	5
Family Dugongidae			O	O	O	O	X	2	2
Trichechidae	X	O		X			O	1	3
Order Proboscidea				X	X			2	2
Family Elephantidae				X	X			2	2
Order Perissodactyla	X		X	X	X			6	18
Family Equidae			X	X	O			1	9
Tapiridae	X				X			1	4
Rhinocerotidae				X	X			4	5
Order Hyracoidea			O	X				3	7
Family Procaviidae			O	X				3	7
Order Tubulidentata				X				1	1
Family Orycteropodidae				X				1	1
Order Artiodactyla	X	X	X	X	X	O		77	187
Family Suidae			X	X	X	O		5	8
Tayassuidae	X	O						2	3
Hippopotamidae			O	X				2	2
Camelidae	X		X					3	6
Tragulidae				X	X			2	4
Cervidae (incl. Moschinae)	X	X	X		X	O		15	38
Giraffidae				X				2	2
Bovidae (incl. Antilocaprinae)		X	X	X	X	O		46	124
Order Pholidota				X	X			1	7
Family Manidae				X	X			1	7

[a] X indicates region of principal distribution; O indicates region of secondary, minor distribution.

families of Insectivora. In this scheme the Tupaiidae and Macroscelididae were grouped together as a suborder, the Menotyphla, with the remaining families forming the suborder Lipotyphla.

Recent attempts to clarify the relationships have produced a variety of classifications (see Findley 1967; McKenna 1975; Van Valen 1967). The classification given here is that presented by Honacki et al. (1982), and includes the families Solenodontidae, Nesophontidae, Tenrecidae, Chrysochloridae, Erinaceidae, Soricidae, and Talpidae. In this system the Tupaiidae and Macroscelididae are treated as separate orders.

The Solenodontidae, or solenodons, are restricted to the islands of Cuba and Hispaniola. A single Recent genus (*Solenodon*) represented by two species is known. A summary of what little is known of the biology of these species is presented by Nowak and Paradiso (1983). The fossil record dates back only to the late Pleistocene epoch.

The Nesophontidae is known from a single Recent genus, *Nesophontes*, and six species, originally found on several West Indian islands; however, it is now extinct. It is thought that their extinction followed shortly after the arrival of the Spanish to the New World. Habitat destruction and the introduction of rats and cats to the islands perhaps led to extinction of these insectivores. Findley (1967) considered the genus *Nesophontes* a member of the family Solenodontidae.

The Tenrecidae, or tenrecs and otter shrews, are an old and diverse group of 12 Recent genera and 33 species in 3 subfamilies. The otter shrews, two genera and three species, are found in equatorial Africa. Many authors regard the otter shrews as a distinct family, the Potamogalidae. The tenrecs are all restricted to the island of Madagascar except for introductions to several islands in the Indian Ocean. Three species of tenrecs are known from the early Miocene epoch of Africa. The only known fossil tenrecs from Madagascar are from the Pleistocene, although the group undoubtedly existed on the island before it separated from the African mainland. Of their relationships Butler (1978:63) stated, "Whatever their relationships may be, it is not disputed that the Tenrecidae are an ancient, isolated group. Though unknown before the Miocene, it is very likely that they formed part of the Paleocene African fauna. In the Miocene there were three very distinct genera, implying an earlier radiation." An excellent study of tenrec behavior was presented by Eisenberg and Gould (1970).

The Chrysochloridae, or golden moles, consists of 7 genera and 18 species of small fossorial insectivores. They are confined to central and southeastern Africa. The oldest known fossils are from the early Miocene of Africa.

The Erinaceidae, or hedgehogs and gymnures, include 9 Recent genera and 18 species. Today the family is found throughout the Ethiopian, Oriental, and Palearctic faunal regions. Fossil hedgehogs are known from the Eocene to early Pliocene epochs of North America, late Eocene of Europe, Miocene of Africa, and Oligocene of Africa.

The Soricidae, or shrews, includes some of the smallest mammals. They are widely distributed throughout most habitat types in the northern hemisphere and Africa. They are found on all continents, but the greatest diversity is in the Nearctic, Ethiopian, and Palearctic faunal regions; there are 21 Recent genera and some 288 species. The systematics of many species in the two largest genera (*Crocidura* and *Sorex*) are poorly understood. The oldest known fossils are from the late Eocene epoch of Europe and early Oligocene of North America.

The Talpidae, or moles and desmans, are fossorial or semiaquatic insectivores of the northern hemispheres. Twelve Recent genera and 27 species are known. The oldest known fossils are from the late Eocene epoch of Europe and the middle Oligocene of North America.

Order Macroscelidea.   The elephant shrews comprise a single family, Macroscelididae, of 15 species in 4 genera restricted to the African continent; one species is found in Morocco and Algeria and the remaining species in central, eastern, and southern Africa. The fossil record is poor and entirely restricted to Africa (Patterson 1965). Fossil elephant shrews are known from the early Oligocene, Miocene, and Pliocene epochs. Elephant shrews have often been considered a subfamily of Insectivora (Le Gros Clark 1932; Patterson 1957); however, recent evidence suggests that the Macroscelidea are best considered a distinct order (Butler 1956; Patterson 1965; McKenna 1975). Romer (1971:211) stated that "They appear to be an isolated African offshoot from some primitive insectivore stock." This family was reviewed by Corbet and Hanks (1968).

Order Scandentia.   The tree shrews comprise a single family, Tupaiidae, of 16 species in five genera, that is restricted to the Oriental region. This family occurs throughout eastern Asia, on the mainland from India and China, through the Malay Peninsula, Borneo, and the Philippines. Tree shrews have been variously classified as a family of Insectivora (Romer 1971; Van Valen 1982), Primates (Le Gros Clark 1932; Simpson 1945), and most recently as a separate order, the Scandentia (Butler 1972; McKenna 1975). The genera within the family Tupaiidae form a coherent grouping, possessing several primitive characters that perhaps place them closest to the ancestral stock of all placental mammals (Campbell 1974). An extensive bibliography of the tree shrews, including parasites, was provided by Elliot (1971). Luckett (1980) reviewed evidence assessing relationships. There is no fossil record for the family Tupaiidae.

In describing a new species of *Psorergates* (Psorergatidae) from *Tupaia dorsalis* and reviewing the mites found on tupaiids, Giesen and Lukoschus (1982:266) stated, "These data on the distribution of parasitic mite taxa confirm neither a relation to Primates nor to Insectivora. Aberrant species and monotypic genera on Tupaiidae seem to support the hypothesis of an unique phylogenetic evolution for the Scandentia."

Order Dermoptera. The Recent flying lemurs, or colugos, comprise a single family with two species in a single genus (*Cynocephalus*) that occurs in Southeast Asia and the Philippines. The only fossils known, tentatively identified as dermopterans, are from the late Paleocene and early Eocene epochs of North America. Van Valen (1967) considered the Dermoptera a suborder of the Insectivora; however, most recent authors consider the group to represent a distinct order. The general consensus, expressed by Patterson (1957:24), is "that dermopterans derive from the general insectivore stock from which chiropterans and primates also arose."

Order Chiroptera. Bats occur on all land masses with the exception of a few isolated oceanic islands and the polar caps. Most remote oceanic islands have endemic populations of bats. Recent bats are divided into two suborders: the Megachiroptera and the Microchiroptera. Megachiropterans are strictly Old World in distribution; microchiropterans are found worldwide. The oldest known fossil bat, *Icaronycteris index* of the early Eocene epoch of Wyoming (Jepsen 1966, 1970), has been assigned to the Microchiroptera, although it and several other early genera may best be considered a distinct suborder, the Eochiroptera. Fossil megachiropterans are known from a few specimens of the Oligocene and Miocene epochs of Europe. Well-developed bats, not dissimilar from modern-day microchiropterans, are known from middle Eocene deposits in both North America and Europe. By the beginning of the Oligocene epoch, at least six extant families of bats and several extant genera were already established. A diphyletic origin to the two major groups of bats, the mega- and microchiropterans, has been postulated (Smith 1976; Smith and Madkour 1980); however, most workers concur that the group is monophyletic (Van Valen 1979). Bats undoubtedly arose from an insectivoran-like ancestor. Excellent recent reviews of the ecology of parasitic insects on bats were provided by Marshall (1981, 1982). A review of the arthropods parasitic on the Neotropical family Phyllostomidae may be found in Webb and Loomis (1977).

Order Primates. The Primates are a diverse group of some 181 species in 54 genera found mostly in tropical and subtropical areas. Currently, 14 families are recognized. The Recent primates are divided into two or three distinct suborders. The more primitive group is the suborder Strepsirhini, which includes the lemurs, indrids, lorises, and galagos, all of which are Old World in distribution. The suborder Haplorhini includes two infraorders: the Platyrrhini, the cebids and marmosets of the New World tropics, and the infraorder Catarrhini, the cercopithecids, hylobatids, pongids, and hominids of the Old World tropics and subtropics. A few species of cercopithecids extend into the southern Palearctic, and the one extant hominid, *Homo sapiens*, is now worldwide in distribution. The tarsiers have been included at various times either in the Strepsirhini (then called the

Prosimi), the Haplorhini, or most recently, as a distinct suborder, the Tarsii.

Precisely where the Primates originated and the routes of dispersal is a subject that has been hotly debated in recent years (see Ciochon and Chiarelli 1980; Luckett and Szalay 1975). *Purgatorius* of the late Cretaceous period of North America has been described as the earliest known primate fossil (Van Valen and Sloan 1965); however, its affinities to the primates is debatable. The earliest unquestioned primate fossils are from the early Paleocene epoch of North America. Late Paleocene primates are abundant in both North American and European faunas. The oldest known primate fossils from South America and Africa are from the Oligocene epoch (Hershkovitz 1974, 1977; Simons et al. 1978; Simons and Kay 1983). The consensus is now that the Primates evolved in Africa, although there is a paucity of early African fossils. Other models for the origin of primates include those of Asian origin (Delson and Rosenberger 1980; Gingerich 1980), North American origin (Simpson 1945), and South American origin. African and Asian models are currently being debated (see Ciochon and Chiarelli 1980).

Concerning the phylogeny of tarsiers, Cartmill (1982:279) wrote: "There are currently three principal schools of thought concerning the phylogenetic affinities of the genus *Tarsius*. One school regards *Tarsius* as the most distant living primate relative of the Simiiformes; the second school regards *Tarsius* plus the extinct Omomyidae as constituting the phyletic sister group of Simiiformes; and the third school holds that *Tarsius* itself is the sister group of the Simiiformes. Oddly, all three schools contend that comparative otic anatomy provides crucial evidence for their favorite phylogenetic schemes." Gingerich (1980:133) concluded that "There is disagreement regarding the major phyletic relationships of Tarsiiformes, Lemuriformes, and Simiiformes, with different results depending on whether one attempts to trace phyletic groups through the fossil record or to infer history from the comparative anatomy of living forms. . . . Parallelisms and reversals are common evolutionary phenomena. . . . This means . . . that our evidence regarding primate phylogeny is still far from complete . . . and that we need to take a more critical look at different methods being used to reconstruct primate history."

Within the suborder Strepsirhini, four families are restricted entirely to Madagascar: the Lemuridae (16 Recent species), the Cheirogaleidae (7 Recent species), the Indriidae (4 Recent species), and the Daubentoniidae (1 Recent species). The Galagidae (8 Recent species) is found only in sub-Saharan Africa; the Lorisidae (5 Recent species) is found in Southeast Asia and west-central Africa; and the Tarsiidae (3 Recent species) is Southeast Asian in distribution and is usually considered a distinct suborder, Tarsii.

Four families comprise the infraorder Catarrhini (suborder Haplorhini), and all were originally Old World in distribution: the Cercopithecidae (76 Recent species) are found in the Ethiopian, Palearctic, and Oriental faunal regions; the Hylobatidae, or gibbons (9 Recent species), are restricted to southeast Asia; the Pongidae, or great apes (4 Recent species), are found in the Ethiopian and Oriental faunal regions; the Hominidae, or man (1 Recent species), are now worldwide in distribution. The earliest known hominid, *Australopithecus afarensis*, is approximately 3.0 million years old. As early as 1945, Simpson (1945:187) stated, "Most students now believe that the gibbons, apes, and man form a natural unit." The Hominidae is recognized as a distinct family, primarily on the basis of intellectual development. There is little biochemical justification for regarding the Hominidae as a family distinct from the Pongidae (see King and Wilson 1975; Yunis et al. 1980). Three families comprise the infraorder Platyrrhini (suborder Haplorhini), all Neotropical in distribution: the Callimiconidae (1 Recent species), the Callithricidae (15 Recent species), and the Cebidae (some 31 + Recent species).

Order Rodentia.   The order Rodentia is the largest and most diverse order of mammals, including some 37 families and 1728 + Recent species. Rodents are the most successful of all living mammals; the order Rodentia has more species, genera, and families than any other mammalian order. Rodents occupy a wide range of niches including arboreal, terrestrial, semiaquatic, and subterranean, and have the widest geographic distribution of any order. They are found worldwide except for Antarctica and some oceanic islands. Rodents often exceed other mammals in local diversity and abundance. Most are small in size (rat sized), although the range is from 5 grams in the Old World harvest mouse, *Micromys minutus*, to 79 kilograms in the South American capybara, *Hydrochoerus hydrochaeris*.

The relationships of the rodents to the other orders of mammals are poorly understood. Historically, the rodents and lagomorphs have been considered together as suborders of the order Rodentia. All recent authors consider the two groups as distinct orders of separate ancestry. It has been suggested that the rodents are most closely related to the Primates (McKenna 1961), Insectivora (Wood 1962), or the Carnivora (Szalay and Decker 1974). McKenna (1975) in a recent classification of the Mammalia, listed the order Rodentia as *incertae sedis*. Romer (1971:303) stated that "The origin of the rodents is obscure. When they first appear, in the late Paleocene, in the genus *Paramys*, we are already dealing with a typical, if rather primitive, true rodent, with the definitive ordinal characters well developed. Presumably, of course, they had arisen from some basal, insectivorous, placental stock; but no transitional forms are known. To perfect the dental and other features of the order, a considerable period of time— perhaps the whole extent of the Paleocene—seems necessary. But in what region or environment this occurred, we do not know." A complex of some five genera of true rodents is now known from the latest Paleocene epoch (Clarkforkian) of the western United States, representing the earliest known rodents (Savage and Russell 1983).

Traditionally, three suborders of Rodentia are recognized, although there is considerable disagreement as to the relationships of the suborders and to the placement of families within the groups: the Sciuromorpha, including the squirrel-like rodents of the families Anomaluridae, Aplodontidae, Castoridae, Geomyidae, Heteromyidae, Pedetidae, and Sciuridae; the Myomorpha or rat and mouse-like rodents of the families Arvicolidae, Cricetidae, Dipodidae, Gliridae, Muridae, Rhizomyidae, Seleviniidae, Spalacidae, and Zapodidae; and the Hystricomorpha or porcupine-like rodents of the families Abrocomidae, Agoutidae, Bathyergidae, Capromyidae, Caviidae, Chinchillidae, Ctenodactylidae, Ctenomyidae, Dasyproctidae, Dinomyidae, Echimyidae, Erethizontidae, Heptaxodontidae, Hydrochaeridae, Hystricidae, Myocastoridae, Octodontidae, Petromyidae, and Thryonomyidae (see Vaughan 1978; Eisenberg 1981; Nowak and Paradiso 1983; Woods 1982; and references therein).

Romer (1971) recognizes three suborders, the Sciuromorpha, Myomorpha, and Caviomorpha, but lists three superfamilies (Castoroidea, Theridomyoidea, and Thryonomyoidea) and four families (Anomaluridae, Ctenodactylidae, Hystricidae, and Pedetidae) as not assigned to either superfamily or suborder. The relict family Aplodontidae was assigned by Wood (1965), along with several extinct families, to the suborder Protrogomorpha. Several recent authors have questioned the value of continuing to recognize suborders within the Rodentia. A review of the problems and classifications of rodent suborders may be found in Anderson (1967).

The Anomaluridae, or scaly-tailed squirrels, are found in central and western Africa in both tropical and subtropical forests; no fossils of this family are known (Kingdon 1974). The Aplodontidae, the mountain beaver or sewellel, is now represented by the single species, *Aplodontia rufa*, which is restricted to the Pacific Northwest. It is the sole surviving member of a once widespread family whose fossils are known from the late Eocene of North America, Oligocene of Europe, and Pliocene epoch of Asia. The Castoridae, or beavers, are Holarctic in distribution, among the largest of rodents (weighing up to 40 kilograms), and semiaquatic. Fossils are known from the early Oligocene of North America and the late Oligocene of Europe.

The Geomyidae, or pocket gophers, are endemic to North America and are found from southern Canada to extreme Northwestern Colombia. Pocket gophers are fossorial; the earliest fossils are from the latest Oligocene. The Heteromyidae, or pocket mice, are the sister group of the geomyids and also are distributed from southern Canada to northern South America; the habitats occupied range from arid deserts to humid tropical forests (Genoways 1973). The earliest fossils are from the early Oligocene epoch. The Pedetidae, or springhaas, are represented by a single living species, *Pedetes capensis*, which occurs in arid areas of the southern Ethiopian region; the earliest fossils are from the Miocene epoch of Africa.

The Sciuridae, or squirrels, marmots, chipmunks, and prairie dogs, are found in all zoogeographic regions except for the Australian; this is a large and diverse family occupying a wide array of habitats. The earliest fossils are from the Oligocene epoch of Europe and North America. The Arvicolidae, or lemmings, muskrats, and voles, are Holarctic and primarily found in forests and prairies; the muskrats are semiaquatic. The earliest fossils are from the Oligocene.

The Cricetidae, or gerbils, hamsters, mice, and rats, are widely distributed in numerous habitats in both the New and Old World; the earliest fossils are Oligocene (Carleton 1980). The Arvicolidae and Cricetidae, along with the murid rodents, are considered by some authors to belong to a single family, the Muridae. The Dipodidae, or jerboas, are arid-adapted jumping rodents of the southern Palearctic; the earliest fossils are from the Pliocene epoch of Asia. The Gliridae, or dormice, are found in the Ethiopian, Oriental, and Palearctic regions; the earliest fossils are from the middle Oligocene epoch of Europe. The Muridae, or rats and mice, are now found worldwide, although they were originally restricted to the Old World; the greatest species diversity is found in southeast Asia; the earliest fossils are from the Pliocene epoch of Europe and Asia.

The Rhizomyidae, or bamboo rats, are found in west-central Africa and Southeast Asia; the earliest fossils are from the late Oligocene of Europe. The Seleviniidae, or desert dormouse, is known from a single species, *Selevinia betpakdalensis*, found in a restricted desert region of the USSR; there are no known fossils (Nowak and Paradiso 1983). The Spalacidae, or mole rats, are found in the southern Palearctic; as the name implies, mole rats are fossorial. The earliest fossils are from the latest Pliocene. The Zapodidae, or jumping mice, are a Holarctic group of small mice generally found in forests, meadows, and marshes; the earliest fossils are from the Oligocene of Europe and the early Miocene of North America (Wrigley 1972).

The Abrocomidae, or chinchilla rats and chinchillones, contain a single genus, *Abrocoma*, with two species that are found at higher elevations in west-central South America; the earliest fossils are from the late Miocene of South America. The family Agoutidae, or pacas, contains a single genus, *Agouti*, with two species, found from east-central Mexico to Paraguay; they are moderate-sized, nocturnal, forest animals (Collett 1981). The pacas are often considered a subfamily of Dasyproctidae. The Dasyproctidae, or agoutis and acouchis, includes two genera of moderate-sized forest rodents that occur from southern Mexico throughout much of South America; the earliest fossils are from the Oligocene of South America (Smythe 1978).

The Bathyergidae, or African mole-rats, occur south of the Sahara Desert in Africa (Kingdon 1974). They are highly adapted to a fossorial existence; fossils are known from the Miocene of Africa and Oligocene of Mongolia. The Capromyidae, or hutias, include several large, nocturnal rodents restricted to the Antilles; the earliest fossils are from the late Pleistocene of the West Indies (Anderson et al. 1983). The Caviidae, or cavies, cuis, and guinea pigs, occur over much of South America; caviids generally are found at higher elevations, rocky outcrops, and savannahs (Rood 1972; Lacher 1981). The earliest fossils are from the middle Miocene of South America. The Chinchillidae, or chinchillas and viscachas, are medium-sized rodents of the South American Andes and pampas; the earliest fossils are from the early Oligocene of South America.

The Ctenodactylidae, or gundis, are restricted to North Africa; they are medium-sized rodents of arid areas (George 1974). The earliest fossils are from the Oligocene of Asia. The Ctenomyidae, or tuco-tucos, are a highly fossorial complex of species belonging to a single genus *Ctenomys*, found in the southern half of South America (Weir 1974). The earliest fossils are from the Pliocene. The Dinomyidae, or pacarana, are represented by a single living species, *Dinomys branickii* from the Andean highlands of central South America. The earliest fossils are from the Miocene. The Echimyidae, or spiny rats, are found in forested regions throughout the northern half of South America and southern Central America. The earliest fossils are from the late Oligocene.

The Erethizontidae, or New World porcupines, include several large-bodied, spiny rodents distributed across the United States and Canada and through much of Central and South America; fossil porcupines are known from the Oligocene of South America and the late Pliocene of North America. The Heptaxodontidae are an extinct family of large-bodied rodents known only from Pleistocene and Recent fossils found in cave deposits in the West Indies. This family undoubtedly became extinct through the activity of man. The Hydrochaeridae, or capybara, is represented by a single living species, *Hydrochaeris hydrochaeris*, the largest of all rodents. Capybaras are found throughout eastern South America and Panama; the earliest fossils are from the early Pliocene of South America. The Hystricidae, or Old World porcupines, are large-bodied, spiny rodents found throughout all of Africa and much of southern Asia. The fossil record dates back to the Oligocene of Europe, middle Pliocene of Asia, and Pleistocene of Africa.

The Myocastoridae, the nutria of coypu, is represented by a single living species, *Myocastor coypus*, a large-bodied, semiaquatic rodent originally found throughout southern South America, and now widely distributed in Europe and the southern and northwestern United States, where they were introduced for the fur industry (Lowery 1974). The earliest fossils are from the early Miocene of South America. The Octodontidae, or octodonts, are rat-sized rodents found in the Andean mountains, foothills, and adjacent coastal areas of South America; the earliest fossils are from the early Oligocene (Woods and Boraker 1975). The Petromyidae, or dassie rat, is known from a single species, *Petromus typicus*, found in arid areas of southwestern Africa; there are no known fossils. The Thryonomyidae, or cane rats, are represented by a single Recent genus, *Thryonomys* with two species, and are found throughout the southern half of Africa; the earliest fossils are from the Miocene (Kingdon 1974).

Order Lagomorpha. The Lagomorpha, the rabbits, hares, and pikas, are an old group, although extant species show little variation in body form. Lagomorphs are native to all continents except Australia and Antarctica; none were endemic to the Australian faunal region, however, they have been introduced onto many of the islands by man in recent years. There are two Recent families in the order: the Leporidae, which includes the rabbits and hares (45 Recent species), is widely distributed and now widely introduced, and the Ochotonidae, or pikas (19 Recent species), which are Holarctic in distribution. McKenna (1982:213) stated that "The order [Lagomorpha] almost certainly originated in Asia, spreading from there to other continents at various times in the Cenozoic." The oldest known lagomorph is *Mimotona* from the Paleocene of Asia. Wilson (1960) suggested that the ochotonids and leporids diverged from a common ancestor during the Oligocene. Europe and Asia seem to have been the center of evolution and diversification for the ochotonids; they reached their greatest diversity and distribution in the Miocene and have declined since. The leporids spread to other continents early on. The order Lagomorpha has no clear relationship to any other mammalian order. Over the years, the systematic position of lagomorphs has varied considerably. Older classifications considered the group a suborder, the Duplicidentata, of the Rodentia. Using amino acid sequence information, Dene et al. (1982) suggested that the Lagomorpha, Scandentia (tree shrews), and Carnivora were a monophyletic group.

Order Carnivora. The Carnivora are a widespread and successful group that includes a diverse array of body forms. Today carnivores are found on all land masses and in the oceans. In recent years several species have been successfully introduced by man onto continents not previously occupied. This is especially true for domestic dogs (*Canis familiaris*), cats (*Felis catus*), and mongooses (*Herpestes*). The Recent Carnivora include some 12 families and 271 species. The number of families and even orders included in this group remains controversial.

The Phocidae, or earless seals, and the Otariidae, the sea lions and fur seals, are often considered together as a distinct order, the Pinnipedia, or as a distinct suborder, the Fissipedia. However, recent evidence indicates that this grouping is artificial and based on convergence; pinnipeds as such are a polyphyletic group, having been derived from separate terrestrial ancestors (McLaren 1960; Tedford 1976). The Phocidae are now thought to have originated from mustelid stock in the North Atlantic; the earliest fossils are mid-Miocene (Ray 1976). The Otariidae are believed to have originated from canoid stock in the north Pacific. Repenning (1976) considers the walruses (Odobenidae) and the Otariidae to have evolved from an extinct aquatic family, the Enaliarctidae, which was derived from the canoid carnivores. He included the two families together in a separate order, the Otarioidea.

Two suborders of Carnivora are often recognized: Caniformia which includes the families Canidae, Ursidae, Odobenidae, Otariidae, Procyonidae, Mustelidae, and Phocidae; and Feliformia which includes the Viverridae, Herpestidae, Protelidae, Hyaenidae, and Felidae. The suborder Caniformia is sometimes grouped into three superfamilies: Canoidea, including only the family Canidae; Arctoidea, including the Ursidae, Otariidae, Odobenidae, and Procyonidae; and Musteloidea, including the Mustelidae and Phocidae.

The Canidae, or wolves, dogs, jackals, and foxes, include some 11 Recent genera and 35 species; they are worldwide in distribution. Canid fossils are known from the Pliocene in Africa (Savage 1978), the late Eocene of North America and Europe, early Oligocene in Asia, and early Pleistocene in South America. A review of the systematics and ecology of the wild canids may be found in Fox (1975).

The Ursidae, or bears and giant panda, includes five Recent genera and nine species. Fossil bears are known from the Pliocene of North America and Asia; they reached the South American and African continents during the Pleistocene (Savage 1978). Although the number of recognized species of bears has remained fairly constant, the number of genera recognized by various authors fluctuates, as does the number of subfamilies (Mondolfi 1983).

The relationships of the giant panda (*Ailuropoda melanoleuca*) and the lesser (or red) panda (*Ailurus fulgens*) to the other carnivores has been problematic for some time and remains so today. Giant pandas and lesser pandas historically have been placed in monotypic families, Ailuropodidae and Ailuridae, respectively, or together as a subfamily of the Procyonidae (Ewer 1973). Excellent reviews of the giant panda which place them as a subfamily of bears may be found in Davis (1964) and Chorn and Hoffmann (1978). Lesser pandas are most often considered members of the procyonids; however, Ginsburg (1982) recently reiterated that *Ailurus* is most closely related to the bears and otariid seals and must be considered a monotypic family, Ailuridae.

The Procyonidae, or raccoons, coatis, and allies, include 7 Recent genera and 19 species. All except the lesser panda are Neotropical or Nearctic in distribution; most are omnivorous, highly opportunistic predators. Fossil procyonids are known from the late Oligocene to Recent in North America, from the late Miocene to the late Pliocene in Europe; they reached South America during the Pliocene. A review of the ecology and systematics of the procyonids can be found in Ewer (1973).

The family Mustelidae is a diverse group of some 23 Recent genera and

63 species, and includes the weasels, martens, skunks, badgers, and otters. Mustelids are found in all faunal regions except the Australian. The distribution of the family is basically Holarctic, and many individual species or species complexes are Holarctic. Ecologically they are replaced in the tropical Old World by the viverrids and herpestids. The geological range of the Mustelidae is early Oligocene to Recent in North America, Europe, and Asia; late Miocene to Recent in Africa; and Pliocene to Recent in South America. Recently, Muizon (1982:259) suggested that the Mustelidae "represents a paraphyletic grouping and the genera *Enhydra* [sea otters] and *Enhydriodon* . . . constitute the sister group of the Phocidae."

The family Viverridae has traditionally included a diverse grouping of six subfamilies; however, C. Wozencraft (in Honacki et al. 1982) recently suggested that the mongooses merit recognition as a distinct family, the Herpestidae. The Viverridae (*sensu stricto*) would thus include the genets, civets, and palm civets, some 19 Recent genera and 34 species restricted to the Ethiopian and Oriental faunal regions. On the origin of the viverrids (*sensu lato*), Savage (1978:257) wrote, "The family is ancient, originating from the Miacidae probably in late Eocene times in Eurasia. On dental characters alone, it is impossible to distinguish late viverravine miacids from early viverrids. . . . There are no known miacids in Africa and the first migrants, probably in earliest Miocene, were already viverrids." Savage also suggested that the viverrids colonized the island of Madagascar "sporadically from Miocene times onward." The viverrids apparently reached Madagascar by rafting, and there were probably at least two separate colonizations.

The family Protelidae contains a single Recent genus and species, *Proteles cristatus*, the aardwolf. Aardwolves are restricted to eastern and southern Africa. They are often considered members of the family Hyaenidae; however, they also have been placed within the Viverridae (Savage 1978). The families Protelidae and Hyaenidae are thought to have evolved from a branch of the Viverridae. There is no fossil record of aardwolves. Aardwolves feed primarily on termites. A review of the biology of *Proteles* may be found in Kingdon (1977).

The Hyaenidae, or hyaenas, contains two Recent genera and three species. Hyaenids are found in the Ethiopian, Oriental, and Palaearctic faunal regions. Hyaenas are highly specialized for carrion feeding, although the spotted hyaena, *Crocuta crocuta* is an effective predator on big game. The oldest known fossils are from the late Miocene of Eurasia, and most are restricted to the Old World; however, one genus crossed the Beringian land bridge and was found in North America during the Pleistocene epoch. Hyaenids reached their peak diversity in the Pleistocene. The family Hyaenidae is thought to be a direct descendent of the Viverridae. A review of hyaenid evolution was provided by Thenius (1966). Reviews of the biology of individual species may be found in Mills (1982) and Kruuk (1972).

The family Felidae, or cats, contains some four or five Recent genera and 37 species. There is little consensus on the number of genera or relationships between species, although there is little disagreement of the number of species recognized. The number of genera recognized by various authors ranges from two to 15 (Hemmer 1978). The family is worldwide in distribution, except for the Australian and Oceanic regions. Cats are the most specialized and predaceous of the carnivores, and are fairly uniform in body structure. Fossils are known from the late Eocene in North America and Eurasia, Miocene of Africa, and Pleistocene of South America. Savage (1977:243) suggested that the "felids and viverrids have independent origins from miacids." Reviews of the Recent species of cats may be found in Eaton (1973) and Guggisberg (1975).

Most Carnivora are flesh eaters, although a few are insectivorous (aardwolves and meerkats) or herbivorous (giant pandas and binturongs). Carnivores are generally medium to large sized and have a complex social behavior. An excellent review of the ecology, systematics, and evolution of the Carnivora was presented by Ewer (1973), and a recent review of behavior in carnivores was presented by Macdonald (1983).

Order Cetacea. The cetaceans, or whales, are found throughout all oceans of the world, and a few species of dolphins are now secondarily found in fresh water. The Recent members of the order Cetacea are generally divided into two distinct suborders: the Mysticeti, or baleen whales, and the Odontoceti, or toothed whales. A third suborder, the Archaeoceti includes the extinct primitive whales. A few authors regard the suborders as separate orders: the Mysticeta, Odontoceta, and Archaeoceta (i.e., DeBlase 1982). This discrepancy results from the problem of whether the whales had a monophyletic or a polyphyletic origin.

The oldest and most primitive cetaceans, the archaeocetes from the Eocene, Oligocene, and Miocene, are thought to have evolved from primitive, carnivorous condylarths. The earliest archaeocete known is *Pakicetus* from the early Eocene of Pakistan (Gingerich et al. 1983); middle Eocene whales are known from formations in North America, Africa, and Asia. The earliest Mysticeti are from the middle Oligocene of Europe, and the oldest Odontoceti are from the late Eocene of North America.

All cetaceans are extremely specialized for an aquatic mode of life; adaptations include fusiform body shape, modification of the anterior limbs into flippers, a lack of external hind limbs, and an essentially hairless body. Ectoparasites of cetaceans include acarids, diatoms, amphipods, ciliates, cirripeds, copepods, isopods, lampreys, and remoras (Arvy 1982). A recent review of the evolution, zoogeography, and ecology of the Cetacea may be found in Gaskin (1982).

Order Sirenia. The order Sirenia includes the manatees, dugongs, and sea cows; all are large bodied, fully aquatic herbivores. The order contains two Recent families: the Dugongidae with two monotypic genera, *Dugong* and *Hydrodamalis*; and Trichechidae with one genus, *Trichechus*, and three species. The extant species are tropical or subtropical in distribution. Stellar's sea cow (*Hydrodamalis gigas*) was confined to the shallow waters around Bering and Copper Islands in the Bering Sea but was hunted to extinction by Russian whaling crews only 27 years after its discovery in the mid-1700s. The most primitive sirenian known is *Prorastomus* of the middle Eocene of Jamaica (Savage 1976). Domning (1982:599) wrote concerning the relationships of *Prorastomus* that "the tooth formula of *Prorastomus* is in fact typical of primitive sirenians, including dugongids, and that its geographic occurrence likewise represents the primeval 'Tethyan' distribution of the order. It may be regarded as a good structural ancestor for all later sirenians. . . ." Late Eocene genera are known from Europe and North Africa. The sirenians reached their peak diversity in the Miocene and Pliocene epochs. It is thought that the Sirenia and Proboscidea evolved from a common condylarth ancestor. Reviews of the biology and evolution of Sirenia may be found in Domning (1976, 1978, 1982), Forsten and Youngman (1982), Hartman (1979), and Husar (1977).

Order Proboscidea.   Recent elephants include only two species: the African elephant, *Loxodonta africana*, and the Indian elephant, *Elephas maximus*. Both belong to the single family Elephantidae. *Loxodonta* was until recently found throughout sub-Saharan Africa, and *Elephas* throughout much of south Asia. The order Proboscidea evolved in Africa, the oldest true proboscidean being *Palaeomastodon*, a tapir-sized animal of the late Eocene epoch of northern Africa. The Proboscidea underwent a rapid radiation during the later Cenozoic era. Fossil elephants include a diverse array of forms now categorized into four families and perhaps representing 100 species. By the late Miocene epoch proboscideans had reached North and South America and Eurasia. Several forms became extinct during the Pleistocene epoch; only the family Elephantidae remains extant today. Reviews of the evolution and origin of the elephants may be found in Aguirre (1969) and Maglio (1973).

Order Perissodactyla.   The Recent members of the order Perissodactyla comprise three distinct families: the Equidae, the horses, zebras, and asses (nine species); the Tapiridae, the tapirs (four species); and the Rhinocerotidae, the rhinoceroses (five species). The Perissodactyla evolved from the order Condylarthra; the initial radiation probably stemmed from the phenacodontid condylarths during the late Paleocene epoch of Eurasia. By the early Eocene, three superfamilies of perissodactyls were already present: the equoids, tapiroids, and chalicotherioids. The rhinocerotoids appeared at the "beginning of the late Eocene, apparently derived from secondary radiations of tapiroids" (Radinsky 1969:308). The order underwent a tremendous adaptive radiation in the Eocene epoch such that shortly after its emergence some 14 families had evolved. The horses and rhinoceroses flourished until the Pleistocene in both the Old and New World; late Pleistocene extinctions drastically reduced their ranges. Before recent introductions by man, the equids were restricted to Africa and Eurasia; rhinoceroses are restricted to the southern half of Africa and south Asia; tapirs are restricted to the northern Neotropics and southeast Asia. The Recent perissodactyls are all herbivorous.

Order Hyracoidea. The Recent Hyracoidea, variously called hyraxes, conies, or dassies, constitute a single family, Procaviidae, with three genera and seven species. All three genera are Ethiopian in distribution, with one species, *Procavia capensis*, extending north along the Nile River, the eastern Mediterranean, and the southeastern Arabian peninsula. The oldest known hyrax fossils are from the upper Eocene-lower Oligocene of Egypt and are little differentiated from modern hyraxes (Meyer 1978). Africa was undoubtedly the center of origin for the order. The relationships of the Hyracoidea to the other orders of mammals are obscure. They have been variously associated with the Perissodactyla (McKenna 1975) and the Proboscidea and Sirenia as subungulates (Romer 1971; Simpson 1945). Dubrovo (1978:380) stated, "they most probably originated from Cretaceous mammals which may also possibly have given rise to the ancestral forms of condylarths, the Proboscidea, the Sirenia, and some other mammalian orders." Hyraxes are herbivorous and live in colonies that are isolated from neighboring colonies by varying distances. The Hyracoidea are unique within the Mammalia in harboring a complex fauna of lice that includes several species of both anoplurans and mallophagans (Hopkins 1949). Reviews of the biology of hyraxes may be found in Olds and Shoshani (1982) and Kingdon (1971).

Order Tubulidentata. The Recent aardvarks include only one family, Orycteropodidae, composed of a single species, *Orycteropus afer*. Aardvarks are now restricted to the Ethiopian faunal region, sub-Sahara Africa. It is believed that the aardvarks evolved in Africa from the extinct order Condylarthra, perhaps during the Paleocene epoch (Patterson 1975). Pliocene *Orycteropus* are known from Asia and Europe. Fossil aardvarks are also known from Madagascar. Aardvarks are insectivorous, semifossorial, solitary, and nocturnal (Melton 1976).

Order Artiodactyla. The Artiodactyla, or even-toed ungulates, are one of the most successful orders of mammals in terms of diversity, adaptive radiation, distribution, and numbers of species. Artiodactyls are found in all biogeographic regions except the Australian and Oceanic. The order includes eight Recent families: the Suidae, or pigs (eight species); Tayassuidae, or peccaries (three species); Hippopotamidae, or hippopotamuses (two species); Camelidae, or camels, llamas, guanacos, and vicuñas (six species); Tragulidae, the chevrotains or mouse deer (four species); Cervidae, or deer (38 species); Giraffidae, giraffes and okapis (two species); and Bovidae, the cattle and antelope (124 species).

The Suidae probably arose in either Europe or Asia during the Oligocene and reached Africa by the early Miocene where they evolved into a wide variety of types. Today they are found throughout the southern half of Eurasia and all of Africa. The Tayassuidae seems to have arisen in North America during the Oligocene epoch; Oligocene and Miocene fossils are known from Europe, Pliocene fossils from Asia, and Pleistocene fossils from Africa and South America (Cooke and Wilkinson 1978). Today they are found throughout the northern three-fourths of South America, all of Central America, Mexico, and the extreme southwestern United States. The Hippopotamidae probably arose in sub-Saharan Africa during the early Miocene epoch and then entered north Africa and Europe in the early Pleistocene (Coryndon 1978). Today they are restricted to sub-Saharan Africa and the Nile River. The Camelidae probably arose during the late Eocene in North America, and dispersed to Asia across the Beringian land bridge during the Pliocene. Today two distinct groups are found, the true camels across north Africa and central Asia and the llamas, guanacos, and vicuñas in the highlands of southern Peru, south through Bolivia and Chile. The latter groups reached South America during the Pleistocene epoch. The Tragulidae are known from the Miocene epoch of Africa, Asia, and Europe. Today the chevrotains are restricted to west Africa, southern India, and Southeast Asia. The Cervidae first appeared in the early Oligocene epoch of Asia, and reached North America across Beringia in the early Miocene. Today the cervids are widely distributed in both the Old and New World. The Giraffidae first appeared in the early Miocene of northern Africa (Churcher 1978), and fossil giraffids are known from India. Today the family is restricted to sub-Saharan Africa.

The Bovidae first appeared in Europe in the early Miocene epoch and underwent a tremendous radiation during the Pliocene. The North American pronghorn, *Antilocapra americana*, was until recently considered a monotypic family, Antilocapridae; however, O'Gara and Matson (1975) demonstrated that it best be considered a subfamily of the Bovidae. Bovids are widespread in the Nearctic, Palearctic, Ethiopian, and Oriental faunal regions.

The enormous Plio-Pleistocene radiation of the artiodactyls corresponds to the decline of the equids, which were perhaps ecological competitors (although see Cifelli 1981). Most artiodactyls are herbivorous and live in groups of varying sizes; the pigs and peccaries are omnivorous. Excellent reviews of the biology of selected species may be found in Clutton-Brock et al. (1982), Gauthier-Pilters and Dagg (1981), Leuthold (1977), and Sinclair (1977).

Order Pholidota. The order Pholidota, the pangolins or scaly anteaters, contains a single family, Manidae, with one extant genus, *Manis*, and seven Recent species. Pangolins are found in tropical and subtropical Africa and Southeast Asia. Four species occur in sub-Saharan Africa and three in Asia, including the islands of the Philippines, Sumatra, Java, and Borneo. The fossil record of the Pholidota is poor, however, pangolins are known from the Oligocene and Miocene epochs of Europe. Historically the pangolins have been included within the order Edentata, however, they are now treated as a distinct order. Of the relationship between the Pholidota and Edentata, Patterson (1978:270) stated that they "had a common ancestry in the later Cretaceous . . . that radiated during the Cenozoic with varying degrees of success in South America, North America, and the Old World. Whether one regards these groups as distinct orders or as suborders of an order Edentata is of minor moment." Pangolins are nocturnal and insectivorous; ants and termites comprise the bulk of the diet for most species. A review of the biology of the African species may be found in Kingdon (1971).

Additional reviews of the Mammalia include Grzimek (1975), Vaughan (1978), Eisenberg (1981), DeBlase (1982), and Anderson and Jones (1984); those with especial reference to parasites include Hoffstetter (1982) and Patterson (1957).

SUMMARY

The symbiotic association of mammals and arthropods perhaps spans more than 190 million years. As mammals evolved to occupy a wide variety of niches on all land masses, arthropods evolved to invade and colonize a wide array of habitats and niches on the mammalian body.

Coevolution between a host and parasite is difficult to demonstrate, and in many host–parasite systems coevolution (*sensu stricto*) may not occur. As the pocket gophers (*Geomys*) and their chewing lice (*Geomydoecus*) show, co-speciation often occurs between host and parasite, where the parasite exhibits an evolutionary response to the host but the host shows no counterevolutionary response (Timm 1983; Timm and Price 1980). Both the host and parasite may also develop coevolved adaptations to maximize reproductive output, as white-footed mice (*Peromyscus leucopus*) evolved a tolerance for cuterebrid parasitism (Timm and Cook 1979).

Mammals provide habitats for parasitic arthropods that include the subcuticular layer, hair, skin, glands, respiratory tract, and various other organs; the parasites may be highly adapted to their specific habitats.

The first mammals appeared on earth during the late Triassic period (Rhaetic) of the Mesozoic era, approximately 190 million years ago. They must have evolved from the mammal-like reptiles of the synapsid Therapsida, although the exact mammalian ancestor within this group of reptiles is unknown. Mammals are commonly considered to have a monophyletic origin, perhaps derived from a single family of cynodonts.

The oldest known mammalian fossils are from the late Triassic period of western Europe (Clemens et al. 1979), although the original continent where mammals evolved is not known. The only earlier fossil is from the early Triassic period of southern Brazil. A period of great mammalian radiation began in the late Cretaceous period (late Mesozoic) and continued on into the Paleocene epoch. During this time many of the Recent mammalian orders first appeared (Table 3.1). This radiation continued throughout the Cenozoic era. By the Oligocene epoch the majority of modern family taxa had developed, and many modern genus taxa appeared as early as the Pliocene epoch. The distribution of extant mammals was affected by continental drift.

Diversity and distribution of mammals are discussed by orders: Monotremata, Marsupialia, Edentata, Insectivora, Macroscelidea, Scandentia, Dermoptera, Chiroptera, Primates, Rodentia, Lagomorpha, Carnivora, Cetacea, Sirenia, Proboscidea, Perissodactyla, Hyracoidea, Tubulidentata, Artiodactyla, and Pholidota (Table 3.2).

ACKNOWLEDGMENTS

We thank John B. Kethley, Barry M. OConnor, and Rupert L. Wenzel for freely sharing their knowledge of and literature on the ectoparasitic arthropods with us. Jack Fooden, Lawrence R. Heaney, L. Henry Kermott, Ke Chung Kim, and Bruce D. Patterson provided helpful discussion and constructive comments on the manuscript. Clara Richardson prepared the illustrations.

REFERENCES

Aguirre, E. 1969. Evolutionary history of the elephant. *Science* **164**:1366–1376.

Anderson, S. 1967. Introduction to the rodents. Pages 206–209 in S. Anderson and J. K. Jones, Jr. (Eds.), *Recent Mammals of the World: A Synopsis of Families.* Ronald Press, New York.

Anderson, S. and J. K. Jones, Jr. 1984. Orders and Families of Recent Mammals of the World. John Wiley, New York.

Anderson, S., C. A. Woods, G. S. Morgan, and W. L. R. Oliver. 1983. *Geocapromys brownii. Mamm. Species* **201**:1–5.

Arvy, L. 1982. Phoresies and parasitism in cetaceans: A review. Pages 233–335 in G. Pilleri (Ed.), *Investigations on Cetacea.* Brain Anatomy Institute, Bern, Switzerland.

Bee, J. W., D. Murariu, and R. S. Hoffmann. 1980. Histology and histochemistry of specialized integumentary glands in eight species of North American shrews (Mammalia, Insectivora). *Trav. Mus. Hist. Nat. Grigore Antipa, Bucuresti* **22**:547–569.

Bonaparte, J. F. and M. C. Barberena. 1975. A possible mammalian ancestor from the middle Triassic of Brazil (Therapsida-Cynodontia). *J. Paleont.***49**:931–936.

Brooks, D. R. 1979. Testing the context and extent of host–parasite coevolution. *Syst. Zool.* **28**:299–307.

Butler, P. M. 1956. The skull of *Ictops* and the classification of the Insectivora. *Proc. Zool. Soc. London* **126**:453–481.

Butler, P. M. 1972. The problem of insectivore classification. Pages 253–265 in K. A. Joysey and T. S. Kemp (Eds.), *Studies in Vertebrate Evolution*. Oliver and Boyd, Edinburgh.

Butler, P. M. 1978. Insectivora and Chiroptera. Pages 56–68 in V. J. Maglio and H. B. S. Cooke (Eds.), *Evolution of African Mammals*. Harvard University Press, Cambridge, MA.

Cain, A. J. 1959. Deductive and inductive methods in post-Linnaean taxonomy. *Proc. Linn. Soc. Lond.* **170**:185–217.

Campbell, C. B. G. 1974. On the phyletic relationships of the tree shrews. *Mammal. Rev.* **4**:125–143.

Carleton, M. D. 1980. Phylogenetic relationships in Neotomine-Peromyscine rodents (Muroidea) and a reappraisal of the dichotomy within New World Cricetinae. *Misc. Publ. Mus. Zool., Univ. Mich.* **157**:1–146.

Cartmill, M. 1982. Assessing tarsier affinities: Is anatomical description phylogenetically neutral? *Geobios, Mem. Spec.* **6**:279–287.

Catts, E. P. 1982. Biology of New World bot flies: Cuterebridae. *Annu. Rev. Entomol.* **27**:313–338.

Chorn, J. and R. S. Hoffmann. 1978. *Ailuropoda melanoleuca*. *Mamm. Species* **110**:1–6.

Churcher, C. S. 1978. Giraffidae. Pages 509–535 in V. J. Maglio and H. B. S. Cooke (Eds.), *Evolution of African Mammals*. Harvard University Press, Cambridge, MA.

Cifelli, R. L. 1981. Patterns of evolution among the Artiodactyla and Perissodactyla (Mammalia). *Evolution* **35**:433–440.

Ciochon, R. L. and A. B. Chiarelli (Eds.). 1980. *Evolutionary Biology of the New World Monkeys and Continental Drift*. Plenum Press, New York.

Clemens, W. A. 1970. Mesozoic mammalian evolution. *Ann. Rev. Ecol. Syst.* **1**:357–390.

Clemens, W. A. 1979. Notes on the Monotremata. Pages 309–311 in J. A. Lillegraven, Z. Kielan-Jaworowska, and W. A. Clemens (Eds.), *Mesozoic Mammals: The First Two-Thirds of Mammalian History*, University of California Press, Berkeley.

Clemens, W. A., J. A. Lillegraven, E. H. Lindsay, and G. G. Simpson. 1979. Where, when, and what—A survey of known Mesozoic mammal distribution. Pages 7–58 in J. A. Lillegraven, Z. Kielan-Jaworowska, and W. A. Clemens, (Eds.), *Mesozoic Mammals: The First Two-Thirds of Mammalian History*, University of California Press, Berkeley.

Clutton-Brock, T. H., F. E. Guinness, and S. D. Albon. 1982. *Red Deer: Behavior and Ecology of Two Sexes*. University of Chicago Press, Chicago, IL.

Collett, S. F. 1981. Population characteristics of *Agouti paca* (Rodentia) in Colombia. *Publ. Mus., Mich. State Univ., Biol. Ser.* **5**:485–602.

Cooke, H. B. S. and A. F. Wilkinson. 1978. Suidae and Tayassuidae. Pages 435–482 in V. J. Maglio and H. B. S. Cooke (Eds.), *Evolution of African Mammals*, Harvard University Press, Cambridge, MA.

Corbet, G. B. and J. Hanks. 1968. A revision of the elephant-shrews, family Macroscelididae. *Bull. Br. Mus. (Nat. Hist.), Zool.* **16**(2):47–111.

Coryndon, S. C. 1978. Hippopotamidae. Pages 483–495 in V. J. Maglio and H. B. S. Cooke (Eds.), *Evolution of African Mammals*, Harvard University Press, Cambridge, MA.

Cracraft, J. 1974. Continental drift and vertebrate distribution. *Ann. Rev. Ecol. Syst.* **5**:215–261.

Crompton, A. W. and F. A. Jenkins, Jr. 1968. Molar occlusion in Late Triassic mammals. *Biol. Rev.* **43**:427–458.

Crompton, A. W. and F. A. Jenkins, Jr. 1973. Mammals from reptiles: A review of mammalian origins. *Ann. Rev. Earth Planet. Sci.* **1:**131–155.

Crompton, A. W. and F. A. Jenkins, Jr. 1979. Origin of mammals. Pages 59–73 in J. A. Lillegraven, Z. Kielan-Jaworowska, and W. A. Clemens (Eds.), *Mesozoic Mammals: The First Two-Thirds of Mammalian History*, University of California Press, Berkeley.

Davis, D. D. 1964. The giant panda: A morphological study of evolutionary mechanisms. *Fieldiana: Zool. Mem.* **3:**1–339.

DeBlase, A. F. 1982. Mammalia. Pages 1015–1061 in S. P. Parker (Ed.), *Synopsis and Classification of Living Organisms*, McGraw-Hill, New York.

Delson, E. and A. L. Rosenberger. 1980. Phyletic perspectives on platyrrhine origins and anthropoid relationships. Pages 445–458 in R. L. Ciochon and A. B. Chiarelli (Eds.), *Evolutionary Biology of the New World Monkeys and Continental Drift.* Plenum, New York.

Dene, H., M. Goodman, M. C. McKenna, and A. E. Romero-Herrera. 1982. *Ochotona princeps* (pika) myoglobin: An appraisal of lagomorph phylogeny. *Proc. Natl. Acad. Sci.* **79:**1917–1920.

Domning, D. P. 1976. An ecological model for Late Tertiary sirenian evolution in the North Pacific Ocean. *Syst. Zool.* **25:**352–362.

Domning, D. P. 1978. Sirenian evolution in the North Pacific Ocean. *Univ. Calif. Publ. Geol. Sci.* **118:**xi + 176.

Domning, D. P. 1982. Evolution of manatees: A speculative history. *J. Paleont.* **56:**599–619.

Dort, W., Jr. and J. K. Jones, Jr. (Eds.). 1970. *Pleistocene and Recent Environments of the Central Great Plains.* University Kansas, Department of Geology, Spec. Publ. No. 3.

Dubrovo, I. A. 1978. New data on fossil Hyracoidea. *Paleont. J.* **12:**375–383.

Eaton, R. L. 1973. *The world's cats. Vol. 1. Ecology and conservation.* World Wildlife Safari, Winston, Oregon.

Eisenberg, J. F. 1981. *The Mammalian Radiations: An Analysis of Trends in Evolution, Adaptation, and Behavior.* University of Chicago Press, Chicago, IL.

Eisenberg, J. F. and E. Gould. 1970. The tenrecs: A study in mammalian behavior and evolution. *Smithson. Contrib. Zool.* **27:**v + 1–137.

Elliot, O. 1971. Bibliography of the tree shrews 1780–1969. *Primates* **12:**323–414.

Ewer, R. F. 1973. *The Carnivores.* Cornell University Press, Ithaca, NY.

Fain, A. 1955. Un acarien remarquable vivant dans l'estomac d'une chauve-souris: *Gastronyssus bakeri* n. g., n. sp. *Ann. Soc. Belg. Med. Trop.* **35:**681–688.

Fain, A. 1962. Les acariens parasites nasicoles des batraciens. Revision des Lawrencarinae Fain, 1957 (Ereynetidae: Trombidiformes). *Bull. Inst. R. Sci. Nat. Belg.* **38**(25):1–69.

Fain, A. 1963. Chaetotaxie et classification des Speleognathinae (Acarina: Trombidiformes). *Bull. Inst. R. Sci. Nat. Belg.* **39**(9):1–80.

Fain, A. 1964a. Chaetotaxie et classification des Gastronyssidae avec description d'un nouveau genre parasite nasicole d'un ecureuil sudafricain (Acarina: Sarcoptiformes). *Rev. Zool. Bot. Afr.* **70:**40–52.

Fain, A. 1964b. Les Lemurnyssidae parasites nasicoles des Lorisidae africains et des Cebidae sud-américains: Description d'une espèce nouvelle (Acarina: Sarcoptiformes). *Ann. Soc. Belg. Med. Trop.* **44:**453–458.

Fain, A. and S. Orts. 1969. *Epimyodex talpae* n. g., n. sp., parasite sous-cutané de la taupe en Belgique (Demodicidae: Trombidiformes). *Acarologia* **11:**65–68.

Fay, F. H. and D. P. Furman. 1982. Nasal mites (Acari:Halarachnidae) in the spotted seal, *Phoca largha* Pallas, and other pinnipeds of Alaskan waters. *J. Wildl. Dis.* **18:**63–67.

Findley, J. S. 1967. Insectivores and dermopterans. Pages 87–108 in S. Anderson and J. K.

Jones, Jr. (Eds.), *Recent Mammals of the World: A Synopsis of Families*. Ronald Press, New York.

Forsten, A. and P. M. Youngman. 1982. *Hydrodamalis gigas*. *Mamm. Species* **165**:1–3.

Fox, M. W. 1975. *The Wild Canids: Their Systematics, Behavioral Ecology and Evolution*. Van Nostrand Reinhold, New York.

Futuyma, D. J. and M. Slatkin. 1983. *Coevolution*. Sinauer Associates, Sunderland, MA.

Gaskin, D. E. 1982. *The Ecology of Whales and Dolphins*. Heinemann, London.

Gauthier-Pilters, H. and A. I. Dagg. 1981. *The Camel: Its Evolution, Ecology, Behavior, and Relationship to Man*. University of Chicago Press, Chicago, IL.

Genoways, H. H. 1973. Systematics and evolutionary relationships of spiny pocket mice, genus Liomys. *Spec. Publ. Mus., Tex. Tech Univ.* **5**:1–368.

George, W. 1974. Notes on the ecology of gundis (f. Ctenodactylidae). Pages 143–160 in I. W. Rowlands and B. J. Weir (Eds.), *The Biology of Hystricomorph Rodents, Symp. Zool. Soc. London*, No. 34.

Giesen, K., M. T. and F. S. Lukoschus. 1982. A new species of the genus *Psorergates* Tyrell, 1883 (Acarina: Prostigmata: Psorergatidae) parasitic on the tree-shrew *Tupaia dorsalis* (Mammalia: Scandentia) from Borneo. *Zool. Meded.* **56**:259–266.

Gillespie, R., D. R. Horton, P. Ladd, P. G. Macumber, T. H. Rich, R. Thorne, and R. V. S. Wright. 1978. Lancefield Swamp and the extinction of the Australian megafauna. *Science* **200**:1044–1048.

Gingerich, P. D. 1980. Eocene Adapidae, paleobiogeography, and the origin of South American Platyrrhini. Pages 123–138 in R. L. Ciochon and A. B. Chiarelli (Eds.), *Evolutionary Biology of the New World Monkeys and Continental Drift*. Plenum Press, New York.

Gingerich, P. D., N. A. Wells, D. E. Russell, and S. M. Ibrahim Shah. 1983. Origin of whales in epicontinental remnant seas: New evidence from the early Eocene of Pakistan. *Science* **220**:403–406.

Ginsburg, L. 1982. Sur la position systématique du petit panda, *Ailurus fulgens* (Carnivora, Mammalia). *Geobios, Mem. Spec.* **6**:247–258.

Griffiths, M. 1968. *Echidnas*. Pergamon Press, New York.

Griffiths, M. 1978. *The Biology of the Monotremes*. Academic Press, New York.

Grzimek, B. (Ed.). 1975. *Grzimek's Animal Life Encyclopedia. Mammals, I–IV*, Vols. 10–13. Van Nostrand Reinhold, New York.

Guggisberg, C. A. W. 1975. *Wild Cats of the World*. Taplinger, New York.

Hartman, D. S. 1979. Ecology and behavior of the manatee (*Trichechus manatus*) in Florida. *Spec. Publ., Am. Soc. Mamm.* **5**:viii + 153.

Heaney, L. R. and R. M. Timm. 1983. Relationships of pocket gophers of the genus *Geomys* from the Central and Northern Great Plains. *Misc. Publ., Mus. Nat. Hist., Univ. Kans.* **74**:1–59.

Hemmer, H. 1978. The evolutionary systematics of living Felidae: Present status and current problems. *Carnivore* **1**:71–79.

Hershkovitz, P. 1974. A new genus of Late Oligocene monkey (Cebidae, Platyrrhini) with notes on postorbital closure and platyrrhine evolution. *Folia Primat.* **21**:1–35.

Hershkovitz, P. 1977. *Living New World Monkeys (Platyrrhini)*. The University of Chicago Press, Chicago, IL.

Hoffstetter, R. 1982. Introduction sur les hôtes. I. Phylogénie des mammifères: méthodes d'étude, résultats, problèmes. Pages 13–20 in *Mem. Mus. Natl. Hist. Nat.*, Nouv. Ser., A, Zool. **123**.

Honacki, J. H., K. E. Kinman, and J. W. Koeppl (Eds.). 1982. *Mammal Species of the World*. Allen Press and The Association of Systematic Collections, Lawrence, KS.

Hopkins, D. M. 1967. *The Bering Land Bridge*. Stanford University Press, Stanford, CA.

Hopkins, D. M., J. V. Matthews, Jr., C. E. Schweger, and S. B. Young (Eds.). 1982. *Paleoecology of Beringia*. Academic Press, New York.

Hopkins, G. H. E. 1949. The host-associations of the lice of mammals. *Proc. Zool. Soc. Lond.* **119**:387–604.

Hopson, J. A. 1967. Mammal-like reptiles and the origin of mammals. *Discovery*. Magazine of the Peabody Museum of Natural History, Yale University **2**(2):25–33.

Hopson, J. A. and A. W. Crompton. 1969. Origin of mammals. Pages 15–72 in T. Dobzhansky, M. K. Hecht, and W. C. Steere (Eds.), *Evolutionary Biology*, Vol. 3. Appleton-Century-Crofts, New York.

Husar, S. L. 1977. *Trichechus inunguis*. *Mamm. Species* **72**:1–4.

Jannett, F. J., Jr. 1975. "Hip glands" of *Microtus pennsylvanicus* and *M. longicaudus* (Rodentia: Muridae), voles "without" hip glands. *Syst. Zool.* **24**:171–175.

Janzen, D. H. 1980. When is it coevolution? *Evolution* **34**:611–612.

Jenkins, F. A., Jr., A. W. Crompton, and W. R. Downs. 1983. Mesozoic mammals from Arizona: New evidence on mammalian evolution. *Science* **222**:1233–1235.

Jepsen, G. L. 1966. Early Eocene bat from Wyoming. *Science* **154**:1333–1339.

Jepsen, G. L. 1970. Bat origins and evolution. Pages 1–64 in W. A. Wimsatt (Ed.), *Biology of Bats*. Academic Press, New York.

Kemp, T. S. 1982. *Mammal-Like Reptiles and the Origin of Mammals*. Academic Press, New York.

Kenyon, K. W., C. E. Yunker, and I. M. Newell. 1965. Nasal mites (Halarachnidae) in the sea otter. *J. Parasit.* **51**:960.

Kim, K. C., V. L. Haas, and M. C. Keyes. 1980. Populations, microhabitat preference and effects of infestation of two species of *Orthohalarachne* (Halarachnidae: Acarina) in the northern fur seal. *J. Wildl. Dis.* **16**:45–51.

King, M.-C. and A. C. Wilson. 1975. Evolution at two levels in humans and chimpanzees. *Science* **188**:107–116.

Kingdon, J. 1971. *East African Mammals: An Atlas of Evolution in Africa. I.* Academic Press, New York.

Kingdon, J. 1974. *East African Mammals: An Atlas of Evolution in Africa. II(B). Hares and Rodents*. Academic Press, New York.

Kingdon, J. 1977. *East African Mammals: An Atlas of Evolution in Africa. III(A). Carnivores*. Academic Press, New York.

Kirsch, J. A. W. 1977a. The comparative serology of Marsupialia, and a classification of marsupials. *Aust. J. Zool., Suppl. Ser.* **52**:1–152.

Kirsch, J. A. W. 1977b. The classification of marsupials, with special reference to karyotypes and serum proteins. Pages 1–50 in D. Hunsaker (Ed.), *The Biology of Marsupials*. Academic Press, New York.

Kirsch, J. A. W. 1977c. Biological aspects of the marsupial-placental dichotomy: A reply to Lillegraven. *Evolution* **31**:898–900.

Kirsch, J. A. W. and M. Archer. 1982. Polythetic cladistics, or, when parsimony's not enough: The relationships of carnivorous marsupials. Pages 595–619 in M. Archer (Ed.), *Carnivorous Marsupials*. Royal Zoological Society of New South Wales, Vol. 2.

Kruuk, H. 1972. *The Spotted Hyena: A Study of Predation and Social Behavior*. University of Chicago Press, Chicago, IL.

Kurtén, B. 1972. *The Age of Mammals*. Columbia University Press, New York.

Kurtén, B. and E. Anderson. 1980. *Pleistocene Mammals of North America*. Columbia University Press, New York.

Lacher, T. E., Jr. 1981. The comparative social behavior of *Kerodon rupestris* and *Galea spixii* and the evolution of behavior in the Caviidae. *Bull. Carnegie Mus. Nat. Hist.* **17**:1–71.

Le Gros Clark, W. E. 1932. The brain of the Insectivora. *Proc. Zool. Soc. Lond.* **1932**:975–1013.

Leuthold, W. 1977. *African Ungulates: A Comparative Review of Their Ethology and Behavioral Ecology.* Springer-Verlag, New York.

Lillegraven, J. A. 1974. Biogeographical considerations of the marsupial-placental dichotomy. *Ann. Rev. Ecol. Syst.* **5**:263–283.

Lillegraven, J. A., M. J. Kraus, and T. M. Bown. 1979. Paleogeography of the world of the Mesozoic. Pages 277–308 in J. A. Lillegraven, Z. Kielan-Jaworowska, and W. A. Clemens (Eds.), *Mesozoic Mammals: The First Two-Thirds of Mammalian History.* University of California Press, Berkeley.

Lombert, H. A. P. M., F. S. Lukoschus, and J. O. Whitaker, Jr. 1983. *Demodex peromysci,* n. sp. (Acari: Prostigmata: Demodicidae), from the Meibomian glands of *Peromyscus leucopus* (Rodentia: Cricetidae). *J. Med. Entomol.* **20**:377–382.

Lowery, G. H., Jr. 1974. *The Mammals of Louisiana and Its Adjacent Waters.* Louisiana State University Press, Baton Rouge, LA.

Luckett, W. P. (Ed.). 1980. *Comparative Biology and Evolutionary Relationships of Tree Shrews.* Plenum Press, New York.

Luckett, W. P. and F. S. Szalay (Eds.). 1975. *Phylogeny of the Primates: A Multidisciplinary Approach.* Plenum Press, New York.

Macdonald, D. W. 1983. The ecology of carnivore social behaviour. *Nature* **301**:379–384.

MacNeish, R. S. 1976. Early man in the New World. *Am. Sci.* **64**:316–327.

Maglio, V. J. 1973. Origin and evolution of the Elephantidae. *Trans. Am. Phil. Soc., n.s.* **63(3)**:1–149.

Marshall, A. G. 1981. *The Ecology of Ectoparasitic Insects.* Academic Press, New York.

Marshall, A. G. 1982. Ecology of insects ectoparasitic on bats. Pages 369–401 in T. H. Kunz (Ed.), *Ecology of Bats.* Plenum Press, New York.

Marshall, L. G. 1979. Evolution of metatherian and eutherian (mammalian) characters: a review based on cladistic methodology. *Zool. J. Linn. Soc.* **66**:369–410.

Marshall, L. G. 1980. The great American interchange—An invasion induced crisis for South American mammals. Pages 133–229 in M. H. Nitecki (Ed.), *Biotic Crises in Ecological and Evolutionary Time.* Academic Press, New York.

Marshall, L. G. 1981. The families and genera of Marsupialia. *Fieldiana: Geol., New Ser.* **8**:vi + 1–65.

Marshall, L. G., R. F. Butler, R. E. Drake, G. H. Curtis, and R. H. Tedford. 1979. Calibration of the Great American Interchange. *Science* **204**:272–279.

Marshall, L. G., S. D. Webb, J. J. Sepkoski, Jr., and D. M. Raup. 1982. Mammalian evolution and the Great American Interchange. *Science* **215**:1351–1357.

Martin, P. S. 1982. The pattern and meaning of Holarctic mammoth extinction. Pages 399–408 in D. M. Hopkins et al. (Eds.), *Paleoecology of Beringia.* Academic Press, New York.

McKenna, M. C. 1961. A note on the origin of rodents. *Am. Mus. Novit.* **2037**:1–5.

McKenna, M. C. 1969. The origin and early differentiation of therian mammals. Ann. N. Y. Acad. Sci. **167**:217–240.

McKenna, M. C. 1975. Toward a phylogenetic classification of the Mammalia. Pages 21–46 in W. P. Luckett and F. S. Szalay, Eds. *Phylogeny of the Primates: A Multidisciplinary Approach.* Plenum Press, New York.

McKenna, M. C. 1982. Lagomorph interrelationships. *Geobios, Mem. Spec.* **6**:213–223.

McLaren, I. A. 1960. Are the Pinnipedia biphyletic? *Syst. Zool.* **9**:18–28.

Melton, D. A. 1976. The biology of aardvark (Tubulidentata-Orycteropodidae). *Mammal Rev.* 6:75–88.

Meyer, G. E. 1978. Hyracoidea. Pages 284–314 in V. J. Maglio and H. B. S. Cooke (Eds.), *Evolution of African Mammals.* Harvard University Press, Cambridge, MA.

Mills, M. G. L. 1982. *Hyaena brunnea. Mamm. Species* **194**:1–5.

Mondolfi, E. 1983. The feet and baculum of the spectacled bear, with comments on ursid phylogeny. *J. Mamm.* **64**:307–310.

Montagna, W. 1974. *The Structure and Function of Skin.* Academic Press, New York.

Montgomery, G. G. (Ed.). in press. *Evolution and Ecology of Sloths, Anteaters and Armadillos.* Smithsonian Institution Press, Washington, D.C.

Mosimann, J. E. and P. S. Martin. 1975. Simulating overkill by Paleoindians. *Am. Sci.* **63**:304–313.

Muizon, C. de. 1982. Les relations phylogénétiques des Lutrinae (Mustelidae, Mammalia). *Geobios, Mem. Spec.* **6**:259–277.

Müller-Schwarze, D. 1983. Scent glands in mammals and their functions. Pages 150–197 in J. F. Eisenberg and D. G. Kleiman (Eds.), *Advances in the Study of Mammalian Behavior.* Spec. Publ. 7, American Society of Mammalogists.

Nitecki, M. H. (Ed.). 1983. *Coevolution.* University of Chicago Press, Chicago, IL.

Nowak, R. M. and J. L. Paradiso. 1983. *Walker's Mammals of the World.* Johns Hopkins University Press, Baltimore, 2 vols.

Nutting, W. B. 1965. Host–parasite relations: Demodicidae. *Acarologia* **7**:301–317.

Nutting, W. B., F. S. Lukoschus, and C. E. Desch. 1980. Parasitic mites of Surinam XXXVII. *Demodex marsupiali* sp. nov. from *Didelphis marsupialis*: Adaptation to glandular habitat. *Zool. Meded.* **56**:83–90.

Nutting, W. B. and P. Woolley. 1965. Pathology in *Antechinus stuartii* (Marsupialia) due to *Demodex* spp. *Parasitology* **55**:383–389.

O'Gara, B. W. and G. Matson. 1975. Growth and casting of horns by pronghorns and exfoliation of horns by bovids. *J. Mamm.* **56**:829–846.

Olds, N. and J. Shoshani. 1982. *Procavia capensis. Mamm. Species* **171**:1–7.

Olson, E. C. 1959. The evolution of mammalian characters. *Evolution* **13**:344–353.

Parrington, F. R. 1967. The origins of mammals. *Adv. Sci.* **24**:165–173.

Patterson, B. 1957. Mammalian phylogeny. Pages 15–48 in *First Symposium on Host Specificity among Parasites of Vertebrates.* Institute of Zoology, University of Neuchatel, Neuchatel, Switz.

Patterson, B. 1965. The fossil elephant shrews (family Macroscelididae). *Bull. Mus. Comp. Zool., Harvard Univ.* **133**:295–335.

Patterson, B. 1975. The fossil aardvarks (Mammalia: Tubulidentata). *Bull. Mus. Comp. Zool., Harvard Univ.* **147**:185–237.

Patterson, B. 1978. Pholidota and Tubulidentata. Pages 268–278 in V. J. Maglio and H. B. S. Cooke (Eds.), *Evolution of African Mammals.* Harvard University Press, Cambridge, MA.

Peck, S. B. 1982. A review of the ectoparasitic *Leptinus* beetles of North America (Coleoptera: Leptinidae). *Can. J. Zool.* **60**:1517–1527.

Phillips, C. J., J. K. Jones, Jr., and F. J. Radovsky. 1969. Macronyssid mites in oral mucosa of long-nosed bats: Occurrence and associated pathology. *Science* **165**:1368–1369.

Price, P. W. 1980. Evolutionary biology of parasites. *Monogr. Pop. Biol.,* Princeton University Press, Princeton, NJ.

Quay, W. B. 1953. Seasonal and sexual differences in the dorsal skin gland of the kangaroo rat (*Dipodomys*). *J. Mamm.* **34**:1–14.

Quay, W. B. 1968. The specialized posterolateral sebaceous glandular regions in microtine rodents. *J. Mamm.* **49**:427–445.

Quay, W. B. 1970. Integument and derivatives. Pages 1–56 in W. A. Wimsatt (Ed.), *Biology of Bats*, Vol. 2. Academic Press, New York.

Radinsky, L. B. 1969. The early evolution of the Perissodactyla. *Evolution* **23**:308–328.

Ray, C. E. 1976. Geography of phocid evolution. *Syst. Zool.* **25**:391–406.

Repenning, C. A. 1976. Adaptive evolution of sea lions and walruses. *Syst. Zool.* **25**:375–390.

Ride, W. D. L. 1964. A review of Australian fossil marsupials. *J. R. Soc. West. Aust.* **47**:97–131.

Romer, A. S. 1971. *Vertebrate Paleontology*. The University of Chicago Press, Chicago, IL.

Rood, J. P. 1972. Ecological and behavioural comparisons of three genera of Argentine cavies. *Anim. Behav. Monogr.* **5**(1):1–83.

Rudnick, A. 1960. A revision of the mites of the family Spinturnicidae (Acarina). *Univ. Calif. Publ. Entomol.* **17**:157–284.

Savage, D. E. and D. E. Russell. 1983. *Mammalian Paleofaunas of the World*. Addison-Wesley, Reading, MA.

Savage, R. J. G. 1976. Review of early Sirenia. *Syst. Zool.* **25**:344–351.

Savage, R. J. G. 1977. Evolution of carnivorous mammals. *Palaeontology* **20**:237–271.

Savage, R. J. G. 1978. Carnivora. Pages 249–267 in V. J. Maglio and H. B. S. Cooke (Eds.), *Evolution of African Mammals*. Harvard University Press, Cambridge, MA.

Savin, S. M. 1977. The history of the earth's surface temperature during the past 100 million years. *Ann. Rev. Earth Planet. Sci.* **5**:319–355.

Simons, E. L., P. Andrews, and D. R. Pilbeam. 1978. Cenozoic apes. Pages 120–146 in V. J. Maglio and H. B. S. Cooke (Eds.), *Evolution of African Mammals*. Harvard University Press, Cambridge, MA.

Simons, E. L. and R. F. Kay. 1983. *Qatrania*, new basal anthropoid primate from the Fayum, Oligocene of Egypt. *Nature* **304**:624–626.

Simpson, G. G. 1945. The principles of classification and a classification of mammals. *Bull. Am. Mus. Nat. Hist.* **85**:xvi + 1–350.

Simpson, G. G. 1980. *Splendid Isolation: The Curious History of South American Mammals*. Yale University Press, New Haven, CT.

Sinclair, A. R. E. 1977. *The African Buffalo: A Study of Resource Limitation of Populations*. University of Chicago Press, Chicago, IL.

Smith, A. G. and J. C. Briden. 1979. *Mesozoic and Cenozoic Paleocontinental Maps*. Earth Sci. Ser., Cambridge University Press.

Smith, D. H. 1975. An ecological analysis of a host–parasite association: *Cuterebra approximata* (Diptera: Cuterebridae) in *Peromyscus maniculatus* (Rodentia: Cricetidae). Unpublished Ph.D. Dissertation, University of Montana, Missoula.

Smith, D. H. 1978. Vulnerability of bot fly (*Cuterebra*) infected *Peromyscus maniculatus* to short-tail weasel predation in the laboratory. *J. Wildl. Dis.* **14**:40–51.

Smith, J. D. 1976. Comments on flight and the evolution of bats. Pages 427–437 in M. K. Hecht, P. C. Goody, and B. M. Hecht (Eds.), *Major Patterns in Vertebrate Evolution*. NATO Advanced Study Institutes Series: Series A: Life Sciences.

Smith, J. D. and G. Madkour. 1980. Penial morphology and the question of chiropteran phylogeny. Pages 347–365 in D. E. Wilson and A. L. Gardner (Eds.), *Proceedings Fifth International Bat Research Conference*. Texas Tech Press, Lubbock, TX.

Smythe, N. 1978. The natural history of the Central American agouti (*Dasyprocta punctata*). *Smithson. Contribution Zool.* **257**:1–52.

Sokolov, V. E. 1982. *Mammal Skin*. University of California Press, Berkeley.

Sweatman, G. K. 1958a. Biology of Otodectes cynotis, the ear canker mite of carnivores. *Can. J. Zool.* **36**:849–862.

Sweatman, G. K. 1958b. On the life history and validity of the species in Psoroptes, a genus of mange mites. *Can. J. Zool.* **36**:905–929.

Szalay, F. S. 1982. A new appraisal of marsupial phylogeny and classification. Pages 621–640 in M. Archer (Ed.), *Carnivorous Marsupials*. Royal Zoological Society of New South Wales, Vol. 2.

Szalay, F. S. and R. L. Decker. 1974. Origins, evolution, and function of the tarsus in Late Cretaceous Eutheria and Paleocene primates. Pages 223–259 in F. A. Jenkins, Jr. (Ed.), *Primate Locomotion*. Academic Press, New York.

Tamarin, R. H. 1981. Hip glands in wild-caught *Microtus pennsylvanicus*. *J. Mamm.* **62**:421.

Tedford, R. H. 1976. Relationship of pinnipeds to other carnivores (Mammalia). *Syst. Zool.* **25**:363–374.

Thenius, E. 1966. Zur Stammesgeschichte der Hyänen (Carnivora, Mammalia). *Z. Saugetierk.* **31**:293–300.

Thompson, J. N. 1982. *Interaction and Coevolution*. John Wiley, New York.

Timm, R. M. 1979. The *Geomydoecus* (Mallophaga: Trichodectidae) parasitizing pocket gophers of the *Geomys* complex (Rodentia: Geomyidae). Unpublished Ph.D. Dissertation, University of Minnesota, St. Paul.

Timm, R. M. 1983. Fahrenholz's Rule and Resource Tracking: A study of host–parasite coevolution. Pages 225–265 in M. H. Nitecki (Ed.), *Coevolution*. University of Chicago Press, Chicago, IL.

Timm, R. M. and E. F. Cook. 1979. The effect of bot fly larvae on reproduction in white-footed mice, Peromyscus leucopus. *Am. Midl. Nat.* **101**:211–217.

Timm, R. M. and R. E. Lee, Jr. 1981. Do bot flies, *Cuterebra* (Diptera: Cuterebridae), emasculate their hosts? *J. Med. Entomol.* **18**:333–336.

Timm, R. M. and R. D. Price. 1980. The taxonomy of *Geomydoecus* (Mallophaga: Trichodectidae) from the *Geomys bursarius* complex (Rodentia: Geomyidae). *J. Med. Entomol.* **17**:126–145.

Van Valen, L. M. 1967. New Paleocene insectivores and insectivore classification. *Bull. Am. Mus. Nat. Hist.* **135**:217–284.

Van Valen, L. M. 1979. The evolution of bats. *Evol. Theory* **4**:103–121.

Van Valen, L. M. 1982. What are treeshrews? *Evol. Theory* **6**:174.

Van Valen, L. M. and R. E. Sloan. 1965. The earliest primates. *Science* **150**:743–745.

Vaughan, T. A. 1978. *Mammalogy*. W. B. Saunders, Philadelphia, PA.

Webb, J. P., Jr. and R. B. Loomis. 1977. Ectoparasites. Pages 57–119 in R. J. Baker, J. K. Jones, Jr., and D. C. Carter (Eds.), *Biology of Bats of the New World Family Phyllostomatidae. Part II.* Special Publ., The Museum, Texas Tech University, Lubbock, TX.

Webb, S. D. in press. On the interrelationships of tree sloths and ground sloths. In G. G. Montgomery (Ed.), *Evolution and Ecology of Sloths, Anteaters and Armadillos*. Smithsonian Institution Press, Washington, D.C.

Weir, B. J. 1974. The tuco-tuco and plains viscacha. Pages 113–130 in I. W. Rowlands and B. J. Weir (Eds.), *The Biology of Hystricomorph Rodents. Symp. Zool. Soc. Lond.* No. **34**.

Wetzel, R. M. 1982. Systematics, distribution, ecology, and conservation of South American edentates. Pages 345–375 in M. A. Mares and H. H. Genoways (Eds.), *Mammalian Biology in South America*. Special Publ. Ser., Pymatuning Laboratory of Ecology, University of Pittsburgh, Pittsburgh, PA.

Wilson, R. W. 1960. Early Miocene rodents and insectivores from northeastern Colorado. *Univ. Kans., Paleontol. Contrib.* **7**:1–92.

Wood, A. E. 1962. The early Tertiary rodents of the family Paramyidae. *Trans. Am. Phil. Soc.*, n.s. **52**(1):1–261.

Wood, A. E. 1965. Grades and clades among rodents. *Evolution* **19**:115–130.

Woodburne, M. O. and W. J. Zinsmeister. 1982. Fossil land mammal from Antarctica. *Science* **218**:284–286.

Woods, C. A. 1982. The history and classification of South American hystricognath rodents: reflections on the far away and long ago. Pages 377–392 in M. A. Mares and H. H. Genoways (Eds.), *Mammalian Biology in South America*. Special Publ. Ser., Pymatuning Laboratory of Ecology, University of Pittsburgh, PA.

Woods, C. A. and D. K. Boraker. 1975. *Octodon degus*. *Mamm. Species* **67**:1–5.

Wrigley, R. E. 1972. Systematics and biology of the woodland jumping mouse, *Napaeozapus insignis*. *Ill. Biol. Monogr.* **47**:1–117.

Yunis, J. J., J. R. Sawyer, and K. Dunham. 1980. The striking resemblance of high-resolution G-banded chromosomes of man and chimpanzee. *Science* **208**:1145–1148.

PART TWO

INSECTA

Chapter 4

Hemimerus deceptus Rehn   *Asidoptera phyllostomatis* (Perty)

Patterns of Insect
Parasitism in Mammals

Ke Chung Kim and Peter H. Adler

Authorized for publication as paper No. 6810, Journal Series of the Pennsylvania Agricultural Experiment Station, The Pennsylvania State University, University Park, PA 16802, U.S.A.

INTRODUCTION

The parasitic insects associated with mammals constitute a minute fraction, less than 0.5%, of the total insect diversity. They belong to only seven orders: Dermaptera, Hemiptera, Anoplura, Mallophaga, Coleoptera, Diptera, and Siphonaptera. All species of Anoplura, most Siphonaptera, and some Mallophaga are parasitic on mammals, whereas only a small number of minor taxa in the Dermaptera, Hemiptera, Coleoptera, and Diptera are mammalian parasites. In this chapter we discuss the adaptations, evolutionary strategies, and host associations of these minor parasitic taxa and compare them with those of other major taxa, toward a synthesis of parasite–host coevolution.

SYMBIOTIC ASSOCIATES AND PARASITIC INSECTS

Many insects of different phylogenetic lines have developed similar symbiotic associations with mammals, although the degree of dependence varies greatly between different taxa. These associations include phoresy, commensalism, inquilinism, and parasitism. However, it is often difficult to separate parasitic associates from those that are commensalistic with or phoretic on mammals.

A small percentage of Lepidoptera are associated with mammals, but their associations are primarily phoretic, commensalistic (Waage and Montgomery 1976; Waage 1979), or opportunistic. The highly specialized pyralid moths, *Cryptoses* (3 spp.) and *Bradypophila* (2 spp.), live as adults in the fur of sloths (Edentata). They are not parasites but, at best, commensals or phoretic symbiotes. Some moths of Geometridae, Noctuidae, Notodontidae, and Pyralidae imbibe lachrymal fluids, usually on a facultative basis, from the eyes of large mammals such as bovids, elephants, and occasionally man (Bänziger and Büttiker 1969; Bänziger 1972). One species of noctuid feeds exclusively on lachrymal secretions (Bänziger 1972), and three species of noctuids (*Calyptra* spp.) are skin-piercing bloodsuckers of mammals (Bänziger 1979). None of these species are parasites as defined in this book.

Some polyphagan Coleoptera associated with mammals are more likely commensals than parasites. These associations include *Loberopsyllus* (Languriidae) with Nearctic rodents, *Catopidius* (Leiodidae) with Palearctic Lagomorpha, and *Myotyphlus* (Staphylinidae) with a Tasmanian murid. Three species of coprophagous scarabaeids live, as adults, on sloths (Marshall 1981).

On the other hand, many mammalian symbiotes are permanent parasites living in the fur: they include Polyctenidae (Hemiptera); Arixeniidae and Hemimeridae (Dermaptera); parasitic Leptinidae, Staphylinidae, and Platypsyllidae (Coleoptera); Hippoboscidae, Nycteribiidae, and Streblidae (pupiparous Diptera); all Anoplura; some Mallophaga; and most Siphonaptera. Permanent parasites such as lice (Anoplura and Mallophaga) spend their entire life on the host, whereas temporary parasites are monostadial. Of the temporary parasites, the myiasis-producing Diptera— Calliphoridae, Sarcophagidae, Gasterophilidae, Cuterebridae, Oestridae, and Hypodermatidae—are parasitic on mammalian hosts as larvae, whereas fleas (Siphonaptera) and louse flies, bat flies, and spider flies (Diptera) parasitize the hosts as adults.

Bats (Chiroptera) are host to the Polyctenidae, Arixeniidae, Nycteribiidae, Streblidae, and Ischnopsyllidae (Siphonaptera). Rodentia and other mammals are host to the Hemimeridae, Leptinidae, Platypsyllidae, Hippoboscidae, myiasis-producing flies, Anoplura, Mallophaga, and most Siphonaptera.

ADAPTATIONS AND EVOLUTIONARY STRATEGIES

Hemiptera

The Hemiptera are a large group of primarily terrestrial insects with distinctive piercing-sucking mouthparts. Many species are plant juice feeders, others are predaceous, and a few attack man and animals. But despite their predaceous habits and piercing-sucking mouthparts, only three taxa— Cimicidae, Polyctenidae, and Reduviidae—are associated with mammals. The Cimicidae and some reduviids like *Triatoma* are blood feeders but are not parasites as defined in this book. The Polyctenidae is the only hemipteran taxon that is wholly parasitic.

Polyctenidae

The 32 species and five genera of Polyctenidae (Fig. 4.1) are found exclusively on the microchiropteran bats of the Old and New World subtropics and tropics. These unique hemipterans defied proper systematic placement for years after their discovery in 1864, variously occupying positions in the Diptera and Anoplura. Current treatment places the family near the Cimicidae and Anthocoridae; features such as the asymmetrical parameres of the male genitalia indicate this affinity (Ferris and Usinger 1939).

The Polyctenidae are remarkably well adapted to an ectoparasitic lifestyle. The anthocorid-type head and the body are dorsoventrally flattened. Eyes and ocelli are lacking, although absence of the latter is characteristic of hemipteran nymphs. The four-segmented antennae are shortened, although the number of segments does not differ from that in closely related families. The wings are reduced to a pair of large lobes. Unlike the cimicids in which the hemelytra narrowly articulate with the body, the rudimentary wings of the polyctenids are broadly joined to the body without evidence

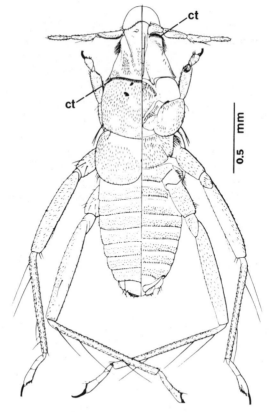

Figure 4.1 *Hesperoctenes longiceps* (Waterhouse) (after Ferris and Usinger 1939); ct, Ctenidia.

of articulation. The forelegs, designed in part for attachment and reminiscent of those of certain Mallophaga, are short, stout, and flattened, while the midlegs and hindlegs are generally long and slender. Such an arrangement allows the insect to pull itself along with the forelegs while using the middle and hindlegs to propel itself. Marshall (1982) points out that this arrangement is clearly designed for existence in the fur of the host; the insect is quite helpless off the host.

A very distinctive feature of the Polyctenidae is the presence of a number of combs (e.g., genal and antennal), resembling those in other ectoparasitic insects such as nycteribiids and fleas. In addition to the combs, polyctenids are clothed in a vestiture of retrose setae.

Polyctenids are characterized by pseudoplacental viviparity (Hagan 1931). The young, born alive, pass through three postnatal instars, little differentiation occurring between molts (Maa 1959).

The close association of polyctenids with their hosts obviates the need for certain structures and behaviors, such as eyes and wings and responsiveness to carbon dioxide (Marshall 1982). The need for frequent blood meals and reliance on contact with the host is so great, in fact, that death results in as few as 6–12 hours if they are removed from the host (Marshall 1982). These facts suggest a unique host ecology, which unfortunately is not well known. Maa (1964) reported that most of the bats that serve as hosts form small assemblages in caves and tree holes. This semicommunal life undoubtedly fosters transference of polyctenids between individuals. A single host may have polyctenids over the entire body, but particularly on the back where the hair is long; host grooming, although limited, is considered a major mortality factor (Marshall 1982).

Dermaptera

Very few orthopteroid genera have evolved a parasitic existence, perhaps owing in part to their generally large size. However, two of the three dermapteran suborders, Arixeniná and Hemimerina, are composed solely of ectoparasitic or mutualistic species. Nakata and Maa (1974) presented a thorough systematic treatment of both suborders.

Hemimeridae

The Hemimeridae (Fig. 4.2) contain *Hemimerus* with nine species and *Araeomerus* with two species. The first species, named in 1871, was placed in the Gryllidae. All species are associated with the pouched rats or the long-tailed rats (Muridae) of tropical Africa (Nakata and Maa 1974).

Members of the Hemimeridae are dorsoventrally flattened and about a centimeter long. The thoracic tergites are broadly expanded, rendering the species somewhat blattoidlike. Eyes and wings are absent. The head is semiprognathous, broad, and triangular, with a pair of large antennae. The legs resemble those of typical earwigs (Forficulina) but are highly specialized, being short with flattened tarsi. *Hemimerus talpoides* Walker clings to the base of the host hairs and moves adroitly through the fur and yet can also walk on an inverted glass surface (Ashford 1970). The abdomen terminates in a pair of elongate, setose cerci.

Popham (1962) suggested that the head, neck, and method of feeding of Hemimerina differs from all other Dermaptera and that they must have been derived from their own common ancestor. *Hemimerus talpoides* feeds on dried skin and exudates from the skin, eyes, ears, mouth, and anus and apparently does not harm the host or elicit self-grooming (Ashford 1970). Perhaps mutualism better defines the relationship. Although hemimerines spend a great deal of time on the host, they leave soon after it dies (Rehn and Rehn 1936).

Females are viviparous. Four nymphal instars occur, and there is some indication of slight maternal care (Ashford 1970).

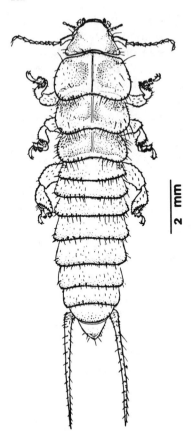

2 mm

Figure 4.2   *Hemimerus deceptus* Rehn (after Deoras 1941).

Arixeniidae

Members of the suborder Arixenina, which consists of the monotypic family Arixeniidae (Fig. 4.3), are intimately associated with *Cheiromeles torquatus*, a colonial roosting molossid bat of the Indo-Malayan subregion. The three species of *Arixenia* are obligate parasites that show somewhat more specialization than the two species of *Xenaria*. The latter species are believed to have an obligatory association with *C. torquatus*, but the nature of the association is unclear; specimens have never been collected from a bat captured in flight (Marshall 1977). *X. jacobsoni* possibly feeds on guano or its associated insect fauna (Popham 1962).

Arixenia is large for an ectoparasite, reaching 25 mm in length. It is dorsoventrally flattened and covered with a vestiture of fine hairs. Wings are lacking, but small eyes and ocelli are present. The 14-segmented antennae are rather long. The cerci are long and unsegmented and more heavily sclerotized than in *Hemimerus*. Selection pressure for forficate cerci (defense type) appears to have been assuaged as the Dermaptera evolved

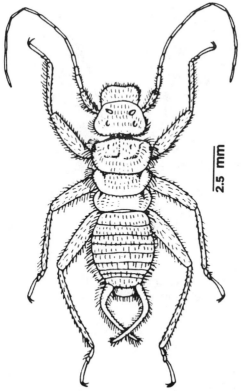

Figure 4.3 *Arixenia esau* Jordan. (From A. G. Marshall, *The Ecology of Ectoparasitic Insects.* Academic Press, London.)

toward a parasitic existence. A greater sensory role was probably assumed by the cerci in the more parasitic forms.

The head is semiprognathous, and the mouthparts are rather unspecialized and are similar to those of typical earwigs (*Forficula*). *Arixenia esau* Jordan feeds on epidermal products and exudates of its host. Marshall (1977) suggested that the host–insect relationship is mutualistic or commensal rather than parasitic, the insect actually cleaning the body of the host. Apparently *A. esau* occurs on the host only to feed; most of its time is spent on the roost. Marshall (1977) indicates that several factors are important in maintaining *A. esau* in the roost: viviparity, gregariousness, strong claws, and negative geotropic and phototropic responses.

Coleoptera

Less than 0.03% (about 72 species) of the Coleoptera, the largest order of insects, are associated with mammals. They belong to six polyphagan families: Leptinidae, Platypsyllidae, Leiodidae, Languriidae, Scarabaeidae,

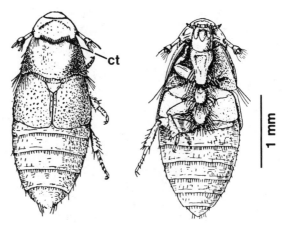

Figure 4.4 *Platypsyllus castoris* Ritsema showing ctenidia (ct) (after Jeannel 1922). Legs removed on one side.

and Staphylinidae. Most symbiotic beetles belong to the amblyopinine Staphylinidae. In this section we discuss three families: Leptinidae, Platypsyllidae, and Staphylinidae.

Leptinidae and Platypsyllidae

Leptinidae and Platypsyllidae are small, louselike beetles measuring only 2–3 mm in length. All four genera, *Leptinillus* (two species, Arnett 1968) and *Leptinus* (five species) (Leptinidae) and *Platypsyllus* (one species) (Fig. 4.4) and *Silphopsyllus* (one species) (Platypsyllidae), are primarily associated with the fur and nests of rodents in the Holarctic region including northern Africa. The head and body are broad and dorsoventrally flattened and covered with a dense vestiture of setae. The head is prognathous, and the antennae have 11 segments. The eyes are reduced in *Leptinillus*, vestigial in *Leptinus*, and absent in *Platypsyllus*. Ocelli are absent. The wings are reduced to tiny pegs, or absent in *Platypsyllus*. The forelegs of all species are short, and the hindlegs are long and slender. In *Platypsyllus*, which shows the greatest degree of specialization and reliance on the host, the tarsal claws of the forelegs are curved for catching the host hair, a cephalic ctenidium is present, the antennae and elytra are shortened, and the mandibles are flat and spatulate. Wood (1965) postulated that the origin of the Leptinidae and Platypsyllidae could be traced to the Catopinae, a small Holarctic subfamily of beetles associated with carrion and mammal burrows.

Much of what is known of the biology of the Leptinidae and Platypsyllidae was worked out by Wood (1965) who studied *Leptinillus validus* (Horn) and *Platypsyllus castoris* Ritsema on beaver, *Castor canadensis*. There are three larval instars in both species. Larvae and adults of *L. validus* feed

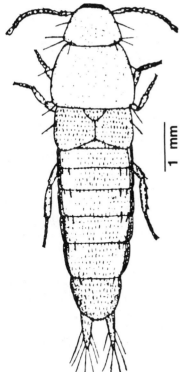

Figure 4.5 *Amblyopinus schmidti* Seevers (after Seevers 1944).

on the epidermis of the beaver, probably by scraping the skin, which may cause superficial abrasions. Larvae of *P. castoris* also feed on the skin of the beaver, and the larval mandibles are capable of creating skin lesions; the adults probably take primarily liquid nourishment from the host.

Oviposition and pupation apparently occur in the lodge of the beaver. Larvae and adults remain in the underfur of the beaver where water does not penetrate. Both adults and larvae of *L. validus* are often found in the lodge in the absence of beaver.

Staphylinidae

Most species of the Staphylinidae (Fig. 4.5) are apparently predaceous. Many occur in the nests of birds and mammals. Perhaps then it is not surprising that some of these beetles are parasitic on mammals, particularly the cricetine and hystricomorph rodents. These beetles represent five closely related genera: *Amblyopinodes, Amblyopinus, Edrabius, Megamblyopinus,* and *Myotyphlus*. Much that is known about these beetles was reviewed by Seevers (1955) who placed them in the tribe Amblyopinini. The information that follows is drawn primarily from that source.

All members of the Amblyopinini are wingless with rather short elytra. These beetles range from 5 to 16 mm in length. The antennae are long and filiform. The legs are rather long and unspecialized, although those of some species bear ctenidia. Species of *Amblyopinodes* are the most highly specialized forms. Their head capsules are like that of other parasitic beetles, especially *P. castoris* Ritsema. They also have rows of movable, clavate setae on several of the sternites. The function of these setae is unknown, but they are not believed to be true ctenidia.

The Amblyopinini probably originated from the quediine stock. The most generalized of the Amblyopinini, *Myotyphlus*, whose parasitic habits are not completely certain, links the Neotropical amblyopinines with the quediines (tribe Quediini, or subtribe Quediina as cataloged by Arnett 1968). The resemblance between the Amblyopinini and the Quediini is manifested particularly in the head capsule and hind coxae of *Myotyphlus* and *Edrabius*. These two taxa have eyes with only one facet. The remaining three taxa have larger, multifaceted eyes, which are reduced nonetheless.

Little is actually known of the biology of these beetles. Most species occur in cool mountain habitats, a few are found in dry or warm areas, and one species is known from a semiaquatic environment (Machado-Allison and Barrera 1972). The adults and larvae occur on the host and are considered obligate parasites. They embed their mandibles deep in the skin, causing irritation. Presumably they feed on skin and body fluids. Barrera (1966) obtained positive results for blood feeding in *Amblyopinus tiptoni* Barrera. The amblyopinines move swiftly on the host and are negatively phototropic.

Diptera

Ten family taxa of the calyptrate Diptera are associated with mammals: Hippoboscidae (48 species), Mystacinobiidae (one species), Nycteribiidae (256 species), Streblidae (221 species), Calliphoridae and Sarcophagidae (16 species), Cuterebridae (83 species), Gasterophilidae (18 species), Oestridae (34 species), and Hypodermatidae (20 species). The Hippoboscidae, Nycteribiidae, and Streblidae are often artificially grouped into the "Pupipara," and their relationships have been the subject of much controversy (Maa 1963; Theodor 1964). Nonetheless, the members of these families share many structural, biological, and ecological features as a result of convergent evolution within the framework of an obligate ectoparasitic existence. The sole species of Mystacinobiidae, *Mystacinobia zelandica* Holloway, is closely associated with a bat in New Zealand (Holloway 1976), but its symbiotic status appears to be primarily commensalistic-phoretic.

Among the Diptera symbiotically associated with mammals, the larvae of approximately 42 genera and about 170 species of Calliphoridae, Sarcophagidae, Cuterebridae, Gasterophilidae, Oestridae, and Hypoder-

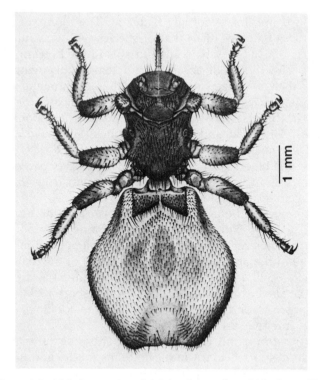

Figure 4.6 *Melophagus ovinus* (L.) (after Schwardt and Matthysse 1948).

matidae are obligate producers of myiasis throughout the world (James 1947; Zumpt 1965). A number of additional dipteran genera are facultative producers of myiasis in mammals. Zumpt (1965) reviewed both the facultative and obligate groups and elucidated the evolution of myiasis. We treat here only the obligate parasites.

Hippoboscidae

The Hippoboscidae (louse flies) (Fig. 4.6) are a well-defined group of cosmopolitan calyptrate muscoid Diptera. Each genus of the Hippoboscidae is composed solely of mammalophilic or ornithophilic species. The most current listing includes 197 species (Marshall 1981). Most species are obligate parasites of birds; only eight genera and 48 species parasitize mammals. The mammalophilic species are the most specialized, particularly the members of the subfamily Lipopteninae. The Hippoboscidae have received exhaustive treatment, notably by Bequaert (1953, 1957) and Maa (1963, 1966, 1969). The following information is drawn from these sources.

Hippoboscids are dorsoventrally flattened. The thorax of winged hipboscids is less flattened than that of species with deciduous wings or apterous species. This may be related in part to the housing of the wing musculature. The thorax of apterous and temporarily winged forms is shorter than that of winged species. The thoracic humeral callosities of mammalophilic species are always rounded and, at most, project anteriorly only mildly. In contrast, most ornithophilic species have pointed, projecting callosities. The head and thorax are robust and have a very tough integument. However, the abdomen is rather soft and distensible for accommodating the blood meal. The body is very setose, particularly in the apterous forms. No true ctenidia are present. Apart from the terminalia, there is no overt sexual dimorphism.

The head is prognathous and wedge shaped, and the mouthparts are modified for piercing skin and sucking fresh blood. Compound eyes are present and vary in size from large and well developed in the fully winged forms to small in the apterous species. About half of the mammalian hipboscids have ocelli, large or rudimentary. The antennae are always flattened, and the small third segment, bearing the short arista, is concealed within the larger second segment. In all mammalophilic species, except *Allobosca*, the antennae are quite small.

The legs are strong and well developed. Those of the mammalophilic species are stoutest, particularly in the apterous species and *Allobosca*. Each leg terminates in a pair of claws, each with a basal thumb. In species infesting mammals, the claws are simple and heavy, whereas most ornithophilic species have deeply cleft claws.

Hippoboscids may be fully winged (temporarily or permanently) or apterous. *Melophagus* has tiny, rudimentary wings and is the only mammalophilic group lacking halteres (Fig. 4.6). All other mammalophilic genera are permanently winged except *Lipoptena* and *Neolipoptena*. After emergence, once a suitable host is located, *Allobosca, Lipoptena,* and *Neolipoptena* lose their wings and their flight muscles atrophy. Calypteres may be present, reduced, or absent.

Intrauterine viviparity occurs in the Hippoboscidae. The female deposits a third-instar larva which soon becomes a puparium. Species of *Melophagus* actually glue the puparium to the host hair. Temporarily winged species also deposit the puparium on the host, but without glue it falls into the duff. Winged forms may larviposit in selected locations apart from the host.

Nycteribiidae

The Nycteribiidae (spider bat flies) (Fig. 4.7) are obligate ectoparasites of bats. They occur throughout the tropics and subtropics, primarily in the

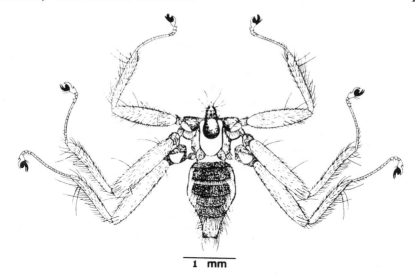

Figure 4.7 *Penicillidia (Eremoctenia) progressa* (Muir), male (after Scott 1917).

Old World. Currently, 256 species and 12 genera are recognized (Marshall 1981). General treatment of the family, plus descriptions of about 200 species can be found in Theodor (1967), upon which much of the following discussion is based.

Nycteribiids are extremely specialized Diptera. Much of their body construction has been modified to such a degree that they have come to resemble spiders superficially. The thoracic sternal plates are fused to form an expanded plate; the dorsum is membranous; the pleura are displaced; and the well-developed legs are attached dorsally. Each leg bears a pair of strongly curved claws. Wings and flight muscles are lacking in all members of the family, but halteres are still present.

The head is small and bent backwards so that it is held in a groove on the mesonotum. When the fly feeds, the head is brought forward. The head may be laterally compressed, dorsoventrally flattened, or broadly rounded. The antennae are small and appear to have two segments. Nonetheless, the characteristic antennal suture of calyptrate muscoid flies is present. Compound eyes, when present, consist of only one or two lenses. They are most developed in *Cyclopodia* and *Archinycteribia* which parasitize fruit bats. The mouthparts are developed for piercing skin and sucking blood.

Ctenidia are present on the thorax between the first and second coxae, except in some species of *Penicillidia*. Abdominal ctenidia occur in many species. Most abdominal sclerites bear rows of posteriorly directed setae. Upright setae, presumably sensory in nature, are also present on the sclerites.

Nycteribiids are intimately associated with their hosts. They take fresh blood meals frequently and die within one or two days if separated from the host. The female, sometimes accompanied by the male, leaves the host for a short while to larviposit in the bat roost on cave walls or leaves (see Marshall 1970). As in all the Pupipara, females are viviparous; the first three instars are nurtured within the female.

Marshall (1981) noted that nycteribiids fall into two sized groups that are ecologically based. Large species generally occupy the surface of the pelage; smaller species occur within the fur. Not surprisingly, the large species are most common in areas on the host that cannot be readily reached for self-grooming, whereas the smaller species are more generally distributed.

Streblidae

The Streblidae (Figs. 4.8, 4.9) are obligate, bloodsucking parasites of bats. They are primarily a tropical group, but, as opposed to the Nycteribiidae, they are well represented in the New World. Marshall (1981) lists a total of 221 species in 31 genera; one-third of the genera are monotypic. Much of the following information is drawn from Wenzel et al. (1966).

Unlike the Nycteribiidae, streblids are far more diverse in structure, hence the much larger number of genera. Members of the subfamily Streblinae are dorsoventrally flattened. Species of *Nycterophilia* are laterally flattened and resemble fleas. The size of streblids varies from less than 1 mm in the louselike *Megistopoda* to more than 5.5 mm in *Joblingia*. The abdomen is almost entirely membranous but is covered with dorsal setae. These setae are longest in apterous and brachypterous species.

Many streblids have normal wings which they fold over the back when moving through the host fur. Reduced wings have evolved in a number of unrelated genera. In *Paradyschiria* the wings are absent. The legs of streblids are also variable. In *Mastoptera* the legs are short and strong. A few genera such as *Megistopoda* have elongate hindlegs. These species also have a large ventral thoracic plate that aids movement through the host fur. They also have ctenidia similar in form and location to those of nycteribiids. A few genera of streblids have greatly elongated tibiae. Elongated legs have also evolved in the Nycteribiidae through convergent evolution; however, it is the first tarsal segment that has been lengthened.

The streblid head has undergone perhaps the greatest modification of any body region. Jobling (1929) provided an overview of the head region and noted the structural diversity, from funnel shaped to dorsoventrally

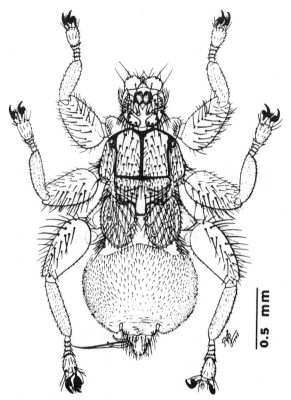

Figure 4.8 *Aspidoptera phyllostomatis* (Perty), male (after Jobling 1949).

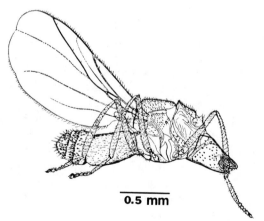

Figure 4.9 *Ascodipteron rhinopomatos* Jobling, female (after Theodor and Moscona 1954).

flattened. Postgenal ctenidia are present in the Streblinae and in *Eldunnia.* Streblids lack ocelli; compound eyes are small or absent. As in the Nycteribiidae, the antennae appear two segmented and are in grooves.

The most unique modifications in the Streblidae are found in species of *Ascodipteron* (Marshall 1981). Upon emergence from the puparium, the female (Fig. 4.9) locates a host and, once settled, sheds its wings and legs. By means of the proboscis it bores into the host and encysts much as do the fleas in the family Tungidae. Only the last three abdominal segments are exposed, enabling the female to extrude the third instar larvae. Males of *Ascodipteron* retain legs and wings and maintain a normal existence.

Certain species of streblids are so dependent on the host that they die within one or two hours after removal, for they feed frequently. On the other hand, the more generalized species such as *Joblingia* do not feed as often and do not stay on the host as much. Most streblids, however, leave the host to deposit a puparium on the walls of their roosting sites. The puparia of *Ascodipteron* simply fall to the ground. Apparently, winged streblids can locate their host by olfaction.

Obligate Myiasis-Producers

Of the six families whose larvae are obligate producers of myiasis, only the Cuterebridae, Gasterophilidae, Hypodermatidae, and Oestridae are exclusively parasites of mammals. The Calliphoridae (blow flies) are mostly scavengers, the larvae living in carrion, excrement, and similar materials. A few species of blow flies, like *Cochliomyia*, are obligate producers of cutaneous or nasal myiasis in large mammals. The larvae of Sarcophagidae (flesh flies) feed mostly on animal material; many are coprophagous, some are parasitoids, and a few are parasitic on mammals.

The larvae of obligate myiasis-producers may occur subdermally or within the internal organs where they feed principally on living tissue and mucus. They are generally robust, subcylindrical, or spindle shaped and possess one or more rows of well-developed spines, microspines, setulae, or similar armature, which may be densely set, on each body segment (Fig. 4.10*A*, *B*). The thoracic spiracles are generally nonfunctional or absent, but in species such as *Cochliomyia* the prothoracic spiracles are conspicuous. The pair of caudal spiracles is well developed and conspicuous but often housed within a cavity as in the Gasterophilidae. The head typically possesses a pair of curved, heavily sclerotized mouth hooks. Antennae and sensory papillae are generally present, but short and inconspicuous.

The adults of many obligate myiasis-producers are large and robust and are strong flyers. Many species resemble different bees. The eyes are large and well developed, and the antennae are small. The mouthparts are small, reduced, or rudimentary but may be functional in some species; all necessary energy for maintenance and reproduction is obtained in the

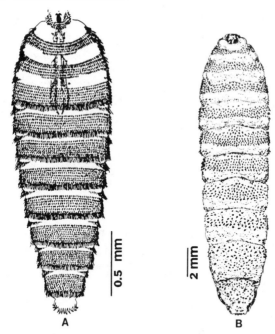

Figure 4.10   Larvae of *Cephenemyia trompe* (Modeer). (*A*) first instar; (*B*) third instar (after Grunin 1966).

larval stage. Some species with functional mouthparts drink fluid probably to maintain internal water balance. Many species form mating aggregations on topographic sites or above a conspicuous marker.

The females of these flies either lay a large number of eggs or larviposit. *Cuterebra* lay a large number of their eggs, 1000–2500 in small batches, in areas frequented by the host (Catts 1982). Larvae generally enter the host by way of moist body openings or occasionally at skin lesions. From the point of entry they migrate through various parts of the body such as the tracheae and abdominal cavity, and thenceforth into a specific subcutaneous position where they cause the formation of a warble. The cuterebrid, *Dermatobia hominis* (L. Jr.), is unique in its egg-laying habits. A female glues her eggs to a host-frequenting dipteran such as a mosquito. After development, the eggs hatch on warming as the transporting host feeds, and the larvae enter the skin.

The Gasterophilidae typically lay their eggs on the body hairs or on vegetation. The host then licks the fur or eats the vegetation, whereupon the larvae hatch and migrate from the mouth to the digestive tract where they attach (James 1947). The Hypodermatidae oviposit on the host hairs and the larvae burrow into the skin and migrate through the body to a subdermal site. The Oestridae (*sensu* Papavero 1977) are usually larviporous, and the female ejects small packets of first instar larvae directly into the nasal and oral portals of the host. The larvae then develop primarily in the nasopharyngeal cavities. The sarcophagid, *Wohlfahrtia*, larviposits in skin lesions.

All myiasis-producing larvae leave the host by way of the warble pore, the anus, the mouth, or the nose. They drop to the ground, burrow shallowly into the soil or duff and form a puparium.

HOST ASSOCIATIONS AND DISTRIBUTIONS

Hemimeridae and Arixeniidae (Dermaptera)

All nine species of *Hemimerus* are found on the giant pouched rats, *Cricetomys*. The taxonomy of these rats has been confused over the years. Many "forms" that differ ecologically and in pelage have been named, but Meester and Setzer (1971) recognize only two species. The fact that nine species of *Hemimerus* are described from only two host species suggests that a reevaluation of host taxonomy is needed.

Little is known of either species of *Araeomerus*. At least one species is associated with the food caches of the long-tailed pouched rats, *Beamys;* it may in fact not occur on the animal (Nakata and Maa 1974).

Hemimerus and *Araeomerus* are restricted to the West and East African subregions, coincident with the distributions of their hosts. The center of distribution is considered to be in the mountains surrounding the Victoria, Tanganyika, and Nyasa Lakes (Nakata and Maa 1974).

The only known true host of *Arixenia esau, A. camura* Maa, and *Xenaria* species is *Cheiromeles torquatus*, a large insectivorous and essentially hairless bat. The skin is very glandular, loose, and wrinkled. Perhaps the lack of hair has permitted the evolution of large ectoparasites, while the loose folds still afford some protection. Association of *A. esau* with other molossid bats is assumed to be accidental (Marshall 1977).

Cheiromeles torquatus is confined to Malaysia, the Philippines, and Indonesia. *A. esau* is known from Malaya, Sumatra, and Borneo. *A. camura* is known only from Mindanao.

Parasitic Beetles

The Leptinidae occur in North America, Europe, Russia, and North Africa. *Leptinillus validus* occurs on beaver in North America, and *L. aplodontiae* Ferris is found on the mountain beaver in western North America. *Leptinus americanus* LeConte, also North American, is associated with small rodents and insectivores. Its European congener, *L. testaceus* Muller, has been found in the nests and fur of mice, shrews, and moles and in the nests of bumblebees and wasps. Three other species of *Leptinus* are parasitic on Holarctic rodents. *Silphopsyllus desmanae* Olsufiev, a Russian species, is known from the water mole, *Desmana maschata*. The platypsyllid *Platypsyllus castoris* occurs solely on *Castor canadensis* in the Nearctic and on *C. fiber* in the Palearctic.

The staphylinid Amblyopinini are believed to have originated in association with primitive rodents during the early Tertiary period. Associations with marsupials are probably secondary, since the infesting beetles are highly specialized. Host specificity is highly developed in the amblyopinines. *Edrabius* (six species) occurs in Chile, western Argentina, and southern Peru on burrowing caviomorph rodents, *Ctenomys*. *Amblyopinus* (34 species) occurs primarily on the sciuromorph, caviomorph, and myomorph rodents, although about five species occur on marsupials and several others on hystricomorph rodents. The distribution records encompass the subtropical and temperate areas of South and Central America. *Amblyopinodes* (15 species) occurs primarily on the cricetine rodents of Argentina, Venezuela, Peru, and Brazil. *Megamblyopinus* (2 species) is not well known but has been found in the nests of *Ctenomys* in *Peru*. The sole species of *Myotyphlus* is known from a Tasmanian murid.

Bats and Their Parasitic Insects

Polyctenidae (Hemiptera)

The 32 species and five genera of polyctenids are distributed among five of the 17 families of Microchiroptera. The largest and probably oldest microchiropteran family, Vespertilionidae, conspicuously lacks polyctenids. However, the bats in this family are distributed primarily in the temperate zones and thus north of the range of the Polyctenidae. The polyctenids probably originated, along with the bats, in the area around the Indian Ocean (Marshall 1982). Only *Hesperoctenes* (16 species), which Maa (1964) suggested is historically the youngest taxon, occurs in the New World and restricted to the New World Molossidae, whereas *Hypoctenes* (five species) occurs on molossids in the Ethiopian and Oriental regions. Other Old World polyctenids include *Adroctenes* (three species) on the Rhinolophidae, and *Eoctenes* (seven species) on the Emballonuridae, Megadermatidae, and Nycteridae (Maa 1964).

Host specificity is high among the polyctenids. Even the New World *Hesperoctenes fumarius* (Westwood), which is recorded from 11 species of Molossidae, may in reality represent a species complex with the siblings expressing host specificity (Marshall 1982). Interhost species transference, of course, rests largely on the occurrence of multiple-species assemblages of bats. Data on this issue are scarce, although *Megaderma spasma* (L.) appears not to rest in contact with other bats (Marshall 1982). Additionally, no bat species harbors more than one species of polyctenid (Maa 1964).

Nycteribiidae (Diptera)

The center of origin of the Nycteribiidae may have been the Malaysian subregion. Certain species are distributed widely within a biogeographic region, but only *Basilia blainvillii* (Leach) occurs in two regions (Ethiopian and Oriental). Many species have a more restricted distribution than their hosts. In general, nycteribiids are quite host specific (Marshall 1981).

Five genera and 65 species occur on the frugivorous Megachiroptera, primarily in the Australian, Ethiopian, and Oriental regions. The remaining seven genera occur on five families of the insectivorous Microchiroptera. The largest genus, *Basilia* (103 species), occurs throughout the subtropics and tropics. *Hershkovitzia* (three species) is the only genus restricted to the New World.

Streblidae (Diptera)

The list of streblid species has increased substantially in recent years and indicates that intense faunistic surveys are likely to yield numerous additional species. Wenzel (1976) found 115 species, including 45 new species, in more than 36,000 specimens collected from Venezuela. The degree of host specificity is considerably high—55% of the Panamanian species are known from only a single host—although some species of bats such as *Phyllostomus hastatus* may have three or four species of streblids at a time (Wenzel et al. 1966). The opportunity for interhost species transfers of streblids has undoubtedly been favorable since many of the hosts roost in mixed-species ensembles (Wenzel et al. 1966). Jobling (1939) noted the congruence of streblid and bat evolution; that is, the most generalized streblids occur on the most generalized bats.

The Streblidae are divided into five subfamilies. The Brachytarsininae (5 genera, 54 species) and the Ascodipterinae (1 genus, 18 species) occur in the Oriental, Ethiopian, and Palearctic regions. Only these two taxa contain species parasitic on the frugivorous Megachiroptera. The Nycterophilinae (2 genera, 6 species), the Trichobiinae (19 genera, 111 species), and the Streblinae (4 genera, 32 species) are strictly Neotropical, although some members of the largest genus, *Trichobius*, occur in the Nearctic region. The Old and New Worlds have no streblid taxa in common (Wenzel 1976; Wenzel et al. 1966).

Louse Flies and Their Mammalian Hosts

The hippoboscids are diverse in terms of host specificity. In general, the mammalophilic species are more host specific than are the ornithophilic species. The first hippoboscids were probably ectoparasites of birds and were similar to present-day forms by the mid-Tertiary. Bequaert (1954) believed that all extant genera were present by that time.

The louse flies of the monotypic genera *Allobosca* and *Proparabosca* are ectoparasites of primates in the Ethiopian region. *Austrolfersia* (one species) and *Ortholfersia* (four species) parasitize marsupials from Queensland to Tasmania. *Hippobosca* (seven species) occurs in the Old World, including the Australian region, on Artiodactyla, Perissodactyla, and Carnivora. The largest genus-taxon, *Lipoptena,* is found worldwide (excluding the sub-polar, polar, and Australian regions) on Artiodactyla. The monotypic *Neolipoptena* occurs on Cervidae in western North America. *Melophagus* (three species) occurs on the Bovidae of the Palearctic region, although one species (*M. ovinus*) that occurs on domestic sheep is now cosmopolitan. Four additional species of mammal-infesting hippoboscids occur exclusively on domestic animals (Maa 1963).

Obligate Myiasis-Producers and Their Hosts

Cuterebridae

Catts (1982) recognized six genera and 83 species of cuterebrids. Cuterebrids are native to the New World, and all species are parasites of rodents, lagomorphs, and primates. They have recently invaded introduced Artiodactyla. They are believed to have coevolved with the Rodentia and secondarily invaded the other hosts.

The largest taxon, *Cuterebra* (72 species), parasitizes the sciuromorphs, myomorphs, and all three genera of North American lagomorphs. The monotypic *Alouattamyia* parasitizes howler monkeys. The Neotropical *Dermatobia* (one species) is not host specific, and its hosts include many wild and domestic mammals and birds, as well as man. Little information is available for *Rogenhofera* (six species), *Pseudogametes* (two species), or *Montemyia* (one species).

Gasterophilidae

The gasterophilids are native to the Old World, although four species of *Gasterophilus* have been introduced into the Nearctic region. The nine species of *Gasterophilus* all parasitize the Equidae of the Ethiopian and Palearctic regions. *Gyrostigma* (three species) parasitizes rhinoceroses. *Rodhainomyia* (one species) and *Platycobboldia* (one species) attack African elephants, whereas *Cobboldia* (one species) parasitizes the Indian elephant. Zumpt (1965) places the nongasterophilous species of *Neocuterebra* (one species) and *Ruttenia* (one species) in this family. They are foot and skin parasites, respectively, of the African elephant.

Oestridae

Zumpt (1965) considered the Oestridae to consist of the Oestrinae and the Hypoderminae, but Papavero (1977) considered only the Oestrinae to be in this family. We follow Papavero's treatment of the Oestridae as a family exclusive of Hypodermatidae. Much of the following information is drawn from Papavero (1977).

The center of oestrid distribution is Eurasia and Africa. *Cephenemyia* (six species), *Acrocomyia* (one species), *Procephenemyia* (one species), and *Paryngomyia* (two species) parasitize the Cervidae and/or the Bovidae of the Holarctic region. *Pharyngobolus* (one species) invades the African elephant of the Congo forest region. *Tracheomyia* (one species) is the only Australian representative of this family and is a parasite of kangaroos. *Cephalopina* (one species) parasitizes camels of the Palearctic region and has recently been introduced with camels into the Oriental region. *Gedoelstia* (two species), *Kirkioestrus* (two species), *Loewioestrus* (one species), *Oestroides* (one species), and *Suinoestrus* (one species) parasitize African Artiodactyla. *Gruninia* (one species) parasitizes the Bovidae of central Asia. *Rhinoestrus* (nine species) invades bovids, giraffes, hippopotamuses, and other artiodactyls of Africa and Eurasia. *Oestrus* (four species) attacks Ethiopian (and recently, Palearctic) Artiodactyla. Papavero (1977) noted that only about 2.5% of the Recent mammalian genera are parasitized by the Oestridae.

Hypodermatidae

The family Hypodermatidae contains parasites of Old World ungulates and rodents. It includes *Hypoderma*, *Pavlovskiata*, *Pallasiomyia*, and *Przhevalskiana* of the Palearctic region, and *Strobiloestrus* of the Ethiopian region. The six species of *Hypoderma* are generally referred to as heel flies or ox warbles. The single species of *Oedemagena* is host specific to reindeer in the Holarctic region. Three taxa (*Oestromyia*, *Oestroderma*, and *Portschinskia*) parasitize rodents of the Palearctic region (Zumpt 1965).

Calliphoridae and Sarcophagidae

Seven genus-taxa of the Calliphoridae and Sarcophagidae are responsible for producing myiasis, primarily in Artiodactyla. These include *Wohlfahrtia* (two species), Holarctic; *Chrysomya* (one species), Ethiopian and Oriental; *Cochliomyia* (one species), New World tropics and subtropics; *Pachychoeromyia* (one species), and *Cordylobia* (three species), Ethiopian; and *Booponus* (four species), Oriental. A single species of *Elephantoloemus* parasitizes the Indian elephant (Zumpt 1965).

DISCUSSION AND CONCLUSIONS

Parasitic Adaptations and Convergence

Nine family-taxa of four orders were compared for their parasitic adaptations: Polyctenidae (Hemiptera); Arixeniidae and Hemimeridae (Dermaptera); Leptinidae, Platypsyllidae, and Amblyopinini–Staphylinidae (Coleoptera); Hippoboscidae, Nycteribiidae, and Streblidae (Diptera). To study these adaptations, the following attributes were examined and their character polarity (indicated by arrow) delineated:

1. Body form: a. normal, b. dorsoventrally flattened, c. laterally compressed; a $\swarrow\!\!\!\!\!\!^{b}_{c}$.
2. Type of head: a. hypognathy, b. semiprognathy, c. prognathy; a → b → c.
3. Compound eyes: a. present, b. reduced, c. absent; a → b → c.
4. Ocelli: a. present, b. reduced, c. absent; a → b → c.
5. Antennae: a. long, normal, b. shortened, modified; a → b.
6. Wings: a. fully developed, b. brachypterous, c. apterous or deciduous; a → b → c.
7. Forelegs: a. long, normal, b. shortened, modified, c. long, modified; a $\swarrow\!\!\!\!\!\!^{b}_{c}$.
8. Combs: a. not developed, b. spiniform setae, c. well developed in head, thorax and/or abdomen; a → b → c.
9. Type of reproduction: a. ovipary, b. ovovivipary–vivipary; a → b.

The results of this analysis are presented in Table 4.1.

These insects, although remotely related phylogenetically, have developed similar adaptations for a parasitic mode of life (Table 4.1). Most of these parasites are dorsoventrally flattened except for some nycteribiids and streblids that are laterally compressed like fleas. In the course of flattening, the head became porrect or semiprognathous. The mouthparts usually moved anteriorly, while retaining the ventral position in those taxa whose ancestors had the hypognathous head. Compound eyes and ocelli are reduced or completely lacking, and the antennae are usually shortened. However, specialized sensory organs are highly developed on the antennae of ectoparasitic insects (Al-Abbasi 1981).

The wings are reduced to a brachypterous condition or completely lacking, or in the case of some hippoboscids they are shed after reaching the host. The forelegs are usually shortened and modified as holdfasts.

Ctenidia or combs (Figs. 4.11, 4.12) are rows of flat, stout spines present

Table 4.1 Character Matrix for Nine Family-Taxa of Parasitic Insects[a]

Order	HEM	DER		COL			DIP		
Family	Polyctenidae	Hemimeridae	Arixeniidae	Leptinidae	Platypsyllidae	Amblyopinini Staphylinidae	Hippoboscidae	Nycteribiidae	Streblidae
1. Body form	b	b	b	b	b	b	b	b, c	b, c
2. Type of head	b	b	b	c	c	c	c	b	a, b
3. Compound eyes	c	c	b	b	c	b	a	b	b
4. Ocelli	c	c	b	c	c	b	a, c	c	c
5. Antennae	b	a	a	a	b	a	b	b	b
6. Wings	b	c	c	b	c	c	a, c	c	a, b, c
7. Forelegs	b	b	b	b	b	a	b	c	a
8. Combs	c	a	a	a	c	b	a, b	c	c
9. Reproduction	b	b	b	a	a	a	b	b	b

[a]HEM, Hemiptera; DER, Dermaptera; COL, Coleoptera; DIP, Diptera; a, plesiomorphy; b, c, apomorphy. Character codes: 1.(a) Normal, (b) dorsoventrally flattened, (c) laterally flattened. 2.(a) Hypognathy, (b) semi-prognathy, (c) prognathy. 3.(a) Present, (b) reduced, (c) absent. 4.(a) Present, (b) reduced, (c) absent. 5.(a) Long, normal, (b) shortened, modified. 6.(a) Present, (b) brachypterous, (c) apterous or deciduous. 7.(a) Long, normal, (b) shortened, modified, (c) long, modified. 8.(a) Not developed. (b) spiniform setae, (c) well developed. 9.(a) Ovipary, (b) ovovivipary–vivipary.

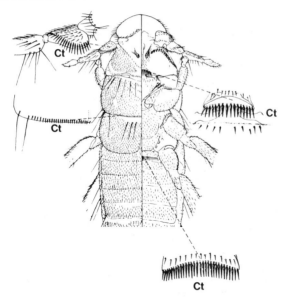

Figure 4.11 Location of ctenidia (Ct) of *Hypoctenes clarus* Jordan (after Ferris and Usinger 1939).

in the Polyctenidae, Platypsyllidae, Nycteribiidae, and Streblidae as well as Siphonaptera. They are found on the antennae, head (genal, occipital, postgenal, cephalic), thorax (pronotal, mesonotal), legs, and/or abdomen. They are well described for Siphonaptera by Traub (1972a, b, 1980) and Amin and Wagner (1983); for Polyctenidae by Ferris and Usinger (1939) and Maa (1959, 1964); and for Nycteribiidae by Theodor (1967). At present, there are two theories concerning the function of ctenidia: (1) they are used for attachment to the host to prevent dislodgment (Humphries 1966; Traub 1972a, b, 1980; Amin and Wagner 1983); (2) they protect the mobile joints (Marshall 1980, 1981). Humphries (1966) and Amin and Wagner (1983)

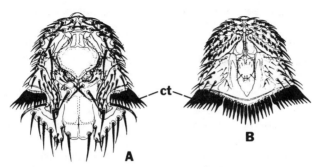

Figure 4.12 Ctenidia (ct) on head of *Metelasmus pseudopterus* Coquillet. (*A*) dorsal view; (*B*) ventral view (after Jobling 1936).

document definite correlations between the spacing of pronotal spines in different fleas and their host hair diameter. Although these correlations seem to support the attachment function of the pronotal ctenidia, other evidence does not discount the protective or other functions of the genal, postgenal, cephalic, mesonotal, and abdominal combs. Marshall (1980) supports the protection theory with his behavioral observations of the Polyctenidae, Nycteribiidae, and Ischnopsyllidae.

The myiasis-producing Diptera that are obligate associates of mammals demonstrate a separate pattern of convergence. The internal or subdermal environment of the larvae, which show site specificity, has selected for a larval body plan emphasizing the mouthparts and body armature. The structural and behavioral design of the adults—strong flight, large eyes, shortened or compact antennae and small or reduced mouthparts, and the formation of mating aggregations—is related to efficient mate and host location. Most species have a short life span and live on the nutritional reserves carried over from the larval stage.

The reproductive strategies of parasitic anthropods are generally related to the degree of host dependence. Ovovivipary and vivipary are common in parasitic insects such as the Polyctenidae, parasitic dermapterans, and pupiparous Diptera (Table 4.1). This reproductive adaptation allows the females to produce fewer progeny and to invest more energy per offspring (*K*-strategy) (Marshall 1981). Permanent parasites like lice produce small numbers (≤30) of eggs rather than young, whereas temporary parasites such as ticks and fleas may produce hundreds or thousands of eggs during a lifetime.

Species of myiasis-producing Diptera, such as those of *Cuterebra,* that oviposit in areas remote from the point of larval entry tend to lay many eggs. In contrast, species like *Wohlfahrtia* and *Cephenemyia* that deposit offspring at the point of larval entry are numerically conservative and typically larviposit early instars.

Patterns of Host Associations

The host associations of parasitic insects are usually limited to specific groups of mammals. Four family taxa—Polyctenidae, Arixeniidae, Nycteribiidae, and Streblidae—are specific ectoparasites of bats, whereas the Hemimeridae and parasitic beetles are primarily associated with rodents (Table 4.2).

Of the bat ectoparasites, the Polyctenidae, Nycteribiidae, and Streblidae are widely distributed throughout all biogeographic regions except the Palearctic for Polyctenidae and the Australian for Streblidae. The Arixeniidae are confined to the Oriental bats.

The Hemimeridae are parasites of the burrowing murids: *Araeomerus* on *Beamys* and *Hemimerus* on *Cricetomys* in Africa. Parasitic beetles (Leptinidae, Platypsyllidae, and Staphylinidae) are primary parasites of rodents: the

Table 4.2 Geographical Distributions and Host Associations of Nine Parasitic Family-Taxa in Hemiptera (HEM), Dermaptera (DER), Coleoptera (COL) and Diptera (DIP)

Order	HEM	DER		COL			DIP		
Family	Polyctenidae	Hemimeridae	Arixeniidae	Leptinidae	Platypsyllidae	Staphylinidae	Hippoboscidae	Nycteribiidae	Streblidae
Geographic regions									
Ethiopian	+	+					+	+	+
Oriental	+		+				+	+	+
Australian	+						+	+	+
Palearctic				+	+		+	+	+
Nearctic	+			+	+		+	+	+
Neotropical	+					+	+	+	+
Host associations									
Marsupialia						+	+		
Insectivora					+	+			
Chiroptera	+		+					+	+
Primates							+		
Rodentia		+		+		+	+		
Carnivora							+		
Artiodactyla							+		
Perissodactyla							+		

Nearctic *Leptinillus* (Leptinidae) on *Castor* (Castoridae) and *Aplodontia* (Aplodontidae); the Holarctic *Platypsyllus* (Platypsyllidae) on *Castor*; the Holarctic *Leptinus* (Leptinidae) on murid rodents; and the Neotropical amblyopinine Staphylinidae including *Amblyopinodes* primarily on Cricetidae, *Amblyopinus* on various rodents and a few marsupials, and *Edrabius* and *Megamblyopinus* on *Ctenomys* (Ctenomyidae). *Leptinillus* and *Leptinus* are primarily nest associates of their respective hosts. The platypsyllid *Silphopsyllus* is found on the Palearctic insectivore *Desmana* (Talpidae).

Members of the Hippoboscidae are primarily parasitic on large mammals—Primates, Marsupialia, Artiodactyla, Perissodactyla, and Carnivora—but are peculiarly missing from rodents (Bequaert 1954–1957; Maa 1963, 1969).

The host associations of the myiasis-producing Diptera, with the exception of the cuterebrids and some hypodermatids, are centered in the large herbivorous mammals, particularly the Artiodactyla and to a lesser extent the Perissodactyla and Proboscidea. The center of host associations for the Cuterebridae is Rodentia; secondary host associations are with the Lagomorpha and Primates. Papavero (1977) considered large size and herbivory to be important factors influencing host selection by oestrids. The abundance of herbivores relative to carnivores also may have been important in centering the host associations of myiasis-producing Diptera.

Looking at other ectoparasites, the Siphonaptera are parasitic on mammalian hosts of almost all orders, although about 74% of the total known species are found on rodents (Marshall 1981). However, Anoplura and Mallophaga do not occur on the Chiroptera, Pholidota, Cetacea, Sirenia, and most Insectivora. Furthermore, the Anoplura, which are ectoparasites of eutherian mammals, are not found on the Edentata and Proboscidea (Kim 1982; Chapter 5), but the Mallophaga, which are ectoparasites of birds and mammals, do not occur on the Monotremata, Insectivora, Dermoptera, Lagomorpha, Pinnipedia, and Tubulidentata (Emerson and Price 1981). The center of host associations for Anoplura is Rodentia, and for Mallophaga is the Carnivora. The host associations of Mallophaga in Rodentia and Artiodactyla are limited to six families: Geomyidae, Caviidae, Ctenomyidae and Echimyidae (Rodentia), Bovidae and Cervidae (Artiodactyla) (Kim, Chapter 7).

Host Specificity

The ectoparasitic insects on mammals are largely host specific and primarily monoxenous. The Polyctenidae are parasitic on the microchiropterans of five families. They are usually monoxenous, but some species are oligoxenous, specific to a genus-taxon of the hosts: for example, *Eoctenes ferrisi* Maa on *Emballonura* spp.; *E. nycterides* (Horvath) on *Nycteris* spp.; and *Androctenes horvathi* Jordan on *Rhinolophus* spp. (Maa 1964).

The dermapteran arixeniids, *Arixenia* and *Xenaria*, are all monoxenous, being associated only with the hairless molossid bat, *Cheiromeles torquatus.* The dermapteran hemimerids, *Hemimerus* and *Araeomerus,* are mostly monoxenous on murine rodents in Africa: nine species of *Hemimerus* on two species of *Cricetomys* and two *Araeomerus* species on two species of *Beamys* (Marshall 1981).

The parasitic beetles, *Platypsyllus* and *Silphopsyllus,* are monotypic and mostly monoxenous. However, the Holarctic Leptinidae and the Neotropical amblyopinine Staphylinidae are generally oligoxenous (Seevers 1955).

Host specificity among hippoboscids is highly variable. A few obligate, perhaps permanent, species are monoxenous; others are primarily oligoxenous or polyxenous, although the mammalian hippoboscids are more host specific than are avian counterparts (Bequaert 1953–1957). The Nycteribiidae and Streblidae, are usually oligoxenous. In Malaysia and the New Hebrides, monoxeny accounts for 66% of the nycteribiids and 75% of the streblids (Marshall 1976, 1981). In the New World 70% of the 94 streblid species studied were monoxenous and 14% oligoxenous (Wenzel et al. 1966).

The majority of Anoplura and Mallophaga are monoxenous or host specific at the species level. For example, monoxeny accounts for 65% of the Anoplura, 75% of the Echinophthiriidae (Kim 1982), and 70% of the carnivore-infesting Mallophaga. Another 24% of the Anoplura are oligoxenous to two to three host species (Kim 1982).

The Siphonaptera are rarely monoxenous and are more often polyxenous or pleioxenous. The host associations may change with altitude and from one locality to another (Marshall 1981). For example, the bat fleas, which constitute the family Ischnopsyllidae with about 107 species and subspecies, are relatively uncommon in Panama. Of the 95 species of bats representing seven families and 40 genera, only four species of two bat families, Vespertilionidae and Molossidae, harbor four species of ischnopsyllids: *Sternopsylla distincta speciosa* Johnson, *Ptilopsylla dunni* Kohls, *Rhynchopsylla megastigmata* Traub and Gammons, and *Hormopsylla kyriophyla* Tipton and Mendez. They are polyxenous (Tipton and Mendez 1966).

Host specificity among the myiasis-producing Diptera is variable. Approximately 38% of the Oestridae, for example, are monoxenous and another 26% are specific to two hosts (based on data from Papavero 1977). In the Cuterebridae, host specificity is high among the native hosts of *Cuterebra,* but "Torsalo," *Dermatobia hominis,* is exceptionally polyxenous (Catts 1982), probably because host selection is mediated by any of almost 50 carriers of the bot fly's eggs (Guimarães and Papavero 1966).

Ectoparasite Assemblages and Associations

A parasite–host association is a multispecies system involving a certain set of different parasite species and the host, although it is generally considered an interspecific interaction between two organisms, one a parasite (or symbiote) and the other a host (or symbiont). A dynamic assemblage ["parasitocenose" of Pavlovsky (1966)] of symbiotes exists on the host, similar to a terrestrial community, where multispecific interactions mold its structure and function. Thus, in discussing parasite–host associations, we must recognize the three primary interactions among the members of the association. These interactions occur between (1) a parasite population and the individual host, (2) a parasite population and populations of other symbiotes on the host, and (3) a parasite assemblage and the host. The patterns of host associations of parasitic arthropods with mammals is molded by all of these interactions.

Many studies on the relationships between parasitic arthropods and mammals have focused primarily on the interactions between a specific parasite taxon and the host (Nelson et al. 1975, 1977). However, no data currently exist on interactions between populations of different symbiote species coexisting on the same host and the effect of the total symbiote assemblage on the host (Waage 1979). Discussions of the parasite–host associations are generally confined to two-species interactions between a parasite taxon and the host, as if no other parasites exist on the host or the parasites do not affect each other or the host as a community.

Coexistence and Disjunction in Parasite–Host Associations

Symbiote assemblages on mammalian hosts show definite patterns of parasite association, and each assemblage appears to be a manifestation of a long evolutionary history of multispecies interactions involving many different symbiotes and the host. The data on the host associations of parasitic arthropods show coexistence of certain parasites on particular hosts or host groups (Tables 4.3, 4.4). In this context, however, the purported host associations in published records must be scrutinized for validity because many records contain stragglers, contaminants, and even misidentifications of parasite species. Furthermore, most records neither include any information on other associated symbiotes nor provide the data on co-occurrence of these symbiotes on the same host. In fact, the records of parasites on mammalian hosts do not necessarily mean a consistent coexistence of any two or more parasite species on the same host simultaneously. For example, the Neotropical Ctenomyidae is recorded to have Anoplura, Mallophaga, and parasitic Coleoptera (Table 4.3), but this record does not imply the coexistence of these taxa on the same host taxon or individual.

Certain species of remotely related parasitic arthropods are regularly found together on a host taxon or individual, whereas some ectoparasites do not consistently occur with certain other taxa on the same host, although they may be sympatric. To examine the compatibility of associations among recorded ectoparasites, the host associations with the New

Table 4.3   Host Associations of Permanent Parasitic Insects with the New World Rodents[a]

Rodent Families (Biogeographical Region)	Parasitic Insects (Orders)		
Aplodontidae (Na)			COL
Sciuridae (W)	ANO		
Geomyidae (Na)		MAL	
Heteromyidae (NW)	ANO		
Castoridae (Ha)			COL
Cricetidae (W)	ANO		COL
Arvicolidae (Ha)	ANO		
Muridae (W)	ANO		COL
Zapodidae (Ha)	ANO		
Erethizontidae (NW)		MAL	
Caviidae (Nt)	ANO	MAL	
Dasyproctidae (Nt)		MAL	
Chinchillidae (Nt)	ANO	MAL	
Capromyidae (Nt)		MAL	
Octodontidae (Nt)	ANO		
Ctenomyidae (Nt)	ANO	MAL	COL
Abrocomidae (Nt)	ANO	MAL	
Echimyidae (Nt)	ANO	MAL	
Agoutidae (Nt)		MAL	

[a]ANO, Anoplura; MAL, Mallophaga; COL, Coleoptera. Na, Nearctic region; Ha, Holarctic region; Nt, Neotropical region; W, World; NW, New World. No records from the Neotropical Hydrochaeridae, Heptaxondontidae, Dinomyidae, Myocastoridae.

World rodents of the Anoplura, Mallophaga, and parasitic Coleoptera are compared (Table 4.3).

Of the 19 host families harboring parasitic insects, 26% have only Mallophaga, 26% only Anoplura, and 11% only Coleoptera. In other words, 63% have only one order of parasitic arthropods. Two rodent families (11%) have Anoplura and Coleoptera together. The remaining families (26%) are infested with both Anoplura and Mallophaga. When the hosts with multitaxa associations are examined, the host taxon at the generic or specific level usually harbors only one of the three ectoparasitic taxa (Kim, Chapter 7).

On the other hand, certain taxa of remotely related parasites are consistently found together on the host (Table 4.4). On the Carnivora, Perissodactyla, and Artiodactyla in Panama, ticks (Ixodoidea) are commonly found with chewing lice (Mallophaga), sucking lice (Anoplura), or fleas

Table 4.4   Host Associations of Parasitic Arthropods with Carnivora, Perissodactyla, and Artiodactyla in Panama

Hosts	Ectoparasites		
	Acari	Insecta	
Order Carnivora			
Canidae, *Canis*	IX	MA,	SI
Procyonidae, *Procyon*	IX	MA,	SI
Nasua	LA, IX, TR	MA,	SI
Potos	IX	MA	
Bassaricyon	IX		SI
Mustelidae, *Mustela*	IX	MA,	SI
Eira	IX		SI
Galictis	IX	MA,	SI
Conepatus	IX		SI
Lutra	IX		
Felidae, *Felis*	IX	MA,	SI
Order Perissodactyla			
Tapiridae, *Tapirus*	IX		
Equidae, *Equus*	IX	MA	
Order Artiodactyla			
Tayassuidae, *Tayassu*	IX, TR	MA	
Suidae, *Sus*	IX		AN, SI
Cervidae, *Odocoileus*	IX	HI	AN
Mazama	IX, TR,	HI	
Bovidae, *Bos*	IX	MA, AN	
Capra	IX	MA	
Ovis		HI	

Source: From Wenzel and Tipton 1966.
[a] AN, Anoplura; HI, Hippoboscidae (Diptera); IX, Ixodoidea; LA, Laelapidae; MA, Mallophaga; SI, Siphonaptera; TR, Trombiculidae.

(Siphonaptera), but chewing lice usually do not occur with sucking lice or hippoboscids (Table 4.4). Close examination of the multitaxa associations, furthermore, reveals that Anoplura and Mallophaga do not occur together on the same host species.

In certain instances a parasite taxon is exclusively parasitic on a particular mammalian taxon, whereas it is completely absent in other closely related mammalian taxa (Table 4.3). The mallophagan *Geomydoecus* (Trichodectidae), for example, is exclusive parasites of the Geomyidae, whereas the sympatric and closely related Heteromyidae harbors the anopluran *Fahrenholzia* but no mallophagans (Kim, Chapter 7). The mallophagan *Eutrichophilus* is found only on the New World porcupines (Erethizontidae) and is absent from the Old World Hystricidae (Emerson and Price 1981). Certain host associations are obviously related to geographical factors when a parasite taxon is found on several unrelated host groups of the same geographical area. For example, among the South American Gyropidae (Mallophaga), *Gliricola* is found on Caviidae, Capromyidae, and Echimyidae; *Macrogyropus* is on Dasyproctidae and Caviidae; and *Phtheiropoios* is on Chinchillidae and Ctenomyidae (Table 4.3).

We here propose the *Association Index* (A.I.) to indicate the degree of compatibility of symbiote associations. The association index may be computed for a member species A of a symbiote assemblage: the number of host taxa that harbors both symbiote A and another symbiote such as B divided by the number of host taxa that harbors symbiote A. The association indices for ectoparasites on Panamanian bats and rodents are shown in Tables 4.5 and 4.6.

Of the symbiotes associated with Panamanian bats, the Streblidae, followed by the Spinturnicidae, are the most common ectoparasites. The least common ectoparasites are Polyctenidae, Ischnopsyllidae, and Spelaeorhynchidae. The Dermanyssidae are most likely to be associated with the Spinturnicidae and Streblidae, and the Spinturnicidae are most likely to be associated with the Streblidae. The Spelaeorhynchidae are commonly associated with all other ectoparasites except the Polyctenidae. Both ticks and chiggers (Trombiculidae) are commonly associated with Spinturnicidae and Streblidae. The Nycteribiidae seem to have a reasonable likelihood of coexistence with other ectoparasites except the Polyctenidae (Table 4.5). The record of host associations shows that the Polyctenidae are confined to molossid bats and that the Ischnopsyllidae are opportunistic species in a specific habitat. In contrast, ticks, although they are opportunistic, have a greater association with Spinturnicidae and Streblidae than others.

The Panamanian rodents of 11 families and 23 genera are host to Laelapidae, Dermanyssidae, Ixodoidea, Trombiculidae (Acarina), *Amblyopinus* (Staphylinidae), Anoplura, Mallophaga, and Siphonaptera (Table 4.6). Fleas and chiggers are the most common ectoparasites, although other taxa, except parasitic beetles, *Amblyopinus*, are also frequently found on rodents. Fleas and chiggers do not readily associate with *Amblyopinus* and chewing lice (Mallophaga), but commonly coexist with other ectoparasites. In fact, parasitic beetles and chewing lice are consistently the least compatible with other ectoparasites. Chewing lice are least compatible with sucking lice. *Amblyopinus* is limited to the cricetid *Oryzomys, Reithrodontomys,* and *Peromyscus* in specific habitats, and, when present, these parasitic beetles coexist reasonably well with other ectoparasites.

As the foregoing discussion demonstrates, two definite patterns exist in the host associations of parasitic arthropods: (1) host specificity and (2)

Table 4.5 Association Indexes (A.I.)[a] of the Ectoparasites on the Panamanian Bats (Chiroptera) of Seven Families and 38 Genera

Ectoparasites	Host Genera Infested		A.I.								
	Number	Percent	DE	SP	IX	SL	TR	NY	ST	PO	IS
Dermanyssidae (DE)	9	0.24	+	0.78	0.44	0.33	0.56	0.22	0.89	0.11	0.33
Spinturnicidae (SP)	21	0.55	0.33	+	0.38	0.10	0.38	0.10	0.81	0	0.05
Ixodoidea (IX)	12	0.32	0.33	0.67	+	0.08	0.33	0.08	0.92	0	0.08
Spelaeorhynchidae (SL)	3	0.08	1.00	0.67	0.33	+	1.00	0.33	1.00	0	0.33
Trombiculidae (TR)	12	0.32	0.42	0.67	0.33	0.25	+	0.17	0.83	0	0.08
Nycteribiidae (NY)	5	0.13	0.40	0.40	0.20	0.20	0.40	+	0.60	0	0.20
Streblidae (ST)	31	0.82	0.26	0.55	0.35	0.10	0.32	0.10	+	0.03	0.06
Polyctenidae (PO)	2	0.05	0.50	0	0	0	0	0	0.50	+	0.50
Ischnopsyllidae (IS)	2	0.05	1.00	0.50	0.50	0	0.50	0.50	1.00	0.50	+

Source: From Wenzel and Tipton 1966.

[a] A.I., number of host genera with ectoparasites A and B divided by the number of host genera with ectoparasite A.

Table 4.6 Association Indexes (A.I.)[a] of the Ectoparasites on the Panamanian Rodents of 11 Families and 23 Genera

Ectoparasites	Host Genera Infested		A.I.							
	Number	Percent	LA	DE	IX	TR	CO	AN	MA	SI
Laelapidae (LA)	16	0.70	+	0.69	0.69	0.88	0.19	0.69	0.19	0.94
Dermanyssidae (DE)	12	0.52	0.92	+	1.00	1.00	0.08	0.83	0.33	1.00
Ixodoidea (IX)	15	0.65	0.80	0.67	+	0.93	0.20	0.00	0.27	0.93
Trombiculidae (TR)	18	0.78	0.72	0.67	0.78	+	0.11	0.61	0.22	0.89
Amblyopinus (CO)	3	0.13	1.00	0.33	1.00	0.66	+	0.66	0	1.00
Anoplura (AN)	11	0.48	1.00	0.91	0.82	1.00	0.27	+	0.18	1.00
Mallophaga (MA)	7	0.30	0.43	0.57	0.57	0.57	0	0.29	+	0.71
Siphonaptera (SI)	20	0.87	0.80	0.60	0.70	0.80	0.15	0.60	0.25	+

Source: Data from Wenzel and Tipton 1966.

[a] A.I., number of host genera with ectoparasites A and B divided by the number of host genera with ectoparasite A.

association specificity. In host specificity, a certain parasite taxon is consistently associated with a particular mammalian host at different taxonomic levels. This association may be parasite species versus host species, parasite species versus host genus, parasite genus versus host family, or parasite family versus host order. In association specificity, a certain parasite taxon is often consistently associated with other parasites on the same host (individual species or genus) whereas it does not coexist with certain other parasites. Both patterns are the results of evolutionary and ecological processes (Kim, Chapter 1).

SUMMARY

Parasitic insects associated with mammals occupy a unique environment, whether on or in the host. Seven orders include insects that are ectoparasites of mammals; only one insect order, Diptera, includes mammalian endoparasites that produce myiasis. This chapter deals primarily with the Dermaptera, Hemiptera, Coleoptera, and Diptera, although general patterns of parasitism are discussed by including lice and fleas as well.

The insects that exploit the external adaptive zone of the host share a number of morphological characters: flattened body; prognathous head; reduced compound eyes, ocelli, antennae, and wings; modified forelegs (as holdfasts) and setae (as combs). The insects inhabiting the internal or subdermal environment of the host do so as larvae that are typically maggotlike with well-developed mouthparts and body armature, whereas the corresponding adults exhibit characters for efficient location of mates and hosts, which include well-developed eyes, strong wing musculature, and reduced mouth parts and antennae. Biological characters such as the reproductive strategy also exhibit patterns related to the degree of association with the host. For example, permanent parasites like the Polyctenidae and Anoplura produce live young or a small number of eggs, whereas parasites such as fleas that are less intimately associated with the host produce many eggs.

The host associations of parasitic insects are generally centered around a particular taxon of mammals. The Polyctenidae, Arixeniidae, Nycteribiidae, and Streblidae are confined to the Chiroptera. The majority of Siphonaptera, Cuterebridae, Anoplura, parasitic Coleoptera, and all hemimerid Dermaptera are parasitic on the Rodentia. The Hippoboscidae, Gasterophilidae, Hypodermatidae, Oestridae, and some opportunistic Calliphoridae and Sarcophagidae are primarily parasitic on the large herbivores, particularly the Artiodactyla. At the species level most parasitic insects are monoxenous or oligoxenous, although opportunistic or less permanent insects such as fleas and the human bot fly tend toward polyxeny.

Certain taxa of parasitic arthropods are regularly found together whereas others are not. These relationships may occur at the family, generic, or species level of the host. We propose the following formula, termed the Association Index, for assessing the degree of coexistence: the number of host genera (or other taxon) with ectoparasites A and B divided by the number of host genera (or other taxon) with ectoparasite A. The relationships of symbiote associations were investigated by computing the A.I. for paired assemblages of ectoparasites on Panamanian bats and rodents. Of the symbiotes associated with Panamanian bats, the Dermanyssidae are most likely to be associated with the Spinturnicidae and Streblidae. The Spelaeorhynchidae and Nycteribiidae are likely to be associated with other ectoparasites except Polyctenidae. In the Panamanian parasitic arthropod–rodent system, fleas and chiggers, although they are most common, do not readily associate with *Amblyopinus* or chewing lice but commonly coexist with other ectoparasites. Parasitic beetles and chewing lice are least compatible with other ectoparasites, and chewing lice usually do not coexist with sucking lice.

ACKNOWLEDGMENTS

We thank Drs. E. Paul Catts, and Charles W. Pitts for kindly reading the manuscript and giving us their valuable comments. We are especially indebted to E. Paul Catts for his expert opinions and additions on parasitic Diptera. Our appreciation also goes to Judi Hicks and Thelma Brodzina, Department of Entomology, Pennsylvania State University for their typing efforts.

REFERENCES

Al-Abbasi, S. H. 1981. *Comparative morphology and phylogenetic significance of antennal sense organs in parasitic Psocodea (Insects).* Ph.D. Thesis. The Pennsylvania State University, University Park, PA. 158 pp.

Amin, O. M. and M. E. Wagner. 1983. Further notes on the function of pronotal combs in fleas (Siphonaptera). *Ann. Entomol. Soc. Am.* **76**:232–234.

Arnett, R. H. 1968. *The Beetles of the United States (A Manual for Identification).* The American Entomological Institute, Ann Arbor, Michigan. 1112 pp.

Ashford, R. W. 1970. Observations on the biology of *Hemimerus talpoides* (Insecta; Dermaptera). *J. Zool. Lond.* **162**:413–418.

Bänziger, H. 1972. Biologie der lacriphagen Lepidopteren in Thailand und Malaya. *Rev. Suisse Zool.* **72**:1381–1469.

Bänziger, H. 1979. Skin-piercing blood-sucking moths II: Studies on a further 3 adult *Calyptra* [*Calpe*] sp. (Lepid., Noctuidae). *Acta Tropica* **36**:23–37.

Bänziger, H. and W. Büttiker. 1969. Records of eye-frequenting Lepidoptera from man. *J. Med. Entomol.* **6**:53–58.

Barrera, A. 1966. Hallazgo de *Amblyopinus tiptoni* Barrera, 1966 en Costa Rica, A.C. (Colo: Staph.). *Acta Zool. Mex.* **8**:1–3.

Bequaert, J. C. 1953–1957. The Hippoboscidae or louse-flies (Diptera) of mammals and birds. Part I. Structure, physiology, and natural history. Part II. Taxonomy, evolution, and revision of American genera and species. *Entomol. Am.* **32**(n.s.):1–209; **33**(n.s.):211–422; **34**(n.s.):1–232; **35**(n.s.):233–416; **36**(n.s.):417–611.

Catts, E. P. 1982. Biology of new world bot flies: Cuterebridae. *Annu. Rev. Entomol.* **27**:313–338.

Deoras, P. J. 1941. Structure of *Hemimerus deceptus* Rehn var. *ovatus*, an external parasite of *Cricetomys gambiense*. *Parasitology* **33**:172–185.

Emerson, K. C. and R. D. Price. 1981. A host–parasite list of the Mallophaga. *Misc. Publ. Entomol. Soc. Am.* **12**:1–72.

Ferris, G. F. and R. L. Usinger. 1939. The family Polyctenidae (Hemiptera: Heteroptera). *Microentomology* **4**:1–50.

Guimarães, J. H. and N. Papavero. 1966. A tentative annotated bibliography of *Dermatobia hominis* (Linnaeus Jr., 1781) (Diptera, Cuterebridae). *Arq. Zool.* (S. Paulo) **14**:223–294.

Grunin, K. Y. 1966. Oestridae (Fam. 64a). Pages 1–96 in E. Lindner (Ed.), *Die Fliegen der paläarktischen Region*. Stuttgart.

Hagan, H. R. 1931. The embryogeny of the polyctenid, *Hesperoctenes fumarius* Westwood, with reference to viviparity in insects. *J. Morphol. Physiol.* **51**:1–117.

Holloway, B. A. 1976. A new bat-fly family from New Zealand (Diptera: Mystacinobiidae). *N. Z. J. Zool.* **3**:279–301.

Humphries, D. A. 1966. The function of combs in fleas. *Entomol. Mon. Mag.* **102**:232–236.

James, M. T. 1947. The flies that cause myiasis in man. *U.S. Dep. Agric. Misc. Publ.* **631**:1–175.

Jeannel, R. 1922. Morphologie camparée due *Leptinus testaceus* Müll. et du *Platypsyllus castoris* Rits. *Arch. Zool. Exp. Gen.* **60**:557–592.

Jobling, B. 1929. A comparative study of the structure of the head and mouthparts in the Streblidae (Diptera: Pupipara). *Parasitology* **21**:417–445.

Jobling, B. 1936. A revision of the subfamilies of the Streblidae and the genera of the Streblinae (Diptera, Acalypterae) including a redescription of *Metelasmus pseudopterus* Coquillet and a description of two new species from Africa. *Parasitology* **28**:355–380.

Jobling, B. 1939. On the African Streblidae (Diptera, Acalypterae) including the morphology of the genus *Ascodipteron* Adens and a description of a new species. *Parasitology* **31**:147–165.

Jobling, B. 1949. A revision of the genus *Aspidoptera* Coquillet, with some notes on the larva and the puparium of *A. clovisi*, and a new synonym (Diptera, Streblidae). *Proc. R. Entomol. Soc. Lond.* (B) **18**:135–144.

Kim, K. C. 1982. Host specificity and phylogeny in Anoplura. Deux. symp. Specif. Parasit. d. parasites Vert. 13–17. Avril 1981. *Mem. Mus. Nat. Hist. Nat.*, Nouv. Ser. A, Zool., **123**:123–127.

Maa, T. C. 1959. The family Polyctenidae in Malaya (Hemiptera). *Pacif. Ins.* **1**:415–422.

Maa, T. C. 1963. Genera and species of Hippoboscidae (Diptera): types, synonymy, habitats, and natural groupings. *Pacif. Inst. Monogr.* **6**:1–186.

Maa, T. C. 1964. A review of the Old World Polyctenidae (Hemiptera: Cimicoidea). *Pacif. Inst.* **6**:494–516.

Maa, T. C. 1966. Studies in Hippoboscidae (Diptera). *Pacif. Inst. Monogr.* **10**:1–148.

Maa, T. C. 1969. Studies in Hippoboscidae (Diptera). Part 2. *Pacif. Inst. Monogr.* **20**:1–312.

Machado-Allison, C. E. and A. Barrera. 1972. Venezuelan Amblyopinini (Inecta: Coleoptera, Staphylinidae). *Brigham Young Univ. Sci. Bull.*, *Biol. Sci.* **17**:1–14.

Marshall, A. G. 1970. The life cycle of *Basilia hispida* Theodor 1967 (Diptera: Nycteribiidae) in Malaysia. *Parasitology* **61**:1–18.

Marshall, A. G. 1976. Host specificity amongst arthropod ectoparasites upon mammals and birds in the New Hebrides. *Ecol. Entomol.* **1**:189–199.

Marshall, A. G. 1977. Interrelationships between *Arixenia esau* (Dermaptera) and molossid bats and their ectoparasites in Malaysia. *Ecol. Entomol.* **2**:285–291.

Marshall, A. G. 1980. The function of combs in ectoparasitic insects. Pages 79–87 in R. Traub and H. Starcke (Eds.), *Fleas—Proceedings of the International Conference on Fleas, June 1977.* Balkema, Rotterdam.

Marshall, A. G. 1981. *The Ecology of Ectoparasitic Insects.* Academic Press, London. 459 pp.

Marshall, A. G. 1982. The ecology of the bat ectoparasite *Eoctenes spasmae* (Hemiptera: Polycteridae) in Malaysia. *Biotropica* **14**:50–55.

Meester, J. and H. W. Setzer. 1971. *The Mammals of Africa: an Identification Manual.* Smithsonian Institution Press, Washington, D.C.

Nakata, S. and T. C. Maa. 1974. A review of the parasitic earwigs (Dermaptera; Arixeniina; Hemimerina). *Pacif. Inst.* **16**:307–374.

Nelson, W. A., J. E. Keirans, J. F. Bell, and C. M. Clifford. 1975. Host–ectoparasite relationships. *J. Med. Entomol.* **12**:143–166.

Nelson, W. A., J. F. Bell, C. M. Clifford, and J. E. Keirans. 1977. Interaction of ectoparasites and their hosts. *J. Med. Entomol.* **13**:389–428.

Popham, E. J. 1962. The anatomy related to the feeding habits of *Arixenia* and *Hemimerus* (Dermaptera). *Proc. Zool. Soc. Lond.* **139**:429–450.

Papavero, N. 1977. *The World Oestridae (Diptera), Mammals and Continental Drift.* Vol. 14, Series Entomologica. (E. Schimitschek and K. A. Spencer, Eds.). W. Junk, The Hague, Netherlands.

Pavlovsky, E. N. 1966. *Natural Nidality of Transmissible Diseases.* University of Illinois Press, Urbana.

Rehn, J. A. G. and J. W. H. Rehn. 1936. A study of the genus *Hemimerus* (Dermaptera, Hemimerina, Hemimeridae). *Proc. Acad. Nat. Sci. Phila.,* **87**:457–508.

Schwardt, H. H. and J. G. Matthysse. 1948. The sheep tick (*Melophagus ovinus* L.), materials and equipment for its control. *Cornell Univ. Agric. Exp. Stn. Bull.* **844**:1–33.

Scott, H. 1917. Notes on Nycteribiidae, with descriptions of two new genera. *Parasitology* **9**:593–610.

Seevers, C. H. 1944. A new subfamily of beetles parasitic on mammals. Staphylinidae, Amblyopininae. *Field Mus. Nat. Hist., Zool. Ser.* **28**:155–172.

Seevers, C. H. 1955. A revision of the tribe Amblyopinini: staphylinid beetles parasitic on mammals. *Fieldiana, Zool.* **37**:211–264.

Theodor, O. 1964. On the relationships between the families of the Pupipara. *Proc. 1st Congr. Parasit.* **1**:999–1000.

Theodor, O. 1967. *An Illustrated Catalogue of the Rothschild Collection of Nycteribiidae (Diptera) in the British Museum (Natural History).* British Museum of Natural History, London. 506 pp.

Theodor, O. and A. Moscoma. 1954. On bat parasites in Palestine. I. Nycteribiidae, Streblidae, Hemiptera, Siphonaptera. *Parasitology* **44**:157–245.

Traub, R. 1972a. The relationship between spines, combs and other skeletal features of fleas (Siphonaptera) and the vesture, affinities and habits of their hosts. *J. Med. Entomol.* **9**:601.

Traub, R. 1972b. Notes on zoogeography, convergent evolution and taxonomy of fleas (Siphonaptera), based on collections from Bunong Benom and elsewhere in South-east Asia. III. Convergent evolution. *Bull. Br. Mus. Nat. Hist. (Zool.)* **23**:307–387.

Traub, R. 1980. Some adaptive modifications in fleas. Pages 33–67 in R. Traub and H. Starke (Eds.), *Fleas—Proceedings of the International Conference on Fleas, June 1977*. Balkema, Rotterdam.

Tipton, V. J. and E. Mendez. 1966. The fleas (Siphonaptera) of Panama. Pages 289–385 in R. L. Wenzel and V. J. Tipton (Eds.), *Ectoparasites of Panama*. Field Museum of Natural History, Chicago.

Waage, J. K. 1979. The evolution of insect/vertebrate associations. *Biol. J. Linn. Soc.* **12**:187–224.

Waage, J. K. and G. G. Montgomery. 1976. *Cryptoses choloepi:* a coprophagous moth that lives on a sloth. *Science* **193**:157–158.

Wenzel, R. L. 1976. The streblid batflies of Venezuela (Diptera: Streblidae). *Brigham Young Univ. Sci. Bull., Biol. Ser.* **20**:1–177.

Wenzel, R. L., V. J. Tipton, and A. Kiewlicz. 1966. The streblid batflies of Panama (Diptera Calypterae: Streblidae). Pages 405–675 in R. L. Wenzel and V. J. Tipton (Eds.), *Ectoparasites of Panama*. Field Museum of Natural History, Chicago.

Wood, D. M. 1965. Studies on the beetles *Leptinillus validus* (Horn) and *Platypsyllus eastoris* Ritsema (Colloptera: Leptinidae) from beaver. *Proc. Entomol. Soc. Ont.* **95**:33–63.

Zumpt, F. 1965. *Myiasis in Man and Animals in the Old World*. Butterworths, London. 267 pp.

Chapter 5

Haematopinus eurysternus Denny

Evolution and Host Associations of Anoplura

Ke Chung Kim

Authorized for publication as Paper No. 6790 in the Journal Series of the Pennsylvania Agricultural Experiment Station.

INTRODUCTION

The Anoplura are closely associated with mammals. These host associations are the manifestation of long evolutionary interactions between Anoplura and mammals. Hence an analysis of the host associations and specificity of Anoplura could provide important clues to the phylogenetic relationships, historical distribution, and perhaps coevolution of the sucking lice and mammals (Hopkins 1957; Vanzolini and Guimaraes 1955; Kim et al. 1975; Traub 1980).

The Anoplura are wingless insects permanently parasitic on eutherian mammals. They inhabit the skin-fur environment of the host body surface and feed exclusively on blood. Their entire life cycle from egg to adult is spent on a single host. All sucking lice glue their eggs on host hairs except *Pediculus humanus* Linn., which attaches its eggs on the fibers of clothing. After eclosion, the sucking louse passes through three nymphal stages. Each nymphal instar usually has distinct morphological characters including definitive chaetotaxy and setal density. The structural parts of the nymphal instars differ in size or proportion. The development of sclerotization and chaetotaxy in successive instars also varies. The second and third instars usually resemble their adults.

The living diversity of sucking lice is estimated to be around 1000 species. However, only about one-half of the diversity, about 500 species, has been described so far, from approximately 32% of the 2600 likely host species in the world. They are grouped into 45 genera and 15 families (Kim and Ludwig 1978a).

Anoplura are distributed throughout the world. They are especially rich in the Ethiopian region, where approximately 35% of all known Anoplura have been reported (Ludwig 1968; Kim 1982). They are found on all major groups of eutherian mammals except the Cetacea, Chiroptera, Edentata, Pholidota, Proboscida, and Sirenia. They are absent from the Monotremata, Marsupialia, and most of the land Carnivora (Fissipedia). The Insectivora are poorly colonized by the Anoplura.

A comprehensive treatise on the host associations of the sucking lice on mammals by Hopkins (1949) provided the basis for subsequent analyses and interpretations of the host associations and specificity of Anoplura (Hopkins 1957; Kim 1982).

Anoplura are highly host specific to particular groups of mammals, being obligate, permanent parasites. Their host specificity is even higher than that of the Mallophaga (Kim 1982; Ludwig 1982), contrary to the commonly accepted view that mallophagans are highly host specific whereas sucking lice are not (Hopkins 1949, 1957). The sucking lice may be specific to certain closely related hosts at the generic, family, or even ordinal level. Thus host specificity may be discussed in terms of host associations for different taxonomic levels. Host specificity for a given taxon is expressed by percent specificity, defined as "the percent of species that utilize only one host"

(Price 1980). However, consideration of Price's "speciation index" is unjustified at this time because only about one-half of the world fauna of Anoplura is currently known and the taxonomy of lice on most host groups is not complete.

This chapter is based on the taxonomic and host data collected for a monograph of the Anoplura through 1978 (Kim and Ludwig 1983). Obvious stragglers and contaminants as well as misidentifications were not included in the analysis. The higher classification of the Anoplura as developed by Kim and Ludwig (1978a), the mammal classification of Anderson and Jones (1967) and Simpson (1945), and the specific and generic names of Honacki et al. (1982) serve as the taxonomic basis for this chapter.

ADAPTATIONS AND EVOLUTIONARY TRENDS

Sucking lice are highly adapted to the skin-fur habitat on mammals, and numerous specialized adaptations maximize their parasitic and bloodsucking modes of life. The characters related to their parasitism in the fur environment on mammals include (1) dorsoventral flattening of the entire body, (2) prognathy of the head, (3) reduction of antennal segments, (4) development and rearrangement of the antennal sensilla, (5) reorganization of the thoracic segments, (6) reduction in the number of thoracic and abdominal spiracles, (7) reduction of the tarsal segments and loss of pulvilli, and (8) shortening of the number or duration of immature stages. However, these adaptations are also common among many other insects parasitic on mammals, such as the Mallophaga, parasitic dipterans (Hippoboscidae, Nycteribiidae, and Streblidae), and parasitic beetles (*Platypsyllus*). The following morphological trends are related to bloodsucking and are unique to the Anoplura: (1) complete lack of the tentorium; (2) development of efficient piercing-sucking mouthparts with three stylets in the trophic sac; (3) reorganization of the thorax, with all thoracic segments fused, the dorsum composed of primarily plural components, and with a pair of dorsal mesothoracic spiracles; (4) reduction of tarsal segments to one-segmented tarsi, each with one claw; (5) reduction of sternites and tergites of the abdomen; and (6) reduction of the number of abdominal spiracles (six pairs or fewer) and paratergal plates.

Body Size

Body size varies greatly among anopluran taxa, ranging from 0.35 mm in *Microphthirus uncinatus* Ferris (Enderleinellidae) to more than 8 mm for *Pecaroecus javalii* Babcock and Ewing (Pecaroecidae). Although its adaptive significance is not known, the size of sucking lice seems to characterize each taxon and to be related to the body size of mammalian hosts (Table 5.1). Large lice are found on relatively large hosts: *Haematopinus* and

Table 5.1   General Morphological Trends in Anoplura

Characters	Large Lice	Small Lice
Body length	> 2 mm	< 2 mm
Eyes or ocular points	Present	Lacking
Midlegs subequal to hindlegs	Yes	No
Paratergites	Lacking, caplike or tuberculiform	Plate-shaped with its apex free
Abdominal setae	Short, dense, or modified	Long, sparse
Abdominal sclerites	Lacking	Highly developed
Hosts	Large mammals	Small mammals
Examples	Haematopinidae, Pecaroecidae, and Microthoracidae on Artiodactyla; Pediculidae on anthropoid Primates; Echinophthiriidae on Carnivora	Enderleinellidae, Hoplopleuridae, Polyplacidae, and Neolinognathidae on Rodentia and Insectivora

Pecaroecus on ungulates, *Pediculus* on anthropoid Primates, *Microthoracius* on camelids, *Hybophthirus* on aardvarks, and Echinophthiriidae on pinnipeds. They usually lack abdominal sclerites with paratergites either caplike or tuberculiform, if present, but have relatively short setae in high density (Kim and Ludwig 1978a). Small lice—Enderleinellidae, Hoplopleuridae, most Polyplacidae, and *Neolinognathus*—are confined to small mammals like rodents and insectivores. They usually have long abdominal setae in regular rows and relatively well-developed abdominal sclerites with platelike paratergites having apices free from the body (Kim and Ludwig 1978a). Other sucking lice of intermediate or medium size are found on diverse host groups: Linognathidae on Artiodactyla, Canidae (Carnivora), and hyraxes; *Ratemia* on horses; *Pedicinus* on Cercopithecidae (Primates); *Pthirus* on Hominidae and Pongidae (Primates); and *Hamophthirius* on flying lemurs (Dermoptera).

Setae and Tubercles

The setae in the Anoplura are usually of the common type, varying in length (Fig. 5.1*a*), but are secondarily modified to serve different purposes

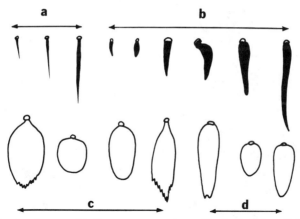

Figure 5.1 Types of setae. (*a*) Common setae (e.g., Hoplopleuridae, Polyplacidae); (*b*) spiniform setae and pegs (e.g., Echinophthiriidae); (*c*) scales (e.g., *Antarctophthirus*); (*d*) scales (e.g., *Lepidophthirus*).

in diverse taxa (Fig. 5.1*a*–*d*) (Kim and Ludwig 1978a). Some thoracic and abdominal setae are modified into scales and spines in *Antarctophthirus* (Fig. 5.1*c*), *Lepidophthirus* (Fig. 5.1*b*), and *Latagophthirus* (all parasitic on aquatic carnivores), as in *Ctenophthirus* (Fig. 5.2*c*) which is found on Echimyidae (Rodentia). Only the body spines are developed in *Echinophthirius* (Fig. 5.1*b*). Strong spines are developed on the head of many taxa, including nymphs of *Proechinophthirus fluctus* (Ferris) and all instars of *P. zumpti* Werneck (Fig. 5.2*a*, *b*).

Strong tubercles, which appear to be holdfast structures, are developed on the head and antennal segments in the sucking lice parasitic on mammals with fine hairs: *Microphthirus* (nymphs and adults) on *Glaucomys* (Sciuridae) (Fig. 5.3*b*); *Mirophthirus* (nymphs and adults) on *Typhlomys* (Cricetidae); *Eulinognathus* (adults) on various rodents; *Phthirpediculus* (adults) on Indriidae (Primates); *Docophthirus* (adults) on Tupaiidae (Scandentia) (Fig. 5.3*c*); and *Hamophthirius* (adults) on Cynocephalidae (Dermoptera). Nymphs of many hoplopleurids and some polyplacids have small, pointed tubercles scattered on the ventral side of the head and antennae (Fig. 5.3), and their primary hosts are rodents: *Hoplopleura* (Fig. 5.3*a*) and *Pterophthirus* (Hoplopleuridae), and *Phthirpediculus, Fahrenholzia, Johnsonpthirus,* and *Linognathoides. Sathrax* has large ventral tubercles on the head in both nymphs and adults (Fig. 5.3*e*).

Neolinognathus (Neolinognathidae), which parasitizes primitive Macroscelididae, has sclerotized tubercles of varying sizes on both sides of the abdomen (Fig. 5.3*f*). In this taxon the number of abdominal spiracles is drastically reduced to a single large pair on segment 8 without any evidence of paratergites.

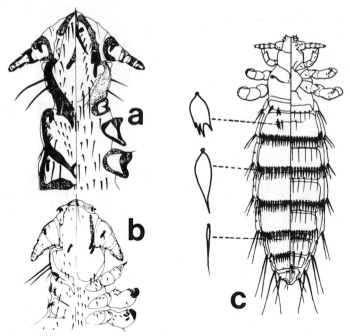

Figure 5.2 Specialized setae. (*a*) Adult head and thorax of *Proechinophthirus zumpti* Werneck; (*b*) head and thorax of *P. zumpti* nymph; (*c*) *Ctenophthirus cercomydis* Ferris, male, showing specialized setae.

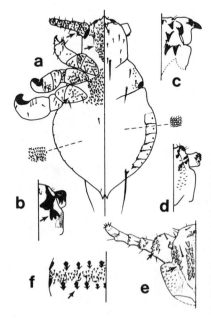

Figure 5.3 Tubercles. (*a*) Nymph 2 of *Hoplopleura johnsonae* Kim showing ventral tubercles on head, antennae, and coxae; (*b*) ventral tubercles on the head and the basal antennal segment of *Microphthirus uncinatus* (Ferris); (*c*) ventral tubercles on the head and the basal antennal segment of *Docophthirus ancinetus* Waterston; (*d*) ventral tubercles on the head and the basal antennal segment of *Ctenophthirus cercomydis* Ferris; (*e*) ventral tubercles on the head and antenna of *Sathrax durus* Johnson; (*f*) abdominal tubercles of *Neolinognathus elephantuli* Beford.

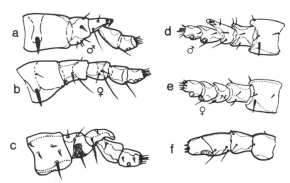

Figure 5.4 Antennae. (*a*) *Neohaematopinus sciuropteri* (Osborn), male; (*b*) *N. sciuropteri*, female; (*c*) *Polyplax brachyuromyis* Kim and Emerson, male; (*d*) *P. spinulosa* (Burmeister), male; (*e*) *P. spinulosa*, female; (*f*) *Ancistroplax crocidurae* Waterston, male.

Antennae

Antennae are largely five-segmented, but the number of segments are occasionally reduced to three or four in the higher taxa (Fig. 5.4). A reduction of antennal segments is found in the highly specialized taxa of Echinophthiriidae, Linognathidae, and Enderleinellidae: *Echinophthirius* on hair seals (Phocidae), *Proechinophthirus* on fur seals (Arctocephalinae, Otariidae), and *Latagophthirus* on river otters (Mustelidae). Furthermore, two hoplopleurids parasitic on Insectivora have reduced antennal segments: *Haematopinoides* on moles (Talpidae) and *Ancistroplax* on shrews (Soricidae) (Fig. 5.4*f*). They are evolutionary transfers from rodent hosts.

Sexual dimorphism in antennae is striking in a number of polyplacid taxa where the third antennal segment of male adults is usually modified with spiniform setae (Fig. 5.4): *Neohaematopinus* (Fig. 5.4*a*, *b*), *Johnsonpthirus*, *Linognathoides* (Kim and Adler 1982), *Sathrax*, *Polyplax* (Fig. 5.4*c*, *d*, *e*), *Lemurphthirus*, and *Mirophthirus* (Chin 1980).

Eyes and Ocular Points

Compound eyes are usually lacking in Anoplura (Fig. 5.5*a*, *b*). However, five taxa parasitic on Artiodactyla and Primates still retain distinct compound eyes (Fig. 5.5*d*): *Microthoracius* on Camelidae and *Pecaroecus* on Tayassuidae (Artiodactyla); and *Pediculus*, *Pthirus*, and *Pedicinus* on anthropoid Primates. The two large taxa *Haematopinus* and *Hybophthirus* have large ocular points instead (Fig. 5.5*c*).

Legs

Anopluran forelegs are usually smaller than other legs, and the midlegs and hindlegs of certain taxa are often modified as holdfast structures. All

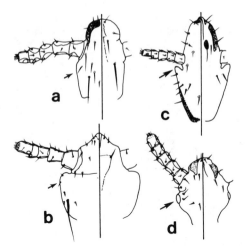

Figure 5.5 Head: (*a, b*) generalized, (*c*) with ocular points, (*d*) with eyes. (*a*) *Linognathus panamensis* Ewing; (*b*) *Hoplopleura johnsonae* Kim; (*c*) *Haematopinus quadripertusus* Fahrenholz; (*d*) *Pediculus schaeffi* Fahrenholz.

legs of the large lice *Haematopinus, Microthoracius, Pecaroecus,* and *Pediculus* are relatively similar in size and shape and have acuminate claws. In the small lice, particularly those parasitic on small mammals, the midlegs and hindlegs are highly modified, as in *Hoplopleura* and *Polyplax* on rodents (Kim and Ludwig 1978a).

ANALYSIS OF THE HOST–PARASITE ASSOCIATIONS

Certain species of the Anoplura are consistently found on particular mammalian taxa, and certain mammalian taxa at the generic, family, and even ordinal level harbor closely related species or particular groups of lice. No Anoplura are known from the Monotremata and Marsupialia, whereas marsupials are infested with the amblyceran Mallophaga. The Insectivora are poorly infested with parasitic Psocodea including Anoplura. The diversity relationship between the Anoplura and their mammalian hosts is summarized in Table 5.2.

Because of the obligate and permanent nature of Anoplura parasitism and the relatively high degree of host specificity, analysis of the associations of Anoplura and mammals reveals significant facts regarding their phylogenetic relationships (Hopkins 1949, 1957; Kim and Ludwig 1978a, b) and historical distribution (Hopkins 1949, 1957; Patterson 1957; Vanzolini and Guimaraes 1955; Traub 1980). However, certain inherent limitations must be recognized and considerable care should be taken in the use of such data because (1) only about one-half of the world Anoplura fauna is presently known, (2) it is often impossible to determine what the lack of reports about sucking lice related to specific hosts in the literature really means (whether it is true absence of lice on the hosts, or simply due to inadequate examination), and (3) many erroneous or unreliable reports exist in the literature owing to improper handling of specimens or contamination. Furthermore, the phylogenetic reasoning could easily become circular when the host associations are used to infer phylogeny of the sucking lice and then the relationships of the lice are employed to make quasiphylogenetic inferences about mammalian evolution (Kim and Ludwig 1978a).

Insectivora and Anoplura

The Insectivora are at present infested with endemic taxa of different phylogenetic origins: *Neolinognathus, Sathrax, Docophthirus, Haematopinoides, Ancistroplax,* and evolutionary emigrants (or "host changes"), such as a few species of *Polyplax.* Of six extant lipotyphlan families, the Talpidae (moles), Chrysochloridae (golden moles), and Soricidae (shrews) are infested with rodent-infesting Hoplopleuridae and Polyplacidae. A North American hoplopleurid, *Haematopinoides squamosus* (Osborn), infests *Scalopus* and *Parascalops* (Talpidae). *Johnsonpthirus chlorotalpae* (Benoit) has been recorded from Central African *Chrysochloris* (Benoit 1961), although a true association is questionable (Kim and Adler 1982). If it is a true association, this represents a recent host change. The multispecies infestation of Soricidae includes five species of *Polyplax* (Polyplacidae) and two species of *Ancistroplax* (Hoplopleuridae) and is confined to *Crocidura, Sorex, Scutisorex,* and *Suncus.* No sucking lice are known from the three remaining lipotyphlans. Taxonomy and distribution of Hoplopleuridae and Polyplacidae suggest that the sucking lice invaded lipotyphlan insectivores secondarily, perhaps after the Tertiary diversification of the hosts. This supports Traub's view (1980) that no Anoplura were present when the Insectivora arose in Asia, perhaps during the Cretaceous period (McKenna 1970).

Two menotyphlan Insectivora, Macroscelididae (elephant shrews) and Tupaiidae (tree shrews), are infested with three highly specialized lice: *Neolinognathus* (Neolinognathidae), which is restricted to the Macroscelididae, primarily *Elephantulus* and *Nasilio; Sathrax,* which is found on *Tupaia,* and *Docophthirus* on *Anathana* are restricted to the Tupaiidae. They are endemic to the menotyphlan Insectivora but show close affinity to the typical forms of Polyplacidae. This suggests that their associations are primary and the progenitors of these lice were associated with rodents.

Dermoptera and Hamophthiriidae

The monotypic family Cynocephalidae ("flying lemurs," colugos) has two extant species and is the host of the monotypic *Hamophthirius,* a highly specialized louse of the linognathid line. Considering the recency of the

Table 5.2 Diversity Relationship beween Anoplura and Mammals (up to 1978).

Family	Anoplura Genus	Species	Order	Infested Mammals[a] Family	Genus	Species	Mammals of World[b] (Genus-Species)
Echinophthiriidae	4	11	Pinnipedia	3	17	17	20–31
	1	1	Carnivora	1	1	1	25–70
Enderleinellidae	5	50	Rodentia	1	24	80	51–261
Haematopinidae	1	16	Artiodactyla	3	11	14	75–171
	(1)[c]	4	Perissodactyla	1	1	3	6–16
Hamophthiriidae	1	1	Dermoptera	1	1	1	1–2
Hoplopleuridae	5	134	Rod., Lag., Ins.	11	75	289	321–1709
(Hoplopleurinae)	2	124	Rodentia	6	65	270	270–1348
	(1)	1	Lagomorpha	1	1	3	1–14
(Haematopinoidinae)	1	7	Rodentia	2	6	14	11–34
	2	3	Insectivora	2	3	4	39–313
Hybophthiriidae	1	1	Tubulidentata	1	1	1	1–1

Family			Order				
Linognathidae	56	3	Artiodactyla	3	31	59	62–152
	4	(1)	Carnivora	1	4	11	15–41
	7	1	Hyracoidea	1	3	11	3–11
Microthoraciidae	4	1	Artiodactyla	1	2	4	2–4
Neolinognathidae	2	1	Insectivora	1	1	4	5–28
Pecaroecidae	1	1	Artiodactyla	1	1	1	1–2
Pedicinidae	14	1	Primates	1	6	27	11–60
Pediculidae	2	1	Primates	3	6	14	16–38
Polyplacidae	157	8	Rodentia	34	89	302	354–1687
	7	2(1)	Insectivora	3	7	19	5–15
	6	1	Lagomorpha	1	3	11	8–49
	7	3	Primates	3	6	8	14–30
Ratemiidae	2	1	Perissodactyla	1	1	4	1–7
Pthiridae	2	1	Primates	2	2	2	5–9

[a] Number of known genera and species is for those families with louse infestations.
[b] After Anderson and Jones (1967).
[c] Number in parentheses indicates the genus already counted for the family.

host, hamophthiriids must have invaded the ancestral cynocephalids, which are perhaps related to the Plagiomenidae (Dawson 1967), before host speciation occurred and then rapidly evolved along with the host. The specialization of *Hamophthirius* supports the idea that the Dermoptera had a long independent history.

Primates and Diverse Groups of Anoplura

Six diverse taxa of Anoplura infest the Primates: *Lemurpediculus* (one species) on Lemuridae (lemurs), *Phthirpediculus* (two species) on Indriidae (indrisoid lemurs), *Lemurphthirus* (two species) on Lorisidae (lorises, galagos), Pedicinidae (*Pedicinus* with 14 species) on Cercopithecidae (Old World monkeys) (Kuhn and Ludwig 1967), and *Pediculus* and *Pthirus* on other anthropoid Primates (Kim and Emerson 1968).

Pediculus (Pediculidae) is found on Cebidae (New World monkeys), Hylobatidae (Gibbons), Pongidae (great apes), and Hominidae (man) with poorly differentiated species, whereas *Pthirus* (Pthiridae) is found on man and gorilla with well-defined species. The first hominoids, *Propliopithecus*, appeared in the early Oligocene epoch in Egypt, and the first ceboids were present in the late Oligocene or early Miocene in South America (Simpson 1945; Dawson 1967). These facts suggest that pediculids were present in the early Oligocene epoch. Furthermore, the fact that *Pthirus pubis* (Linnaeus) is found on man (Hominidae) and *P. gorillae* Ewing on gorillas (Pongidae) suggests the infestation of Eocene hominoids with pthirids.

There is no record of Anoplura from the two prosimian families, Daubentoniidae (aye-ayes) and Tarsiidae (tarsiers). *Pedicinus*, a unique parasite with some similarity to the forms of Hoplopleuridae and Polyplacidae, is endemic to the Old World Cercopithecidae. This fact suggests that *Pedicinus* is of polyplacid ancestry and that the invasion and radiation of pedicinid ancestors on the Old World monkeys must have been recent, perhaps as late as the Pliocene. The three genera, *Lemurpediculus*, *Phthirpediculus*, and *Lemurphthirus*, represent the polyplacid line, and *Pedicinus*, *Pediculus*, and *Pthirus* each has an independent phylogenetic line. Multiple infestations of these phylogenetic lines and the taxonomic status of these taxa suggest that the initial invasion of the Primates by each taxon must have occurred at different times.

Lagomorpha with *Haemodipsus*, and a Hoplopleuran Emigrant

The earliest lagomorphs are known from the Paleocene epoch of Asia, and the Leporidae (rabbits and hares) and the Ochotonidae (pikas) diverged from a common ancestor during the Oligocene epoch (Wilson 1960). The leporids are infested with endemic *Haemodipsus* (Polyplacidae), but pikas (*Ochotona*) are infested with a single emigrant, *Hoplopleura ochotonae* Ferris. Considering its isolated taxonomic relationship and its present diversity,

Haemodipsus seems to have infested the early leporids and evolved along with them since the Oligocene. The ever-expanding *Hoplopleura* also colonized *Ochotona* long after it spread to Europe and North America during the Pliocene epoch (pikas were most abundant in the Miocene), perhaps as late as the Pleistocene when it reached as far as Great Britain and eastern North America (Dawson 1967).

Rodentia and Their Lice

The Rodentia harbor 70% of the Anoplura diversity represented by three major families: Enderleinellidae (50 species), Hoplopleuridae (131 species), and Polyplacidae (157 species). Louse infestation is concentrated on the three families of rodents as follows: Sciuridae with seven genera and 96 species, Cricetidae with five genera and 94 species, and Muridae with three genera and 89 species. Additionally, the Heteromyiidae harbors *Fahrenholzia* with 13 species, and the Dipodidae is infested with 19 species of *Eulinognathus*. All other taxa are infested with one to five species of diverse genera.

Of the seven families of the sciuromorph rodents, only the Sciuridae (squirrels), Heteromyidae (kangaroo rats), and Pedetidae (spring haas) are found to harbor the Anoplura. The Geomyidae (pocket gophers) are exclusively infested with the Mallophagan *Geomydoecus* (Trichodectidae) (Price and Emerson 1971). The Apolodontidae (mountain beavers) and Anomaluridae (scaly tailed squirrels) lack louse infestations. The monotypic *Apolodontia*, the most primitive rodent restricted to the west coast of North America (McLaughlin 1967), does not harbor anoplurans despite its long geological history and eight extinct genera from the Upper Eocene epoch. The Anomaluridae, a recent family with four genera and 14 species found in Western and Central Africa, has no known louse infestations.

The Sciuridae harbors three major taxa, Enderleinellidae with five genera and 50 species, Polyplacidae (*Neohaematopinus*, *Johnsonpthirus*, and *Linognathoides*) (Kim and Adler 1982), and *Hoplopleura* (Hoplopleuridae). The Enderleinellidae are restricted to the Sciuridae; *Enderleinellus* is cosmopolitan and widely distributed throughout the squirrel family. A peculiar *Microphthirus* is restricted to the North American flying squirrel *Glaucomys*, *Werneckia* to the African bush squirrels (*Paraxerus*), and *Phthirunculus* and *Atopophthirus* to the Oriental *Petaurista*. The polyplacid *Neohaematopinus* is a primary parasite of the Sciuridae and is found on most genera examined. In more recent geological time, *Neohaematopinus* s.l. began to invade other hosts closely associated with squirrels, perhaps in the same habitat, like the North American wood rat *Neotoma* (Cricetidae), and the African golden moles *Chrysochloris* (Chrysochloridae) (Benoit 1961, 1969). The hoplopleuran distribution on Sciuridae is patchy in different regions: *H. erismata* Ferris, *H. mitsuii* Kaneko, *H. thurmonae* Johnson, *H. distorta* Ferris, *H. emarginata* Ferris on the Oriental *Callosciurus, Funambulus*,

Sciurotamias, Tamiops, Rhinosciurus, and *Menetes; H. arboricola* Kellogg and Ferris, *H. sciuricola* Ferris, and *H. trispinosa* Kellogg and Ferris on the Holarctic *Eutamias, Sciurus, Tamiasciurus,* and *Glaucomys,* respectively. Considering its patchy distribution, the invasion of the Sciuridae by *Hoplopleura* must have been relatively recent and have taken place in different regions.

African tree squirrels (Tribe Funambulini) and flying squirrels in the Holarctic and Oriental regions appear to provide diverse niches for the Anoplura. African tree squirrels (*Funambulus, Paraxerus, Funisciurus*) are infested with *Enderleinellus, Werneckia, Johnsonpthirus* (Benoit 1961, 1969) and *Hoplopleura,* and flying squirrels (Petauristinae) are infested with *Neohaematopinus, Enderleinellus, Microphthirus, Atopophthirus* (Kim 1977), *Phthirunculus,* and *Hoplopleura.*

The Heteromyidae are uniquely infested with *Fahrenholzia* (Polyplacidae), but no other parasitic Psocodea, whereas the Pedetidae (*Pedetes* with two species) are infested with *Eulinognathus* (Polyplacidae). *Eulinognathus* is parasitic on many different rodents, but primarily on jerboas (Dipodidae), *Pedetes* (Pedetidae), and Bathyergidae (mole rats) in central and southern Africa, on *Hypogeomys* in Madagascar, on *Lophiomys* in Africa (both in the Cricetidae), and on *Ctenomys* (tuco tucos) (Ctenomyidae), and Chinchillidae (chinchillas) in South America. The Dipodidae are distributed in northern Africa, Asia Minor, and northern Arabia eastward through southern Russia and Turkestan to Mongolia and northeastern China. Within the Dipodidae *Eulinognathus* is presently restricted to the Dipodinae (*Allactagulus, Allactaga, Dipus, Jaculus,* and *Pygeretmus*). *Lagidiophthirus* and *Cuyana,* close relatives of *Eulinognathus,* are known from *Lagidium* (Chinchillidae); *Galeophthirus* is known from *Galea* (cavies) (Caviidae); and all are from South America. The taxonomic status of these three genera remains to be clarified.

The Cricetidae (New World rats and mice) and Muridae (Old World rats and mice) are the two largest families of Rodentia, having 97 genera and about 567 species and 98 genera and 457 species, respectively. They are primarily infested with *Hoplopleura* (Hoplopleuridae), *Polyplax* (Polyplacidae), and their minor relatives. Approximately 40% of the known Anoplura fauna is found in the families Cricetidae and Muridae. *Polyplax* also infests the Rhyzomyidae. No lice are known from the Spalacidae. In addition to *Hoplopleura* and *Polyplax,* the Cricetidae, which are distributed throughout the continents except in the Austro-Malayan area, are infested with other Anoplura: *Mirophthirus* on the unusual host *Typhlomys* in Guizhou, China; *Neohaematopinus* is found on *Neotoma; Fahrenholzia* on Central American *Zygodontomys;* and *Eulinognathus* on the Madagascar *Hypogeomys* and African *Lophiomys.* The monotypic *Proenderleinellus* also infests *Cricetomys* (Muridae).

Among the gliroid and dipodoid rodents there is no record of Anoplura from Platacanthomyidae (spiny dormice) and Seleviniidae (dzhalmans). Whereas the Dipodidae harbors *Eulinognathus,* the Gliridae (dormice) and

Zapodidae (jumping mice) of the Palearctic and Ethiopian regions are exclusively parasitized by *Schizophthirus* (Hoplopleuridae), which are clearly endemic to the glirids and zapodids.

The Anoplura are curiously absent in the hystricoid and erethizontoid rodents (porcupines). The New World porcupines (Erethizontidae) are infested with several species of the mallophagan *Eutrichophilus* (Trichodectidae). The cavioid and chinchilloid rodents are host to few sucking lice but are abundantly infested with Mallophaga. The Caviidae harbors three species of *Pterophthirus* and the monotypic *Galeophthirus*. *Pterophthirus* is also found on the South American Echimyidae. *Galeophthirus* and *Lagidiophthirus* found on Chinchillidae are specialized relatives of *Eulinognathus*. No records of Anoplura are known from Hydrochoeridae (capybaras), Dinomyidae (pacarana), Heptaxodontidae, or Dasyproctidae (agoutis).

The distribution of Anoplura on the octodontoid rodents provides a curious picture, closely related to geographical patterns. Although no Anoplura are recorded from Capromyidae (hutias), Myocastoridae (nutria), or Ctenodactylidae (gundis), the mallophagan Gyropidae are well established on Capromyidae. The African families Thryonomyidae (cane rats) and Petromyidae (dassie rat), both monotypic, and the Bathyergidae (mole rats) are solely parasitized by sucking lice: *Scipio* on *Thryonomys* and *Petromus*; and *Eulinognathus* on Bathyergidae. The South American families, Octodontidae (hedge rat), Ctenomyidae (tuco tucos), Abrocomidae (chinchilla rats), and Echimyidae (spiny rats) have been invaded by numerous diverse taxa of Anoplura and Mallophaga. The Ctenomyidae are infested with both Gyropidae (Mallophaga) and *Eulinognathus*. The Octodontidae is infested by *Hoplopleura*, and Abrocomidae by both Gyropidae and *Polyplax*. The Echimyidae harbors both Mallophaga and Anoplura: Trimenoponidae, Gyropidae, and the hoplopleurid *Pterophthirus* on *Proechymys*, and the monotypic *Ctenophthirus* (Polyplacidae) on *Thrichomys (Cercomys)*. *Pterophthirus*, already mentioned, is also found on Caviidae. The *Eulinognathus* infestation of the South American hystricomorph Rodentia is significant in that *Eulinognathus* also is a primary parasite of Palearctic and Ethiopian rodents.

Carnivora and Echinophthiriidae

Fissipedia are primarily infested with Mallophaga, but Pinnipedia are infested with Anoplura. The aquatic carnivores, the Pinnipedia and aquatic Mustelidae (otters), are universally parasitized by the Echinophthiriidae (Kim et al. 1975). All land carnivores harbor the ischnoceran Trichodectidae and an amblyceran *Heterodoxus spiniger* (Enderlein), except for aquatic Mustelidae (*Lutra* and perhaps *Enhydra*), which are infested with *Latagophthirus* (Echinophthiriidae). The Canidae have a multiple infestation of four taxa: *Linognathus* (Linognathidae), *Trichodectes*, *Suricatoecus* (Ischnocera), and *Heterodoxus* (Amblycera). *Linognathus*, which is characteristic of the Artiodactyla, must have secondarily invaded the Canidae. The mode of changeover is not certain as yet but must have taken place quite early (Hopkins 1957). The presence of *Latagophthirus* on Mustelidae strongly suggests that the absence of endemic Anoplura on Fissipedia is secondary (Kim, Chapter 7).

Tubulidentata, Hyracoidea, and Their Anoplura

The monotypic *Orycteropus* (Orycteropodidae, Tubulidentata), which may be traced back to the Miocene epoch in Africa, is parasitized by the monotypic *Hybophthirus* (Hybophthiridae). No other parasitic Psocodea is known from aardvarks. The hyraxes are known from the early Oligocene epoch to Recent in Africa from which six genera of Geniohyidae and Procaviidae and the Pliocene *Pliohyrax* are recorded (Dawson 1967), but they have never spread beyond Africa and the Mediterranean region. The Hyracoidea are heavily infested with *Prolinognathus* (Linognathidae) and members of the mallophagan family Trichodectidae (Emerson and Price 1981). The apparent center of the *Prolinognathus* distribution on *Procavia* may be the result of mutual incompatibility of the heavy mallophagan infestation with the Anoplura (Hopkins 1949). The Linognathidae are considered monophyletic despite the remote relationships of their hosts. Protolinognathids might have colonized ancestral hyraxes secondarily through bodily contact with hoofed neighbors, since the Artiodactyla used to be mostly small forms frequently in close contact with the Procaviidae during their evolution in Africa (Weisser 1975).

Perissodactyla and Their Lice

The Perissodactyla, which first appeared in the early Eocene epoch, developed and diversified rapidly into five superfamilies and 12 families. Their diversity diminished during the Oligocene; only four families persisted after the early Miocene, and only three families were extant beyond the Pleistocene: Equidae, Tapiridae, and Rhinocerotidae (Dawson 1967). No lice have ever been recorded from Tapiridae or Rhinocerotidae. Equidae is the only taxon infested with lice: two genera of Anoplura, *Ratemia* (Ratemiidae) and *Haematopinus* (Haematopinidae), and the trichodectid Mallophaga, *Bovicola*. *Ratemia* is endemic to *Equus* (Equidae) and possesses many linognathid characters. *Haematopinus* is a generalized taxon with many primitive characters and is widely distributed on the ungulates including Equidae. The distribution of *Haematopinus* in infested families is relatively even: five species on Equidae, six species on Suidae, three species on Cervidae, and eight species on Bovidae. Considering the present distribution and relatively generalized morphology, the Haematopinidae must have been widely distributed on the ungulates during the Tertiary period when the ungulate diversity was greater.

Artiodactyla and Diverse Groups of Lice

The Artiodactyla are infested with both Anoplura and Mallophaga, but Hippopotamidae has no lice. No Anoplura are recorded from Tragulidae (mouse deer) or Antilocapridae (pronghorn). Many families have a multiple infestation of both Anoplura and Mallophaga: Tayassuidae (peccaries), Camelidae (camels and llamas), Cervidae (deer), Giraffidae, and Bovidae (antelopes, gazelles, cattle, goats, sheep). The Suidae (pigs) are infested with *Haematopinus*. The family Tayassuidae harbors the monotypic *Pecaroecus* (Pecaroecidae) and *Macrogyropus* of the amblyceran Gyropidae. *Pecaroecus* is endemic to the Tayassuidae.

The Camelidae are infested with *Microthoracius* (Microthoraciidae)—one species on the African *Camelus* and three species on the South American *Lama*—and *Bovicola* (Trichodectidae, Mallophaga) which is found on the *Lama*. While *Microthoracius* is endemic to the Camelidae, *Bovicola* is also found on the Equidae, Cervidae, and Bovidae. The Traguilidae are also infested with another ischnoceran, *Damalinia*, which again occurs on Cervidae and Bovidae.

The family Cervidae harbors three Anoplura genera—*Solenopotes*, *Linognathus*, and *Haematopinus*—and three genera of Trichodectidae, and the family Giraffidae is infested with a species of *Linognathus*. The primary distribution of *Solenopotes* is in the Cervidae: *Odocoileus*, *Mazama*, *Rangifer*, *Cervus*, and *Platyceros* (Kim and Weisser 1974). *Linognathus* primarily occurs on the Bovidae, which also harbor *Haematopinus* and *Solenopotes* and large numbers of Trichodectidae (Weisser 1975). *Haematopinus* seems to infest larger members of the host families (Hopkins 1949).

HOST SPECIFICITY

Anoplura are closely associated with major groups of mammals and are uniquely adapted to skin and fur habitats on the mammalian body surface. The life cycle and propagation of the sucking lice is to a large extent determined by the biology and behavior of the host (Kim and Ludwig 1978a; Kim 1982). Likewise, the mammalian hosts have evolved with sucking lice. The present diversity and host distribution of Anoplura are the result of long interactive evolutionary processes (Table 5.2).

The Anoplura have developed a high degree of host specificity with mammals (Table 5.3). Considering the entire Anoplura diversity, 63% of the known species is specific to a single host species, and 24% is specific to two or three closely related species. The percent specificity ranges from 100% of the monotypic families Hamophthiriidae, Pecaroecidae, and Pthiridae, to 0% of Neolinognathidae. The rodent-infesting sucking lice show that the mean percent specificity is about 62% for the Anoplura: that is, 66% for Enderleinellidae, 62% for Hoplopleuridae, and 58% for Poly-

Table 5.3 Host Specificity of Anoplura on Mammals

Anoplura (Families)	Total Number of Species	Percent Specificity	Number of Louse Species with Hosts of				
			1 Species	2 Species	3 Species	4 Species	5 or More Species
Echinophthiriidae	12	75	9	1	—	—	2
Enderleinellidae	50	66	33	9	4	1	3
Haematopinidae	20	95	19	1	—	—	—
Hamophthiriidae	1	100	1	—	—	—	—
Hoplopleuridae	134	62	83	19	17	7	8
(Hoplopleurinae)	(124)	(61)	(76)	(18)	(15)	(7)	(8)
(Haematopinoidinae)	(10)	(70)	(7)	(1)	(2)	(—)	(—)
Hybophthiriidae	1	100	1	—	—	—	—
Linognathidae	67	66	44	15	3	4	1
Microthoraciidae	4	25	1	2	1	—	—
Neolinognathidae	2	0	—	1	1	—	—
Pecaroecidae	1	100	1	—	—	—	—
Pedicinidae	13	54	7	1	4	—	1
Pediculidae	2	50	1	—	—	—	1
Polyplacidae	173	58	101	21	15	8	28
Ratemiidae	2	50	1	—	1	—	—
Pthiridae	2	100	2	—	—	—	—
Anoplura (Total)	484	63	304	70	46	20	44

placidae. The ungulate-infesting Haematopinidae and Linognathidae have a high host specificity, 95% and 66%, respectively.

The distribution of Anoplura on mammals at the generic level provides some insights into the specificity characteristics of many taxa. Ludwig (1968) documented the associations of the Anoplura genera with families of mammals: 29 genera of Anoplura are associated with a single host family, 6 genera with 2 host families, 3 genera with 3 families, and 4 genera with 4–6 families; and 32 host families are infested with a single genus of Anoplura, 8 families with 2 genera, 4 families with 3 genera, and 2 families with 4–7 genera.

Of the five genera of the Echinophthiriidae, *Proechinophthirus*, *Lepidophthirus*, and *Latagophthirus* have 100% host specificity involving one or two host genera. *Proechinophthirus* is restricted to fur seals (*Arctocephalinae*, *Otariidae*), *Lepidophthirus* to monk seals (Monachinae, Phocidae). *Latagophthirus* exclusively infests *Lutra* (Mustelidae, Carnivora). The monotypic *Echinophthirius* is the parasite of the Phocidae, infesting six genera and seven species. The distribution of *Antarctophthirus* involves the entire Pinnipedia, infesting 10 host species of three families and nine genera; 67% of the distribution is specific to a single host species and the remainder to two or more hosts. These polyhospitalic species are either polytypic species of complexes or sibling species.

Four genera of Enderleinellidae—*Werneckia*, *Microphthirus*, *Phthirunculus*, and *Atopophthirus*—are parasitic on a single host genus. The monotypic *Microphthirus* infests two species of *Glaucomys* in North America, while *Werneckia* (three species) is found on four species of the African *Paraxerus* (12 species). *Enderleinellus* is widely distributed in the Sciuridae, infesting 17 genera and about 80 species; 66% of the distribution is specific to single host species, 16% to two hosts, and the remainder to three or more host species.

Hoplopleura, the largest genus of the family Hoplopleuridae (five genera, 134 species), has a broad host distribution with 9 families, 87 genera, and more than 200 species of rodents, of which 61% infests single host species and 14% two hosts. The distribution of *Pterophthirus* includes two families, four genera, and nine species of rodents, and *Schizophthirus* spreads over two host families, including five genera and 11 species. However, those infesting the Insectivora are narrowly distributed within the host family; *Ancistroplax* (two species) is adapted to *Crocidura* (Soricidae), and *Haematopinoides* is confined to two of the 15 talpid genera.

Linognathus is the most diversified taxon within the family Linognathidae, with 50 described species infesting 75 host species of 6 families and 34 genera in Artiodactyla, Perissodactyla, and Canidae (Carnivora). The majority of *Linognathus* infests the Bovidae (44 genera, 111 species) and is widely distributed throughout the family, with 72% host specificity. Considering the diversity of these hosts, we expect to discover many more species of *Linognathus*. The primary distribution of *Solenopotes* (10 species) is found in the Cervidae (16 genera, 37 species), infesting 15 species of 2 families and 8 genera. All species are specific to one or two host species.

The Polyplacidae, which are broadly adapted to diverse groups of mammals, include *Polyplax*, *Neohaematopinus*, and *Eulinognathus*. The distribution of *Polyplax* (72 species) involves 2 orders, 7 families, 54 genera, and about 170 species of mammals, but primarily centers in the Cricetidae and Muridae. About 54% of *Polyplax* is specific to a single host species. *Neohaematopinus* s.l. infests the mammalian hosts of 2 orders, including 3 families, 31 genera, and more than 85 species; its primary distribution is in the family Sciuridae, and 55% is restricted to a single host species. *Eulinognathus* is widely distributed throughout the order Rodentia, infesting 12 genera in 6 families with 75% host specificity. *Fahrenholzia* with 62% host specificity infests the Heteromyidae (75 species), of which 5 genera and about 30 species are known hosts. *Scipio* (four species) is distributed in the two African rodent families Thryonomyidae and Petromyidae, each having one to five host species. *Haemodipsus*, with 57% host specificity, is specific to the Leporidae (eight genera, 49 species), infesting three host genera and more than 15 species. The polyplacids infesting the prosimian Primates—*Lemurpediculus*, *Lemurphthirus*, and *Phthirpediculus*—are widely distributed with high host specificity within their respective families. *Neolinognathus* (two species) broadly infests the Macroscelididae (five genera, 28 species).

DISCUSSION AND CONCLUSIONS

Origin and Evolution of Anoplura

The Anoplura arose from the Protanoplura of the late Cretaceous period, but they probably were facultative and temporary parasites on primordial mammals (Kim and Ludwig 1978a, 1982). The Protanoplura must have made two major evolutionary shifts: (1) the invasion of a new habitat on the body surface (skin and fur) of mammals, perhaps between the Jurassic and Cretaceous period, and (2) the exploitation of new food sources (blood) after they were well established on the mammalian body surface in the first shift, perhaps during the late Cretaceous or early Paleocene (Fig. 5.6) (Kim and Ludwig 1982).

The preliminary cladogram of the Anoplura was constructed on the basis of morphological analysis (Kim and Ludwig 1978a) and shows the late Eocene-Oligocene dates for the origin of Anoplura. These dates are much later than the data show, as Traub (1980) correctly questioned the lateness of the late Eocene-Oligocene dates. The data on mammals, fossils, distribution, and endemism indicate that the Anoplura were present on primitive rodents, carnivores (perhaps Creodonta), and prosimian Primates in the North American Paleocene, as suggested by Kim and Ludwig (1978b, 1982) and Traub (1980). If this inference is accepted, the origin of

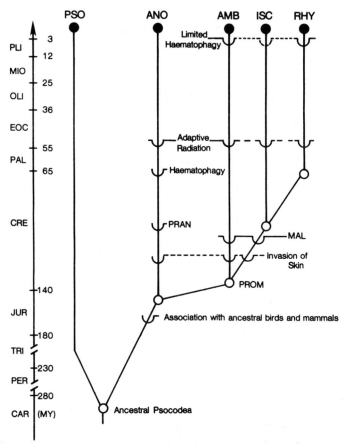

Figure 5.6 Phylogenetic tree of parasitic psocodea showing the origin of Anoplura. PSO, Psocodea; ANO, Anoplura; AMB, Amblycera; ISC, Ischnocera; RHY, Rhynchophthirina; MAL, Mallophaga; PRAN, Protanoplura; PROM, Protomallophaga; CAR, Carboniferous; PER, Permian; TRI, Triassic; JUR, Jurassic; CRE, Cretaceous; PAL, Paleocene; EOC, Eocene; OLI, Oligocene; MIO, Miocene; PLI, Pliocene. (From Kim and Ludwig, 1982. Ann. Entomol. Soc. Am. 75:537–548. By permission of the Entomological Society of America.)

the Protanoplura should be dated back to the Cretaceous period as Kim and Ludwig (1982) suggested. This dating is further corroborated by the fact that the earliest known mammals appeared during the Jurassic and Cretaceous periods and were very small insectivorous animals (Kemp 1982). Mammals began to radiate at the late Cretaceous period; they included the Marsupialia and prosimian Primates of North America and Insectivora of North America and Asia. The Carnivora and Rodentia appeared during the Paleocene epoch in North America. The Perissodactyla and Artiodactyla appeared in the early Eocene epoch of the Northern Hemisphere (Kurtén 1969, 1971). However, the theory proposed by Traub

(1980) that suggests the North American Paleocene origin of the Anoplura on the Rodentia is somewhat too premature to accept because of the still inadequate data concerning the associations of sucking lice and mammals.

Preliminary analysis shows that in early evolutionary time the Anoplura diverged into the two major phyletic lines, linognathoids and haematopinoids. The linognathoid lines include Linognathidae, Polyplacidae, Hoplopleuridae, Pedicinidae, and Enderleinellidae; the haematopinoid lines are Haematopinidae, Microthoraciidae, Echinophthiriidae, and Pecaroecidae. Along with or within these phyletic lines many taxa evolved independently for a long time, and their phylogenetic relationships are not clear as yet. These taxa are Hybophthiriidae, Hamophthiridae, Pthiridae, Ratemiidae, Neolinognathidae, and Pediculidae (Kim and Ludwig 1978a, 1983).

Taxonomic analysis of the sucking lice and their host associations suggests that the large number of Anoplura were associated with archaic mammals and evolved along the line of mammalian cladogenesis (Kim and Ludwig 1978a; Kim 1982; Ludwig 1982). The species diversity of certain taxa like *Microthoracius* and *Ratemia*, parasitic on those mammalian hosts that once thrived in the Neogene, must have been much greater than at present. The diversity of Primates, Artiodactyla, Perissodactyla, and Carnivora was apparently much greater in the pre-Pleistocene periods than it is today (Dawson 1967). If this thesis is accepted, large sections of anopluran taxa once very successful disappeared as their host groups became extinct (Table 5.4). Some of the monotypic or small taxa, like *Hybophthirus* on aardvarks, *Microthoracius* on camels, and *Latagophthirus* on river otters, perhaps represent the remnants of once very large taxa. At the same time, certain taxa like the Polyplacidae and Hoplopleuridae may have speciated rapidly by invading new hosts, or they may have radiated along with the mammalian cladogenesis (Fig. 5.7). Through this process few or none of the evolutionary connecting links have been left in the present diversity of Anoplura, and the original structure–function complex in primitive sucking lice has been modified greatly by reduction or loss of many characters.

Origin of Host Associations and Specificity

The survival of Anoplura depends largely upon the well-being of the host animals because they spend their entire life on an individual host. Each host must provide acceptable ecological settings for a particular anopluran taxon to survive, including habitat (skin-fur) and environmental conditions determined by its physiology and surrounding climate. Also propagation and transmission of sucking lice are to a large extent determined by the biology and behavior of the host (Kim 1972, 1975; Kim and Ludwig 1978a). These facts support the concept that the Anoplura have evolved closely with mammals and that the present host associations and distributions are

Table 5.4 Diversity of Some Anoplura and Their Eutherian Hosts[a]

Anoplura ex Mammalian Hosts	Fossils		Recent	
Pecaroecidae	(1 g/10 sp)		1 g/1 sp	
ex Tayassuidae		10 g		1 g/2 sp
Ratamiidae	(1 g/18 sp)		1 g/2 sp	
ex Equidae		18 g		1 g/7 sp
Microthoraciidae	(1 g/40 sp)		1 g/4 sp	
ex Camelidae		21 g		2 g/4 sp
Hybophthiridae	(1 g/2 sp)		1 g/1 sp	
ex Orycteropodidae		2 g/2+ sp		1 g/1 sp
Linognathidae (part)	(2 g/130 sp)		2 g/61 sp	
ex Bovidae;		89 g		44 g/111 sp
Cervidae;		39 g		16 g/37 sp
Giraffidae		16 g		2 g/2 sp
Enderleinellidae	(4 g/10 sp)		5 g/49 sp	
ex Sciuridae		4 g		51 g/261 sp

Source: Anderson and Jones (1967) for extant mammals; Simpson (1945), Dawson (1967) for fossils; Kim and Ludwig (1978a) for Anoplura.
[a]Number of genera/number of species; g, genera, sp, species; number (genus or genera/ species) in parenthesis is an estimate based on species diversity of extant and fossil mammals and recent Anoplura.

the result of long interactive evolutionary processes. Furthermore, host–parasite specificity is the height of such obligatory associations.

Origin of Anoplura Associations with Mammals

Host associations and specificity must be considered in terms of multispecies community and species diversity because they are the result of interspecific interactions in both ontogenetic (or ecological) and phylogenetic (or evolutionary) context (Kim, Chapter 1). The occurrence of particular species on a host organism at a given ecological time is determined by the contingent of ectoparasites on the parent host and an existing biotic makeup surrounding the host in a given habitat. This community structure is shaped by extinction and speciation in the evolutionary time scale (MacArthur and Wilson 1967; Whittaker 1972, 1977; Whittaker et al. 1973; Price 1980).

The present host associations of Anoplura are the manifestation of both ecological and evolutionary processes. The development of these host–parasite associations in Anoplura must have followed one of the following models (Fig. 5.7):

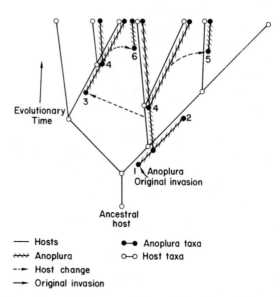

Figure 5.7 Evolution of parasite–host associations in Anoplura. 1. Original invasion of Pro-tanoplura. 2. Extinction. 3. Primary host change. 4. Cladogenesis. 5. Secondary host (vacant) change. 6. Secondary host change, causing multiple infestation.

1. A certain host species was colonized by a specific parasite species before host cladogenesis, and then these parasites coevolved and speciated along with the adaptive radiation of the hosts throughout the entire history of their associations (1, 4 in Fig. 5.7). This process resulted in obligatory associations of parasites and hosts, and thus certain hosts harbor closely related species of parasites ("primary infestation" of Hopkins 1949, 1957; "phylogenetic specificity" of Sprent 1969). In other words, divergence and speciation of the host species leads to the divergence of parasites; these types of parasites may be referred to as "natives" or "heirloom parasites" (Sprent 1969). This model supports several parasitophyletic rules (Stammer 1957): Fahrenholz' rule (Eichler 1941a), Szidat's rule (Szidat 1956), and Eichler's rule (Eichler 1941a, b, 1973). Most of the host–parasite associations of Enderleinellidae and Sciuridae, *Prolinognathus* and Hyracoidea, Echinophthiriidae and aquatic Carnivora, *Haemodipsus* and Leporidae, *Hamophthirius* and Dermoptera, *Hybophthirus* and Tubulidentata, *Microthoracius* and Camelidae, *Ratemia* and *Equus* (Equidae), *Pthirus* and Primates, *Pedicinus* and Cercopithecidae, and *Pediculus* and Primates are examples of this model (Table 5.3).

2. A certain host was invaded and colonized by a specific parasite species from distantly related hosts which had coinhabited the same range during or after the height of host cladogenesis ("host-changes," Wirte-wechsel of Weisser 1975) (3, 5, 6 in Fig. 5.7). The parasites of this model may be called "emigrants" or "souvenir parasites" (Sprent 1969). Then these parasites coevolved with the hosts without much diversification or speciated along with adaptive radiation of the hosts. These processes led to disparity in the distribution of parasites in the host groups at lower taxonomic levels, usually at the generic level. The distributions of *Linognathus* in Canidae, *Hoplopleura* in Ochotonidae, *Polyplax* in Insectivora, *Neohaematopinus* in Cricetidae (*Neotoma*), and *Hoplopleura* in Sciuridae, and the colonization of and adaptation to *Tupaia* by the polyplacid *Sathrax* are good examples.

The result of such evolutionary processes is that a certain host taxon becomes closely associated with a constant set of parasites. However, large sections of once successful groups of Anoplura could have disappeared along with the extinction of the host group ("secondary absence" of Hopkins 1949; 2 in Fig. 5.7). Such unique taxa as *Pecaroecus* on peccaries, *Microthoracius* on camels and llamas, and *Ratemia* on horses, represent a remnant of what once was a large group (Kim and Ludwig 1978a).

Evolution of Parasite–Host Associations

Although parasite cladogenesis is closely related to host evolution, the rate of evolution often differs between parasites and hosts (Brooks 1979). Certain groups of Anoplura might have speciated rapidly along with greatly diversifying mammalian taxa such as Rodentia. The associations of rodents with Polyplacidae and Hoplopleuridae is an example of this process. However, sometimes adaptive radiation of the parasites may have been much slower than that of the hosts. As a result, a single parasite species infests a group of host species. This process lowers host–parasite specificity. The Hoplopleuridae and Polyplacidae associated with Sciuridae, Cricetidae, and Muridae provide many examples for this phenomenon. *Enderleinellus suturalis* (Osborn), infesting many species and subspecies of *Spermophilus* (*Citellus* J. A. Allen) in the Holarctic region, is a polytypic species. The populations on each host species are distinct but not differentiated enough to be considered separate species (Kim et al. 1963; Kim 1966a). As Werneck (1948) suggested, many other species of *Enderleinellus*, such as *E. osborni* Kellogg and Ferris (Seeley 1972) and *E. marmotae* Ferris, are species complexes (Kim 1966b). Similar situations are also found in *Hoplopleura*, exemplified by *H. hesperomydis* (Osborn) (Kim et al. 1966), *H. ferrisi* Cook and Beer, *H. emphereia* Kim (Kim 1965; Johnson 1972), and *H. captiosa* Johnson (Johnson 1960; Kim 1966c), and in *Polyplax* such as *P. serrata* (Burmeister) and *P. reclinata* (Nitzsch) (Johnson 1960), among many others.

Sometimes parasite cladogenesis might have outrun the host, and consequently many similar species of parasites are found on one host species within the same habitat (Sprent 1969). The mode of parasite speciation may be either synhospitalic or allohospitalic. Synhospitalic speciation involves a single host species, while allohospitalic speciation occurs when parasites on one host species originated on different host species (Sprent 1969; Brooks 1979) (Fig. 5.7). Synhospitalic speciation, relatively uncommon to Anoplura, is observed in the Linognathidae and Polyplacidae. Speciation must have been intense in *Prolinognathus* and some of *Linognathus* (Linognathidae) and in the mallophagan family Trichodectidae, causing unusually severe competition among ectoparasites (Hopkins 1949, 1957; Kim and Ludwig 1983). While the Trichodectidae infests all three genera of Hyracoidea—*Dendrohyrax*, *Heterohyrax*, and *Procavia*—*Prolinognathus* primarily infests *Procavia*. Many species of hyraxes harbor multiple congeneric species. For example, *Dendrohyrax arboreus adolfifriedrici* (Brauer) harbors four genera and eight species of *Procavicola* and *Prolinognathus faini* Benoit; and *Procavia capensis* (Pallas) has three species of *Procaviphilus* and two species of *Prolinognathus*. Likewise, in the Bovidae, *Connochaetes taurinus* (Burchell) harbors two species of *Linognathus* and one species each of *Solenopotes*, *Haematopinus*, and the ischnoceran *Damalinia* (Hopkins 1949; Weisser 1975). A similar phenomenon is also found in *Neohaematopinus* parasitic on the African bush squirrels *Paraxerus* and *Polyplax* from the African Muridae (e.g., *Acomys* and *Tatera*) (Johnson 1960; Kim and Ludwig 1983). The data from the Linognathidae and Polyplacidae seem to support the model of synhospitalic speciation, contrary to the contention by Brooks (1979).

Phylogenetic Relationships and Historical Biogeography

Because of the obligate and permanent nature of Anoplura parasitism and the relatively high degree of their host specificity, analysis of the associations of Anoplura and mammals reveals significant facts regarding their phylogenetic relationships (Hopkins 1949, 1957; Kim and Ludwig 1978a, b) and historical distribution (Hopkins 1949, 1957; Patterson 1957; Vanzolini and Guimaraes 1955; Traub 1980). More definitive phylogenetic inferences on the evolution and historical distribution of Anoplura and mammals are discussed in this section.

Anopluran Relics and Mammals (Table 5.4)

The monotypic *Orycteropus* (Orycteropodidae, Tubulidentata), which may be traced back to the Miocene epoch in Africa, is parasitized by the monotypic *Hybophthirus* (Hybophthiridae), but no other parasitic Psocodea.

The monotypic family Cynocephalidae ("flying lemurs," colugos), which has two extant species, is infested with another monotypic genus, *Hamophthirius*, a highly specialized louse of the linognathid line. The morphological uniqueness of *Hamophthirius* suggests a long association with the dermopterans, modern Cynocephalidae, and Plagiomenidae of the North American late Paleocene and early Eocene epochs (Dawson 1967).

Such an association supports the view that the Dermoptera had a long independent history, and that the Plagiomenidae may have had primitive hamophthiriids (Traub 1980).

The hyraxes, of which six genera of Geniohyidae and Procaviidae are known (Dawson 1967), existed in the early Oligocene epoch in Africa but never spread beyond Africa and the Mediterranean region. The Hyracoidea are heavily infested with *Prolinognathus* (Linognathidae) and members of the mallophagan family Trichodectidae. The apparent center of the *Prolinognathus* distribution on *Procavia* may be the result of the mutual incompatibility of the heavy mallophagan infestation with the Anoplura (Hopkins 1949). The Linognathidae are considered monophyletic despite the remote relationships of their hosts. Protolinognathids might have colonized ancestral hyraxes secondarily through bodily contact with hoofed neighbors, since the Artiodactyla used to be mostly small forms frequently in close contact with the Procaviidae during evolution in Africa (Weisser 1975).

The taxonomy of *Pecaroecus* suggests that the monotypic Pecaroecidae was once widely distributed among the diverse tayassuids (Dolichoerinae with three genera and Tayassuinae with seven genera) in both Old and New Worlds during the mid-Tertiary period and that most species became extinct along with most of the peccaries in the late Tertiary (Dawson 1967).

Similarly, considering the taxonomy and distribution of lice as well as the geological history of camelids, it can be inferred that camelines were infested with *Microthoracius* long before their Quarternary expansion from North America to South America, Eurasia, and Africa, and that many species of microthoraciids became extinct along with the North American camels from the late Eocene through the Pleiocene epoch. Extant species of *Ratemia* are remnants of once much larger Ratemiidae parasitic on equids in the Tertiary period.

Echinophthiriidae and Pinnipedia

From the data on morphological and ecological adaptations, host specificity, and modes of dispersal (Kim 1971, 1972, 1975) I infer that ancestral carnivores must have had primitive echinophthiriids before they ventured into aquatic habitats—perhaps during the middle to late Oligocene epoch (Sarich 1969)—and that the echinophthiriids have coevolved and radiated along with the pinnipeds and aquatic fissipeds (Kim et al. 1975). The common distribution of *Antarctophthirus* supports the view that the pinnipeds are a monophyletic group derived from a canoid stock whose divergence took place in the middle to late Oligocene epoch (Sarich 1969). The discovery of *Latagophthirus rauschi* Kim and Emerson on the North American river otters further supports that the echinophthiriids are the ancient primary parasites of the canoid (or arctoid-musteloid) Carnivora, and that the absence of Anoplura on land fissipeds is secondary—the cause of which is still a matter of conjecture (Hopkins 1949, 1957; Kim and Emerson 1974b; Traub 1980).

Living sea lions of the subfamily Otariinae (Otariidae) include the three southern genera, *Phocarctos*, *Otaria*, and *Neophoca*, and two northern, *Eumetopias* and *Zalophus*. Both northern and southern sea lions harbor a single polytypic species of sucking louse, *Antarctophthirus microchir* (Troussart and Neuman) (Kim et al. 1975). The fossil records suggest that the dispersal of the sea lions to the Southern Hemisphere occurred perhaps two million years ago (Repenning et al. 1971). These facts require the interpretation that *A. microchir* existed on the Otariinae prior to the southern dispersal of sea lions, that the southern sea lions did not evolve separately from some fur seals in the Southern Hemisphere, and that the Otariinae is monophyletic (Kim et al. 1975). The evolution of *A. microchir* was exceedingly slow relative to that of the sea lions. Likewise, the presence of *Echinophthirius horridus* (von Olfers) on diverse phocines such as *Cystophora*, *Erignathus*, *Halichoerus*, *Pagophilus*, *Phoca*, and *Pusa*, suggests the monophyly of Phocinae.

The fact that two distinct species of *Proechinophthirus* have evolved on fur seals—*P. fluctus* (Ferris) on the northern *Callorhinus* and *P. zumpti* (Werneck) on the southern *Arctocephalus*—suggests that the two fur seals evolved along separate lineages much more ancient than the sea lion lineage, if the evolutionary rate of *Proechinophthirus* is comparable to that of *Antarctophthirus* (Kim et al. 1975).

Insectivora and Anoplura

Taxonomy of Hoplopleuridae and Polyplacidae associated with lipotyphlan insectivores suggests that their associations are secondary and thus that they must have developed after the Tertiary diversification of the hosts. This further supports the view of Traub (1980) that no Anoplura were present when the Insectivora arose in Asia, perhaps during the Cretaceous period (McKenna 1970). On the other hand, the infestations of the menotyphlan insectivores with polyplacids—*Neolinognathus* on Macroscelididae, and *Sathrax* and *Docophthirus* on Tupaiidae—are clearly primary, suggesting that the progenitors of these lice from rodents must have invaded the menotyphlans rather early, perhaps in the Miocene epoch, and they evolved rapidly.

Anthropoid Primates and Anoplura

The presence of *Pediculus* on the South American Cebidae, Oriental Hylobatidae, African Pongidae, and Hominidae represents primary infestations contrary to the view of Hopkins (1957) that they were secondary infestations from humans. The first hominoids, *Propliopithecus*, appeared in the early Oligocene epoch of Egypt, and the first ceboids were present in the late Oligocene or early Miocene epoch of South America (Simpson 1945; Dawson 1967). These facts suggest that pediculids were present in the early Oligocene epoch. Furthermore, the fact that *Pthirus pubis* (Linnaeus) on man (Hominidae) and *P. gorillae* Ewing on gorillas (Pongidae) suggests that the Eocene hominoids had pthirids before they radiated into Hominidae and Pongidae in the early Oligocene epoch. On the other hand, *Pedicinus*, unique lice with characters of hoplopleurids and polyplacids, which are primary parasites of rodents, are endemic to the Old World Cercopithecidae. This fact supports the theory of a more recent invasion of pedicinds on the Old World monkeys, perhaps as late as the Pliocene epoch.

Anoplura and South American Mammals

The distribution of *Eulinognathus* and *Pediculus* gives considerable support to the idea that the South American hystricomorph rodents and ceboid monkeys with their lice have African roots (Hoffstetter 1972; Lavocat 1974; Raven and Axelrod 1975; Cracraft 1975; Wood 1977; Traub 1980). Fossil records and the parasite–host associations suggest that pediculids and pthirids were present on hominoids in the early Oligocene epoch and on ceboids in the South American late Oligocene. Likewise, *Eulinognathus* must have already invaded the African Pedetidae and Bathyergidae by the early Miocene epoch, the South American chinchilloids and cavioids by the mid-Miocene or even the early Oligocene, and the Dipodidae in the late Miocene in Asia.

Anoplura and Madagascar Isolation

Madagascar was isolated from Africa throughout the Tertiary period, and no pre-Pleistocene mammals are known from there. The exclusively Malagasy mammal fauna includes the three primates, Lemuridae, Indriidae, and Daubentoniidae, and the cricetid Nesomyinae (Coryndon and Savage 1973). The Malagasy primates are infested with polyplacids: Lemuridae with *Lemurpediculus* (one species) and Indridae with *Phthirpediculus* (two species) (Kim and Ludwig 1978a).

The known autochthonous Anoplura on the Malagasy rats (Nesomyinae) include *Polyplax nesomydis* Paulian and *P. brachyuromyis* Kim and Emerson, both of which belong to the *P. jonesi* group, and *Eulinognathus hypogeomydis* Paulian on another nesomyine, *Hypogeomys* (Kim and Emerson 1974a). Although the present taxonomy is not completely satisfactory, the inclusion of the Ethiopian species in the *jonesi* group suggests the African connections of endemic Malagasian *Polyplax: P. kaiseri* Johnson on *Gerbillus*, *P. plesia* Johnson on *Mystromys*, and *P. jonesi* Kellogg and Ferris on *Saccostomus* (murid). These observations also support the theory that the invasion of Madagascar by African mammals occurred perhaps by chance dispersal over the wide stretches of sea in the post-Early Miocene epoch (Coryndon and Savage 1973) and that the ancestral Malagasy mammals already had polyplacids at the time of invasion.

SUMMARY

Analysis of host–parasite associations offers many significant clues regarding taxonomic relationships, the historical biogeography, and the evolution of sucking lice and mammals. The present pattern of host associations in Anoplura has been molded by adaptation and evolution involving multispecies interactions of sucking lice and mammals, and host–parasite specificity is the height of such obligatory associations.

Most of the associations between sucking lice and mammals followed the model of primary infestation. Certain mammalian species were colonized by specific lice before host cladogenesis; these lice then coevolved with the hosts and speciated along with adaptive radiation of the hosts throughout the entire history of their associations. The distribution of Anoplura has been modified by emigration and extinction (secondary absence). Although the cladogenesis of Anoplura is closely related to mammalian evolution, the evolutionary rate often differs between sucking lice and mammals as well as between two parasite taxa within the same genus.

Infestations of the lipotyphlan insectivores by hoplopleurid and polyplacid lice are secondary, but sucking lice on the menotyphlans represent primary infestations for each host family. The specialization of *Hamophthirius* supports the idea that the Dermoptera had a long independent history (Dawson 1967) and that ancestral dermopterans must have had louse infestations. The three genera of prosimian Primates, *Lemurphthirus, Lemurpediculus,* and *Phthirpediculus,* represent the polyplacid line, and their associations are secondary. However, infestations of anthropoid Primates with *Pediculus, Pthirus,* and *Pedicinus* are primary. It is likely that pediculids were present in the early Oligocene, and that the Eocene hominoids had pthirids before they radiated into Hominidae and Pongidae in the early Oligocene.

The Rodentia harbor 70% of the known Anoplura diversity represented by Enderleinellidae, Hoplopleuridae, and Polyplacidae, and louse infestation is concentrated on the three families of rodents, Sciuridae, Cricetidae, and Muridae. *Fahrenholzia* solely infests the Heteromyidae, and the primary host of *Eulinognathus* is the Dipodidae. The Cricetidae and Muridae are primarily infested with *Hoplopleura* (Hoplopleuridae), *Polyplax* (Polyplacidae), and their minor relatives which account for more than 40% of the known Anoplura fauna. The infestation of *Eulinognathus* on the South American hystricomorph rodents supports the African connection (Traub 1980). The early leporids were invaded by *Haemodipsus,* but other lagomorphs such as *Ochotona* were secondarily infested by rapidly evolving *Hoplopleura* during the Pliocene epoch.

The absence of endemic Anoplura and infestations of *Linognathus* on fissiped carnivores are secondary. Ancestral carnivores had primitive echinophthiriids before they ventured into aquatic habitats, perhaps during the middle to late Oligocene epoch, and the echinophthiriids coevolved and radiated along with the pinnipeds. The three taxa, Pecaroecidae, Microthoraciidae, and Ratemiidae, represent the remnants of once much larger groups in early evolutionary time. Cervidae and Bovidae are primary distributions of *Solenopotes* and *Linognathus*, respectively. Early camelids were infested with microthoraciids long before their Quarternary expansion from North America to South America, Eurasia, and Africa, and many species of microthoraciids became extinct along with the North American camels.

Infestations of the Malagasy Primates with the polyplacids, *Lemurpediculus* and *Phthirpediculus*, and the inclusion of the Ethiopian species in the *Jonesi* group suggest the African connection of endemic Malagasian lice. These observations also support the theory that the invasion of Madagascar by African mammals occurred perhaps by chance dispersal over the side stretches of sea in the post-Early Miocene epoch (Coryndon and Savage 1973) and that the ancestral Malagasy mammals already had polyplacids at the time of invasion (Traub 1980).

Anoplura were present on primitive rodents, carnivores (perhaps Creodonta), and prosimian Primates in the North American Paleocene epoch, as suggested by Kim and Ludwig (1978b) and Traub (1980). If such dates are accepted, the origin of the Protanoplura should be dated back to the late Cretaceous period, which gave rise to the Anoplura, when they may not as yet have been permanent parasites on primordial mammals (Kim and Ludwig 1982). However, the proposed North American Paleocene origin of the Anoplura on the Rodentia (Traub 1980) is somewhat too premature at this time considering the still inadequate data for associations of sucking lice and mammals.

ACKNOWLEDGMENTS

Much of the synthesis on the associations of sucking lice and mammals including many ideas presented in this chapter could not have been made without the base developed through daily discussions I had with my dear colleague, Dr. Herbert W. Ludwig, while I was in residence as a visiting professor at the Universität Heidelberg, West Germany, during the spring and summer of 1976, for which I am much indebted and express my gratitude. I also thank the participants of the Ecosystematics Group of The Frost Entomological Museum, whose discussions during weekly meetings have contributed significantly to the refinement of my conceptualization of and inferences on host–parasite specificity. Much appreciation goes to my colleagues, Charles W. Pitts, Jr., and Clarence H. Collison, and my students, Peter H. Adler, Jae C. Choe, and Allen L. Norrbom, for reading the manuscript critically and making valuable suggestions for improvement. The opinions and inferences presented are, however, my responsibility and do not reflect their views in any way.

REFERENCES

Anderson, S. and J. K. Jones. 1967. *Recent Mammals of the World: A Synopsis of Families*. Ronald Press, New York.

Benoit, P. L. G. 1969. Anoplura recueillis par le Dr. A. Elbl au Ruwanda et au Kivu (Congo). *Rev. Zool. Bot. Afr.* **80**:97–120.

Benoit, P. L. G. 1961. Anoploures du Centre Africain. *Rev. Zool. Bot. Afr.* **63**:231–241.

Brooks, D. R. 1979. Testing the context and extent of host–parasite coevolution. *Syst. Zool.* **28**:299–307.

Chin, Ta-hsiung. 1980. Studies on Chinese Anoplura IV. The description of two new species and proposal of new families and new suborders for the lice of *Typhlomys cinereus* Milne-Edwards. *Acta Acad. Med. Guiyang* **5**(2):91–100.

Coryndon, J. C. and R. J. G. Savage. 1973. The origin and affinities of African mammal faunas. Pages 121–135 in N. F. Hughes (Ed.), *Organisms and Continents Through Time*. Special Paper in Palaeontology No. 12, The Palaeontological Association, London.

Cracraft, J. 1975. Historical biogeography and earth history: perspective for a future synthesis. *Ann. Mo. Bot. Gard.* **62**:227–250.

Dawson, M. R. 1967. Chapter 2. Fossil history of the families of recent mammals. Pages 12–53 in S. Anderson and J. K. Jones (Eds.), *Recent Mammals of the World: A Synopsis of Families*. Ronald Press, New York.

Eichler, W. 1941a. Wintsspezifität und stammesgeschichtliche Gleichlaufigkeit (Fahrenholzsche Regel) bei Parasiten im allgemeinen und bei Mallophagen im besondern. *Zool. Anz.* **132**:254–262.

Eichler, W. 1941b. Korrelation in der Stammesentwicklung von Wirten und Parasiten. *Zeitschr. Parasitenkd.* **12**:94.

Eichler, W. 1973. Neuere überlegungen zu den parasitophyletischen Regeln. *Helminthologia* **14**:1–4.

Emerson, K. C. and R. D. Price. 1981. A host–parasite list of the Mallophaga on mammals. *Misc. Publ. Entomol. Soc. Am.* **12**:1–72.

Hoffstetter, R. 1972. Relationships, origins and history of the ceboid monkeys and caviomorph rodents: a modern reinterpretation. *Evol. Biol.* **6**:323–347.

Honacki, J. H., K. E. Kinman, and J. W. Koeppl. 1982. *Mammal Species of the World*. Allen Press and The Association of Systematics Collections, Lawrence, KS.

Hopkins, G. H. E. 1949. The host-associations of the lice on mammals. *Proc. Zool. Soc. Londor.* **119**:387–604.

Hopkins, G. H. E. 1957. The distribution of Phthiraptera on mammals. *Prem. Symp. Specif. Parasitol. Parasites Verteb. Internatl. Union Biol. Sci. Ser. B* **32**:7–14.

Johnson, P. T. 1960. The Anoplura of African rodents and insectivora. *Tech. Bull. U.S. Dept. Agr.* **1211**:1–116.

Johnson, P. T. 1972. On the rodent-infesting Anoplura of Panama. *Great Basin Nat.* **32**:121–136.

Kemp, T. S. 1982. *Mammal-like Reptiles and the Origin of Mammals.* Academic Press, London.

Kim, K. C. 1965. A review of the *Hoplopleura hesperomydis* complex (Anoplura, Hoplopleuridae). *J. Parasitol.* **51**:871–887.

Kim, K. C. 1966a. The nymphal stages of three North American species of the genus *Enderleinellus* Fahrenholz (Anoplura: Hoplopleuridae). *J. Med. Entomol.* **2**:327–330.

Kim, K. C. 1966b. The species of *Enderleinellus* (Anoplura, Hoplopleuridae) parasitic on the Sciurini and Tamiasciurini. *J. Parasitol.* **52**:988–1024.

Kim, K. C. 1966c. New species of *Hoplopleura* from Thailand, with notes and descriptions of nymphal stages of *Hoplopleura captiosa* Johnson (Anoplura). *Parasitology* **56**:603–612.

Kim, K. C. 1971. The sucking lice (Anoplura: Echinophthiriidae) of the northern fur seal: descriptions and morphological adaptation. *Ann. Entomol. Soc. Am.* **64**:280–292.

Kim, K. C. 1972. Louse populations of the northern fur seal (*Callorhinus ursinus*). *Am. J. Vet. Res.* **33**:2027–2036.

Kim, K. C. 1975. Ecology and morphological adaptation of the sucking lice (Anoplura: Echinophthiriidae) on the northern fur seal. *Rapp. Reun. Cons. Int. Explor. Mer* **169**:504–515.

Kim, K. C. 1977. *Atopophthirus emersoni*, new genus and new species (Anoplura: Hoplopleuridae) from *Petaurista elegans* (Sciuridae, Rodentia), with a key to the genera of Enderleinellinae. *J. Med. Entomol.* **14**:417–420.

Kim, K. C. 1982. Host specificity and phylogeny of Anoplura. Deux. Symp. Specif. Parasit. Parasites Vertebr., 13–17. April 1980. Paris. *Mem. Mus. Nat. Hist. Nat.*, N.S., Ser. A., Zool. **123**:123–127.

Kim, K. C. and P. H. Adler. 1982. Taxonomic relationships of *Neohaematopinus* to *Johnsonpthirus* and *Linognathoides* (Polyplacidae: Anoplura). *J. Med. Entomol.* **19**:615–627.

Kim, K. C. and K. C. Emerson. 1968. Descriptions of two species of Pediculidae (Anoplura) from great apes (Primates, Pongidae). *J. Parasitol.* **54**:690–695.

Kim, K. C. and K. C. Emerson. 1974a. A New *Polyplax* and records of sucking lice (Anoplura) from Madagascar. *J. Med. Entomol.* **11**:107–111.

Kim, K. C. and K. C. Emerson. 1974b. *Latagophthirus rauschi*, a new genus and new species (Anoplura: Echinophthiriidae) from the river otter (Carnivora: Mustelidae). *J. Med. Entomol.* **11**:442–446.

Kim, K. C. and H. W. Ludwig. 1978a. The family classification of the Anoplura. *Syst. Entomol.* **3**:249–284.

Kim, K. C. and H. W. Ludwig. 1978b. Phylogenetic relationships of parasitic Psocodea and taxonomic position of the Anoplura. *Ann. Entomol. Soc. Am.* **71**:910–922.

Kim, K. C. and H. W. Ludwig. 1982. Parallel evolution, cladistics, and classification of Parasitic Psocodea. *Ann. Entomol. Soc. Am.* **75**:537–548.

Kim, K. C. and H. W. Ludwig. 1985. *The Sucking Lice of the World: Biology, Systematics, and Host Associations.* A monograph (in preparation).

Kim, K. C. and C. F. Weisser. 1974. Taxonomy of *Solenopotes* Enderlein, 1904, with redescription of *Linognathus panamensis* Ewing (Linognathidae: Anoplura). *Parasitology* **69**:107–195.

Kim, K. C., B. W. Brown, Jr., and E. F. Cook. 1963. A quantitative taxonomic study of the *Enderleinellus suturalis* complex (Anoplura: Hoplopleuridae). *Syst. Zool.* **12**:134–148.

Kim, K. C., B. W. Brown, and E. H. Cook. 1966. A quantitative taxonomic study of the *Hoplopleura hesperomydis* complex (Anoplura, Hoplopleuridae), with notes on *a posteriori* taxonomic characters. *Syst. Zool.* **15**:24–45.

Kim, K. C., C. A. Repenning, and G. V. Moorejohn. 1975. Specific antiquity of the sucking lice and evolution of otariid seals. *Rapp. Reun. Cons. Int. Explor. Mer* **169**:544–549.

Kuhn, H. J. and H. W. Ludwig. 1967. Die Affenläuse des Gattung *Pedicinus*. *Zeitschr. Zool. Syst. Evol.* **5**:144–297.

Kurtên, B. 1969. Continental drift and evolution. *Sci. Am.* **220**:54–64.

Kurtén, B. 1971. *The Age of Mammals*. Weidenfield and Nicholson, London. 250 pp.

Lavocat, R. 1974. The evolution of the old world and new world hystrichomorphs. *Symp. Zool. Soc. London* **34**:21–60.

Ludwig, H. W. 1968. Zahl, Vorkommen und Verbreitung der Anoplura. *Zeitschr. Parasitenkd.* **31**:254–256.

Ludwig, H. W. 1982. Host specificity in evolution and coevolution of Anoplura and Mammalia. Deux. Symp. Specif. Parasit. Parasites Vertebr., 13–17. April 1980, Paris. *Mem. Mus. Nat. Hist. Nat.*, N.S., Ser. A., Zool. **123**:145–151.

McKenna, M. C. 1970. The origin and early differentiation of therian mammals. *Ann. N.Y. Acad. Sci.* **167**:217–240.

McLaughlin, C. A. 1967. Chapter 11. Aplodontoid, Sciuroid, Geomyoid, Castoroid, and Anomaluroid Rodents. Pages 210–225 in S. Anderson and J. K. Jones (Eds.), *Recent Mammals of the World: A Synopsis of Families*. Ronald Press, New York.

MacArthur, R. H. and E. O. Wilson. 1967. *The Theory of Island Biogeography*. Princeton University Press, Princeton, NJ.

Patterson, B. 1957. Mammalian Phylogeny. *Prem. Symp. Specif. Parasit. Parasites* Vertebr. Internatl. Union Biol. Sci. Ser. B **32**:15–49.

Price, P. W. 1980. *Evolutionary Biology of Parasites*. Princeton University Press, Princeton, NJ.

Price, R. D. and K. C. Emerson. 1971. A revision of the genus *Geomydoecus* (Mallophaga: Trichodectidae) of the New World pocket gophers (Rodentia: Geomyidae). *J. Med. Entomol.* **8**:228–257.

Raven, P. H. and D. I. Axelrod. 1975. History of the flora and fauna of Latin America. *Am. Sci.* **63**:420–429.

Repenning, C. A., R. S. Peterson, and R. S. Hubbs. 1971. Contributions to the systematics of the southern fur seals, with particular reference to the Juan Fernandez and Guadalupe species. Pages 1–34 in W. H. Burt (Ed.), *Antarctic Pinnipedia*, Antarct. Res. Ser. 18. Am. Geophys. Union, 226 pp.

Sarich, V. M. 1969. Pinniped phylogeny. *Syst. Zool.* **18**:416–422.

Seeley, N. J. 1972. A taxonomic study of the *Enderleinellus osborni* complex (Hoplopleuridae, Anoplura). A Paper for Master of Education Degree in Biological Sciences, Pennsylvania State University (unpublished).

Simpson, G. G. 1945. The principles of classification and a classification of mammals. *Bull. Am. Mus. Nat. Hist.* **85**:1–350.

Simpson, G. G. 1950. History of the fauna of Latin America. *Am. Sci.* **38**:361–389.

Sprent, J. F. A. 1969. Evolutionary aspects of immunity in zooparasitic infections. Pages 3–62 in G. J. Jackson, R. Herman, and I. Singer (Eds.), *Immunity to Parasitic Animals*, Vol. 1. Appleton-Century-Crofts, New York.

Stammer, H. L. 1957. Gedanken zu den parasitophyletischen Regeln und zur Evolution der parasiten. *Zool. Anz.* **159**:225–267.

Szidat, L. 1956. Geschichte, Anwendung und einige Folgerungen aus den parasitogenetischen Regeln. *Zeitschr. Parasitenkd.* **17**:237–268.

Traub, R. 1980. The zoogeography and evolution of some fleas, lice and mammals. Pages 93–172 in R. Traub and H. Starck (Eds.), *Fleas*. Proc. Int. Conf. Fleas, Ashton Wold/Peterborough/UK, 21–25 June 1977, A. A. Bolkema, Publ., Rotterdam.

Vanzolini, P. E. and L. Guimaraes. 1955. Lice and the history of South American land mammals. *Rev. Bras. Ent.* (Sao Paulo) **3**:13–46.

Weisser, C. F. 1975. *A monograph of the Linognathidae, Anoplura, Insecta* (excluding the genus *Prolinognathus*). 2 Vols. Doctoral Dissertation, University of Heidelberg, Zool. Inst., Heidelberg, Germany.

Werneck, F. L. 1948. Notas sobre o genero *Enderleinellus* (Anoplura). *Mem. Inst. Osw. Cruz* (Rio de Janeiro) **45**:281–305.

Whittaker, R. H. 1972. Evolution and measurement of species diversity. *Taxon* **21**:213–251.

Whittaker, R. H. 1977. Evolution of species diversity in land communities. *Evol. Biol.* **10**:1–67.

Whittaker, R. H., S. A. Levin, and R. B. Root. 1973. Niche, habitat, and ecotopes. *Am. Nat.* **107**:321–338.

Wilson, R. W. 1960. Early Miocene rodents and insectivores from northeastern Colorado. *Univ. Kans. Paleontol. Contrib.* **7**:1–92.

Wood, A. E. 1977. Paleontology and plate tectonics with special reference to the history of the Atlantic Ocean. R. T. West (Ed.), Proc. Symp. N. Am. Paleont. Conv. II, Lawrence, KS. *Milwaukee Publ. Mus. Spec. Publ.* **2**:95–109.

Chapter 6

Bovicola bovis (Linnaeus)

Evolution of Mallophaga on Mammals

K. C. Emerson and Roger D. Price

INTRODUCTION

Mallophaga (chewing lice) are small wingless obligatory external parasites that live on certain species of land mammals and birds. Although they are more numerous in genera and species on birds than on mammals, this chapter discusses only the Mallophaga found on mammals.

Mallophaga spend the entire life cycle on the host. Their eggs are attached to the hair or fur of the host. Following hatching, there are three nymphal or preadult stages. Adults and nymphs, depending on the species, feed on fur, hair, blood, serum, and secretions of sebaceous glands. The microclimate, composition of food, and other ecological conditions found near the skin of the mammal and their tolerance of these factors likely are the greatest influences in determining host specificity of the Mallophaga. Most species of chewing lice cannot tolerate a temperature much higher than that of their normal host, however, they are more tolerant of a lower temperature. The humidity of the microclimate next to the skin appears to effect Mallophaga. Chewing lice depend on an intimate and continuous association with their host, living at most only a few days when deprived of their normal food.

A few species of Mallophaga are found on several species of related mammals, however, many are limited in distribution to a single species or subspecies of host. Therefore we believe that, as mammalian species evolved, their Mallophaga evolved too, but at a slower rate, because the host's environment sometimes changed more drastically than the microhabitat near the skin of the host where the Mallophaga live. When mammals could not adapt to changing ecological conditions, Mallophaga unique to those hosts probably became extinct with their hosts. To date, the remains of many extinct species of mammals are known, but no fossils of any Mallophaga have ever been found, or have any specimens been found in amber.

Hopkins (1949) has presented an excellent summary of mammal–louse associations, and Werneck (1948, 1950) has treated the taxonomy of mammalian Mallophaga described up to that time.

The classification and scientific names of mammals used in this chapter are those of Honacki et al. (1982). We published a host–parasite list of the Mallophaga on mammals (Emerson and Price 1981) and here follow taxonomic treatments and scientific names used in that list.

DIVERSITY AND HOST DISTRIBUTION OF MALLOPHAGA

The current classification, up to the generic level, of the Mallophaga found on living mammals and the probable faunal region of origin for each mallophagan family are listed as follows:

Suborder AMBLYCERA

> Family Boopiidae—Australasian
> > Genera: *Boopia, Heterodoxus, Latumcephalum, Macropophila, Paraboopia, Paraheterodoxus,* and *Phacogalia*
> Family Trimenoponidae—Neotropical
> > Genera: *Chinchillophaga, Cummingsia, Harrisonia, Hoplomyophilus, Philandesia,* and *Trimenopon*
> Family Abrocomophagidae—Neotropical
> > Genus: *Abrocomophaga*
> Family Gyropidae—Neotropical
> > Subfamily Protogyropinae
> > Genus: *Protogyropus*
> > Subfamily Gliricolinae
> > Genera: *Gliricola, Monothoracius,* and *Pitrufquenia*
> > Subfamily Gyropinae
> > Genera: *Aotiella, Gyropus, Macrogyropus,* and *Phtheiropoios*

Suborder ISCHNOCERA

Family Trichodectidae—Holarctic and/or Ethiopian
   Genera: *Bovicola, Cebidicola, Damalinia, Dasyonyx, Eurytrichodectes, Eutrichophilus, Felicola, Geomydoecus, Lorisicola, Lutridia, Lymeon, Neofelicola, Neotrichodectes, Parafelicola, Procavicola, Procaviphilus, Stachiella, Suricatoecus, Trichodectes,* and *Tricholipeurus*

Family Philopteridae—Madagascar (for genus listed below)
   Genus: *Trichophilopterus*

Suborder RHYNCHOPHTHIRINA

Family Haematomyzidae—Ethiopian and/or Oriental
   Genus: *Haematomyzus*

Kim and Ludwig (1978) believe the suborder Amblycera developed in the late Cretaceous period and early Paleocene epoch, the suborder Ischnocera and Rhynchophthirina in the late Paleocene and early Eocene epochs. However, there are no fossil records to prove or disprove these conclusions. As we later discuss, species in the suborder Amblycera are found only on mammals considered to be the more primitive of the living mammals.

When Simpson completed the manuscript of *The Principles of Classification and a Classification of Mammals* in 1942 (published in 1945), he noted that 54% of the mammalian families and 67% of the genera were extinct. He also stated that ". . . of the 18 surviving orders, 15 include known extinct families." The numbers of extinct genera and species of mammals have increased significantly since 1942, while the number of new species of living mammals found has been almost nil. To date, Mallophaga have been found on only nine orders of living mammals. The number of extinct and living genera in these orders, as given by Simpson, are as follows:

Order	Genera	
	Extinct	Living
Marsupialia	81	57
Primates	99	59
Edentata	113	19
Rodentia	275	344
Carnivora	261	144
Proboscidea	22	2
Hyracoidea	10	3
Perissodactyla	152	6
Artiodactyla	333	86

We believe there are at least 4268 species of living mammals, and a majority of these (3400) do not normally have Mallophaga (Emerson and Price 1981). No chewing lice are known from Monotremata, Insectivora, Scandentia, Dermoptera, Chiroptera, Pholidota, Lagomorpha, Cetacea, Pinnipedia, Sirenia, and a majority of Rodentia. Of the remaining 868 species of living mammals, Mallophaga have been recorded from 356. So it is believed that, to date, less than half of the existing species of Mallophaga on living mammals have been found and described. Many of the endangered or rare and seldom-collected mammals have not been examined for Mallophaga; current laws protecting the larger mammals, especially the Carnivora, Perissodactyla, and Artiodactyla, have reduced opportunities to collect Mallophaga from wild hosts. We hope that, within the next 20 years, Mallophaga can be obtained from all known living species of Carnivora, Perissodactyla, and Artiodactyla.

Here, we summarize the Mallophaga recorded on species of each mammalian family. We omit the mammalian families in each order for which no species of Mallophaga have been recorded, even though the possibility exists of their having the parasites.

Marsupialia

Of 16 families of the Marsupialia, only six families have been recorded to harbor mallophagans.

Family	Number of Living Species	Number of Species Known with Mallophaga
Didelphidae	65	4
Dasyuridae	58	10
Peramelidae	19	3
Caenolestidae	7	1
Vombatidae	2	2
Macropodidae	54	23

One species of *Cummingsia* (Trimenoponidae) is found on the didelphid *Monodelphis*, another on the didelphid *Marmosa*, and a third on the caenolestid *Lestoros*. The other species of *Cummingsia* is found on a Neotropical rodent and probably was acquired from a Neotropical marsupial. All of the mallophagans recorded for the Boopiidae are found on the marsupials of Australia and New Guinea, except for one which originated there on the Dingo and is now found on *Canis* species worldwide. Recent work by Kéler (1971) and Clay (1976) provides excellent data on the Mallophaga found on marsupials of Australia and New Guinea. No comprehensive study of the Mallophaga found on marsupials of the Neotropics has been published.

Primates

Of 14 families of the Primates, chewing lice are found on three prosimian families and one anthropoid family. They are as follows:

Family	Number of Living Species	Number of Species Known with Mallophaga
Lemuridae	16	2
Indriidae	4	2
Lorisidae	11	1
Cebidae	37	7

The philopterid species *Trichophilopterus babakotophilus* (Stobbe) has been recorded from two species of Lemuridae and two species of Indriidae, both of these families found on the island of Madagascar. All records are from specimens taken from museum skins, so some host records are suspect. However, this species is unique, as are many organisms from Madagascar.

The slow loris (Lorisidae) of southeastern Asia has the ischnoceran species *Lorisicola mjobergi* (Stobbe) (Trichodectidae).

Within the Cebidae, douroucouli (*Aotus trivirgatus*) have the gyropid *Aotiella aotophilus* (Ewing). Gyropid Mallophaga are known only from Neotropical mammals. Howler monkeys (*Alouatta* sp.) have an ischnoceran species of *Cebidicola* (Trichodectidae), which has not been found on other hosts. Trichodectid Mallophaga are found worldwide on a variety of hosts, as discussed later.

Edentata

Of seven living species of Bradypodidae only two species harbor the trichodectid mallophagans (Ischnocera): *Lymeon cummingsi* Eichler on the three-toed sloth (*Bradypus tridactylus*) and *Lymeon gastrodes* (Cummings) on the two-toed sloth (*Choloepus didactylus*).

Rodentia

The occurrence of Mallophaga on rodents is sporadic, although the order Rodentia represents the largest diversity among living mammals. The following shows their diversity relationship:

Family	Number of Living Species	Number of Species Known with Mallophaga
Geomyidae	31	31
Cricetidae	623	2
Erethizontidae	9	8
Caviidae	15	10
Dasyproctidae	17	8
Chinchillidae	6	3
Capromyidae	11	4
Ctenomyidae	27	10
Abrocomidae	2	2
Echimyidae	43	21

During the last 35 years, mammalogists have conducted extensive research on the taxonomy and evolution of the pocket gophers (Geomyidae). In the last 12 years, Price and his co-workers, Emerson, Hellenthal, and Timm, have examined Mallophaga from all 31 species and more than 300 subspecies of Geomyidae. Before these studies began, only 11 specific taxa of Mallophaga were known from pocket gophers. Now 102 species and subspecies of *Geomydoecus* (Trichodectidae) are known, and there are undoubtedly others still to be described. The trichodectid *Geomydoecus* is confined to hosts of Geomyidae. Detailed results of these research efforts, dealing with each species of Geomyidae and their Mallophaga, have been published in a series of papers (e.g. Price 1972; Price and Emerson, 1971, 1972; Price and Hellenthal 1975, 1976, 1979, 1980a–c; Timm and Price 1980; and Hellenthal and Price 1976, 1980). This research is continuing; however, it already represents the most exhaustive study done to date of a family of mammals and their Mallophaga. These results indicate that the Mallophaga exhibit varying degrees of host specificity at the host generic, specific, and subspecific levels and that many hosts have more than one species of *Geomydoecus*. These studies are providing new clues to mammalogists on the taxonomy and evolution of the pocket gophers, and it is hoped that in the future comparable research efforts can be expended on other groups of mammals and their Mallophaga.

Cummingsia inopinata Mendez (Trimenoponidae) has been recorded from the cricetid *Thomasomys cinereiventer;* it probably represents a relatively recent transfer from a Neotropical marsupial. Another cricetid, *Scapteromys gnambiquarae,* now has *Gyropus ribeiroi* Werneck (Gryopidae), which it probably acquired from another Neotropical rodent, perhaps Echimyidae.

The ischnoceran *Eutrichophilus* (Trichodectidae) is found only on the New World porcupines (Erethizontidae). The number of *Eutrichophilus* species found on each species of porcupine varies from one to three, and they are host specific. *Eutrichophilus* is unique in that some species have asymmetrical heads, a condition not found in other mammalian Mallophaga.

Five species of *Cavia* (Caviidae) have Mallophaga in the genera *Gliricola* and *Gyropus* (Gyropidae), and *Trimenopon* (Trimenoponidae). *Trimenopon hispidum* (Burmeister) has been found on guinea pigs. The species of *Gliricola* and *Gyropus* on guinea pigs do not appear to be host specific, but are found only on *Cavia*. Other species of each taxon are found on other Neotropical hosts.

Mallophaga found on rodent hosts in the Dasyproctidae, Chinchillidae, Capromyidae, Ctenomyidae, Abrocomidae, and Echimyidae are all in the families Trimenoponidae, Gyropidae, and Abrocomophagidae; all are restricted to Neotropical mammalian hosts. These mallophagan families originated in the Neotropical region and have not become established elsewhere on other hosts. Mammalogists seem to be having difficulty with the taxonomy of the spiny rats of the genus *Proechimys* (Echimyidae). To date, we (Emerson and Price 1975) have seen Mallophaga from 10 species of *Proechimys*, and, based on our studies of these specimens, the Mallophaga will provide clues to help separate the populations of *Proechimys*.

Carnivora

The Mallophaga occur exclusively on land carnivores (Fissipedia) and are completely absent on marine carnivores (Pinnipedia) and some aquatic fissipeds like river otters, which harbor the anopluran Echinophthiriidae. The major diversity of mallophagans is found on Mustelidae and Viverridae. The breakdown of this diversity is the following:

Family	Number of Living Species	Number of Species Known with Mallophaga
Canidae	38	14
Ursidae	7	3
Procyonidae	18	5
Mustelidae	68	36
Viverridae	82	36
Protelidae	1	1
Hyaenidae	3	1
Felidae	37	13

Trichodectes canis (De Geer) (Trichodectidae), the common mallophagan on the domestic dog, has now been collected from other wild hosts in the genus *Canis*. The other chewing louse found on these hosts is *Heterodoxus spiniger* (Enderlein) (Boopiidae), originally found only on the dingo. These two species of Mallophaga have been found occasionally on some of the foxes; however, *Suricatoecus* (Trichodectidae) is the mallophagan normally found on foxes. *Suricatoecus* species apparently are more host specific than are those of *Trichodectes* or *Heterodoxus*. In the future, with more intermingling between domestic dogs and wild canids, the Mallophaga common on the dog may replace other species on wild foxes.

Trichodectes ferrisi Werneck (Trichodectidae) is found on the spectacled bear (*Tremarctos ornatus*) of South America, *T. pinguis pinguis* Burmeister on the European brown bear (*Ursus arctos*), and *T. pinguis euarctidos* Hopkins on the North American black bear (*Ursus americanus*). No Mallophaga have been seen from the Asiatic black bear (*Selenarctos thibetanus*), the North American brown or grizzly bears (*Ursus arctos*), the polar bear (*Thalarctos maritimus*), the sun bear (*Helarctos malayanus*), or the sloth bear (*Melursus ursinus*). If Mallophaga from the bears just listed could be studied, relationships of the living species of bears might be answered.

The North American cacomistle, *Bassariscus astutus*, has the ischnoceran *Neotrichodectes thoracicus* (Osborn) (Trichodectidae). No Mallophaga have been examined from the other species of *Bassariscus*. The North American raccoon, *Procyon lotor*, has *Trichodectes octomaculatus* Paine (Trichodectidae); the crab-eating raccoon, *P. cancrivorus*, has *T. fallax* Werneck. No Mallophaga have been seen on the five insular species of *Procyon*, so it cannot be stated at this time whether or not they might be subspecies of the North American raccoon. The common coati, *Nasua nasua*, has another ischnoceran *Neotrichodectes pallidus* (Piaget); no Mallophaga have been examined from the other two species of coatis. The kinkajou, *Potos flavus*, has *Trichodectes potus* Werneck. No Mallophaga have been seen from the three species of olingos, *Bassaricyon* species. We hope that in the near future Mallophaga can be obtained from the two species of pandas, as they should prove to be interesting to study.

Mallophaga have been collected from slightly more than half of the living species of mustelids, from which the mallophagans collected are *Lutridia*, *Neotrichodectes*, *Stachiella*, and *Trichodectes* (Trichodectidae). The genus *Lutridia* has been found only on four species of otters (Lutrinae). No specimens have been seen to date on the North American or sea otters. The *Neotrichodectes* has been found on some New World species of Mustelinae, the American badger (*Taxidea taxus*), and New World skunks (*Mephitis, Spilogale,* and *Conepatus*), where they are mostly host specific. *Stachiella* has been found on mammalian species of *Mustela, Martes, Galictis, Ictonyx,* and *Poecilictis*, where they are host specific. The genus *Trichodectes* has been found on species of *Martes* (Asian), *Eira, Galictis, Grisonella, Mellivor, Meles,* and *Melogale*. The species of Mallophaga found on the mustelids are generically diverse. Perhaps, when Mallophaga from other species of mustelids are collected, a revision of the mallophagan species will be required.

Less than half of the species of the mammalian family Viverridae have been examined for Mallophaga; the genera found to date are *Felicola, Neofelicola, Parafelicola, Suricatoecus,* and *Trichodectes* (Trichodectidae). Most of the species of these Mallophaga are host specific, and several hosts have more than one species of louse. The species on *Genetta* ought to be reexamined when better identifications of the hosts are available; there is confusion over the early records off *Genetta genetta* and *G. tigrina*. We need new material from properly identified hosts to clarify these records. The ischnoceran *Neofelicola* and *Parafelicola* are found only on species of Viverridae, while *Felicola*, *Suricatoecus*, and *Trichodectes* are also found on other mammalian families. The water mongoose (*Atilax paludinosus*) has the distinction of having seven species of Mallophaga that are host specific to it.

The aardwolf (*Proteles cristatus*) has a *Felicola* species that is only subspecific from the form found on the brown hyaena (*Hyaena brunnea*). Mallophaga have not been examined from the other two species of hyaenas.

Felicola has been the only genus found to date on the cats (Felidae). Mallophaga have been examined from only a third of the species of Felidae, but those studied show a good degree of host specificity.

Proboscidea

There are two species of living elephants (Elephantidae); *Elephas maximus* and *Loxodonta africana*. The rhynchophthirinan species *Haematomyzus elephantis* Piaget (Haematomyzidae) has been recorded on both species of living elephants. We have seen specimens collected from wild Asian elephants but have not seen any from wild African elephants. The only other species of *Haematomyzus* known is a closely related species, *H. hopkinsi* Clay, found on wart hogs in Africa. There is little doubt that the original hosts for this genus were elephants.

Hyracoidea

The living hyraxes consist of three genera, *Dendrohyrax*, *Heterohyrax*, and *Procavia*. Three species each are recognized for *Dendrohyrax* and *Heterohyrax*. There is a controversy over the taxonomy of *Procavia*: previously six species were recognized for this group, but recently it was considered a single species (Honacki et al. 1982). Of the 11 species previously recognized, nine species of hyraxes are recorded to harbor mallophagans.

The ischnoceran genera *Dasyonyx*, *Eurytrichodectes*, *Procavicola*, and *Procaviphilus* (Trichodectidae) are found only on hyraxes. They exhibit host specificity ranging from host genus to host subspecies, and in many cases more than one species of a mallophagan genus will be found on a host subspecies. The host–parasite relationships are similar to those found on the pocket gophers (Geomyidae) and the spiny rats (Echimyidae: *Proechimys*). The present data show that the Mallophaga are specific to various hyrax populations (probably subspecies) and thus could be used to identify more accurately the host of Procaviidae. Cooperation between mammalogists and Mallophaga taxonomists probably can produce a new and useful classification of the Procaviidae.

Perissodactyla

Nine species of extant *Equus* (Equidae) are recognized at present, and most of these equids harbor chewing lice. Mallophagan species of the genus *Bovicola* (subgenus *Werneckiella*) (Trichodectidae) have been recorded from all living species of horses, asses, and zebras except Grevy's zebra (*Equus grevyi*). Moreby (1978) found one species of *Bovicola* on each of the other six species of Equidae, except for the common zebra (*Equus burchelli*) which has two species, one on the subspecies in southern Africa and one on the subspecies of central Africa.

Artiodactyla

Of the eight families of Artiodactyla recognized (Honacki et al. 1982), the following six families have infestation records of Mallophaga:

Family	Number of Living Species	Number of Species Known with Mallophaga
Suidae	8	1
Tayassuidae	2	2
Camelidae	6	3
Tragulidae	4	2
Cervidae	41	14
Bovidae	126	52

Haematomyzus hopkinsi Clay (Haematomyzidae), a close relative of the mallophagan from elephants, has been taken from wart hogs in Africa (Clay 1963).

Macrogyropus dicotylis (Macalister) (Gyropidae) has been recorded on both species of living peccaries. The genus *Macrogyropus*, as noted later, is restricted to larger Neotropical mammals.

Bovicola breviceps (Rudow) (Trichodectidae) has been recorded from the llama (*Lama glama*), the guanaco (*Lama guanicoe*), and the alpaca (*Lama pacos*) (Camelidae). No Mallophaga have been seen on the vicuna (*Vicugna vicugna*) or the camels.

Damalinia traguli Werneck (Trichodectidae) has been taken from the larger Malay chevrotain (*Tragulus napu*) and the lesser Malay chevrotain (*Tragulus javanicus*) (Tragulidae).

Mallophaga have been recorded on only about a third of the species of living Cervidae, those being in the genera *Bovicola*, *Damalinia*, and *Tricholipeurus* (Trichodectidae). The red deer (*Cervus elaphus*) of Europe and Asia and the wapiti (*Cervus canadensis*) of North America have the same mallophagan species, *Bovicola longicornis* (Nitzsch) and *B. concavifrons* (Hopkins). The mule deer (*Odocoileus hemionus*) and the white-tail deer

(*Odocoileus virginianus*), both found in North America, have *Tricholipeurus lipeuroides* (Megnin) and *T. parallelus* (Osborn). Since so many of the other species of Cervidae have not been examined for Mallophaga, what the host specificity of their lice is not known.

The situation concerning Mallophaga on species of Bovidae is not much different from that of the Cervidae, with the same three genera of Mallophaga found to date on less than half of the known living species of hosts. A few hosts have two species of Mallophaga, and a few species of hosts in the same genus share the same mallophagan species. There are, however, some unique examples. The aoudad (*Ammotragus lervia*) has two species of *Bovicola*, subgenus *Werneckiella*, which are not typical of the species of *Bovicola* found on other species of Cervidae or Bovidae; they are typical of the type found on horses, asses, and zebras. Furthermore, the domestic Angora goat has *Bovicola crassipes* (Rudow), not found on domestic short-haired goats. This species is closely related to the species found on the Himalayan tahr (*Hemitragus jemlaica*) and the bharal (*Pseudois nayaur*). Three species of *Damalinia* have been recorded from the brindled gnu (*Connochaetes taurinus*), which constitutes the only record we have of three species of Mallophaga from a species of Bovidae. As in the Cervidae, Mallophaga from the hosts that have no present collection records would be very interesting to study.

The foregoing brief discussion provides some data on the presently known distribution of Mallophaga on living land mammals. A complete list of mammals and their mallophagan parasites has recently been published (Emerson and Price 1981). The classification of Mallophaga currently used has, for several reasons, not been accepted by some entomologists. It is based upon a few morphological similarities and a belief that all Mallophaga, including those found on birds, evolved from a common ancestral stock. Also, it does not recognize the fact that many recent mammals, now extinct, probably also had chewing lice and that Mallophaga have been examined from less than half of the living mammals that probably have these parasites. Although there are no known fossil records of Mallophaga, it is postulated that the evolution of Mallophaga must have occurred later and more slowly than that of their hosts. The following discussion concerns all of these factors and their impact upon the present classification of Mallophaga.

EVOLUTION OF MAMMALIAN MALLOPHAGA

Amblycera

The amblyceran Boopiidae, commonly found on marsupials of New Guinea and Australia, most likely originated in Australia. Marsupials are known from fossils in the Cretaceous period in both South and North

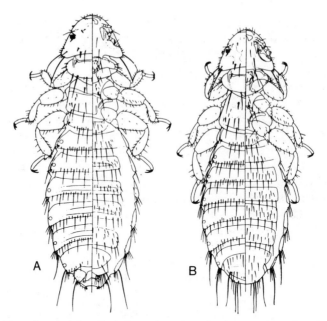

Figure 6.1 *Heterodoxus spiniger* (Enderlein), from *Canis familiaris*. (*A*) Female; (*B*) male. (From K. C. Emerson and R. D. Price. 1975. Mallophaga of Venezuelan mammals. *Brigham Young Univ. Sci. Bull.*, *Biol. Sci.* **20**:1–77. By permission of the Brigham Young University Press.)

America, but not in Australia. They probably were in Australia in the late Cretaceous period, even though no fossils have been found. Those three continents have not been part of a single land mass since mammals evolved (Kurtén 1969; Simpson 1980). Only two species of Boopiidae are not found on marsupials. *Heterodoxus spiniger* (Fig. 6.1) is found on the dingo and has spread in modern times to canids worldwide. This louse is similar to other *Heterodoxus* found on marsupials and must have evolved recently from one of those species as it became established on the feral or semidomesticated dogs introduced by the aborigines. *Therodoxus oweni* Clay is a monotypic species recently described from specimens collected from a cassowary (Casuariiformes) in New Guinea (Clay 1971). This species is properly placed in the Boopiidae and probably originated from a form found on marsupials in the area. Because no other species of this genus, or any closely related to it, have been found on other species of cassowaries, it may be assumed that the species did not evolve from another avian mallophagan form. We believe the Boopiidae originated in Australia from a stock not found elsewhere and became parasitic as the marsupials evolved on that land mass. The family Boopiidae is a logical grouping of related genera and species.

    The Trimenoponidae, found only on certain land mammals in the Neotropics, contains diverse genera of Mallophaga (Ferris 1922; Mendez 1967).

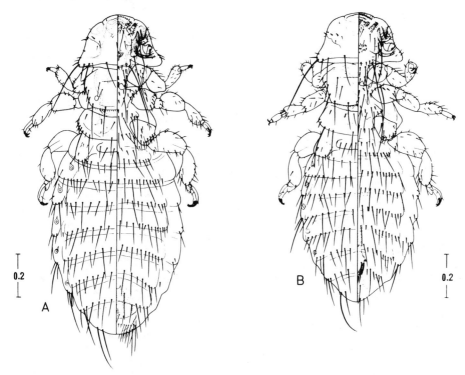

Figure 6.2 *Cummingsia peramydis* Ferris, from *Monodelphis brevicaudata*. (*A*) Female; (*B*) male. (From K. C. Emerson and R. D. Price. 1975. Mallophaga of Venezuelan mammals. *Brigham Young Univ. Sci. Bull., Biol. Sci.* **20**:1–77. By permission of the Brigham Young University Press.)

The genus *Cummingsia* (Fig. 6.2) contains four species at present; three are known from South American marsupials and a fourth is found on a rodent (*Thomasomys cinereiventer:* Cricetidae). These four species are closely related, each having prominent spinelike projections on the ventral side of the head. These structures differ in shape and size from those found on species of Boopiidae, but they are an indication that the species are very old. The species found on the rodent apparently represents a recent transfer from a marsupial. Despite many specimens examined, no Mallophaga have been found to date on the common opossum (*Didelphis marsupialis*); we can offer no explanation for this. We predict that, with better collecting, more species of *Cummingsia* will be discovered. The genus *Cummingsia* is not found outside the Neotropics, so it is likely to have evolved there. It has morphological structures that suggest that it could have originated on marsupials or some early form of Cavimorpha, but we believe the prominent ventral spinelike projections on the head (Fig. 6.2) indicate that it may have originated on the marsupials. We believe it would be logical to remove the genus *Cummingsia* from Trimenoponidae and place it in a new monotypic family.

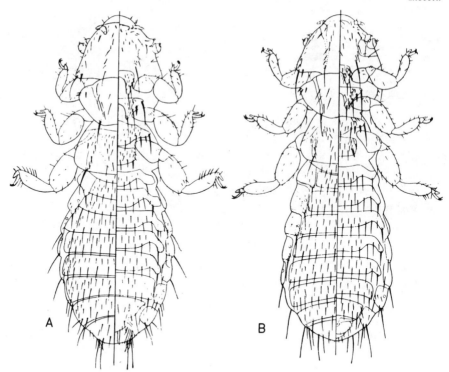

Figure 6.3 *Trimenopon hispidum* (Burmeister), from *Cavia porcellus*. (*A*) Female; (*B*) male. (From K. C. Emerson and R. D. Price. 1975. Mallophaga of Venezuelan mammals. *Brigham Young Univ. Sci. Bull., Biol. Sci.* **20**:1–77. By permission of the Brigham Young University Press.)

The earliest rodents in South America were the Caviomorpha, descendants of some early stock which reached South America from Africa or North America before the early Oligocene epoch (Simpson 1980). The mallophagan genus *Philandesia* is restricted to hosts of the family Chinchillidae, so it probably evolved in South America. The genus *Chinchillophaga* is found only on the mara (*Dolichotis patagonum*), which is probably improperly placed in the family Caviidae. It, too, likely had its origin in South America. *Trimenopon* (Fig. 6.3) is found only on guinea pigs (*Cavia:* Caviidae) and is another of probable South American origin. These genera, we believe, are properly placed in the family Trimenoponidae. The other two genera now in the family Trimenoponidae, *Harrisonia* and *Hoplomyophilus*, are found only on spiny rats (*Proechimys* and *Hoplomys:* Echimyidae) in the Neotropics, so their origin may also have been in South America. It is possible that *Harrisonia, Hoplomyophilus, Philandesia, Chinchillophaga,* and *Trimenopon* had their origin from a common stock that was originally parasitic on an early species of Caviomorpha that is now extinct. These taxa probably evolved later than *Cummingsia,* the genus found on South American marsupials.

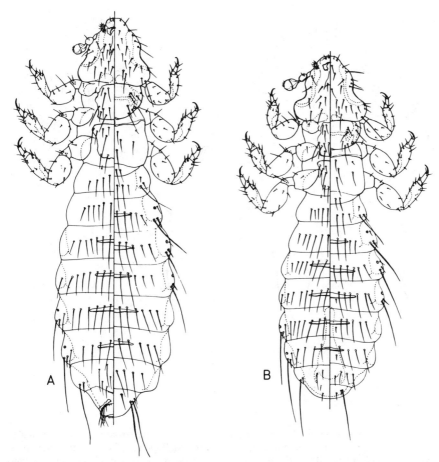

Figure 6.4 *Abrocomophaga chilensis* Emerson and Price, from *Abrocoma bennetti*. (*A*) Female; (*B*) male. (From K. C. Emerson and R. D. Price. 1976. Abrocomophagidae (Mallophaga), a new family from Chile. *Fla. Entomol.* **59**:425–428. By permission of the Florida Entomological Society.)

Abrocomophaga chilensis Emerson and Price (Fig. 6.4), the only species in the amblyceran family Abrocomophagidae, has been found only on the rat chinchilla (*Abrocoma bennetti*: Abrocomidae) in Chile (Emerson and Price 1976). This taxon also is likely to have originated in South America and probably evolved early from the ancestor that produced most of the taxa in the family Trimenoponidae.

The family Gyropidae (Figs. 6.5, 6.6), containing three subfamilies and eight genera, is found only on Neotropical mammals. Diversity within the family is properly represented by this classification. Species of Gyropidae are found on hosts of the rodent Caviidae, Dasyproctidae, Capromyidae, Ctenomyidae, Abrocomidae, Echimyidae, and Cricetidae, and the an-

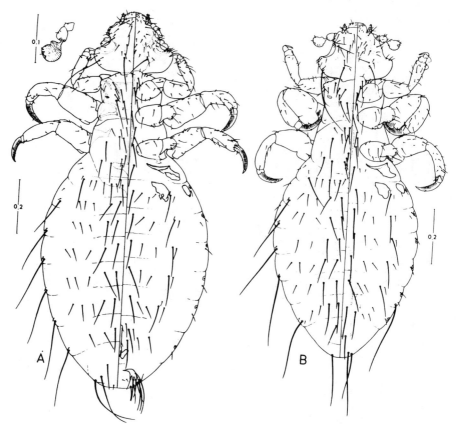

Figure 6.5 *Gyropus wernecki* Emerson and Price, from *Proechimys semispinosus*. (*A*) Female; (*B*) male. (From K. C. Emerson and R. D. Price. 1975. Mallophaga of Venezuelan mammals. *Brigham Young Univ. Sci. Bull., Biol. Sci.* **20**:1–77. By permission of the Brigham Young University Press.)

thropoid Cebidae. The species *Aotiella aotophilus* (Ewing) found only on *Aotus trivirgatus* (Cebidae) and *Gyropus ribeiroi* on *Scapteromys gnambiquarae* (Cricetidae) are recent transfers from other hosts. All species of Gyropidae probably evolved from a common stock that became parasitic on early forms of Cavimorpha. Since no species of Gyropidae are known from Recent native mammals in North America or Africa, it might be assumed the common stock from which these Mallophaga evolved was found only in South America. Differences in morphology and feeding habits between the gyropid species and those of Trimenoponidae and Abrocomophagidae are great enough to suggest that the gyropids evolved from an ancestral form different from those of the other two families and that the evolution of Gyropidae occurred at a later period. The common stock from which the gyropids evolved probably became parasitic on one or more species of Cavimorpha.

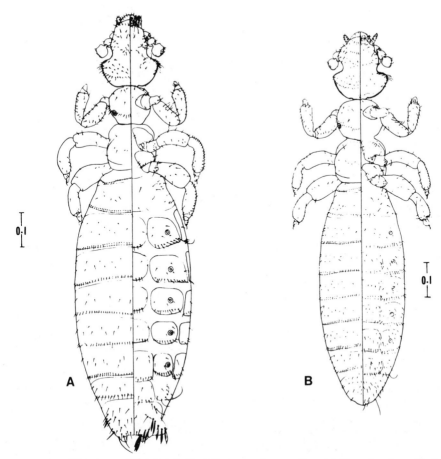

Figure 6.6 *Gliricola porcelli* (Schrank) and *Cavia porcellus*. (*A*) Female; (*B*) male. (From K. C. Emerson and R. D. Price. 1975. Mallophaga of Venezuelan mammals. *Brigham Young Univ. Sci. Bull., Biol. Sci.* **20**:1–77. By permission of the Brigham Young University Press.)

Ischnocera

Species of the ischnoceran family Trichodectidae (Fig. 6.7) are now found on the following mammalian hosts: Primates (Lorisidae and Cebidae), Edentata (Bradypodidae), Rodentia (Geomyidae and Erethizontidae), Carnivora (Canidae, Ursidae, Procyonidae, Mustelidae, Viverridae, Protelidae, Hyaenidae, and Felidae), Hyracoidea (Procaviidae), Perissodactyla (Equidae), and Artiodactyla (Camelidae, Tragulidae, Cervidae, and Bovidae). The Trichodectidae do not occur on any native wild land mammals of Australia or on mammals of North or South America, except those which appeared after the "great American interchange" which started in the early Pliocene (Vanzolini and Guimaraes 1955). This means the taxon evolved from a stock found in Africa, Asia, Europe, or North America. We

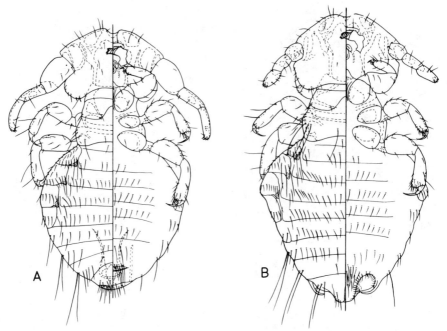

Figure 6.7 *Geomydoecus mexicanus* Price and Emerson, from *Pappogeomys merriami saccharalis.* (*A*) Male; (*B*) female. (From R. D. Price and C. Emerson. 1971. A revision of the genus Geomydoecus (Mallophaga: Trichodectidae) of the New World pocket gophers (Rodentia: Geomidae). (*J. Med. Entomol. Bishop Mus.* **8:**228–257. By permission of the Bernice D. Bishop Museum.)

believe it occurred after Gondwanaland divided into the various future continental land masses and before Laurasia was completely divided into future continental land masses. By the time the "great American interchange" occurred, Mallophaga in the suborder Amblycera were well established in South America. Species of the mallophagan suborder Ischnocera, which include those of Trichodectidae, are obviously of more recent origin than those of Amblycera. This means that species of Trichodectidae now found in South America were obtained during the "great American interchange" and, in so doing, may have displaced some species of the suborder Amblycera on the native mammalian species. The mammals introduced to South America during this interchange took their mallophagan parasites with them. This suggests that the mallophagan family evolved from stock in Africa (less Madagascar) or Europe and Asia. We believe it was probably on the Europe-Asia landmass.

Trichophilopterus babakotophilus (Fig. 6.8) is a philopteran parasite found only on certain primates (Lemuridae and Indriidae) restricted to the island of Madagascar. Currently this species is placed in the family Philopteridae even though all other philopterids occur only on birds. When the species and genus were described by Stobbe (1913), they were placed in the family

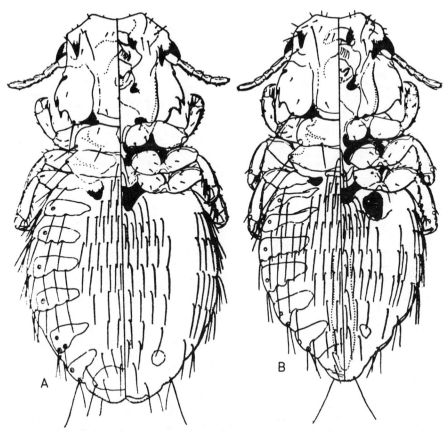

Figure 6.8 *Trichophilopterus babakotophilus* Stobbe, from *Lemur cronatus*. (A) Female; (B) male (from Ferris 1933).

Philopteridae because each leg had two tarsal claws. Mjöberg (1919) and Ewing (1929) considered the species to be sufficiently different so that it should be placed in a monotypic family—Trichophilopteridae. Ferris (1933) moved the genus and species back to the family Philopteridae. Opinions are still divided as to which family it should be assigned. We believe that we have now examined more species of mammalian and avian Mallophaga than any other workers in the field and herewith offer our opinion on the matter. The head, thorax, and abdomen of *T. babakotophilus* are similar to species of trichodectids with these exceptions: (1) both sexes possess spinelike projections on the head such as those found elsewhere on mammal Mallophaga in the genus *Cummingsia* and the family Boopiidae, this suggesting that it is as old as the Mallophaga found on marsupials; (2) the female does not have gonopophyses on the terminal abdominal segment; and (3) each leg has two tarsal claws. It should be noted that the mouth-

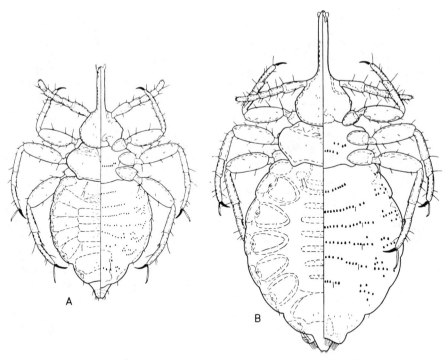

Figure 6.9 *Haematomyzus elephantis* Piaget, from *Elephas maximus*. (*A*) Male; (*B*) female. (From Werneck 1950.)

parts are typical of those found in trichodectid species. The absence of gonopophyses (used to hold the egg while it is being cemented to the hair of the host) and the presence of two claws on each leg are the only features typical of philopterid species. Shape and size of the head, thorax, and abdomen and their chaetotaxy are essentially the same as those found on both trichodectids and philopterids. The male genitalia are unique, differing considerably from any found on trichodectids and philopterids. We believe *Trichophilopterus babakotophilus* is older than any living species of Trichodectidae or Philopteridae, that it evolved from a stock found only on Madagascar, and that it is sufficiently unique to warrant assignment to the monotypic family Trichophilopteridae.

Rhynchophthirina

The mallophagan species *Haematomyzus elephantis* Piaget (Haematomyzidae) (Fig. 6.9) has been found on both species of living elephants (Ferris 1931). It is common on the hairy ears of young elephants. The second species in this family, *H. hopkinsi*, is found on the wart hog. Distinctions between the two species of *Haematomyzus* are so slight that they may prove to be only of a subspecific nature. Since the wart hog and the African elephant share common watering and wallowing sites, it is probable that the species (or subspecies) on the wart hog evolved recently from the species found on the elephant. The Haematomyzidae is the only family in the suborder Rhynchophthirina, and it differs greatly from members of the other two suborders, Amblycera and Ischnocera. Some workers have considered it to represent a link between the Mallophaga and Anoplura (sucking lice); however, it is assigned to the Mallophaga because the mouthparts are mandibulate rather than piercing. We agree with that assignment. The Haematomyzidae probably evolved from stock found in the Old World tropics—either the Ethiopian or Oriental region, or both.

CONCLUSION AND SUMMARY

1. Mallophaga (chewing lice) are obligatory parasitic insects which live on birds and some mammals.

2. Mallophaga have evolved from several stocks after Laurasia and Gondwanaland broke up into the present continents. As the birds and mammals evolved, mallophagan species evolved also, but at a slower rate.

3. The more species and genera of Mallophaga found on a living host, the better the data will be concerning evolution of the host and its relationships with other hosts.

4. There are no fossil records of Mallophaga.

5. Birds have more species and genera of Mallophaga than do mammals.

6. In two groups of mammals—pocket gophers (Geomyidae) and hyraxes (Procaviidae)—each host species has almost as many species of Mallophaga as do species of birds. Extensive research on the Mallophaga found on pocket gophers has contributed greatly to the understanding of the taxonomy and distribution of pocket gophers. For other groups, the degree of assistance to mammalogists may not be as great.

7. Mallophaga have been collected to date from less than half of the living mammals that are expected to have these parasites. Collections of Mallophaga from these hosts would help our understanding of the relationships within the Mallophaga and the mammals.

There are insects, other than Mallophaga, that are ectoparasites on mammals. The Anoplura (sucking lice) are also wingless obligatory external parasites and exhibit varying degrees of host specificity. Some mammals have both Anoplura and Mallophaga, some may have only one of the orders, and some may have neither (Kim and Adler, Chapter 4; Kim,

Chapters 5, 7). Since the Mallophaga and Anoplura developed in different geological periods, the host specificity of each parasite should be considered by mammalogists in their reviews of mammal relationships. There are parasitic mites, ticks, and insects in other orders, each with varying degrees of specificity, that can also assist the mammalogists. We believe that some mammalogists are overlooking very useful data—Mammal–Mallophaga relationships—in their research.

REFERENCES

Clay, T. 1963. A new species of *Haematomyzus* Piaget (Phthiraptera, Insecta). *Proc. Zool. Soc. London* **141**:153–161.

Clay, T. 1971. A new genus and two new species of Boopidae (Phthiraptera: Amblycera). *Pac. Insects* **13**:519–529.

Clay, T. 1976. The *spinosa* species-group, genus *Boopia* Piaget (Phthiraptera: Boopiidae). *J. Aust. Entomol. Soc.* **15**:333–338.

Emerson, K. C. and R. D. Price. 1975. Mallophaga of Venezuelan mammals. *Brigham Young Univ. Sci. Bull., Biol. Ser.* **20**:1–77.

Emerson, K. C. and R. D. Price. 1976. Abrocomophagidae (Mallophaga: Amblycera), a new family from Chile. *Fla. Entomol.* **59**:425–428.

Emerson, K. C. and R. D. Price. 1981. A host–parasite list of the Mallophaga on mammals. *Misc. Publ. Entomol. Soc. Am.* **12**:1–72.

Ewing, H. E. 1929. *A manual of external parasites*. C. C. Thomas, Springfield, IL.

Ferris, G. F. 1922. The mallophagan family Trimenoponidae. *Parasitology* **14**:75–86.

Ferris, G. F. 1931. The louse of elephants. *Haematomyzus elephantis* Piaget (Mallophaga: Haematomyzidae). *Parasitology* **23**:112–127.

Ferris, G. F. 1933. The mallophagan genus *Trichophilopterus*. *Parasitology* **25**:468–471.

Hellenthal, R. A. and R. D. Price. 1976. Louse–host associations of *Geomydoecus* (Mallophaga: Trichodectidae) with the yellow-faced pocket gopher, *Pappogeomys castanops* (Rodentia: Geomyidae). *J. Med. Entomol.* **13**:331–336.

Hellenthal, R. A. and R. D. Price. 1980. A review of the *Geomydoecus subcalifornicus* complex (Mallophaga: Trichodectidae) from *Thomomys* pocket gophers (Rodentia: Geomyidae), with a discussion of quantitative techniques and automated taxonomic procedures. *Ann. Entomol. Soc. Am.* **73**:495–503.

Honacki, J. H., K. E. Kinman, and J. W. Koeppl (Eds.). 1982. *Mammal Species of the World*. Allen Press and Assoc. Syst. Coll., Lawrence, KS.

Hopkins, G. H. E. 1949. The host-associations of the lice of mammals. *Proc. Zool. Soc. Lond.* **119**:387–604.

Kéler, S. von. 1971. A revision of the Australasian Boopiidae (Insecta: Phthiraptera), with notes on the Trimenoponidae. *Aust. J. Zool. Suppl. Ser., Suppl.* **6**:1–126.

Kim, K. C. and H. W. Ludwig. 1978. Phylogenetic relationships of parasitic Psocodea and taxonomic position of the Anoplura. *Ann. Entomol. Soc. Am.* **71**:910–922.

Kurtén, B. 1969. Continental drift and evolution. *Sci. Am.* **220**:54–64.

Mendez, E. 1967. Description of a new genus and species of Trimenoponidae from Panama (Mallophaga). *Proc. Entomol. Soc. Wash.* **69**:287–291.

Mjöberg, E. 1919. Preliminary description of a new family and three new genera of Mallophaga. *Entomol. Tidskr.* **40**:93–96.

Moreby, C. 1978. The biting louse genus *Werneckiella* (Phthiraptera: Trichodectidae) ectoparasitic on the horse family Equidae (Mammalia: Perissodactyla). *J. Nat. Hist.* **12**:395–412.

Price, R. D. 1972. Host records for *Geomydoecus* (Mallophaga: Trichodectidae) from the *Thomomys bottae-umbrinus* complex (Rodentia: Geomyidae). *J. Med. Entomol.* **9**:537–544.

Price, R. D. and K. C. Emerson. 1971. A revision of the genus *Geomydoecus* (Mallophaga: Trichodectidae) of the New World pocket gophers (Rodentia: Geomyidae). *J. Med. Entomol.* **8**:228–257.

Price, R. D. and K. C. Emerson. 1972. A new subgenus and three new species of *Geomydoecus* (Mallophaga: Trichodectidae) from *Thomomys* (Rodentia: Geomyidae). *J. Med. Entomol.* **9**:463–467.

Price, R. D. and R. A. Hellenthal. 1975. A review of the *Geomydoecus texanus* complex (Mallophaga: Trichodectidae) from *Geomys* and *Pappogeomys* (Rodentia: Geomyidae). *J. Med. Entomol.* **12**:401–408.

Price, R. D. and R. A. Hellenthal. 1976. The *Geomydoecus* (Mallophaga: Trichodectidae) from the hispid pocket gopher (Rodentia: Geomyidae). *J. Med. Entomol.* **12**:695–700.

Price, R. D. and R. A. Hellenthal. 1979. A review of the *Geomydoecus tolucae* complex (Mallophaga: Trichodectidae) from *Thomomys* (Rodentia: Geomyidae) based on qualitative and quantitative characters. *J. Med. Entomol.* **16**:265–274.

Price, R. D. and R. A. Hellenthal. 1980. The *Geomydoecus oregonus* complex (Mallophaga: Trichodectidae) of the western United States pocket gophers (Rodentia: Geomyidae). *Proc. Entomol. Soc. Wash.* **82**:25–38.

Price, R. D. and R. A. Hellenthal. 1980. The *Geomydoecus neocopei* complex (Mallophaga: Trichodectidae) of the *Thomomys umbrinus* pocket gophers (Rodentia: Geomyidae) of Mexico. *J. Kans. Entomol. Soc.* **53**:567–580.

Price, R. D. and R. A. Hellenthal. 1980. A review of the *Geomydoecus minor* complex (Mallophaga: Trichodectidae) from *Thomomys* (Rodentia: Geomyidae). *J. Med. Entomol.* **17**:298–313.

Simpson, G. G. 1945. The principles of classification and a classification of mammals. *Bull. Am. Mus. Nat. Hist.* **85**:1–350.

Simpson, G. G. 1980. *Splendid isolation. The curious history of South American mammals.* Yale University Press, New Haven, CT.

Stobbe, R. 1913. Mallophagen. I. Beitrag: Neue Formen von Saugetieren (*Trichophilopterus* und *Eurytrichodectes* nn. gen.). *Ent. Rundsch.* **30**:105–106, 111–112.

Timm, R. M. and R. D. Price. 1980. The taxonomy of *Geomydoecus* (Mallophaga: Trichodectidae) from the *Geomys bursarius* complex (Rodentia: Geomyidae). *J. Med. Entomol.* **17**:126–145.

Vanzolini, P. E. and L. R. Guimaraes. 1955. Lice and the history of South American land mammals. *Rev. Bras. Entomol.* **3**:13–46.

Werneck, F. L. 1948. *Os malofagos de mamiferos. Parte I: Amblycera e Ischnocera (Philopteridae e parte de Trichodectidae).* Revista Brasileira de Biologia, 243 pp.

Werneck, F.L. 1950. *Os malofagos de mamiferos. Parte II: Ischnocera (continuacao de Trichodectidae) e Rhyncophthirina.* Inst. Oswaldo Cruz, 207 pp.

Chapter 7

Linognathus setosus (von Olfers)

Trichodectes canis (DeGeer)

Evolutionary Aspects of the Disjunct Distribution of Lice on Carnivora

Ke Chung Kim

INTRODUCTION

The peculiar absence of endemic Anoplura on land carnivores (Fissipedia) and of Mallophaga on marine carnivores (Pinnipedia) was first noted by

Authorized for publication as paper No. 6758 in the Journal Series of the Pennsylvania Agricultural Experiment Station. A contribution from the Frost Entomological Museum (PaAES Proj. No. 2594).

Hopkins (1949, 1957). He and others after him have sought the cause of such anomaly and disjunction in the host associations (Jellison 1942; Hopkins, 1949, 1957; Vanzolini and Guimaraes 1955; Traub 1980). Here, I describe the host associations and distributions of Anoplura and Mallophaga on Carnivora, and discuss the origin and evolution of the disjunct distributions and host associations.

Discussions of host associations and specificity have been complicated and often confused by the misidentification of both parasites and hosts, the occurrence of erroneous published records, and the poor state of louse taxonomy. However, during the last 30 years, the Smithsonian Institution and other major museums (e.g., Wenzel and Tipton 1966) have made large mammalian ectoparasite collections from around the world. These collections have provided the basis for needed taxonomic studies of Anoplura and Mallophaga that ascertain true parasite–host relationships. The present analysis of the host associations and specificity of mammalian lice is based on the data accumulated from these studies.

The collection records and diversity data, both published and stored in my file, were critically evaluated, and the inaccurate records, including stragglers, contamination, and erroneous identification, were excluded from analysis. I followed the basic classification of mammals by McKenna (1975) and taxonomic treatment of families, genera, and species by Honacki et al. (1982).

Definition of the Problem

The land carnivores, Fissipedia, are infested primarily with chewing lice of the ischnoceran Trichodectidae and three aberrant taxa, *Heterodoxus spiniger* (Enderlein), *Linognathus* (four species), and *Latagophthirus rauschi* Kim and Emerson, whereas the marine carnivores, Pinnipedia, harbor exclusively the sucking lice of Echinophthiriidae (Kim 1982). Considering the host associations and specificity of parasitic Psocodea on Carnivora, we can pose the following questions:

1. Were both the Anoplura and the Mallophaga permanent parasites of the Paleogene Carnivora?
2. Why do chewing lice not occur on aquatic carnivores?
3. Is *Linognathus* endemic to Fissipedia?
4. Is *Latagophthirus rauschi* an endemic species of *Lutra*?

To answer these questions, Hopkins (1949, 1957) and Traub (1980) offered the following two alternative hypotheses:

1. Parasitic Psocodea were originally associated with and coevolved with carnivores since the Paleocene epoch, but some secondarily became extinct on certain host groups (Hopkins 1949, 1957).

2. All the chewing lice of mammals arose in the Southern Hemisphere, whereas the sucking lice originated in the Northern Hemisphere. Thus the mammals of austral origin lack primary Anoplura, and those that arose in North America have no primary Mallophaga (Traub 1980).

ADAPTIVE STRATEGIES OF ANOPLURA AND MALLOPHAGA

Anoplura (sucking lice) and Mallophaga (chewing lice) constitute what are commonly called "lice," referring specifically to the parasitic Psocodea. Lice are obligate, permanent ectoparasites of birds and mammals, and spend the entire life cycle on the host body surface.

Both lice probably have a common psocodean origin, but they have evolved independently as ectoparasites of mammals from their specific preparasitic ancestors in the early Cretaceous period (Kim and Ludwig 1978b, 1982). As a result they have similar adaptations in those traits related to the ectoparasitic mode of life. The common adaptations include (1) dorsoventral flattening of the body, (2) prognathy of the head, (3) reduction in the number of antennal segments, (4) development and rearrangement of the antennal sensilla, (5) reorganization of the thoracic segments, and (7) reduction in the number of molts and shortening of immature stages to three nymphal instars (Kim and Ludwig 1982).

As both lice are closely adapted to the microhabitat and its microclimatic characteristic of the host body covering, their biology is essentially the same except for those parameters involving anatomical and physiological constraints. The eggs are glued to the base of the host hairs and hatch in one to two weeks. After hatching, the first-instar nymphs go through three molts to reach the adult stage, which takes two to three weeks. It usually takes 28–30 days to complete a life cycle (from egg to oviposition) at an optimal temperature (usually 35°C) and relative humidity (75%) (Craufurd-Benson 1941; Matthysse 1946).

Some aspects of the life cycle and reproductive strategy are clearly different between Anoplura and Mallophaga. For example, the short-nosed cattle louse, *Haematopinus eurysternus* Denny, lays 30–35 eggs during the period of the female's adult life (the maximum adult longevity is 16 days), usually one to two eggs per female per day. On the same host, however, the cattle chewing louse, *Bovicola bovis* (Linnaeus), usually produces three to five eggs per female per week for a maximum longevity of 42 days, with an average of 20 eggs for a female's reproductive life. *B. bovis* is primarily parthenogenetic. No parthenogenesis is reported for Anoplura (Craufurd-Benson 1941; Matthysse 1946). Chewing lice may survive off the host for four to five weeks, and as long as 11 weeks (Craufurd-Benson 1941; Matthysse 1946; Mock 1974). They successfully reproduce in low relative humidity, and their reproduction seems to be slowed at high relative humidity (Mock 1974). Sucking lice cannot survive off the host longer than a day.

Body structures in Anoplura are highly adapted for bloodsucking, as they feed exclusively on blood. The head, mouthparts, and thorax are completely modified to maximize bloodsucking, and the tibio-tarsus complex, particularly of the mid- and hindlegs, is well adapted for grasping host hairs (Kim and Ludwig 1978a). The tentorium in the head is completely lacking. The piercing-sucking mouthparts, which are elaborately modified for vessel feeding, consist of the snoutlike proboscis armed with small internal teeth, the sucking pump provided with dilater muscles, and the trophic sac fitted with three long stylets (Stojanovich 1945; Ramcke 1965). The three thoracic segments are completely fused, and the notal components are reduced and concentrated in the center of the dorsum (Kim and Ludwig 1978a).

The mouthparts of Mallophaga consist of paired mandibles for scraping and chewing (Symmons 1952; Weber 1969). However, a few Amblycera like *Trochiloecestes* have the hypopharynx modified into three styletlike structures adapted for piercing (Clay 1949). Mallophaga feed on sloughed epidermal tissues, parts of feathers and hairs, sebum and other epidermal secretions, and serum or rarely blood, depending on the species (Clay 1949, Hopkins 1949, 1957; Emerson and Price 1981). The head is widened posteriorly and with the tentorium, except *Haematomyzus* in which the forehead is prolonged into a long, slender rostrum with only tentorial pits apparent and with apical mandibles (Symmons 1952). The thorax still retains segmentation with the lateral spiracles, although it is flattened and sometimes partly fused (Kim and Ludwig 1978a). The legs are relatively unmodified (Cope 1940, 1941). The ventral surface of the head is often grooved, as in *Bovicola*, to receive host hair. The chewing louse usually has its mandibles encircling a hair lying in the groove (Matthysse 1946).

ASSOCIATIONS OF ANOPLURA AND MALLOPHAGA WITH MAMMALS

Neither Anoplura nor Mallophaga are found on the Monotremata, Cetacea, Sirenia, and Pholidota. Anoplura infest all other major groups of eutherian mammals except Edentata and Proboscida, and their major diversity is found on Rodentia and Artiodactyla. At present, about 500 species are known from about 850 species of mammals (Kim and Ludwig 1978a). On the other hand, Mallophaga are predominantly parasitic on birds (85% of the total known diversity) and are also found on limited groups of mammals including those of the Metatheria like marsupials. There are 467 species and subspecies of Mallophaga from about 360 species of mammals (Emerson and Price 1981 and Chapter 6 of this book; Emerson 1982).

Parasitic Psocodea are closely associated with eutherian mammals, and they show a high degree of host specificity. The majority of both Anoplura and Mallophaga is monoxenous. Monoxeny accounts for 65% of Anoplura (Kim 1982), 79% of the carnivore-infesting Mallophaga, 75% of Echinophthiriidae, and 100% of *Neofelicola*, *Proechinophthirus*, and *Lepidophthirus*.

In certain instances a psocodean taxon is exclusively parasitic on a particular mammal taxon, while it is completely absent in other closely related host taxa. The trichodectid *Geomydoecus*, for example, is found only on Geomyidae, whereas Heteromyidae, its sympatric relative, does not harbor them but instead hosts the anopluran *Fahrenholzia*. *Eutrichophilus* is found only on the New World porcupines (Erethizontidae) and is absent from the Old World Hystricidae (Emerson and Price 1981). Certain host associations are obviously related to geographical factors; a parasite taxon is found on several unrelated host groups of the same geographical area. For example, among the South American Gyropidae, *Gliricola* is found on Caviidae, Capromyidae, and Echimyidae, whereas *Macrogyropus* is on Dasyproctidae and Caviidae and *Phtheiropoios* is on Chinchillidae and Ctenomyidae.

The multispecies associations of parasitic Psocodea occur in limited groups of mammals, although a host species usually has a single species of either Anoplura or Mallophaga (Table 7.1). It is relatively common for the species of some rodent taxa like Sciuridae and Muridae, certain Primates,

Table 7.1 Multispecies Associations of Parasitic Psocodea on Mammals[a]

A + A	M + M	A + M
Primates	Rodentia	Primates
Cercopithecidae	Geomyidae	Cebidae (*Alouatta*)
Hominidae	Erethizontidae	Rodentia
Pongidae	Capromyidae	Echimyidae
Rodentia	Caviidae	Carnivora
Cricetidae	Chinchillidae	Canidae
Muridae	Ctenomyidae	Mustelidae
Sciuridae	Dasyproctidae	Hyracoidea
Thryonomyidae	Carnivora	Procaviidae
Carnivora	Herpestidae	Perissodactyla
Otariidae	Viverridae	Equidae
	Felidae	Artiodactyla
		Suidae
		Tayassuidae
		Camelidae
		Cervidae
		Bovidae

[a] A, Anoplura; M, Mallophaga.

and the pinniped Otariidae to have two or more species of congeneric or heterogeneric Anoplura. Man harbors *Pthirus pubis* (Linnaeus) and two subspecies of *Pediculus humanus* Linnaeus, each confined to a distinct part of the body. Northern fur seals have two species of sucking lice, *Proechinophthirus fluctus* (Ferris) and *Antarctophthirus callorhini* (Osborn) (Kim 1971). It is relatively common for the rodents, Geomyidae, Erethizontidae and several South American taxa, and the feloid carnivores to have a set of mallophagan guilds. For example, the geomyid *Thomomys umbrinus intermedius* Mearns is reported to have six species of *Geomydoecus* (Emerson and Price 1981).

Co-occurrence of both Anoplura and Mallophaga on a host species is primarily limited to Artiodactyla, Perissodactyla, Hyracoidea, the South American Echimyidae, and occasional hosts of Canidae, Mustelidae, and the cebid *Alouatta* (Table 7.2) (Hopkins 1949; Emerson and Price 1981; Kim 1982). The South African springbuck, *Antidorcas marsupialis* (Zimmermann), harbors six species of Linognathidae (five species of *Linognathus* and one *Solenopotes*) and sometimes the trichodectid *Tricholipeurus antidorcas* Bedford (Weisser 1975; Emerson and Price 1981). Similarly, the hyrax, *Procavia capensis* (Pallas), is known to harbor as many as 17 species of Trichodectidae (five species of *Dasyonyx*, 10 *Procavicola*, and 2 *Procaviphilus*) (Emerson and Price 1981) and two species of the anopluran *Prolinognathus* (Kim 1982). Likewise, many host taxa harbor diverse lice: Bovidae with three trichodectids, *Tricholipeurus*, *Damalinia*, and *Bovicola*, and three anoplurans, *Haematopinus*, *Linognathus*, and *Solenopotes*.

Individual hosts rarely harbor both Anoplura and Mallophaga simultaneously, although both lice are reported from the host species. This is particularly obvious when individual hosts are examined. Of the skins of

Table 7.2   Occurrence[a] of Anoplura and Mallophaga on Carnivora, Pinnipedia, Artiodactyla, and Perissodactyla

Mammals	Taxonomic Rank	Exclusive Occurrence		Co-occurrence of Both Taxa
		Anoplura	Mallophaga	
Carnivora	Genus	0.02	0.85	0.13
	Species	0.01	0.86	0.13
Pinnipedia	Genus	1.00	0	0
	Species	1.00	0	0
Artiodactyla	Genus	0.14	0.28	0.58
	Species	0.11	0.19	0.70
Perissodactyla	Genus	0	0	1.00
	Species	0	0	1.00

[a] Percent of occurrence equals the total number of host taxa with known infestation of lice divided by the number of host taxa exclusively infested with Anoplura, Mallophaga, or both.

Table 7.3   Proportions of Anoplura and Mallophaga Found on Individual
Skins of Mammals

Species of Mammals	Number of Skins	Proportion (%)	
		Mallophaga	Anoplura
Hyracoidea, Procaviidae:			
Procavia capensis[a]	16	100	0
Procavia capensis	40	71	29
Heterohyrax brucei[a]	6	100	0
Heterohyrax brucei	3	77	23
Carnivora, Felidae:			
Felis silvestris[a]	3	100	0
Primates, Cercopithecidae:			
Colobus polykomos[a]	18	0	100
Cercopithecus nictitans[a]	6	0	100
Artiodactyla, Bovidae:			
Kobus kob[a]	3	99	1
Ovis dalli	2	100	0
Raphicerus campestris[a]	1	100	0
Raphicerus campestris	2	68	32
Sylvicapra grimmia[a]	10	0	100
Cervidae			
Cervus canadensis	1	100	0
Odocoileus virginianus	3	100	0
Odocoileus virginianus	1	0	100
Odocoileus hemionus	1	0	100
Rangifer tarandus granti	2	0	100

[a]Data based on Hopkins (1949).

Odocoileus virginianus examined, three had only mallophagans and one had
anoplurans. However, some mammals do occasionally harbor both lice
simultaneously. Such a condition is primarily limited to Hyracoidea and
Artiodactyla (Table 7.3). A hyrax of *Procavia capensis* (Pallas) may be infested simultaneously with one or more species of the trichodectid *Dasyonyx, Procavicola,* and *Procaviphilus,* and a species of the anopluran *Prolinognathus.*

EVOLUTION AND CLASSIFICATION OF CARNIVORA

In the early Cretaceous period, primitive mammals such as the multituberculates, marsupials, and the earliest eutherians, were already well estab-

lished in the continent of Euramerica and perhaps in the contiguous super-continent of Gondwanaland (Eisenberg 1981). The Ferae, Insectivora, Archonta, and Ungulata, represented by four tokotheres, *Cimolestes*, *Batodon*, *Purgatorius*, and *Protungulatum*, respectively, were present in North America (McKenna 1975).

The most primitive carnivores, the Creodonta, flourished in North America, Europe, and Asia during the Paleocene and Eocene epochs. By the Oligocene epoch they were found in Africa. Although phylogenetic relationships between creodonts and recent carnivores are uncertain, modern fissipeds can be traced back to the middle Paleocene Miacidae which were apparently the predominant predators in North America and Eurasia during the Paleocene and Eocene (Romer 1966; Repenning and Tedford 1977). By this time, 50 million years before the present, the North American continent was already separated from Eurasia, and South America was totally isolated. Africa and India were approaching Asia, and a filter bridge still existed between Africa and Madagascar. Australia was isolated and on its way south. About 12 million years ago, in Pliocene, the closure of the isthmus took place between North and South America (Smith et al. 1973; Hallam 1973), but the Atlantic Northeast Current was passing through between two continents so that Caribbean seals were washed all the way to Hawaii (Repenning et al. 1979). North America became connected to South America by dry land only about 3 million years ago (Repenning, personal communication).

In the early Oligocene the five modern fissipeds, Canidae, Mustelidae, Procyonidae, Viverridae, and Felidae, were present in North America and Eurasia, and the Ursidae and Hyaenidae appeared later in the Miocene (Dawson 1967). The primitive pinnipeds, the arctoid fissipeds, appeared in the northern Pacific and Atlantic areas during the late Oligocene and the early Miocene (Dawson 1967; Repenning et al. 1976; Repenning and Tedford 1977; Muizon 1982). The first otarioid Enaliarctidae was found in California during the early Miocene. Enaliarctids were derived from primitive ursid land carnivores, perhaps *Cephalogale* in the North Pacific region, during the Oligocene, and they were diverse in the early Miocene (Repenning and Tedford 1977). The phocoid ancestors, the Mustelida, which are a sister group of the lutrines, appeared in the northern hemisphere during the Oligocene. The Semantoridae (*Potamotherium* and *Semantor*), a sister group of the Phocidae, was present in the early and middle Eocene of Europe and western Asia, and the oldest fossil Phocidae, about 15 million years old, is known from the middle Miocene of Europe and North America. The phocoid seals are a sister group of the lutrine mustelids (Muizon 1982).

The modern carnivores include the terrestrial Fissipedia and the marine Pinnipedia, each of which is often treated as an independent order. They are grouped into nine families of Fissipedia and three families of Pinnipedia (Honacki et al. 1982):

FISSIPEDIA

Canoidea: Canidae, Ursidae (including *Ailuropoda*), Procyonidae (including *Ailurus*), Mustelidae

Feloidea: Viverridae, Herpestidae, Protelidae, Hyaenidae, Felidae

PINNIPEDIA

Otarioidea: Otariidae, Odobenidae

Phocoidea: Phocidae

The Pinnipedia are marine carnivores with two major groups: Otarioidea—Otariidae and Odobenidae; and Phocoidea—Phocidae (true seals). The otarioid seals first appeared in the North Pacific region during the middle or late Miocene. By the early Pliocene the otariids had dispersed to the South Pacific Ocean, and during the Pleistocene they spread to their present circumantarctic distribution, never reaching the North Atlantic (Repenning and Tedford 1977). The living Otariidae include the sea lions (Otariinae) and the fur seals (Arctocephalinae). The Odobenidae (walrus) are distributed in the North Atlantic and North Pacific.

The Phocidae originated in northern Europe and North America. The first phocids appeared in the middle Miocene of Atlantic North America and Europe (Dawson 1967; Muizon 1982). The Phocinae (hair seals), well established in the North Atlantic, spread to the Pacific Ocean by way of the Arctic sea during the Pleistocene (Ray 1976), while the Monachinae migrated southward (Muizon 1982). The distribution of extant phocine seals is primarily polar and circumpolar, with three species of *Phoca* that are landlocked in freshwater lakes like Lake Baikal and the Caspian Sea, whereas *Cystophora* is found in North Atlantic and Arctic Oceans, and elephant seals are in the North and South Pacific and in the South Atlantic north to the Falklands and the Patagonia coast (Ray 1976). Of the monachine seals that migrated southward, the Monachini stayed in warm subtropical and tropical waters, but the Lobodontini and Miroungini reached the Antarctic waters during the Pliocene (Muizon 1982). At present *Monachus* is tropical in its distribution, and other taxa, *Leptonychotes, Lobodon, Hydrurga*, and *Ommatophoca*, are distributed in the Antarctic seas.

HOST ASSOCIATIONS AND DISTRIBUTIONS OF LICE ON CARNIVORA

The Fissipedia are host to 91 species of the ischnoceran Trichodectidae and the amblyceran boopiid *Heterodoxus spiniger* whose sister species are primary parasites of marsupials, and four species of the anopluran *Lino-*

MALLOPHAGA

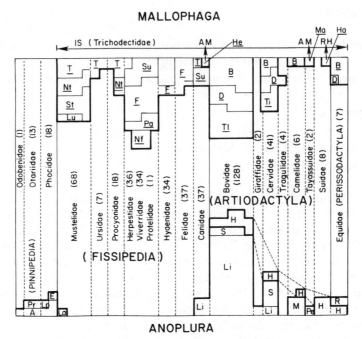

ANOPLURA

Figure 7.1 Host associations and distribution of Anoplura and Mallophaga on Carnivora, Artiodactyla, and Perissodactyla. The number after each mammalian family indicates the number of extant species. The bold line indicates the family boundary. A, *Antarctophthirus*; B, *Bovicola*; D, *Damalinia*; E, *Echinophthirius*; F, *Felicola*; H. *Haematopinus*; Ha, *Haematomyzus*; He, *Heterodoxus*; Li, *Linognathus*; Lp, *Lepidophthirus*; La, *Latagophthirus*; Lu, *Lutridia*; M, *Microthoracius*; Ma, *Macrogyropus*; Nt, *Neotrichodectes*; Nf, *Neofelicola*; Pa, *Parafelicola*; Pe, *Pecaroecus*; Pr, *Proechinophthirus*; R, *Ratemia*; S, *Solenopotes*; St, *Stachiella*; Su, *Suricatoecus*; T, *Trichodectes*; Ti, *Tricholipeurus*; AM, *Amblycera*; IS, Ischnocera; RH, Rhynchophthirina.

gnathus and *Latagophthirus rauschi* (Fig. 7.1). The Pinnipedia are parasitized exclusively by 11 species (four genera) of the anopluran Echinophthiriidae and no chewing lice (Table 7.4). Of 107 species of Carnivora-infesting lice, only three species, which are parasitic on Canidae, are cosmopolitan and oligoxenous: *H. spiniger, Trichodectes canis* (DeGeer), and *Linognathus setosus* (Von Olfers).

Parasitic Psocodea on fissipeds, except for *Heterodoxus*, have the following geographical distribution:

Palearctic Region

Anoplura:	*Linognathus* (2 Holarctic; 1 cosmopolitan species)
Mallophaga, Ischnocera:	Trichodectidae (6 endemic species; 6 Holarctic, 6 Ethiopian/Palearctic, 5 Palearctic/Oriental)

Table 7.4 Diversity and Host Associations of Lice on Carnivora: The Number of Species (sp.) per Genus (g.) by Host Family[a]

Host (Family, Genera/Species)	Heterodoxus	Suricatoecus	Trichodectes	Neotrichodectes	Stachiella	Lutridia	Felicola	Neofelicola	Parafelicola	Mallophaga	Linognathus	Latagophthirus	Antarctophthirus	Proechinophthirus	Lepidophthirus	Echinophthirus	Anoplura	Total Diversity (g./sp.)
Canidae (14/37)	1	6	1	—	—	—	—	—	—	8	4	—	—	—	—	—	4	(4/12)
Ursidae (7/7)	—	—	2	—	—	—	—	—	—	2	—	—	—	—	—	—	0	(1/2)
Procyonidae (9/18)	—	—	3	2	—	—	—	—	—	5	—	—	—	—	—	—	0	(2/5)
Mustelidae (26/68)	—	—	7	8	9	3	—	4	6	27	—	1	—	—	—	—	1	(5/27)
Viverridae (19/34)	—	1	1	—	—	—	3	4	6	15	—	—	—	—	—	—	0	(5/15)
Herpestidae (17/36)	—	9	—	—	—	—	14	—	—	23	—	—	—	—	—	—	0	(2/23)
Protelidae (1/1)	—	—	—	—	—	—	1	—	—	1	—	—	—	—	—	—	0	(1/1)
Hyaenidae (2/3)	—	—	—	—	—	—	1	—	—	1	—	—	—	—	—	—	0	(1/1)
Felidae (5/37)	—	—	—	—	—	—	9	—	—	9	—	—	—	—	—	—	0	(1/9)
Otariidae (6/13)	—	—	—	—	—	—	—	—	—	0	—	—	2	2	—	—	4	(2/4)
Odobenidae (1/1)	—	—	—	—	—	—	—	—	—	0	—	—	1	—	—	—	1	(1/1)
Phocidae (13/18)	—	—	—	—	—	—	—	—	—	0	—	—	3	—	2	1	6	(3/6)
Total Carnivora (120/393)	1	16	14	10	9	3	28	4	6	91	4	1	6	2	2	1	16	(15/107)

[a]Numbers in parentheses are the number of genera/number of species.

Nearctic Region

Anoplura:	*Linognathus* (2 Holarctic; 1 cosmopolitan)
	Latagophthirus (1 endemic)
Mallophaga, Ischnocera:	Trichodectidae (7 endemic; 6 Holarctic; 8 New World)

Neotropical Region

Anoplura:	*Linognathus* (1 endemic)
Mallophaga, Ischnocera:	Trichodectidae (12 endemic; 8 New World)

Ethiopian Region

Mallophaga, Ischnocera:	Trichodectidae (29 endemic; 6 Ethiopian/ Palearctic)

Oriental Region

Mallophaga, Ischnocera:	Trichodectidae (10 endemic; 6 Palearctic/ Oriental)

Trichodectidae are found widely on the Perissodactyla, Artiodactyla, Hyracoidea, Fissipedia, Lorisidae and Cebidae of Primates, and an unusual trichodectid *Lymeon* on Bradypodidae (Edentata) (Fig. 7.1). Two families of rodents are also infested with trichodectids, Erethizontidae and Geomyidae, each of which is parasitized by their unique chewing lice, *Eutrichophilus* and *Geomydoecus* respectively.

Of the Carnivora-infesting trichodectids, *Trichodectes* is found throughout the world wherever the Canidae, Ursidae, Procyonidae, Mustelidae, and Viverridae are present. *Neotrichodectes* on Mustelidae and Procyonidae has a Holarctic and Neotropical distribution (Fig. 7.2*a*). *Stachiella* is a parasite of the Mustelidae in the Holarctic, Neotropical, and Ethiopian regions. *Lutridia* occurs on the Palearctic *Lutra*, the Neotropical *Pteronura*, and the Ethiopian *Aonyx*. *Suricatoecus* occurs on Canidae, Viverridae, and Herpestidae in the Holarctic, Ethiopian, and Neotropical regions. *Felicola*, *Parafelicola*, and *Neofelicola* are parasites of the Viverridae, Herpestidae, Protelidae, Hyaenidae, and Felidae of the world, except of the Australian region (Fig. 7.2*b*). The largest fauna of Trichodectidae is found in the Ethio-

---

Figure 7.2 Geographical distribution (%) of the Carnivora-infesting Trichodectidae. A. *Trichodectes* group: T, *Trichodectes*; St, *Stachiella*; Lu, *Lutridia*; Nt, *Neotrichodectes*. One cosmopolitan *Trichodectes* species is not included. B. *Felicola* group and *Suricatoecus*: F, *Felicola*; Pa, *Parafelicola*; Nf, *Neofelicola*; Su, *Suricatoecus*. Number indicates the number of species.

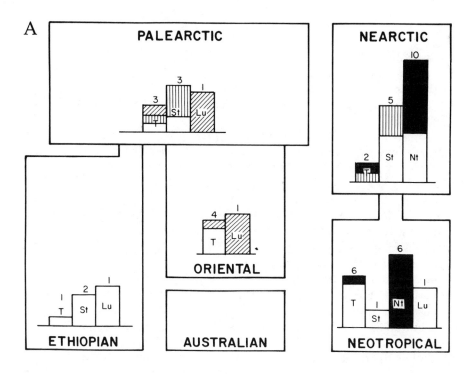

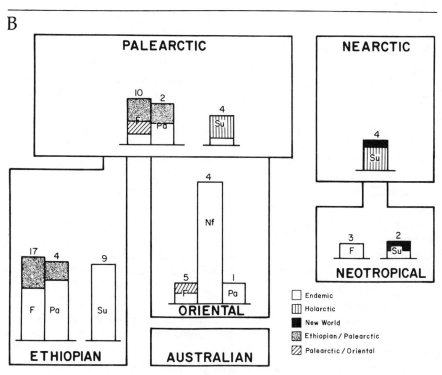

pian region with 28 species (six genera) or 32% of the total known diversity. The Neotropical region is represented by 13% of the total diversity, with 11 species (five genera). The Nearctic fauna is relatively small with seven species represented by two genera, *Neotrichodectes* and *Stachiella* (Table 7.5).

Approximately 44% of the Trichodectidae, primarily *Felicola*, *Parafelicola*, and *Suricatoecus*, is found in the Ethiopian and Palearctic regions. *Felicola* is a primary parasite of feloid hosts, Viverridae (with 3 species), Herpestidae (14 species), Protelidae (1 species), Hyaenidae (1 species), and Felidae (9 species). *Felicola subrostratus* (Burmeister) is widely distributed among feloids. Almost 85% of *Felicola* is found in the Ethiopian (12 species), Palearctic (2 species), and Oriental (2 species) regions, and on the Neotropical *Felis* (3 species), of which *F. felis* (Werneck) is a common oligoxenous species (Tables 7.4, 7.5).

Trichodectes is a primary parasite of canoid hosts, primarily in South America and the Oriental region, with one species on Canidae, two species on Ursidae, three species on Procyonidae, seven species on Mustelidae, and one species on Viverridae. *T. canis* (DeGeer) is a cosmopolitan species found on *Canis*, *Vulpes*, *Dusicyon* (Canidae), and *Viverra* (Viverridae). On the other hand, *Suricatoecus* is a parasite of two major taxa, the Holarctic Canidae and the African Herpesteidae, with one species, *S. hopkinsi* (Bedford), recorded from the African *Nandinia binotata* (Gray). *S. quadraticeps* (Chapman) is an oligoxenous species of the Nearctic *Vulpes velox* (Say) and the New World *Urocyon cinereoargenteus* (Schreber) (Fig. 7.2; Table 7.5).

Stachiella (nine species) is a parasite of the Mustelidae, distributed widely throughout the range of the Holarctic, Neotropical, and Ethiopian regions. With a similar geographical distribution, the Mustelidae are also host to *Lutridia* (three species) and *Neotrichodectes* (10 species), of which two species are found on the New World procyonids, *Bassariscus* and *Nasua*, along with seven species of *Trichodectes*. The Oriental viverrids, *Priondon* and *Paradoxurus*, are the sole hosts of *Neofelicola* (four species), and the African *Genetta* and the Oriental *Viverricula* are the hosts of *Parafelicola* (six species), of which *P. acuticeps* (Neumann) is an oligoxenous species of *Genetta* (Table 7.5) (Emerson and Price 1981).

Of the two aberrant anopluran taxa found on fissipeds, four species of *Linognathus*, primary parasites of Bovidae but with one species on Giraffidae (Artiodactyla), are found on the Canidae. *Linognathus setosus* is an oligoxenous species. This species is found on most species of *Canis*, the circumpolar *Alopex lagopus* (Linnaeus), and the Holarctic *Vulpes vulpes* (Linnaeus); *L. vulpes* Werneck on the Holarctic *Vulpes*; *L. fenneci* Fiedler and Stampa on the Arabian *Vulpes* (*Fennecus*) *zerda* (Zimmermann); and *L. taeniotrichis* Werneck on the South American *Dusicyon* and *Chrysocyon* (Weisser 1975).

Latagophthirus is the sole known member of Echinophthiriidae on Mustelidae. It is in many aspects similar to *Antarctophthirus*, perhaps repre-

Table 7.5 Geographical distribution of Trichodectidae on Carnivora[a]

Chewing Lice	P	O	E	Nt	Na	HA[b]	NW[b]	PO[b]	EP[b]	EO[b]	Total
Felicola group	3	7	15	3	0	3	3	13	23	31	37
Felicola	2	2	12	3	0	2	3	7	19	22	27
Neofelicola	0	4	0	0	0	0	0	4	0	4	4
Parafelicola	1	1	3	0	0	1	0	2	4	5	6
Distribution (%)	0.08	0.18	0.42	0.08	0	0.08	0.08	0.35	0.62	0.84	1.00
Trichodectes group	2	3	4	7	7	12	21	7	6	9	36
Trichodectes	1	3	1	5	0	2	6	5	2	5	14
Neotrichodectes	0	0	0	0	4	4	10	0	0	0	10
Stachiella	1	0	2	1	3	6	4	1	3	2	9
Lutridia	0	0	1	1	0	0	1	1	1	2	3
Distribution (%)	0.06	0.09	0.11	0.19	0.19	0.33	0.58	0.19	0.17	0.25	1.00
Suricatoecus	1	0	9	1	0	4	2	1	10	9	15
Distribution (%)	0.07	0	0.60	0.07	0	0.27	0.13	0.07	0.67	0.60	1.00
Trichodectidae	6	10	28	11	7	19	26	21	39	49	88
Distribution (%)	0.07	0.11	0.32	0.13	0.08	0.22	0.30	0.24	0.44	0.56	1.00

[a]Regions: E, Ethiopian; EO, Ethiopian/Oriental; EP, Ethiopian/Palearctic; Ha, Holarctic; Na, Nearctic; NW, New World (Na + Nt); O, Oriental; Nt, Neotropical; P, Palearctic; PO, Palearctic/Oriental; W, World.
[b]Include the species common to both regions plus species of each region.

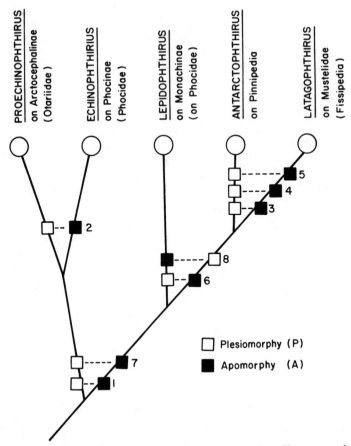

Figure 7.3 Cladogram for the generic taxa of Echinophthiriidae. Character and character states: 1. Thorax: (P) about one-half as wide as abdomen, (A) about three-fourths as wide as abdomen. 2. Forelegs: (P) small and slender with acuminate claw, (A) subequal to fore and midlegs. 3. Number of antennal segments: (P) five, (A) four or three. 4. Notal pit: (P) distinct, (A) indistinct. 5. Dorsal principal thoracic setae: (P) single, (A) multiple. 6. Thorax with: (P) spiniform setae or pegs, (A) scales. 7. Abdomen with: (P) no scales, (A) scales. 8. Male pseudopenis: (P) connected at apex, (A) separated.

senting an offshoot of the primitive *Antarctophthirus* line (Fig. 7.3). *Latagophthirus rauschi* is a primary parasite of the Nearctic *Lutra canadensis*. Other lutrines are infested with the ischnoceran *Lutridia*: *L. exilis* on the Palearctic *Lutra lutra* (Linn.), *L. matschiei* Strobbe on the African *L. maculicollis* Lichtenstein and *Aonyx*, and *L. lutrae* Werneck on the Neotropical *Pteronura brasiliensis* (Blumenbach).

Of the Echinophthiriidae, the most successful *Antarctophthirus* occurs on all pinnipeds, but other taxa are confined to particular groups. *Proechinophthirus* is confined to the fur seals (Arctocephalinae), and a monotypic *Echinophthirius* is an oligoxenous species widely parasitic on the phocine

Phocidae. They represent a primitive phylogenetic line, evidenced by retention of a typical anopluran body shape without scales. *Lepidophthirus* is found exclusively on the monachine Phocidae of the southern hemisphere (Fig. 7.3). In the main echinophthiriid line, *Lepidophthirus*, *Antarctophthirus*, and *Latagophthirus*, represent highly specialized lice with compact bodies and scales on thorax and abdomen. Their body structures and spiracles are adapted to aquatic habitats and microhabitats on the host skin (Kim 1975).

DISCUSSION

Diversity and Host Specificity

The diversity of extant Carnivora is approximately 270 species representing 12 families and 120 genera (Table 7.6), whereas 107 species of parasitic Psocodea found on Carnivora include two suborders (Amblycera and Ischnocera), two families (Boopiidae and Trichodectidae), and nine genera of Mallophaga, plus two families (Echinophthiriidae and Linognathidae) and six genera of Anoplura (Table 7.4). So far only 137 species of Carnivora, representing one-half of the total diversity, are recorded as harboring parasitic Psocodea (Emerson and Price, Chapter 6) (Table 7.6). The remaining carnivores need to be closely examined for louse infestations.

Table 7.6   Carnivora Diversity and Hosts with Recorded Infestation of Parasitic Psocodea

Carnivora (Family)	Diversity		Carnivora with Lice		Percent Infested[a] (Genera/Species)
	Genera	Species	Genera	Species	
Canidae	14	37	7	20	0.50/0.54
Ursidae	7	7	2	3	0.29/0.42
Procyonidae	9	18	4	5	0.44/0.28
Mustelidae	26	68	17	36	0.65/0.53
Viverridae	19	34	8	13	0.42/0.38
Herpestidae	17	36	13	24	0.76/0.67
Protelidae	1	1	1	1	1.00/1.00
Hyaenidae	2	3	2	1	1.00/0.33
Felidae	5	37	1	13	0.20/0.35
Otariidae	6	13	6	7	1.00/0.54
Odobenidae	1	1	1	1	1.00/1.00
Phocidae	13	18	11	13	0.85/0.72
Total	120	273	74	137	0.62/0.50

[a] Percent of the total Carnivora diversity with louse infestation records.

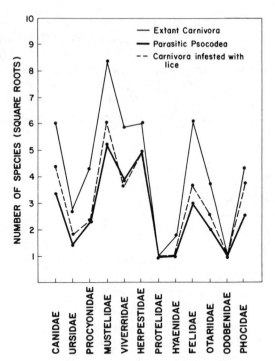

Figure 7.4   Diversity relationships between parasitic Psocodea and their hosts.

The diversity patterns of parasitic Psocodea and their Carnivora hosts are closely congruent (Fig. 7.4). The similar patterns are maintained whether the diversity of parasitic Psocodea is compared with the total Carnivora diversity or with the number of carnivores infested with lice. When those Carnivora not recorded to have louse infestations are carefully examined, the result will be an increase of 30–40% of the known diversity, or approximately 50 new species totaling 190 species, primarily in Mustelidae, Procyonidae, Viverridae, Felidae, and Otariidae. However, this increase should not affect the basic diversity patterns (Fig. 7.4). This observation suggests that adaptive radiation of the ischnoceran Trichodectidae has followed closely the evolution of Fissipedia and that echinophthiriid evolution was closely parallel with the adaptive radiation of Pinnipedia.

Along with Anoplura and Mallophaga, some other parasitic arthropods are also found on Carnivora. Marine carnivores harbor the mesostigmatid Halarachnidae (Acarina) in their respiratory organs, while they are infested with the anopluran Echinophthiriidae on their body surfaces (Dailey and Brownell 1972; Kim 1975; Kim et al. 1980). The fissipeds are infested with numerous other parasitic arthropods in addition to parasitic Psocodea. Ectoparasites include Acarina: Ixodidae, Demodicidae, Trombiculidae, Cheyletidae, Sarcoptidae, and Psoroptidae; Insecta: Siphonaptera, Ano-

Table 7.7 The host specificity[a] of Anoplura and Mallophaga on Carnivora

Parasitic Psocodea	Number of Species	Number of Hosts				
		1	2	3	4	> 5
MALLOPHAGA	91[c]	0.79	0.11	0.03	0.03	0.03
Boopiidae (Am)[b]	1	0	0	0	0	1.00
Heterodoxus	1	0	0	0	0	1.00
Trichodectidae (Is)[b]	90[c]	0.86	0.11	0.03	0.03	0.02
Felicola	28	0.87	0.04	0.04	0	0.04
Lutridia	3	0.66	0.33	0	0	0
Neofelicola	4	1.00	0	0	0	0
Neotrichodectes	4	1.00	0	0	0	0
Parafelicola	6	0.70	0.20	0.10	0	0
Stachiella	9	0.78	0.11	0.11	0	0
Suricatoecus	16	0.87	13	0	0	0
Trichodectes	14	0.87	0.06	0	0.06	0
ANOPLURA	16[c]	0.56	0.13	0.06	0	0.25
Echinophthiriidae	12[c]	0.75	0.08	0	0	0.17
Antarctophthirus	6	0.66	0.16	0	0	0.16
Proechinophthirus	2	1.00	0	0	0	0
Lepidophthirus	2	1.00	0	0	0	0
Echinophthirius	1	0	0	—	0	1.00
Latagophthirus	1	1.00	0	0	0	0
Linognathidae	4	0.25	0	0.25	0	0.50
Linognathus	4	0.25	0	0.25	0	0.50

[a] Percent specificity.
[b] Am, Amblycera; Is, Ischnocera.
[c] Total.

plura, Mallophaga. Endoparasites include Acarina: Halarachnidae (Wenzel and Tipton 1966; Flynn 1973, Index-Catalog 1968–1981). Among the most common ectoparasitic guilds are ticks and fleas, along with sucking lice and chewing lice (Amin 1973; Coultrip et al. 1973; Rogers 1975; Stone and Pence 1977).

The host associations of Anoplura and Mallophaga with Carnivora show a high degree of host specificity (Table 7.7). On the fissipeds the species of Trichodectidae are 86% monoxenous and 11% oligoxenous, with two host species. Two heterophyletic guilds of parasitic Psocodea also are found on the Fissipedia: the amblyceran *Heterodoxus* and the anopluran *Linognathus*. They are broadly oligoxenous and have a cosmopolitan distribution (Table 7.7, Fig. 7.2). Echinophthiriidae are mostly monoxenous, with the exception of two oligoxenous species, *Echinophthirius horridus* (von Olfers) and

Antarctophthirus microchir (Trouessart and Neumann), which may turn out to be a species complex.

Individual hosts rarely harbor both Anoplura and Mallophaga simultaneously, although both lice are reported from the host species (Table 7.3). For example, three species, the anopluran *Linognathus setosus*, the amblyceran *Heterodoxus spiniger*, and the ischnoceran *Trichodectes canis*, are reported from dogs, but a domestic dog most likely harbors a single species.

The facts on diversity, host specificity, ectoparasitic guilds, and infestations on Carnivora suggest that the species diversity of parasitic Psocodea on fissipeds and pinnipeds must have reached its height and must have maintained the equilibrium with other guilds. Therefore the ectoparasite community of Carnivora is in a steady state as conceptualized by MacArthur and Wilson (1967) and Wilson (1969).

Antiquity of Lice and Their Host Associations

Hopkins (1949) formalized the concept of primary and secondary infestation of parasite–host relationships. Primary infestation is defined as the presence of a parasite taxon on the progenitor of extant hosts dating back to the time when the host groups diverged from the ancestral stock. In a primary infestation individual members of a host taxon are infested with closely related species of a parasite taxon. Secondary infestation describes an infestation of hosts by a parasite taxon after the divergence of a host taxon from its parent stock. Noting a high rate of extinction among parasitic Psocodea and hosts, we expect the distribution of parasites in a given group of hosts to be discontinuous although they were once widespread ("secondary absence") (Kim, Chapters 1 and 5). Extinction could have occurred on some or a majority of its host ranges in such a way that disjunction could be observed in host–parasite associations and geographical distribution. The type of infestation is usually revealed by studying the host associations, the geographical distributions of parasites, and the historical biogeography of mammals. Such observations can be further supported by findings in ecology, behavior, and in the biology of parasites and their hosts.

The survival of parasitic Psocodea as obligate and permanent ectoparasites of mammals depends largely on the ecological and evolutionary success of their hosts. The pattern of close host association of Anoplura and Mallophaga with Carnivora represents a manifestation of such interactive evolutionary processes (Tables 7.4, 7.5, Fig. 7.1, 7.2).

The Echinophthiriidae are highly specialized and thus have developed unique morphological and ecological adaptations not found in other Anoplura (Kim 1971, 1972, 1975; Kim and Emerson 1974). Such specialization is a culmination of continuous evolutionary interactions between ectoparasites and carnivores and between parasitic guilds on hosts, and of adaptations by sucking lice to selection pressures of two contrasting environments, sea and land.

Louse transmission rarely occurs between two different hosts owing to the peculiar behavior of pinnipeds: there is little opportunity for two seals or two different species to come in contact at sea. On land, each pinniped species usually occupies a distinct territory—although occasional interspecies contacts are possible, which often produce contamination or straggling among syntopic species (Peterson 1968; Ridgway 1972). Studies on the ecology and biology of several species of Echinophthiriidae support the observation that the echinophthiriids are primarily transmitted from nursing cows to newborn pups (Kim 1975), and that they must have associated and evolved with pinnipeds from the time ancestral pinnipeds ventured into marine life—and also with aquatic mustelids from the day of primitive mustelids. These studies include Murray and Nichols (1965) on *Lepidophthirus macrorhini* Enderlein; Murray, Smith, and Soucek (1965) on *Antarctophthirus ogmorhini;* Kim (1971, 1972, 1975) on *Proechinophthirus fluctus* and *Antarctophthirus callorhini* on the northern fur seal; and Kim and Emerson (1974) on *Latagophthirus rauschi.*

The following Carnivora associations of Anoplura and Mallophaga represent primary infestations: *Trichodectes* and its sister taxa on Canoidea (Canidae, Procyonidae, Mustelidae); *Felicola* and its sister taxa on Feloidea (Herpestidae, Viverridae, Hyaenidae, Felidae); Echinophthiriidae on Pinnipedia; *Latagophthirus* and *Lutridia* on the lutrine Mustelidae. *Linognathus* and *Solenopotes* on Artiodactyla (Cervidae, Giraffidae, Bovidae) also represent primary infestations (Tables 7.4, 7.5). The presence of the heterophyletic marsupial-infesting *Heterodoxus* and the ungulate lice *Linognathus* on Canidae is considered secondary.

The pattern of host associations in parasitic Psocodea is shaped by (1) original association of the progenitors of extant taxa and their hosts ("primary infestations"), (2) host and parasite extinction ("secondary absence"), (3) dispersal and host transfer ("secondary infestation"), (4) host dispersal and migration, and (5) evolutionary differentiation (Hopkins 1949, 1957; Ludwig 1982; Kim, Chapters 1 and 5). If the parasite progenitors survived on the original host and evolved along with the host through adaptive radiation, the extant host association of a parasitic psocodean taxon would be limited to the particular host taxon that is the descendant of the original host. This is exemplified by the echinophthiriid associations of pinnipeds (Eichler 1948; Inglis 1971). However, the host associations of certain parasitic Psocodea may become discontinuous by the extinction of a parasite population or of the original hosts. The causes of such extinction might include climatic or geomorphological changes. Some "generalist" species may transfer to new heterophyletic hosts just as *Linognathus* and *Heterodoxus* invaded Canidae ("secondary infestation"). Host associations may also become disjunct by host dispersal and speciation.

The specific associations of parasitic Psocodea with their respective host taxa also suggest a phylogenetic history. The chewing lice of the *Felicola* group, including *Felicola, Neofelicola,* and *Parafelicola,* are primary feloid parasites, whereas the *Trichodectes* group include the canoid ectoparasites *Trichodectes, Neotrichodectes, Stachiella,* and *Lutridia* (Table 7.4). *Suricatoecus* is closely related to *Felicola* but transitional to *Trichodectes* (Hopkins and Clay 1952), and this taxonomic status is supported by its host associations with both the canoid Canidae and the feloid Herpestidae and Viverridae (Table 7.4). These facts suggest that the progenitors of the *Trichodectes* group were present on canoid ancestors and that those of the *Felicola* group were already associated with feloid ancestors well before their adaptive radiation led to the divergence of modern taxa. *Suricatoecus* must have originated on Herpestidae in the Ethiopian region and secondarily invaded Canidae and Viverridae. The radiation and movement of Canidae through the Holarctic pathway to the Neotropics increased the diversity of *Suricatoecus.*

The Otarioidea, including fur seals, sea lions, and walrus, originated in the North Pacific, whereas the Phocoidea (seals) originated in the North Atlantic (Repenning et al. 1976; Muizon 1982). The echinophthiriid *Antarctophthirus* is, so far, found on all of the extant pinniped taxa except the Phocinae, which is parasitized by *Echinophthirius* and *Lepidophthirus* (Table 7.2, Fig. 7.3). This suggests that the echinophthiriids were widely parasitic on pinnipeds of both lineages in the early Miocene epoch.

The first pinniped Enaliarctidae were present in California during the early Miocene. The arctoid carnivore, *Enaliarctos mealsi,* is considered to be a transitional species between ancestral ursids and modern seals. This species departed in structure from terrestrial ursids (Hemicyonidae) and evolved in the direction of aquatic pinnipeds (Otarininae) in the early Miocene (Mitchell and Tedford 1973). Their ursid and mustelid ancestors were already prevalent in the early Oligocene of North America (Repenning and Tedford 1977). The presence of echinophthiriids on modern pinnipeds suggests that the Enaliarctidae and their ursid ancestors must have been infested with the echinophthiriid progenitors.

If we accept that the echinophthiriid associations of pinnipeds and aquatic mustelids are primary infestations, we can conclude that the forerunners of the Echinophthiriidae were associated with the ursid and mustelid ancestors of modern seals before they adopted marine life. The phocoid seals are considered a sister group of the lutrine mustelids (Muizon 1982). The occurrence of *Lutridia* on the Old World, African, and South American lutrines (Emerson and Price 1981) and the finding of *Latagophthirus* on the North American *Lutra canadensis* (Kim et al. 1975) further support that the early lutrids and arctoid-mustelid ancestors of pinnipeds were infested with both trichodectids and echinophthiriids during the early Oligocene epoch, 35 million years before the present. These data may even push further back to the Paleocene, 55–65 million years before the present, if the middle Paleocene Miacidae is considered to be ancestral to the arctoid pinnipeds.

The exclusive presence of *Proechinophthirus* on Arctocephalinae (fur seals) of both northern and southern hemispheres suggests that this lineage is of great antiquity. If the evolutionary rates between *Proechinophthirus* and *Antarctophthirus* are comparable, the fact that two species, *P. fluctus* and *P. zumpti,* have evolved on the northern *Callorhinus ursinus* and on the southern *Arctocephalus pusillus*, respectively, suggests that the two fur seals have evolved along separate lineages much more ancient than the sea lion lineages. These lineages can be traced back to the early Pliocene epoch, about 10 million years before the present (Kim et al. 1975). On the other hand, the presence of *Antarctophthirus microchir* on the northern and southern sea lions (Otariinae), *Zalophus* and *Eumetopias* (northern), and on *Neophoca, Otaria*, and *Phocarctos* (southern), suggests that a monospecific lineage of the sucking lice existed for at least two million years and also suggests a monophyletic origin of the sea lions out of an ancestral arctocephaline (Kim et al. 1975).

Accordingly, the specific antiquity of the Echinophthiriidae on pinnipeds and aquatic mustelids puts the beginning of Echinophthiriidae at not later than the middle Tertiary period, or even further back to the Eocene epoch, about 50 million years before the present. Furthermore, the antiquity of trichodectids puts the presence of ischnocerans earlier than the Paleocene epoch, and ancestral trichodectids were probably present on the middle Paleocene Miacidae.

Distribution and Paleogeography

The present geographical distribution of parasitic Psocodea has been shaped by a number of factors: (1) the center of origin where the progenitors of extant taxa arose, (2) successive host availability, (3) geomorphological changes, (4) climatic changes, (5) host dispersal and migration, (6) host transfer, and (7) the extinction of the original host (Hopkins 1957; Pielou 1979). The geographic ranges of a parasite taxon may also become disjunct by host speciation.

The distribution of parasitic Psocodea on Carnivora shows distinct patterns (Fig. 7.2, Table 7.5). The Anoplura are centered in the Holarctic and Oceanic regions, whereas the Mallophaga are concentrated in the Paleotropical and Neotropical realms. The distribution of Mallophaga suggests that the Amblycera have an austral origin, as Traub (1980) theorized, and the Protomallophaga must have existed in the early Cretaceous period, about 120 million years before the present or even earlier, when the continents of Australia, Africa, and South America were still joined together as a supercontinent Gondwanaland (Kurtén 1969; Smith et al. 1981). Kim and Ludwig (1978b, 1982) arrive at the similar conclusion by cladistic analysis of parasitic Psocodea: "The forerunners of the parasitic psocodeans were common in the early Triassic Period, and associated with ancestral birds and mammals during the Jurassic Period. Later, they further exploited their associations by invading new habitats and niches, the skin surface and dermal derivatives of primitive birds and mammals, perhaps during the Cretaceous." Traub (1980) also supported the presence of Anoplura in the Cretaceous.

The Protoanoplura invaded a new habitat (the body surface) on mammals between the Jurassic and Cretaceous periods, and made a second major evolutionary shift by exploiting a new food source (blood), perhaps during the late Cretaceous period or the early Paleocene epoch (Kim and Ludwig 1982).

Of 90 known species of the ischnoceran Mallophaga, 29 species (33% of the total known diversity) are endemic to the Ethiopian region, followed by the Neotropical fauna of 12 species (13%). The distribution of the *Felicola* group and *Suricatoecus* is centered in the Ethiopian region with *Felicola* (46%) on Herpestidae, Protelidae, Hyaenidae, and Felidae; *Suricatoecus* (63%) on Viverridae and Herpestidae; and *Parafelicola* (50%) on Viverridae. The species of *Trichodectes* (36%) on Canidae, Ursidae, Procyonidae, and Mustelidae are centered in the Neotropical region with "spill overs" to the Nearctic or other regions, and *Neotrichodectes* (100%) on Procyonidae and Mustelidae is clearly centered in the Nearctic region. *Stachiella* (60%) is primarily distributed in the Holarctic region, with 25% found in the Ethiopian region (Table 7.5). The anopluran *Latagophthirus* is limited to North America.

The geographical regions are represented by the following taxa of the ischnoceran Trichodectidae (Fig. 7.2):

Palearctic Region:	*Felicola, Parafelicola, Stachiella, Suricatoecus, Trichodectes*
Oriental Region:	*Felicola, Neofelicola, Parafelicola, Trichodectes*
Ethiopian Region:	*Felicola, Parafelicola, Stachiella, Suricatoecus, Trichodectes, Lutridia*
Neotropical Region:	*Felicola, Stachiella, Suricatoecus, Trichodectes, Lutridia*
Nearctic Region:	*Stachiella, Neotrichodectes*

The Trichodectidae are widely parasitic on many different mammals throughout the world, except in the Australian region, although particular taxa are confined to specific host taxa in a particular geographic region. The Trichodectidae infestation of fissiped Carnivora, Hyracoidea,Perissodactyla, and Artiodactyla is so universal that it is considered primary, whereas their associations with Rodentia (Erethizontidae and Geomyidae), Edentata, and Primates (Lorisidae and Cebidae) are sporadic and hence represent secondary infestations. Hyracoidea have been exclusively endemic to

Africa since the Oligocene epoch, and they evolved from Eocene ancestors (Coryndon and Savage 1973). The rhinocerotids, suids, tragulids, and bovids colonized the continent of Africa at the end of the Oligocene, and the equiids and giraffids at the end of the Miocene (Cooke 1968). On the other hand, the Neotropical ungulates are Pliocene immigrants (Keast 1969). During the Eocene epoch the mammalian faunas of the African, Eurasian, and North American continents were already well established with immigrants and endemic taxa (Simpson 1947; Cooke 1968; Keast 1969). In view of these facts, the trichodectid associations with primary hosts suggest that ancestral trichodectids were already widely present on these mammals before the Eocene epoch in Africa.

The taxonomy, diversity, and host associations of Trichodectidae on Carnivora suggest that the *Felicola* group originated in the Paleotropical realm along the line of feloid adaptive radiation where radiation was intense, particularly in the Ethiopian region, and that a few species moved into the Palearctic and even spread to the entire Holarctic region (Fig. 7.2, Tables 7.5, 7.8). On the other hand, the *Trichodectes* group appears to have originated in the New World (Fig. 7.2, Table 7.8), although its canoid hosts including canids and mustelids were already common in both Eurasia and North America during the early Tertiary period (Ewer 1973). *Suricatoecus* is centered in the Ethiopian region (Fig. 7.2, Table 7.5). If all the Carnivora-infesting Trichodectidae are considered, their range is again centered in the Paleotropical realm with 56% of the distribution.

The *Felicola* group is parasitic on the feloids Viverridae, Herpestidae, Protelidae, Hyaenidae, and Felidae. Of the Feloidea, Viverridae (mongooses, civets, and genets) first appeared in the Oligocene epoch of Europe and entered Africa at the end of the Oligocene. Living viverrids are basically the Paleotropical carnivores (Keast 1969). The mongooses or Herpestidae are distributed throughout Africa, the Near East, and Southeast Asia. The monotypic Protelidae are found in southern Africa. The earliest Hyaenidae appeared in the late Miocene epoch of Eurasia; living hyaenids currently occupy Africa, Near East, and central and south Asia (Ewer 1973; Eisenberg 1981). The Felidae, descendants of the Nimravinae known from the early Oligocene to middle Pliocene epoch in North America, Eurasia, and Africa, first appeared in the late Miocene or Pliocene epoch in Eurasia (Dawson 1967). The Felidae appeared in South America during the Uquian of the Pleistocene epoch (Simpson 1980), and the African felids appeared in the early Miocene (Keast 1969). The extant species are found in all regions except for Australia and Madagascar (Ewer 1973; Eisenberg 1981).

Felicola and its sister groups *Neofelicola* and *Parafelicola* are associated with Herpestidae, Viverridae, and other feloids in the Ethiopian, Palearctic, and Oriental regions. This suggests that the *Felicola* group is of Ethiopian origin following the feloid dispersal to the Palearctic and Oriental regions and eventually to North America. The introduction of *Felicola* to South America from North America along with *Felis* must have taken place

Table 7.8 Number of Endemic Species of the Mallophagan Trichodectidae (Ischnocera) on Carnivora with Percent Distribution and Diversity

Mallophaga Trichodectidae (Genera)	Nearctic	Neotropical	Ethiopian	Palearctic	Oriental	Total	Diversity (%)
Felicola	0	3	12	2	2	19	0.37
Parafelicola	0	0	3	1	1	5	0.10
Neofelicola	0	0	0	0	4	4	0.08
Felicola group	0	3	15	3	7	28	0.55
Distribution (%)	0	0.11	0.54	0.11	0.25	—	—
Suricatoecus	0	1	9	1	0	11	0.22
Distribution (%)	0	0.09	0.82	0.09	0	—	—
Trichodectes	0	5	1	1	3	10	0.20
Stachiella	3	1	2	1	0	7	0.14
Neotrichodectes	4	0	0	0	0	4	0.08
Lutridia	0	1	1	0	0	2	0.04
Trichodectes group	7	7	4	2	3	23	0.45
Distribution (%)	0.30	0.30	0.17	0.09	0.13	—	—
Trichodectidae	7	11	28	6	10	51	1.00
Distribution (%)	0.14	0.22	0.55	0.12	0.20	—	—

during the early Pleistocene interchange by way of the Holarctic pathway (Simpson 1947, 1980; Ewer 1973).

The canoid carnivores are parasitized by the *Trichodectes* group, *Suricatoecus*, and the Anoplurans, *Linognathus*, and *Latagophthirus*. The *Trichodectes* group is centered in Mustelidae with 27 species belonging to *Trichodectes*, *Neotrichodectes*, *Stachiella*, and *Lutridia* in all the geographical regions except the Australian (Table 7.4). The host associations of the *Trichodectes* group portray a rather confusing faunal picture. Although the *Trichodectes* group is generally centered in the New World, *Stachiella* is centered (67% distribution) in the Holarctic region, and its distribution is extended to the Ethiopian region (Table 7.5).

The modern canids can be traced to Oligocene ancestors through a series of extinct tertiary taxa of the Miocene *Vulpes* and the Pliocene *Canis* of North America. Except for the Pliocene Palearctic *Nyctereutes*, the North American *Alopex* and *Urocyon*, the Palearctic *Canis*, *Cuon*, and *Vulpes*, and the African *Vulpes* (*Fennecus*) and *Lycaon* all appeared in the Pleistocene epoch (Stains 1975). *Otocyon* is exclusively of the Ethiopian region (Ewer 1973). These canids are primarily parasitized by *Suricatoecus*, which are also primary parasites of the Ethiopian Herpestidae.

The earliest ursid, *Ursavus*, was found in the European middle Miocene epoch (Dawson 1967), and the ursids entered North America in the middle Pliocene (Simpson 1947). Extant bears, *Ursus*, are distributed in the Holarctic region, and the Andean *Tremarctos ornatus* is the descendant of the Pliocene immigrants. No bears occur in Africa; *Ursus* is host to *Trichodectes pinguis* Burmeister and *Tremarctos* to *T. ferrisi* Werneck.

The Procyonidae were present in the early Oligocene epoch of North America and Europe, and their invasion of South America occurred somewhat earlier than that of other northern mammals, around the end of the Miocene epoch (Dawson 1967; Keast 1969). The living Procyonidae are confined entirely to the New World, except for the pandas, confined to the montane region of South China (Eisenberg 1981). Only the New World procyonids are host to *Trichodectes* and *Neotrichodectes*.

The Mustelidae are host to the largest diversity of trichodectids, *Stachiella*, *Lutridia*, *Neotrichodectes*, and *Trichodectes*, two of which are also found on Procyonidae. The earliest Mustelidae appeared in the early Oligocene epoch in North America and Eurasia. The African mustelids appeared in the middle Pliocene epoch; the Neotropical mustelids appeared in the late Pliocene (Keast 1969). The European lutrines appeared in the late Oligocene, and the North American otters existed in the middle Miocene; *Lutra* appeared in the Pliocene in North America and Eurasia (Dawson 1967). The living Mustelidae are distributed throughout the world except for Madagascar and Australasia. The Mustelinae are parasitized by *Stachiella*, *Neotrichodectes*, and *Trichodectes* in the Holarctic, Oriental, Ethiopian, and Neotropical regions. *Trichodectes* is centered in the Mustelinae but has secondarily invaded Procyonidae (three species), Ursidae (two species), Canidae, and Viverridae (one species each). The African *Mellivora capensis* (Schreber) is the host of *T. vosseleri* (Stobbe), and the Old World Melinae, *Meles* and *Melogale,* are also parasitized by *Trichodectes.* On the other hand, the New World meline *Taxidea* and Mephitinae are the hosts of *Neotrichodectes.* The Old World lutrines, *Lutra lutra,* the African *Lutra maculicallis* and *Aonyx,* and the South American *Pteronura* are infested with *Lutridia,* but the North American *Lutra canadensis* is parasitized by the anopluran *Latagophthirus.* The Eurasian procyonids and mustelines moved into North America in the early Oligocene epoch, while the Melinae, Mephitinae, and Lutrinae were already present in North America. Some Eurasian lutrines appear to have already moved into North America by the early Miocene epoch and some into Africa in the early Pliocene. *Lutra* appeared in North America in the mid-Pliocene (Simpson 1947; Keast 1969), but the progenitors of the North American *Lutra* were present much earlier, perhaps in the late Oligocene or early Miocene epoch. Thus the lutrines that entered North America during the Pliocene epoch or earlier must have had *Lutridia,* and the absence of *Lutridia* on *L. canadensis* must have been secondary. These data suggest that the *Trichodectes* group represents the primary parasites of Mustelidae, which perhaps originated in Africa but radiated extensively in the New World. Since the arctoid ancestors of pinnipeds, which date back to the early Miocene (Repenning et al. 1976), harbored echinophthiriids, the progenitors of the North American *Lutra* in the Miocene must have harbored both *Latagophthirus* and *Lutridia.*

The occurrence of *Linognathus traeniotrichis* on the South American *Dusicyon* and *Chrysocyon* suggests that the canids involved in the southern migration during the early Pliocene interchange, about 3 or 4 million years before the present, harbored *Linognathus* (Vanzolini and Guimaraes 1955; Simpson 1980). *Linognathus* includes the primary ectoparasites of the advanced artiodactyles like Bovidae and Giraffidae. Giraffids are late Old World descendants of palaeomerycines that first appeared in the late Miocene epoch, and the bovids can be traced to the late Miocene *Eotragus* in Eurasia and Africa. The host associations and fossil data suggest that the host transfer of *Linognathus* to canids took place sometime during the late Miocene and early Pliocene, perhaps 12 million years before the present in Eurasia (Dawson 1967).

The pinnipeds originated in the temperate waters of the North Atlantic and North Pacific Oceans, the Otarioidea in the North Pacific and the Phocoidea in the North Atlantic (Repenning et al. 1976). During the early Miocene epoch the Phocidae had already differentiated into the Phocinae and Monachinae present in Europe and North America. Considering the host specificity, *Lepidophthirus* must have evolved on the monachine seals after they separated from the main phocid line and moved southward, perhaps during the late Miocene, and similarly, the monotypic *Echinophthirius* developed on the phocine phocids after their separation from the main phylogenetic line in the North Atlantic.

The Phocoidea are considered a sister group of the lutrids (Lutrinae, Mustelidae), and the primitive phocoid Semantoridae is closely related to the lutrids (Muizon 1982). Considering the phylogenetic relationships of phocids and lutrids, the finding of *Latagophthirus* on the North American *L. canadensis* is expected, and *Latagophthirus* must have evolved along with aquatic habitation of lutrids. Furthermore, *Latagophthirus* is a sister group of *Antarctophthirus* which widely infest pinnipeds (Fig. 7.3). These facts suggest that primitive echinophthiriids were present on the arctoid progenitors of pinnipeds as well as on the mustelids, perhaps the Mustelida (Muizon 1982), in the northern hemisphere long before the lutrid cladogenesis, perhaps as early as the late Oligocene (Kim and Emerson 1974; Kim et al. 1975). Accordingly, the arctoid-musteloid ancestors of recent pinnipedia and lutrids must have harbored both the anopluran Echinophthiriidae and the ischnoceran Trichodectidae in the Oligocene epoch.

Archaic Infestation

The Cretaceous radiation of archaic mammals involved the Monotremata, Metatheria, Edentata, and Pholidota (Eisenberg 1981), but the parasitic Psocodea apparently were not involved in this early radiation. Neither Anoplura nor Mallophaga are associated with Monotremata and Pholidota, and the chewing lice found on Edentata appear to be evolutionary emigrants from other hosts ("host-changes" or secondary infestations). In this period existed the progenitors of parasitic Psocodea, which must have frequented archaic mammals (Kim and Ludwig 1978b, 1982) but had not developed any specific associations.

The mallophagan association with Marsupialia is limited to the Amblycera, the Australian Boopiidae, and the South American Trimenoponidae (Emerson and Price 1981). Analysis of the host association suggests that the North American marsupials of the Cretaceous period were not associated with mallophagans. Permanent association of the amblyceran Mallophaga with marsupials had already been made before the final breakup of Gondwanaland during the late Cretaceous period and early Tertiary. The Boopiidae in Australia and Trimenoponidae in South America radiated in the late Paleocene or early Eocene epoch (Smith et al. 1973; Eisenberg 1981).

No mallophagans are associated with Insectivora. The Anoplura associated with Insectivora are Hoplopleuridae and Polyplacidae, the primary parasites of Rodentia. *Neolinognathus* of the Macroscelididae and *Sathrax* and *Docophthirus* of Tupaiidae are endemic to the menotyphlan Insectivora, and other taxa associated with the lipotyphlan Insectivora are typical members of Hoplopleuridae and Polyplacidae. These facts suggest that the progenitors of the Hoplopleuridae on the lipotyphlans such as soricids and talpids made host changes during the Eocene or early Oligocene epoch and that ancestral polyplacids likely invaded the menotyphlans during the Miocene (Kim, Chapter 5). Although various primitive insectivores were present in the Paleocene (Dawson 1967), the associations of Anoplura with Insectivora were not established until the late Eocene and early Oligocene epochs when recent insectivores like soricoids appeared. Likewise, the close associations of Anoplura as permanent ectoparasites with other mammals appear to have developed during the Eocene and Oligocene epochs.

If we consider the distribution of trichodectids and fossil records of the arctoid (canoid) Fissipedia (Repenning et al. 1976), the occurrence of trichodectids ancestral to *Lutridia* on the Mustelidae during the Oligocene and Miocene movements from Eurasia to North America seems tenable. Furthermore, it appears that the arctoid ancestors of recent pinnipeds and the primitive lutrine mustelids in North America were infested with ancestral echinophthiriids in the late Eocene or early Oligocene epoch. In other words, the North American arctoid fissipeds of the late Eocene and early Oligocene must have had primary infestations of both the anopluran Echinophthiriidae and ischnoceran Trichodectidae. If this pattern of host associations has persisted throughout the Cenozoic era, then extant fissipeds should harbor echinophthiriids or their relatives. Likewise, some trichodectids should be found on pinnipeds, unlike the present pattern of their host associations.

Colonization and Extinction

During the Paleocene epoch ancestral Anoplura frequented and began to be more closely associated with certain archaic mammals after successfully exploiting a new food source (blood) (Kim and Ludwig 1982). However, major host associations and rapid radiation of Anoplura appear to have taken place much later during the late Eocene and early Oligocene epochs when Tertiary mammals reached their highest diversity (at ordinal and familial levels) (Lillegraven 1972), and the Anoplura must have succeeded in occupying some arctoid fissipeds, Musteloidea. Throughout this period the North American fissipeds were already occupied by primitive trichodectids.

The Eocene anoplurans and ischnocerans were likely to be more generalized in their biological and ecological requirements than extant taxa, and yet they must have had adaptive strategies to survive and speciate under rigorous selection pressure generated by adaptive radiation of mammals. Biological strategies for the Eocene Anoplura and Ischnocera may be determined by inference from the biological and ecological requirements of extant taxa such as *Haematopinus eurysternus* (Craufurd-Benson 1941) and *Bovicola bovis* (Matthysse 1946; Mock 1974). They include (1) a higher reproductive potential, (2) broader environmental tolerance, (3) longer generation time, (4) greater adult longevity, (5) longer nymphal survival under starvation, (6) oligoxenous or pleioxenous host specificity (Marshall 1981),

Table 7.9   Inferred Biological Attributes of the Eocene Anoplura
and Ischnoceran Mallophaga

Category	Anoplura	Ischnocera
Host associations	Temporary	Permanent
Mouthparts	Complex, specialized: three stylets in a sac	Simple, generalized: a pair of mandibles
Food	Blood, plasma	Epidermal derivatives
Feeding mechanism	Piercing-sucking	Scraping-chewing
Reproduction	Bisexual	Bisexual, parthenogenesis
Environmental conditions	Fluctuating	Stable
Effects on host	Acute	Minimal

and (7) overdispersed population distribution (Crofton 1971). Food was not a limiting factor because food supplies were sufficient for both lice, and each fed on different resources. However, space became a limiting factor in host associations because both lice were tracking the same resource, the fur habitat (Kethley and Johnston 1975).

The ischnoceran Mallophaga already occupied the fur habitats of musteloid fissipeds where microhabitats were stable and food (sloughed epidermal tissues) was readily available. Their population density on hosts must have been relatively high although distribution was overdispersed. They have developed parthenogenesis which increases population size (Table 7.9). However, their effect on the host and the host reactions to their presence must have been minimal because they are basically detritivores feeding on dermal products.

The Anoplura, on the other hand, were colonizers, actively tracking food (blood) and space (skin and fur). They were at a considerable disadvantage to colonize the host in the presence of resident ischnocerans and other ectoparasitic guilds because the process of sanguinivorous feeding takes more time and energy. The sucking louse first must find a microhabitat and position itself in a suitable site. Then it pierces the skin and finally sucks the blood. Furthermore, this process induces acute reactions of the host that include behavioral and immune responses (Table 7.9). Grooming and scratching behavior of the host are major factors in reducing louse infestations and controlling louse distribution on the host (Murray 1955, 1957a, b, c, d, 1961). Sucking lice had to compete with established resident guilds for suitable microhabitat free from host predation.

Invasion of the arctoid fissipeds by anoplurans could have resulted in one of two conditions (MacArthur and Wilson 1967): (1) In the event that the host was heavily infested with ischnocerans and perhaps other guilds, the anoplurans could not succeed in colonization. (2) On the other hand, if the host was barely infested with the resident guilds, the sucking lice could have colonized it. To permanently maintain such colonization, however, this condition would have had to be repeated for many generations. Otherwise, the sucking lice must have become extinct. It is difficult to speculate as to whether sucking lice first became established on fissipeds and later became extinct or whether they never succeeded in colonizing them. Nevertheless, some arctoid fissipeds ancestral to recent pinnipeds and the North American lutrines, which were already infested with ischnocerans, must have been colonized by sucking lice; therefore, they were infested by both ischnocerans and anoplurans until they adopted aquatic life during the middle Oligocene.

Once the Anoplura colonized the habitats on certain arctoid fissipeds, they must have developed adaptations to hold on to hairs effectively, to minimize the time and energy required for feeding, and to avoid competition and predation. They must have optimized their fitness continuously by specialization which warranted residence in the microhabitat on the host. Since the arctoid fissipeds frequented and exploited aquatic habitats, the Anoplura with their adaptations and specialized biological attributes (Table 7.9) were better prepared for survival against flooding and aquatic conditions (Kim and Emerson 1974). On the other hand, the generalized ischnocerans were more likely to be washed off and die in water when their hosts were invading new aquatic (primarily marine) habitats.

CONCLUSIONS AND SUMMARY

No parasitic Psocodea are found on the Monotremata, Cetacea, Sirenia, and Pholidota. Anoplura infest all other major groups of eutherian mammals except Edentata and Proboscida. The majority of Anoplura are found on Rodentia and Artiodactyla, and they are centered in the Ethiopian region. The Mallophaga are predominantly parasitic on birds (85% of the total diversity), but some mallophagans are found on limited groups of mammals. The Amblycera are primarily parasitic on the Australian and Neotropical marsupials and Neotropical rodents. The Ischnocera are permanent parasites of various mammals, and the Rhynchophthirina are parasitic on African and Oriental elephants and African wart hogs. Most lice are monoxenous or host specific at the species level (Table 7.7). The occurrence of multispecies associations in parasitic Psocodea is reported in limited groups of mammals, like Sciuridae and Muridae (Rodentia), certain Primates, and the pinniped Otariidae. Both Anoplura and Mallophaga occasionally coinfest a host species in Artiodactyla, Perissodactyla, Hyracoidea, the South American Echimyidae, and some hosts of Canidae, Mustelidae, and the cebids (Tables 7.1, 7.2). However, individual hosts rarely harbor both Anoplura and Mallophaga simultaneously, although both lice may be reported from the host species (Table 7.3).

The land carnivores (Fissipedia) are host to 90 species of Trichodectidae (Ischnocera) and *Heterodoxus spiniger*, an amblyceran boopiid, and to the species of the Anopluran *Linognathus* and *Latagophthirus* (Table 7.4). The marine carnivores (Pinnipedia) are parasitized solely by the anopluran Echinophthiriidae (currently 11 species known) (Table 7.4). Of 107 species of lice infesting Carnivora, only three species that are parasitic on Canidae are oligoxenous and cosmopolitan: *Heterodoxus spiniger*, *Trichodectes canis*, and *Linognathus setosus* (Table 7.7).

Of the Carnivora-infesting trichodectids (Table 7.4), *Trichodectes* is found on the Canidae, Ursidae, Procyonidae, Mustelidae, and Viverridae throughout the world. *Neotrichodectes* is found on the Holarctic and Neotropical Mustelidae and Procyonidae. *Stachiella* is a parasite of the Holarctic, Neotropical, and Ethiopian Mustelidae. *Lutridia* is found on the lutrines, the Palearctic *Lutra*, the Neotropical *Pteronura*, and the Ethiopian *Aonyx*. *Suricatoecus* is parasitic on Canidae, Viverridae, and Herpestidae in the Holarctic, Ethiopian, and Neotropical regions. *Felicola*, *Parafelicola*, and *Neofelicola* are parasites of the Viverridae, Herpestidae, Protelidae, Hyaenidae, and Felidae of the world (except in the Australian region). *Latagophthirus* is the single endemic anopluran taxon found on the Nearctic lutrine *Lutra canadensis* (Table 7.5). *Linognathus* and *Heterodoxus* are recent invaders of the Canidae through host changes.

A close congruence exists between the diversity patterns of the parasitic Psocodea and their Carnivora hosts (Table 7.6, Fig. 7.4). This similar diversity pattern suggests that the trichodectids radiated closely following the evolution of Fissipedia and that the echinophthiriids evolved in parallel with the adaptive radiation of Pinnipedia.

The following host associations represent primary infestations: *Trichodectes* and its sister taxa on Canoidea, *Felicola* and its sister taxa on Feloidea; Echinophthiriidae on Pinnipedia; *Latagophthirus* and *Lutridia* on the lutrine Mustelidae. The infestations of the marsupial-infesting *Heterodoxus* and the ungulate lice *Linognathus* on Canidae are secondary (Fig. 7.2; Table 7.6).

The progenitors of parasitic Psocodea began to frequent archaic mammals during the Cretaceous period but had not established any specific associations. The marsupial associations of the amblyceran Mallophaga have an austral origin, perhaps developed before the final breakup of the Gondwanaland during the late Cretaceous and early Tertiary periods. However, the North American marsupials were not associated with mallophagans at that time. Although primitive Insectivora were present in the Paleocene epoch, no parasitic Psocodea developed specific associations with them. The associations of Anoplura with insectivores were not established until the late Eocene and early Oligocene epochs.

The Eocene and Oligocene fissipeds in North America were already infested with the ischnoceran Mallophaga. During the late Eocene and early Oligocene epochs the major host associations of Anoplura were established with the North American arctoid fissipeds. The North American arctoid fissipeds ancestral to recent pinnipeds and *Lutra* (or lutrines) were infested by both ischnocerans and anoplurans before they adopted aquatic (primarily marine) life. The ischnoceran Mallophaga on these hosts became extinct as the hosts adopted aquatic life.

At the beginning of the Eocene epoch ancestral trichodectids were already widely present on archaic mammals in Africa. Of the Carnivora-infesting trichodectids, the *Felicola* group (*Felicola*, *Neofelicola*, and *Parafelicola*) is primary parasites of the feloid carnivores, whereas the *Trichodectes* group (*Trichodectes*, *Neotrichodectes*, *Stachiella*) is closely associated with the canoid fissipeds. *Suricatoecus*, which is closely related to *Felicola* but transitional to *Trichodectes*, is parasitic on both the canoid Canidae and the feloid Herpestidae and Viverridae (Table 7.4). These facts suggest that the progenitors of the *Trichodectes* group were present on canoid ancestors and that those of the *Felicola* group were already associated with the feloid ancestors well before they diverged from a common ancestor. *Suricatoecus* must have originated on Herpestidae in the Ethiopian region and secondarily invaded Canidae and Viverridae, and its diversity increased as the Canidae radiated and moved through the Holarctic pathway to South America.

The *Felicola* group originated on Herpestidae in the Paleotropica, most likely in the Ethiopian region, whereas the *Trichodectes* group radiated in the New World, perhaps with a Paleotropical origin, and was associated with the adaptive radiation of Mustelidae. Modern pinnipeds are descendants of the arctoid-musteloid ancestors. The Otarioidea have an arctoid origin (Mitchell and Tedford 1973), whereas the Phocoidea are a sister group of the lutrine mustelids (Muizon 1982). The presence of *Antarctophthirus* on all of the extant pinnipeds (Table 7.4) suggests a wide occurrence of echinophthiriids on the Miocene pinnipeds of both lineages. The progenitors of the Echinophthiriidae must have been associated with the arctoid-musteloid ancestors of modern seals before they adopted marine life. The host transfer of *Linognathus* from Artiodactyla to canids took place sometime during the late Miocene and the early Pliocene epochs in Eurasia. Following the model by Traub (1980) that advocates the North American origin of Anoplura, probably on the Paleocene rodents, the North American origin of the echinophthiriids in the Miocene seems reasonable!

ACKNOWLEDGMENTS

Many colleagues and my students extended their kind assistance to me in numerous ways in preparing this chapter. Of these, I am most grateful to K. C. Emerson for providing me with a list of species of the mammalian Mallophaga and to Charles A. Repenning for his enlightenment and guidance in pinniped evolution. Thanks are due to Peter H. Adler, K. C. Emerson, Roger D. Price, and Charles A. Repenning for reading the manuscript and making valuable suggestions, and to Thelma Brodzina, Judi Hicks, and Jackie Wolfe for patiently typing the manuscript many times.

REFERENCES

Amin, O. M. 1973. A preliminary survey of vertebrate ectoparasites in Southeastern Wisconsin. *J. Med. Entomol.* **10:**110–111.

Clay, T. 1949. Piercing-mouthparts in the biting lice (Mallophaga). *Nature* **164:**617.

Cooke, H. B. S. 1968. The fossil mammal fauna of Africa. *Q. Rev. Biol.* **43:**234–264.

Cope, O. B. 1940. The morphology of *Esthiopterum diomedeae* (Fabricius) (Mallophaga). *Microentomology* **5:**117–142.

Cope, O. B. 1941. The morphology of a species of the genus *Tetrophthalmus* (Mallophaga: Menoponidae). *Microentomology* **6:**71–92.

Coryndon, S. C. and R. J. G. Savage. 1973. The origin and affinities of African mammal faunas. In N. F. Hughes (Ed.), *Organisms and Continents Through Time. Spec. Pap. Paleontol., Paleontol. Assoc.* **12:**121–135.

Coultrip, R. L., R. W. Emmons, L. J. Legters, J. D. Marshall, Jr., and K. F. Murray. 1973. Survey for the arthropod vectors and mammalian hosts of Rocky Mountain spotted fever and plague at Fort Ord, California. *J. Med. Entomol.* **10:**303–309.

Craufurd-Benson, H. J. 1941. The cattle lice of Great Britain. Part I. Biology, with special reference to *Haematopinus eurysternus*; Part II. Louse populations. *Parasitology* **33:**331–358.

Crofton, H. D. 1971. A quantitative approach to parasitism. *Parasitology* **62:**179–194.

Daily, M. D. and R. L. Brownell, Jr. 1972. Chapter 9. A checklist of marine mammal parasites. Pages 528–529 in S. H. Ridgway (Ed.), *Mammals of the Sea: Biology and Medicine.* Charles C. Thomas, Springfield, IL.

Dawson, M. R. 1967. Fossil History of the Families of Recent Mammals. Pages 12–53 in S. Anderson and J. K. Jones, Jr. (Eds.), *Recent Mammals of the World: A Synopsis of Families.* Ronald Press, New York.

Eichler, W. 1948. Some rules in ectoparasitism. *Ann. Mag. Nat. Hist.* (12) **1:**588–598.

Eisenberg, J. E. 1981. *The Mammalian Radiations: An Analysis of Trends in Evolution, Adaptation, and Behavior.* University of Chicago Press, Chicago.

Emerson, K. C. 1982. List of the Genera and Species of the Mammalian Mallophaga (unpublished).

Emerson, K. C. and R. D. Price. 1981. A host–parasite list of the Mallophaga on mammals. *Misc. Publ. Entomol. Soc. Am.* **12:**1–72.

Ewer, R. F. 1973. *The Carnivores.* Cornell University Press, Ithaca, N.Y.

Flynn, R. J. 1973. *Parasites of Laboratory Animals.* Iowa State University Press, Ames.

Hallam, A. 1973. Distributional patterns in contemporary terrestrial and marine animals. In N. F. Hughes, (Ed.), *Organisms and Continents through Time. Spec. Pap. Paleont. Paleontol. Assoc.,* **12:**1–42.

Honacki, J. H., K. E. Kinman, and J. W. Koeppl (Eds.). 1982. *Mammal Species of the World: A Taxonomic and Geographic Reference.* Allen Press and Assoc. Syst. Coll., Lawrence, KS.

Hopkins, G. H. E. 1949. The host associations of the lice of mammals. *Proc. Zool. Soc. Lond.* **119:**387–604.

Hopkins, G. H. E. 1957. The distribution of Phthiraptera on mammals. Prem. Symp. Specif. parasit. parasites Vertebr., *Int. Union Biol. Sci.*, Ser. B. **32**:88–119.

Hopkins, G. H. E. and T. H. Clay. 1952. *A Check List of the Genera and Species of Mallophaga.* British Museum (N. H.), London.

Index Catalog of Medical and Veterinary Zoology. U.S. Dept. Agr., Agr. Res. Serv. Parasite–Subject Catalogue: Hosts. Suppl. **16**:7 (1968); **17**:7 (1971); **18**:7 (1974); **19**:7 (1975); **20**:7 (1976); **21**:7 (1978); **22**:7 (1981).

Inglis, W. G. 1971. Speciation in parasitic nematodes. In B. Davies (Ed.). *Adv. Parasitol.* **9**:185–223.

Jellison, W. L. 1942. Host distribution of lice on native North American rodents north of Mexico. *J. Mammal.* **23**:245–250.

Keast, A. 1969. Evolution of mammals on southern continents. VII. Comparisons of the contemporary mammalian faunas of the southern continents. *Q. Rev. Biol.* **44**:121–167.

Kethley, J. B. and D. E. Johnston. 1975. Resource tracking patterns in bird and mammal ectoparasites. *Misc. Publ. Entomol. Soc. Am.* **9**:231–236.

Kim, K. C. 1971. The sucking lice (Anoplura: Echinophthiriidae) of the northern fur seal; descriptions and morphological adaptation. *Ann. Entomol. Soc. Am.* **64**:280–292.

Kim, K. C. 1972. The louse population of the northern fur seal. *Am. J. Vet. Res.* **33**:2027–2036.

Kim, K. C. 1975. Ecology and morphological adaptation of the sucking lice (Anoplura, Echinophthiriidae) on the northern fur seal. *Rapp. P. -V. Réun. Cons. Int. Explor. Mer* **169**:504–514.

Kim, K. C. 1982. Host specificity and phylogeny in Anoplura. Deux. Symp. Specif. Parasit. parasites Vertebr. 13–17 Avril 1981, *Mem. Mus. Nat. Hist. Nat.*, Nouv. Ser. A, Zool., **123**:123–127.

Kim, K. C. and K. C. Emerson. 1974. *Latagophthirus rauschi,* new genus and new species (Anoplura: Echinophthiriidae) from the river otter (Carnivora: Mustellidae). *J. Med. Entomol.* **11**:442–446.

Kim, K. C. and H. W. Ludwig. 1978a. The family classification of the Anoplura. *Syst. Entomol.* **3**:249–284.

Kim, K. C. and H. W. Ludwig. 1978b. Phylogenetic relationships of parasitic Psocodea and taxonomic position of the Anoplura. *Ann. Entomol. Soc. Am.* **71**:910–922.

Kim, K. C. and H. W. Ludwig. 1982. Parallel evolution, cladistics and classification of parasitic Psocodea. *Ann. Entomol. Soc. Am.* **75**:537–548.

Kim, K. C., C. A. Repenning, and G. V. Morejohn. 1975. Specific antiquity of the sucking lice and evolution of otariid seals. *Rapp. P. -V. Réun. Cons. Int. Explor. Mer* **169**:544–549.

Kim, K. C., V. L. Haas, and M. C. Keyes. 1980. Populations, microhabitat preference and effects of infestation of two species of *Orthohalarachne* (Halarachnidae: Acarina) in the northern fur seal. *J. Wildl. Dis.* **16**:45–51.

Kurtén, B. 1969. Continental drift and evolution. *Sci. Am.* **220**:54–64.

Lillegraven, J. A. 1972. Ordinal and familial diversity of Cenozoic mammals. *Taxon* **21**:261–274.

Ludwig, H. W. 1982. Host specificity in Anoplura and coevolution of Anoplura and Mammalia. *Deux. Symp. Specif. parasit. parasites Vertebr.* 13–17 Avril 1981, *Mem. Mus. Nat. Hist. Nat.*, Nouv. Ser. A, Zool. **123**:145–152.

MacArthur, R. H. and E. O. Wilson. 1967. *The Theory of Island Biogeography.* Princeton University Press, Princeton, N.J.

Marshall, B. G. 1981. *The Ecology of Ectoparasitic Insects.* Academic Press, London.

Matthysse, J. G. 1946. Cattle lice: Their biology and control. *Cornell Univ. Agric. Exp. Stn. Bull.* **832**:1–67.

McKenna, M. C. 1975. Toward a phylogenetic classification of the Mammalia. Pages 21–46 in W. P. Luckett and F. S. Szalay (Eds.), *Phylogeny of the Primates*. Plenum Press, New York.

Mitchell, E. and R. H. Tedford. 1973. The Enaliarctinae: a new group of extinct aquatic Carnivora and a consideration of the origin of the Otariidae. *Bull. Am. Mus. Nat. Hist.* **151**:203–284.

Mock, D. E. 1974. The cattle biting louse, *Bovicola bovis* (Linn.). I. In vitro culturing, seasonal population fluctuations, and role of the male. II. Immune response of cattle. Ph.D. Thesis, Graduate School, Cornell University, Ithaca, NY. 193 pp.

Muizon, Ch. de. 1982. Phocid phylogeny and dispersal. *Ann. Afr. Mus.* **82**:175–213.

Murray, M. D. 1955. Infestation of sheep with the face louse (*Linognathus ovillus*). *Aust. Vet. J.* **31**:22–26.

Murray, M. D. 1957a, b, c, d. The distribution of the eggs of mammalian lice on their hosts. I. Description of the oviposition behavior. II. Analysis of the oviposition behavior of *Damalinia ovis* (L.). III. The distribution of the eggs of *Damalinia ovis* (L.) on the sheep. IV. The distribution of the eggs of *Damalinia equi* (Denny) and *Haematopinnus asini* (L.) on the horse. *Aust. J. Zool.* **5**:13–18, 19–29, 173–182, 183–187.

Murray, M. D. 1961. The ecology of the louse *Polyplax serrata* (Curm.) on the mouse *Mus musculus* L. *Aust. J. Zool.* **9**:1–13.

Murray, M. D. and D. G. Nichols. 1965. Studies on the ectoparasites of seals and penguins. I. The ecology of the louse *Lepidophthirus macrorhini* Enderlein on the southern elephant seal, *Mirounga leonina* (L.). *Aust. J. Zool.* **13**:437–454.

Murray, M. D., M. S. R. Smith, and Z. Soucek. 1965. Studies on the Ectoparasites of seals and penguins. II. The ecology of the louse *Antarctophthirus ogmorhini* Enderlein on the Weddel Seal, *Leptonychotes weddeli* Lesson. *Aust. J. Zool.* **13**:761–71.

Peterson, R. S. 1968. Social behavior in pinnipeds with particular reference to the northern fur seal. Pages 3–53 in R. J. Harrison et al. (Eds.), *The Behavior and Physiology of Pinnipeds*. Appleton-Century-Crofts, New York.

Pielou, E. C. 1979. *Biogeography*. John Wiley, New York.

Ramcke, J. 1965. Kopf der Schweinlaus (*Haematopinus suis* L., Anoplura). *Zool. Jb. Anat.* **82**:547–663.

Ray, C. S. 1976. Geography of phocid evolution. *Syst. Zool.* **25**:391–406.

Repenning, C. A., C. E. Ray, and D. Grigorescu. 1976. Pinniped biogeography. Pages 357–369 in J. Gray and A. J. Boucot (Eds.), *Historical Biogeography, Plate Tectonics, and the Changing Environments*. Proc. 37th Ann. Biol. Colloq. Oregon State University Press, Corvallis.

Repenning, C. A. and R. H. Tedford. 1977. Otarioid seals of the Neogene. *Geol. Surv. Prof. Pap.* **992**:1–93.

Ridgway, S. H. (Ed.). 1972. *Mammals of the Sea: Biology and Medicine*. Charles C. Thomas, Springfield, IL.

Rogers, L. L. 1975. Parasites of black bears of the Lake Superior region. *J. Wildl. Dis.* **11**:189–192.

Romer, A. S. 1966. *Vertebrate Paleontology*, 3rd ed. University of Chicago Press, Chicago.

Simpson, G. G. 1947. Holarctic mammalian faunas and continental relationship during the Cenozoic. *Bull. Geol. Soc. Am.* **58**:613–618.

Simpson, G. G. 1980. *Splendid Isolation: The Curious History of South American Mammals*. Yale University Press, New Haven, CT.

Smith, A. G., J. C. Briden, and G. E. Drewry. 1973. Phanerozoic World Maps. In N. F. Hughes (Ed.), *Organisms and Continents Through Time, Spec. Publ. Paleont., Lond. Paleontol. Assoc.* **12**:1–43.

Smith, A. G., A. M. Hurley, and J. C. Briden. 1981. *Phanerozoic Paleocontinental World Maps.* Cambridge University Press, Cambridge.

Stains, H. J. 1975. 1. Distribution and taxonomy of the Canidae. Pages 3–26 in M. W. Fox (Ed.), *The Wild Canids: Their Systematics, Behavioral Ecology and Evolution.* Van Nostrand Reinhold, New York.

Stojanovich, C. J. 1945. The head and mouthparts of the sucking lice (Insecta: Anoplura). *Microentomology* **10**:1–46.

Stone, J. E. and D. B. Pence. 1977. Ectoparasites of the bobcat from West Texas. *J. Parasit.* **63**:463.

Symmons, S. 1952. Comparative anatomy of the Mallophagan head. *Proc. Zool. Soc. Lond.* **27**:349–436.

Traub, R. 1980. The zoogeography and evolution of some fleas, lice and mammals. Pages 93–172 in R. Traub and H. Starcke (Eds.), *Fleas.* Proc. Int. Conf. Fleas, Ashton Wold/Peterborough/UK 21–25 June 1977. A. A. Balkema, Rotterdam.

Vanzolini, P. E. and L. R. Guimaraes. 1955. Lice and the history of South American land mammals. *Rev. Bras. Entomol.* **3**:13–46.

Weber, H. 1969. Die Elefantenlaus *Haematomyzus elefantis* Piaget 1869. *Zoologica* **41**(1)(116):1–155.

Weisser, C. F. 1975. A Monograph of the Linognathidae, Anoplura, Insecta (excluding the genus *Prolinognathus*). Unpublished doctoral dissertation, University of Heidelberg, Heidelberg.

Wenzel, R. L. and V. J. Tipton. 1966. *Ectoparasites of Panama.* Field Mus. Nat. Hist., Chicago.

Wilson, E. O. 1969. The Species equilibrium. *Brookhaven Symp. Biol.* **22**:38–47.

Chapter 8

Ctenocephalides felis (Bouché)

Coevolution of
Fleas and Mammals

Robert Traub

This study was supported by Grant No. AI-04242 of the National Institutes of Health, Bethesda, Maryland, with the Department of Microbiology, University of Maryland School of Medicine, Baltimore, Maryland.

INTRODUCTION

Data on the classification, distribution, historical biogeography, and host relationships of fleas and other ectoparasites have contributed significantly to our knowledge of the ecology and distribution of diseases like scrub typhus, tick typhus, and other zoonoses and have helped resolve questions about the phylogeny and evolution of certain mammals and birds (Traub et al. 1967; Traub 1980c; Traub and Jellison 1981). Such findings have been possible only because of the manifest interrelationships between ectoparasites and the hosts with which they have been so intimately associated for untold eons. The conclusion that most taxa of Siphonaptera (fleas) must have coevolved with their mammalian hosts is not surprising because their aboriginal hosts were mammals and this order of insects is monophyletic (Smit 1972; Traub 1972d; Rothschild and Schlein 1975; Traub 1980c). The host associations of stephanocircid and hystrichopsyllid fleas are ancient, and their infestation of marsupials may date back to the Cretaceous period. They must have accompanied marsupials from South America across Antarctica to Australia more than 100 million years ago (Traub 1972d; Traub and Dunnet 1973; Traub 1980c). The relationship between flea and mammal has been so intimate that many species bear physical marks imposed by the vestiture or habits of their hosts (Traub 1972b; Traub 1980b). In other instances coevolution with contemporary hosts has occurred for a much more limited period, for instance, after some species left their congeners and typical hosts and switched to other kinds of mammals. Nevertheless, in most such cases the close association has persisted long enough for the fleas to exhibit structural modifications resulting from the change. Some fleas exhibit little host specificity. They can feed on a gamut of mammals and show no signs of coevolution with any particular host group. Others are adapted to live as denizens of the nests of the hosts rather than on the bodies of the mammals and are modified accordingly.

Coevolution of fleas and mammals is revealed primarily by the physiology, host relations, historical biogeography, and the morphology of fleas. The intensely close relationship between parasite and host over a long time has resulted in structural adaptive changes in the fleas as illustrated in Figures 8.1–8.126.

When additional pertinent information about certain points is presented in another section of the chapter, an asterisk * has been inserted after the key term, name, or phrase to indicate that the appropriate cross-reference can be located in the index.

The siphonapteran modifications occur generally in response to some physical or behavioral attribute of the host, and thus coevolution does not necessarily mean simultaneous evolution or comparable rates of change by both flea and host. The changes in vestiture must have first occurred in the mammal, and the spines and bristles of the flea became tailored to fit the hairs of the host.

The Order Siphonaptera is remarkably homogeneous and distinctive.

All the adults are bloodsucking ectoparasites of mammals or birds, are built on the same general body plan, and cannot be confused with any other insect. Fleas are wingless, laterally compressed, and are nearly always equipped with one or more rows of caudad-directed stiff bristles (Figs. 8.112–8.114). Frequently they have a comb of spines on the head or thorax, and occasionally on one or more of the abdominal terga. These morphological features facilitate rapid passage through the hairs of the host. The chaetotactic specializations also enhance the general ability of the flea to remain on the host and impede the attempts of the mammal or bird to dislodge it by pulling it backwards or sideways (Traub 1980b). All fleas have the hind legs long, narrow, and modified for jumping, although a few species—as an adaptive measure—have a reduced capacity to do so (Rothschild and Schlein 1975; Traub 1980b). Despite these similarities, there is considerable variation in morphology and behavior among the fleas. Thus some species have anchoring mouthparts and stay attached to the host for days; others are virtually completely buried within the superficial layers of the skin of the host. Certain fleas are highly modified for living within the nest of the host and feed while the mammal is sleeping (Traub 1972b; Traub 1980b). The majority, however, are free moving within the fur or feathers of the host.

Association of Fleas with Mammals

Certain groups of mammals are not known to have any characteristic fleas: the Cetacea (whales), Sirenia (manatees), Proboscidea (elephants), Dermoptera (flying lemurs), Tubulidentata (aardvarks), and the Primates (sensu excluding tupaiids). Marine, aquatic, and semiaquatic mammals lack specific fleas: the platypus, potamogalid otter-shrews, soricid water shrews (if at least semiaquatic), beaver, muskrats, fish-mice (*Ichthyomys*), water mice (*Rheomys*), river-dwelling hydromyines, otters, seals, walruses, hippopotami, and so on. Supposedly, this is primarily because Siphonaptera cannot survive immersion in water for more than a few minutes (unless surrounded by an air bubble, as in fleas of some aquatic birds).

Mammals that have vast home ranges and do not inhabit dens for rearing their young almost invariably lack major fleas of their own. Grazing or browsing animals like artiodactyls, perissodactyls, proboscideans, and kangaroos are in this category, perhaps because there is no suitable place for larval development where these hosts are likely to return and become infested. A few species of vermipsyllid fleas do infest hoofed animals in Central Asia, but these are especially adapted for their particular life-style (Traub 1972b). Sticktight fleas may also be found on such hosts, but they are nonspecific. Since they remain attached for weeks and the females lay large numbers of eggs over a long period of time, these fleas can often overcome the perils such hosts present to the perpetuation of the siphonapteran species. Fleas of the small family Ancistropsyllidae* infest cervids but are appropriately highly specialized.

The agility of the true primates in grooming themselves and one another probably accounts for their lack of specific fleas. *Pulex irritans** Linnaeus, the so-called human flea, in my opinion, had porcines as its original host. Spiny* hosts, such as porcupines, hedgehogs, and tenrecs, harbor only few species of fleas—or none at all—a condition that has been ascribed to the difficulties posed to Siphonaptera by the coat of spines (Traub 1980b), but the fleas of spiny hosts are highly modified. In general, the remaining groups of mammals have a relatively rich fauna of fleas (except in localized habitats, as in hot, humid areas). Bird fleas occur in at least six families of fleas, but always as a small minority, and in each case are believed to have been secondarily derived from mammalian fleas.

Host Responses to Infestation by Fleas

There have been few overt responses by the host to Siphonaptera in the course of their joint evolutionary path, and as fleas are parasites and not symbionts, any host responses at all serve to discommode rather than accommodate them. The few that have been observed are physiological and behavioral rather than morphological, and little studied. There is nothing like the agile tail with a terminal whisk or the faculty of twitching the skin, both so useful in the ruminant for shooing bloodsucking flies. Observations on trombiculid mites (chiggers) suggest that mammals have come into immunological equilibrium with their own particular siphonapteran fauna although data are lacking. Trombiculid mites which normally feed on birds or reptiles may cause intense itching and dermatitis in man, whereas mammalian mites like *Leptotrombidium* cause virtually no noticeable reaction in man (Traub and Wisseman 1974). The obvious inference is that these mammalian chiggers have become attuned to their particular hosts and hence do not provoke the scratching reflex or some other reaction that might endanger their survival.

The host tissue response is a major factor in the production of the cyst in which the tungids are ensconced as hypodermal species, for these fleas do not really "burrow" into the skin. In our Ethiopian studies, *Echidnophaga* frequently were embedded in the skin of the muzzle of *Rattus* to the level where only 10–20% of the flea body was exposed. Some powerful immunological responses must have taken place at the site of attachment of sticktight fleas as shown by the signs of healing in foci traumatized by tooth and claw in efforts to dislodge the large number of fleas.

Dogs and other animals gnash their front teeth in a characteristic manner when "defleaing." Baby rats and mice, observed in our laboratory, scratch vigorously at heavy infestations of *Leptopsylla* and *Xenopsylla* fleas well before the scratching reflex is supposed to have developed. Such host grooming behavior against ectoparasites may have arisen in response to infestation by ticks or mites before the fleas arose. At present it cannot be documented that the shape of the claw or vestiture of the ear or tail developed in response to infestation by a particular group of arthropod. Certain fleas of *Hylopetes* are generally found at the base of long hairs on the tail of these flying squirrels. This is probably the result of the grooming habits of the host, just as *Leptopsylla segnis* (Schönherr) characteristically is found in patches on the vertex of the skull or midline of the back of *Mus* and *Rattus*, sites that the host cannot reach readily when trying to scratch or bite fleas (Traub 1980b).

Physiological Adaptations of Fleas

Fleas must have been adapted physiologically in innumerable ways to survive as bloodsucking parasites of mammals and probably also adapted to their specific hosts, although sufficient data are unavailable. Most species must ingest host blood as adults before they can lay viable eggs, and larvae may require host blood indirectly through ingestion of flea feces, which consist mainly of blood (Rothschild and Clay 1952). For example, host blood obtained in this way is essential for proper development of larval cat fleas (Strenger 1973).

One of the best examples of coevolution of fleas and mammals is found in the spilopsylline fleas on rabbits. The sex life of these fleas is dependent on the sexual hormonal status of the rabbits they infest, and, in consequence, flea larvae and adults emerge while and where the baby rabbits are born and housed (Rothschild 1965a, b; Rothschild et al. 1970). In this way both the larvae and adults escape the danger of emerging in some completely unfavorable habitat of the extensive host territory. However, the related genus *Euhoplopsyllus*, on hares, has apparently solved this problem in another way. The larvae live in the fur of the host (Freeman and Madsen 1949) where warmth and food in the form of scurf are assured. The Australian pygiopsyllid *Uropsylla tasmanica* Rothschild, a flea of the far-ranging marsupial "cat," *Dasyurus*, has gone further, for the larvae are parasitic in the superficial tissues of the host (Dunnet and Mardon 1974; Traub 1980b; Pearse 1981). Pearse (1981) suggested that sexual maturation in *U. tasmanica* is dependent on the hormonal status of its host. Whether this is true or not, her studies clearly indicate that the life cycle of this flea is so intimately associated with that of *Dasyurus* that the eggs develop and hatch, the parasitic larvae emerge, and the adults emerge from the cocoon (the following autumn) at a time when the host is occupying its breeding nest. Temperature and humidity are also important determinants in the synchronization of the life cycles of fleas with those of their hosts, as in the case of bird fleas which greet the birds upon their return to the nests in the spring (Rothschild and Clay 1952). These factors, as well as the exceptional abundance of food for larvae and adults, presumably account for large populations of fleas in the nests of lactating or breeding mammals (Ioff 1953; Traub 1972b).

Legends for Figures 8.1–8.126

Figures 8.1–8.4 Lateral aspect of ♀ head and pronotum:
1. *Palaeopsylla laxata;* **2.** *P. similis;* **3.** *P. sinica;* **4.** *P. setzeri.*

Figures 8.5–8.8 Lateral aspect of ♀ head and pronotum:
5. *Ctenophthalmus (Alloctenus) cryptotis;* **6.** *C. (C.) bisoctodentatus;* **7.** *Ctenophthalmus* n. sp.; **8.** *C. (Palaeoct.) fissurus.*

Figures 8.9–8.11 Lateral aspect of ♀ head and pronotum:
9. *Ctenocephalides felis;* **10.** *C. arabicus;* **11.** *C. rosmarus.*

Figures 8.12–8.15 Lateral aspect of ♀ head and pronotum:
12. *Sigmactenus cavifrons;* **13.** *Leptopsylla (Pectinoctenus) pectiniceps;* **14.** *Epirimia aganippes;* **15.** *Ctenophthalmus (Ethioct.) calceatus.*

Figures 8.16, 8.17 ♀ Head and pronotum (lateral).

Figures 8.18, 8.19 ♀ Terga I-III:
16. *Dinopsyllus (D.) longifrons;* **17.** *D. (Cryptoctenophyllus) ingens;* **18.** *D. (D.) longifrons;* **19.** *D. (C.) ingens.*

Figures 8.20–8.22, 8.24 *Stephanocircus dasyuri.*

Figures 8.23, 8.25, 8.26 *Dinopsyllus (D.) lypusus:*
20. ♂ Head and pronotum; **21.** ♂ head and prothorax (dorsal); **22.** ♂ head (ventral); **23.** ♂ head and pronotum (dorsal); **24.** ♂ head (frontal); **25.** ♂ head (frontal); **26.** ♂ head, pronotum, and procoxa (lateral).

Figures 8.27–8.29 *Peromyscopsylla h. hesperomys.*

Figures 8.30–8.32 *Frontopsylla (Profrontia) ambigua:*
27. ♂ Head (frontal); **28.** ♂ head, pronotum, and procoxa (lateral); **29.** ♂ head (dorsal); **30.** ♀ head and pronotum (dorsal); **31.** ♀ head, pronotum, and procoxa (lateral); **32.** ♀ head (frontal).

Figures 8.33, 8.34, 8.36 *Meringis altipecten.*

Figures 8.35, 8.37, 8.38 *Migrastivalius (Gryphopsylla) hopkinsi:*
33. ♀ Head, pronotum, and procoxa (lateral); **34.** ♀ head and pronotum (dorsal); **35.** ♀ head, pronotum, and procoxa (lateral); **36.** ♀ caudal aspect of pronotum; **37.** ♀ head (frontal); **38.** ♀ head and pronotum (dorsal).

Figures 8.39–8.41 *Mesopsylla tuschkan propinacta.*

Figures 8.42–8.44 *Aviostivalius klossi:*
39. ♀ Head, pronotum, and procoxa (lateral); **40.** ♀ head (frontal); **41.** ♀ head and pronotum (dorsal); **42.** ♀ head, pronotum, and procoxa (lateral); **43.** ♀ head (frontal); **44.** ♀ head (dorsal).

Figures 8.45–8.49 *Macropsylla hercules:*
45. ♀ Head, pronotum, and procoxa (lateral); **46.** ♀ entire abdomen (dorsal); **47.** ♀ head (ventral); **48.** ♀ head and pronotum (frontal); **49.** ♀ head and pronotum (dorsal).

Figures 8.50–8.53 *Choristopsylla tristis:*
50. ♀ Head, pronotum, and procoxa; **51.** ♀ metatarsus 5; **52.** ♀ metatibia and tarsus; **53.** ♀ meso- and metathorax, tergum I.

Figures 8.54–8.56 *Choristopsylla ochi:*
54. ♀ Head, pronotum, and procoxa; **55.** ♀ metatibia and tarsus; **56.** meso- and metathorax, tergum I.

Figures 8.57–8.60 *Opisodasys pseudarctomys:*
57. ♀ Head, pronotum, and procoxa; **58.** ♀ metatarsus 5; **59.** ♀ metatibia and tarsus; **60.** ♀ meso- and metathorax, tergum I.

Figures 8.61–8.63 *Tarsopsylla octodecimdentata coloradensis:*
61. ♀ head, pronotum, and procoxa; **62.** ♀ metatibia and tarsus; **63.** ♀ meso- and metathorax, tergum I.

Figures 8.64, 8.67 *Foxella ignota.*

Figures 8.65, 8.66 *Bradiopsylla echidnae:*
64. ♀ Head, pronotum, and procoxa; **65.** ♀ metatarsus 5; **66.** ♀ metatibia and tarsus; **67.** ♀ terga I-III.

Figures 8.68–8.70 *Parastivalius novaeguineae.*
68.   ♀ Head, pronotum, and procoxa; **69.** ♀ metatibia and tarsus; **70.** ♀ metatarsus 5.

Figures 8.71–8.73 *Thrassis francisi:*
71.   ♀ Head, pronotum, and procoxa; **72.** ♀ metatibia and tarsus; **73.** ♀ meso- and metathorax, terga I-II.

Figures 8.74–8.76   ♀ Terga I–III:
74. *Opisocrostis hirsutus;* **75.** *Thrassis francisi;* **76.** *Orchopeas leucopus.*

Figures 8.77–8.79   ♀ Terga I–III:
77. *Ctenophthalmus* n. sp.; **78.** *C. (Palaeoct.) fissurus;* **79.** *C. (Ethioct.) calceatus.*

Figures 8.80, 8.81 *Palaeopsylla similis.*

Figures 8.82, 8.83 *P. setzeri.*

Figures 8.84, 8.85 *P. laxata:*
80.   ♀ Terga I–III; **81.** ♀ protibia; **82.** ♀ terga I-III; **83.** ♀ protibia; **84.** ♀ terga I–III; **85.** ♀ protibia.

Figures 8.86–8.90   ♀ Metatibia:
86. *Dinopsyllus (D.) longifrons;* **87.** *D. (C.) ingens;* **88.** *Palaeopsylla laxata;* **89.** *P. setzeri;* **90.** *P. similis.*

Figures 8.91–8.93 *Macropsylla grossiventris:*
91.   ♀ Protibia; **92.** ♀ metafemur, tibia, and tarsus; **93.** ♀ metatarsus 5.

Figures 8.94, 8.95, 8.97 *Phthiropsylla agenoris.*

Figures 8.96, 8.98 *Polygenis gwyni:*
94.   ♀ Head and prothorax; **95.** ♀ metafemur, tibia, and tarsus; **96.** ♀ metatibia and tarsus; **97.** ♀ metatarsus 5; **98.** ♀ metatarsus 5.

Figures 8.99, 8.102 *Chaetopsylla (C.) globiceps.*

Figures 8.100, 8.101 *Vermipsylla alakurt:*
99.   ♀ Metafemur, tibia, and tarsus; **100.** ♀ metatarsus 5; **101.** ♀ metafemur, tibia, and tarsus; **102.** ♀ metatarsus 5.

Figures 8.103–8.105 *Bradiopsylla echidnae:*
103. ♀ Head, pronotum, and procoxa (lateral); **104.** ♀ head, thoracic nota, terga I–II (dorsal); **105.** ♀ head (frontal).

Figures 8.106–8.111   ♀ Metatibia:
106. *Centetipsylla madagascariensis;* **107.** *Archaeopsylla erinacei;* **108.** *Parapulex echinatus;* **109.** *Pariodontis riggenbachi;* **110.** *Xenopsylla cheopis;* **111.** *Ctenocephalides felis.*

Figures 8.112, 8.113   ♂ Habitus:
112. *Polygenis roberti beebei;* **113.** *Porribius caminae.*

Figure 8.114 *Coronapsylla jarvisi.* ♀ Habitus.

Figure 8.115 *Nycteridopsylla eusarca.* ♂ Pronotal comb. (Courtesy of R. Traub. 1968. Smitella. *J. Med. Entomol.* **5:** 375–404.)

Figure 8.116 *N. chapini.* ♂ Pronotal comb. (Courtesy of R. Traub. 1968. Smitella. *J. Med. Entomol.* **5:** 375–404.)

Figure 8.117 *Ernestinia eximia.* ♀ Head. (Courtesy of R. Traub. 1968. Smitella. *J. Med. Entomol.* **5:** 375–404.)

Figure 8.118 *Chiropteropsylla brockmani.* ♀ Metepimere with false comb.

Figure 8.119 *Myodopsylla insignis.* ♂ Meso- and metathorax, terga I–II.

Figure 8.120 *Scolopsyllus colombianus.* ♀ Pronotum.

Figures 8.121, 8.123, 8.125 *Pulex sinoculus.*

Figures 8.122, 8.124, 8.126 *P. irritans:*
121.   ♀ Head and prothorax; **122.** ♀ head and prothorax; **123.** ♀ meso- and metathorax, terga I–II; **124.** ♀ meso- and metathorax, terga I–II; **125.** ♀ modified abdominal segments; **126.** ♀ modified abdominal segments.

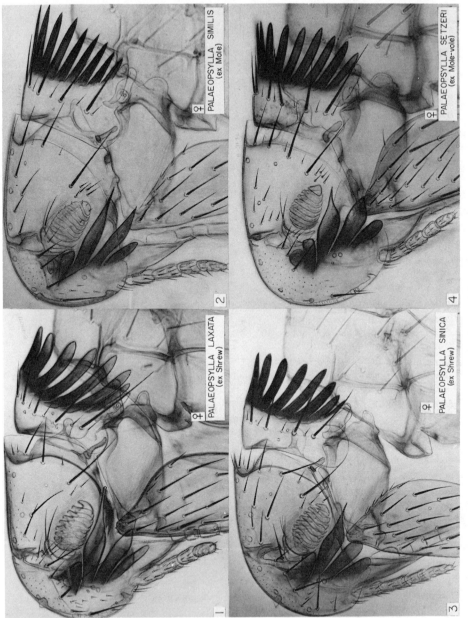

PALAEOPSYLLA LAXATA
(ex Shrew) ♀

PALAEOPSYLLA SIMILIS
(ex Mole) ♀

PALAEOPSYLLA SINICA
(ex Shrew) ♀

PALAEOPSYLLA SETZERI
(ex Mole-vole) ♀

Figures 8.1–8.4

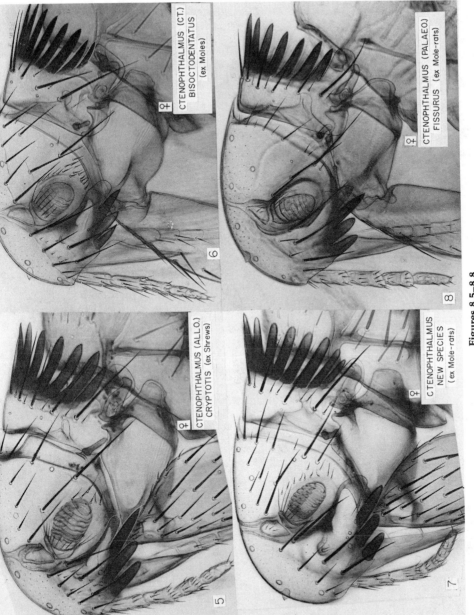

Figures 8.5–8.8

303

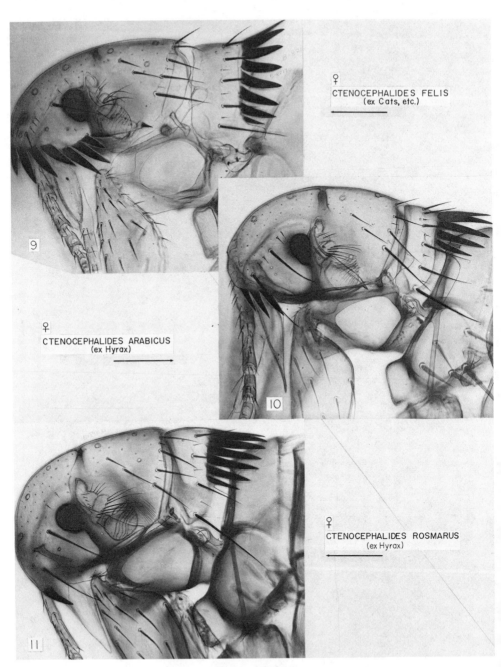

♀
CTENOCEPHALIDES FELIS
(ex Cats, etc.)

♀
CTENOCEPHALIDES ARABICUS
(ex Hyrax)

♀
CTENOCEPHALIDES ROSMARUS
(ex Hyrax)

Figures 8.9–8.11

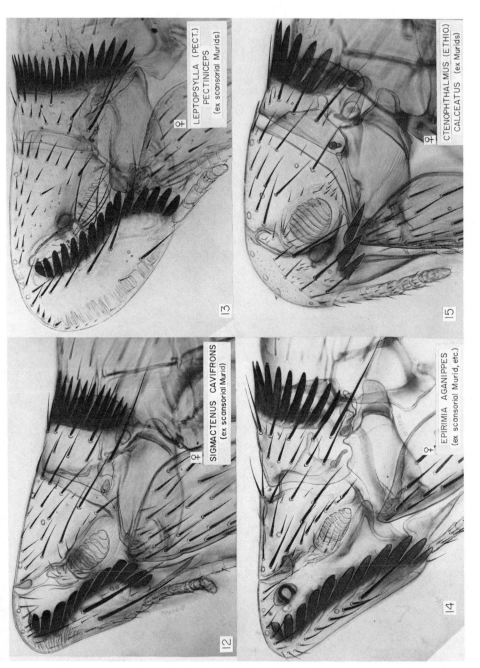

12 SIGMACTENUS CAVIFRONS ♀ (ex scansorial Murid)

13 LEPTOPSYLLA (PECT.) PECTINICEPS ♀ (ex scansorial Murids)

14 EPIRIMIA AGANIPPES ♀ (ex scansorial Murid, etc.)

15 CTENOPHTHALMUS (ETHIO) CALCEATUS ♀ (ex Murids)

Figures 8.12–8.15

305

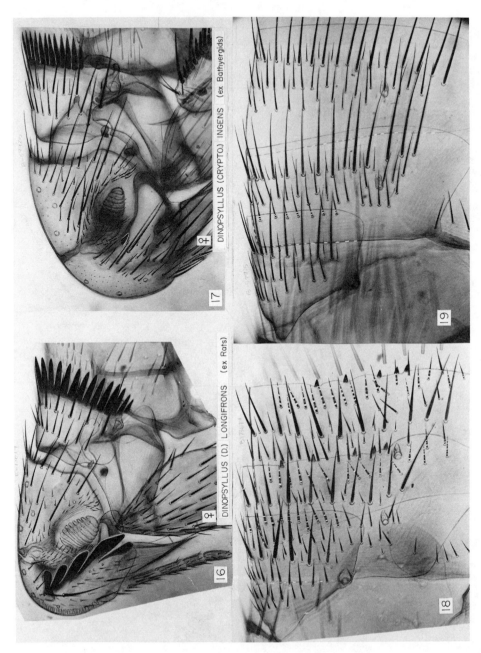

DINOPSYLLUS (CRYPTO.) INGENS (ex Bathyergids)

DINOPSYLLUS (D.) LONGIFRONS (ex Rats)

Figures 8.16–8.19

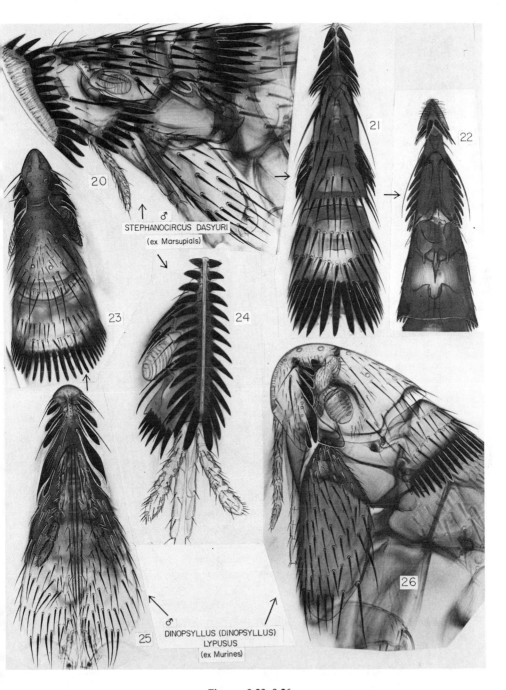

STEPHANOCIRCUS DASYURI
(ex Marsupials)

DINOPSYLLUS (DINOPSYLLUS)
LYPUSUS
(ex Murines)

Figures 8.20–8.26

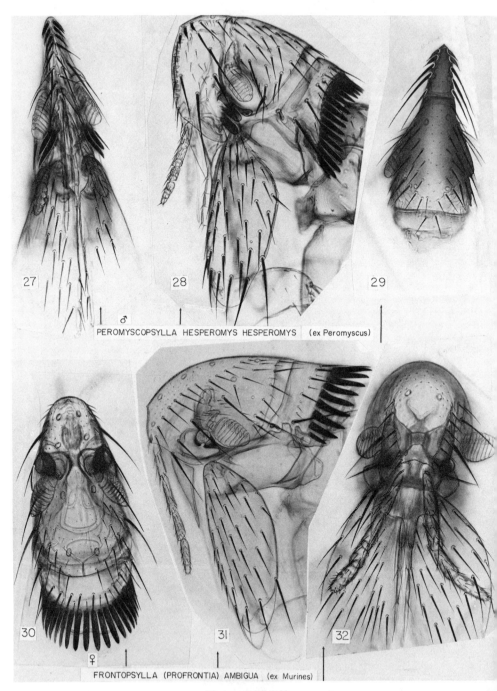

27   28   PEROMYSCOPSYLLA HESPEROMYS HESPEROMYS   (ex Peromyscus)   29
♂

30   31   FRONTOPSYLLA (PROFRONTIA) AMBIGUA   (ex Murines)   32
♀

Figures 8.27–8.32

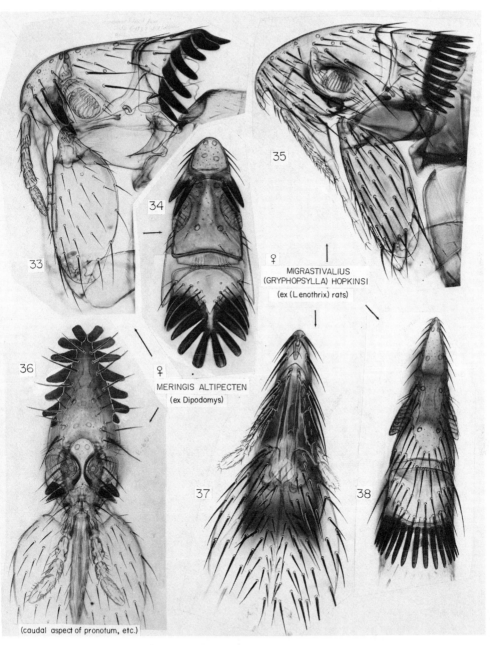

34

35

♀ MIGRASTIVALIUS
(GRYPHOPSYLLA) HOPKINSI
(ex (Lenothrix) rats)

33

36

♀ MERINGIS ALTIPECTEN
(ex Dipodomys)

37

38

(caudal aspect of pronotum, etc.)

Figures 8.33–8.38

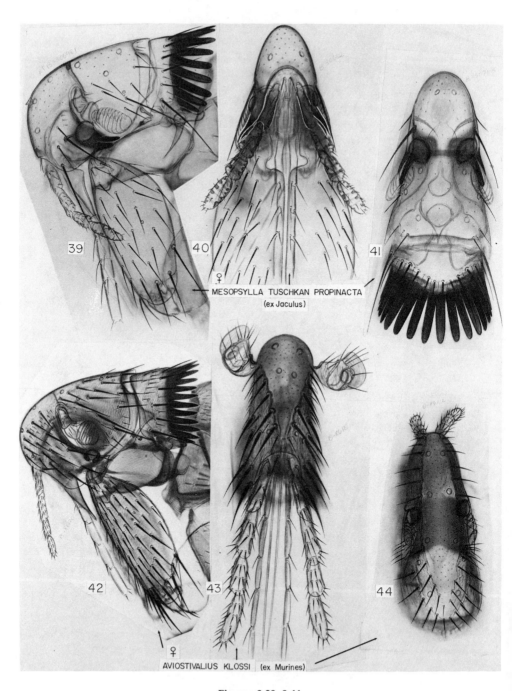

39 40 41

♀

MESOPSYLLA TUSCHKAN PROPINACTA
(ex Jaculus)

42 43 44

♀   ♀

AVIOSTIVALIUS KLOSSI (ex Murines)

Figures 8.39–8.44

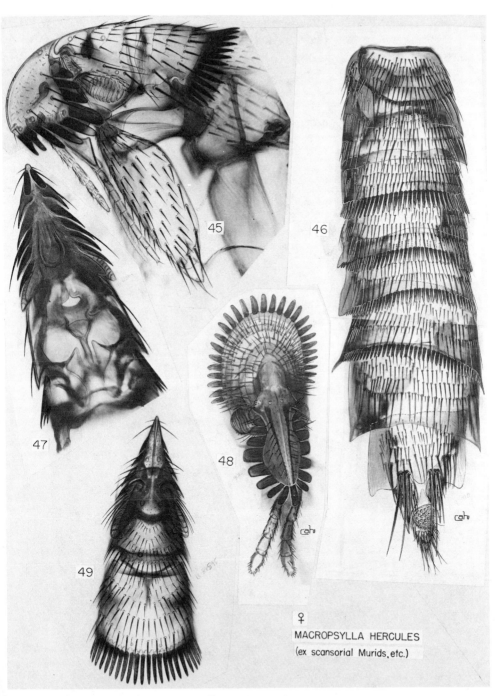

45   46

47

48

49

♀
MACROPSYLLA HERCULES
(ex scansorial Murids, etc.)

Figures 8.45–8.49

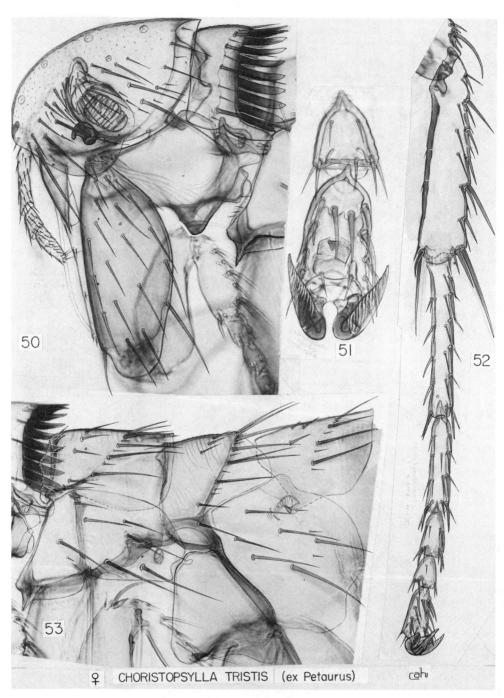

50

51

52

53

♀ CHORISTOPSYLLA TRISTIS (ex Petaurus) cah

Figures 8.50–8.53

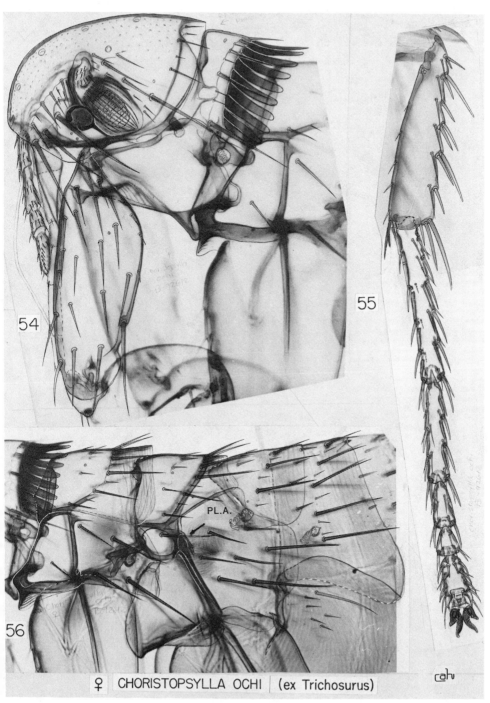

54

55

PL.A.

56

♀ CHORISTOPSYLLA OCHI (ex Trichosurus)

cah

Figures 8.54–8.56

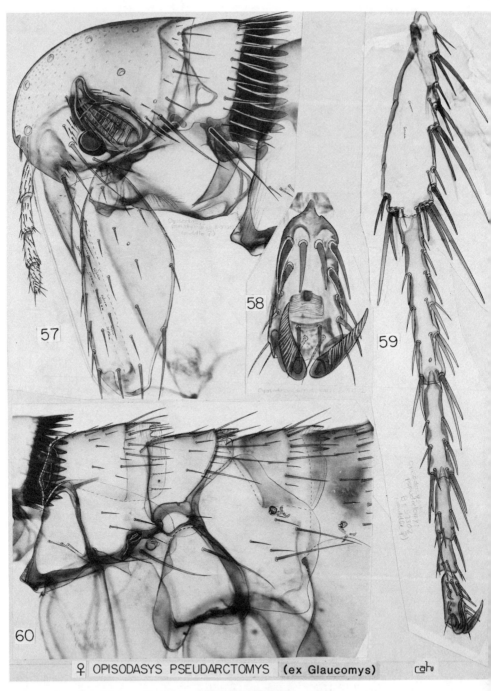

57

58

59

60

♀ OPISODASYS PSEUDARCTOMYS (ex Glaucomys)

Figures 8.57–8.60

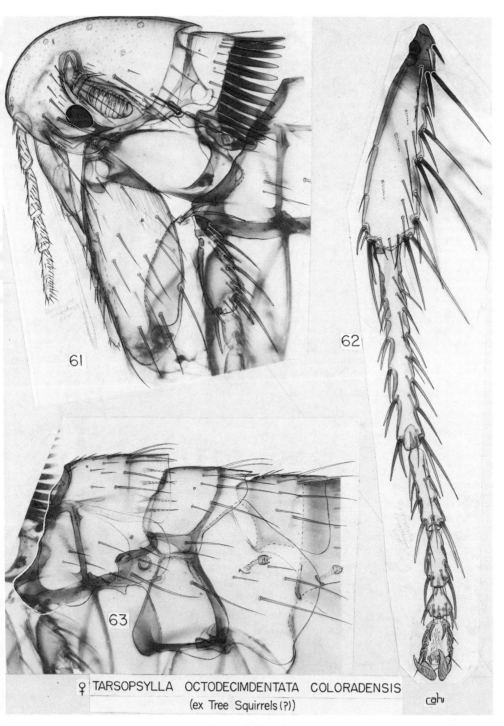

61

62

63

♀ TARSOPSYLLA OCTODECIMDENTATA COLORADENSIS
(ex Tree Squirrels(?))

cah

Figures 8.61–8.63

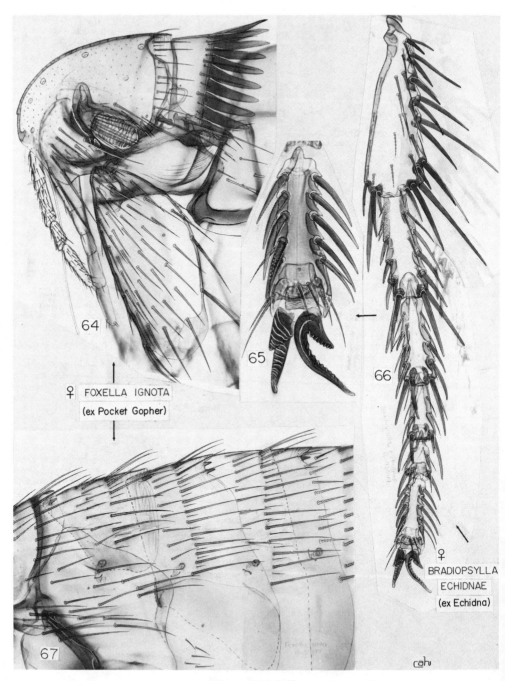

64

65

66

67

♀ FOXELLA IGNOTA
(ex Pocket Gopher)

♀ BRADIOPSYLLA
ECHIDNAE
(ex Echidna)

Figures 8.64–8.67

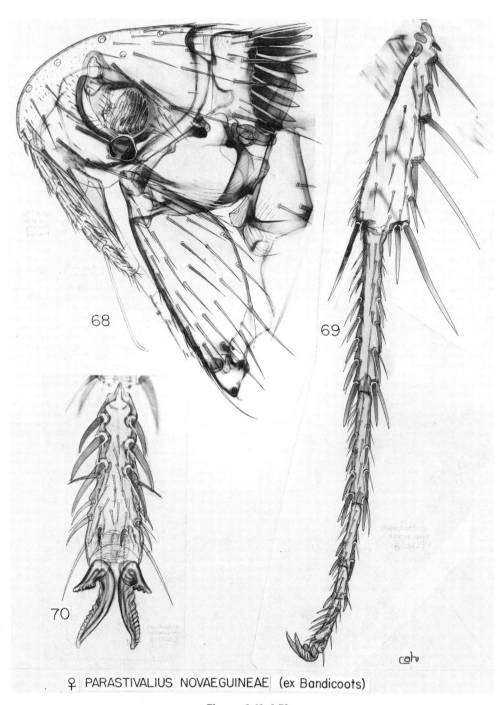

68

69

70

♀ PARASTIVALIUS NOVAEGUINEAE (ex Bandicoots)

Figures 8.68–8.70

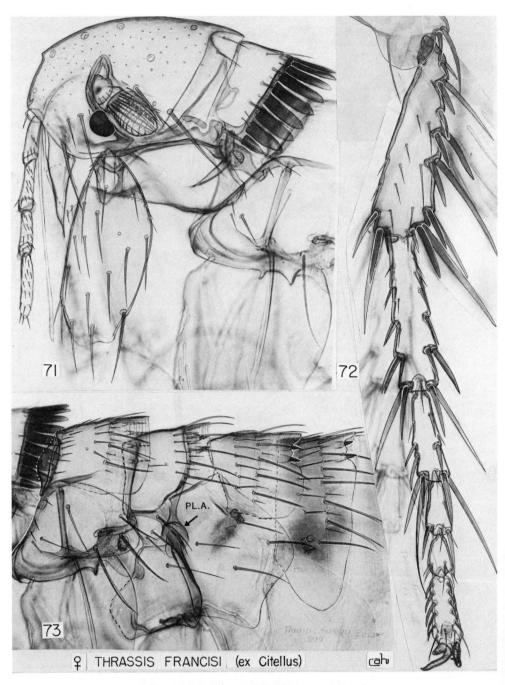

71

72

73

PL.A.

♀ THRASSIS FRANCISI (ex Citellus)

Figures 8.71–8.73

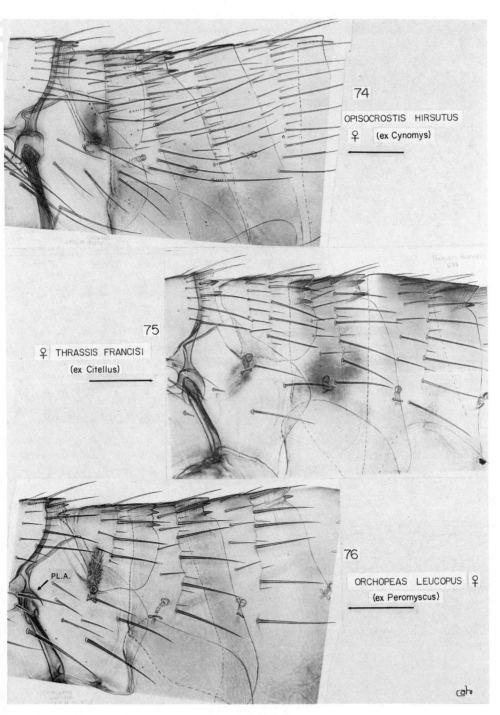

74

OPISOCROSTIS HIRSUTUS

♀ (ex Cynomys)

♀ THRASSIS FRANCISI

(ex Citellus)

75

76

ORCHOPEAS LEUCOPUS ♀

(ex Peromyscus)

PL.A.

Figures 8.74–8.76

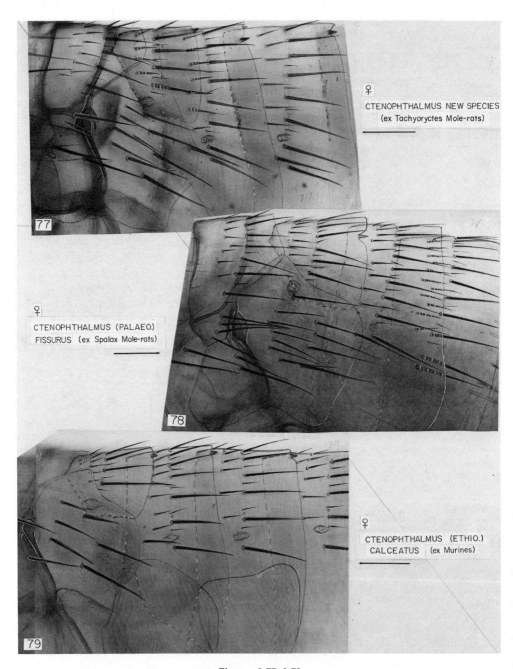

♀

CTENOPHTHALMUS NEW SPECIES
(ex Tachyoryctes Mole-rats)

♀

CTENOPHTHALMUS (PALAEO.)
FISSURUS (ex Spalax Mole-rats)

♀

CTENOPHTHALMUS (ETHIO.)
CALCEATUS (ex Murines)

Figures 8.77–8.79

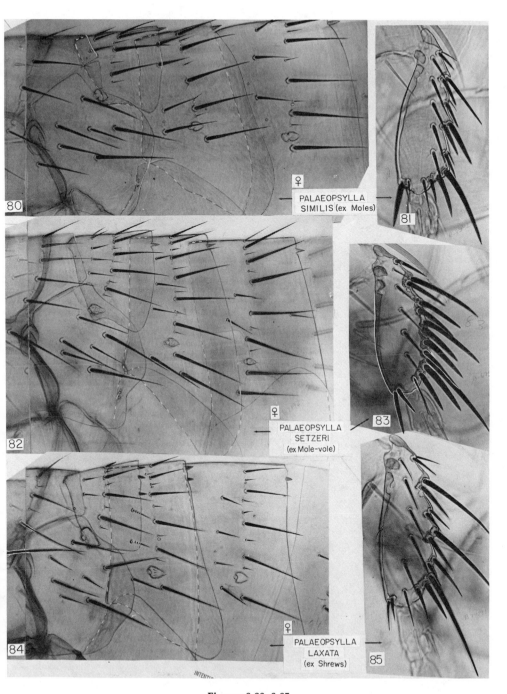

PALAEOPSYLLA
SIMILIS (ex Moles) ♀

PALAEOPSYLLA
SETZERI
(ex Mole-vole) ♀

PALAEOPSYLLA
LAXATA
(ex Shrews) ♀

Figures 8.80–8.85

86 ♀ DINOPSYLLUS (D.) LONGIFRONS (ex Murines)

♀ DINOPSYLLUS (C.) INGENS (ex Bathyergids) →

87

88

♀ 89 PALAEOPSYLLA SETZERI (ex Mole-vole)

♀ PALAEOPSYLLA LAXATA (ex Shrews)

90 ♀ PALAEOPSYLLA SIMILIS (ex Moles)

Figures 8.86–8.90

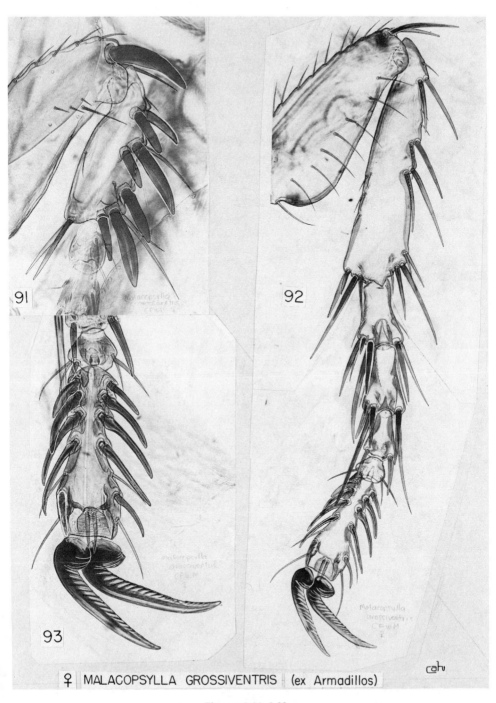

91

92

93

♀ MALACOPSYLLA GROSSIVENTRIS (ex Armadillos)

Figures 8.91–8.93

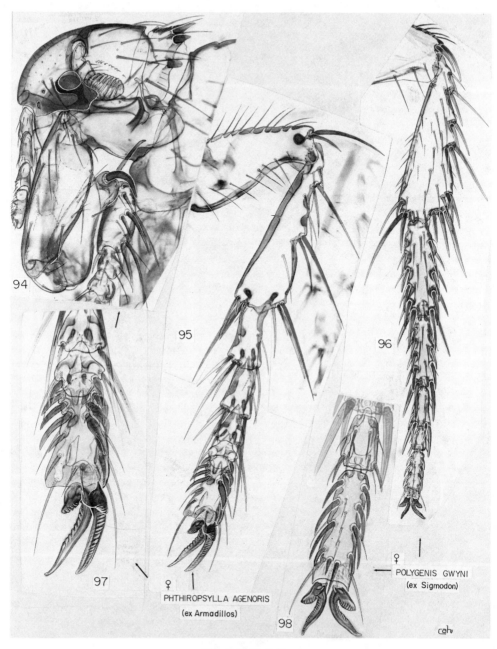

94

95

96

97

♀
PHTHIROPSYLLA AGENORIS
(ex Armadillos)

98

♀
POLYGENIS GWYNI
(ex Sigmodon)

Figures 8.94–8.98

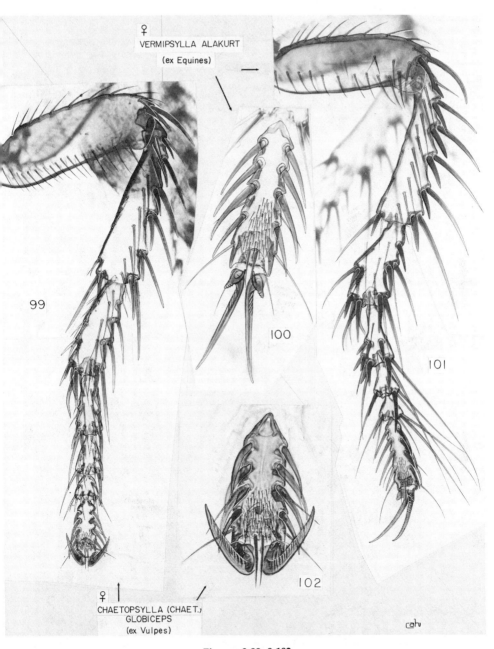

♀
VERMIPSYLLA ALAKURT
(ex Equines)

99

100

101

102

♀
CHAETOPSYLLA (CHAET.)
GLOBICEPS
(ex Vulpes)

cahn

Figures 8.99–8.102

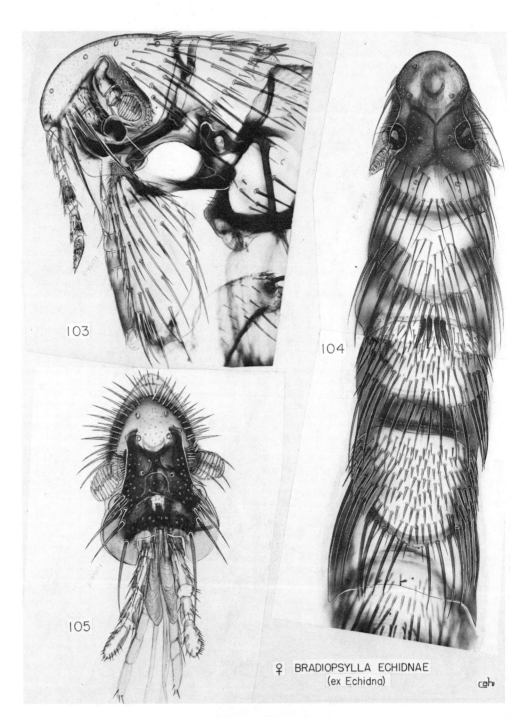

103

104

105

♀ BRADIOPSYLLA ECHIDNAE
(ex Echidna)

Figures 8.103–8.105

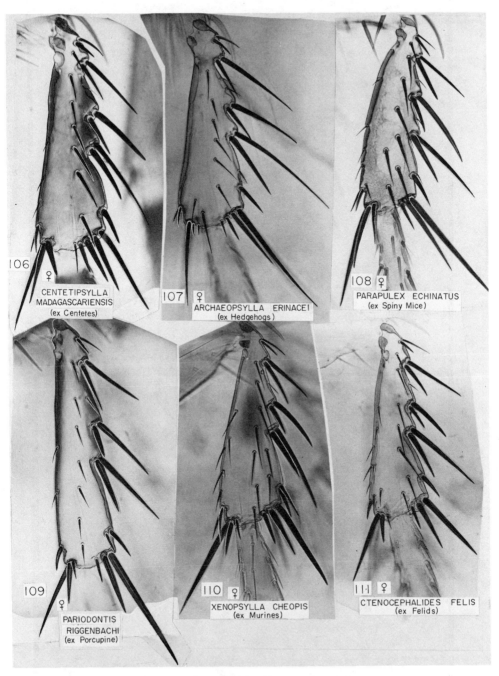

106 ♀
CENTETIPSYLLA
MADAGASCARIENSIS
(ex Centetes)

107 ♀
ARCHAEOPSYLLA ERINACEI
(ex Hedgehogs)

108 ♀
PARAPULEX ECHINATUS
(ex Spiny Mice)

109 ♀
PARIODONTIS
RIGGENBACHI
(ex Porcupine)

110 ♀
XENOPSYLLA CHEOPIS
(ex Murines)

111 ♀
CTENOCEPHALIDES FELIS
(ex Felids)

Figures 8.106–8.111

327

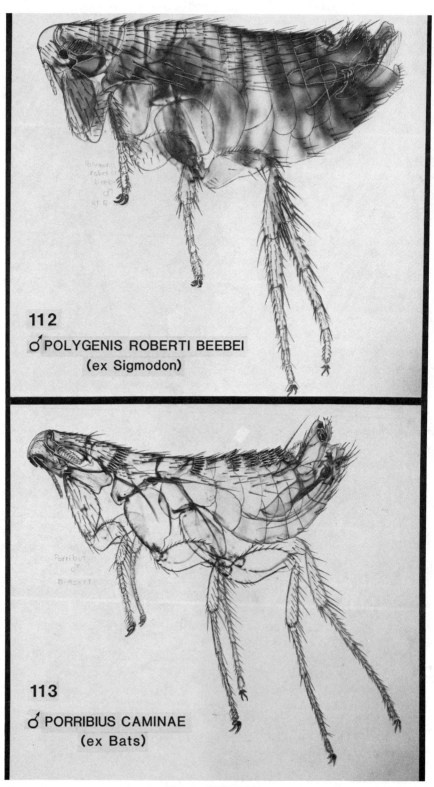

112

♂ POLYGENIS ROBERTI BEEBEI
(ex Sigmodon)

113

♂ PORRIBIUS CAMINAE
(ex Bats)

Figures 8.112–8.113

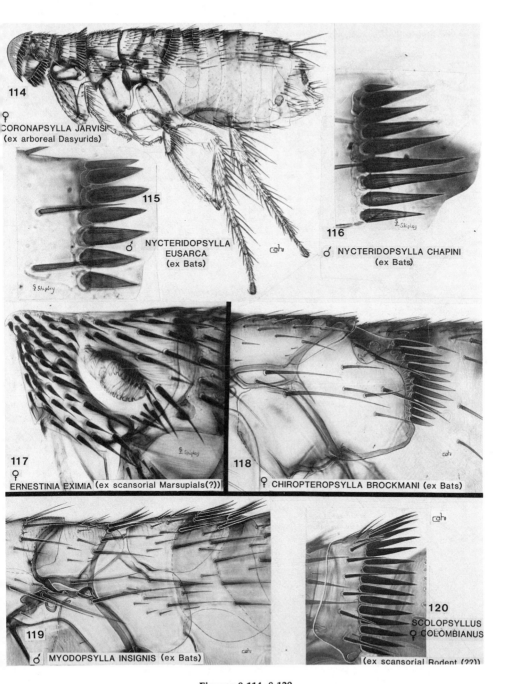

114
♀
CORONAPSYLLA JARVISI
(ex arboreal Dasyurids)

115
♂
NYCTERIDOPSYLLA
EUSARCA
(ex Bats)

116
♂ NYCTERIDOPSYLLA CHAPINI
(ex Bats)

117
♀
ERNESTINIA EXIMIA (ex scansorial Marsupials(?))

118
♀ CHIROPTEROPSYLLA BROCKMANI (ex Bats)

119
♂ MYODOPSYLLA INSIGNIS (ex Bats)

120
SCOLOPSYLLUS
♀ COLOMBIANUS
(ex scansorial Rodent (??))

Figures 8.114–8.120

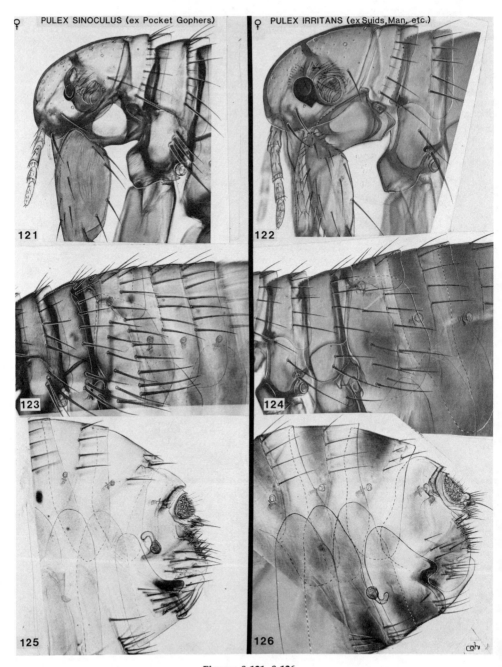

PULEX SINOCULUS (ex Pocket Gophers)

PULEX IRRITANS (ex Suids, Man, etc.)

121

122

123

124

125

126

Figures 8.121–8.126

HOST SPECIFICITY AND HOST DISTRIBUTION

Host specificity and host distribution constitute some of the most important evidence for the thesis of coevolution of fleas and mammals. The
correlation of fleas and host is so close that in many instances it seems
axiomatic that they must have been intimately associated in locality and
habitat for countless eons. Before presenting such data, however, it is
necessary to provide some generalizations and to define or propose terms
that facilitate analysis. The data on fleas have proven so useful in studies
on medical ecology, evolution, and zoogeography (Traub 1980b, c; Traub
and Jellison 1981), because the phylogeny of the obligate parasite to a great
extent reflects that of the host—a principle that Brooks (1979) refers to as
Fahrenholz's rule. Primitive hosts such as shrews, moles, and marsupials
not only tend to be infested with primitive fleas (Traub 1980c), but they
also almost invariably lack highly evolved siphonapterans, a point made
for fish parasites (Szidat 1956). Conversely, the more specialized and more
recently developed groups of mammals at the ordinal and familial levels
lack primitive fleas, and, if infested, have the more advanced types. The
principle that primitive ectoparasites are rarely, if ever, found on a highly
evolved host has also been noted in myobiid mites by Fain (1979). Moreover, most taxa of fleas tend to remain with the hosts with which they
evolved. Although a few fleas have switched and then adapted to higher
groups of mammals, as those hosts evolved or penetrated into an area,
only rarely have they reverted and parasitized hosts of lower evolutionary
standing. When they have done so, they generally show morphological
signs of secondary adaptations to them (Traub 1980c).

Many of the host records in the literature are misleading or nigh useless
because they list all the reported hosts without comment, and, accordingly, the reader cannot determine which constitute the true or important
hosts and which represent accidental or meaningless records. Also, many
of the terms used to describe the degree of host affinity have been too
indefinite to permit critical evaluation of the data, and many have been
employed in different senses by various authors. There are serious and
inherent problems in attempting to study or classify host specificity of
fleas, and some cannot be resolved because of inadequate information on
the host relationships of a surprisingly large number of fleas. This is especially true regarding South American Siphonaptera, particularly the
Rhopalopsyllidae. What follows is an attempt to obtain the maximum
amount of information possible within such limitations. However,
throughout it should be realized that specificity at the generic level may be
different from that exhibited by species or species groups.

There is a significantly greater degree of host specificity among
Siphonaptera than the remarks of Keast (1977) and a casual reading of
Hopkins (1957) would indicate. Holland (1964) provides excellent examples
along these lines. Also, data on host specificity were cardinal features in

the evidence on faunal connections among the southern continents long before such theories became fashionable (Traub 1968b, 1972c, d, 1980c). It is true, however, that only a minority of species or genera of fleas are what is termed here *ultraspecific*, that is, limited to infestation of a single *species* of host (either as fur fleas, etc., or confined to the nest of the host), but many are *generispecific*, that is, essentially or primarily restricted to a genus of mammal or to a group of closely related hosts. There are fleas that exhibit a broader host range than do the generispecific Siphonaptera and infest a fairly large assortment of mammalian genera (e.g., rodents) belonging to a single tribe or subfamily. We term these *ethnoxenous* fleas. Provided the genera of hosts are not closely allied, the term *ethnoxenous* is also applicable to fleas of host families that are small and have not been divided into subfamilies. Other species or genera of Siphonaptera are less specific and are commonly found on some hosts that are not closely related, such as microtines and peromyscines, but these fleas nevertheless have a somewhat limited host range, that is, they occur within a large family of mammals. Such fairly facultative fleas are herein referred to as *intrafamilial* fleas. Species of fleas that infest several families of hosts within an order are called *extrafamilial* parasites. Several genera of fleas are extrafamilial. Fleas that may regularly infest more than one order of mammals are termed *extraordinal*. Certain fleas can successfully parasitize various mammals of an appropriate size that are encountered within their particular habitat. These are here designated as *biotopical* fleas in order to stress the ecological aspect, but the term is not necessarily exclusive, since such fleas can also be extrafamilial, and so forth. Other fleas, while tending to infest mammals of a certain order, are so adaptable that they can parasitize almost any kind of mammal they meet or find anywhere; they are termed *indiscriminate* or *catholic* fleas.

Hopkins (1957) used *monoxenous* and *oligoxenous* to denote fleas which infest one host, or two hosts, respectively. These terms are useful provided that they refer to the numbers of species or genera of hosts, and not to individuals, because fleas feed repeatedly, and most species feed on more than one mammal in their lifetime. *Polyxenous* can then be used as a general term regarding fleas that have a broad host range. Hopkins (1957) used the term *promiscuous* to indicate fleas that can "reproduce indefinitely on unrelated hosts," but such precise information on breeding is unavailable for the vast majority of fleas. It is treated here as a synonym of catholic. *Polyhaemophagous* was restricted by Hopkins (1957) to apply to fleas having the facility "to feed on the blood of hosts other than the one normal to the flea in question." Even though we frequently are ignorant about which species is the "normal" host, the term is useful to generally denote any flea that can feed on more than a single species of host.

A genus or species of mammal may be host to more than one species of ultraspecific or generispecific flea, and some species of mammals are infested with 4–14 such species of fleas. Where multiple species of ultraspecific fleas exist within a siphonapteran genus, there generally is only one such species on a particular species of host in any one area (*Phaenopsylla*, a parasite of *Calomyscus*, is an exception). It follows, then, that the ultraspecific fleas may have a much more restricted geographic range than do the hosts. The same is true for the generispecific Siphonaptera. Another important point is that the highly specific fleas may not necessarily be the most prevalent component of the siphonapteran fauna of a species (or individual) of host in any one area. Rodents may be infested with 12–25 species of fleas, and mathematically it is unlikely that any one of them would outnumber all or most of the others five- or tenfold. Thus estimating specificity by calculating percent infestation of various flea species on a host could be misleading. Primitive mammals such as the insectivores, marsupials, and bats, however, tend to be infested with a much more limited number of species of Siphonaptera than are cricetids.

It is impossible to express satisfactorily specificity in mathematical terms. Nevertheless, if 90–95% of the specimens of an ultraspecific species come from the particular host, this host can be considered the *ultraspecific host* for that flea species. About 80–95% of the records of generispecific species or genera of fleas are collected from the appropriate genus or group of hosts, and such mammals may therefore be designated as *prime* hosts of those fleas. The situation becomes complex regarding the ethnoxenous and facile fleas because of the numbers of different hosts that can be involved, and because each such host may entertain a rich fauna of fleas. Where 30% or more of these fleas are found on any one group of mammals, they are referred to as *main* hosts. When only 5–29% of the fleas are associated with certain mammals, the hosts are *occasional*, whereas those that are very infrequently infested (less than 5%) are termed *accidental*. For example, fleas that are parasites of rodents are often collected from carnivores preying upon those hosts. Indeed, exceptionally, some rodent fleas have transferred to the predators and eventually produced characteristic taxa like *Ceratophyllus lunatus* Jordan and Rothschild and *Nearctopsylla brooksi* (Rothschild). *C. lunatus*, a flea of martens and fishers (*Martes*), was placed with bird fleas because of the superficial similarity of the pronotal comb with those of *Ceratophyllus*. Actually the comb is fundamentally different and is typical of the pattern seen in fleas of other such nocturnal,* arboreal hosts (Traub 1969, 1972b, 1980b).

Ultraspecific Fleas

The species of fleas that are restricted to a certain species of host, ultraspecific species, are listed in Tables 8.1–8.4. The data are arranged in phylogenetic sequence of mammals (Traub and Jellison 1981), commencing with the monotremes, the primitive orders, and four families of rodents (Table 8.1). Other families of rodents are listed in Tables 8.2 and 8.3, and the highest flea-bearing orders, Carnivora, Hyracoidea, and Artiodactyla,

Table 8.1 Species of Mammals Parasitized by Specific Species of Fleas (Ultraspecific Fleas). I. Nonplacentals, Insectivora, Chiroptera, Pholidota and Aplodontid, Ctenodactylid, Rhizomyid, and Pedetid Rodents[a]

Mammals		Siphonaptera
MONOTREMATA		
1. *Tachyglossus aculeatus*	M	*Bradiosylla echidnae* (M)
MARSUPIALIA		
2. *Acrobates pygmaeus*	M	*Choristopsylla thomasi*
3. *Antechinus swainsoni*	U	*Acanthopsylla dunneti*
4. *Cercartetus nanus*	U	*Acanthopsylla scintilla*
5. *Eudromecia caudata*	U	*Acanthopsylla bella, A. eudromicia*
6. *Gymnobelideus leadbeateri*	M	*Stephanocircus domrowi, Wurunjerria warnekei* (M)
7. *Lasiorhinus latifrons*	U	*Echinophaga calabyi, E. octotricha, Lycopsylla lasiorhini*
INSECTIVORA		
8. *Hemiechinus auritus*	U	*Synosternus longispinus*
9. *Hylomys suillus*	M	*Cratynius bartelsi, C. crypticus, C. audyi*
10. *Neurotrichus gibbsi*	M	*Corypsylla jordani, Nearctopsylla hamata*
11. *Talpa altaica*	U	*Callopsylla (T.) semenovi* (M)
12. *Talpa caeca*	U	*Palaeopsylla oxygonia, P. cisalpina, P. iberica*
13. *Talpa europea*	U	*P. atlantica, P. hamata, P. kohauti, P. minor, P. similis, P. steini, Ctenophthalmus (C.) bisoctodentatus*
14. *Urotrichus talpoides*	M	*Stenischia fujisania*
15. *Anourosorex squamipes*	M	*Palaeopsylla remota*
16. *Cryptotis parva*	U	*Corrodopsylla hamiltoni*
17. *Cryptotis species* (2)	U	*Ctenophthalmus (A.) expansus, C. (A.) cryptotis*
18. *Diplomesodon pulchellum*	M	*Leptopsylla putoraki*
19. *Sorex trowbridgei*	U	*Nearctopsylla princei*
20. *Soriculus caudatus*	U	*Palaeopsylla recava*
21. *Tenrec ecaudatus*	M	*Centetipsylla madagascariensis*
CHIROPTERA		
22. *Eptesicus fuscus*	U	*Nycteridopsylla chapini*
23. *Plecotus auritus*	M	*Ischnopsyllus hexactenus, Nycteridopsylla pentactena*
PHOLIDOTA		
24. *Manis temmincki*	U	*Neotunga euloidea*
LAGOMORPHA		
25. *Ochotona alpina*	U	*Ctenophyllus subarmatus*
26. *Ochotona cansa*	U	*Geusibia torosa*

Table 8.1 (Continued)

Mammals		Siphonaptera
27. *Ochotona dahurica*	U	*Ochotonobius hirticrus*
28. *Ochotona hyperborea*	U	*Ctenophyllus rigidus*
29. *Ochotona roylei*	U	*Aconothobius martensi, A. orientalis, Geusibia triangularis, Chaetopsylla gracilis*
30. *Nesolagus netscheri*	M	*Nesolagobius callosus* (M)
31. *Oryctolagus cuniculus*	M	*Spilopsylla cuniculus, Odontopsyllus quirosi*
32. *Romerolagus diazi*	M	*Cediopsylla tepolita, Hoplopsyllus pectinatus*
33. *Sylvilagus brasiliensis*	U	*Cediopsylla spillmani, Euhoplopsyllus andensis, E. manconis, E. exoticus*
34. *Sylvilagus floridanus*	U	*Cediopsylla simplex, Odontopsyllus multispinosus*
RODENTIA		
35. *Aplodontia rufa*	M	*Paratyphloceras oregonensis* (M), *Trichopsylloides oregonensis* (M), *Dolichopsyllus stylosus* (M), *Hystrichopsylla schefferi*
36. *Ctenodactylus gundi*	M	*Caenopsylla mira*
37. *Tachyoryctes splendens*	U	*Ctenophthalmus* (G.) *audax, C.* (G.) *edwardsi, C.* (G.) new spp. (2)
38. *Pedetes capensis*	U	*Synosternus caffer*

[a]M, monotypic genus (or subgenus, if indicated); U, no other member of the host genus occurs in area or habitat.

are treated in Table 8.4. A total of 92 genera and 233 species of fleas and 122 species of mammals are listed in these tables. Since there are approximately 1790 species of fleas (including 123 bird fleas) described to date, and there is access to about 60 new species from mammals, about 14% (233/1667) of the known siphonapteran fauna of mammals can be regarded as essentially ultraspecific. As the precise hosts are unknown for quite a few fleas, the number of such fleas may actually be somewhat higher.

A striking and previously unreported conclusion is immediately discernible from the tabulated data. Virtually all of these ultraspecific fleas parasitize hosts that either lack close relatives altogether, that is, they belong to monotypic genera or else have no close kin (congeners) in the immediate area or habitat. Thus 31 (25%) of the 122 mammals listed are monotypic (designated by "M" in the tables) and 89 (73%) are marked "U" since they are unique in the particular foci. The 31 monotypic hosts are infested by a total of 73 fleas—all ultraspecific, and amounting to 31% of the 233 species in this category. The unique hosts carry 158 such species (68%). These figures exclude the newly discovered, monotypic, leptopsyllid *Typhlomyopsyllus** *cavaticus* Li and Huang. This sole species is an

Table 8.2 Species of Mammals Parasitized by Specific Species of Fleas (Ultraspecific Fleas). II. Sciurid, Geomyid, Heteromyid, Spalacid, Glirid, Zapodid, Dipodid, and Bathyergid Rodents[a]

Mammal		Siphonaptera
SCIURIDAE		
39. *Callosciurus nigrovittatus*	—	*Macrostylophora* new sp. "A"
40. *Callosciurus notatus*	—	*Macrostylophora* new sp. "B"
41. *Citellus adocetus*	U	*Polygenis adocetus*
42. *Citellus citellus*	U	*Ctenophthalmus (Eu.) orientalis*
43. *Citellus pygmaeus*	U	*Ctenophthalmus (Eu.) breviatus, C. (Eu.) pollex*
44. *Citellus undulatus*	U	*Oropsylla alaskensis*
45. *Dremomys lokriah*	U	*Rowleyella arborea* (M)
46. *Eutamias sibiricus*	U	*Ctenophthalmus (Eu.) pisticus*
47. *Lariscus insignis*	U	*Medwayella batibacula, M. calcarata*
48. *Marmota marmota*	U	*Frontopsylla tjanshanica*
49. *Marmota olympus*	U	*Oropsylla eatoni*
50. *Rhinosciurus laticaudatus*	M	*Medwayella limi*
51. *Sciurus griseus*	U	*Opisodasys enoplus, Orchopeas dieteri*
52. *Sciurus vulgaris*	U	*Aenigmopsylla grodekovi* (M)
53. *Spermophilopsis leptodactylus*	M	*Rostropsylla daca* (M)
54. *Tamias striatus*	M	*Tamiophila grandis, Monopsyllus acerbus*
55. *Tamiasciurus douglasii*	U	*Orchopeas nepos*
56. *Tamiasciurus hudsonicus*	U	*Rhadinopsylla media, Monopsyllus vison, Orchopeas caedens*
57. *Glaucomys sabrinus*	U[b]	*Opisodasys vesperalis*
58. *Glaucomys volans*	U[b]	*Opisodasys pseudarctomys, Conorhinopsylla stanfordi, Epitedia faceta*
59. *Petaurista elegans*	U	*Smitipsylla maseri*
60. *Petaurista leucogenys*	U	*Callopsylla (C.) petaurista*
61. *Petaurista magnifica*	U	*Smitipsylla prodigiosa*
GEOMYIDAE		
62. *Orthogeomys grandis*	U	*Pulex sinoculus*
HETEROMYIDAE		
63. *Heteromys desmarestianus*	U	*Wenzella obscura, W. yunkeri*
64. *Perognathus californicus*	U	*Carteretta carteri*
SPALACIDAE		
65. *Spalax ehrenbergi*	U	*Ctenophthalmus (M.) costai, C. (Eu.) levanticus, Rhadinopsylla golana*
66. *Spalax leucodon*	U	*C. (Eu.) gratus, C. (Sp.) monticola, C. (P.) fissurus, C. (P.) turcicus*

Table 8.2 (Continued)

Mammals		Siphonaptera
67. *Spalax microphthalmus*	U	*C. (Sp.) spalacis, C. (Sp.) jeanneli, C. (Sp.) auris, C. (Sp.) gigantospalacis*
GLIRIDAE		
68. *Dryomys nitedula*	U	*Myoxopsylla dryomydis*
69. *Graphiurus murinus*	U	*Dinopsyllus kempi*
70. *Glirulus japonicus*	M	*Monopsyllus yamane*
ZAPODIDAE		
71. *Sicista subtilis*	U	*Leptopsylla sicistae*
DIPODIDAE		
72. *Allactaga tetradactyla*	U	*Hopkinsipsylla o. occulta* (M)
73. *Dipus sagitta*	M	*Ophthalmopsylla (C.) karakum*
BATHYERGIDAE		
74. *Bathyergus suillus*	U	*Dinopsyllus (Cr.) ingens* (M)
75. *Cryptomys hottenttotus*	U	*Cryptopsylla ingrami* (M)

[a]M, monotypic genus (or subgenus, if indicated); U, no other member of the host genus occurs in area or habitat.
[b]Both species of flying squirrel may occur together in periphery.

ultraspecific flea of the monotypic, pygmy dormouse, *Typhlomys* (Platacanthomyidae).

The only clear-cut instance where two congeneric mammals with ultraspecific fleas occur together is represented by items 39 and 40 in Table 8.2. This exception, which occurs on Mt. Kinabalu in Borneo, is informative since the two species of *Callosciurus* coexist only in an extreme and limited part of their ranges. *C. notatus* (Boddaert) is a squirrel of the lowlands and foothills, whereas *C. nigrovittatus* (Horsfield) is a montane species, and the two are found together only at an elevation of about 1250–1300 m. Each species of squirrel carries only its particular species of *Macrostylophora* in its allopatric, as well as in the sympatric range, even when on the same tree. Two hosts listed as unique, *Melomys lorentzi* (Jentink) and *Pogonomelomys ruemmleri* Tate and Archbold (numbers 104 and 106 in Table 8.3) may really be monotypic. *M. lorentzi*, the specific host of the highly modified, monotypic genus *Smitella*, differs so greatly from the other rats classified as *Melomys* that A. Ziegler (in litt.) believes that it will be placed in a genus by itself. Two of its distinctive characters seem pertinent here: (1) the unusually large hind feet, which suggest this rat may be somewhat saltatorial, and (2) the shaggy fur whose hairs are of such size and so spaced that the unique "helmet" of *Smitella** can readily slip between them and latch onto them by the crown of false spines. Such vestiture must

Table 8.3   Species of Mammals Parasitized by Specific Species of Fleas (Ultraspecific Fleas). III. Cricetid and Murid Rodents[a]

Mammals		Siphonaptera
CRICETIDAE		
76. *Baiomys taylori*	M	*Jellisonia ironsi*
77. *Neotoma cinerea*	U	*Phalacropsylla allos, Anomiopsyllus montanus*
78. *Neotoma floridana*	U	*Conorhinopsylla nidicola, Epitedia cavernicola, E. neotomae*
79. *Neotoma lepida*	U	*Anomiopsyllus amphibolus*
80. *Neotomodon alstoni*	M	*Strepsylla mina, S. taluna, Pleochaetis paramundus*
81. *Onychomys leucogaster*	U	*Pleochaetis exilis*
82. *Alticola roylei*	U	*Callopsylla (C.) mygala*
83. *Dolomys bogdanovi*	U	*Rhadinopsylla dolomydis, Ctenophthalmus (M.) nifetodes, C. (M.) dolomydis*
84. *Ellobius talpinus*	U	*Ctenophthalmus (D.) dux, Amphipsylla dumalis, Neopsylla bactriana, Xenopsylla magdalinae*
85. *Eothenomys custos*	U	*Ctenophthalmus (Si.) parcus, C. (Si.) dinormus*
86. *Lagurus curtatus*	U	*Amphipsylla s. washingtona, Megabothris clantoni, Thrassis bacchi johnsoni*
87. *Microtus mexicanus*	U	*Ctenophthalmus (Ne.) haagi*
88. *Microtus nivalis*	U	*Rhadinopsylla mesa*
89. *Pitymys sikimensis*	U	*Stenoponia himalayana, Amphipsylla t. gregorii, Neopsylla pageae, Callopsylla (C.) dolabella*
90. *Calomyscus bailwardi*	M	*Phaenopsylla hopkinsi, P. jordani, P. kopetdag, P. mustersi, P. newelli, P. tiflovi, Ph.* new spp. (2), *Peromyscopsylla tikhomerovae, Amphipsylla argoi, A. montium, A. transcaucasica, Paradoxopsyllus microphthalmus,* New Genus, n.sp.
91. *Mesocricetus auratus*	M	*Ctenophthalmus (P.) rettigi, C. (P.) acuminatus*
92. *Myospalax myospalax*	U	*Amphipsylla daea, A. aspalacis, Ctenophthalmus (N.) dilatatus, Stenoponia singularis, Brachyctenonotus myospalacis* (M)
93. *Myospalax fontanieri*	U	*Amphipsylla casis, Neopsylla anoma, N. aliena*
94. *Hypogeomys antinema*	M	*Xenopsylla petteri*
95. *Rhombomys opimus*	M	*Coptopsylla olgae, C. bondari, C. mesghali, Nosopsyllus (G.) ziarus, N. (G.) vlasovi*
96. *Tatera brantsi*	U	*Xenopsylla goldenhuysi*
97. *Tatera indica*	U	*Xenopsylla hussaini*
98. *Tatera leucogaster*	U	*Xenopsylla frayi*
99. *Tatera robusta*	U	*Xenopsylla humilis*

Table 8.3   (Continued)

Mammals		Siphonaptera
MURIDAE		
100. *Aethomys namaquensis*	U	*Praopsylla powelli* (M)
101. *Cricetomys gambianus*	M	*Xenopsylla crinita, X. torta*
102. *Lorentzimys nouhuysi*	M	Five species of *Astivalius* (4 new), 3 new spp. of *Metastivalius*
103. *Malacomys longipes*	U	*Afristivalius rahmi*
104. *Melomys lorentzi*	U	*Smitella thambetosa* (M)
105. *Nesokia indica*	M	*Xenopsylla nesokiae*
106. *Pogonomelomys ruemmleri*	U	*Tiflovia pachnopoata, T. stellalpestris*
107. *Rattus baluensis*	U	*Bibikovana tiptoni*
108. *Rattus (S.) bowersi*	U	*Neopsylla dispar*
109. *Rattus (S.) infraluteus*	U	*Neopsylla luma*
110. *Saccostomus campestris*	U	*Xenopsylla sarodes, X. bechuanae, X. scopulifer*
111. *Otomys sloggitti*	U	*Chiastopsylla monticola*
112. *Otomys sulcata*	U	*Chiastopsylla quadrisetis*
113. *Otomys unisulcatus*	U	*Chiastopsylla mulleri*
114. *Parotomys littledalei*	U	*Xenopsylla occidentalis*

[a]M, monotypic genus (or subgenus, if indicated); U, no other member of the host genus occurs in area or habitat.

provide a very different microhabitat for a flea than that of the very fine fur of *Melomys platyops* (Thomas), which is the only other *Melomys* within the range of *Smitella*. The remarkable helmet of *Smitella* seems to be an adaptation to the shaggy fur of a host with specialized habits. *P. ruemmleri*, the specific host of the two species of *Tiflovia,* is another example of a rodent that differs markedly from the other species currently listed in the same genus. Apparently a mammalogist is in the process of placing it in a separate, monotypic genus (A. Ziegler, in litt.).

The number of species of mammals that have ultraspecific fleas and that belong to monotypic genera is disproportionately great. For example, in the Cricetidae there are 107 genera; 16 of these (15%) carry ultraspecific fleas. Six of these genera of hosts are monotypic. Of the 64 ultraspecific fleas on cricetids, 24 (37%) are on monotypic hosts. Many of the other hosts with ultraspecific fleas belong to genera consisting of only two or three species. Moreover, whether monotypic or not, the hosts in Tables 8.1–8.4 are often highly distinctive and have no close relatives at the generic level. Forty-five species, 37% of the 122 hosts listed belong to this category (see items 2, 6, 9, 10, 14, 18, 21, 24–32, 57, 58, 65–67, 71–75, 90, 92–94, 101–103, 105, 106, 111–113, 120–122).

Table 8.4   Species of Mammals Parasitized by Specific Species of Fleas (Ultraspecific Fleas). IV. Carnivora, Hyracoidea, and Artiodactyla[a]

Mammals		Siphonaptera
CARNIVORA		
115. *Hyaena hyaena*	U	*Chaetopsylla hyaena*
116. *Meles meles*	M	*Chaetopsylla trichosa, Paraceras melis*
117. *Melogale orientale*	U	*Paraceras pendleburyi*
118. *Procyon lotor*	U	*Chaetopsylla lotoris*
119. *Vulpes vulpes*	U	*Chaetopsylla globiceps, C. korobkovi, Caenopsylla laptevi*
HYRACOIDEA		
120. *Procavia capensis*	U	*Ctenocephalides rosmarus,* 4 spp. of *Procaviopsylla*
121. *Dendrohyrax brucei*	U	*Procaviopsylla spinifex, Ctenocephalides craterus*
ARTIODACTYLA		
122. *Phacochoerus aethiopicus*	M	*Delopsylla crassipes* (M), *Moeopsylla sjoestedti* (M), *Echidnophaga inexpectata*

[a]M, monotypic genus (or subgenus, if indicated); U, no other member of the host genus occurs in area or habitat.

Another noteworthy feature in Tables 8.1–8.4 is that 69 (57%) of these hosts are specialized in habit or structure. For example, 32 (26%) are scandent or arboreal (items 2–6, 39, 40, 45–47, 50–52, 54–61, 68–70, 76–78, 101, 102, 104, 106, and 114); 18 (15%) are fossorial (items 10–14, 37, 48, 49, 62, 65–67, 74, 75, 84, 93, and 105); four (3%) are spiny or scaly in vestiture (items 1, 8, 21, 24); four (3%) have spinose or coarse fur (items 63, 64, 108, 109); and 11 (9%) (items 9, 30–34, 38, 71–73, 94) are saltatorial (*M. lorentzi* is probably saltatorial as well). *Calomyscus* (90) and *Lorentzimys* (102) also have the hindlegs modified as in jumping mammals, but the latter is actively scansorial and nests in *Pandanus* (screw pines), and *Calomyscus* climbs well on rocks and bushes. All the fleas of these unusual mammals have special, adaptive morphological modifications.

Ultraspecific fleas occur in all the mammalian orders which include representatives parasitized by fleas. However, many of the constituent families lack them entirely. Thus among the marsupials the species of didelphids, dasyurids, peramelids, and coenolestids are not known to have ultraspecific fleas, and the same is true regarding the macroscelids and tupaiids, which are often considered as insectivorans. The only family of bats noted is the Vespertilionidae. Among the rodents, all subfamilies of the Cricetidae and only one geomyid *(Orthogeomys)* are represented. Among the murids there are neither dendromurines, otomyines, nor hydromyines. No hystricids are cited, nor are any of the Neotropical rodents on the list. Among the Carnivora, the ursids, viverrids, and felids all lack ultraspecific fleas.

Monotypic species are among the mammals infested by fleas that are generispecific or less specific in nature, and generally one or more kindred hosts also occur locally. All the foregoing observations lead to the conclusion that there is no suitable alternate host available for ultraspecific fleas to infest. The absence of alternate hosts *ipso facto* supports the idea that coevolution must have occurred among the ultraspecific fleas and their particular hosts.

Since only about 14% of the siphonapteran fauna of mammals is known to be specific to a single species of host, it would be expected that hosts with more than one species of such ultraspecific fleas would be uncommon. However, 39% of the 122 hosts listed in Tables 8.1–8.4 have more than one such flea, and, indeed, 22% have three or more. For example, *Aplodontia rufa* Rafinisque (35) has four species belonging to four different genera; *Myospalax myospalax* Laxmann (92), *Rhombomys opimus* Lichtenstein (95), and *Procavia capensis* (Pallas) (120) each are infested with five such highly specific species; and *Lorentzimys nouhuysi* Jentink (102) is host to eight such fleas. *Talpa europea* Linnaeus (13) has seven, and *Calomyscus bailwardi* Thomas has the incredible total of 14 (but only a few of these occur in any one part of its range). Even more striking is that, by convergence,* the specific but unrelated fleas shared by a host often resemble one another more in certain fundamental morphological features (within phylogenetic constraints) than they do their congeners that occur on other hosts.

For analysis of the ultraspecific fleas, the data from Tables 8.1–8.4 are further summarized by family of flea and order or family of host in Table 8.5, which, for editorial reasons, excludes the sole flea of monotremes, *Bradiopsylla echidna* (Denny). The majority of ultraspecific fleas (143 of 232 = 62%) are parasites of rodents. Of these, the greatest proportion occurs on the Cricetidae, that is, 25% of the 92 listed genera and 28% of the species are found on cricetids. The Sciuridae are second, with 20% of the genera and 13% of the ultraspecific species. In contrast, such examples occur significantly less often on the Marsupialia (7% of genera and 4% of species) and Insectivora (12% of genera and 11% of species). Although 11% of the specific genera and 11% of the species are murid fleas, it must be noted that none of these is a parasite of *Rattus* (*Rattus*), and there are only three species listed for the genus *Rattus* at all. In fact, *R. baluensis* (Thomas) is a distinctive rat that probably does not belong in the subgenus *Rattus* and perhaps not even in that genus, and the other two are in the subgenus *Stenomys*. All of the murids listed (items 100–114) are specialized in one way or another, and the absence of ultraspecific fleas on the subgenus *Rattus** is in accord with the concept that these rats are a recently developed, rapidly diversifying group (Traub 1972c, 1980c).

The maximum representation of ultraspecificity occurs in the Hystrichopsyllidae, which include 23% of the pertinent genera and 33% of the

Table 8.5 Numbers and Percentages of 233 Ultraspecific Species of Fleas, Arranged by Family of Flea and Groups of Hosts[a,b]

Hosts		Hystricho-psyllidae	Pygio-psyllidae	Cerato-phyllidae	Lepto-psyllidae	Pulicidae	Others	Sub-Total	Percent of Total
Marsupialia	g	0	4 = 67%	0	0	1 = 17%	1 = 17%	6	7%
	s	0	7 = 70%	0	0	2 = 20%	1 = 10%	10	4%
Insectivora	g	6 = 55%	0	1 = 9%	2 = 18%	2 = 18%	0	11	12%
	s	19 = 73%	0	1 = 4%	4 = 15%	2 = 8%	0	26	11%
Lagomorpha	g	0	0	0	5 = 45%	5 = 45%	1 = 9%	11	12%
	s	0	0	0	9 = 47%	9 = 47%	1 = 5%	19	8%
Aplodontidae	g	3 = 75%	0	0	1 = 25%	0	0	4	4%
	s	3 = 75%	0	0	1 = 25%	0	0	4	2%
Rhizomyidae	g	1 = 100%	0	0	0	0	0	1	1%
	s	4 = 100%	0	0	0	0	0	4	2%
Sciuridae	g	5 = 28%	1 = 6%	10 = 55%	1 = 6%	0	1 = 6%	18	20%
	s	8 = 26%	3 = 10%	18 = 58%	1 = 3%	0	1 = 3%	31	13%
Spalacidae	g	2 = 100%	0	0	0	0	0	2	2%
	s	11 = 100%	0	0	0	0	0	11	8%
Gliridae	g	1 = 33%	0	2 = 67%	0	0	0	3	3%
	s	1 = 33%	0	2 = 67%	0	0	0	3	1%
Cricetidae	g	9 = 39%	0	6 = 26%	6 = 26%	1 = 4%	1 = 4%	23	25%
	s	25 = 39%	0	9 = 14%	21 = 33%	6 = 9%	3 = 5%	64	28%

Muridae	g	2 = 20%	5 = 50%	0	0	1 = 10%	2 = 20%	10	11%
	s	2 = 8%	13 = 50%	0	0	7 = 27%	4 = 27%	26	11%
Carnivora	g	0	0	1 = 50%	0	0	1 = 50%	2	2%
	s	0	0	2 = 29%	0	0	5 = 71%	7	3%
Hyracoidea	g	0	0	0	0	2 = 100%	0	2	2%
	s	0	0	0	0	7 = 100%	0	7	3%
Artiodactyla	g	0	0	0	0	3 = 100%	0	3	3%
	s	0	0	0	0	3 = 100%	0	3	1%
Other	g	3 = 21%	1 = 7%	0	4 = 29%	3 = 21%	3 = 21%	14	15%
	s	4 = 25%	1 = 6%	0	5 = 29%	3 = 18%	4 = 24%	17	7%
Totals	g[c]	21 = 23%	12 = 13%	17 = 18%	18 = 20%	15 = 16%	9 = 10%	92[c]	
	s	77 = 33%	24 = 10%	32 = 14%	41 = 18%	39 = 17%	19 = 8%	232	
Total Number in family[d]	g	45	46	39	24	22	—	—	
	s	460	243	323	184	148	—	—	
Total in family (%)	g	47%	26%	44%	75%	68%	—	—	
	s	17%	10%	10%	22%	26%	—	—	

[a] *Bradiopsylla echidnae*, on monotremes, is excluded.
[b] g, genera; s, species.
[c] Genera parasitizing more than one group are only counted once.
[d] Nonvolant hosts only.

total number of ultraspecific species, either as fur fleas or nest fleas. Moreover, 17% of the species of Hystrichopsyllidae are specific to a single species of host. The Leptopsyllidae and Pulicidae also include a notable number (and proportion) of such fleas, and the Pygiopsyllidae also rank fairly high in this respect. Although 24% of the 1358 species of fleas under consideration belong to the Ceratophyllidae, only 14% of the ultraspecific fleas belong to this family. The remaining unspecified families of fleas, combined, account for 10% or less of the total number of ultraspecific fleas.

Of the 77 species of such highly specific hystrichopsyllids, 25 (32%) occur on the Cricetidae and 19 (25%) on the Insectivora. The low number on murids (2 = 3%) and the absence on marsupials are noteworthy. With only one genus and three species in the family Spalacidae, it is impressive that a total of 11 species of hystrichopsyllids is restricted to this group of fossorial rodents. Thus the large family Sciuridae includes only eight hystrichopsyllids in this category, many of which are nest fleas, such as *Conorhinopsylla, Tamiophila,* and *Rhadinopsylla.*

Among the Pygiopsyllidae 10% of the species, representing 26% of the genera, are ultraspecific, and the vast majority are found on the marsupials and murids (13 and 7, respectively, of 24). However, this distribution reflects the origin and main radiation of this taxon in the Australian region (Traub 1972c, 1980c). The majority of the species of ultraspecific Ceratophyllidae (58% of 32) are sciurid fleas. Since the ceratophyllids are believed to have arisen on Sciuridae (Traub 1983b), this is to be expected. The dearth of ultraspecific ceratophyllids* on insectivorans and their absence on marsupials and certain other groups of mammals can apparently be explained by their Nearctic provenance and/or their relatively recent origin (Oligocene) (Traub 1983a; Traub and Rothschild 1983).

More than 50% (21/41) of the ultraspecific leptopsyllids occur on cricetids, in contrast to the single such species on sciurids, and 21% of such leptopsyllids are found on lagomorphs. Four are associated with Insectivora; three of these are parasites of the monotypic gymnuran *Hylomys,* and the remaining species is from a talpid. The absence of ultraspecific leptopsyllids on species of soricids, murids, and carnivorans is conspicuous since the ranges of these fleas and hosts coincide. In contrast, there are few or no marsupials where leptopsyllids are found, and hence the lack of association is not surprising.

Among the ultraspecific Pulicidae, 23% of 39 occur on the Lagomorpha, and the next highest are on the Muridae (18%) and Cricetidae (15%). The absence of such pulicids* on sciurids, insectivorans, and marsupials is noteworthy and provides support for some of the conclusions presented regarding zoogeography. In the category of "other" fleas are the families Stephanocircidae (one genus, one species on marsupials); Ischnopsyllidae (two genera, three species on bats); Rhopalopsyllidae (one genus, one species on *Citellus*); Vermipsyllidae (one genus, five species on Carnivora, one genus, one species on Lagomorpha); Chimaeropsyllidae (two genera, four species on murids; one genus, one species on Bathyergidae); and Coptopsyllidae (one genus, three species on gerbillines).

Specificity to a single species of mammalian host therefore occurs in at least 10 of the 15 families of Siphonaptera, including all of the major groups. The degree to which this occurs varies, ranging from 10 to 26% of the five main families (listed in Table 8.5), but the phenomenon is rarely seen in large groups like the Stephanocircidae, Ischnopsyllidae, and Rhopalopsyllidae. To a limited extent, the sparsity is due to lack of information, but there is no doubt that fleas of these three families tend to infest more than one species of host.

Generispecific Fleas

There are 183 species and species groups of fleas that infest a genus or a group of related hosts (Tables 8.6–8.9). Although lack of detailed information about some taxa preclude calculating the precise number of species of Siphonaptera that are generispecific, at least 102 genera of fleas infest a single genus, or closely related groups of hosts. It is estimated that a minimum of 500 species of fleas, equivalent to 30% of the known fauna of mammalian fleas, do so. This is a very conservative estimate since, for example, *Phalacropsylla allos* Wagner and *P. paradisea* Rothschild (almost undoubtedly *Neotoma* fleas), and *Procaviopsylla* spp. (on hyracoids) and *Pulex porcinus* Jordan and Rothschild (on peccaries) are excluded here because of the sparsity of unequivocal records or other limitations in the data. Similarly we have excluded the highly specialized *Hollandipsylla neali* Traub, which may prove to be generispecific or perhaps even ultraspecific. This species is known from only a single record, from the sole specimen from *Hylopetes* flying squirrel (presumably *H. lepidus* Horsfield) which we examined at Bundu Tuhan on Mt. Kinabalu in N. Borneo. No fleas of any kind were collected from a series of the giant flying-squirrels (*Petaurista*) in the same locality. Two species of *Hylopetes* occur in the vicinity of Bundu Tuhan, but their precise distribution and habits are unknown.

Two-thirds (122/183) of the species or species groups of generispecific fleas infest rodents; 10% (19/183) Insectivora; 9% (17 entries) marsupials; 5% (10) lagomorphs; and 4% (7/183) Carnivora. The genera in these tables are generispecific, but in some instances the taxon as a whole is ethnoxenous, intrafamilial, or extrafamilial. Significantly, only three genera of hystrichopsyllids, *Ctenophthalmus*, *Nearctopsylla*, and *Palaeopsylla*, infest more than one order of host (Table 8.6); in each case there is only one such aberrant species, such as *Ctenophthalmus particularis*, *Nearctopsylla brooksi*, and *Palaeopsylla setzeri*. Some hystrichopsyllids are ethnoxenous genera infesting more than one genus of host within a single subfamily of rodent, such as *Rhadinopsylla heiseri* (McCoy) and *R. japonica* Sakaguti and Jameson on petauristines (Table 8.6). However, several genera are extrafamilial and parasitize members of more than one family within an order, for example,

Table 8.6   Species of Fleas that Infest a Genus or Related Groups of Mammals (Generispecific Fleas). I. Hystrichopsyllid Fleas[a]

Hystrichopsyllidae	Mammals
F1 *Acedestia chera*	Bandicoots, *Potorous*
F2 *Adoratopsylla (A.) bisetosa*, etc.	Opossums
F3 *Adoratopsylla (T.) intermedia*	*Didelphis, Marmota*, etc.
F4 *Anomiopsyllus nudatus*, etc.	*Neotoma*
F5 *Atyphloceras nuperus*, etc.	Microtines
F6 *Callistopsyllus terinus*	*Peromyscus*
F7 *Catallagia ioffi*, etc.	Microtines
F8 *Catallagia mathesoni*, etc.	*Peromyscus*
F9 *Corrodopsylla curvata*, etc.	Soricids
F10 *Corypsylla kohlsi*	*Sorex*
F11 *Corypsylla ornata*	*Scapanus*
F12 *Ctenophthalmus (C.) egregius*, etc.	Microtines
F13 *Ctenophthalmus (Et.) eximius*, etc.	Murines
F14 *Ctenophthalmus (Eu.) congener*, etc.	Microtines
F15 *Ctenophthalmus (Id.) particularis*	*Crocidura*
F16 *Ctenophthalmus (M.) lydiae*	Murines
F17 *Ctenophthalmus (M.) nivalis*, etc.	Microtines
F18 *Ctenophthalmus (P.) dolichus*	Gerbillines
F19 *Ctenophthalmus (Sp.) caucasicus*	*Spalax*
F20 *Ctenophthalmus (Si.) parcus*	Microtines
F21 *Delotelis telegoni, D. hollandi*	*Peromyscus*
F22 *Epitedia faceta*	Tree squirrels, flying squirrels
F23 *Idilla caelebs*	*Antechinus*
F24 *Megarthroglossus becki*, etc.	*Neotoma*
F25 *Megarthroglossus procus, M. divisus*	*Sciurus, Tamiasciurus*
F26 *Meringis altipecten*, etc.	*Dipodomys*
F27 *Nearctopsylla (B.) jordani*, etc.	Talpids
F28 *Nearctopsylla (B.) genalis*	Soricids
F29 *Nearctopsylla (N.) brooksi*	Mustelids
F30 *Neopsylla acanthina*, etc.	Microtines
F31 *Neopsylla inopina*	*Citellus*
F32 *Neopsylla setosa*	*Citellus*
F33 *Neopsylla stevensi*, etc.	*Rattus, Apodemus*
F34 *Palaeopsylla miyama*	*Urotrichus*
F35 *Palaeopsylla mogura*	*Mogera*
F36 *Palaeopsylla setzeri*	*Hyperacrius*
F37 *Palaeopsylla sinica, P. laxata*	Shrews
F38 *Palaeopsylla soricis*	Shrews

Table 8.6 (Continued)

Hystrichopsyllidae	Mammals
F39 *Palaeopsylla tauberi*, etc.	*Soriculus*
F40 *Paraneopsylla ioffi*, etc.	Microtines
F41 *Rhadinopsylla (A.) heiseri*, etc.	*Citellus*, etc.
F42 *Rhadinopsylla (A.) isacantha*, etc.	Microtines
F43 *Rhadinopsylla (A.) japonica*, etc.	Petauristines
F44 *Rhadinopsylla (M.) sectilis*	*Peromyscus*
F45 *Stenistomera (S.) alpina*	*Neotoma*
F46 *Stenistomera (M.) macrodactyla*	*Peromyscus*
F47 *Stenoponia montana*, etc.	Microtinae
F48 *Stenoponia ponera*	*Peromyscus*
F49 *Stenoponia tripectinata*, etc.	Gerbillines
F50 *Strepsylla davisae*	*Peromyscus*
F51 *Xenodaeria telios*	*Crocidura*

a Etc., and allied taxa.

Corypsylla, Megarthroglossus, Neopsylla, Palaeopsylla, and *Rhadinopsylla. Ctenophthalmus* and *Nearctopsylla* are extraordinal. Some taxa in the Ceratophyllidae on Sciuridae and Peromyscinae are biotopical (Jellison 1945; Holland 1964). The hystrichopsyllids that infest certain or distinctive groups of hosts usually consist of specific species groups or subgenera, as observed in *Ctenophthalmus* as well as *Adoratopsylla, Rhadinopsylla,* and *Stenistomera.* (In many other taxa of fleas the taxonomic distinctions based upon morphological grounds likewise parallel grouping by host, suggesting that the scheme of systematics is correct.)

The Pygiopsyllidae include generispecific species or groups for the major part of the genera, although some are extraordinal; for example, in *Lentistivalius* some generispecific species infest murids, and others are on sciurids. *Bibikovana* and *Metastivalius* are also extraordinal at the generic level: some species are restricted to certain murids, and others are found on marsupials like bandicoots. Some of the genera of leptopsyllids are intrafamilial or extrafamilial, that is, there is some crossing of subfamily or family lines of hosts: *Amphipsylla* on microtines and Cricetini; *Paradoxopsyllus* and *Peromyscopsylla* among murids and various cricetids. In the Leptopsyllidae, the constituents are either found on lagomorphs or rodents, but none of the genera are found on both. Ochotonid and leporid fleas do not seem to have switched from one family of lagomorph to the other, and this statement applies to the fleas belonging to other families as well.

In the ceratophyllids most of the genera are also limited to a genus or allied group of hosts, but some of the genera are associated with two families or even two orders of hosts (Table 8.8). *Monopsyllus*, for example,

Table 8.7 Species of Fleas that Infest a Genus or Related Groups of Mammals (Generispecific Fleas). II. Pygiopsyllid and Leptopsyllid Fleas[a]

Fleas	Mammals
PYGIOPSYLLIDAE	
F52 *Acanthopsylla bisinuata*, etc.	*Melomys*, scansorial rats
F53 *Acanthopsylla incerta*, etc.	*Antechinus*, etc.
F54 *Aviostivalius klossi*	Native *Rattus*
F55 *Bibikovana rainbowi*, etc.	Native *Rattus*
F56 *Bibikovana traubi*	*Peroryctes*
F57 *Choristopsylla ochi*	*Trichosurus, Pseudocheirus*, etc.
F58 *Choristopsylla tristis*	*Petaurus*, etc.
F59 *Idiochaetis illustris*	*Hyomys goliath* (M), *Mallomys rothschildi* (M)
F60 *Lentistivalius alienus*	*Crocidura*
F61 *Lentistivalius aestivalis*	*Apodemus*
F62 *Lentistivalius ferinus*	*Suncus*
F63 *Lycopsylla nova*	*Phascolomys, Vombatus*
F64 *Medwayella dryadosa*	Ground callosciurines
F65 *Medwayella robinsoni*, etc.	Callosciurines
F66 *Metastivalius anaxilas*, etc.	*Melomys*, scansorial *Rattus*
F67 *Metastivalius rectus*, etc.	Native rats
F68 *Metastivalius shawmayeri, M. tamberan*	*Peroryctes*
F69 *Migrastivalius (G.) hopkinsi*	*R. (Lenothrix) whiteheadi; R.(L.) alticola*
F70 *Muesebeckella mannae, M. nadi*	*Pseudocheirus*
F71 *Nestivalius pomerantzi*	Native *Rattus*
F72 *Obtusifrontia simplex*	Native *Rattus*, etc.
F73 *Orthopsylla abacetus*, etc.	*Rattus*
F74 *Papuapsylla barretti*, etc.	*Rattus, Melomys*, etc.
F75 *Parastivalius novaeguineae*, etc.	*Peroryctes*
F76 *Pygiopsylla spinata*	*Peroryctes*, etc.
F77 *Stivalius cognatus*, etc.	*Rattus (Rattus)*
F78 *Striopsylla vandeuseni*, etc.	*Peroryctes*
F79 *Uropsylla tasmanica* (M)	*Dasyurus, Sarcophilus*
LEPTOPSYLLIDAE	
F80 *Amphipsylla anceps*, etc.	Cricetini
F81 *Amphipsylla orthogonia*	*Alticola*
F82 *Amphipsylla rossica*, etc.; *A. tuta*, etc.	Microtines
F83 *Acropsylla* (3 spp.)	*Mus*
F84 *Conothobius conothoae*	*Ochotona*
F85 *Ctenophyllus tarasovi*	*Ochotona*

Table 8.7 (Continued)

	Fleas	Mammals
F86	*Desertopsylla rothschildae* (M)	Dipodids
F87	*Geusibia ashcrafti, G. lacertosa*	*Ochotona*
F88	*Leptopsylla (L.) taschenbergi*, etc.	*Apodemus*
F89	*Leptopsylla (L.) segnis*, etc.	Murines
F90	*Leptopsylla (L.) nana*	Cricetini
F91	*Leptopsylla (P.) pamirensis*, etc.	Murines
F92	*Mesopsylla tuschkan*, etc.	*Jaculus, Allactaga*
F93	*Ochotonobius rufescens*, etc.	*Ochotona*
F94	*Odontopsyllus dentatus*	Hares, rabbits
F95	*Paradoxopsyllus custodis*	*Rattus*
F96	*Paradoxopsyllus hesperius*	Microtines
F97	*Paradoxopsyllus oribatus*, etc.	*Rattus, Mus*
F98	*Paradoxopsyllus repandus*, etc.	Gerbillines
F99	*Peromyscopsylla himalaica*, etc.	Murines
F100	*Perom. bidentata*, etc.; *P. silvatica*, etc.	Microtines
F101	*Peromyscopsylla hesperomys*, etc.	*Peromyscus*
F102	*Sigmactenus cavifrons*	*Pogonomelomys*
F103	*Sigmactenus* (6 spp.)	*Rattus*, etc.

[a]Etc., and allied taxa; (M), monotypic.

is extraordinal, with some species on sciurids, others on cricetids, and *M. lunatus* on carnivores. *Callopsylla* also parasitize two orders, Lagomorpha and Rodentia like sciurids and microtines; there also is a subgenus on birds. The subgenus *Nosopsyllus* infests sciurids, cricetids, and murids. Most *Thrassis* are found on sciurids, although *T. bacchi johnsoni* occurs on a microtine, and *T. aridis* parasitizes the heteromyid *Dipodomys*. The generispecific species infesting the smaller orders generally belong to genera found on only one group of hosts, but, as shown in Table 8.9, some stephanocircids parasitize marsupials and others, murids. *Synosternus* is extraordinal, occurring on hedgehogs and gerbillines, and *Xenopsylla* spans murines and gerbillines.

Genera with Generispecific Species

As shown in Table 8.10, the major families of Siphonaptera all include a substantial number (8–25% of the totals) of genera with species that infest closely allied hosts or only a single genus of mammal. However, the groups of hosts infested by these genera vary significantly, depending on the family of flea. Nine (39%) of 23 such genera of hystrichopsyllids infest

Table 8.8   Species of Fleas that Infest a Genus or Related Groups of Mammals (Generispecific Fleas). III. Ceratophyllid Fleas[a]

Ceratophyllidae	Mammals
F104 *Amalaraeus arvicolae*, etc.	Microtines
F105 *Amphalius runatus*, etc.	*Ochotona*
F106 *Callopsylla (C.) lagomys, C. caspia tiflovi*	*Ochotona*
F107 *C.(C.) sparsilis*, etc.; *C. caspia* spp.	Microtines
F108 *Callopsylla (C.) dolabris*	*Marmota*
F109 *Citellophilus tesquorum*, etc.	*Citellus*
F110 *Citellophilus lebedewi, C. menzbieri*	*Marmota*
F111 *Dactylopsylla comis*, etc.	Geomyids
F112 *Diamanus montanus*	*Citellus*
F113 *Foxella ignota*, etc.	Geomyids
F114 *Jellisonia hayesi*, etc.	*Peromyscus*
F115 *Jellisonia klotsi*, etc.	*Reithrodontomys*
F116 *Kohlsia graphis*	*Sciurus*
F117 *Kohlsia osgoodi*, etc.	*Peromyscus*
F118 *Libyastus hopkinsi*, etc.	African tree squirrels
F119 *Macrostylophora fimbriata*	Hylopetes, *Petaurista*
F120 *Macrostylophora hastata*, etc.	Callosciurines
F121 *Malaraeus sinomus*, etc.	*Peromyscus*
F122 *Megabothris asio*, etc.	Microtines
F123 *Monopsyllus ciliatus*	*Eutamias, Tamiasciurus*
F124 *Monopsyllus eumolpi*, etc.	*Eutamias*
F125 *Monopsyllus lunatus*	*Martes, Mustela*
F126 *Monopsyllus tolli*	*Ochotona*
F127 *Monopsyllus wagneri, M. thambus*	*Peromyscus*
F128 *Myoxopsylla laverani, M. jordani*	*Eliomys, Dryomys, Glis*
F129 *Nosopsyllus (N.) alladinis*, etc.	Funambulini, Xerini
F130 *Nosopsyllus (N.) barbarus*, etc.	Murines
F131 *Nosopsyllus (N.) durii, N.(N.) farahae*	*Microtus*
F132 *Nosopsyllus (Gerbillophilus)* spp.	Gerbillines
F133 *Nosopsyllus (Penicus) geneatus* (M)	*Acomys*
F134 *Opisocrostis oregonensis*, etc.	*Citellus*
F135 *O.t. tuberculatus, O. labis*, etc.	*Citellus*
F136 *O.t. cynomuris, O. hirsutus*	*Cynomys*
F137 *Opisodasys hollandi*, etc.	*Sciurus*
F138 *Opisodasys keeni, O. nesiotus*	*Peromyscus*
F139 *Orchopeas howardii*	*Sciurus*
F140 *Orchopeas leucopus*	*Peromyscus*
F141 *Orchopeas sexdentatus, O. neotomae*	*Neotoma*

Table 8.8 (Continued)

Ceratophyllidae	Mammals
F142 *Oropsylla alaskensis*	*Citellus undulatus*, etc.
F143 *Oropsylla arctomys*, etc.	*Marmota*
F144 *Oropsylla idahoensis*, etc.	*Citellus*
F145 *Paraceras javanicum*	Badgers, etc.
F146 *Paraceras sauteri*	Viverrids
F147 *Paramonopsyllus desertus*	*Ochotona*
F148 *Pleochaetis asetus*	*Microtus*
F149 *Pleochaetis mundus, P. mathesoni*	*Peromyscus*
F150 *Syngenopsyllus calceatus*	*Callosciurus*
F151 *Thrassis acamantis*, etc.	*Marmota*
F152 *Thrassis aridis*, etc.	*Dipodomys*
F153 *Thrassis bacchi johnsoni*	*Lagurus*
F154 *Thrassis fotus, T. bacchi* spp.	*Citellus*

a Etc., and allied taxa; (M), monotypic.

peromyscine rodents, eight occur on voles (Microtinae), and there are only four such genera on squirrels (Sciurinae). Although many insectivorans coexist with these rodents, only six hystrichopsyllid genera with generispecific fleas infest talpids, soricids, or other members of the order. Among ceratophyllids, 22 (88%) of the 25 genera include generispecific species on rodents, and the only nonrodents with such fleas are the ochotonids (with four) and carnivorans (with two). Most (56%) of these ceratophyllid genera are found on sciurids. In the Leptopsyllidae most of the generispecific species belong to genera parasitizing murids (5/13 = 38%) or ochotonids (31%).

If limitation of infestation to closely kindred hosts is indeed an index of coevolution, then the major families of fleas evidently had different evolutionary histories. Except for one pygiopsyllid (*Lentistivalius*), all the genera with generispecific fleas on shrews and moles belong to the Hystrichopsyllidae. Nevertheless, the fleas of this family are also intimately associated with mice belonging to the Peromyscini and Microtinae—much more so than with the squirrels. A noteworthy number (15) of genera of this type are found on marsupials, but these are primarily (67%) pygiopsyllids. Of the 20 such genera on squirrels, 70% are ceratophyllids. The majority of the 22 such genera on murids are pygiopsyllids (50%) or leptopsyllids (23%).

Specificity in Families of Fleas

Consolidation of the data on ultraspecific and generispecific fleas reinforces the belief that, at the familial level, fleas tend to exhibit characteristic

Table 8.9 Species of Fleas that Infest a Genus or Related Groups of Mammals (Generispecific Fleas). IV. Ischnopsyllid, Stephanocircid, Pulicid, Vermipsyllid, Chimaeropsyllid, Ancistropsyllid, Coptopsyllid and Malacopsyllid Fleas[a]

Fleas	Mammals
ISCHNOPSYLLIDAE	
F155 *Lagaropsylla mera*, etc.	*Tadarida*
STEPHANOCIRCIDAE	
F156 *Coronapsylla jarvisi* (M)	*Antechinus*
F157 *Stephanocircus connatus*, etc.	Native *Rattus*
F158 *Stephanocircus greeni*	*Antechinus*
F159 *Stephanocircus harrisoni*	Bandicoots
PULICIDAE	
F160 *Archaeopsylla erinacea, A. sinensis*	*Erinaceus*
F161 *Echidnophaga ambulans*	*Tachyglossus*, Echidna
F162 *Euhoplopsyllus glacialis*	Hares, rabbits
F163 *Hoplopsyllus anomalus*	*Citellus*
F164 *Parapulex chephrenis, P. echinatus*	*Acomys*
F165 *Pariodontis riggenbachi, P. subjugis*	Hystricids
F166 *Synosternus cleopatrae*	Gerbillines
F167 *Synosternus longispinus*	Erinaceids
F168 *Xenopsylla debilis, X. difficilis*	*Tatera*
F169 *Xenopsylla conformis*	*Meriones*
F170 *Xenopsylla nilotica*, etc.	Gerbillines
F171 *Xenopsylla papuensis*	Ground-nesting *Pogonomys*
VERMIPSYLLIDAE	
F172 *Chaetopsylla matina*, etc.	*Martes, Mustela*, etc.
F173 *Chaetopsylla mikado, C. homoea*	Carnivora
F174 *Chaetopsylla ursi, C. tuberculaticeps*	Ursids
F175 *Dorcadia ioffi, D. dorcadia*	Artiodactylids
F176 *Vermipsylla alakurt*	Equines
CHIMAEROPSYLLIDAE	
F177 *Chimaeropsylla haddowi, C. potis*	Macroscelids
F178 *Demeillonia granti, D. miriamae*	Macroscelids
F179 *Macroscelidopsylla albertyni*	Macroscledis
ANCISTROPSYLLIDAE	
F180 *Ancistropsylla nepalensis*	*Axis, Muntiacus*
COPTOPSYLLIDAE	
F181 *Coptopsylla lamellifer*, etc.	Gerbillines
MALACOPSYLLIDAE	
F182 *Malacopsylla grossiventris* (M)	Armadillos
F183 *Phthiropsylla agenoris* (M)	Armadillos

[a]Etc., and allied taxa; (M), monotypic.

Table 8.10 Numbers of Genera of Fleas that Include Species Restricted to a Single Host Genus or to Closely Allied Taxa

Hosts	Hystricho-psyllidae	Pygio-phyllidae	Cerato-psyllidae	Lepto-psyllidae	Pulicidae	Others	Totals
Monotremes	0	0	0	0	1 = 100%	0	1 = 1%
Marsupials	3 = 20%	10 = 67%	0	0	0	2 = 13%	15 = 15%
Talpids	3 = 100%	0	0	0	0	0	3 = 3%
Soricids	6 = 86%	1 = 14%	0	0	0	0	7 = 7%
Other insectivora	0	0	0	0	2 = 40%	3 = 60%	5 = 5%
Chiroptera	0	0	0	0	0	1 = 100%	1 = 1%
Edentates	0	0	0	0	0	2 = 100%	2 = 2%
Ochotonids	0	0	4 = 50%	4 = 50%	0	0	8 = 8%
Leporids	0	0	0	1 = 50%	1 = 50%	0	2 = 2%
Sciurinae	4 = 20%	1 = 5%	14 = 70%	0	1 = 5%	0	20 = 20%
Petauristinae	1 = 50%	0	1 = 50%	0	0	0	2 = 2%
Spalacidae	1 = 100%	0	0	0	0	0	1 = 1%
Gliridae	0	0	1 = 100%	0	0	0	1 = 1%
Geomyidae	0	0	2 = 100%	0	0	0	2 = 2%
Heteromyids	1 = 50%	0	1 = 50%	0	0	0	2 = 2%
Dipodids	0	0	0	2 = 100%	0	0	2 = 2%
Cricetini	0	0	0	2 = 100%	0	0	2 = 2%
Peromyscini	9 = 53%	0	7 = 42%	1 = 6%	0	0	17 = 17%
Gerbillinae	2 = 29%	0	1 = 14%	1 = 14%	2 = 29%	1 = 14%	7 = 7%
Microtinae	8 = 47%	0	6 = 35%	3 = 18%	0	0	17 = 17%
Hystricidae	0	0	0	0	1 = 100%	0	1 = 1%
Muridae	2 = 9%	11 = 50%	1 = 5%	5 = 23%	2 = 9%	1 = 5%	22 = 22%
Carnivora	1 = 33%	0	2 = 67%	0	0	0	3 = 3%
Equines	0	0	0	0	0	1 = 100%	1 = 1%
Cervids	0	0	0	0	0	2 = 100%	2 = 2%
Totals (nominal)[a]	41	23	40	19	10	14	—
Totals (actual)	23	20	25	13	8	13	102
Percent of 102	23%	20%	25%	13%	8%	13%	—
Percent of family[b]	51% of 45	43% of 46	64% of 39	54% of 24	36% of 22	—	—

[a]Some genera infest more than one group of hosts. These "nominal" figures are *not* used in calculating percentages.
[b]Total numbers of genera of fleas of nonvolant hosts per family.

353

host associations. As shown in Table 8.11, which deals with nonrodent hosts, only six of the 23 groups of mammals listed are parasitized with any of the 50 hystrichopsyllid genera or subgenera (hereafter designated as "g./sg.") that include ultraspecific or generispecific fleas. Of these, all but one genus are found on marsupials or insectivorans, and the highest rate of infestation on these nonrodents (7 of 50 = 14%) occurs among shrews (Soricidae). There are two such g./sg. on the Australian marsupials and two on Latin American marsupials. Except for one g./sg. on echidna and another on shrews, the pygiopsyllids listed are completely restricted to the marsupials. (The fleas of the sole Neotropical genus of pygiopsyllid are not cited because of insufficient records, but marsupials are probably the primary hosts.) Since the ceratophyllids are mainly fleas of rodents, only six (18%) g./sg. are listed in this table. All but one of these infest ochotonids or carnivorans. Leptopsyllids do not occur within the range of marsupials, and among the other groups of nonrodents, the genera with ultraspecific and generispecific fleas are best represented (5/24 = 21%) on ochotonids. The dearth on insectivorans (only one g./sg. each on soricids, erinaceids, and tenricids) is noteworthy. The pulicids are remarkable in including members that are highly specific to groups of mammals that rarely, if ever, have other fleas, such as the monotremes, wombats, erinaceids, pholidotids, hyracoids, and suids. Most of these unusual hosts are found in, or have roots in, the Southern Hemisphere. These data support the idea that the pulicid* fleas are an ancient group that arose in Africa (Traub 1972c, 1980c). Five of the pulicid genera with ultraspecific or generispecific species are found on leporids, but there are none on ochotonids. These, and the other marked differences in host specificity in the fleas of leporids as compared to ochotonids, suggest that the families of lagomorphs originated in different parts of the world.

Among the nonrodent hosts, the maximum percentage of the 171 specific g./sg. occurs on the ochotonids (6%). Among the soricids and talpids, the great majority of these fleas (78% and 83%) are hystrichopsyllids. Altogether, 18 (11%) of these 171 g./sg. occur on marsupials, 19 (11%) on insectivorans, and 16 (9%) on lagomorphs.

Comparable data on rodent fleas contribute additional support to the concept that certain siphonapteran taxa evolved together with their hosts. Of the 171 g./sg. of fleas that include ultraspecific or generispecific fleas, 116 (68%) infest some kind of rodents. Only seven of the 22 families and subfamilies of rodents listed in Table 8.12 lack specific hystrichopsyllids. The greatest number of these occur among the Peromyscini (22% of 50), Microtinae (22%), and Sciurinae (14%). Rodent taxa that consist of only one genus of host account for a disproportionate number of the ultraspecific or generispecific hystrichopsyllid fleas, that is, Spalacidae (10%), Aplodontidae (6%), Myospalacini (4%), and Ellobiini (2%). Rodent pygiopsyllids in this category are reported only for Sciurinae (1 g./sg.) and Murinae (62% of 26), but the pygiopsyllids are rare or completely absent where the other groups of rodents occur. In the Ceratophyllidae 30 of the 34 g./sg. on rodents (88%) include ultraspecific or generispecific fleas, and by far the greatest representation is in the Sciurinae (56% of the total), with 21% each on Peromyscini and Microtinae. Notably, however, only seven of the 22 listed groups of rodents (32%) have genera of ceratophyllids with specific fleas. The absence of ceratophyllids on aplodontids, dipodids, zapodids, ctenodactylids, heteromyids, and other geologically more ancient groups of rodents has been ascribed to the relatively youthful age of the Ceratophyllidae (Traub 1980c, 1983a). Zoogeographical factors help account for some of the others; for example, the Nearctic provenance of the family and its scarcity in Africa, where the pedetids and bathyergids are found (Traub 1980c, 1983a).

In the Leptopsyllidae 19 of the 24 g./sg. (79%) include ultraspecific or generispecific species, and 25% of those on Murinae do so, as do 21% of the Cricetini and 17% of the Dipodidae. Twelve of the groups of rodents have leptopsyllids, and these include Ctenodactylidae, Dipodidae, and Zapodidae which are not infested by the ceratophyllids, hystrichopsyllids, and others. This point, and the high rating (22%) for Cricetini no doubt reflect in part the Palaearctic origin of the Leptopsyllidae, just as the dearth of leptopsyllids on sciurids is explainable by the Nearctic origin and Oligocene dating of the squirrels (Traub 1983a). Although seven rodent taxa are infested with pulicids, a total of only seven specific g./sg. are involved, and none of the families or subfamilies carry more than two such g./sg. of fleas. The low numbers (or complete absence) of specific pulicids* on such large taxa as Muridae, Sciurinae, Microtinae, and Peromyscini is striking.

Thirty g./sg. are listed in Table 8.12 for Sciurinae, of which 56% are ceratophyllids. However, there is only one pygiopsyllid. The five g./sg. on petauristines are also found on the tree squirrels. The Muridae also have 30 ultraspecific or generispecific g./sg., but the distribution is markedly different than that for the sciurids. There are no ceratophyllids represented, and 53% are in the Pygiopsyllidae. None of the host-specific g./sg. associated with the Microtinae is in the Pulicidae or Pygiopsyllidae. There are no voles in the Ethiopian or Neotropical regions where the various subfamilies of Pulicidae arose, and none in the Australian Region, the homeland of the pygiopsyllids (Traub 1972c, d, 1980c).

In summary, the greatest number of the genera and subgenera with ultraspecific or generispecific species of fleas occur on rodents (116 of 171 = 68%) (Table 8.13). Of these, 28% are found on the Cricetidae, whereas the totals for all the marsupials and insectivorans are 11% and 12%, respectively. However, the numbers of genera of rodents (343) outnumber those of the marsupials (80) by a factor of 4.3, and those of insectivorans (53) by a factor of 6.4 (Traub 1983a). Accordingly, the actual proportion of highly specific fleas among these nonrodents is no doubt much greater than the figures seem to suggest. It is significant that for each of the families of fleas

Table 8.11 Numbers of Genera and Subgenera with Ultraspecific or Generispecific Fleas Infesting Mammals other than Rodents

	Hystricho-psyllidae	Pygio-psyllidae	Cerato-phyllidae	Lepto-psyllidae	Pulicidae	Others	Totals
Number specific in family[a]	50	26	34	24	18	19	171
Monotremes	0	1 = 50% 4%[b]	0	0	1 = 50% 6%[b]	0	2 = 100% 1%
Didelphidae	2 = 100% 4%[b]	0	0	0	0	0	2 = 100% 1%
Dasyuridae	1 = 25% 2%	2 = 50% 8%	0	0	0	1 = 25% 5%	4 = 100% 2%
Peramelidae	0	5 = 83% 19%	0	0	0	1 = 17% 5%	6 = 100% 4%
Phalangeridae	1 = 17% 2%	4 = 67% 15%	0	0	0	1 = 17% 5%	6 = 100% 4%
Vombatidae	0	1 = 50% 4%	0	0	1 = 50% 6%	0	2 = 100% 1%
Tenrecidae	0	0	0	1 = 100% 4%	0	0	1 = 100% 1%
Erinaceidae	0	0	0	1 = 33% 4%	2 = 67% 11%	0	3 = 100% 2%
Macroscelidae	0	0	0	0	0	3 = 100% 16%	3 = 100% 2%
Soricidae	7 = 78% 14%	1 = 11% 4%	0	1 = 11% 4%	0	0	9 = 100% 5%

Talpidae	5 = 83% 8%	0	0	1 = 20% 2%	0	6 = 100% 4%
Chiroptera	0	0	0	0	3 = 100% 16%	3 = 100% 2%
Pholidota	0	0	0	1 = 100% 6%	0	1 = 100% 1%
Edentata	0	0	0	0	2 = 100% 11%	2 = 100% 1%
Ochotonidae	0	4 = 40% 12%	5 = 50% 21%	0	1 = 10% 5%	10 = 100% 6%
Leporidae	0	0	1 = 17% 4%	5 = 83% 28%	0	6 = 100% 4%
Carnivora	1 = 25% 2%	2 = 50% 6%	0	0	1 = 25% 5%	4 = 100% 2%
Hyracoidea	0	0	0	1 = 100% 6%	0	1 = 100% 1%
Equines	0	0	0	0	1 = 100% 5%	1 = 100% 1%
Suidae	0	0	0	3 = 100% 17%	0	3 = 100% 2%
Other Artiodactylids	0	0	0	0	2 = 100% 11%	2 = 100% 1%

[a]Totals for family regarding genera/subgenera with specific fleas, including those on rodents.
[b]Percent of total number of such taxa of fleas in family or group. Some genera occur on more than one group of host.

Table 8.12 Numbers of Genera and Subgenera with Ultraspecific or Generispecific Fleas of Rodents

Specific genera and subgenera[a]	Hystricho-psyllidae	Pygio-psyllidae	Cerato-phyllidae	Lepto-psyllidae	Pulicidae	Others	Totals
	50	26	34	24	18	19	171
Aplodontidae	3 = 75% 6%[b]	0	0	1 = 25% 4%[b]	0	0	4 = 100% 1%[b]
Sciurinae	7 = 23% 14%	1 = 3% 4%[b]	19 = 63% 56%[b]	1 = 3% 4%	1 = 3% 6%[b]	1 = 3% 5%[b]	30 = 100% 18%
Petauristinae	1 = 20% 2%	0	4 = 80% 12%	0	0	0	5 = 100% 3%
Geomyidae	0	0	2 = 67% 6%	0	1 = 33% 6%	0	3 = 100% 2%
Heteromyidae	3 = 100% 6%	0	0	0	0	0	3 = 100% 2%
Pedetidae	0	0	0	0	1 = 100% 6%	0	1 = 100% 1%
Cricetini	1 = 17% 2%	0	0	5 = 33% 21%	0	0	6 = 100% 4%
Peromyscini	11 = 58% 22%	0	7 = 37% 21%	1 = 5% 4%	0	0	19 = 100% 11%
Myospalacini	2 = 67% 4%	0	0	1 = 33% 4%	0	0	3 = 100% 2%
Nesomyinae	0	0	0	0	1 = 100% 6%	0	1 = 100% 1%

Host							Total
Ellobiini	1 = 50% 2%	0	0	1 = 50% 4%	0	0	2 = 100% 1%
Other Microtinae	11 = 52% 22%	0	7 = 33% 21%	3 = 14% 13%	0	0	21 = 100% 12%
Gerbillinae	1 = 17% 2%	0	1 = 17% 2%	1 = 17% 4%	2 = 34% 11%	1 = 17% 5%	6 = 100% 4%
Spalacidae	5 = 100% 10%	0	0	0	0	0	5 = 100% 3%
Rhizomyidae	1 = 100% 2%	0	0	0	0	0	1 = 100% 1%
Muridae	3 = 10% 6%	16 = 53% 62%	0	6 = 20% 25%	2 = 7% 11%	3 = 10% 16%	30 = 100% 18%
Gliridae	1 = 33% 2%	0	2 = 67% 6%	0	0	0	3 = 100% 2%
Zapodidae	0	0	0	1 = 100% 4%	0	0	1 = 100% 1%
Dipodidae	0	0	0	4 = 100% 17%	0	0	4 = 100% 2%
Hystricidae	0	0	0	0	1 = 100% 6%	0	1 = 100% 1%
Bathyergidae	1 = 50% 2%	0	0	0	0	1 = 50% 5%	2 = 100% 1%
Ctenodactylidae	0	0	0	1 = 100% 4%	0	0	1 = 100% 1%

[a]Totals for family regarding genera/subgenera with specific fleas, including those on nonrodent hosts.
[b]Percent of total number of such taxa of fleas in family or group. Some genera occur on more than one group of host.

359

Table 8.13 Summary of Numbers of Genera and Subgenera with Ultraspecific or Generispecific Fleas, Arranged by Major Groups of Mammalian Hosts

	Hystricho-psyllidae	Pygio-psyllidae	Cerato-phyllidae	Lepto-psyllidae	Pulicidae	Others	Totals
Number specific in family[a]	50	26	34	24	18	19	171
Marsupials	4 = 22% 8%[a]	11 = 61% 42%[a]	0	0	1 = 6% 6%[a]	2 = 11% 11%[a]	18 = 100% 11%[a]
Insectivora	11 = 52% 22%	1 = 5% 4%	1 = 5% 3%[a]	2 = 10% 8%[a]	3 = 14% 17%	3 = 14% 16%	21 = 100% 12%
Lagomorpha	0	0	4 = 25% 12%	6 = 38% 25%	5 = 31% 28%	1 = 6% 5%	16 = 100% 9%
Sciuridae	7 = 23% 14%	1 = 3% 4%	20 = 65% 59%	1 = 3% 4%	1 = 3% 6%	1 = 3% 5%	31 = 100% 18%
Cricetidae	24 = 50% 48%	0	14 = 29% 41%	7 = 15% 29%	2 = 4% 11%	1 = 2% 5%	48 = 100% 28%
Muridae	3 = 9% 6%	16 = 50% 62%	2 = 6% 6%	6 = 19% 25%	2 = 6% 11%	3 = 9% 16%	32 = 100% 19%
Other rodents	14 = 48% 28%	0	4 = 13% 12%	7 = 24% 29%	3 = 10% 17%	1 = 3% 5%	29 = 100% 17%
All rodents	40 = 30% 80%	17 = 13% 65%	30 = 23% 88%	17 = 13% 71%	6 = 5% 33%	6 = 5% 32%	116 = 100% 68%
Carnivora	1 = 25% 2%	0	2 = 50% 6%	0	0	1 = 25% 5%	4 = 100% 2%
Hoofed animals	0	0	0	0	3 = 50% 17%	3 = 50% 16%	6 = 100% 4%
Other	0	0	0	0	3 = 38% 17%	5 = 63% 26%	8 = 100% 5%

[a]Percent of total number of such taxa of fleas in family or group. Some genera occur on more than one group of host.

360

listed, the highest rates of specificity are for the Rodentia. It is also striking that 48% of the ultraspecific and generispecific entries for the Hystricho-psyllidae and 41% for the Ceratophyllidae involve the Cricetidae. The specificity of the Pulicidae and Leptopsyllidae infesting Lagomorpha, a very small group of hosts, and the predilection of the ceratophyllids for sciurids are impressive.

Intermediate Degrees of Specificity

Many fleas are less host specific than the ultraspecific and generispecific species mentioned previously but still are more restricted regarding host range than are the indiscriminate fleas, which can infest almost any sort of host they encounter. Among the fleas of intermediate specificity are the ethnoxenous species, which are limited to infestation of a single tribe or subfamily of host, and the intrafamilial fleas, which are capable of para-sitizing various members within a mammal family. As shown in Table 8.14, 33 of the 36 listed ethnoxenous groups of fleas, or 92% of the total, infest rodents, as do 100% of the 19 groups of intrafamilial groups. One reason for the dearth of insectivoran fleas in this table is that the overwhelming majority of fleas on shrews, moles, and their kin are either ultraspecific (28 species, Table 8.5) or generispecific (19 species or species groups). There are an additional eight species found on insectivorans (excluding tupaiids), but seven of these are really rodent fleas with an unusually broad host range, and the eighth is found on both shrews and moles. Carnivoran fleas are either ultraspecific (items 115–119 in Table 8.4) or generispecific (seven species or species groups), or else they infest a variety of hosts. There are 31 entries recorded for lagomorphs in the tables, and 29 pertain to ultraspecific or generispecific species of fleas. The same is true for 31 of the 40 citations for marsupials.

Even though many kinds of rodents are parasitized by ethnoxenous fleas, there is only one such sciurid flea known, namely *Medwayella limi* Traub. Most sciurid fleas are either ultraspecific or generispecific, but there are five biotopical species of intrafamilial specificity, and these infest both flying squirrels and tree squirrels. Microtines and peromyscines often live in the same microhabitat, but there are only three species of fleas listed as commonly infesting both groups. Among the 35 ethnoxenous fleas listed, there are 11 hystrichopsyllids, 7 ceratophyllids, and 7 leptopsyllids, whereas among the intrafamilial fleas, 17 of the 18 are either ceratophyllids or hystrichopsyllids.

Fleas with a Broad Host Range

The extrafamilial and extraordinal fleas, as well as the major indiscriminate Siphonaptera, are listed in Table 8.15. Among the species that infest hosts of several families within an order of mammals, 17 of 20 parasitize rodents.

Table 8.14 **Ethnoxenous and Intrafamilial Fleas of Mammals**

Fleas	Mammals
Ethnoxenous Fleas Within a Subfamily of Host	
Amphipsylla schelkovnikovi	Cricetini
Atyphloceras bishopi	Voles
Atyphloceras echis	*Peromyscus, Neotoma*
Cleopsylla townsendi	Sigmodontines
Ctenophthalmus agyrtes, C. assimilis	Voles
Ctenophthalmus congener, C. golovi	Voles
Dinopsyllus echinus	Murines
Frontopsylla elata	Voles
Frontopsylla spadix	Murines
Jellisonia johnsonae	Peromyscines
Juxtapulex porcinus	Edentates
Kohlsia keenani	Sigmodontines, peromyscines
Leptopsylla aethiopica	*Mus*, etc.
Leptopsylla algira	*Mus*, etc.
Malaraeus arvicolae, M. penicilliger	Microtines
Malaraeus euphorbi	*Peromyscus, Neotoma*
Medwayella limi	Ground callosciurines
Meringis cummingi	*Dipodomys, Perognathus*
Metastivalius mordax, etc.	Murines
Neotyphloceras crassispina	Sigmodontines
Nosopsyllus (N.) punjabensis	Murines
Nosopsyllus (G.) henleyi	Gerbillines
Ophthalmopsylla volgensis	*Allactaga, Jaculus*
Papuapsylla corrugis	Giant rats
Papuapsylla luluai	Murines
Peromyscopsylla draco	*Peromyscus, Reithrodontomys*
Phalacropsylla spp.	*Neotoma, Peromyscus* (nests)
Procaviopsylla spp.	Procaviids
Pulex porcinus	Peccaries
Rothschildiana spp.	Murines
Sphinctopsylla inca, etc.	Sigmodontines
Traubia spp.	Murines
Xenopsylla vexabilis	Murines
Intrafamilial Fleas Within a Family of Host	
Amphipsylla rossica	Voles, gerbillines
Callopsylla fragilis	Hamsters, voles
Catallagia decipiens	*Peromyscus*, voles

Table 8.14   (Continued)

Fleas	Mammals
Conorhinopsylla stanfordi	Tree, flying squirrels (nests) (BIO.)[a]
Ctenophthalmus cabirus, etc.	Murids
Ctenophthalmus calceatus, etc.	Murids
Dinopsyllus ellobius	Murids
Epitedia stanfordi	*Peromyscus*, voles
Lentistivalius (D.) mjoebergi	Tupaiids
Listropsylla spp.	Murids
Malaraeus telchinum	*Peromyscus*, voles
Monopsyllus indages	Tree, flying squirrels (BIO.)
Monopsyllus vison	Tree, flying squirrels (BIO.)
Neopsylla pleskei	Hamsters, voles
Nosopsyllus consimilis, N. mikulini	Microtines, cricetines
Nosopsyllus incisus	Scansorial murids (BIO.)
Syngenopsyllus calceatus	Tree, flying squirrels (BIO.)
Tarsopsylla octodecimdentata	Tree, flying squirrels (BIO.)

[a]BIO., biotopical, that is, dependent on habitat.

Thrassis bacchi (Rothschild) is noteworthy in that one subspecies is restricted to *Lagurus*, while another subspecies is found on *Citellus*. In some instances there has been a transfer from the aboriginal host to others living in the same environment or niche. Thus *Pleochaetis dolens* (Jordan and Rothschild) and *Epirimia aganippes* (Rothschild) occur on various scansorial and arboreal rodents. Such biotopical fleas are common among the extraordinal fleas. *Tunga penetrans* (Linnaeus) is the only extraordinal flea that does not regularly include rodents as one of its hosts. Generally the aboriginal hosts were rodents, but *Acanthopsylla, Macropsylla, Pygiopsylla,* and *Stephanocircus* include exceptions. The pygiopsyllid *Migrastivalius jacobsoni* (Jordan and Rothschild) infests a variety of nocturnal scansorial or semiarboreal hosts such as murids (*Chiropodomys, Pithecheir,* and rats of sundry native subgenera of *Rattus*) and tupaiids (*Ptilocercus* and *Tupaia*) and even occasionally is found on climbing sciurids such as *Dremomys* and *Sundasciurus tenuis* (Horsfield), which are essentially diurnal.

Although there are no pulicids cited for the extrafamilial fleas, there are nine (30%) extraordinal ones, including six species of *Xenopsylla*. Most of the extrafamilial fleas are either hystrichopsyllids (11/20 = 55%) or ceratophyllids (25%). The two families are not nearly so well represented among the extraordinal fleas, 17% and 10%, respectively, whereas there

Table 8.15 Examples of Extrafamilial, Extraordinal, and Indiscriminate Fleas of Mammals

Fleas	Mammals
Extrafamilial Fleas within an Order of Host	
Acanthopsylla eudromiciae	Small scansorial marsupials (BIO.)[a]
Carteretta clavata	*Perognathus, Peromyscus*
Catallagia decipiens	*Peromyscus*, voles, sciurids
Ctenophthalmus (C.) solutus	*Apodemus*, voles, etc.
Epirimia aganippes	Scansorial murids, *Graphiurus* (BIO.)
Epitedia wenmanni	*Peromyscus*, voles, sciurids
Megarthroglossus divisus	*Neotoma*, sciurids (nests)
Meringis shannoni, M. hubbardi	*Dipodomys, Onychomys*, etc.
Monopsyllus sciurorum	Tree squirrels, glirids
Ophthalmopsylla (C.) jettmari	Cricetini, sciurids (BIO.)
Palaeopsylla nippon	Soricids, talpids
Paraceras javanicum, P. sauteri	Badgers, civets
Paradoxopsyllus acanthus, etc.	Microtines, murines
Pleochaetis dolens	Scansorial cricetids, squirrels (BIO.)
Rhadinopsylla heiseri, R. sectilis	Cricetids, sciurids
Stenoponia tripectinata, etc.	Murines, gerbillines, dipodids
Thrassis bacchi	*Citellus, Lagurus*
Extraordinal Fleas within Two or More Orders of Host	
Acanthopsylla enderleini, etc.	Scansorial murids, marsupials (BIO.)
Ctenocephalides connatus	*Xerus* and its predators (BIO.)
Ctenophthalmus (C.) agyrtes, etc.	Microtines, murids, soricids, etc.
Ctenophthalmus (Eu.) assimilis, etc.	Microtines, murids, soricids
Dinopsyllus lypusus, etc.	Rodents, shrews
Echidnophaga larina, etc.	Rodents, carnivores, warthogs, etc.
Hystrichopsylla ozeana, etc.	Rodents, insectivores (nests)
Lentistivalius vomerus	Tupaiids, *Dremomys*, scansorial murids (BIO.)
Macropsylla hercules	Scansorial marsupials, murids (BIO.)
Malaraeus bitterrootensis	Peromyscines, *Ochotona* (BIO.)
Medwayella loncha, etc.	Tupaiids, callosciurines
Migrastivalius (M.) jacobsoni	Scansorial murids, tupaiids (BIO.)
Monopsyllus vison	Tree sciurids and their predators (BIO.)
Neotyphloceras rosenbergi	Rodents, marsupials
Papuapsylla alticola	Giant rats, scansorial marsupials
Papuapsylla corrugis	Giant rats, *Peroryctes*
Paradoxopsyllus integer, etc.	*Ochotona*, murids

Table 8.15 (Continued)

Fleas	Mammals
Pleochaetis soberoni	Flying squirrels, scansorial procyonids (BIO.)
Polygenis gwyni	*Sigmodon, Didelphis*
Pygiopsylla hoplia	Bandicoots, murids
Rectodigitus traubi, etc.	Rats, bandicoots
Stephanocircus dasyurus, etc.	Marsupials, murines
Synopsyllus spp.	Tenrecs, Neosomyinae
Tunga penetrans	Suids, man, etc.
Xenopsylla cheopis, X. bantorum	Murids, insectivores
Xenopsylla conformis, X. astia	Gerbils, murids, insectivores
Xenopsylla erilli, X. cryptonella	*Xerus* and its predators
Indiscriminate or Catholic Fleas—Very Broad Host Range	
Echidnophaga gallinacea and *E. myrmecobii,* etc.	
Nosopsyllus fasciatus and *N. londiniensis*	
Ctenocephalides felis and *Megabothris turbidus*	
Pulex irritans, P. simulans, and *Synosternus pallidus*	

*a*BIO., biotopical, that is, dependent on habitat.

are eight (27%) pygiopsyllids. However, most of the pygiopsyllids listed are biotopical forms that have capitalized on the environment. At least nine of the extraordinal fleas include species that parasitize predators of rodents, but in most instances the rodents constitute the typical host of the flea. Among such predators are procyonids, bandicoots, martens, and foxes. There are only seven entries for insectivorans as partial hosts for extraordinal fleas, and in each case rodents constitute the prime host for those species, as well as for the genus as a whole.

Actually, there are only a limited number of fleas that are truly catholic or indiscriminate under normal circumstances, and even some of the most notorious and cosmopolitan fleas such as *Nosopsyllus fasciatus* (Bosc) and *Xenopsylla cheopis* (Rothschild) are far more prone to infest rats rather than man or carnivores. These fleas attack man only when rats are not immediately available. *Synosternus pallidus* (Taschenberg) is primarily a hedgehog flea, and many of the earlier reports of it on odd hosts and in unusual areas are due to misidentifications of *Xenopsylla* (Hopkins and Rothschild 1953). However, *S. pallidus* can and does feed on rodents, and apparently on carnivores and man as well, and is found on the floor of huts in West Africa. *Ctenocephalides felis* (Bouche)* has an extremely broad host range—it is no longer believed that "forms" or "subspecies" have adapted to goats or cattle in Asia. Seven of the nine indiscriminate species listed are pulicids, and the remainder are ceratophyllids. This may be explained by a correlation between the jumping ability of fleas and the size of the host they infest. Good jumpers tend to parasitize large animals (Rothschild et al. 1973; Traub 1972e), and often overleap small hosts. This is also true for the catholic species that are powerful jumpers; there are relatively few records of *Ctenocephalides* or *Pulex* from rats and mice.

ZOOGEOGRAPHY AND EVOLUTION

As suggested by foregoing allusions, the data on zoogeography of mammals and fleas also strongly favor the concept of their coevolution. The distribution of the Stephanocircidae and marsupial-infesting Doratopsyllinae can only by adequately explained by continental drift, specifically the passage of the primordial marsupial hosts and fleas by island-hopping from South America to Australia by way of Antarctica, by at least the Cretaceous period (Traub 1972c, d, 1980c). This means that these host–flea relationships must have been in existence since that period, if not earlier. The distribution of the Pygiopsyllidae likewise suggests antiquity and dispersal among the austral continents, since the provenance and center of development of this family is the Australian region: (1) there is one genus (*Ctenidiosomus*) in South America and two in Africa; (2) the family is unknown in Europe, southwestern Asia, America north of Central America, and is represented by only very few species in eastern Asia (Traub 1972c, 1980c). Moreover, the pygiopsyllids have a basic association with marsupials even though the later derivatives of these fleas accommodated themselves to other hosts, especially murids in New Guinea. If the ancestral *Ctenidiosomus* indeed accompanied marsupials on a return trip to South America from Australia, the Pygiopsyllidae must date to the Cretaceous period, although it is more youthful than the stephanocircids and hystrichopsyllids.

As murids* retraced the route of their ancestors and moved from the Australian region toward Wallacea and the Oriental region, and as marsupials dispersed from Australia and New Guinea to the nearby islands, they carried pygiopsyllids with them. Eventually, as in Sulawesi and Borneo, some of the pygiopsyllids encountered and transferred to tree squirrels of Palearctic origin, culminating in *Medwayella* and *Farhangia* (and perhaps other genera awaiting discovery) (Traub 1972a, c, 1980a, c). *Medwayella* is associated with callosciurines in the lower montane habitats in various areas from Sulawesi and the Philippines to the Southeast Asian mainland, including Indochina. Twenty-eight species are known to me, including undescribed species. However, in dispersing from continental Asia to the upper montane parts of these Pacific islands, the callosciurine and petauristine squirrels carried Malayan genera of ceratophyllid fleas with them, for example, *Syngenopsyllus* to Java and *Macrostylophora* at least to Borneo and the Philippines. Flying squirrels in Borneo are parasitized by another ceratophyllid, *Hollandipsylla*, which may be a bioendemic taxon. *Macrostylophora*, like *Medwayella* and their hosts, has speciated prolifically, and I know of 28 species. Thus these pygiopsyllid and ceratophyllid fleas have coevolved with their hosts as the squirrels dispersed, even though, in the case of the pygiopsyllids, the squirrels are not the original hosts. The Ceratophyllidae, however, are believed to have arisen on sciurids in North America (Traub and Rothschild 1983; Traub et al. 1983) and have accompanied these hosts to the limits of their ranges in South America, Europe, continental Asia (except for certain tropical humid lowlands), and to the near extremes of the distribution of squirrels in Africa. In the archipelagos of the Oriental region there are ceratophyllids on montane or submontane squirrels wherever these rodents occur, with the possible exception of Sulawesi, where such fleas have not been reported but may very well occur. There are no squirrels in either the Australian region or Luzon nor are there native ceratophyllids or other squirrel fleas.

Among the murid pygiopsyllids that have dispersed with their hosts from the homeland in the Australian region is *Bibikovana*, known from Australia, New Guinea, and Borneo. A new genus which is well represented on rats in Sulawesi (by new species), is also found in extreme western New Guinea. Murids have pygiopsyllids, often characteristic genera or species, throughout Australia, New Guinea, and the Malfilindo archipelagos (Indonesia west of New Guinea, Philippines, and Borneo) all the way to Southeast Asia, and one such species also occurs in Japan. For example, *Migrastivalius jacobsoni** is intrafamilial and biotopical and restricted to scansorial rodents and tupaiids. *Stivalius* (s. str., *sensu* Traub, 1972a) is characteristic of *Rattus* (*Rattus*) in parts of the Malfilindo archipelagos, India, and Sri Lanka.

The same principles are exhibited by other zoogeographic data involving rats and shrews in these parts of the world. Simpson (1961) hypothesized that the penetration of murids* into Australia/New Guinea from Indonesian islands occurred in several waves over millions of years (Miocene to late Pliocene or Pleistocene) and that the last invasion was by *Rattus*. Observations on Siphonaptera support this general concept of distribution, especially concerning *Rattus* (Traub 1972c, d, 1980c). Thus rat fleas of the leptopsyllid *Sigmactenus,** a derivative of the Palearctic *Leptopsylla*, are known from Borneo, various Philippine islands, Sulawesi, Timor, and New Guinea (with a total of six species, including two new ones). Similarly, murids from the Asian mainland have carried two hystrichopsyllid genera of Palearctic origin or affinity (*Neopsylla* and *Rothschildiana*) to Java and/or Borneo; endemic species thereof occur on those islands. *Xenopsylla** have probably also traversed this route on rats. A wealth of *Palaeopsylla* are found on shrews and moles in southern Asia (and Europe). Soricids like *Crocidura* have penetrated across the islands to the Philippines and

Sulawesi, but also exist in Africa and elsewhere. *Palaeopsylla* are known from Java and are expected to occur on some of the other islands as well.

Data on transatlantic faunal connections (Traub 1980c) also suggest a long-standing association between certain groups of fleas and mammals. For example, spilopsylline fleas (Pulicidae) and the leptopsyllid *Odontopsyllus*, both infesting rabbits, show a North American/European distribution that seems best explained by Greenland connections which became inoperable about 50 million years ago in the mid Eocene epoch (Traub 1980c). Such a route for mammalian dispersal was discussed by McKenna (1975). The distribution of *Atyphloceras bishopi* Jordan and *A. nuperus* Jordan (Hystrichopsyllidae) may also be attributed to such North Atlantic dispersal, but their current hosts are microtines whose known fossil history dates only from the Upper Miocene. All indications are that the Ceratophyllidae is too young a group to have utilized North Atlantic connections, and, significantly, there are no members of that family with such a distribution (Traub and Rothschild 1983). Some mammalogists who opposed the idea that the South American "hystricomorph" rodents and ceboid monkeys had African roots now believe that such a dispersal actually occurred, by rafting, in the South Atlantic* in the Eocene when the two continents were much closer together than they are now. The siphonapteran data suggest that the ctenophthalmines, and perhaps the stenoponiine fleas (Hystrichopsyllidae), and ancestral pulicines entered South America in the early Eocene by that means (Traub 1980c).

Fleas must have traversed, with their mammalian hosts, the Beringian route between North America and Asia on many occasions. There are 22 Holarctic genera among the Ceratophyllidae, Hystrichopsyllidae, and Leptopsyllidae, and 13 Holarctic species of fleas are known (Holland 1958, 1963, 1964; Haddow et al. 1982; Hopla 1980b; Traub 1980c, 1983a). The usual hosts are microtines and sciurids. Faunal relationships between Africa and Madagascar are of course well established. However, the siphonapteran data further suggest that Madagascar was originally closer to Somalia than is generally believed, and received a filtered fauna from southwest Asia, including elemental cricetid stocks and leptopsyllid fleas (Traub 1980c). This hypothesis would also help explain the remarkable absence of bioendemic murids in Madagascar.

It appears that both fleas and hosts could not have undertaken vast journeys together during the eons without undergoing evolutionary changes as they encountered new conditions. It is therefore expected that the noncosmopolitan Siphonaptera that are far removed from their aboriginal homeland tend to be much more specialized morphologically than are their more conservative kin at home. For example, *Migrastivalius** (Malaya, Borneo, Java, Sumatra, Philippines, on various subgenera of *Rattus*) has lost the aedeagal crochet, and the aedeagus has compensated for the loss of this clasping device by developing (1) a hooklike Ford's sclerite and (2) an unusually long sclerotized inner tube (S.I.T.). Moreover, in the female, the bursa copulatrix (B.C.) is specially lengthened and thickened (Traub 1980a). Likewise, *Stivalius s. str.* (Sulawesi, Timor, Philippines, India, etc. on *Rattus (Rattus)*) has no crochet but their S.I.T. and B.C. are highly developed (Traub 1972a). Squirrel fleas in the genus *Medwayella* (Indonesian islands to Thailand or Indochina) have a complex crochet process and a unique patch of spiniform bristles on the male eighth sternum. Abdominal combs are unknown on New Guinean pygiopsyllids but often occur on those fleas, such as *Migrastivalius (Gryphopsylla)*, *M. (Migrastivalius)* and *Lentistivalius (Destivalius)* in Borneo, Malaya, and elsewhere. The same trend is also seen in other murine fleas. Thus the leptopsyllid *Sigmactenus* has more head spines and a better-developed tibial comb than the *Leptopsylla* stock from which it is derived. In *Nosopsyllus* (Asia and Africa, on murimes or sciurids) and *Libyastus* (Africa, on sciurids) the male eighth sternum is much more reduced (or lost entirely) than in the Nearctic squirrel fleas from which these genera ultimately descended.

Thus the evidence shows that many fleas have been associated with their particular hosts (or groups thereof) for millions of years and have accompanied these animals as they gradually penetrated new territory and eventually reached areas remote from their nativity. Like their hosts, the fleas adapted to new conditions during the course of dispersal and of time, resulting in the evolution of new species and genera or even higher taxa.

EVIDENCE BASED ON THE MORPHOLOGY OF FLEAS

Several workers have stressed that the external morphology of fleas reflects the attributes of their particular hosts (Ioff 1953; Humphries 1966, 1967; Traub 1953, 1968b, 1969, 1972b, 1972c, 1980b; Traub and Barrera 1966; Traub and Dunnet 1973; Traub and Evans 1967). To a greater degree than is still generally appreciated, the number, shape, position, and arrangement of the siphonapteran combs and spines can be explained by the vestiture and habits of the host; for example, whether it is volant, spiny, fossorial, arboreal, or otherwise specialized. In other words, the amount of selective pressure exerted, directly or indirectly, upon the flea by the host is an important factor in determining some of the characteristic morphological features of the flea. Simple observation of a siphonapteran specimen often suffices to ascertain whether the host is subterranean, scansorial, volant, or another type. There are certainly other predisposing factors, such as the life-style or phylogenetic history of the flea, and here too the structure of the insect frequently provides potent clues to whether the parasite is a permanent denizen of the nest of the host, feeding while it is asleep, or whether the flea stays attached to the host for long periods.

Convergence* and Divergence* in Fleas

Even under equally intensive selective pressure, all taxa are not expected to respond in the same manner regarding combs and bristles that may serve as holdfasts or for protection. Some species have developed supernumerary spines or combs, but some other fleas have evolved substitute devices such as spiniform bristles and false combs, which apparently serve the same purpose. Certainly, as per Dollo's law,* we cannot expect fleas that have lost combs untold millions of years ago to suddenly develop a ctenidium of typical spines (Traub 1968a, 1972b, 1980b). Thus fleas whose relatives all lack genal combs and spines on the head, may infest a host whose habits differ from those parasitized by their kin and develop a cephalic crown of spiniform bristles in consequence. Some completely combless fleas have even developed supernumerary bristles arranged in a special way to compensate, in effect, for the absence of ctenidia, as in the vermipsyllids.*

If it is correct that many fleas coevolved with or became adapted to their hosts, it would theoretically follow that (1) such fleas would differ significantly from their relatives (even close) that infest other kinds of mammals; and (2) insofar as adaptive features, they would come to resemble other, unrelated fleas that likewise parasitize those particular hosts. There are many examples illustrating these principles of (1) divergence and (2) convergence. For instance, concerning divergence, the first two genal spines of the *Ctenophthalmus** that are shrew* fleas are quite rounded apically, the pronotal spines tend to be bluntly rounded at the tips, and the longitudinal axis is somewhat concave (Fig. 8.1), whereas in the *Ctenophthalmus* mole* fleas (Fig. 8.6) and murid* fleas (Fig. 8.15) these genal spines are acute, and the pronotal spines are straight and pointed. Significantly, the spines of the mole *Ctenophthalmus* tend to be even narrower and more stilettolike than those of the fleas of rats and microtines. The *Ctenophthalmus* of "mole-rats" exhibit both divergence and convergence since their hosts actually belong to three different families (Rhizomyidae, Spalacidae, and two subfamilies of Cricetidae, including the nominate tribe Myospalacini) and yet the fleas (like their hosts) share some pertinent features. Of these *Ctenophthalmus, C. (Ducictenophthalmus) dux* Jordan and Rothschild, the unique species infesting the microtine *Ellobius*, cannot be discussed here because specimens are unavailable and the pertinent features have not been illustrated. However, in the other "mole-rat"* fleas: *C. (Geoctenophthalmus)* (Fig. 8.7) on *Tachyoryctes* (Rhizomyidae), the *fissurus* group of *C. (Palaeoctenophthalmus)* (Fig. 8.8), the *spalacis* group of *C. (Spalacoctenophthalmus)* on *Spalax*, and the monotypic *C. (Neoctenophthalmus)* on *Myospalax*, the first two genal spines are blunt or rounded as in the shrew fleas, but the pronotal spines are straight. The pronotal comb of the *Ctenophthalmus* infesting squirrels differs from that of their congeners in that the subventral spines are distinctly broader than the others in the comb. Marked variations also exist in this taxon in the degree of development of the eye (cf. Figs. 8.5–8.8 and 8.15) (Traub and Barrera 1966; Traub and Evans 1967; Traub 1972).

The hystrichopsyllid *Palaeopsylla*, *Corypsylla*, and *Nearctopsylla* primarily infest Insectivora, and each includes some species that are shrew* fleas and others that are found on moles, as well as one or two species associated with a carnivoran or murid. However, they not only exhibit divergence in that the combs on the fleas of the three kinds differ significantly from one another, but in each genus the differences are consistent, according to the host and are seen in all the shrew fleas, mole fleas, or others. Moreover, there also is a remarkable convergence, for the same general modifications occur in the shrew fleas of all three genera, and the mole fleas also resemble one another. Furthermore, the trends are the same as those noted for *Ctenophthalmus*. Thus in the shrew fleas some of the genal spines tend to be apically blunt or rounded (Figs. 8.1 and 8.3), and the pronotal spines are up-curved, with ovate tips. In contrast, in the mole fleas all but the first genal spine are long and acutely pointed, and the pronotal spines are stiletto shaped (Fig. 8.2). *Palaeopsylla setzeri* Traub and Evans, the only member of the genus not found on Insectivora, is parasitic on the burrowing vole, *Hyperacrius*, which is quite molelike in behavior. It is fitting that the spines of the pronotal comb are much more like those of mole fleas than the *Palaeopsylla* infesting shrews and that the second genal spine is apically triangular, but not acutely so (cf. Figs. 8.2–8.4). The same evolutionary trend toward apically rounded, somewhat concave pronotal spines is seen in shrew fleas of *Lentistivalius*, *Hypsophthalmus*, *Xenodaeria*, *Doratopsylla*, and *Corrodopsylla* (Traub and Barrera 1966; Traub and Evans 1967). The first two also include species found on other hosts, such as murids, which have other patterns of spines.

The pulicid *Ctenocephalides** also exhibits divergence in the pattern of the genal and pronotal combs. At least six of the ten species that infest carnivores, particularly scansorials, have well-developed combs. A large genal comb (seven spines) and a full pronotal comb (eight or nine spines per side) is typical for the taxon (Fig. 8.9, *C. felis*). However, greatly reduced combs are seen in two species infesting hyrax: *C. rosmarus* (Rothschild) (Fig. 8.10; one genal spine; six to seven pronotal spines per side) and *C. arabicus* (Jordan) (Fig. 8.11; three genal spines, etc.). Their spines are straighter and narrower than those in the carnivoran *Ctenocephalides*. However, the third species, *C. craterus* (Jordan and Rothschild), that infests hyrax has the typical number of spines, although they are the same shape as in the other hyrax *Ctenocephalides*. The differences in the numbers of spines is explainable in terms of the scansorial* habits of the host, as indicated by *C. craterus*. *C. crataepus* (Jordan) has unusually short spines in the genal ctenidium, but its true host is unknown.

It has escaped notice that a remarkable variety of murine* fleas (i.e., of *Rattus s. lat.*) in widely separated areas have either full, vertical, genal combs of true spines (combined with false combs on the tibia) or else an anteromarginal, vertical, false comb of spiniform bristles. Such modifications occur in rat fleas of five families. Among the fleas with true combs are the leptopsyllids *Sigmactenus* (Borneo, New Guinea, etc.) (Fig. 8.12) and *Leptopsylla* (*Pectinoctenus*) (Asia) (Fig. 8.13), the hypsophthalmid *Epirimia aganippes* (Ethiopian region, also found on dormice) (Fig. 8.14), and the typical species of the large, African hystrichopsyllid *Dinopsyllus* (Fig. 8.16). The murid-infesting subgenus *Leptopsylla* has a somewhat shorter comb but also has a group of frontomarginal spiniforms. A false helmet of "spines," which are in reality bristles, is seen in the murid flea *Smitella*,* and a frontomarginal row of spiniforms occurs in other pygiopsyllids of murids, such as *Idiochaetis* and certain *Metastivalius* (Traub 1980b). The stephanocircid* murid fleas also have a vertical genal comb (as well as a helmet of spines), but these fleas are derived from marsupial-infesting taxa which already possessed a vertical comb and a helmet. However, the shape of their spines consistently differs from that in the marsupial fleas. Upright genal combs that are wholly or partially frontomarginal in position, special helmets with spines or spiniforms, and frontal spiniforms have been termed *crowns of thorns** (Traub 1980b). Not all murid fleas have such combs. In South America, where there are no murids, *Chiliopsylla* has a vertical ctenidium. The immediate question regarding the presence or absence of upright combs is "Why the variation?" The answer may lie in the divergence in congeners already mentioned, for example, where species of fleas on shrews have different types of combs than do their close relatives on moles. Another marked example of variations within a siphonapteran genus that correspond to the host is shown in Fig. 8.17. Here, no genal comb exists in the only species of *Dinopsyllus* that infests bathyergid moles, namely *D. ingens** (Rothschild), in contrast to the well-developed comb seen in other members of the genus (Fig. 8.16). The murids infested with fleas with such combs differ in behavior from the rats whose fleas lack them.

Adaptive Value of the Modifications in Fleas

Modifications of the Head and Crowns of Thorns

The vesture of the host is one of the factors that may account for the shape of the spines of its fleas (Humphries 1966, 1967; Traub 1972b; Traub and Dunnet 1973). Mammals with coarse or somewhat spinose fur, such as bandicoot* marsupials, tend to have fleas with sharply pointed pronotal spines, as in Figure 8.68, and in the case of spiny fleas of spiny* mammals, even the bristles are tailored to fit the spines of the host (Traub 1980b). First we treat the question of modification and hyperdevelopment of spines and bristles in fleas in response to selection pressure exerted by the habits of the host, commencing with a comparison of ordinary fur fleas and those that show the extraordinary development of modified bristles or spines on the front part of the head that has been designated a crown of thorns. These remarkable structures have been treated at length elsewhere (Traub 1980b), and only a few salient points need be mentioned here. In my opinion, such structures are clearly adaptive and functional, serving the flea to hold onto the hairs of the host or to reduce the likelihood of dislodgement. Second, the modifications are complex in nature and involve "coordinated evolution" of at least four different morphological changes: the development of spiniforms from bristles (either by evolution, or, in some species, during the lifetime of the flea), movement of the spiniforms toward the anterior margin of the head, the assumption of a vertical pattern of barbs, and actual narrowing of the head itself. The net result, which also may be accomplished by the development of a true helmet (Figs. 8.20, 8.114), is a device that can be slipped between two relatively closely set hairs of the host (or even grasp a single one) and effectively hook on to them. The false helmet of *Smitella* functions in the same way. To free itself, however, the flea need only move forward and disengage the barbs or spines. Fleas with crowns of thorns often are *semisessile* in behavior and remain fixed in position for hours or days, as in *Leptopsylla* (Farhang-Azad et al. 1983). To demonstrate the narrow flea head with crowns of thorns and the uniformly broad head of fleas that lack anterior cephalic spines or spiniforms, photographs of microdissections from various aspects are presented. In the helmet flea, *Stephanocircus*, the acute nature of the preantennal region is immediately apparent from the dorsal (Fig. 8.21) and ventral (Fig. 8.22) aspects, and the proximity of the bases of the spines on the helmet, seen from the anterior (Fig. 8.24), emphasizes the keellike shape of the helmet. In the leptopsyllid *Peromyscopsylla* the crown consists of only about three anteromarginal spiniforms (Fig. 8.28), but the head, again, is narrow (Figs. 8.27, 8.29).

In contrast, ordinary fur fleas, which feed for a few minutes and then move about in the pelage of the host, or else leave entirely, lack a crown of thorns, even if they are closely related to those that possess one. In such fleas, the head is broad anteriorly; as in the leptopsyllid *Frontopsylla* (Figs. 8.30–8.32), *Mesopsylla* (Figs. 8.39–8.41), and the pygiopsyllid *Aviostivalius* (Figs. 8.42–8.44). In *Metastivalius*,* some species have a marginal row of spiniforms and have all the other attributes of fleas with crowns of thorns, including the narrowed head; some have no spiniforms at all and have a broad head, and some are intermediate in both respects (Traub 1980b). Fleas with an upright genal comb on the posterior part of the frons (Figs. 8.1–8.4) do not have a narrowed head and do not remain fixed for long periods in one position. In fleas like *Sigmactenus*, in which a portion of the vertical genal comb is near the frontal margin and the rest is caudal in position, only the section of the head closest to the comb is narrowed. This condition is also found in the hystrichopsyllid *Dinopsyllus* (Figs. 8.23, 8.25, 8.26) in which the upper part of the comb is closer to the frontal margin than are the rest of the spines. The vertex is somewhat narrowed, and the lower part of the head is broad.

The monotypic, Bornean pygiopsyllid *Migrastivalius* (*Gryphopsylla*) is unique in possessing a distinct subapical notch along the ventral margin of the head (Fig. 8.35). This has been presumed to be an adaptive mechanism to enable the flea to cling to the spiny bristles of its host, rats of the genus *Lenothrix* (Traub 1980b). As shown in Figures 8.37 and 8.38, the front end of the flea is narrowed, and the "hook" is so placed that it seems it could readily grasp a spiny bristle near its base. If this idea is correct, then it would be expected that other such "hook-bearing" fleas, not necessarily belonging to *Migrastivalius* (*Gryphopsylla*), exist on spiny rats elsewhere.

The majority of mammalian fleas have a pronotal comb and lack a genal comb and hence superficially resemble the ceratophyllid shown in Figure 8.71, although the eye may be reduced, and the details of the spine shapes may differ. Most fleas also lack full combs on any other segment. All ceratophyllids (save the unique Antarctic bird flea), the majority of pygiopsyllids, and many leptopsyllids meet this description. As previously pointed out (Traub 1980b), 11% of the 241 genera known to me include species with crowns of thorns, and 130 (87%) of the 149 species in those genera are so adorned. The crown of thorns is the acme of one line of evolution of hyperdevelopment of spines, combs, and bristles, but there are other manifestations of a trend toward supernumerary spines and bristles, such as fleas with upright genal combs that are caudal in position (Figs. 8.1–8.4) or horizontal genal combs (Figs. 8.5–8.9), along with many other examples of combed fleas in Pulicidae, Leptopsyllidae, and Hypsophthalmidae. All ischnopsyllids (bat fleas) have a ventromarginal comb of two (rarely three) spines at the front of the head (Fig. 8.113). We now discuss these and other variations from the common mode of chaetotaxy, and then give a possible explanation for the hyperdevelopment of spines, combs and bristles, and for certain other structural modifications.

Pronotal Combs

The degree of elaboration in ctenidia varies from slight to extreme. In the first category are the pronotal combs of fleas of tree squirrels and semi-arboreal, diurnal squirrels. These ctenidia well illustrate the point that the fleas of a specific kind of host tend to converge toward a particular configuration. A minimum of 94 species of such squirrel* fleas, representing 13 genera and three families of Siphonaptera occurring in various parts of the world, all possess pronotal spines that are broader and proportionately shorter than those found on related fleas (often congeneric) found on hosts like insectivores, murids, and peromyscines (Traub 1972b). This trend is particularly pronounced in the subventral spines 2–4. Also, usually there is some degree of overlapping at the base of the subventral spines. Even fleas that are found on ground squirrels that are somewhat scansorial demonstrate this trend toward broad spines (Fig. 8.71), whereas the pronotal spines of fleas of flying squirrels (Fig. 8.57) tend to be narrower, greater in number, and to extend further ventrad, adjacent to the vinculum. *Tarsopsylla octodecimdentata* (Kolenati) (Fig. 8.61), which is found on arboreal or scandent hosts like tree squirrels, flying squirrels, and martens, has a comb of this latter type. In contrast, pocket gopher fleas have combs with narrower spines (Fig. 8.64), as do most murine fleas (Figs. 8.12, 8.13, 8.31, 8.42). In some instances the comb of the murine fleas covers the vinculum (Figs. 8.26, 8.35). Scattered throughout the order are fleas with well-developed combs on one or more of the abdominal segments (Jordan 1947, 1950; Traub 1972b, 1983b): *Ctenidiosomus, Lentistivalius* (*Destivalius*), *Migrastivalius* (all pygiopsyllids), *Stenoponia* (a hystrichopsyllid), some stephanocircids (Fig. 8.114), and many bat parasites (ischnopsyllids) (Fig. 8.113). One of the most heavily combed fleas is *Macropsylla hercules* Rothschild (Fig. 8.46).

There is an opposite evolutionary trend, namely, a reduction of the pronotal comb, as seen in true nest* fleas, and discussed at length elsewhere (Traub 1972b), and in fleas of spiny* hosts (Figs. 8.103, 8.104) (Traub 1980b). Reduction in the number and length of the spines in such fleas is clearly adaptive and is associated with circumstances where either a full comb is disadvantageous, for example, where there is a need to disengage rapidly from the host and avoid being carried away from the nest, or else where the vestiture of the host would cause a complete comb to malfunction, as in spiny animals. Such associations also reflect geologically ancient associations between flea and host.

Specialized Bristles

Not only spines, but bristles may also be highly developed and modified in fleas. For example, fleas generally have five to six dorsomarginal notches bearing one to three bristles on the hind tibia (Figs. 8.52, 8.69, 8.72), whereas some fleas have a false comb on this structure due to a vertical alignment of these and supernumerary bristles (Figs. 8.86, 8.87). Such tibial combs are found in many of the families of Siphonaptera, including the Ceratophyllidae (certain *Jellisonia* and *Kohlsia*); Leptopsyllidae (*Peromyscopsylla, Leptopsylla, Amphipsylla, Sigmactenus,* etc.); Hystrichopsyllidae (*Dinopsyllus, Stenistomera alpina* (Baker)); Pygiopsyllidae (some *Metastivalius, Astivalius*); and Chimaeropsyllidae (*Epirimia*). There may be a false comb on the protibia (Fig. 8.83) and/or mesotibia as well. The protibial comb in *Palaeopsylla setzeri* is due to vertical displacement of some of the dorsolateral bristles, not to the addition of supernumerary ones. Tibial combs are generally associated with other features, such as a foreshortened head with a flattened anterior margin (Traub 1972b). The tibiae may be specialized in other ways. A few additional bristles or groups may occur on the hind tibia (Fig. 8.66), which conversely may have a reduced number of notches and bristles (Figs. 8.92, 8.95) as compared with the more typical pattern (Fig. 8.96). There may be unusually deep notches (Fig. 8.106) and/ or the lack of a subapical pair (Figs. 8.107, 8.108), as in some fleas of spiny hosts, or else one of the dorsomarginal pairs may be exceptionally long (Fig. 8.101). Even the protibia can be extremely specialized, as in *Malacopsylla grossiventris* (Weyenbergh) (Fig. 8.91), in which the dorsomarginal bristles are exceedingly stout, appearing like spiniforms. As a general rule, the hindfemur bears only about three to four subapical ventral bristles and lacks lateral bristles except for perhaps two to four subapical ones (Fig. 8.96). However, in some cases the hindfemur is liberally clothed, as with a full subventral row (Figs. 8.92, 8.95, 8.99), perhaps reinforced with a few subapical laterals (Figs. 8.101).

Modifications in the chaetotaxy of the tarsi frequently occur. For example, some of these segments may bear very long thin bristles on the dorsal margin (Fig. 8.101) (cf. Figs. 8.96 and 8.99), and the apical bristles on some of the segments may extend far beyond the apex of the succeeding segment (Fig. 8.92). The fifth (apical) segment of the metatarsus also varies from one species or group to another. Many lack median, tiny plantar bristles (Figs. 8.65, 8.93, 8.97); others have just a few (Figs. 8.70, 8.98) or bear a dense patch of them (Figs. 8.100, 8.102). The number, size, and arrangement of the lateral plantar bristles on metatarsus 5 exhibit a broad range of differences. Most fleas possess five pairs (Figs. 8.65, 8.72, 8.93), although the first pair may be displaced toward the midline (Figs. 8.51, 8.55, 8.58, 8.62). Pygiopsyllids tend to have six pairs of lateral plantar bristles, with the first three pairs generally stouter than the others, and the third overlapping the fourth (Fig. 8.70). Rhopalopsyllids have only four pairs on hindtarsus 5, and they are all lateral (Fig. 8.98), a pattern common among neopsyllids and standard in the rhadinopsyllines. The species of hypodermal *Tunga* have only two (rarely three) such pairs of plantar bristles, and they are weakly developed, as are the bristles of the sticktight pulicid *Echidnophaga*, where the number varies from one to five. In some taxa these are very large (Figs. 8.93, 8.97). New data on displacement of some of the lateral plantar* bristles is discussed later. The adaptive value and structure of the tarsal claws have been well treated by Smit (1972).

Supernumerary* or modified bristles may occur on the head and abdomen as well. In some fleas the gena is clothed with spinose* bristles (Fig. 8.117) (Traub 1980b). In many murine pygiopsyllids there are four or more rows of bristles on the preantennal region and three full rows on the occiput (Figs. 8.35, 8.42), whereas most fleas have far fewer bristles on at least the front part of the head (Figs. 8.54, 8.57, 8.61). As a rule, ceratophyllid fleas have only one full row of occipital caudal bristles. However, *Jellisonia, Kohlsia, Pleochaetis,* and *Nosopsyllus* (*Penicus*) have three complete rows of bristles.

Commonly, fleas have two rows of bristles on the abdominal terga: the first generally incomplete and consisting of small bristles, the second row consisting of five to eight larger bristles per side (Figs. 8.76, 8.79, 8.112). Frequently the caudal row extends downward only to slightly below the spiracular fossa. Pulicids have only one row of such tergal bristles, but in other families there are fleas that have three, four, or even more rows of abdominal bristles (Figs. 8.18, 8.19, 8.46). Even where two tergal rows occur, there are species that have supernumerary bristles, and the rows extend more ventrad than usual, and/or the bristles are thinner, longer, and placed closer together (cf. Figs. 8.74, 8.75 with 8.76; Figs. 8.80, 8.82 with 8.84). Also, the bristles may be longer and thinner than in related fleas: for example, *Dinopsyllus ingens* (Fig. 8.19) versus *D. longifrons* Jordan and Rothschild (Fig. 8.18). The pocket gopher fleas like *Foxella** (Fig. 8.67) likewise have supernumerary* and longer, thinner, abdominal bristles than do most other ceratophyllids (Fig. 8.73).

False combs occur on parts of the body of fleas other than the limbs (Traub 1968a, 1972b, 1980b). The comb on the helmet of *Smitella* consists of spiniform bristles. False combs are known on the pronotum (*Scolopsyllus*, Fig. 8.120); on the metepimere, as in the bat flea *Chiropteropsylla* (Fig. 8.118); and on the seventh tergum (Figs. 8.116, 8.120); or there may be comblike clusters of flattened or spinelike bristles on the metathoracic and anterior abdominal terga, as in *Myodopsylla* bat fleas (Fig. 8.119). Another type of unique specialization is seen in the echidna flea, *Bradiopsylla echidnae*, where the meso- and metanota are clothed with stubby thornlike bristles (Fig. 8.104).

Nonchaetotactic Changes

There are some other pertinent adaptive morphological changes besides chaetotaxy. First, the reduction of the pleural arch of the metathorax and concomitant modification of the adjacent area are seen in fleas that have lost the ability to jump effectively. These fascinating developments, discussed elsewhere (Smit 1972; Traub 1972b, 1980b; Rothschild et al. 1975), are seen in true nest fleas and in some fleas of nocturnal/crepuscular animals that are volant (bats) (Fig. 8.113) or gliding (flying squirrels and phalangers) (Figs. 8.53, 8.60), where virtually complete reduction of the arch has been effected. In contrast, the pleural arch (PL.A.), which houses the resilin that provides the motive power for jumping, is well developed in the great majority of fleas (Figs. 8.56, 8.73, 8.76). Second, there are interesting and instructive differences regarding the tarsi. On the hindleg, the first tarsal* segment may be much shorter than the tibia (Fig. 8.92), or else slightly shorter (Fig. 8.66), subequal (Figs. 8.69, 8.72), or much longer (Fig. 8.62). The proportionate length of this first segment as compared to the second also varies considerably, ranging from twice as long (Fig. 8.62) to subequal (Figs. 8.66, 8.92, 8.93).

Correlation of Morphology with the Habits of the Hosts

Throughout most of the Order Siphonaptera there is an apparently haphazard and bewildering variety of bristles, spines, combs, and other morphological features. However, one factor is common to all the exceptions to the basic morphological model of the fur-infesting flea, namely, the modifications are all adaptive and serve to enhance the flea's chance of survival in its specific life-style. The variations are not casual, but form a pattern, as seen in the spines of the combs of shrew* fleas. In the case of the sticktight fleas, the trend is toward the development or perfection of anchoring mouthparts, increased reproductive capacity, and the loss of spines and bristles (which serve no purpose in a flea that is permanently attached and which interfere with gross enlargement of the abdomen upon engorgement or when gravid, etc.). This particular evolutionary tendency reaches its peak in hypodermal fleas like *Tunga* or *Rhynchopsyllus*, where the flea is wholly or partly ensconced within the tissues of the host. In the specialized nest fleas, the direction is toward structural changes that tend to improve the flea's ability to remain in the nest and feed while the host is sleeping, or else lessen the possibility of its remaining on the host when the mammal leaves the nest and enters an environment hostile to this kind of flea (or its larvae). Thus the mouthparts are elongated; the combs, spines, and bristles are reduced in number and size; and the ability to jump is lost.

In the case of the fur-inhabiting fleas, hyperdevelopment of spines, combs, and bristles regarding number or size is invariably associated with infestation of a host whose vestiture or habits adversely affects the likelihood of survival of the flea as an individual or as a species. Moreover, the greater the concomitant need for the parasite to remain on the host, the greater is the degree of development of its combs and bristles (Traub 1980b). As pointed out previously, in *Leptopsylla*, semisessile fleas that depend on frontomarginal spiniforms to cling to the hairs of the host, the modified bristles and the genal and pronotal spines are fewest in number in those species that infest ground-dwelling murines. They are best developed in the *Leptopsylla* that infest the desert shrew *Diplomesodon* and parasitize scansorial murines.

This same principle applies throughout the order and is illustrated by the following review. This review proceeds from those hosts that present the least hazard to their fleas to those whose habits or pelage pose the greatest problems to the fleas. However, we are dealing with qualitative traits. There is no purely linear progression in either the degree of risk or the morphological response of the flea. Therefore, to minimize duplication in names and examples, certain groups of hosts or evolutionary developments are treated at some length, without full regard for the actual sequence of degree of hazard or hyperdevelopment in chaetotaxy.

Fleas of Relatively Unspecialized Hosts

Fleas of diurnal, ground-dwelling rodents that have unspecialized fur, a small home range, and an established nest lack combs and specialized bristles on the body and limbs. They are good or fair jumpers: for example, *Polygenis* (Fig. 8.112) and other rhopalopsyllids on rodents like *Sigmodon* and *Oryzomys*, and most *Xenopsylla* on hosts like *Arvicanthis*. These fleas do not spend much time on the host and feed relatively rapidly. They abound in the nests, burrows, and protected runways of the host. Such fleas have a reasonable chance for survival if they leave the host, and their larvae are presumably relatively tolerant of variations in environmental conditions. Some of the combless fleas like *Pulex** have the capability of feeding on a large variety of hosts and hence are not under evolutionary stress to remain on a single individual.

The survival pressure is somewhat greater on fleas whose hosts are diurnal but live and breed in well-established, long burrows and that are either colonial, like prairie dogs, or somewhat communal, like marmots and ground squirrels. Among the other hosts that are somewhat more hazardous to fleas are ground-dwelling or slightly scansorial mammals that are active much of the day (sometimes at night too) or occupy a small territory that may be somewhat restrictive (mossy, rocky, or a fringe nature, etc.). This is true of many cricetines, especially peromyscines and certain voles, chipmunks, some murids, and others. Fleas of marmots, cricetines, and the like have well-developed eyes and pronotal combs of about 16–20 spines (but no other combs). Many fleas, including many ceratophyllids and leptopsyllids and some pygiopsyllids, share these traits. When the mammals are partly scandent, or even when they nest in tree holes, there is a tendency for the fleas to have supernumerary bristles, perhaps rows of them on the occiput, as in *Orchopeas leucopus* (Baker) and *Pleochaetis*, and perhaps tibial combs, as in some *Jellisonia*. Although *Amphipsylla* usually shares these traits, it faces exacerbated hazardous conditions in arid climes or because of extremely long winters. These fleas are characterized by having full occipital rows of bristles, full pronotal ctenidia, and false tibial combs.

Diurnal mammals, like tree squirrels, which not only are largely arboreal but also nest in tree holes or among the branches, provide an environment for flea larvae that is no doubt rather specialized (Traub 1972a). Such fleas tend to mate in the nest and fix their eggs to the leaves, rather than to mate on the host and lay eggs loosely among the hairs as *Xenopsylla* does (Traub 1972b, 1980b; Traub and Rothschild 1983). These fleas possess one comb on the pronotum, relatively broad spines, and the subventral spines partially overlapping at the base. The fleas of squirrels that live in extremely tall trees in the equatorial rain forest, where a leap from the nest might be fatal, tend to be extremely poor jumpers and lack a pleural arch (e.g., *Syngenopsyllus*, *Libyastus*, etc.).

There are small mammals that are active largely at ground level but whose habits make survival somewhat difficult for fleas. These include mammals that nest in holes under rocks or on the ground or among roots of trees and may be restricted to meadows or grassy margins of streams, as are some *Microtus*, or to damp, mossy terrain, as *Clethrionomys*, or to rocky or sandy soil or to the environs of certain bushes or other vegetation upon which these rodents feed or scamper. Some murines, ground squirrels, and certain xerophilic rodents also inhabit restricted environments. Fixed habitats for a host generally mean unfavorable conditions for its fur fleas unless they are on the mammal and the larvae hatch in its home territory. Fleas with these types of hosts tend to have a pronotal comb, plus a short genal comb of two or three spines: for example, some ctenophthalmines, neopsyllines, and many chimaeropsyllids. As expected, some of these fleas show morphological signs of living primarily in the nests of their hosts. Those fleas that infest hosts that are somewhat scansorial tend to have false combs on some of the tibiae, as in some *Neopsylla*.

Fleas of Nocturnal, Gliding Mammals

Among the nocturnal* arboreal hosts that obviously pose more problems to fleas than do those of ordinary diurnal squirrels are the flying squirrels (petauristines) that glide from tree to tree, perhaps over distances of 100 m. Their characteristic fleas, such as the Nearctic ceratophyllid, *Opisodasys pseudarctomys* (Baker), have a somewhat reduced eye and a pronotal comb that extends further down than usual, overlapping the base of the vinculum, consisting of about 24 spines (Fig. 8.57).

Just as the marsupial flying phalangers of Australia resemble the Holarctic flying squirrels in appearance and glissant behavior to an extraordinary degree, so do their respective fleas exhibit remarkable signs of convergence in the reduction of the eye and of the pleural arch, in an increase in the number of pronotal spines, and in certain features of the hindleg, as shown by *Choristopsylla tristis* (Rothschild) (Figs. 8.50, 8.53) and *Opisodasys pseudarctomys* (Figs. 8.57, 8.59). The first and second tarsal segments are unusually long as compared to the length of the tibia and other tarsal segments, as shown in Table 8.16. This pattern regarding eye, spines, bristles, and tarsus also occurs in the Holarctic ceratophyllid *Tarsopsylla octodecimdentata* (Figs. 8.61–8.63), which often infests scansorial crepuscular/nocturnal hosts like martens and flying squirrels (as well as tree squirrels or their nests). The Bornean *Hollandipsylla neali** is another ceratophyllid that infests flying squirrels and is similarly modified, although here the eye is completely vestigial. Significantly, this same series of features is seen in *Myoxopsylla jordani* Ioff and Argyropulo, another flea whose host (a dormouse) is nocturnal and virtually completely arboreal. It too has a reduced eye, a comb with supernumerary spines that extends down over the vinculum, exceptionally long segments on metatarsus 1 and 2, and it has

(TIB.) and Other Tarsal Segments (3–5) in Certain Fleas and the Displacement of the First Pair of Lateral Plantar Bristles (L.P.B.) on 5

Figures	Fleas	Hosts	Hind Leg Ratios				Hind Tarsus 5 First L.P.B. Displaced
			1/3–5	TIB./1	2/3–5	TIB./2	
57–60	*Opisodasys pseudarctomys*	Flying squirrels U.S.A.	1.3	1.2	0.7	2.2	+
50–53	*Choristopsylla tristis*	Flying phalanger Australia	1.5	1.1	0.8	1.8	+
54–56	*Choristopsylla ochi*	Arboreal marsupial Australia	1.3	1.2	0.8	1.9	+
61–63	*Tarsopsylla o. coloradensis*	Flying squirrels and martens (?) Holarctic	1.5	1.1	0.7	2.4	+
65–66	*Bradiopsylla echidnae*	Echidna Australia	0.6	1.7	0.5	2.0	0
68–70	*Parastivalius novaeguineae*	Bandicoots New Guinea	1.0	1.4	0.7	2.1	0
71–73	*Thrassis francisi*	Ground squirrels U.S.A.	1.0	1.3	0.6	2.2	0
92	*Malacopsylla grossiventris*	Armadillos South America	0.4	3.5	0.4	3.9	0
94, 95,97	*Phthiropsylla agenoris*	Armadillos South America	0.5	2.9	0.4	3.5	0
96,98	*Polygenis gwyni*	Sigmodon Mexico, etc.	0.9	1.7	0.6	2.4	(Nil)
	Myoxopsylla jordani	Dormouse SW Asia	1.3	1.1	0.7	2.0	+
	Hollandipsylla neali	Flying squirrel Borneo	1.2	1.1	0.7	1.7	+ (and 3rd)
	Orchopeas howardii	Sciurus Nearctic	0.9	1.3	0.5	2.2	+

lost the pleural arch. Loss of the eye and lessened ability to jump are clearly adaptive responses in fleas whose gliding or arboreal hosts are active only in the dark, and where dislodgement or an active leap could be fatal to the flea (Rothschild et al. 1973; Traub 1972b, 1980b). The pygiopsyllid *Choristopsylla ochi* (Rothschild) which parasitizes nocturnal scansorial phalangers is an intermediate type regarding modifications of legs and pleural arch (Figs. 8.54–8.56). Such long tarsal segments are highly unusual in other kinds of fleas. This is shown by the examples listed in Table 8.16. In the above-mentioned fleas of glissant mammals and dormice, the first tarsal segment of the hindleg is at least 1.2 times the length of segments 3–5 combined and is nearly as long as the tibia, whereas segment 2 is 0.7–0.8 times the length of segments 3–5. In contrast, segment 1 in the other species is generally shorter than the combined length of segments 3–5, and the tibia is at least 1.3 (or even 3.5) times as long as segment 1, whereas segment 2 is generally 0.4–0.6 times the length of segments 3–5. *Parastivalius novaeguineae* (Rothschild) is an exception with an unusually long second tarsal segment on the hindleg, but its host is somewhat saltatorial,* or galloping, and such fleas are specially modified.

Displacement of Plantar Bristles

Displacement of one of the lateral plantar pairs of distotarsal bristles (generally the first pair) toward the midline is very frequently associated with infestation of hosts that exert unusual survival pressure upon the fleas. This displacement occurs in those fleas in Table 8.16 that infest arboreal hosts but not on the other fleas. This displacement is widespread in fleas of arboreal, semiarboreal, and scandent mammals, as in the ceratophyllids of sciurids such as *Macrostylophora* and *Syngenopsyllus*; leptopsyllids like *Sigmactenus* and *Leptopsylla* on murines; in the hystrichopsyllid *Dinopsyllus* on murines, dormice, and others. In the phalanger pygiopsyllids *Muesebeckella* and *Wurunjerria*, the third pair of lateral plantar bristles is displaced toward the midline (and in the latter, the first pair as well). The monotypic genus *Hollandipsylla** is unusual among mammal ceratophyllids in that it possesses two pairs of displaced plantar bristles, the basal and third pairs (Traub 1953). The two species of *Smitipsylla*, described from flying squirrels in Nepal, have both of these pairs somewhat displaced. In the ischnopsyllid bat fleas the basal pair is in the midline, and some species have two pairs displaced. Bat* fleas naturally face more hazards than do other fleas. The basal pair alone is displaced in many shrew fleas, including the Doratopsyllinae, the phalacropsylline fleas, the doratopsylline fleas of marsupials, stenoponiine species, all of which face handicaps because of the need to live in a special environment or to develop at a certain season. In contrast, the fleas that have a relatively simple or unhazardous mode of life and a more readily available food source lack median plantar bristles. These bristles are all lateral, as in the Siphonaptera of ground squirrels; for example, *Opisocrostis, Oropsylla* and some *Medwayella,* all the Hystricho-psyllinae, the Neopsyllini that infest ground-dwelling rodents, and the fleas of pocket gophers. Fleas that live on mammals that are partly scanso-rial tend to have the first pair of plantar bristles displaced somewhat, as in *Monopsyllus.* In the African ceratophyllid *Libyastus,* which has been consid-ered a nest flea of squirrels (and lacks a pleural arch, for example), the first pair of plantars is only slightly displaced. The habits of *Libyastus* (and of their hosts) are inadequately known. Remarkable examples in *Meta-stivalius** are discussed later.

This shift of plantar bristles toward the midline is also seen in the vari-ous bird fleas even though many genera and several families of avian fleas include or consist of taxa that are derived from mammalian fleas. Bird fleas, of course, are under tremendous selection pressure because of the volant nature and other attributes of the host (Jordan 1926; Holland and Losh-baugh 1958; Smit 1972; Traub 1972b; Pearse 1981).

Desert Fleas

Some siphonapterans, like their hosts, have succeeded in coping with environmental* hazards such as aridity or cold. In the desert environment the basic survival strategy for fleas is either to remain in the fur of the host or, when unattached, to spend the time wholly in the burrow. Exposure to the external hostile elements could be rapidly fatal to the flea or completely unsuitable for development of eggs or larvae. Many mammals in hot, xeric terrain are active only at night, when the temperature is more equable. They often plug the openings of the burrows during the day. Both of these traits help the fleas to cope as well. Fleas may swarm at the entrance of the burrows, as Darskaya (1955) noted for the xerophilic rodent *Rhombomys.* A large proportion of the most highly modified nest fleas, such as the hys-trichopsyllids *Megarthroglossus, Callistopsyllus,* and *Anomiopsyllus,* are asso-ciated with desert rodents (Traub 1972b), and this is true for the leptopsyl-lid *Jordanopsylla* as well. These are adapted for feeding while the host is sleeping. The unmodified *Xenopsylla* on desert rodents presumably feed rapidly like *X. cheopis,* which may leave the host after a few minutes (Farhang-Azad et al. 1983). Some arenicolous fleas like *Xenopsylla* have exceptionally long tarsal bristles which they can use to burrow rapidly in the sand to escape the dangers of heat and desiccation (Wagner 1932; Ioff 1947). Other fleas of desert rodents have some of the features of nest fleas, such as a reduced eye, but have a well-developed pronotal comb (e.g., *Phaenopsylla*) which is used to remain on the host if it leaves the burrow. In certain desert fleas, like *Mesopsylla* and most *Ophthalmopsylla,* the eye is exceptionally large, as it is in their nocturnal dipodid hosts (Traub and Evans 1967). Some fleas of desert rodents, like *Mesopsylla* and *Carteretta,* have both genal and pronotal combs. These are well developed in some macroscelid fleas, and particularly so in the bristly genus *Stenoponia,* which also has an abdominal comb. Furthermore, *Stenoponia** is found only in cool or cold seasons or areas.

Fleas of Fossorial Hosts

Fleas of fossorial mammals like moles, mole rats, and pocket gophers and/ or fleas that spend virtually all of their lives in the dark tend to be essentially eyeless (Jordan 1948; Traub and Evans 1967; Traub 1972b) (Figs. 8.2, 8.64, 8.113). Degrees of eye development as associated with the habits of the host are well illustrated in the genus *Pulex*, where most species infest diurnal mammals like porcines, the natural hosts of *P. irritans* (Fig. 8.122). *P. sinoculus* Traub is highly unusual because it parasitizes pocket gophers and is completely eyeless (Fig. 8.121). The monotypic *Callopsylla* (*Typhlocallopsylla*) *semenovi* (Ioff) is the only ceratophyllid to infest insectivorans; its host is a mole. This flea has practically no trace of an eye. In contrast, the other *Callopsylla* parasitize rodents or birds and have well-developed or large eyes, corresponding with the diurnal activity of the host (Traub 1972b, 1980b). In *Dinopsyllus* most of the species in the subgenus *Dinopsyllus* have a reduced eye, as befits the fleas of a nocturnal/crepuscular host like *Rattus*. In Figures 8.16 and 8.26 the eye is visible as an indiscrete dark patch near the dorsalmost spine of the upright genal comb. However, in *D.* (*Cryptoctenopsyllus*) *ingens*, the flea of the bathyergid "mole" (a rodent), there is no trace of the eye (Fig. 8.17). The same trend is exhibited in *Ctenophthalmus*, where, although there is never a fully developed eye, the largest ones occur in squirrel fleas. The fleas infesting the various fossorial mole rats (which are highly disparate phylogenetically) are essentially blind (Figs. 8.7, 8.8), as are the mole fleas (Fig. 8.6). The murid *Ctenophthalmus* (Fig. 8.15) are intermediate in this regard.

Fleas of fossorial mammals are relatively well protected against environmental changes while in the burrows of their hosts and have reasonably ready access to their food source since these mammals occupy such a small territory. In such a habitat there is little need for devices to facilitate attachment to the host, but particles of soil and sand in that environment may have a traumatic effect upon the fleas. Thus it is expected that reduction of spines and combs might ensue in fleas switching to or coevolving with hosts that are, or are becoming, fossorial in certain types of soil. *D.* (*C.*) *ingens* is outstanding in this connection, since the genal comb, so conspicuous in other members of the genus (Figs. 8.16, 8.26), is represented only by a single small spine (Fig. 8.17). Furthermore, the pronotal comb is definitely shorter, with fewer spines, and the apical spinelets (vestigial combs) on the abdominal terga that are present in the other *Dinopsyllus* (Fig. 8.18) are virtually lost (Fig. 8.19). The same but modified tendency is seen in *Brachyctenonotus myospalacis* Wagner, a flea of the zokor *Myospalax*, which is fossorial. Here the individual spines of the pronotal comb are much shorter than in other leptopsyllids. Significantly, the pronotal spines are pale and short in the unique *Ctenophthalmus* that infests the zokor. The four subgenera of *Ctenophthalmus* that parasitize the diverse mole rats have a smaller number of pronotal spines than do other *Ctenophthalmus*, and their pleural arch is greatly reduced or absent.

Nevertheless, there are reasons why certain fleas of burrowing animals do not lose their combs, and in fact undergo hyperdevelopment of spines and bristles. That the hystrichopsyllid mole fleas are extremely host specific suggests that they cannot survive on other mammals, although shrews and mice often enter the burrows of moles. Furthermore, there are also the critical factors of temperature, humidity, and larval food in the tunnels. Thus these fleas need comblike devices for host attachment and survival. Sometimes there is a need to remain attached to a fossorial host, as when the young disperse, or the mammal must seek a new supply of food. In certain types of soil, combs may help protect fleas against particulate matter in the burrows. The overall size of the genal and pronotal combs differs little in the congeneric mole and shrew fleas (cf. Figs. 8.1 and 8.2; 8.5 and 8.6). However, the shape of the spines and other features differ according to the hosts.

Life on burrowing hosts seems to have brought about certain other adaptive modifications regarding spines and bristles. Ioff (1929) ascribed the rounded nature of the first two genal spines of *Ctenophthalmus* of *Spalax* mole-rats to the extremely thick, soft, silky fur of the host and also remarked on the frequent occurrence of damaged genal spines on these fleas. Such naturally blunt or rounded genal spines, and injured spines as well, are found on the *Ctenophthalmus* of rhizomyid, spalacid, and myospalacine mole-rats (Hopkins and Rothschild 1966; Smit 1976; Traub 1980b). Smit (1963, 1967) commented that the fleas of fossorial rodents, such as the various kinds of mole-rats, tend to have broken genal and/or pronotal spines and wondered why this was not the case for fleas of fossorial insectivores. Smit (1967) did not specifically suggest any cause for this phenomenon but regarding *Brachyctenonotus*, went on to state that "it looks as if the forebear of [the characteristic zokor flea] *Brachyctenonotus myospalacis* . . . became wise to the potential damage of pronotal spines after it had transferred from presumably nonfossorial rodents to zokors, for these spines are now reduced to extremely short little stumps." However, I believe that particles of sand and soil traumatize the spines of fleas on fossorial rodents in their particular microhabitats, and that the blunt or rounded apices characteristic of many of these fleas (cf. Figs. 8.7 and 8.8 with Fig. 8.15) are the result of an adaptive trend to eliminate sharp apices, which seem much more likely to break off than obtuse tips. The same processes of selection have led to a reduction in the numbers of ctenidial spines or their foreshortening in some cases. An alternate mechanism to reduce the hazards posed by certain types of soil particles would be the development of a special type of pronotal comb in which the spines are close-knit, parallel sided, apically rounded, and sufficiently numerous to completely extend over the sides. This is precisely the type of comb in *Amphipsylla dumalis* Jordan and Rothschild, a flea of the fossorial microtine *Ellobius*. Its congeners on ordinary voles possess the usual type of comb, with fewer, more widely spaced spines.

That evolution has indeed resulted in such protective measures is suggested by another feature that is even more characteristic of fleas of subterranean mammals, namely, the possession of supernumerary bristles on the thorax and abdomen. Pocket gopher fleas such as *Foxella* (Fig. 8.67) are conspicuously more densely clothed than are fleas of ground squirrels (Figs. 8.73, 8.75) or of *Peromyscus* (Fig. 8.76), and their bristles are finer, more closely set, and longer than in the other fleas. The four ultraspecific fleas (in four genera) of another fossorial rodent, *Aplondontia rufa*,* also bear unusually long, approximated bristles. Prairie dogs live in extensive underground burrows, although they are active on the surface during the day, and *Opisocrostis hirsutus* (Baker), a generispecific flea of this host, is truly hirsute, as the name suggests (Fig. 8.74). Jellison (1947) believed that *Opisocrostis* was originally a *Cynomys* flea but later switched in part to *Citellus*, on which most species occur today. He suggested that in the process of adaptation there was "a reduction in the number of bristles." *Dinopsyllus* are very bristly fleas, but this is particularly true of *D. (C.) ingens*, a parasite of the bathyergid "mole" (Fig. 8.19), where the bristles are also more attenuated and approximated than in the murine species (Fig. 8.18). Another bathyergid flea, the chimaeropsyllid *Cryptopsylla ingrami* (de Meillon) of *Cryptomys*, also has lost its combs and eye and possesses close-knit supernumerary bristles and false tibial combs. The pulicid *Xenopsylla georychi* (C. Fox), likewise a combless bathyergid flea, is very unusual in the genus by being completely eyeless and in possessing extra abdominal bristles which are approximated.

The same chaetotactic trend occurs in the *Ctenophthalmus* of various kinds of mole rats. There are twice as many bristles on the metepimere (Figs. 8.77, 8.78) in these fleas as there are in the murine *Ctenophthalmus* (Fig. 8.79). They also have two or more bristles inserted below the spiracular fossa on the unmodified abdominal terga instead of just one. These extra bristles have been noted at taxonomic characters (Smit 1963; Hopkins and Rothschild 1966), but also they probably offer more protection against sand and soil (in the burrow or in the fur of the host) than does the customary setal clothing of fleas of ordinary hosts. The chaetotaxy of the myospalacine *Ctenophthalmus* is similar, and here the segments covering the genitalia are adorned with an unusually large number of bristles, some of them exceptionally long and thin. This tendency toward supernumerary, longer and thinner bristles in the fleas of fossorial hosts is also seen in *Amphipsylla daea* (Dampf) of *Myospalax* and *A. dumalis* of *Ellobius*. Another flea of *Ellobius*, *Xenopsylla magdalinae* Ioff, has unusually long, thin bristles. The basic association between such unusual chaetotaxy and soil habitats even extend to bird fleas. Rothschild and Clay (1952) sagely remarked "It has been claimed that certain mammal fleas develop finer and longer bristles on their legs if they parasitize rodents living in sandy soil. *C. styx* (a bird flea) has finer and more numerous bristles than any other bird flea and this may be a direct result of the type of soil in which birds make their nest." The close-set abdominal and thoracic bristles of *C. styx* are indeed striking when compared to those on fleas that infest birds nesting aboveground.

All of the mammalian fleas cited previously infest rodents, whereas in the fleas of fossorial insectivores hyperdevelopment of bristles is exceptional. There are no clear-cut differences in the number, size, and arrangement of the thoracic and abdominal bristles in mole fleas and shrew fleas of the genus *Palaeopsylla* (cf. Figs. 8.80 and 8.84). It is noteworthy that in *Palaeopsylla setzeri*, the sole species not found on insectivores, the tergal row of abdominal bristles extends down below the spiracular fossa (Fig. 8.82) and that this flea parasitizes a molelike vole, *Hyperacrius*. Before attempting to explain the dearth or absence of supernumerary bristles in insectivoran fleas, let us further probe the basic premise that chaetal hyperdevelopment is characteristic of fleas of fossorial rodents. A good test case exists in the Pulicidae, a family that possesses only a single row of abdominal bristles and is sparsely clothed in general. *Pulex sinoculus*, the unusual, blind species that infests pocket gophers, definitely has more bristles on the thorax and abdomen and on the female genital segments (Figs. 8.123, 8.125) than do the other species of *Pulex* (e.g., Figs. 8.124, 8.126). *P. sinoculus* is anomalous in that these bristles are relatively stout, but this may represent an adaptive response. Since pulicids have so few bristles in general, there is a special need for strength to fend off the particles of soil. If this correlation between supernumerary bristles and flea infestations of fossorial rodents is valid, then it would be expected that these fleas have supernumerary bristles on the legs. Since all the other *Dinopsyllus* have tibial combs (Fig. 8.86), it is not very surprising that the *Bathyergus-Dinopsyllus* has one (Fig. 8.87). However, its retention is significant because this flea exhibits a marked reduction in ctenidial combs. There is a false comb on the protibia of *P. setzeri* (Fig. 8.83), unlike the situation in the *Palaeopsylla* of shrews and moles (Figs. 8.81, 8.85). This *Hyperacrius* flea also has somewhat of a comblike arrangement on the metatibia (Fig. 8.89; cf. Figs. 8.88 and 8.90). There are no data as to whether fleas can use tibial combs to groom themselves, but such combs could presumably serve to brush away soil in the fur as the flea progresses through the hairs. In the latter events, a comb on the mesotibia would probably be more useful than one on the hindleg.

Since mole fleas lack supernumerary bristles and have unmutilated spines, it seems reasonable to assume that they face less risk from the grains of soil in their particular habitats than do the fleas of fossorial rodents: thus there has been less selection pressure to round off the apices of the spines and to generate protective clothing. I believe that the pocket gophers, bathyergids, and sundry mole-rats inhabit soil that is definitely more sandy, rocky, or particulate than do the flea-bearing species of moles, which are associated with humus and fine rich soil. Although precise data are unavailable, it seems that flea-bearing moles and fossorial rodents only rarely occur together in the same microhabitat. Moles are very broadly distributed throughout Hungary but are not found in the extremely limited foci inhabited by *Spalax* (I. Szabo, in litt.). There seem to be no records of geomyid fleas from moles, and reports of mole fleas from pocket gophers are highly exceptional. According to E. W. Jameson, Jr. (in litt.), these hosts do occur together in a few parts of the Pacific Northwest, but I wonder if these represent a recent development due to ecological changes wrought by man. Environmental factors, and not host phylogeny, may perhaps also account for the patterns of attenuated, sharp, pointed ctenidial and genal spines of talpid fleas as compared to congeners on soricids (cf. Figs. 8.2 and 8.6 with Figs. 8.1, 8.3 and 8.5) (Traub and Barrera 1966; Traub and Evans 1967). In *Callopsylla (T.) semenovi* the sole ceratophyllid that infests insectivorans (a mole), the pronotal spines are narrower and more numerous than in other members of the family (Traub 1980b), which primarily parasitize rodents. The characteristic shapes of the spines in the mole fleas perhaps may be associated with the type of hairs of the host, but it is also possible that they are modified to help ward off particulate matter in the fur or burrows of the host. Such debris must be a problem to all fleas of underground-dwelling mammals, even though the humus and rich soil where these moles burrow may be less traumatic to the fleas than the more sandy and pebbly environment of the fossorial rodents. This line of reasoning is reinforced by the case of *Palaeopsylla remota* Jordan. The pronotal and genal combs of this species resemble those of talpid *Palaeopsylla*, but its host is a soricid. However, the host *Anourosorex* is aptly called a moleshrew because of its habits and habitus. The menace of soil debris in the fur or burrows of fossorial hosts is also suggested by the pattern of the pronotal comb in the pocket gopher fleas. The spines are more numerous, narrower, and closer set than in the ceratophyllids of nonfossorial hosts, recalling the case in the *Ellobius* flea, *A. dumalis.**

Fleas of shrews must face more hazards than do those of fossorial rodents or even mole fleas. As in the case of other insectivoran fleas, they are ultraspecific and are rarely collected on other hosts. Thus those fleas recorded are clearly trespassers. Shrews have a highly circumscribed home range and seldom leave their short burrows and runways. Because of their small size, shrews are believed to be more sensitive to temperature and humidity than other mammals (Walker et al. 1975). Their requirement for vast amounts of food in proportion to their size undoubtedly helps restrict their activity to established, satisfactory foci. Circumstantial evidence and collection data strongly suggest that adult and larval fleas of soricids are as sensitive to ambient conditions as their hosts and are limited to the haunts of the shrews. For all of these reasons it is to be expected that these siphonapterans are adapted to remain within the fur of their hosts, with a well-developed pronotal comb and often a genal comb as well (Figs. 8.1, 8.2, 8.5). The potency of the evolutionary forces molding the detailed composition of the comb is illustrated by the marked convergence in design exhibited by the shrew fleas in four families and six subfamilies on four continents (Traub and Barrera 1966; Traub and Evans 1967).

Fleas of Saltatorial Hosts

Saltatorial hosts, if infested by fur fleas at all, place their siphonapterans at special risk (Traub 1980b). This is not so much because of the hopping or erratic gait of the mammals, but presumably because the frightened or food-seeking host may rapidly enter an environment that is hazardous for the various stages of fleas. Moreover some of these hosts have very large home ranges and/or lack a permanent den. They either have no specific fur fleas (e.g., macropodid kangaroos) or have highly specialized types, as do rabbits, hares, and bandicoots. Even some of the small saltatorial rodents such as zapodids either lack characteristic fleas (e.g., *Zapus*) or have them rarely (*Sicista*). The fur fleas of all saltatorial hosts are invariably specialized, and the degree of evolutionary response in hyperdevelopment of bristles, spines, and combs seems to be determined by the proportionate hazard to the flea. Among the least specialized of the fleas in this category, the leptopsyllid *Cratynius* (four species), which infests the monotypic gymnuran *Hylomys*, has a small tiara of frontomarginal spiniforms and a full pronotal comb (which is also found in all of the following fleas). The dipodid fleas *Mesopsylla* (Figs. 8.39–8.41) and *Desertopsylla* have a short genal comb, and the latter has spiniforms on the head. The kangaroo rats (Heteromyidae) are infested with *Meringis*, which have a genal comb of two spines. In *M. altipecten* Traub and Hoff the pronotal ctenidium is arched and flares high over the mesonotum (Figs. 8.33, 8.34, 8.36). This strange condition is seen in some other desert fleas and in certain shrew fleas (Fig. 8.1). Although this was formerly regarded as inexplicable (Traub and Evans 1967), I now believe it is an adaptive modification that discourages dislodgement by tooth or claw when the flea stands on its head to feed; the comb is then in position to grasp hairs if the flea is threatened. The horizontal genal comb also functions in that manner when the flea is in the feeding position (Traub 1980b).

Some bandicoot marsupials gallop, if they are not saltatorial, and their behavior and extended range also pose survival problems to their fleas, further complicated by their coarse-haired fur. Several kinds of pygiopsyllids infesting bandicoots, such as some *Striopsylla*, have the preantennal region clothed with thorny bristles or subspiniforms, and *Ernestinia* (Fig. 8.117) is even more elaborately adorned with thorns. These fleas have been observed or are presumed to abut their heads against the body of the host with the assistance of unusually broad forecoxae, facilitated by exceptionally short mouthparts, and cling to the thickened hairs of the bandicoots by means of the thorny bristles or subspiniforms (Traub 1968a, 1972b, 1980b). The elephant shrews (macroscelids) are excellent jumpers, and they carry chimaeropsyllids with a conspicuous genal comb, or a *Caenopsylla* with a shorter comb on the gena. (They also have sticktight fleas.) The acme of adaptations of fleas of saltatorial wide-ranging hosts is seen in fleas of rabbits, where *Spilopsyllus, Cediopsylla, Nesolagobius,* and *Hoplopsyllus pectinatus* Barrera have a relatively well-developed, upright or inclined genal comb. *Nesolagobius* is an archaeopsylline pulicid, but its convergence with the other genera on rabbits, which are spilopsyllines, is quite remarkable and includes changes in the shape of the head (Barrera 1967). The leptopsyllid rabbit fleas, in the genus *Odontopsyllus,* which lack a genal comb, have an exceptionally large pronotal ctenidium and are very bristly. In *Cediopsylla* and *Spilopsyllus* (and probably in *Hoplopsyllus pectinatus* on *Romerolagus,* if not the other rabbit fleas as well), the sexual activity of the fleas is determined by the hormonal* status of the host so that the larvae hatch out in the temporary nests of the rabbits where the young rabbits are available as food for the adult fleas as they emerge (Rothschild 1965a, b; Rothschild and Ford 1973; Rothschild et al. 1970).

Fleas of Nocturnal and Ground-Dwelling Hosts

Mammals that are nocturnal and live at ground level tend to have well-combed fleas, like *Ctenocephalides* on canids and felids, and some *Nearctopsylla* on mustelids, especially when the hosts are wide-ranging carnivorans. As illustrated in Figure 8.9, versus Figures 8.10 and 8.11, the *Ctenocephalides** on Carnivora have conspicuously more spines in the genal comb than do those on hyrax. *Caenopsylla,* which has two combs including a small genal one, is found on foxes as well as on macroscelids and the gundi. The species on *Ctenodactylus* has highly developed tibial combs, and it is pertinent that gundis are notorious for their ability to rapidly climb sheer, almost vertical, rock faces. Lodgement is critical on such hosts. However, what about combless fleas that are found on hosts with a vast territory? Consider the vermipsyllids which even lack vestigial spines or spinelets and yet parasitize carnivorans and ungulates. In the species on foxes and weasels, there are supernumerary bristles on the caudal margin of the head and on other strategic places on the body, and the long bristles are arranged in unusually straight, vertical rows so that they are pectinate. Vermipsyllids are also specialized for their particular mode of life in other respects (Lewis 1971; Traub 1972, 1980b). The ancistropsyllids,* on cervids, are unique in possessing hornlike spiniforms on the gena. Most species of *Pulex* infest porcines, deer, and carnivorans, and their ability to feed on such a variety of hosts helps compensate for the absence of combs. (Some of the combless fleas with large eyes infest diurnal hosts.)

Fleas of Nocturnal*/Scandent Hosts

Hosts that are both crepuscular/nocturnal and climbing (scansorial or arboreal) provide the most evidence concerning the setal response to selective pressure. Not only do all the fleas of such hosts show one or more adaptive modifications, but also a gradation in the hyperdevelopment of their bristles and spines corresponds to the degree of risk imposed upon them by their hosts. At the bottom of the scale among fleas of nocturnal/scansorial hosts are species in which the existing structures have undergone a slight change, such as the addition of a few spines to the pronotal comb (20–24 spines instead of 16–18), with a tendency for them to encroach upon the link-plate and, in the case of fleas on nonmurines, to slope somewhat downward. Examples of such fleas include the Nearctic ceratophyllids: *Pleochaetis ponsi* Barrera on certain tree-climbing *Peromyscus*, and *P. smiti* Johnson and *Kohlsia keenani* Tipton and Mendez on some *Oryzomys* and other cricetids; and the pygiopsyllids: *Lentistivalius vomerus* Traub on tree shrews and native rats in Borneo and some *Stivalius s. str.* and *Aviostivalius klossi* (Jordan and Rothschild) on climbing *Rattus* in the Asiatic/Pacific region. Here the hosts are only somewhat scansorial and are mainly crepuscular or nocturnal. The *Neopsylla* with tibial combs, which occur on rats with these characteristics, are in this general category. The ultraspecific fleas (*Astivalius* and some *Metastivalius*) of the murid *Lorentzimys*,* which is primarily arboreal and nests in *Pandanus* trees, are at a somewhat higher level. These fleas have a similar type of pronotal comb but in addition have tibial combs and a reduced eye. They resemble one another closely, regardless of generic assignment. The fleas of flying squirrels,* flying phalangers,* and dormice* likewise have more spines in the pronotal comb than do their close relatives on diurnal or terrestrial sciurids, even in the case of *Macrostylophora fimbriata* (Jordan and Rothschild) and *Monopsyllus argus* (Rothschild), which presumably are relatively recent converts to petauristines, but which also show a degree of reduction of the pleural arch (Traub 1972b).

As expected under this hypothesis, the ceratophyllid fleas of nocturnal*/scansorial carnivorans, which face additional perils posed by the large territory covered by the hosts, have supernumerary spines in the pronotal comb: for example, *Ceratophyllus lunatus* on martens and *Paraceras sauteri* (Rothschild) and *P. pendleburyi* Jordan on the ferret badger (which climbs trees). Interestingly enough, the tendency toward chaetotactic hyperdevelopment in such fleas is even seen in combless taxa. The *Chaetopsylla* of martens have more bristles in the subpectinate rows (back of the head, etc.) than do their congeners on terrestrial carnivorans. *Scolopsyllus*, the only rhopalopsyllid with a pronotal comb, may be an excellent case in point, for the ctenidium consists of modified bristles, not true spines (Fig. 8.120). If these concepts are correct, its primary host will turn out to be a nocturnal scandent mammal.

Fleas of nocturnal arboreal or semiarboreal hosts such as some *Rattus* possess 26 or more spines in the pronotal comb, and it extends down over the vinculum, as in certain *Stivalius s. str.* In the recently described, mono-typic *Typhlomyopsyllus* (Li and Huang 1981), the comb also covers the vin-culum. This eyeless leptopsyllid from China is a parasite of *Typhlomys cinereus* Milne-Edwards, the unique or "blind" dormouse (Platacanthomy-idae) whose habits have not yet been reported. The type of pronotal comb and other features of this flea, including the absence of an eye and the possession of a displaced third pair of plantar setae, lead me to believe that the host is nocturnal and scansorial or semiarboreal, like the other member of the family, *Platacanthomys* (whose fleas are unknown). The trend toward multiple spines and downward extension of the pronotal comb, supple-mented by the development of a crown of thorns and, at times, by an accentuated downward slope of the pronotal comb, is seen in pygiopsyl-lids whose hosts spend a great deal of time aloft in trees: for example, *Muesebeckella* (ex arboreal phalangers), where the frons also is leathery and otherwise specialized (Traub 1969); *Idiochaetis* on giant rats; *Smitella* on a unique kind of *Melomys; Acanthopsylla* on *Antechinus* and *Melomys*. Both a pronotal and vertical genal comb are found in some fleas in this category like *Sigmactenus** (Fig. 8.12) and the chimaeropsyllid *Epirimia* (Fig. 8.14), both of which also have well-developed combs of spiniforms on the tibiae. In *Leptopsylla* (Fig. 8.13) and allies, and *Acanthopsylla* there are frontomar-ginal spiniforms as well. Since the murid host of *Smitella** has a shaggy fur in addition to being scandent or semiarboreal, it is no surprise that this flea has also developed a helmet with false spines (Traub 1968a, 1980b).

In my opinion, a pronotal comb that contains groups of spines whose axes are clearly divergent, as when the upper spines are horizontal and the lower spines are inclined obliquely ventrad (Fig. 8.13), serves in a dual capacity. The horizontal spines operate when the flea is moving or resting, and the lower, oblique spines function when the flea assumes an angle in the feeding position. Many murid fleas have spines that are almost uni-formly horizontal and subparallel (Figs. 8.12, 8.26), but the "dual-purpose" type is seen in those at greatest risk, such as *Idiochaetis, Smitella, Macropsylla* (Fig. 8.45) and *Epirimia* (Fig. 8.14), and as in fleas of tree marsupials (e.g., *Muesebeckella*) and coarse-furred bandicoots (Fig. 8.68).

Well-developed, vertical genal combs are seen in many fleas whose hosts are nocturnal and in scansorial/semiarboreal murids, such as in *Sig-mactenus* (Fig. 8.12) (Asiatic-Pacific islands), *Epirimia* (Fig. 8.14) (Africa), while in *Leptopsylla* (Fig. 8.13) and allies (Asia and Africa), there are also frontomarginal spiniforms. All of these fleas have very large pronotal combs. Not only does *Dinopsyllus* correspond in general regarding these combs (Figs. 8.16, 8.26) and the traits of the host, but also in the case of *D. kempi* Jordan and Rothschild, which is an ultraspecific flea of the dormouse *Graphiurus*, an arboreal host, the pronotal comb is even better developed than in the murid *Dinopsyllus*.

Many of the fleas of nocturnal, climbing, or arboreal hosts have abdominal combs of true spines, particularly among the Pygiopsyllidae and Stephanocircidae, which are primitive families. Here again there is a correlation between the presence or degree of development of the supernumerary* combs and the extent to which the host lives in trees. A comb is present on tergum 2 in *Migrastivalius* (*Gryphopsylla*) (monotypic) and *M.* (*Migrastivalius*) (three species). Both occur on noctural hosts; the former (Figs. 8.35, 8.37, 8.38) on spinose, scansorial rats of the genus *Lenothrix* and *M.* (*M.*) *jacobsoni*,* for example, on *Chiropodomys, Pithecheir*, and other scandent or semiarboreal rats. *Lentistivalius* (*Destivalius*) *mjoebergi* (Jordan) has short combs (comblets) on terga 2 and 3 and is a characteristic flea of the Bornean tree shrew, *Tupaia baluensis*, which spends most of its time in trees. Some *Afristivalius* have three abdominal comblets, and the South American *Ctenidiosomus* also has several; the hosts of these fleas climb trees regularly.

Some of the pygiopsyllids of nocturnal scansorial/arboreal hosts have another set of pertinent, unusual characters that are reported here for the first time. In all of the *Migrastivalius*,* in *Acanthopsylla richardsoni* Smit, and in *Stivalius s. str.*, the sclerotized inner tube (S.I.T.) of the aedeagus is remarkably elongate and is believed to function as an intromittent organ (Traub 1972a, 1980a), whereas in the majority of fleas (i.e., at least in the other families), it is the penis rods, not the S.I.T., that penetrate deeply into the bursa copulatrix of the female (Holland 1955). In *Migrastivalius* the female has the bursa copulatrix correspondingly modified, greatly enlarged and thickened to accommodate the S.I.T., and the male has unique or exceptional holdfast mechanisms to more than compensate for the loss of the customary crochet processes: fanglike extensions of the hood in *Migrastivalius*, the very large, hooklike Ford's sclerite in *Stivalius*, and so forth. In my opinion, these devices heighten the likelihood that the pair of fleas remain together on the host while in copula: that is, the female is securely held by the male at a time when she would otherwise be particularly susceptible to dislodgement. The selective value of the modifications seems apparent, since (1) the elongate S.I.T. occurs only in fleas whose hosts place them at exceptional risk, (2) they are examples of convergence, since *Acanthopsylla* is not closely related to the other genera, and (3) these species mate on the host (Traub, unpublished observations). Furthermore, in *L.* (*D.*) *mjoebergi*, the S.I.T. and the other structures are only partially modified along these lines, although the tendency thereto is evident, and this species is under less survival pressure than are the others.

In the Stephanocircidae the head is extremely modified (Figs. 8.20, 8.22, 8.24) to form the so-called helmet, which is greatly narrowed and bears a full tiara of true spines so that the flea can readily latch on to the hairs of the host (Traub and Dunnet 1973; Traub 1980b). In addition they have full genal and pronotal combs and other specialized features to facilitate remaining on the hosts. At least in Australia, the shape of the spines indicates whether the infested mammal is a rodent (murid) or a bandicoot* marsupial. Insufficient data are available to permit discussion of the South American helmet fleas other than to point out that scansorial hosts are often represented. In Australia all of these fleas are under selection pressure regarding vestiture and habits of the hosts (Traub and Dunnet 1973). In the stephanocircids, the species on nocturnal arboreal hosts are clearly the best armed regarding numbers of spines and combs. For example, *Stephanocircus domrowi* Traub and Dunnet, the flea of Leadbeater's opossum, has vestigial combs on most of the terga; the pronotum extends well down over the link-plate; and there are more spines in the various combs than on the *Stephanocircus* of murids and ground-dwelling marsupials. *Coronapsylla jarvisi* (Rothschild) (Fig. 8.114) surpasses *S. domrowi* in all of these respects and in addition has a full comb on terga 1 and 2 and a comblet on tergum 3, as well as pronounced tibial combs. This is presumably a flea of *Antechinus*, which is an active climber in the forest.

The macropsyllids are related to the stephanocircids and are generally regarded as helmet fleas in which the helmet became fused with the genal comb (Hopkins and Rothschild 1956; Traub 1980b). The constituent two genera are monotypic, and *Macropsylla** is one of the most heavily combed and bristly fleas in the order (Figs. 8.45–8.49). There are supernumerary bristles on the tibiae, the protibia is subpectinate, and the dorsomarginals on the others are exceptionally long. With these traits, it is not surprising that *Macropsylla* infests a number of scansorial hosts like certain native *Rattus*, *Pseudomys*, and *Melomys*, and the climbing marsupial *Antechinus*. However, *M. hercules* is commonly found on other native rats that are mainly terrestrial.

Fleas in Unusually Cold Climes

To survive, fleas whose hosts live in subarctic or arctic climes must either be active as adults in the warm parts of the year or else remain on the body of the host or in its nest and burrows. It is expected that pikas (*Ochotona*) living in alpine terrain high in the mountains have leptopsyllid fleas which possess a crown of thorns, a large pronotal ctenidium, and false tibial combs, as in some *Ctenophyllus* (*s. lat.*). In another ochotonid flea, the ceratophyllid *Amphalius*, the pronotal comb has so many narrow spines that it superficially resembles that of a bird flea, and the component species are clothed with long bristles on the body and legs. *Amphalius* is also unusual in possessing highly modified genitalia (i.e., a stout bursa copulatrix and a heavily tanned fistula that probably serves as an intromittent organ) which must provide an especially strong grip in copulation, suggesting (in function) devices seen in *Migrastivalius** and certain other fleas at unusual risk. The basic need to reinforce mechanisms favoring host retention also occurs in fleas that are pronouncedly seasonal. Most species of *Hystrichopsylla** and *Stenoponia* are fleas limited to cool or cold climes, areas, or times. *Stenoponia*, which includes montane, mesic species as well as xerophilic desert forms, has three well-developed combs. Some *Hystrichopsylla* have even larger genal and pronotal combs, and abdominal combs or comblets as well. Both taxa are densely clothed with long bristles and tend to be extrafamilial or extraordinal regarding host specificity, as are a few of the species—another asset for fleas that are handicapped by temperature limitations.

The ceratophyllid *Oropsylla alaskensis* (Baker) is a good example of the hyperdevelopment of spines apparently resulting from selection pressure to remain on the host in adverse climates. This is a Holarctic flea of ground squirrels, which in North America, at least, are limited to tundra of the subarctic or arctic areas. In the southern parts of its range, *O. alaskensis* has about 22–24 spines in the pronotal comb, a total that exceeds by two to six spines the number found on the other, more southern, species of *Oropsylla* in both Asia and North America. Moreover, near Barrow, the most northern point in Alaska, the *O. alaskensis* frequently have combs with 26 spines, approximating the number often ascribed to "bird fleas." The far northern specimens are also more bristly than the southern representatives of the genus.

Bat Fleas

Bats are erratic and swift fliers, often covering large areas in flight. If hyperdevelopment of spines, combs, and bristles is associated with the hazards posed to the flea by the host, then bat fleas should be the most affected. This is obviously the case, since the Ischnopsyllidae, all of which infest bats, represent the acme concerning the development of true and false combs within the order. There are sticktight fleas that parasitize bats, namely, some hectopsyllids and pulicids, but they are not pertinent to this discussion since they have evolved other means to solve the problems of infesting bats: remaining attached by anchoring mouthparts for long periods, distension of the abdomen to permit gross engorgement, and production of huge numbers of eggs. All ischnopsyllids have two or three ventral spines on the front of the head; a well-developed pronotal comb which frequently consists of many close-set narrow spines and extends down to, or over, the vinculum; a vestigial eye; and they lack a pleural arch (Fig. 8.113). Many species have one or more combs on the abdomen, and some even have a true ctenidium on the metanotum and on the fourth to sixth abdominal terga (e.g., *Hormopsylla* and some *Ischnopsyllus*). A false comb of spiniform bristles, fully simulating a regular comb, may occur on the seventh tergum, replacing antepygidial bristles (Figs. 8.115, 8.116). The dorsalmost bristles on the metanotum and some of the abdominal terga may be pectinate in arrangement (Fig. 8.119). Tibial combs are present at times, as in the thaumapsyllines, in which the pronotal comb is unique in the extent to which it overlaps the propleuron.

In all probability, when we know more about bat fleas and their hosts, we will be able to associate the degree of hyperdevelopment of spines and combs in ischnopsyllids with the relative amount of selection pressure exerted by specific kinds of bats. For example, the highly modified *Thaumapsylla* are found on bats that fly for miles to feed on a particular fruit that is in season. At the other extreme, certain bats seem to lack specific fleas altogether, perhaps because the microenvironments in their haunts are unsuitable for larval development, or because the bats are so solitary in habit and scarce in nature that there is little chance for fleas to find them. Among the bats that do have fleas, there are various forms of selection pressure: leaving a site of hibernation in winter to fly to another, as in *Pipistrellus*; or inhabiting small crevices in rocks or tombs in desert areas, as does the host of the *Chiropteropsylla* which has an extraordinary false comb of spiniforms on the metepimere (Fig. 8.118). The seasonal cycles of certain bat fleas, whereby the adults emerge in numbers at the time the baby bats appear, suggest that the breeding activity of these fleas may be bound to the hormones of the host. If so, that would be another factor leading to the natural selection of structures aiding a flea's attachment to the host.

Fleas of Spiny* Hosts

If my ideas are correct concerning the functions of combs, spines, and bristles, then spiny mammals, by virtue of their vestiture, present special problems to fleas. The dearth of fleas that parasitize such hosts indicate that this is indeed the case (Traub 1980b). For example, the New World porcupine and the *Perognathus* with spinose fur lack specific fleas, and some spiny hosts that are infested have fleas only in a limited portion of their range. The fleas that have managed to parasitize hosts like hedgehogs and echidnas all have highly modified spines and bristles that reflect the influence of the environment provided by the spines of the host and serve as further evidence of coevolution. Among Siphonaptera the spines of the pronotal comb are often tailored to fit the hairs of the host, and the space between the apices of two adjacent spines closely corresponds to the diameter of some of the major hairs (Humphries 1966, 1967; Traub 1972b; Traub and Dunnet 1973). Bandicoot fleas (Fig. 8.68), for instance, tend to have sharply pointed pronotal spines (Holland 1969; Traub 1968a, 1972b), and these are divergent, with the gap fitting the coarse hairs of the host (Traub 1972b, 1980b). Not only are the pronotal spines of the *Stephanocircus* of bandicoots stilettolike, but also so are those of the helmet comb and genal comb (Fig. 8.20). In the helmet fleas of murids, all of these spines are parallel sided and bluntly rounded at the tip and are fairly closely set, thereby corresponding with the fine hairs of the host (Traub and Dunnet 1973). According to these principles, a flea infesting a porcupine or echidna should possess pronotal and other spines that diverge even more from a broad base than do the spines of bandicoot fleas, because they would have to lock onto much stouter hairs, that is, spines or quills. Obviously if such a flea possessed a full pronotal comb of divergent spines of regular length, the comb could not function because some of the spines would simply overlap. Such factors seem to have caused the fleas of hedgehogs, tenrecs, and echidnas to lose their combs, or to have led to a striking reduction in the number and size of the spines. The characteristic pronotal comb in such fleas consists of short dorsal or subdorsal spines that are broad at the base and apically pointed (Holland 1964; Traub 1972b, 1980b). The genal comb, if present, is similarly reduced and modified, as in *Archaeopsylla** and *Centetipsylla*. In the echidna flea, *Bradiopsylla*, there are only four to six spines in the pronotal comb, and these are short, pale, and dorsal in position (Figs. 8.103, 8.104). In some individuals there is a tiny spine along the antennal groove near the remnant of the eye. This is believed to represent a vestigial* genal comb (Jordan 1947). In the combless fleas of spiny hosts, some of the *bristles* of the body are greatly modified and are spinelike in appearance and are set wide apart to compensate for the girth of the hosts' spines, as in *Pariodontis* and *Parapulex*. In the key sites on the body of these fleas, these specialized bristles are arranged in short vertical rows so that they can ostensibly act like false combs, and this is also true in the tenrec flea *Centetipsylla*, which has reduced combs (Traub 1980c). The abdominal bristles are also widely spaced in *Archaeopsylla.** The Old World porcupine fleas (*Pariodontis*) have a hooklike genal process which presumably serves to grasp a spine if the claws or teeth of the host attempt to pull the flea backwards.

In *B. echidnae* the metanotum is dorsally clothed with stubby, spinose bristles anterior to the caudal fringe of long bristles (Fig. 8.104). In other fleas the thoracic notal bristles are merely shorter versions of the caudal row (Fig. 8.46). It is easy to envisage how the *Bradiopsylla* metanotum can serve as a defense against backward displacement on a spiny host, in contrast to the notal vestiture of ordinary fleas. Also, the hind tibiae of this group of fleas are unusual. In the tenrec flea (Fig. 8.106), the hedgehog flea *Archaeopsylla* (Fig. 8.107), the *Acomys* flea *Parapulex echinatus* Smit (Fig. 8.108), and the porcupine flea *Pariodontis* (Fig. 8.109) the dorsal margin is conspicuously incised at the notches where the dorsomarginal bristles are inserted. It therefore seems likely that by means of this cleft the fleas can latch onto the spines of the host under certain conditions, since the gap is notably smaller in fleas on hosts with normal vestiture, as shown in Figures 8.62, 8.110, and 8.111. In *B. echidnae* (Fig. 8.66) the sinus at the insertion of the dorsomarginals is not as pronounced as in the other fleas of spiny hosts, but the notch is more mesal in position and is further removed from the lateral margin of the tibiae so there is a gap that may be able to function regarding the hosts' spines. Moreover, the smaller member of each pair of dorsomarginals is proportionately larger than in most other fleas, but the significance of this is unknown. Another unfathomable but possibly pertinent oddity is the straight, relatively long and narrow tibia in *Pariodontis* (Fig. 8.109).

An intriguing point concerns whether the fleas with specialized spines, combs, and bristles coevolved with their hosts as the mammals started to develop modified features such as fossorial habits and habitus, or whether the fleas adapted to these hosts *after* the mammals had already become spiny, or arboreal, or volant. Certainly there can be no question that these fleas had been living on those hosts for long periods, for the structural changes exhibited by the fleas are not only clearly adaptive but also often striking or complex (e.g., the crowns of thorns). The fleas of spiny hosts are pertinent here. *Pariodontis* and *Bradiopsylla*, or their ancestors, already may have been associated with their porcupine and echidna hosts while the mammals were still in the process of developing spines. The other fleas of spiny hosts may have switched after these mammals had already developed spines. For example, *Parapulex*, *Archaeopsylla*, and *Centetipsylla* (all pulicids) resemble fairly closely the other members of their respective subfamilies (on other hosts) except for the adaptive features such as spinose bristles. *Synosternus pallidus*, another pulicid, which infests hedgehogs in certain areas, is similar to its congener on gerbils except that its thoracic and abdominal notal bristles are conspicuously spaced further apart to fit the diameter of the spines of the host (Traub 1980b). More to the point, some of the adaptive features are highly variable and differ in number or intensity from individual to individual. Some specimens of *A. erinacei* (Bouche) have no pronotal spines and others have as many as a total of nine, with the majority possessing around four or five. In this species, the small genal comb is also variable, ranging from one to three spines per side. Such instability in a somatic character suggests a recent evolutionary change that is still in a state of flux, as is often seen regarding vestigial structures. *Pulex irritans*, which is generally viewed as a typical combless flea, occasionally has a genal spine, albeit it is short, pale, and difficult to see. *Pulex porcinus* may have three such vestigial genal spinelets. The hystrichopsyllid *Agastopsylla* is essentially a *Ctenophthalmus* that is en route to becoming a nest flea, and its genal spines are short, only lightly tanned, and variable in number, at times differing on the opposite sides of an individual (Traub 1952, 1980c). Also, in some specimens of *Archaeopsylla* and *Parapulex* the cleft by the dorsomarginal pairs of bristles on the hind tibia is much larger than those depicted in Figs. 8.107 and 8.108. This instability is again suggestive of evolutionary change, but this time in a rudimentary structure, not a vestigial one.

On the other hand, not only does *Pariodontis* lack close relatives but also it is more specialized in its modifications than are the other pulicids of spiny hosts. The unique, hooklike genal lobe and comblike groups of subspiniforms on various parts of its body are quite remarkable. The two species in the genus are also highly distinctive in possessing a comblike set of two or three stout bristles on the ventral portion of the mid- and hind-coxae, above the insertion of the trochanters, suggestive of the pectinate groups on other parts of the body. The association between *Pariodontis** and hystricids is most likely ancient. The same may be true for *Bradiopsylla* and the echidna. This monotypic genus was placed in a distinctive tribe in the Lycopsyllinae by Mardon (1978), along with some other tribes of Australian fleas that are generally regarded as being primitive. *Bradiopsylla* is greatly specialized in features that I believe are adaptive to life on the echidna, but there has been no evidence cited that the taxon or its forebears actually arose with the monotremes, whose origin, in my opinion, must have been a far earlier occurrence.

Survival Pressure on Combless Fleas

Although the vast majority of the combless fleas are active, presumably rapid-feeding fur fleas, true nest fleas, or hypodermal or sticktight fleas, there are a few others that are free moving and subject to selective pressure to remain on the host. The last type of combless flea should exhibit modified chaetotaxy if the hypothesis advanced here is correct; this is apparently the case. Such adaptive responses have already been discussed for some *Chaetopsylla* and ancistropsyllids* and for the fleas of spiny* hosts. The pulicids* of fossorial hosts have been shown to have supernumerary, relatively close-knit bristles. There are additional chaetotactic and structural responses in fleas without ctenidia, particularly on the legs (cf. Figs. 8.93–8.102).

Smit (1972) has emphasized how the tarsal claws (ungues) and the plantar bristles on the fifth segment are used for clinging to the hairs of the host and has shown how their degree of development is correlated with the need for maintaining a hold. Although Smit stated that *Malacopsylla grossiventris* has legs that are among the most powerful in fleas, he restricted his valuable discussion to the distal tarsal segment, pointing out that this combless armadillo flea is found only on the venter of its host and must cope with the problem of being brushed against the substrate. However, I believe that the huge apical spiniform on the forefemur and the stout spinelike dorsomarginal bristles on the tibia of the foreleg (Fig. 8.91) likewise function as attachment devices to forestall dislodgement. The hindleg (Fig. 8.92) is modified in other ways to achieve this: the very long tarsal bristles; the shape of the tibia resembles that seen in fleas of spiny hosts (Fig. 8.109). All armadillos are covered with horny plates at least dorsally, so that in most instances only the venter is readily available as an environment for fleas. In some dasypodids, hairs project between the scutes, but it is not known if fleas attach there. The only other true armadillo flea reported is the related *Phthiropsylla agenoris* (Rothschild) (Figs. 8.94, 8.95,

8.97), which also has highly modified ungues, plantar bristles, tibial and femoral bristles, although only the forefemur has a large spiniform. This species has a greatly reduced pronotal comb.

In the Vermipsyllidae, a combless family of fleas on wide-ranging carnivores or ungulates, the legs are also very heavily bristled and otherwise specialized (Figs. 8.99–8.102), with unusually well-developed lateral plantar bristles, plantar microsetae, a full row of dorsomarginal and subventral bristles on the femora, and in some instances, very long tarsal bristles. Similarly, the Coptopsyllidae, which parasitize desert rodents and which also lack combs, have strikingly stout, short subspiniform bristles on the protibia. The legs and general chaetotaxy of a combless fur flea of the ordinary type, as exemplified by *Polygenis** (Figs. 8.96, 8.98, 8.112) are in marked contrast to the fleas under special selection pressure.

DISCUSSION

It now appears from the preceding data that most groups of Siphonaptera have had a long-standing and intimate association with their particular hosts. In some instances it seems that the fleas and these mammals were connascent. In others the relationship has been of shorter duration but of sufficient length to permit the fleas to develop clear-cut adaptive modifications.

Host Relationships

Several features of the host distribution of ultraspecific species of fleas merit emphasis and further comment. Not only are such fleas found essentially only on those mammals that lack close relatives or else are unique in their homelands, but also their hosts are generally of a specialized nature. At least 103 of the 122 (84%) hosts listed in Tables 8.1–8.4 are unusual in that they are fossorial, spiny, nocturnal/arboreal, gliding or volant, or else in that they live in restrictive microhabitats. In other words, they are in the category that places their fleas at extra risk for survival. Some of the most abundant species and groups of mammals that are infested with fleas completely lack fleas that are ultraspecific. Examples include opossums and ground-dwelling marsupials, *Peromyscus* and *Reithrodontomys*, *Sigmodon*, *Cricetulus*, and most gerbillines. In general, ultraspecific fleas are also conspicuously absent on the more recently evolved, highly adaptable and successful, rapidly diversifying and morphologically unspecialized groups like *Rattus* (*Rattus*)* and *Oryzomys*. These hosts are also relatively generalized in vesture and habits. Certain other mammals, such as the tree squirrels and ground squirrels, are seldom parasitized by ultraspecific fleas, but when they are, the siphonapterans frequently are nest fleas like *Tamiophila* rather than fur fleas like *Monopsyllus*. Moreover, the ultraspecific fleas tend to be morphologically modified to a greater degree than are the fleas of *Arvicanthis* and other relatively unspecialized hosts. Eighty-eight percent (204) of the 233 fleas listed in Tables 8.1–8.4 are more specialized than even their kin regarding such features as the size of the eye, the state of the pleural arch, and chaetotaxy, if the latter species are less specific. For example, *Xenopsylla magdalinae,* an ultraspecific flea of the fossorial *Ellobius talpinus* Pallas, which belongs to a combless and otherwise unspectacular genus, differs considerably from the rat flea *X. vexabilis* Jordan: the eye is smaller; the thoracic and abdominal bristles are longer and thinner; more spiniforms are present on the inner side of the hindcoxa; and more very long bristles are present on the hindtarsus. As in the case of many of the sciurid-infesting ceratophyllids like *Orchopeas caedens* (Jordan), there are no noticeable differences from what is seen in the less specific species in *Orchopeas.* Such fleas have not been considered as "specially modified" and hence are included in the minority of 12% (along with some species not available for study). Nevertheless, the unmodified fleas actually may differ greatly regarding taxonomic characters. Only adaptive features have been considered here.

Another noteworthy point is that the highly modified fleas constitute a small part of the siphonapteran fauna. For example, only 11% of the genera have species with crowns of thorns, and in any one area a host is generally parasitized by only one such species (Traub 1980b).

Genera and subgenera with ultraspecific or generispecific fleas occur among all the major siphonapteran families infesting rodents and the other groups of mammals. Although there is great variation among the taxa as to numbers and kinds of hosts so infested, actually this is a reflection of the overall pattern of host relationships exhibited by the sundry taxa. There are only two groups of hosts (Sciurinae and Muridae) listed for Pygiopsyllidae in Table 8.12, but because of the geographic distribution of the family, these are the only rodents with which these fleas come into real contact (excluding South America, where we lack adequate data). Moreover, while only one g./sg. of pygiopsyllid is shown as highly specific on Sciurinae, there are only two pygiopsyllid genera, *Medwayella* and *Farhangia* that include purely squirrel fleas. Too little is known about *Farhangia* to permit discussion. The important generalization is that wherever a group of fleas is fairly extensively associated with a group of hosts, a high degree of specificity occurs, provided the hosts are distinctive in some respect: for example, they may possess unusual habits or vesture, or they may be restricted in geographic or ecological distribution.

The data also clearly indicate that the vast majority of species of mammal fleas are highly specific: that is, they are restricted to infestation of a single species of host (ultraspecific) or else to closely related taxa (generispecific). All but a small proportion are limited to a single order or even family of mammal. The true shrew fleas parasitize no other host, and the mole fleas only infest moles. Not only are rodent fleas almost invariably found on rodents (and are even seldom collected on the predators), but also pocket gopher fleas and *Aplondontia* fleas are limited to their respective hosts. Sciurid fleas almost always infest squirrels. Since we are necessarily often dealing with species groups in the tables of generispecific fleas, the actual numbers of species in that category is far greater than the figures suggest, that is, some groups include 3–10 species of fleas.

The high degree of specificity among the fleas of Insectivora and Marsupialia is strikingly manifest (Tables 8.1, 8.5, 8.10, 8.11, 8.13). There are 55 species or species groups of insectivoran fleas cited in the tables, and 45 are either ultraspecific (i.e., 26 species = 47%) or generispecific (i.e., 19 species = 35%). Moreover, 46 (84%) of these are wholly restricted to the Insectivora. The remaining nine (16%) are basically rodent fleas that have accommodated themselves to shrews and tenrecs living in their biotope. Of the 39 species or species groups listed as infesting marsupials, 30 (77%) are ultraspecific or generispecific, one is extrafamilial, and eight (21%) are extraordinal (but are limited to two orders).

The lagomorph fleas are even more specific in their host relationships (Tables 8.1, 8.5, 8.10, 8.11, 8.13). There are 24 species or species groups listed, and 22 of these (92%) are restricted to species or closely allied taxa of lagomorphs. The two exceptions (*Malaraeus bitterrootensis* (Dunn) and the group including *Paradoxopsyllus integer* Ioff) infest both *Ochotona* and a rodent. Even more remarkable is the consistency with which the lagomorph fleas parasitize only one suborder of host. No species, or even any genus, is found both on pika and on leporids even though in Holarctica there are many areas where the two coexist. Such host specificity is in accord with the complex hormonal interrelationships seen in the rabbit fleas. Fifteen (68%) of the 22 entries for fleas of carnivorans are either ultraspecific or generispecific, two (species of *Paraceras*) are extrafamilial and five (23%) are extraordinal, parasitizing both predator and prey. It is impressive that so few fleas have adapted to both of these kinds of hosts even though carnivorans often come into contact with flea-infested burrows and nests when seeking food. The fleas of 376 species/species groups are rodent fleas, and of these, 143 (38%) are ultraspecific, and the same number are generispecific (Tables 8.1–8.3, 8.5, 8.10, 8.12, 8.13). Unlike the fleas of the other orders, about 24% of the rodent fleas are less specific; of these 95 taxa, 50 (13% of the total on rodents) are found only within a family of rodents; 14 (4%) are extrafamilial; and 26 (7%) are found on hosts of another order as well as the rodents. It is clear, however, that only a small proportion of fleas in general have a host range that extends to two or more families.

The fleas that have adapted to new hosts provide additional evidence for coevolution. Such transfers have occurred when opportunistic species have switched to other hosts that either already existed in their habitat or else entered their area after dispersal from elsewhere. For example, *Pleochaetis dolens* not only parasitizes tree-climbing cricetids, but also infests local tree squirrels. Fleas like *Ctenocephalides connatus* (Jordan) and *Monopsyllus vison* (Baker) may infest both their regular rodent hosts and the carnivorans that prey upon those mammals. Other fleas, although clearly descended from rodent parasites, as in the case of *Paraceras sauteri*, or from insectivoran fleas, as with *Nearctopsylla brooksi*, are specific fleas of carnivorans. The stephanocircids on murids in Australia and on cricetids in South America illustrate the transfer to emigrant mammals, as does the massive radiation of pygiopsyllids on murids in New Guinea. The ancestral hosts for both these taxa were marsupials (Traub 1972c, 1980c). Notably, in such transfers the secondary association has generally been extensive or pervasive enough, and of sufficiently long duration, for the fleas to have developed adaptive features such as crowns of thorns, supernumerary spines, and the like. Even the occasional species of *Metastivalius* and *Papuapsylla* (Pygiopsyllidae) that have departed murids to revert to bandicoot* marsupials bear the distinctive stiletto pronotal spines typical of fleas living in the coarse bristly fur of bandicoots (Traub 1980b).

The data on host relationships further support the idea of coevolution. The high degree of specificity to marsupials exhibited by certain doratopsylline, stephanocircid, and pygiopsyllid fleas, as well as the observations on zoogeography, strongly intimate that these fleas and hosts have been associated for untold eons. Since the majority of fleas of all kinds are highly specific (ultraspecific or generispecific), and since these fleas, and their hosts, tend to be specialized or distinctive, they probably evolved in close association. In the process, the fleas adapted to the changes wrought in their environment by their hosts. Some of these fleas parasitize hosts that belong to relatively primitive orders such as the marsupials and insectivorans, but many of the mammal hosts involved are murids, widely recognized as one of the most recently evolved groups. The Siphonaptera, then, do not fully fit the concept that Brooks (1979) referred to as Manter's second rule, namely, that the longer the existence of the relationship between host and parasite groups, the greater the degree of specificity exhibited by the parasite (Manter 1955, 1966).

The bond between host and flea often transcends the ties of kinship and the opportunities presented by close physical contact. For example, even though the great majority of mole fleas have congeners on shrews, there is only one mole flea (*Palaeopsylla nippon* Jameson and Kumada) that infests shrews as well. Among lagomorph fleas, 29 of 31 are either ultraspecific or generispecific. Although peromyscine and microtine rodents frequently live side by side, there are only three species known to regularly infest both groups of mammals, namely, *Catallagia decipiens* Rothschild, *Epitedia wenmanni* (Rothschild), and *Megarthroglossus divisus* (Baker). The potency of the ties to particular hosts strongly suggests coevolution.

If the basic premise that ctenidia are adaptive in nature is correct, and the combs indeed are often geared to the vestiture or habits of a certain kind of host, then it would be expected that fleas that have a very broad and diverse host range would either have nondescript combs or lack them altogether. The evidence strongly indicates that this is so, that is, such fleas lack structural modifications that point to a specific host. Five of the nine species designated as "indiscriminate" in Table 8.15 are combless pulicids, and the three ceratophyllids have an unmodified pronotal ctenidium. Although *C. felis* has both a genal and pronotal comb, the spines are not noticeably attuned to the fur of any particular host. The strategic need for devices offering protection against dislodgement in fleas of wide-ranging carnivorans* has already been stressed. Fleas of the genus *Coptopsylla*, the sole member of the Coptopsyllidae, are a case in point. These have no trace of spines or combs, and where data are available, the species clearly infest a great variety of desert rodents which tend to be abundant and diurnal in activity. In view of the hazards facing desert* fleas, it is not surprising that the coptopsyllids are specialized in other respects that are pertinent to feeding on, or remaining on, a host with generalized habits. Thus they possess structures like a feathery frontal region; an exceptionally well-developed clypeus; unusually long mouthparts; and supernumerary and/or displaced bristles on the distotarsal segments.

Zoogeography and Evolution

Evolutionary Status of Fleas and Hosts

To a significant degree, the composition of the flea fauna of any particular group of hosts depends on (1) the provenance of the family of fleas and (2) the geological time of their origin (Traub 1972c, d, 1980c, 1983; Traub and Rothschild 1983). Also, it seems that fleas tend to continue parasitizing the groups with which they evolved (Fain 1965; Traub 1980c), and while they at times may shift to new kinds of hosts that are higher on the phylogenetic scale, only rarely do they (or other ectoparasites) adapt to mammals that are more primitive or of greater antiquity than their own hosts (Traub 1980c). Because of their Nearctic nativity and relatively late origin (Oligocene), ceratophyllids are sparsely represented (only one species) on Insectivora, and are absent on ancient and/or austral groups such as marsupials, hyracoids, lagomorphs, and rodents like aplodontids, rhizomyids, spalacids, dipodids, zapodids, ctenodactylids, and heteromyids (Traub 1983; Traub and Rothschild 1983). In contrast, the hystrichopsyllids, pulicids, and stephanocircids evolved at a much earlier date (at least the Cretaceous) and are either austral (the stephanocircids) or occur in both the southern and northern hemispheres.

Comparison of the evolutionary status of sundry fleas with that of their hosts also supports the idea of coevolution since there is a definite correlation between fleas and their hosts for phylogenetic advancement. The oldest and most primitive groups of hosts have the most elemental types of fleas, and the more recently evolved or highly specialized groups of fleas occur on equally advanced types of mammals (Traub 1980c). However, I reiterate that primitive animals can be highly specialized in certain respects for their particular mode of life. Moles, for example, are adapted for a fossorial life, but nevertheless still are insectivorans. The Stephanocircidae and the Acedestiini and Tritopsyllini in the hystrichopsyllid subfamily Doratopsyllinae are among the most ancient and generalized groups of Siphonaptera extant and are primarily associated with marsupials in Latin America and Australia; this distribution suggests continental drift and a geological age of the early Cretaceous or late Jurassic (Traub and Wisseman 1968; Traub 1972c, d; Traub and Dunnet 1973). Holland (1964) also regarded the association of these particular fleas with marsupials as basic to phylogenetic inference. Other doratopsyllines infest shrews, and certain hystrichopsyllids, equally generalized in morphology, characteristically parasitize talpids. These insectivorans date at least to the Lower Oligocene and Eocene respectively (Simpson 1945), but the primitive insectivorans are known from the Cretaceous (Traub 1980c). Both hystrichopsyllids and stephanocircids, then, have components associated with primitive hosts (and as indicated before, these fleas are usually highly specific). Other hystrichopsyllids are ultraspecific or generispecific parasites of the more primitive rodents like *Aplodontia* and the heteromyids. The Ctenophthalminae and Stenoponiinae (Hystrichopsyllidae) have been cited as evidence for south Atlantic* faunal connections (by way of rafting) between Africa and South America in the early Eocene (Traub 1980c).

The Leptopsyllidae are more advanced and more recently evolved than are the Hystrichopsyllidae and Stephanocircidae, but the transatlantic (Europe and North America) leporid fleas suggest that this family dates back to at least mid Eocene, at which time the North Atlantic* route ceased to operate. The leptopsyllids are intermediate in morphological complexity between the more primitive families and the highly evolved Ceratophyllidae. Their hosts have a similar evolutionary status. Ochotonids, leporids, and some rodents like Palaearctic cricetids are the important hosts of the leptopsyllids and there is only one species found on shrews (and very few on macroscelids or erinaceids). There are few or no leptopsyllid fleas on sciurids and the higher groups of rodents (save for some murids), nor are there any on carnivorans or artiodactylids (Traub 1983a). In many ways ceratophyllid fleas are more advanced than those of the other families mentioned previously and are much more recently evolved, but they presumably arose with their own particular hosts, namely sciurids (in North America) in the Oligocene epoch (Traub 1983; Traub & Rothschild 1983).

There is only one ceratophyllid on insectivorans, and none infest the older rodents. Presumably, they early adapted to cricetids and some transferred to murids, which are the most youthful and most highly advanced rodents, much later. The most complex forms of ceratophyllids occur on the most highly evolved taxa of rodents. The most primitive pygiopsyllids are associated with the most primitive marsupials, but more advanced groups are found on the higher families of marsupials. The most recently evolved pygiopsyllids are those on murids or those infesting callosciurine squirrels. Infestation of murids and callosciurines by pygiopsyllids must represent relatively recent events (Traub 1972c, 1980c). The Pulicidae* and Tungidae are ancient groups, judged from the zoogeographic data (Traub 1972d, 1980c). Although, certain pulicids are very host specific and infest phylogenetically old mammals, there has been an evolutionary tendency toward the development of broad host ranges and the capacity to infest the most advanced mammals, including murids and artiodactylids. Such evolutionarily advanced mammals tend to have highly evolved fleas. The ancistropsyllids, one of the few groups of specific artiodactylid fleas, are really modified leptopsyllids, and *Dorcadia* and *Vermipsylla*, parasites of hoofed animals in the mountains of Central Asia, are, in effect, extreme morphological and physiological forms of the vermipsyllids of carnivores.

Historical Biogeography

The geographical distribution of fleas and their mammalian hosts, their interrelationships, and the adaptive modifications of the fleas, all suggest that the fleas are of ancient lineage and, in general, coevolved with their hosts. Marsupials* obviously carried stephanocircid and doratopsylline fleas with them when they penetrated Australia from South America by way of Antarctica (Traub 1980c). These groups of fleas must therefore date back at least 125 million years. The metatherian pygiopsyllids likewise must be extremely ancient, probably dating to the Cretaceous period, since their current distribution, morphology, and taxonomy provide evidence of faunal connections between Australia and South America (Traub 1972c, 1980c). For example, unlike the representatives on murids, the pygiopsyllids of marsupials have diverged considerably at the subfamily and tribal level. The transfer to murids in New Guinea occurred at a later date, and the pygiopsyllids radiated extensively there, along with their hosts, and then presumably accompanied some rats to Australia. The high degree of morphological variation and specialization of the murid pygiopsyllids in New Guinea, their intense host specificity, and their infestation of some of the oldest stocks of rats, suggest a long joint history.

If, indeed, aboriginal hystricomorphs or other rodents carrying ancestral ctenophthalmine fleas (Hystrichopsyllidae) entered South America after rafting from Africa across the South Atlantic* in the Eocene epoch (Traub 1980c), then the *Ctenophthalmus** of Middle America and Africa have roots that are at least 46 million years old. Data on the related genus *Palaeopsylla* support the estimate for the age of the ctenophthalmines, along with an ancient association with Insectivora. The only documented and recognized fossil fleas are two Baltic Amber species of *Palaeopsylla*,* believed to date back to the Eocene epoch, about 35–50 million years ago (Peus 1968; Traub 1980c). There is a wealth of *Palaeopsylla* extant, nearly all infesting shrews and moles. Since *Palaeopsylla* is essentially a modified *Ctenophthalmus*,* the latter genus may be even more ancient (Traub 1980c). However, at present it is difficult to say how long *Ctenophthalmus* has been associated with its current hosts. The shrew-infesting *Ctenophthalmus* in Mexico have the same upturned rounded spines as in the species found on African soricids. This similarity is, however, likely to be due to convergence. The Nearctic soricids are of Asian origin and entered North America by way of Bering connections and are unlikely to have direct African roots (Traub 1980c). However, regardless of how and where the shrews encountered Nearctic *Ctenophthalmus*, this association of flea and soricid must date back many millions of years, since (1) the characters separating these fleas from the microtine *Ctenophthalmus* are distinct at the subgeneric level; (2) taxonomic changes at the higher level proceed slowly in this genus, as in other primitive fleas (Traub 1980c); (3) half (three of six) of the described New World *Ctenophthalmus* are found on shrews; and (4) the remainder are basically *Microtus* fleas. Since no other hosts are involved, their associations seem primary.

Zoogeographical data also suggest the antiquity of host–parasite relationships. The distribution of spilopsylline (Pulicidae) and leptopsyllid rabbit fleas suggest dispersal by way of the North Atlantic prior to the closing of the route in mid Eocene. Accordingly, even the close association of reproductive cycles between the spilopsyllines and rabbits must date back to 50 million years ago, as do morphological features like the patterns of ctenidia in the fleas.

The Pulicidae presumably originated in Africa (Traub 1980c), which is not only its center of development but also the center of dispersal as demonstrated by the representatives in other parts of the world. Also, ultraspecific species of pulicids* are unknown among sciurids and insectivorans—hosts of boreal origin. Marsupials likewise lack ultraspecific pulicids and probably arose in the New World, well after land connections between South America and Africa had been severed. The African provenance of pulicids no doubt helps account for the absence of genera and subgenera with ultraspecific and generispecific fleas of Microtini and Peromyscini, and their dearth on Sciurinae (Table 8.12). The relatively low incidence of such fleas on murids lends support to the idea that the rats did not originate in Africa even though murids are well represented there today. The Pulicidae is the only family besides Hystrichopsyllidae that has native representatives on all the continents. Whereas some of the genera may have entered a continent in relatively recent times, as *Xenopsylla** seems to have done with *Rattus* in Australia (by way of the Pacific islands), other associations must be far more ancient. The pulicine *Echidnophaga* in Australia may have descended from an avian introduction from Africa (where the genus is well known). However, since it has speciated fairly extensively on Australian marsupials, entry from Africa by Antarctic connections cannot be ruled out even in the absence of known fossil marsupials in Africa. Of the 22 genera of Pulicidae, 14 are found in Africa, but one genus (*Pulex*) probably arose in Middle America or South America (Hopla 1980a, Traub 1980c). Jordan (1948) suggested that "cave-dwelling man" acquired the human flea, *P. irritans,* from the fox and badger. This statement seems even more logical today because of its subsequent pre-Columbian record on man in Europe (Hopla 1980a; Traub 1980c). This would still not exclude provenance of the species (or genus) in the New World because there are many examples of holarctic distribution ascribed to Beringian connections.* Regardless, *Pulex irritans** seems to be basically and aboriginally a parasite of the Suoidea (pigs and peccaries). In fact, African suids (e.g., warthogs) are parasitized by at least two other pulicines: the monotypic *Moeopsylla* and a new species of *Neotunga.* Many of the zoogeographic features of the Pulicidae suggest dispersal by austral routes (i.e., the Southern Hemisphere), but the situation is complicated (Traub 1972c, 1980c). Thus, whereas the living peccaries are restricted to the New World, they formerly existed in Europe (Oligocene, Miocene) and Asia (Miocene) as well, but the recently discovered South African representative is the youngest known, dating only to the Pliocene (Hendey 1976). There is no known evidence indicating that suoids have African roots, and it is premature to discuss whether the ancestor of *Pulex* could have accompanied other mammals rafting across the South Atlantic or whether it was derived from leporid pulicines that utilized North Atlantic connections in the mid Eocene epoch. More germane, however, is the obviously long association between pulicids and certain groups of hosts. For example, pulicids of distinctive groups of leporids structurally resemble one another, either by parallel evolution (spilopsyllines) or convergence (the archaeopsylline *Nesolagobius*). The many species of murid *Xenopsylla* scattered over most of Africa and much of Asia, and those found in the Pacific islands and Australia, have the same basic morphological pattern.

The idea that there were faunal connections (by rafting) between Africa and South America is startling and controversial, although it is gaining credence (Ciochan and Chiarelli 1979; Traub 1980c). Data on Anoplura also support this concept (Traub 1980c). Fain (1965) noted that the highly specialized and unusual mites in the sarcoptiform family Rhyncoptidae, undoubtedly a very ancient group, are known only from an African porcupine (a hystricomorph) and from African and South American monkeys. He therefore suggested that there must be some relationship between the two groups of hosts, as indicated by the infestation of these mammals (and a few others) by two other highly evolved mites, *Psorergates* (Trombidiformes) and *Rhinophaga* (Mesostigmata), in both of these continents. Fain hypothesized that there was a "true affinity" of an unknown nature ". . . between the porcupines and the monkeys" (and perhaps some of the others). He stressed that the affinities were "not morphological or ecological" but perhaps were chemical—as if a common protein were required for the metabolism of the mites. I believe that the queer distribution and association immediately suggest South Atlantic connections because the South

American ceboid monkeys and hystricomorph rodents are the very groups that have been claimed to have descended from stocks crossing from Africa by rafting in the Eocene (Ciochan and Chiarelli 1979; Traub 1980c). If the basic tenet is correct, these unusual mites (like ancestral ctenophthalmines on rodents) probably accompanied their hosts en route and occasionally adapted to other local mammals. Thus the land connections between Africa and South America and South America and Australia, deduced by Manter (1963, 1966) on the basis of trematode parasites of fish and supported by geological and other evidence (Traub 1980c) antedated the appearance of mammals by millions of years.

In conclusion, the zoogeography and evolution of both Siphonaptera and host provide overwhelming evidence of long-term, even geologically ancient, relationships between them. Various groups of fleas have accompanied their hosts as they penetrated new areas of the world, both groups speciating en route and as they settled. It is because of such coevolution that ceratophyllids are found nearly everywhere squirrels occur, and *Stivalius s. lat.* is found on murids in New Guinea and the archipelagos all the way to southeastern Asia.

Coevolution and Morphological Adaptations of Fleas

Since all species of fleas are bloodsucking ectoparasites which are laterally compressed, one may ask why there is such a dazzling and varied array of spines, bristles, and other modifications in the Siphonaptera. Why is there not a standard model or "universal flea"? The answer is that there is a broad range of variations among fleas, and the fleas respond differently to the various vestitures and habits of their hosts. Any adaptive modifications developed by the flea in the course of evolution to meet these challenges constitute evidence for intimate coexistence over an extremely long time, that is, coevolution of parasite and host. A discussion follows of some of the structures and finer points in more detail, before attempting a synthesis of the observations on the morphological adaptations.

The Length of the Tarsal* Segments

The first segment of the metatarsus of the fleas of flying squirrels, flying phalangers, and dormice is unusually long compared to that in fleas of diurnal tree squirrels and ground squirrels (Table 8.16). The first group lacks a pleural arch (and hence are very poor jumpers), and the hosts are all nocturnal and arboreal. This convergence seems adaptive and associated with an enhanced capacity to run and climb. The tarsal segments are not particularly lengthened in any species of *Libyastus* or in *Syngenopsyllus*, also archless fleas of arboreal sciurids, but here the hosts are diurnal, and hence the fleas are under less selection pressure. Also, in jumping, the flea takes off from the trochanter and femur, and not the tarsus, which remains parallel to the substratum (Rothschild and Schlein 1975; Rothschild et al.

1972, 1973, 1975). All bat fleas lack a pleural arch, and in *Sternopsylla,* metatarsus 1 is exceptionally long, three times the length of segment 5. I believe that these and some other fleas of cave-dwelling bats can not only climb the walls, but run about on the floor of the cave and find their hosts when the bats are feeding or drinking water on the ground.

Since the total mass of an arthropod structure tends to remain a relative constant even though its dimensions may change in the course of evolution (Traub 1969), the unusual length of the tibia or first tarsal segment in these fleas of glissant hosts and bats may be due to both narrowing and lengthening of the segment, or one segment may become lengthened at the cost of the subsequent joint. The key segments of the legs of *Choristopsylla tristis* (Fig. 8.52) and *Tarsopsylla* (Fig. 8.62) are long and thin, in contrast to *Malacopsylla grossiventris,* where they are relatively short and stout (Fig. 8.92). The flying squirrel fleas *Smitipsylla maseri* Lewis and *Macrostylophora* do not have elongate tarsi, but these species are much less modified regarding the eye, comb, and pleural arch than are other petauristine fleas. Presumably they have only relatively recently become parasites of flying squirrels.

Supernumerary Combs

The evidence presented strongly indicates that hyperdevelopment of combs and bristles in fleas is an adaptive development following infestation of hosts whose habits or vesture place the fleas at special risk. However, the multicombed condition has also been regarded as a primitive feature. It behooves us to try to reconcile these apparently contradictory points. Oudemans (1909) believed that the highest number of any somatic character (such as combs, segments) represents the primitive condition in development, even though it was assumed that the ancestral flea was originally combless and acquired ctenidia later (presumably in response to ectoparasitism). Jordan (1947) likewise regarded the many-combed condition as the primal form among comb-bearing fleas. I agree with the general principle that the Order Siphonaptera arose from an aboriginal combless insect (such as a mecopteran) and that this ancestral flea gave rise to forms with combs in certain lines of descent, some of which later lost combs in whole or in part. Further, I believe that although some primitive fleas have retained multiple combs, they also still have the capacity to develop combs or comblets in response to selection pressure. Such supernumerary ctenidia of true spines appear on segments where such combs had occurred in the past, and where vestigial spinelets had probably remained. (The power of regeneration provides an analogy. Amphibians can regenerate limbs; some reptiles can regenerate an inferior type of tail, but not feet, and the birds and mammals can at best regenerate only certain tissues.)

For example, the only pygiopsyllids known to possess a full genal comb are species of the genus *Hoogstraalia,* which are bird fleas. However, the single spine behind the eye of *Bradiopsylla echidnae* (Figs. 8.103, 8.105) apparently is the sole vestige of a genal ctenidium (Jordan 1947), and a species of *Hoogstraalia* has recently been described (*H. imberbis* Smit, 1979, host unknown) in which the genal comb is represented by only a single spine. Moreover, the pygiopsyllid *Pagipsylla galliralli* (Smit, 1965) also possesses a single postocular spine but no comb. Therefore I assume that while the comb of *Hoogstraalia* arose as a result of the pressure exerted by the volant host, it does not represent a development *de novo*. *Macropsylla hercules* is another instructive example. This, the only representative of the family, has four abdominal combs, and has accordingly been deemed a primitive flea (Jordan 1947). However, I believe that this flea has become adaptively modified to sundry scansorial hosts. The head is remarkably specialized: the helmet and genal combs are fused (Fig. 8.45); the frontal margin is narrowed (as in fleas with crowns of thorns), but the head is broad near the postantennal region (Figs. 8.47–8.49), and the remnant of the eye has moved dorsad, just above the comb. Moreover, *M. hercules* is primarily a murid flea, as per our own records and those of Dunnet and Mardon (1974), who listed more than 90 reports from murids and only three from marsupials (*Antechinus*). Yet the marsupials presumably arrived in Australia, carrying stephanocircid fleas, at least by the Cretaceous period (Traub 1980c), whereas the murids reached that continent in the Miocene epoch (Simpson 1961; Lidicker 1968). The primitive bioendemic Australian fleas infest marsupials, not rats. Also, *Macropsylla* is a helmet flea (Hopkins and Rothschild 1956), but morphologically it is very different from the stephanocircids. It therefore seems that while *Macropsylla* has ancient roots, and one or more of its abdominal ctenidia may perhaps be archetypical, at least some of the combs, like many of its other modifications, are more recent developments, resulting from selective host pressure.

Hystrichopsylla talpae (Curtis) is pertinent here since it is another large flea with some primitive features and also is extremely bristly, multispined, and multicombed. It too is extraordinal in host relationships (rodents and moles). Like most other *Hystrichopsylla*, this species is definitely associated with cold seasons or climes, and undoubtedly is subject to special environmental pressure to remain safely on the host. In both *Macropsylla* and *H. talpae*, the ability to feed on a variety of hosts compensates in part for the environmental or climatological hazards, but the need to remain attached still exists.

Spines, Spiniforms, and Dollo's Law

The structure known as a crown of thorns is a complex adaptive response to unusual selection pressure of dislodgement, permitting the flea to anchor when feeding or resting on a host. Such adaptation involves not only a modification of the genal comb but also a marked narrowing of the apical region of the head, and other changes. Similar conditions are also found in some fleas that lack genal combs and where bristles have become modified as spiniforms to perform this function. The latter group of fleas are of special interest because they clearly indicate that their crown of thorns is indeed a distinctive, new, evolutionary development based upon modifications of existing structures and not the resurrection of devices possessed by their forebears. Fleas that have tiaras of spiniforms but lack genal combs invariably are devoid of close relatives that possess genal ctenidia. This is true for all the leptopsyllids, pygiopsyllids, and hystrichopsyllids. The hystrichopsyllid *Stenistomera alpina* has frontomarginal spiniforms, while no other members of the Anomiopsyllinae are so equipped. Some of the most elaborate tiaras of spiniforms are found in the Pygiopsyllidae, where the only member with a full genal comb is *Hoogstraalia*.

Jordan (1947) stressed that, according to Dollo's law,* structures lost in the process of evolution cannot be regenerated. Many fleas have developed substitute mechanisms that serve the same purpose as true ctenidia: false helmets with a comb of spiniforms as in *Smitella* or a complete, marginal tiara of spiniforms as in *Idiochaetis* (Traub 1968a, 1969). Crowns of thorns, which are composed solely of modified bristles, then, reconfirm Dollo's law.

Supernumerary* abdominal combs which seem to have developed in response to special host pressure are found in certain members of the Pygiopsyllidae, Macropsyllidae (Fig. 8.46), Stephanocircidae (Fig. 8.114), and Ischnopsyllidae (Fig. 8.113). The Leptopsyllidae and Ceratophyllidae apparently are incapable of producing supernumerary ctenidia of true spines, but the species that infest hosts that impose unusual hazards upon their fleas are the species that have developed alternative mechanisms for enhancing attachment, such as additional pronotal spines (Fig. 8.57), tibial combs, and, in the case of leptopsyllids, frontal spiniforms (Figs. 8.27, 8.28). All of these modifications are in accord with Dollo's law, as are the false combs of the ischnopsyllids *Nycteridopsylla* (Figs. 8.115, 8.116) and *Chiropteropsylla* (Fig. 8.118), and the unique comb of spiniform bristles in *Scolopsyllus* (Fig. 8.120), the only rhopalopsyllid with a comb of any sort. The species of *Ancistropsylla*, the sole genus in Ancistropsyllidae,* are among the very few fleas that infest deer despite their vast home range and lack of a den. I believe that in these combless fleas, the two remarkable, large hooklike bristles on the head serve as an attachment device. The presence of vestigial comblets on the metanotum and some abdominal terga indicate that the ancistropsyllids were derived from combed ancestors.

Adaptive Nature of Morphological Variations

There is a significant common factor among the disparate fleas possessing a crown of thorns, namely, their hosts pose severe survival problems to fleas that become dislodged. The elaborate modifications seen in these fleas surely are adaptations resulting from survival pressure to cope with the unusual habits or vestiture of their hosts. This is illustrated by the specializations and variations that have later developed in the modified structures themselves. The ctenidial spines that are stilettolike in *Stephanocircus harrisoni* Traub and Dunnet, which infests coarse-haired bandicoot* marsupials, are broad and apically rounded in *S. concinnus* Rothschild, which parasitizes fine-furred murines. A marginal row of spiniforms, as seen on the head of *Idiochaetis,* or an apical group of thorns, as in *Acanthopsylla* and *Leptopsylla,* would be unable to grasp a pair of hairs as the keel of the flea's head was thrust between them, unless the barbs were finely adapted to the shape, diameter, and spacing of the hairs of the host in question. This tailoring of spines and spiniforms to fit the hairs of the host, and special structures like a crown of thorns, therefore strongly support the idea of a long and intimate association between mammal and flea, if not actual coevolution dating back to the aboriginal forebears. If the unique notch on the head of *Migrastivalius (G.) hopkinsi* (Traub) (Fig. 8.35) is indeed a device for grasping the spinose hairs of its specialized host, then here again the flea has clearly adapted to its host.

The marked convergence in structural features of certain fleas lead to the same conclusions regarding coevolution and the development of adaptive modifications that are attuned to a particular host. The species of *Metastivalius* on the semiarboreal nocturnal murid *Lorentzimys** have come to resemble the *Astivalius* on that host so closely (including reduction of the eye, development of a chunky body, with tibial combs and displacement of the first pair of tarsal plantar bristles) that they are difficult to recognize as *Metastivalius.* A new species of *Metastivalius* on the glissant nocturnal *Petaurus* has a pronotal comb and tarsal segments that are unusual in the genus but agree with those of other fleas on the arboreal marsupials. In contrast, most of the species of *Metastivalius* infest rats and have yet another type of comb and tarsal segments.

Since the shrew-infesting fleas belonging to three families and to four hystrichopsyllid subfamilies all have the same type of spines in their comb(s), whether in Africa, Eurasia, or North America (Figs. 8.1, 8.3, 8.5), and since their relatives on moles and other hosts have different patterns of spines (Figs. 8.4, 8.6), it seems that the modifications must be adaptive and are functional. These comments also apply to the various murid fleas that have developed vertical genal combs (Figs. 8.12–8.14, 8.16). The markedly reduced eye, unusually large size of body, and vestiture consisting of long, thin, close-set bristles (Figs. 8.17, 8.19, 8.64, 8.67, 8.77, 8.78) so characteristic of fleas of fossorial rodents must also be adaptive. These features are seen in the four species of ultraspecific fleas on the mountain beaver, *Aplondontia rufa,** namely, *Paratyphloceras oregonensis* Ewing, *Trichopsylloides oregonensis* Ewing, *Dolichopsyllus stylosus* (Baker), and *Hystrichopsylla schefferi* Chapin (Table 8.1). Their physiognomic similarity is quite striking, considering that they belong to widely separated genera in two families. The modifications seen in some rabbit fleas reinforce these points. *Nesolagobius callosus* Jordan and Rothschild, the flea of the monotypic Sumatran rabbit *Nesolagus* is an archaeopsylline pulicid that in general appearance, the shape of the genal ctenidium, and with heavy tanning of the cuticle resembles the spilopsylline rabbit fleas *Cediopsylla* and *Spilopsyllus* (Hopkins and Rothschild 1953; Barrera 1967). The host is nocturnal and is restricted to a specialized forest habitat in limited areas. *Hoplopsyllus pectinatus* is unique in the *Euhoplopsyllus/Hoplopsyllus* complex in possessing a genal comb. Its host, the relict *Romerolagus* is found only in the alpine meadows of some of the volcanoes of central Mexico. All these examples of convergence imply development of functional modifications resulting from the coevolution of fleas and their particular kinds of hosts.

The divergence* in chaetotaxy exhibited by a single Siphonapteran genus likewise seems adaptive and associated with the degree of survival pressure exerted by the host. The hosts of the *Ctenocephalides,** which possess only one to three spines in the genal comb (Figs. 8.10, 8.11), are diurnal, live in groups or colonies, and are ground dwelling. These fleas therefore can relatively easily locate a host. In contrast, *C. craterus*, which has a full genal comb, infests tree or bush hyraxes (*Dendrohyrax*), which are primarily arboreal, nocturnal, and are less gregarious. The other species of *Ctenocephalides* with well-developed combs (Fig. 8.9) parasitize carnivorans which have a large home range and are often scansorial. Such examples, and likewise the striking correlation between the hyperdevelopment of spines, combs, or bristles in fleas and the degree of selection pressure imposed by the hosts, suggest adaptive changes following a long history of joint evolution of flea and host. These conclusions are supported by the observations that (1) there are no many-combed, spinose, or bristly fleas on hosts that are gregarious, diurnal, and surface dwelling and hence pose no unusual survival problems for their fleas; (2) conversely, hosts that are nocturnal, arboreal, or otherwise specialized, lack generalized types of fleas, namely, combless species with unmodified bristles.

It might seem that the adaptive value of the combs and highly modified bristles of fleas is evident beyond question. Nevertheless, some authorities (Smit 1972; Schlein, as quoted by M. Rothschild 1976) do not believe that the pronotal comb serves to reduce the chances of dislodgement and aver that its main function is to protect an underlying membrane. Marshall (1980) suggested that the various combs in all ectoparasites, including fleas, merely protect mobile joints and the membrane beneath. Inasmuch as one of the major arguments supporting the concept of coevolution of fleas and mammals is that the sundry chaetotaxal modifications are adaptive, it is necessary to further discuss the question of the function of combs and bristles, supplementing points made elsewhere (Traub 1977, 1980b).

1.   If the combs do not function in enhancing the possibility of remaining on the host and interfering with attempts by the host to pull the flea

backwards out of the fur, then why are there so many firm examples of convergence in pattern and design of spines and combs such as ctenidia of sundry shrew fleas, mole fleas, and murid fleas? Why should fleas of coarse-furred hosts like bandicoots have dagger-shaped spines, and why should the combs of fleas of spiny mammals like *Echidna* be greatly reduced (Figs. 8.103, 8.104)?

2. If the main service of combs is to overlap membranous joints, why would not some sort of standardized, universal model of comb suffice? Why should there be so many intriguing variations, and why should they fall into patterns of arrangements instead of being haphazard?

3. None of these authors mentioned modified *bristles,* and it seems inconceivable that the peculiar bristles in the spiny fleas of spiny* hosts are not adaptations to facilitate remaining on the host. Certainly there are no membranes or joints for these widely spaced, unusually stout, short bristles to protect. If bristles can be functional in this way, why are not the spines equally adaptive?

4. At least in the case of the pronotum, fleas are able to raise or open the comb away from the body to a certain extent (Traub and Evans 1967). In that position the comb is much more apt to snag hairs if the flea is pulled backwards. Sliding a dead flea through the hairs (Marshall 1980) will not test that faculty.

5. Since the combs are so uniformly associated with the mobile joints, they are located in the best possible positions for the ectoparasites to control and retard movement in any direction except forward.

6. If the combs are not adaptive in the ways suggested herein, then there is no apparent explanation for the impressive correlation between hyperdevelopment of spines and ctenidia and the selection pressure exerted by the host. Not only is the correlation nigh perfect but it even applies to the degree of ctenidial development and the amount of risk.

7. In my opinion, the membrane under a dorsal comb is the result of the development of the ctenidium, not vice versa. The comb is not an *extension* to the notum, overlapping the next segment. Instead, the comb is in effect derived from the body of the notum itself, and if one drew a line connecting the apices of the spines, it would outline the original margin of the segment. It is as if a series of triangles were excised from the caudal margin of the segment and the projections between the interstices were rendered tanned and hardened, with the tips blunt or acutely pointed, or otherwise modified systematically. The length of the notum plus the length of the spine adds up to a constant, as per the principle of totality of mass (Traub and Evans 1967; Traub 1969); when the ctenidial spines are long, the notum is correspondingly short. Thus in nest fleas that have lost the comb, like *Wenzella* and *Jordanopsylla,* the pronotum seems to be inordinately long. As indicated in Figure 8.33, the comb cannot adequately "protect" the membrane in species possessing a "flared" ctenidium.

8. Smit (1972) did not discuss the function of the genal comb, but as shown here and elsewhere, the spines can also be modified adaptively,

and presumably operate when the flea is standing in an oblique position (Traub 1980b), as when feeding.

9.   The relative numbers and kinds of fleas infesting various categories of hosts support the idea that the combs and bristles facilitate survival along the lines indicated herein. Hosts with highly specialized vestiture (such as spinose or covered with plates) or habits (volant or glissant) tend to be parasitized by relatively few species of fleas in any one area as compared to those that are generalized in nature (e.g., *Peromyscus, Rhombomys*). Moreover, the fleas on the first group are greatly modified in chaetotaxy as compared to the second and are much more host specific. The obvious implication is that only a few species could adjust to life on these difficult hosts and that the structural changes exhibited by the fleas are truly adaptive.

For our thesis of coevolution, however, the precise function of the combs is perhaps an extraneous issue. The fact that the shape of the spines varies considerably throughout the taxon and is consistently associated with certain groups of hosts is in itself an argument for coevolution.

Duration of the Period of Coevolution

Synchronous Evolution

Although the majority of fleas seem to have coevolved with their hosts, the length of their associations vary considerably among different groups of mammals and siphonapterans. In some cases the fleas and hosts may have been connascent, with the coevolution essentially synchronous as the fleas adapted to any new conditions posed by the mammals. In other instances the fleas switched to, and adapted to, new hosts with characteristics like coarse hair or fossorial habits.

It is necessary to differentiate between coevolution at a high taxonomic level (family, subfamily, or tribe of flea) and at the level of genera and species. For instance, the doratopsylline (Hystrichopsyllidae), stephanocircid, and pygiopsyllid fleas which characteristically infest marsupials* have roots that go back scores of millions of years; for the first two taxa, at least to the early Cretaceous period, if not to the Jurassic period (Traub and Dunnet 1973; Traub 1980c). Since the Australian stephanocircids differ from the South American at the subfamily level and the doratopsyllines differ similarly at the tribal level, we can assume the distinctions arose sometime after the emigration of the marsupials from South America to Australia by way of the Antarctic, more than 130 million years ago. It does not automatically ensue, however, that any of the contemporary species (or perhaps even genera) of these fleas have been infesting their particular kinds of hosts since then. None of the marsupials extant is known to have such a long history as genera. There undoubtedly has been too much extinction, speciation, and evolution of the hosts (if not of the fleas) to render such taxonomic longevity likely. However, since only two genera of stephanocircines that are closely allied are known and *Coronapsylla* is monotypic, it seems possible that *Stephanocircus* has been in existence since the Cretaceous period. Certainly, as families and subfamilies at least, these fleas have been coevolving with marsupials for a long time.

These same higher taxa of fleas are also associated with hosts that evolved much later than the marsupials: the soricids, whose earliest known representatives date from the Oligocene (Patterson 1957; Kurten and Anderson 1960); the talpids, recorded from late Eocene (Simpson 1945; Kurten and Anderson 1980); cricetids, dating from Upper Eocene (Wood 1977); and the murids, whose history commences in the Upper Oligocene or Lower Miocene (Simpson 1961; Chaline and Mein 1979). The two fossil Baltic Amber species of *Palaeopsylla** so closely resemble the current members of the genus, virtually all of which infest shrews or moles, that it has been concluded that (1) *Palaeopsylla* has remained essentially unchanged since mid or Upper Eocene (35–50 m.y.a.) and (2) the fossil species likewise parasitized Insectivora (Hopkins and Rothschild 1966; Traub 1980c). Although these fossil species of fleas antedate the known records of talpids and soricids, the Insectivora are recorded for the Cretaceous (Kurtén and Anderson 1980). It seems reasonable to believe that the Eocene *Palaeopsylla* infested precursors of the moles and shrews, especially since the basic pattern of the *Palaeopsylla* ctenidia is seen in some other genera parasitizing Insectivora. Furthermore, there are inherent drawbacks in paleontological evidence because earlier fossils may exist that have not yet been discovered. If one were to interpret the fossil record literally, he might conclude that moles antedate shrews, yet moles are much more highly specialized morphologically than are soricids, and must have evolved from some non-burrowing form.

The fossil record of the Soricini and Blarini shrews begins in the Miocene epoch (Kurten and Anderson 1980), and both groups are parasitized by ultraspecific fleas. *Cryptotis parva* (Say) dates from the Upper Pliocene, and its ultraspecific flea, *Corrodopsylla hamiltoni* (Traub), may be equally ancient. The aplodontids have a rich fossil history in North America and perhaps Asia (Simpson 1945; Chaline and Mein 1979; Wood 1977; Colbert 1980), but today are represented by only one species, *Aplodontia rufa*,* the mountain beaver of the Pacific Coast, which is considered to be the most primitive living rodent. Its four ultraspecific fleas (*Paratyphloceras oregonensis*, *Trichopsylloides oregonensis*, *Dolichopsyllus stylosus*, and *Hystrichopsylla schefferi*) combine primitive features with convergent modifications typical of fleas of fossorial rodents. The association between these fleas and the mountain beaver must be a long one even if some of these species are descended from fleas whose original aplodontid hosts became extinct. For example, *Dolichopsyllus* is a monotypic leptopsyllid belonging to a unique tribe in a family whose roots are Palaearctic (Traub 1983a).

Since the overwhelming majority of ultraspecific fleas are adaptively modified in obvious ways and their hosts also tend to be unusual, coevolution of flea and mammal seems implicit. It seems likely that the profound changes that characterize a crown of thorns developed gradually as the host became scandent or otherwise specialized. For example, fleas of the genus *Metastivalius* primarily infest murids, and the species that, like *M. mordax* (Rothschild), occur on rats that spend most of their time on the ground lack spiniforms on the head and possess preantennal bristles that in general are of the ordinary type. Others, like *M. anaxilas* (M. Rothschild) and *M. rothschildae* Holland, infest scandent rats and have a crown of thorns, whereas *M. molestus* (Jordan and Rothschild), which parasitizes murines that are intermediate in habit, has frontal bristles that are somewhat thorny. Since the complex genitalia are essentially similar in all of these species, they are all closely related despite the variations in cephalic spines. It therefore seems that the modifications occurred in response to changes in habits and vestiture of the hosts of these fleas.

Coevolution after Switching to New Hosts

In the *Metastivalius*,* which transferred from infesting murids to parasitizing marsupial bandicoots and "mice," not only did the pronotal spines become acutely pointed to accommodate the coarse fur of their hosts, but there have also been changes in the chaetotaxy of the head analogous to those just cited for the murid species. *M. huonensis* Holland is a flea of nonscansorial bandicoot* marsupials (*Peroryctes*) in New Guinea and lacks a crown of thorns, but the frontal/genal bristles in general are subspiniform. Such cephalic bristles are characteristic of bandicoot fleas, presumably in response to the coarse hairs of the host (Traub 1980b). However, in an undescribed *Metastivalius* of nocturnal, scandent *Antechinus*, there is a crown of thorns, and this species so closely resembles *M. anaxilas* that only its stiletto type pronotal spines serve for rapid differentiation. The new *Petaurus-Metastivalius*, previously mentioned, is not related to the *M. anaxilas* group and is particularly pertinent since it represents another switch from murids to an arboreal marsupial, but in this case the host is fine furred, and the spines of the flea's comb are broadly rounded and the ctenidium is specialized in a different manner.

The *Ctenocephalides** on hyrax and *Dinopsyllus ingens** on bathyergid moles, already cited regarding reduction in the genal comb, also serve as examples of a switch to a radically different type host, with subsequent coevolution. There can be little doubt that these species have departed from the norm in their respective genera and hence do not represent the primitive condition regarding numbers of genal spines. A glance at Figures 8.9, 8.10 and 8.17 suffices to refute such a possibility and to reinforce the widely held belief that the highest number of any somatic feature is the most primitive state. In *C. felis*, which has a large genal comb, there is an additional small spine at the tip of the genal lobe near the ventrocaudal angle of the head, indicating that in the past the ctenidium extended to that point. Such a vestigial spine is also visible in *C. arabicus* (but not in *C. rosmarus*). In *D. ingens*, the comb is represented by only a tiny spine by the antennal groove, in line with the middle placoid. As Jordan (1947) pointed out, this isolated spine also occurs in *Listropsylla*, where none of the species have a genal ctenidium, the helmet fleas *Stephanocircus*, in *Bradiopsylla echidna* (Fig. 8.103), and *Typhloceras poppei* Wagner, and can be regarded as a part of a comb that became reduced. In the first two genera this odd spine is actually enlarged and may be functional (Jordan 1947; Traub 1972c). In the hyrax *Ctenocephalides* the numbers of spines in the abbreviated genal comb may vary somewhat even on the two sides of an individual, recalling the case in *Archaeopsylla** hedgehog fleas. Such instability is common in structures undergoing pronounced evolutionary change.

Among the fleas that clearly adapted to hosts with specialized features are *Synosternus pallidus* and *S. longispinus* (Wagner), which moved to hedgehogs from rodents and consequently acquired widely spaced body bristles (Traub 1980b). The transfer is apparently a relatively recent one: (1) the degree of morphological modification is relatively slight as compared to *Archaeopsylla.** (2) *S. pallidus*, at least, is a catholic flea and can survive on other hosts, even in the presumed absence of hedgehogs. (3) The geographic range of erinaceids significantly exceeds that of *Synosternus*. (4) There are at least four *Synosternus* infesting rodents. (5) The genera allied to *Synosternus* are primarily rodent fleas.

Pulex sinoculus surely accommodated itself to life on pocket gophers after those rodents became fossorial, since it closely resembles its congeners on suoids, except for loss of the eye and a few pertinent changes in chaetotaxy.

Ioff (1953) was impressed with the speciation and morphological changes that occurred in fleas following transfer to a new host. He regarded *Leptopsylla putoraki* Ioff as having evolved from *L. sexdentata* (Wagner) that had switched from *Mus* to shrews, and as a result, eventually differed "by the shape of its frons, which is unusually pointed, cone-shaped and oblong." He further stated that it "is plausible to assume that the dense and very fine fur of the shrew required such an adaptation."

The Rate of Evolution in Fleas

The actual rate of evolution in the various taxa of fleas is obviously germane to the duration of the period of coevolution of flea and mammal. For this subject there is virtually no direct evidence and very little specific discussion in the siphonapteran literature. The point is frequently made that the evolution of parasites in general lags a step or more behind that of the hosts; this was referred to by Brooks (1979) as Manter's first rule (1955, 1966). However, this fundamental observation has been recognized and applied for a much longer period in parasitology. The same is true for its corollary, namely, that as a result of the slower rate of evolution, it is much easier to recognize kinship in parasites than among the more widely divergent birds and mammals serving as hosts. For example, Harrison used data on lice as long ago as 1914 to argue that the emus, cassowaries, and other ratite birds were actually related, and that their resemblances were not due solely to convergence (Harrison 1914; Traub 1980c). In 1916 he presented analogous evidence on Mallophaga to contend that the Australian marsupials originated in South America (Harrison 1928). Wagner (1932) and Jordan (1942) advanced similar arguments about the rate of evolution in fleas. Holland (1950b) also stated that this slower rate of evolution in fleas explains why more than one subspecies is apparently never found on a single subspecies of host and why one subspecies of flea "occupies the same range of several subspecies of hosts" (Holland 1950a). Smit (1972) and Barrera (1967) accepted this premise of a time lag, as did Dubinin (1947), who cited the distribution of *Pariodontis*, the Old World porcupine flea, as an example. Dubinin believed that there was only one species of *Pariodontis* and that the forms occurring in South Africa, North Africa, Turkmenia, India, and Southeast Asia (Indochina and Malaya) each represented subspecies of *P. riggenbachi*. He regarded Africa as the center of origin for porcupines in the Oligocene epoch, with the move to the Middle East [complete with *P. riggenbachi* (Rothschild)] occurring in the Miocene epoch. According to that account, then, the evolution and distribution of the Old World porcupines in about 30 million years culminated in four contemporary genera and 12 species of porcupines, as listed by Corbet and Hill (1980), and in one genus, one species, and five subspecies of pulicid flea, as per Dubinin's scheme of classification. Today, the North African *Pariodontis* is considered a synonym of *P. r. riggenbachi* (Rothschild), and the Southeast Asian representatives are regarded as a full species, *P. subjugis* Jordan (which is also known to me from Borneo). Nevertheless, Dubinin was correct in citing this case to illustrate the slower rate of speciation in fleas as compared to that of the hosts.

However, this phenomenon does not apply to all fleas, for there are striking examples where the siphonapterans must have speciated more rapidly than the host, as in the case of the five species of *Astivalius* and three species of *Metastivalius* (both pygiopsyllid genera) on the monotypic murid *Lorentzimys nouhuysi* and the eight *Phaenopsylla* and three *Amphipsylla* (both leptopsyllid genera) on the monotypic cricetid *Calomyscus bailwardi*. Except perhaps for some *Phaenopsylla*, no two species in any one of these genera are sympatric. It therefore seems unlikely that some represent species who transferred to the current host after the original host became extinct. On the contrary, it would appear that speciation occurred in the flea in isolated portions of the ranges of the rodent. Another example of a rapidly evolving flea is the leptopsyllid *Cratynius* on the monotypic gymnuran *Hylomys suillus* Müller, where there are species known from Java (1),

China (1), and Borneo (2). The Bornean species appear to be sympatric, including elevation on the same mountain, and perhaps may also be syn-topic. There are other examples of genera of fleas with species that are limited to only part of the range of the host, indicating that speciation was occurring at a slower rate in the mammal than in the parasite. For example, *Callosciurus nigrovittatus* and *C. notatus* carry different *Medwayella* in Borneo than they do in Malaya, as do some of the same species of ground squir-rels, but these insular and peninsular forms are sibling species (Traub 1972a). In other instances (Sarawak versus Malaya), a single species of host carried different subspecies of fleas.

In continental and insular Southeast Asia, speciation in general has proceeded at a faster rate in sciurids and murids than in their fleas. There are many more kinds of rodents than there are species of fleas infesting them. *Syngenopsyllus* is a monotypic ceratophyllid on various squirrels, ranging from at least Thailand to Java, with subspecies occurring in some of the areas. The leptopsyllid, ceratophyllid, and hystrichopsyllid fleas in the Malfilindo Archipelago all have Palaearctic ancestry and are usually bioendemic; that is, they are restricted to one island. Among these, there is only one genus, the leptopsyllid *Sigmactenus,** that does not occur on the Asian mainland, and it has speciated on various islands. Although *Rothschildiana, Neopsylla, Macrostylophora,* and *Paraceras* apparently traveled with their hosts from continental Asia to some of the islands, and *Avio-stivalius, Medwayella,* and *Lentistivalius* dispersed in the opposite direction, differentiation has not progressed higher than the species level. These are primarily fleas of callosciurines or murids. It is interesting that the rate of evolutionary change seems much more accelerated in the fleas of the petauristines *Hylopetes* and *Petaurista,* assuming that the generic concepts for these fleas and their hosts are correct. Because the changes in the fleas of continental and insular Southeast Asian callosciurines and murids mainly stop at the level of species, it appears that dispersal into these areas occurred in relatively recent times, geologically speaking.

Other evidence bolsters this impression concerning how long murids and their fleas have had an opportunity to coevolve. It has been postulated that the murids* arose in Asia in the Miocene or late Oligocene epoch and that, in a series of waves, these rodents worked their way to Australia or New Guinea by island-hopping through the Indonesian islands, arriving in various periods in the Miocene or Pliocene epoch (Simpson 1961; Tate 1951). A main wave in the late Pliocene or Pleistocene epoch included *Rattus.* Much later, commensal rats entered Australia and New Guinea through the agency of man. Data on Siphonaptera support the general idea that the ultimate origin of murids was in Asia and that *Rattus* arrived at a relatively late date (Traub 1972c, 1972d, 1980c). Lidicker (1968) accepted Simpson's general concepts but believed that there had been only two waves and suggested the first invasion, in the Miocene, "may have been composed of a single species of primitive murid . . . [which] could have differentiated into all of the extant New Guinea rodents except for *Rattus* and *Mus*." The observations of Siphonaptera lend support to Lidicker's hypothesis.

Unlike the primitive species of pygiopsyllids of New Guinean marsupials, which are a somewhat heterogeneous group of fleas, the murid-infesting members of the family are remarkably uniform in basic genitalic morphology, so much so that prior to Holland's monographic treatise (1969), the vast majority of them were all called *Stivalius*. Moreover, the few that had been placed in other genera owed their assignment to elaborate modifications on the head (e.g., *Idiochaetis* and *Ernestinia*) or specializations like the loss of the eye and development of tibial combs (*Astivalius*). They too have the fundamental *Stivalius* genitalic apparatus. In other words, the differences are in features that I regard as adaptive, and the *Stivalius* (*s. lat.*) could all have had a common ancestor derived from a marsupial flea after the first species of murid appeared on the scene. Similarly, the hydromyine and "phloemyine" murids do not have any characteristic fleas at the generic level; the fleas generally found on those rats are essentially *Stivalius* (*s. lat.*). In their spectacular nongenitalic modifications, the pygiopsyllids of the New Guinean murids mirror the accelerated rate of evolution exhibited by their hosts, which are generally deemed as a modern, rapidly diversifying, and hence a taxonomically confusing group. *Sigmactenus*,* a leptopsyllid, indeed must have been a later arrival in New Guinea, probably on the first *Rattus*, resulting in today's *S. toxopeusi* Smit, from which *S. cavifrons* Smit and an undescribed subspecies later evolved after some of the stock switched to scandent hosts like *Melomys* or *Pogonomelomys*.

Xenopsylla probably traveled to the Australian region on some of the later waves of *Rattus* from Asia. In the entire Malfilindo Archipelago and the Australian region, *Xenopsylla* is known only from a total of four or five species. One of these, *X. vexabilis*, ranges from Indochina and Thailand to New Guinea and Australia. Notably, two bioendemic species, one in New Guinea and one undescribed species on Luzon, are eyeless and have a reduced pleural arch (Traub 1972c, 1980c), suggesting a relatively long local history and adaptations to hosts that nest underground. The dearth of *Xenopsylla* in Australia, where aridity and the presence of indigenous murines recall conditions in parts of Africa and southwest Asia where there is a rich *Xenopsylla* fauna, indicates that *Xenopsylla* is a relatively recent emigrant to Australia.

It was suggested that the specialized pygiopsyllid *Tiflovia*, a parasite of the primitive murid *Pogonomelomys ruemmleri* found in the frost grass above timberline in New Guinea, must have evolved in the late Pliocene at the earliest (3–6 million years ago), and that the two known species differentiated since the Pleistocene (1–3 million years ago) (Traub 1977, 1980c). The bizarre adaptive specializations in the pygiopsyllids mentioned before could therefore indeed have evolved in the indicated time frame (since the late Oligocene or Miocene) and the various New Guinean species of *Sigmactenus** could very well have evolved in the one to four million years since the arrival of *Rattus* in the late Pliocene or Pleistocene. However, the origin and development of *Sigmactenus* from taxa like the Asian *Leptopsylla* must of course have taken much longer.

If the murids actually arose in the late Oligocene, then their characteristic genera of fleas cannot be older than 25–30 million years. The genera of fleas specific to *Rattus* would then have a maximum age of about four million years. In geological terms, then, the murids and their fleas are youthful. This situation is in marked contrast to that in *Palaeopsylla** where the fossil evidence indicates a minimum age of 35–50 million years and where the association with the Insectivora is inferred to be at least that old. The related *Ctenophthalmus** may be more ancient since it is more primitive. The spilopsylline rabbit fleas also suggest a slow rate of evolution. There are only three genera on rabbits: *Cediopsylla* (with four species), in the New World; the closely related, monotypic *Spilopsyllus* in the Palaearctic region; and *Hoplopsyllus* (two species, one of which parasitizes ground squirrels). *Cediopsylla tepolita* Barrera, is an ultraspecific parasite of the unique Mexican volcano rabbit (*Romerolagus*), and there are three species of *Cediopsylla* on the only other genus of New World rabbit, *Sylvilagus* (11 species). *Hoplopsyllus pectinatus* is a *Romerolagus* flea. Yet the North American spilopsyllines probably date back to mid Eocene, following dispersal by northern transatlantic connections (Traub 1980c). Moreover, the spilopsyllines must be even much older than that, since *Spilopsyllus cuniculi* (Dale) and *Cediopsylla* have complex hormonal interrelationships with their hosts (Rothschild and Ford 1973), and hence this physiological adaptation must have arisen before the dispersal of the ancestral spilopsyllines.

In most instances, the families and tribes of fleas have been primarily associated with only certain groups of mammals. The length of time that the higher taxa have been in existence is therefore pertinent to coevolution. The stephanocircids and the doratopsyllines of marsupials* probably date back to the beginning of the Cretaceous period, and the pygiopsyllids and the pulicids* probably arose in that period (Traub 1980c). Riek (1970) reported, but did not name or describe, a Lower Cretaceous fossil flea from Australia that he regarded as an ancestral pulicid. The Leptopsyllidae are a minimum of 50 million years of age and the Ceratophyllidae, the most recently evolved family in which differences at the subfamily or tribal level do not seem to have been formed, is at least 35 million years old (Traub and Rothschild 1983).

There is very little in the literature regarding the rate of evolution at the species and subspecies level in the Siphonaptera. Paleontology is of little assistance since only two true fossil fleas have been described, and since the main hosts of fleas, the rodents, are inadequately known as fossils, particularly in the tropics. Deductive reasoning has provided some clues, as with *Tiflovia* (Traub 1977), and remains the most feasible tool. For example, it is surprising how frequently the same species of mammal, and species or subspecies of flea or other ectoparasite, occur on a series of mountains that are completely isolated from one another by desert, semidesert, eternal peaks of snow, lowland rainforest, or other barriers. In Pakistan, ecological islands with much the same faunal and floral components were found in segregated areas under such conditions in the Himalayan and Karakorum mountains (Traub and Evans 1967; Traub et al. 1967; Traub and Wisseman 1968). Despite the awesome barriers, certain murids, microtines, and petauristines were found in all these various isolated foci, as were various trombiculid mites, and fleas like *Palaeopsylla setzeri*, *Stenischia* new sp., *Frontopsylla* (*Profrontia*) *ambigua* Fedina, *Leptopsylla* (*Pectinoctenus*) *pamirensis* (Ioff), and *Macrostylophora fimbriata*. In general the species of fleas showed little differentiation whether from Swat, Dir, Gilgit, or the Kaghan Valley, although subspeciation seemed to have occurred in some areas.

Analogous findings have been made in other countries where we have studied ectoparasites on isolated mountains, such as in Mexico, the United States, Ethiopia, Borneo, and Malaya. Generally there is a slight degree of endemism in the fleas on a particular mountain, where the local form is often a sibling of a species on another mountain, and the remainder of the fauna are shared in common at the level of species, or even subspecies. Where distinctive subspecies occur, the association is not with a particular mountain, but rather with another, broader, geographical category. For example, *Jellisonia klotsi* Traub, *J. hayesi* Traub, and at least five other species of fleas are found on five mountains in central Mexico, ranging from Michoacan to Guerrero. Here again speciation and subspeciation may antedate the barriers that today separate the mountains. At least four competent investigators have studied the fleas of Utah, but none have reported local endemism on the isolated mountains. *Ctenophyllus armatus* has been found on pika on mountain peaks in western North America from Alaska to California, including mountains in the middle of the desert. Further, this flea is widely distributed in eastern Asia as well, and alleged subspecific differences are no longer acknowledged.

It is not known how long these various mountains have been surrounded by desert, but in Gilgit, Pakistan, the rain shadow that accounts for the aridity, caused by the upthrust of the Himalayas, was probably cast a million or more years ago (Traub et al. 1967). Intermittent periods of glaciation may have resulted in confluence of habitats and dispersal of mammals in these various areas. Nevertheless, it seems probable that many thousands of years, if not much longer periods of time, have elapsed since such major glaciation took place.

If these evolutionary processes were taking place at an accelerated rate among Siphonaptera, the evidence should be most readily apparent in the murids, which are the most recently evolved of the flea-bearing mammals and which themselves are speciating and changing at a quickened rate. Diversification at the generic and specific level of murid fleas has taken place since late Oligocene or the Miocene. Even among parasites of *Rattus*, several million years have been available for evolution. Although some *Rattus* (*s. lat.*) are referred to subgenera like *Lenothrix, Maxomys* or *Stenomys* or *Praomys* (names regarded as full genera by some workers, and deemed without validity at all by others), there are very few fleas restricted to such hosts. The same is true regarding fleas of murine genera like *Pithecheir* and *Chiropodomys*, which nest in trees, bamboos, or tree ferns. If there are any fleas found on these, they are biotopical in nature, like *Migrastivalius jacobsoni* on a variety of small semiarboreal or scandent hosts. The subgenus *Rattus** has no known ultraspecific fleas—a fact that is in accord with the youthfulness of the group.

Because of their potential medical importance, *Nosopsyllus fasciatus* and its allies in India were carefully studied by Jordan and Rothschild (1921), who recognized eight indigenous species, along with the cosmopolitan *N. fasciatus*. This is not a surprisingly large total considering that the material studied was from various parts of the subcontinent and that sciurids (*Funambulus*), as well as rats, were represented as hosts. The variations noted in *N. fasciatus* were deemed "individual" and not geographic. There is no evidence in India, Europe, or elsewhere that *N. fasciatus* has evolved into subspecies since its introduction with black rats into various parts of the world.

Although *N. londiniensis* (Rothschild) infests commensal rats, particularly *Rattus rattus*, in widespread parts of the world, subspeciation has not been noted in the fleas on these hosts. However, a distinct subspecies, *N. l. declivus* Traub, is found on the Nile grass rat, *Arvicanthis niloticus* in Egypt. The author suggested that this was the ancestral form of the species and that *N. londiniensis* subsequently transferred and adapted to *Rattus*. Lewis (1967) contended that the morphological uniformity of *N. l. londiniensis* on rats indicated that *N. l. declivus* "is an isolated off-shoot," and not the aboriginal form. Lewis presumably believes that the rate of evolution in fleas is quite rapid, for he recommended "rearing studies" to determine "whether the choice of host could, in some way, effect variations in the morphology of the ectoparasite" (Lewis 1967). As amply attested in this chapter, I believe that fleas adapt structurally to their host, but nevertheless I doubt exceedingly that the changes occur rapidly enough to be induced in the laboratory within the lifetime of an observer. The lack of significant or consistent variations in specimens of *N. l. londiniensis* from *Rattus* in different parts of the world indicates to me that the dispersal of these commensal rats is a relatively recent phenomenon, and that it commenced long after the original transfer of *N. londiniensis* to rats.

My disbelief in the evolution of siphonapteran subspecies within a time frame of decades or centuries is heightened by the data on the species of *Xenopsylla* that have been introduced by commensal rats into various countries. *X. cheopis*, which I regard as originally a flea of *Arvicanthis* in northern Africa (Traub 1963, 1972f), does not exhibit any variations of subspecific rank, even when specimens from areas as disparate as Europe, Asia, Africa, North America, South America, and Australia are critically studied. The synonyms that arose in the literature were due to nomenclatorial confusion, not because of purported morphological differences. Subspecies of *X. astia* Rothschild and *X. brasiliensis* (Baker) are not recognized, and those reported for *X. vexabilis* are now regarded as invalid. *Echidnophaga gallinacea* (Westwood) is another widely introduced flea of rats (and poultry, etc.) that has not produced subspecies either. The widespread and often intense association of these fleas and cosmopolitan rats apparently represents recent events, not coevolution.

Other observations likewise suggest that fleas evolve at a relatively slow rate. The deserts and mountains of Central Asia separate *Ctenophthalmus golovi* Ioff and Tiflov in the subarctic Himalayan peaks from its kin in the Caucasus, yet there is no doubt that both groups belong to the same species. *Ctenocephalides caprae* Ioff was supposed to represent a transfer from carnivorans to goats in historical times, with subsequent changes in morphology, but this name is now considered a synonym. There is no convincing evidence that the *C. felis** on cattle are subspecifically distinct from those on carnivorans. Originally deemed an Oriental form attributable to introduction of the cat flea through human agency, *C. felis orientis* Jordan is now regarded as a full species (Hopkins and Rothschild 1966), as befits a parasite of the rich fauna of viverrid and felids in the Pacific islands and mainland Asia. Significantly, no subspecies of *C. orientis* have been designated in all that huge territory. Although presumably valid names for subspecies of *C. felis* exist, it has not been claimed that they are the result of introductions effected through the agency of man.

The available data on the duration of the association between mammals and their fleas, and on the rates of evolution in these insects, all seem to indicate that whenever there is a degree of specificity, flea and host have coevolved for countless eons. Since the primitive hosts tend to have primitive fleas, evolution in these groups must have proceeded slowly, but the association is manifestly an ancient one. Since the specialized mammals have highly modified fleas, here too the parasites have progressed with their hosts in evolution. The observations on specificity herein reported, with the ultraspecific fleas almost invariably being associated with unique or specialized hosts, lead to the same conclusions about joint evolution. So does the evidence that the impressive morphological modifications characteristic of so many groups of fleas are adaptive in nature and can generally be ascribed to the vestiture or habits of their particular hosts.

CONCLUSIONS

1.   Most fleas have clearly evolved with their hosts, although the intensity and history of the association varies throughout the Order Siphonaptera. The relationships between these fleas and their hosts have existed for

millions of years as indicated by the data and observations on the physiology, host relationships, zoogeography and evolution, and morphology of fleas. These conclusions are based upon the principle that the phylogeny of the fleas, as obligate monophyletic parasites, usually reflects that of the host. A complex physiological interrelationship exists between fleas and mammalian hosts, as illustrated by the hormonal or other synchronization of their reproductive cycles whereby the fleas and their larvae emerge at the critical time when the host is in its breeding den.

   2.   The vast majority of mammalian fleas are either ultraspecific, wholly restricted to a single species of mammal, or generispecific, found only on members of a single genus of host or on closely related taxa. About 14% of the known fauna of mammal-infesting fleas (1667 species) are ultraspecific, and at least 122 species of mammals are parasitized by such fleas. Virtually all of these hosts are either monotypic or else lack close relatives in the foci where the ultraspecific fleas occur. If a kindred mammal were available, probably it would be infested as well. Approximately 84% of the hosts of the ultraspecific fleas are specialized either in habit or structure: they may be scandent, fossorial, spiny, or else live in restrictive habitats. Ultraspecific fleas occur in all the orders that are infested by Siphonaptera, but some of the constituent families lack them. About 88% of the ultraspecific fleas are structurally more modified than are their less specific kin. These points all suggest that these fleas must have coevolved with their hosts.

   3.   About 500 species of fleas, in at least 102 genera, are generispecific. About two-thirds of these are found on rodents, the group of mammals that harbors the overwhelming majority of species and species groups of fleas that are restricted to hosts of a single tribe or family. However, the vast majority of the fleas on Insectivora, Lagomorpha, and Carnivora are either ultraspecific or generispecific, and the marsupial fleas are also highly specific as a rule. The fleas found on leporids do not occur on ochotonids, and vice versa.

   4.   Most species of fleas infest only a single order, or even family, of mammals. For example, shrew fleas are found only on shrews, even though congeners may infest moles. Some fleas have left their original hosts and adapted to other kinds of hosts that reside in their particular environment. Such biotopical Siphonaptera include fleas that have switched to carnivorans, or fleas of tree-nesting cricetids that have transferred to squirrels. Other fleas have transferred to hosts with already-established peculiarities, such as spiny vesture or fossorial habits, and as a result, they become morphologically modified: they represent a coevolution of relatively short duration. Such opportunistic change is unusual; when it occurs, the adaptation of fleas tends to be quite specific, as in the case of modified combs.

   5.   There are only three species of fleas known to infest *both* peromyscines and microtines, even though these rodents commonly coexist. Host specificity can transcend close kinship and frequent contact with different mammals. Within a genus, fleas of some subgenera or groups of species may be limited to different kinds of hosts. This point also illustrates the strong bond between flea and host and implicitly suggests coevolution.

6.   Zoogeographic data indicate that the stephanocircid and doratopsylline fleas of marsupials have been associated with these hosts for more than 125 million years and probably were connascent with the groups of metatherians they infest. The relationships between certain pygiopsyllids and marsupials probably also date back to the Cretaceous period. Some pulicids have parasitized their hosts since the Eocene epoch, if not longer. The most recently evolved family, the Ceratophyllidae, probably arose in the Oligocene epoch about 35 million years ago, with their sciurid hosts. In general, the various families of Siphonaptera have been associated with their aboriginal hosts and descendants ever since they arose. Fleas have often accompanied their hosts as they dispersed to sundry parts of the world over the eons, speciating and radiating in the process. However, fleas have usually lagged somewhat behind their hosts in evolution, as shown by the taxonomy and nomenclature of both groups and by the fleas' adaptations, such as spines and bristles, which are often tailored to fit the hairs or the life-styles of the mammals the fleas infest.

7.   There is a clear-cut correlation between the level of evolutionary development of the flea and that of the host. The ancient, primitive hosts have the most primitive fleas, and the more recently evolved hosts have the more specialized fleas. This again suggests that host and parasite evolved in tandem. Some primitive fleas have remained unchanged at the generic level for millions of years, but even in the ancient Pygiopsyllidae, a rich flea fauna at the generic level has developed in New Guinea since the Miocene or Pliocene epoch, paralleling the situation in their murid hosts.

8.   In fleas, subspeciation has generally occurred slowly through geological times. Barrierlike deserts or isolated mountains have frequently failed to result in speciation. Even subspeciation under such conditions has not been common.

9.   There is a bewildering array of bristles, spines, and combs in the Siphonaptera. These are adaptive, functional, and evolutionary responses to the problem of dislodgement posed by the vestiture and habits of the hosts. Hence these modifications indicate intimate, age-old coexistence between parasite and host. Fleas that have coevolved with, or adapted to, a certain group of hosts, consistently differ in major respects from their close relatives that infest other kinds of mammals (divergence). Fleas that parasitize the same hosts have come to resemble each other by adaptive modifications, although they are of different phylogenetic lines (convergence).

10.   The chaetotactic modifications may be so diagnostic that infestation of a shrew can be recognized by merely examining the spines of the flea, irrespective of the taxonomic standing or geographic location of the flea. Representatives of five families of murine fleas have developed vertical

combs of spines or spiniforms on the head, and most of these also possess false combs on the tibiae. Similarly, among the taxa within a genus the shape and numbers of spines in combs may diverge greatly but consistently depending on the specific hosts they infest, such as moles, shrews, mole-rats, squirrels, or rats. The shape of the spines and bristles often appear to be obviously molded to the vestiture of the host. Thus spiny hosts tend to have spiny fleas with reduced combs and pectinate rows of widely spaced bristles, and mammals with coarse fur have fleas with acutely pointed ctenidial spines with appropriate gaps between the apices.

11. The degrees of development of spines, spiniforms, and combs seem to be closely correlated with the amount of selection pressure exerted upon the flea by the traits of its particular host. Mammals that are surface dwelling, diurnal, and somewhat communal tend to have fleas that are unspecialized in structure. They lack fleas with multiple combs or elaborate structures like crowns of thorns. In contrast, mammals that live in or climb trees and operate at night, are fossorial, or are otherwise specialized, have fleas with devices that militate against dislodgement. Such hosts lack generalized types of fleas. Volant hosts like bats have fleas with highly specialized devices for preventing dislodgement. Such fleas have also lost the power to leap, a faculty hazardous to fleas seeking or infesting a host on the roof of a cave or roosting by itself in a tree. Fleas whose hosts are intermediate between the two extremes in dangers to their fleas are themselves correspondingly intermediate in development of combs and hypertrophied bristles.

12. The number of spines in a comb may represent reduction in some instances, as in the loss of spines in fleas of fossorial rodents. However, in other cases there is hyperdevelopment: the number of spines is so increased that the pronotal comb extends down over the vinculum, as in the fleas of scandent, nocturnal mammals. The morphological features are thus determined by the type of host. The presence of multiple abdominal combs in *Macropsylla hercules* and other species is an adaptive character and does not represent purely plesiomorphy.

13. The more highly evolved fleas like ceratophyllids and some leptopsyllids, which have lost combs on the abdomen and head, may have false combs of spiniform bristles instead of developing supernumerary combs, or else supplement the spines of the pronotal comb, in response to the unusual risks posed by the habits of their particular hosts. Other fleas in such circumstances may produce false combs on segments where true ctenidia do not occur, or else develop unique hooklike structures to enhance their chances of remaining on their hosts.

14. Even the combless fleas may exhibit chaetotactic hyperdevelopment in response to undue selection pressure exerted by the host. Displacement of one or more pairs of the lateral plantar bristles of the fifth tarsal segment in fleas is also associated with arboreal mammals, for example. Elongate tarsal segments are seen in a variety of fleas infesting

some nocturnal glissant or arboreal mammals and certain bats. In addition to the loss of the eye and modifications in the shape and number of ctenidial spines, the highly diverse group of fleas that infests fossorial hosts tends to have supernumerary, unusually long and thin bristles. These bristles presumably serve to protect the flea from particles of soil in the burrow.

15. The morphological evidence and the data on zoogeography, evolution, and host relationships all indicate the intensely close association of fleas and mammalian hosts and of their coevolution.

SUMMARY

Most mammalian fleas have evolved with their hosts over a period of millions of years, as indicated by data and observations on the host relationships, zoogeography and evolution, and the morphology of fleas. The majority of fleas are either ultraspecific and are wholly restricted to a single species of host, or they are generispecific and are limited to infestation of closely related taxa of mammals. The ultraspecific fleas, and their hosts as well, are almost invariably specialized or unique in a major respect. Some fleas have switched to new hosts and have developed characteristic adaptive morphological modifications in consequence.

In general, the various families of Siphonaptera have been associated with the aboriginal hosts and descendants ever since they arose and have often accompanied the mammals as they dispersed to various parts of the world. The stephanocircids, doratopsylline fleas, and certain pygiopsyllids of marsupials, like their hosts, date back to at least the Cretaceous period. Some pulicids have parasitized certain groups of hosts since at least the Eocene epoch, and the most recently evolved family, the Ceratophyllidae, probably arose in the Oligocene epoch about 35 million years ago, along with their sciurid hosts. Inasmuch as the fleas have adapted to the vestiture of the mammals and to the survival problems posed by the special activities of their hosts, in general, evolution obviously proceeded more slowly in the fleas than in the animals they infest.

There is a well-established correlation between the level of evolutionary development of the flea and that of the host. The ancient, primitive hosts have the most primitive fleas, and the more recently evolved mammals have the more specialized fleas. On the whole, the rate of evolution has been relatively slow in the Siphonaptera. The same species and even subspecies of fleas often occur in areas separated by barriers such as desert and mountain ranges.

The diversity of bristles, spines, and combs in the Siphonaptera generally represents adaptive, functional, and evolutionary responses associated with the prevention of dislodgement from the host. Some fleas have demonstrated marked divergence as they adapted to moles rather than shrews, for example, and such fleas have come to resemble, to a remarkable degree, unrelated fleas occurring on those same kinds of hosts (convergence). The shape of the spines and bristles often appear to be molded to the vestiture of the hosts. Thus spiny mammals tend to have spiny fleas. The degree of development of spines, spiniforms, and combs appears to be correlated with the amount of selection pressure exerted upon the flea by the traits of its particular host. Chaetotaxal development is minimal in fleas of mammals that are diurnal, surface dwelling, and somewhat communal. Supernumerary bristles and combs of spines or spiniform bristles are characteristic of fleas whose hosts are nocturnal and arboreal or volant.

All these points indicate that for countless eons fleas that are host specific have been in highly intimate association with the mammals they infest.

ACKNOWLEDGMENTS

Dr. K. C. Kim organized and directed the Symposium on "Coevolution of Parasitic Arthropods and Mammals," and made arrangements for the publication of these papers. All the participants are greatly indebted to him. E. W. Jameson, Jr. and K. C. Kim read the text and kindly made many extremely helpful comments. Information on the habits of certain hosts was helpfully supplied by Alan Ziegler, E. W. Jameson, Jr., Istvan Szabo, and Waclaw Skuratowicz. R. E. Lewis kindly provided some specimens for examination. Fruitful discussions on evolution and host distribution were held with Miriam Rothschild; with Harry Hoogstraal on specificity; and with Abdu Farhang-Azad on various significant points. Jytte Pedersen helped check bibliographic references, and, as usual, Helle Starcke provided much editorial assistance. Dissections were prepared by the writer and mounted by Phuangthong Malikul. The illustrations are photomicrographs retouched by the preparator with the assistance of the author. Figures 8.115–8.117 are from my 1968 article in the *Journal of Medical Entomology* and are reproduced with permission of Frank J. Radovsky, Editor. Of the 126 figures, the first 45 (excluding Fig. 8.22) and Figs. 8.45, 8.47, 8.49 and 8.77–8.90 were prepared by J. Navarro. The remaining 57, including 14 complete plates, were by Caroline A. Herbert of our Department. To these friends and colleagues I extend many and sincere thanks.

REFERENCES

Barrera, A. 1967. Redefinicion de *Cediopsylla* Jordan y *Hoplopsyllus* Baker. *Rev. Soc. Mex. Hist. Nat.* **27**:67–83.

Black, C. C. 1972. Holarctic evolution and dispersal of squirrels (Rodentia: Sciuridae). *Evol. Biol.* **6**:305–322.

Brooks, D. R. 1979. Testing the context and extent of host–parasite coevolution. *Syst. Zool.* **28**:299–307.

Chaline, J. and P. Mein. 1979. *Les rongeurs et l'évolution.* Doin Editeurs, Paris. 235 pp.

Chin, T.-h. 1980. Studies on Chinese Anoplura IV. The description of two new species and proposal of new families and new suborder for the lice of *Typhlomys cinareus* Milne-Edwards. *Acta Acad. Med. Guiyang* **5**:91–100.

Ciochan, R. L. and A. B. Chiarelli. 1979. *Evolutionary Biology of the New World Monkeys and Continental Drift.* Proc. Symp. 7th Congr. Int. Primatol. Soc., Bangalore, India, Jan. 1979, Plenum Press, New York and London. 528 pp.

Colbert, E. H. 1980. *Evolution of the Vertebrates.* John Wiley, New York. 510 pp.

Corbet, G. B. and J. E. Hill. 1980. *A World List of Mammalian Species.* Brit. Mus. (Nat. Hist.), Comstock Publ. Assoc., a division of Cornell University Press, London and Ithaca. 226 pp.

Darskaya, N. F. 1955. On counting fleas at the entrances to burrows of *Rhombomys opimus* in northern and western Kyzul-Kum. *Vop. kraev. obshch. eksp. Parazit. med. Zool.* **9**:87–95.

Dubinin, V. 1947. Geographic distribution and probable direction of migration of fleas of the genus *Pariodontis* in connection with the history of their hosts (porcupines). *Dokl. Akad. Nauk* **58**:1557–1560.

Dunnet, G. M. and D. K. Mardon. 1974. A monograph of Australian fleas (Siphonaptera). *Aust. J. Zool.*, Suppl. Ser. No. **30, 271** pp.

Fain, A. 1965. A review of the family Rhyncoptidae Lawrence parasitic on porcupines and monkeys (Acarina: Sarcoptiformes). Pages 135–158 in J. A. Naegele (Ed.), *Advances in Acarology*, Vol. 2. Comstock Publ. Assoc., Cornell University Press, London and Ithaca.

Fain, A. 1979. Specificity, adaptation and parallel host–parasite evolution in acarines, especially Myobiidae, with a tentative explanation for the regressive evolution caused by the immunological reactions of the host. Pages 321–327 in J. G. Rodriguez (Ed.), *Recent Advances in Acarology*, Vol. 2. Academic Press, New York.

Farhang-Azad, A., R. Traub, and C. L. Wisseman, Jr. 1983. *Rickettsia mooseri* infection in the fleas *Leptopsylla segnis* and *Xenopsylla cheopis*. *Am. J. Trop. Med. Hyg.* **32**(6):1392–1400.

Freeman, R. B. and H. Madsen. 1949. A parasitic flea larva. *Nature* **164**(4161):187–188.

Haddow, J., R. Traub, and M. Rothschild. 1983. Distribution of Ceratophyllid fleas and their hosts. Pages 42–163 in R. Traub, M. Rothschild, and J. Haddow (Eds.), *The Rothschild Collection of Fleas. The Ceratophyllidae: Key to the genera and host relationships. With notes on evolution, zoogeography and medical importance.* Cambridge University Press/Academic Press, Cambridge and London.

Harrison, L. 1914. The Mallophaga as a possible clue to bird phylogeny. *Aust. Zool.* **1**:7–11.

Harrison, L. 1916. Bird parasites and bird-phylogeny. *Aust. Zool.* (10)**4**:254–263.

Harrison, L. 1928. Host and parasite. Presidential Address. *Proc. Linn. Soc. N.S. Wales* **53**:IX–XXXI.

Hendey, Q. B. 1976. Fossil peccary from the Pliocene of South Africa. *Science* **192**:787–789.

Holland, G. P. 1950a. Notes on *Megabothris asio* (Baker) and *M. calcarifer* (Wagner) with the description of a new subspecies (Siphonaptera: Ceratophyllidae). *Can. Entomol.* **82**:126–133.

Holland, G. P. 1950b. Notes on some British Columbian fleas, with remarks on their relationship and distribution. *Proc. Entomol. Soc. Br. Col.* **46**:1–9.

Holland, G. P. 1955. Primary and secondary sexual characteristics of some Ceratophyllinae, with notes on the mechanism of copulation (Siphonaptera). *Trans. R. Entomol. Soc. Lond.* **107**:233–248.

Holland, G. P. 1958. Distribution patterns of northern fleas (Siphonaptera). *Proc. 10. Int. Congr. Entomol.* (1956) **1**:645–658.

Holland, G. P. 1963. Faunal affinities of the fleas (Siphonaptera) of Alaska, with an annotated list of species. Pages 45–63 in *Pacific Basin Biogeography, A Symposium.* Symp. 10th Pacif. Sci. Congr., 21 Aug.–6 Sept. 1961, Bishop Mus. Press, Honolulu.

Holland, G. P. 1964. Evolution, classification, and host relationships of Siphonaptera. *Annu. Rev. Entomol.* **9**:123–146.

Holland, G. P. 1969. Contribution towards a monograph of the fleas of New Guinea. *Entomol. Soc. Can. Mem.* **66**:1–77.

Holland, G. P. and G. Loshbaugh, Jr. 1958. Two new species of fleas from Utah with notes on the genus *Ornithophaga* Mikulin (Siphonaptera). *Can. Entomol.* **90**:486–493.

Hopkins, G. H. E. 1957. Host-associations of Siphonaptera. Pages 64–87 in *First Symposium on Host Specificity among Parasites of Vertebrates.* Inst. Zool., University of Neuchatel.

Hopkins, G. H. E. and M. Rothschild. 1953. *An Illustrated Catalogue of the Rothschild Collection of Fleas (Siphonaptera) in the British Museum,* Vol. 1. British Museum (N.H.). London. 361 pp.

Hopkins, G. H. E. and M. Rothschild. 1956. *An Illustrated Catalogue of the Rothschild Collection of Fleas (Siphonaptera) in the British Museum,* Vol. 2. British Museum (N.H.). London. 445 pp.

Hopkins, G. H. E. and M. Rothschild. 1966. *An Illustrated Catalogue of the Rothschild Collection of Fleas (Siphonaptera in the British Museum,* Vol. 4. British Museum (N.H.). London. 549 pp.

Hopla, C. E. 1980a. A study of the host associations and zoogeography of *Pulex.* Pages 185–208 in R. Traub and H. Starcke (Eds.). *Fleas.* Proc. Int. Conf. Fleas, Ashton, England, June, 1977. A. A. Balkema, Rotterdam.

Hopla, C. E. 1980b. Fleas as vectors of tularemia in Alaska. Pages 287–300 in R. Traub and H. Starcke (Eds.). *Fleas.* Proc. Int. Conf. Fleas, Ashton, England, June, 1977. A. A. Balkema, Rotterdam.

Humphries, D. A. 1966. The function of combs in fleas. *Entomol. Mon. Mag.* **102**:232–236.

Humphries, D. A. 1967. Function of combs in ectoparasites. *Nature* **215**:319.

Ioff, I. G. 1929. Material for the study of the ectoparasite fauna of southeast USSR. VI. Fleas of mole-rats (Spalacidae). *Izv. gos. mikrobiol. Inst. Rostov.* **8**:29–43. German summary pp. 56–59.

Ioff, I. G. 1947. *Problems in the Ecology of Fleas in Relation to Their Epidemiological Importance.* Ordzhonikidze Regional Publishing House, Piatigorsk. 116 pp.

Ioff, I. G. 1953. New cases of species formation in fleas in the case of change of hosts. *Akad. Nauk. SSSR (Dok.)* **89**:189–192.

Jellison, W. L. 1945. Siphonaptera: A new species of *Conorhinopsylla* from Kansas. *J. Kans. Entomol. Soc.* **18**:109–111.

Jellison, W. L. 1947. Siphonaptera: Host distribution of the genus *Opisocrostis* Jordan. *Trans. Am. Micros. Soc.* **66**:64–69.

Jordan, K. 1926. On *Xenopsylla* and allied genera of Siphonaptera. *3rd. Int. Ent.-Kongr. Verh.* (1925) **2**:593–627.

Jordan, K. 1942. On *Parapsyllus* and some closely related genera of Siphonaptera. *Eos* **18**:7–29.

Jordan, K. 1947. On some phylogenetic problems within the order of Siphonaptera (= Suctoria). *Tijdschr. Entomol.* **88**:79–93.

Jordan, K. 1948. Suctoria. Chapter on fleas. Pages 211–245 in J. Smart (Ed.), *A Handbook for the Identification of Insects of Medical Importance,* 2nd ed. British Museum (N.H.), London.

Jordan, K. 1950. On characteristics common to all known species of Suctoria and some trends of evolution in this order of insects. *8th Int. Congr. Entomol.* Stockholm: 87–95.

Jordan, K. and N. C. Rothschild. 1912. On Siphonaptera collected in Algeria. *Novit. Zool.* **19**:357–372.

Jordan, K. and N. C. Rothschild. 1921. On *Ceratophyllus fasciatus* and some allied Indian species of fleas. *Ectoparasites* **1**:178–198.

Keast, A. 1977. Historical biogeography of the marsupials. Pages 69–95 in B. Stonehouse and D. Gilmore (Eds.), *The Biology of Marsupials.* University Park Press, Baltimore, London, Tokyo. 486 pp.

Kurtén, B. and E. Anderson. 1980. *Pleistocene mammals of North America.* Columbia University Press, New York. 442 pp.

Lewis, R. E. 1967. Contributions to a taxonomic revision of the genus *Nosopsyllus* Jordan, 1933 (Siphonaptera: Ceratophyllidae). I. African species. *J. Med. Entomol.* **4**:123–142.

Lewis, R. E. 1971. A new species of *Chaetopsylla* Kohaut, 1903, infesting pikas in Nepal (Siphonaptera: Vermipsyllidae). *J. Parasitol.* **57**:1344–1348.

Li, K.-c. and G.-p. Huang. 1981. A new genus and three new species of fleas from Kuan-kuoshi Natural Reserve in Guizhov province. *Acta Zootaxon. Sinica* **6**:291–297.

Lidicker, W. Z., Jr. 1968. A phylogeny of New Guinea rodent genera based on phallic morphology. *J. Mammal.* **49**:609–643.

Manter, H. W. 1955. The zoogeography of trematodes of marine fishes. *Exp. Parasitol.* **4**:62–86.

Manter, H. W. 1963. The zoogeographical affinities of trematodes of South American freshwater fishes. *Syst. Zool.* **12**:45–70.

Manter, H. W. 1966. Parasites of fishes as biological indicators of recent and ancient conditions. Pages 59–71 in J. E. McCauley (Ed), *Host–Parasite Relationships,* Proc. 26th Ann. Biol. Colloq., Oregon State University.

Mardon, D. K. 1978. On the relationships, classification, aedeagal morphology and zoogeography of the genera of Pygiopsyllidae (Insecta: Siphonaptera). *Aust. J. Zool.*, Suppl. Ser. No. **64**. 69 pp.

Marshall, A. G. 1980. The function of combs in ectoparasitic insects. Pages 79–88 in R. Traub and H. Starcke (Eds.), *Fleas.* Proc. Int. Conf. Fleas, Ashton, England, June, 1977. A. A. Balkema, Rotterdam.

McKenna, M. C. 1975. Fossil mammals and early Eocene North Atlantic land continuity. *Ann. Mo. Bot. Gard.* **62**:335–353.

Oudemans, A. C. 1909. Klassification der Suctoria. *Novit. Zool.* **16**:133–158.

Patterson, B. 1957. Mammalian phylogeny. Pages 15–49 in *First Symposium on Host Specificity among Parasites of Vertebrates.* Inst. Zool., University of Neuchatel.

Pearse, A. M. 1981. *Aspects of the Biology of Uropsylla tasmanica Rothschild (Siphonaptera).* M.Sc. Thesis. University of Tasmania, Hobart, Tasmania.

Peus, F. 1968. Uber die beiden Bernstein-Flohe (Insecta: Siphonaptera). *Palaontol. Zeitschr.* **42**:62–72.

Riek, E. F. 1970. Lower Cretaceous fleas. *Nature* **227**:746–747.

Rothschild, M. 1965a. Fleas. *Sci. Am.* **216**:44–52 and p. 126.

Rothschild, M. 1965b. The rabbit flea and hormones. *Endeavour* **24**:162–168.

Rothschild, M. 1976. Notes on fleas (part II): The internal organs: can they throw any light on relationships within the Order? *Proc. Trans. Br. Entomol. Nat. Hist. Soc.* **9**:97–110.

Rothschild, M. and T. Clay. 1952. *Fleas, Flukes and Cuckoos.* Collins, London. 304 pp.

Rothschild, M. and B. Ford. 1973. Factors influencing the breeding of the rabbit flea (*Spilopsyllus cuniculi*): A spring-time accelerator and a kairomone in nestling rabbit urine with notes on *Cediopsylla simplex*, another "hormone bound" species. *J. Zool., Lond.* **170**:87–137.

Rothschild, M., B. Ford, and M. Hughes. 1970. Maturation of the male rabbit flea (*Spilopsyllus cuniculi*) and the oriental rat flea (*Xenopsylla cheopis*): some effects of mammalian hormones on development and impregnation. *Trans. Zool. Soc. Lond.* **32**:105–188.

Rothschild, M. and J. Schlein. 1975. The jumping mechanism of *Xenopsylla cheopis*. I. Exoskeletal structures and musculature. *Phil. Trans. R. Soc. Lond.* **(B):271**:457–490.

Rothschild, M., Y. Schlein, K. Parker, and S. Sternberg. 1972. Jump of the oriental rat flea *Xenopsylla cheopis* (Roths.). *Nature* **239**(5366):45–48.

Rothschild, M., Y. Schlein, K. Parker, C. Neville, and S. Sternberg. 1975. The flying leap of the flea. *Sci. Am.* **229**(5):92–100, 136.

Rothschild, M., J. Schlein, K. Parker, C. Neville, and S. Sternberg. 1975. The jumping mechanism of *Xenopsylla cheopis*. III. Execution of the jump and activity. *Phil. Trans. R. Soc. Lond.* **(B)271**:499–515.

Simpson, G. G. 1945. The principles of classification and a classification of mammals. *Bull. Am. Mus. Nat. Hist.* **85**. 350 pp.

Simpson, G. G. 1961. Historical zoogeography of Australian mammals. *Evolution* **15**:431–446.

Smit, F. G. A. M. 1963. Species-groups in *Ctenophthalmus* (Siphonaptera: Hystrichopsyllidae). *Br. Mus. (Nat. Hist.) Ent. Bull.* **14**(3):105–152.

Smit, F. G. A. M. 1967. Siphonaptera of Mongolia. Results of the Mongolian-German Biological Expeditions since 1962, No. 23. *Mitt. Zool. Mus. Berl.* **43**(1):77–115.

Smit, F. G. A. M. 1972. On some adaptive structures in Siphonaptera. *Folia Parasitol.* (Praha) **19**:5–17.

Smit, F. G. A. M. 1976. Two new east African mole-rat fleas (Siphonaptera: Hystrichopsyllidae). *Rev. Zool. Afr.* **90**(1):46–52.

Strenger, A. 1973. Zur Ernarungsbiologie der larve von *Ctenocephalides felis felis* B. *Zool. J. Syst.* **100**(1):64–80.

Szidat, L. 1956. Der marine charakter der Parasitenfauna der Susswasserfische des Stromsystems des Rio de la Plata und ihre Deutung als Reliktfauna des Tierarten Tethys-Meeres. *Proc. 14th. Int. Congr. Zool.* **1953**:128–138.

Tate, G. H. H. 1951. Results of the Archbold expeditions. No. 65. The rodents of Australia and New Guinea. *Bull. Am. Mus. Nat. Hist.* **97**(Art. 4):187–430.

Traub, R. 1952. Records and descriptions of fleas from Peru (Siphonaptera). *Proc. Entomol. Soc. Wash.* **54**:1–22.

Traub, R. 1953. *Hollandipsylla neali*, a new genus and new species of flea from North Borneo, with comments on eyeless fleas (Siphonaptera). *J. Wash. Acad. Sci.* **43**:346–353.

Traub, R. 1963. The fleas of Egypt. Two new fleas of the genus *Nosopsyllus* Jordan, 1933 (Siphonaptera: Ceratophyllidae). *Proc. Entomol. Soc. Wash.* **65**:81–97.

Traub, R. 1968a. *Smitella thambetosa*, n.gen. and n.sp., a remarkable "helmeted" flea from New Guinea (Siphonaptera, Pygiopsyllidae) with notes on convergent evolution. *J. Med. Entomol.* **5**:375–404.

Traub, R. 1968b. Book review: R. L. Wenzel and V. J. Tipton (Eds.), *Ectoparasites of Panama*, Field Museum of Natural History, Chicago, Ill. pp. xii + 861, illus., 1966. *Bull. Entomol. Soc. Am.* **14**:143–145.

Traub, R. 1969. *Muesebeckella*, a new genus of flea from New Guinea, with notes on convergent evolution (Siphonaptera: Pygiopsyllidae). *Proc. Entomol. Soc. Wash.* (Muesebeck Jubilee Issue) **71**:374–396.

Traub, R. 1972a. Notes on zoogeography, convergent evolution and taxonomy of fleas (Siphonaptera), based on collections from Gunong Benom and elsewhere in Southeast Asia. I. New Taxa (Pygiopsyllidae, Pygiopsyllinae). *Bull. Br. Mus. Nat. Hist.* (*Zool.*) **23**:201–305.

Traub, R. 1972b. Notes on zoogeography, convergent evolution and taxonomy of fleas (Siphonaptera), based on collections from Gunong Benom and elsewhere in Southeast Asia. II. Convergent evolution. *Bull. Br. Mus. Nat. Hist.* (*Zool.*) **23**:307–387.

Traub, R. 1972c. Notes on zoogeography, convergent evolution and taxonomy of fleas (Siphonaptera), based on collections from Gunong Benom and elsewhere in Southeast Asia. III. Zoogeography. _Bull. Br. Mus. Nat. Hist._ (_Zool._) **23**:389–450.

Traub, R. 1972d. The colloquium on the zoogeography and ecology of ectoparasites, their hosts and related infections at the Second International Congress of Parasitology, Washington, D.C., 1970. 2. The zoogeography of fleas (Siphonaptera) as supporting the theory of continental drift. _J. Med. Entomol._ **9**:584–589.

Traub, R. 1972e. 23. The relationship between the spines, combs and other skeletal features of fleas (Siphonaptera) and the vestiture, affinities and habits of their hosts. _J. Med. Entomol._ **9**:601.

Traub, R. 1972f. 27. Notes on fleas and the ecology of plague. _J. Med. Entomol._ **9**:603.

Traub, R. 1977. _Tiflovia_, a new genus of pygiopsyllid fleas from New Guinea, with notes on convergent evolution and zoogeography (Siphonaptera). _J. Med. Entomol._ **13**:653–685.

Traub, R. 1980a. New genera and subgenera of pygiopsyllid fleas (Siphonaptera). Pages 13–29 in R. Traub and H. Starcke (Eds.), _Fleas_. Proc. Int. Conf. Fleas, Ashton, England, June 1977. A. A. Balkema, Rotterdam.

Traub, R. 1980b. Some adaptive modifications in fleas. Pages 44–69 in R. Traub and H. Starcke (Eds.), _Fleas_. Proc. Int. Conf. Fleas, Ashton, England, June 1977. A. A. Balkema, Rotterdam.

Traub, R. 1980c. The zoogeography and evolution of some fleas, lice and mammals. Pages 93–172 in R. Traub and H. Starcke (Eds.), _Fleas_, Proc. Int. Conf. Fleas, Ashton, England, June 1977. A. A. Balkema, Rotterdam.

Traub, R. 1983a. The hosts of the Ceratophyllid fleas. Pages 164–187 in R. Traub, M. Rothschild, and J. Haddow. _The Rothschild Collection of Fleas._ The Ceratophyllidae: Key to the genera and host relationships. With notes of evolution, zoogeography and medical importance. Cambridge University Press/Academic Press, Cambridge and London.

Traub, R. 1983b. Medical importance of the Ceratophyllidae. Pages 202–228 in R. Traub, M. Rothschild and J. Haddow. _The Rothschild Collection of Fleas._ The Ceratophyllidae: Key to the genera and host relationships. With notes on evolution, zoogeography and medical importance. Cambridge University Press/Academic Press, Cambridge and London.

Traub, R. and A. Barrera. 1966. New species of _Ctenophthalmus_ from Mexico, with notes on the ctenidia of shrew-fleas (Siphonaptera) as examples of convergent evolution. _J. Med. Entomol._ **3**:127–145.

Traub, R. and G. M. Dunnet. 1973. Revision of the siphonapteran genus _Stephanocircus_ Skuse, 1893 (Stephanocircidae). _Aust. J. Zool._, Suppl. Ser. (1973) **20**:41–128.

Traub, R. and T. M. Evans. 1967a. Notes and descriptions on some leptopsyllid fleas (Siphonaptera). _J. Med. Entomol._ **4**(3):340–359.

Traub, R. and T. M. Evans. 1967b. Descriptions of new species of hystrichopsyllid fleas, with notes on arched pronotal combs, convergent evolution and zoogeography (Siphonaptera). _Pac. Insects_ **9**(4):603–677.

Traub, R. and W. L. Jellison. 1981. Evolutionary and biogeographic history and the phylogeny of vectors and reservoirs as factors in the transmission of diseases from other animals to man. Pages 517–546 in W. Burgdorfer and R. L. Anacker (Eds.), _Rickettsiae and Rickettsial Diseases._ Proc. RML Conf. Rickettsiae and Rickettsial Diseases, 1980. Academic Press, New York.

Traub, R. and M. Rothschild. 1983. Evolution of the Ceratophyllidae. Pages 188–201 in R. Traub, M. Rothschild, and J. Haddow. _The Rothschild Collection of Fleas._ The Ceratophyllidae: Key to the genera and host relationships. With notes on evolution, zoogeography and medical importance. Cambridge University Press/Academic Press, Cambridge and London.

Traub, R., M. Rothschild, and J. Haddow. 1983. *The Rothschild Collection of Fleas. The Ceratophyllidae: Key to the genera and host relationships. With notes on evolution, zoogeography and medical importance.* Cambridge University Press/Academic Press, Cambridge and London. 288 pp.

Traub, R. and C. L. Wisseman, Jr. 1968. Ecological considerations in scrub typhus. 2. Vector species. *Bull. W.H.O.* **39**(2):219–230.

Traub, R. and C. L. Wisseman, Jr. 1974. The ecology of chigger-borne rickettsiosis (scrub typhus). *J. Med. Entomol.* **11**(3):237–303.

Traub, R., C. L. Wisseman Jr., and Nur Ahmad. 1967. The occurrence of scrub typhus infection in unusual habitats in West Pakistan. *Trans. R. Soc. Trop. Med. Hyg.* **61**(1):23–57.

Wagner, J. 1932. Die bedeutung der Flohe fur die Frage nach der genesis der Saugetierfauna. *Zoogeographica* **1**(2):263–268.

Walker, E. P. 1968. *Mammals of the World.* Vol. 1. Johns Hopkins Press, Baltimore. 646 pp.

Wood, A. E. 1977. *Paleontology and Plate Tectonics with Special Reference to the History of the Atlantic Ocean.* In R. M. West (Ed.). Proc. Symp. N. Am. Paleont. Conv. II, Lawrence, Kans., Milwaukee Publ. Mus. Spec. Publ. **2**:95–109.

PART THREE

ACARI

Chapter 9

Evolution of Mammalian Mesostigmate Mites

Frank J. Radovsky

INTRODUCTION

The suborder Mesostigmata is one of the major categories of the arachnid subclass Acari. The mesostigmate parasites exhibit considerable diversity in their relationships with hosts, including mammals. Here, the several groups of mammalian-associated Mesostigmata are analyzed and discussed separately. Most of the ecologically diverse taxa that occur on or in vertebrates appear to have evolved from one ancestral group, and their relationships can be stated with a reasonable degree of confidence. This chapter expands the system in which some of these relationships were previously outlined (Radovsky 1966, 1967, 1969).

Concerning insect associates of vertebrates, Waage (1979) wrote that "Unfortunately, some of the better known groups are highly evolved, parasitic forms whose evolutionary origins are obscure . . . groups with clearer links to free-living ancestors, such as commensal Coleoptera and Lepidoptera, have received little study." The Mesostigmata associated with vertebrates present a very different picture. These mites have evolved to parasitism through first becoming commensal nidicoles, what Waage (1979) called macroevolutionary pathway I, and their various stages can be traced through living species. Within one genus, there are polyphagous facultative nidicoles with no indication of parasitic feeding, facultative parasites, and obligatory parasites. In some cases transitional forms also exist to indicate the stages in evolution from one type of parasitism to another, such as ectoparasitism to endoparasitism (cavity parasitism).

The principal group of mesostigmate associates of vertebrates, Dermanyssoidea, also includes many free-living forms (e.g., in soil and leaf litter) and a number of lines that developed associations with invertebrates. Investigators have generally concentrated on one or another of these three major ecological groupings, and more work is still needed to unify the classification across ecological lines (Evans 1955; Evans and Till 1962, 1966). However, there appear to be only two major lines that led to the diverse vertebrate associations in the Dermanyssoidea; the development of these pathways is the emphasis of this chapter.

I restrict my discussion to mammalian associates in a group that also includes symbionts of reptiles and birds. This poses no problem because the principal associations in the lines that led to nest dwelling and then to parasitism were with mammals, and the major exploitation of reptilian and avian hosts was secondary. In most instances the dermanyssoid taxa on vertebrates other than mammals are sufficiently distinctive to be treated as separate families. In the two wholly parasitic families, Dermanyssidae and Macronyssidae, with vertebrate hosts of two or more classes, mammals were apparently the primary hosts in the evolution of each group. Consequently, it has been relatively easy to exclude discussion of the nonmammalian associates, for the most part. However, I have treated the avian respiratory mites (Rhinonyssidae) in discussing other dermanyssoid families because of their origin in a family that occurs principally on mammals, their convergent adaptations with some mammalian parasites, and the instructive nature of the patterns of their host associations.

Our knowledge of the parasitic Mesostigmata is uneven, and the gaps in the most elementary sorts of information about some groups, for example, where and how they attach to the host, are frustrating. Therefore, I have been opportunistic in presenting data. In the discussion of a family or subfamily, particular phenomena relating to parasitism and coevolution are brought out where there is the most information to develop a theme. For example, the application of parasite taxonomy to the interpretation of host taxonomy is particularly developed for the Spinturnicidae.

General Description of the Mesostigmata

The free-living Mesostigmata, about 0.2–2.0 mm in length and found in a wide range of habitats but particularly leaf litter and soil, typically are active, predatory organisms with long legs adapted for running. Some of the parasitic forms retain this habitus whereas others are extremely modified. The tarsi in the free-living forms are provided with strong paired claws and a suckerlike caruncle. The first of the four pairs of legs may be particularly long and slender and specialized as sensory organs. The mouthparts are complex but include two paired structures most significant in food acquisition: the sensory and sometimes manipulative palps and, especially important, the chelicerae. The latter are retractable supraoral structures, provided with distal toothed jaws composed of a dorsal fixed chela and a ventral movable chela. The chelicerae function in grasping, tearing, and puncturing prey or other food. The legs and the gnathosoma, which bears the mouthparts, are attached to a saclike, unsegmented idiosoma, which comprises the body of the mite. The idiosoma bears a number of sclerotized plates that provide protection and relatively rigid loci for muscle attachment. A usually forked tritosternum, arising near the anterior ventral margin of the idiosoma, interacts with a ventral gnathosomal groove (deutosternum) to control overflow of fluid food during feeding (Wernz and Krantz 1976). Respiration is by means of tracheae branching from the single pair of spiracular openings (stigmata) located lateroventrally, most often in the region of coxae III and IV. A peritreme extends from each spiracle, generally anteriad; it is a tube of uncertain function, with a narrow slit-opening that may be under a degree of muscular control.

The mesostigmate life cycle includes egg, larva, protonymph, deutonymph, and adult female and male. The larva is hexapod, lacking the fourth pair of legs, in contrast to the subsequent octopod stages. Each of the immature stages typically has distinctive plates and setation. The plates may be suppressed totally in the larva. The larva is frequently nonfeeding, with mouthparts relatively well formed but clearly nonfunctional, even in some free-living groups. Variations in the life cycle, with suppression or specialization of various stages, are very important in the evolution of vertebrate relationships. All stages are uniformly present, but they may be completed inside the female or, rarely, one stage may remain within the ecdysed skin of the preceding stage.

Preadaptation for Parasitism

The free-living Mesostigmata are significantly preadapted for parasitism. Morphological changes that have occurred in some parasites are so subtle that it occasionally may not be possible to tell by morphology alone whether a species is an obligatory parasite. Several phyletic lines appear to have progressed independently either from the free-living or the commensal stage to obligate parasitism. Some important preadaptive features are the small size; the ambulacral apparatus, particularly the stout claws; the mouthparts, particularly the chelicerae, which even in their unmodified form are highly versatile in feeding on secretions, scales, and so forth, and may be used in picking into the skin to reach the capillary bed; and, developmentally, the ability of nonfeeding stages to be passed in the host habitat, such as the nest, or to be completed within the female.

The chelicerae are fundamentally important in the adaptation of Mesostigmata to parasitism. The heavier-jawed, toothed, grasping chelicerae of the free-living mesostigmate can be used to abrade the skin and cause a flow of blood. However, repeatedly, in separate lines of evolutionary descent, the chelae have become slender, essentially edentate structures that can operate together as the tip of an otherwise modified chelicera to penetrate the vertebrate skin and cause a flow of blood or tissue fluids, and most often to act as a conduit for these fluids. Thus the "generalized" chelicera in free-living Mesostigmata is an elongate pistonlike structure, with a musculature that can be used for penetration, and it is already advanced toward the form of a piercing stylet, able with little modification to achieve the same functions as stylets in bloodsucking insects.

It has been indicated that vertebrate associations, and particularly parasites of mammals, are largely restricted to the Dermanyssoidea. It is not clear why this is so, since the foregoing preadaptive features apply to other higher taxa in the suborder. This question is discussed further in the concluding section.

DIVERSITY, DISTRIBUTION, AND PHYLOGENY

Classification

The classification of the Mesostigmata has been relatively stable over the last decade, although continued changes in higher category names, generally without change in value, and the continued use of different levels for the same groups (e.g., family versus subfamily) by various workers may make it appear superficially to be in flux. In this chapter, I follow Krantz (1978) for the most part, which is essentially in agreement with Radovsky (1969) for the groups to be treated. However, Mesostigmata is retained as a subordinal name rather than "Gamisida" used by Krantz (following van der Hammen 1972).

The Acari or mites are increasingly being recognized as a subclass of the Arachnida with two orders, which may be diphyletic in origin. The order Parasitiformes includes the suborder Mesostigmata, as well as the vertebrate-parasitizing ticks (Metastigmata or Ixodida) and two smaller and free-living suborders. Within the Mesostigmata, most vertebrate associations are in the superfamily Dermanyssoidea of the cohort Gamasina, super-cohort Monogynaspides. Further discussion is concerned almost entirely with classification in the Dermanyssoidea among the 20 superfamilies of Mesostigmata.

Krantz (1978) includes 14 families in the Dermanyssoidea, all of which include a significant proportion of associates of various animals, 13 exclusively so. Of the latter, one (Varroidae) is found with bees; three (Ixodorhynchidae, Omentolaelapidae, Entonyssidae—the last endoparasitic in the lungs) with snakes; one (Rhinonyssidae) in the respiratory tracts of birds; one (Dermanyssidae) primarily on birds but with several species on mammals; one (Macronyssidae) primarily on mammals but also on reptiles and birds; and the remaining six (Spinturnicidae, Halarachnidae, Dasyponyssidae, Manitherionyssidae, Hystrichonyssidae, and Spelaeorhynchidae) are associated only with mammals. The other family, Laelapidae, ranges from free-living predators to obligatory parasites.

Krantz (1978) recognizes nine subfamilies in Laelapidae: Hypoaspidinae includes many free-living species, many associated with arthropods, and a few regularly found in the nests or on the bodies of vertebrates; Haemogamasinae is primarily associated with mammals, occasionally birds, but some species facultatively so; Laelapinae is associated with mammals, occasionally birds, and ranges from polyphagous nidicoles to obligate parasites; Alphalaelapinae, Myonyssinae, and Hirstionyssinae are parasitic on mammals; Iphiopsinae and Melittiphinae are associated with arthropods; and Pseudolaelapinae is free-living. Thus three subfamilies are entirely associated with mammals, two are largely so, and one has some mammalian associations. A tenth subfamily, also found entirely on mammals, was added to the Laelapidae by splitting the monogeneric Mesolaelapinae from the Laelapinae (Tenorio and Radovsky 1974).

Do the Dermanyssoidea represent a natural grouping? The bat-parasitizing Spelaeorhynchidae may belong to the cohort Sejina rather than the Gamasina (Radovsky 1969). Sejina includes *Asternolaelaps* (Ichthyostomatogasteridae), which is found sufficiently often on small mammals to indicate a more than casual association. Most important, the morphology of Spelaeorhynchidae, at least suggestive of a sejine relationship, does not appear to bear out dermanyssoid placement. I treat the Spelaeorhynchidae here as a nondermanyssoid family of uncertain position. The remainder of the dermanyssoid families are all reasonably placed together on morphological grounds, and they appear to constitute a monophyletic group.

Myonyssoides, with three species on burrowing mammals in Africa, has been placed in the Ascidae (Gamasina: Ascoidea) but needs reexamination. Nothing except the host associations is known about its biology, and it is not treated further here.

To reduce confusion, it should be noted that many workers (mostly British and in Commonwealth countries) continue to use a broad interpretation of Dermanyssidae, so that it is approximately equivalent to Dermanyssoidea as applied here. In that system most of the dermanyssoid families given are treated as subfamilies of Dermanyssidae. This organization goes back to Vitzthum (1940–1943) who, however, used Laelapidae as the name of the family in which most dermanyssoids were placed. Evans (1957 and later papers), Evans and Till (1966), and Domrow (1969, etc.) are among the productive researchers who treat Dermanyssidae in a broad sense, and reference to their work here involves conversion of the subfamily categories to families equivalent in content.

There is no single recent work that treats the classification of all or any major part of the Mesostigmata of mammals. Useful references for views on the higher classification and some relationships of higher categories are Evans and Till (1966), Radovsky (1967, 1969), and Krantz (1978). Evans and Till (1966) treat the British fauna, and consequently a representation of the western Palearctic species, in detail. Baker et al. (1956) and Strandtmann and Wharton (1958) are still very useful references, although the classifications at the generic level and above are considerably outmoded; the second of these references is the most recent attempt to summarize information and sources on all Mesostigmata parasitic on vertebrates.

Some useful references for the classification within families and subfamilies follow: Laelapinae—Tipton (1960); Till (1963), Ethiopian *Androlaelaps*; Furman (1972), Venezuela; Herrin and Tipton (1976), Ethiopian *Laelaps*. Haemogamasinae—Keegan (1951); Bregetova (1956), USSR; Williams et al. (1978), North American *Haemogamasus*. Mesolaelapinae—Tenorio and Radovsky (1974). Alphalaelapinae—Tipton (1960). Myonyssinae—Strandtmann and Garrett (1970). Hirstionyssinae—Herrin (1970), Nearctic *Echinonyssus*; Herrin and Yunker (1975), Neotropical *Echinonyssus*; Tenorio and Radovsky (1979). Macronyssidae—Radovsky (1967); Saunders (1975), Venezuela; Micherdzinski (1980). Dermanyssidae—Evans and Till (1962); Moss (1967, 1968, 1978). Halarachnidae—Furman (1979). Raillietiinae—Potter and Johnston (1978); Domrow (1981). Halarachninae—Newell (1947); Furman (1979). Spinturnicidae—Rudnick (1960); Furman (1966a), Panama; Domrow (1972), Australia, New Guinea; Herrin and Tipton (1975), Venezuela; Deunff (1977), Europe. Dasyponyssidae and Manitherionyssidae—Radovsky and Yunker (1971). Hystrichonyssidae—Keegan et al. (1960). Spelaeorhynchidae—Fain et al. (1967); Dušbábek (1970).

Diversity and Distribution

Host relationships and parasitopes in the parasitic Dermanyssoidea are quite diverse. Most mammalian parasites are nest or roost inhabitants, including those in the Laelapidae (with several important subfamilies), the Dermanyssidae, and the majority of the Macronyssidae. Among the obligate parasites, Spinturnicidae is the only large suprageneric grouping of external parasites that is essentially entirely host restricted. The

Spelaeorhynchidae, Dasyponyssidae, and Manitherionyssidae, each with one to several species, and a few species scattered among the genera of Macronyssidae are known or believed to complete their life cycles on the surface of the host. Finally, the endoparasitic (or cavity-parasitic) family Halarachnidae, found in the respiratory passages (Halarachninae) and the ears (Raillietiinae) of mammals, are host-restricted parasites.

Many of the larger groups are cosmopolitan, such as Laelapinae, Hirstionyssinae, Macronyssidae, Spinturnicidae, and Halarachnidae. However, some families and subfamilies are geographically restricted or have major areas of distribution: for example, Spelaeorhynchidae is restricted to the Neotropical Region, and Haemogamasinae is concentrated in the Holarctic Region.

Associations other than of an incidental and transient nature have been recorded for 12 mammalian orders. Two of these stand out as central to the evolution of dermanyssoid–mammal associations: the Rodentia and the Chiroptera, which also are the two largest mammalian orders in numbers of species. The Chiroptera are the central group in the evolution of the Macronyssidae (particularly Macronyssinae but also for a major evolutionary burst in the Ornithonyssinae), Spinturnicidae, and Spelaeorhynchidae. The last two families are restricted to bats. The rodents have had a central position in the evolution of all the laelapine subfamilies and are important hosts for ornithonyssine Macronyssidae and for genera in some other families.

The continuing existence of forms intergradient between free-living mites and parasites and between parasites of various ecological types has been noted. The survival of these "links" does not indicate that the vertebrate associations are recent. On the contrary, the broad host and geographic distributions, the high degree of specialization in many groups, and especially the evidences for congruence between the phylogenies of parasites and some host groups point to the great antiquity of many of the parasitic groups. The intergradient forms themselves appear to have survived over long periods because they continue to occupy niches in which they are able to compete successfully.

Dermanyssoid Phylogeny

Much of the following section, which is the core of this chapter, is concerned with the adaptations of dermanyssoid mites to various associations with mammals, and it is organized by families and subfamilies. A brief treatment of the relationships between these categories is given here, primarily to assist the reader in placing any group from the narrative that follows.

The first dermanyssoid radiation was within the family Laelapidae. The associates of vertebrates and invertebrates had their beginnings in the *Hypoaspis* complex (Evans 1957). This group of genera and subgenera in the subfamily Hypoaspidinae largely comprises free-living soil and litter mites, but it also includes a number of arthropod symbionts and species that have developed at least a facultative commensal, and often phoretic, association with mammals and/or birds. At least three lines lead from the Hypoaspidinae toward vertebrate associations that extend to parasitism. Haemogamasinae includes forms that range from free-living mites to obligatory parasites; however, they are mostly polyphagous nidicoles, and even the most advanced parasites in the subfamily are at a crude level of hematophagous specialization. The Haemogamasinae do not appear to have given rise to any other parasitic groups. The second line is through the Laelapinae. Again, there is a broad range of relationships from polyphagous nidicoles to obligatory parasites. Laelapinae is a large and widely distributed subfamily from which all of the other dermanyssoid associates of vertebrates may be primarily or secondarily derived, that is, the subfamilies of Laelapidae (excepting Haemogamasinae and Mesolaelapinae) and the other families in the complex.

The small subfamily Mesolaelapinae shares some features of both Laelapinae and Haemogamasinae, but it appears to have independently separated from the hypoaspidine stock to develop a relationship with mammals in a restricted geographic area.

The laelapid subfamilies Hirstionyssinae, Myonyssinae, and Alphalaelapinae are direct offshoots from the small mammal (primarily rodent) associated Laelapinae, and all have retained these host associations.

Macronyssidae evolved initially on bats, and this family is derived from a small group of laelapines that are bat parasites. Lines within the Macronyssidae went onto other mammals, birds, and reptiles. One of the macronyssid groups on birds gave rise to the Rhinonyssidae, a large family of avian respiratory-tract mites.

Dermanyssidae stemmed from the Laelapinae and appears to have developed first as rodent parasites, but they transferred to birds, on which most species now occur.

Halarachnidae currently includes two subfamilies of mammalian cavity parasites: the Raillietiinae in the ears, and the Halarachninae in the respiratory passages. There are some indications that the Raillientiinae may have branched first from the laelapines and have given rise to the Halarachninae.

The Spinturnicidae are highly specialized bat parasites, but they too appear to have a laelapine origin.

Dasyponyssidae possibly derived from the Macronyssidae.

Hystrichonyssidae may be related to *Ancoranyssus*, a genus currently placed in the Hirstionyssinae.

Manitherionyssidae, a monotypic family, is poorly known and has not been associated with any other group.

The only other group treated here is Spelaeorhynchidae, which is apparently not dermanyssoid. The three dermanyssoid families entirely associated with reptiles (Ixodorhynchidae, Omentolaelapidae, and Entonyssidae) also appear to stem from the Laelapinae, possibly as a single line (Radovsky 1969).

EVOLUTIONARY STRATEGIES AND ADAPTATIONS

Origin of Vertebrate Associations

The Hypoaspidinae include the most generalized dermanyssoids and most of the free-living mites of the Dermanyssoidea. The *Hypoaspis* complex comprises a number of subgenera in the genus *Hypoaspis*, most of which have been or sometimes are treated as separate genera, and several other related genera.

Evans and Till (1966) characterized a basic dermanyssoid type ("dermanyssid" in their classification) based on features of predatory soil mites in the *Hypoaspis* complex. The detailed description includes characters of each postegg immature stage as well as the adults, thus characterizing the developmental pattern. The adult characteristics include moderately well-sclerotized mites; an undivided dorsal plate with 39 setal pairs; well-developed legs with a chaetotaxial pattern specified as "normal"; jawlike, toothed chelicerae (chelate-dentate); spermadactyl, or sperm transfer organ, of the male arising ventrally and proximally from the movable chela and free distally (so that the male chelicerae appear to be functional both as feeding structures and in sperm transfer). The larvae are nonfeeding, whereas both nymphal stages are active and require food to complete development.

The Hypoaspidinae appear ecologically as well as morphologically similar to many other soil- and litter-inhabiting Gamasina, but hypoaspidines are exceptional in the variety of niches occupied and associations developed by the subfamily, and even within *Hypoaspis*. *Hypoaspis* in the broad sense is found in soil, litter, various decaying materials, the nests of such social Hymenoptera as bumble bees and ants, on beetles, and in the nests and on the bodies of birds and mammals. Their association with mammals does not appear to involve even facultative parasitism, and none of the species with this association is entirely restricted to mammals. However, such species as *Hypoaspis sardoa* (Berlese) and *H. miles* (Berlese) are frequently recovered from the nests of rodents and other small mammals, suggesting a degree of dependency. Such species are able to survive and reproduce continuously as predators on free-living arthropods and/or on decaying organic material under laboratory conditions, and they may be found in leaf litter or other decaying materials in nature. We do not know if they will maintain themselves indefinitely in nature without the nest association. The more consistently that such species occur in nests or on mammals, the more likely it is both that they are adapted to locate and remain in this environment and that they depend on it for long-term survival.

The significance of this mite–mammal association is suggested by the recovery of *H. sardoa* from 21% of 263 *Rattus rattus* examined from a midmontane rain forest, and from 15% of the same host collected in a variety of habitats from 840 to 2135 m, on Mauna Loa in the Hawaiian Islands (Radovsky et al. 1979; Radovsky and Tenorio 1981). In the same study, *Pseudoparasitus* (*Gymnolaelaps*) *annectans* (Womersley) [*Hypoaspis nidicorva* Evans and Till, following Domrow (1973)] (Fig. 9.1) was found on 9% of the *R. rattus* in the forest but only rarely on rodents at any point on the transect. (*Pseudoparasitus* is closely related to *Hypoaspis*; *P. annectans* was cited under the synonym *Hypoaspis nidicorva* in all previous publications on this mite in Hawaii. *P. annectans*, as illustrated in Fig. 9.1, shows the well-developed plates and setation and the strong chelate-dentate chelicerae that are typical of the *Hypoaspis* group.) *H. sardoa* did not appear in any of the soil or litter samples from the same localities (regular sampling over a 2-year period, with both pitfall traps and Berlese extraction), whereas *P. annectans* was present only in four litter samples from as many sites (one, two, two, and eight specimens) (F. J. Radovsky and J. M. Tenorio, unpublished). The rodents and most likely the mites are recent introductions to the island ecosystems that were studied. Although this is not an entirely "natural" situation, the data indicate reliance on association with small mammals for long-term survival in nature by *H. sardoa* and *P. annectans*. Rodent nests were not collected in the cited study, but *P. annectans* has been identified from nests of *Rattus* and *Mus* in Hawaii (J. M. Tenorio 1982); this species is now known from many parts of the world, primarily from the nests or bodies of birds and mammals.

These mites apparently do not have any direct trophic relation to the host, and thus other factors must be sought as adaptive, and hence causal, in the association. The nest provides a concentration of food aside from the host itself. Active nests of small mammals have a rich fauna of arthropods, both free-living and nest parasites, suitable in size as food for predatory dermanyssoid mites. Nests are also enriched in comparison to the general litter and soil habitats by a variety of nonliving organic materials resulting from the host's presence: feces, blood and bits of skin, remains of food, which all may serve as food both directly and as a substrate for fungal growth. Most *Hypoaspis* appear to be predators (Karg 1961; Nelzina et al. 1967) to the extent that they have been studied, but more investigations are needed on their food range. In any case, the varied nonliving organic materials in the nest are important food sources for a number of saprophagous and polyphagous mites, including facultative parasites, that are significant in the evolution of dermanyssoid parasitism, which is discussed further in the chapter.

Phoresy on the mammalian nest occupant is obviously adaptive as a

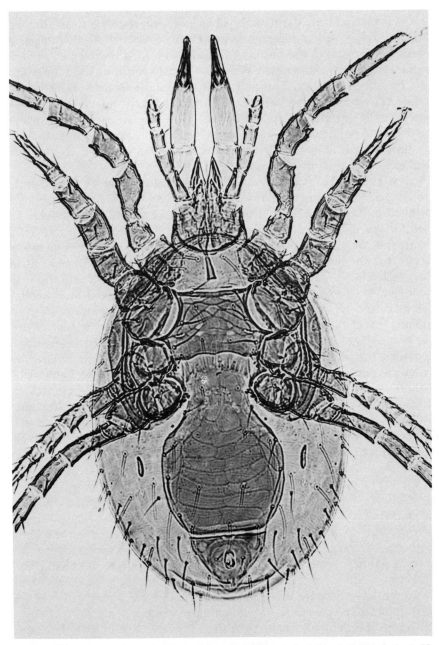

Figure 9.1 *Pseudoparasitus (Gymnolaelaps) annectans* (Womersley) female (ventral view) (Hypoaspidinae, Laelapidae). (Brightfield photomicrograph by J. M. Tenorio.)

dispersal mechanism, particularly since nests represent a discontinuous habitat that is usually temporary in nature. In fact, the advantages of specific adaptation to the nest as a concentrated food source may depend on the availability of the nest builder to effect dispersal, at least for organisms of limited vagility. The high incidence of *Hypoaspis* species on the bodies of rodents and other small mammals taken outside of the nest appears to be directly related to dispersal. However, we must also consider the dispersal requirement as well as the feeding relationship in evaluating the host associations and the proportion of time spent on the host by facultative and obligatory parasites.

A less evident but possibly very important factor in the evolution of the nest association among dermanyssoid Mesostigmata is that of aggregation for mating (Radovsky 1969). The mouthparts of male dermanyssoids are used for sperm transfer as well as for feeding (Fig. 9.2G–K). In free-living forms, the basic structure of the chelicerae is quite similar in both sexes, except that the male has a spermadactyl mounted on the movable chela. Such male chelicerae are functional in predaceous feeding, and the male must feed to survive until it encounters a female in the soil/litter habitat. *Hypoaspis* and its close relatives that are found in mammal nests have retained this cheliceral type (Fig. 9.2G). However, in males of *Androlaelaps*, the least derived laelapine genus and very close in most characteristics to *Hypoaspis*, the chelicera is entirely restructured to assist sperm transfer: the chelae are without teeth and not operable as opposable jaws; the movable chela is reduced to a weak remnant except for the highly developed and often complex spermadactyl; the fixed chela is small and weak. This functional loss in the male chelicerae as feeding organs is evident in nearly all of the more highly specialized vertebrate-associated Dermanyssoidea; in the haemogamasine line leading separately from the Laelapinae to parasitism on vertebrates; and in *Hypoaspis*-related associates of ant nests, such as *Laelaspis*. The parallel in development of "nonchelate chelicerae" between *Androlaelaps* and more derived laelapines and such myrmecophiles as *Laelaspis* was pointed out by Evans and Till (1966), who also noted a similar development in *Julolaelaps*, associated with myriapods. It appears that there are several lines deriving from the free-living *Hypoaspis* type that have developed symbiotic associations of several kinds, involving both vertebrates and invertebrates, in which, at the same time, the male chelicerae have become more highly specialized for sperm transfer and apparently less or no longer functional in feeding.

Perhaps the change in male chelicerae may be related in part to a change from predation to saprophagy on soft or liquid materials, not requiring the chelate-dentate type of chelicerae. However, the main factor appears to be a loss in selective pressure to retain a feeding capability comparable to that of the female. The male in the nest (or on a host) can quickly locate any unmated females and copulate with them. Subsequently, a single mating being sufficient for reproduction during the life of the female, male sur-

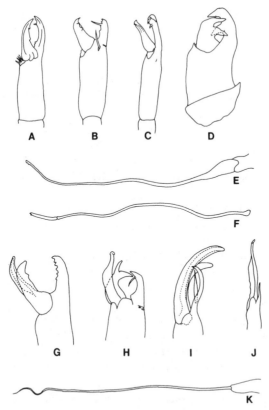

Figure 9.2 Chelicerae of some Mesostigmata (not to scale): (*A–F*) females; (*G–K*) males (*G–J* showing only distal portion of chelicerae with chelae). (*A*) *Hypoaspis aculefer* (Canestrini); (*B*) *Haemogamasus ambulans* (Thorell); (*C*) *Macronyssus longisetosus* (Furman); (*D*) *Spelaeorhynchus praecursor* Neumann; (*E*) *Dermanyssus* sp.; (*F*) *Hystrichonyssus turneri* Keegan, Yunker, and Baker; (*G*) *Hypoaspis aculeifer*; (*H*) *Haemogamasus ambulans*; (*I*) *Androlaelaps hirsti* (Keegan); (*J*) *Dermanyssus gallinae* (De Geer); (*K*) *Alphalaelaps aplodontiae* (Jellison).

vival becomes superfluous; in fact, survival by feeding males under these circumstances may be detrimental in drawing on the available food supply without contributing to the reproductive process. The nest association allows a significant change in reproductive strategy by dermanyssoids: from one requiring similar adaptations and comparable utilization of the food supply by both sexes in a diffuse habitat to one where the food supply is in a concentrated habitat and the male no longer needs to have cheliceral adaptations for both the functions of mating and feeding.

The following two sections concern the transitional stages in two genera from polyphagous feeding and a more casual, even facultative, nest association to obligatory parasitism: *Androlaelaps* in the Laelapinae and *Haemogamasus* in the Haemogamasinae.

Transitions to Parasitism

Androlaelaps

The laelapine line from *Hypoaspis* begins with *Androlaelaps* (= *Haemolaelaps*); in fact, these two genera are not consistently distinguishable except by the specialization of the male chelicerae in the latter, although there are differing trends with regard to a few other characters (Till 1963; Evans and Till 1966). *Androlaelaps* is a nidicolous taxon. A few species, such as *A. casalis* (Berlese), may be found in environments influenced by man, to which they are presumably carried by commensal rodents or by birds, such as straw, hay, stored grain, and poultry litter. *A. fahrenholzi* (Berlese) is more typical of the genus in being found consistently in nests or on hosts but not elsewhere, the hosts including a great variety of small mammals and occasionally birds, but primarily rodents. Although the feeding habits of relatively few of the many *Androlaelaps* species have been studied, available data reveal interesting information concerning the evolution of ectoparasitism through the nest habitat.

Reytblat (1965) compared the feeding behavior of four *Androlaelaps* species as studied by various Russian authors. *A. longipes* (Bregetova), *A. casalis*, *A. fahrenholzi* [= *Haemolaelaps glasgowi* (Ewing)], and *A. semidesertus* (Bregetova) were all described as being able to feed on mammal hosts with varying degrees of facility. *A. longipes* was rated as the least and *A. semidesertus* as the most adapted for parasitism, based on reproduction with a blood diet versus predation on arthropods as well as observed feeding from a host. It is necessary to examine the lesion that they produce to appreciate the nature of feeding on a host by these mites. The wounds made on suckling mice were described for *A. fahrenholzi* (Furman 1959a) and for *A. centrocarpus* (Berlese) (Furman 1966b). Feeding appears to be similar in these two species. The former produced "relatively large wounds, having a punched out appearance . . . The tissues appeared to have been rasped or cut away until a flow of lymph was obtained. Apparently several mites fed at a single lesion . . . Many . . . appeared to be well distended, but only one showed a dark, blood-like residue in its body." *A. centrocarpus* "readily penetrated the intact host skin. Several mites usually fed at a given skin puncture, gradually eroding the skin to produce a large pit-like crater." *A. longipes,* was kept in mass culture on suckling mice in the same way as *A. centrocarpus* (Furman 1966), and it may feed in a similar way.

A. fahrenholzi can be used as a model for the biology of a typical obligatory nidicole in the genus *Androlaelaps*. The female and nymphs have chelate-dentate chelicerae of the *Hypoaspis* type. They are highly generalized in their feeding behavior, although they are adapted to utilize a mammalian host when available. They penetrate the skin and feed on lymph of suckling rodents, sometimes reaching the capillary bed and engorging on blood. They will feed at preexisting wounds on adult rodents (dissertation of R. G. Kozlova, cited by Reytblat 1965) including those made by other parasites. However, they also feed readily on both free-living and parasitic acarines and small insects, on flea feces, and presumably (since they feed readily on dried blood) on any blood or scabs from the host that may drop in the nest.

Androlaelaps species may differ in the effect of diet on reproduction (Reytblat 1965). *A. fahrenholzi* and *A. semidesertus* were unable to reproduce on a diet of arthropods alone; *A. longipes* and *A. casalis* had comparable numbers of offspring on arthropods or blood, and both of these species seemed to reproduce best on a mixed diet of these foods.

Androlaelaps females are regularly found on trapped mammals, whereas males and immatures are infrequently taken except in the nest. *A. fahrenholzi* recovered from rodents in California often has a female:male ratio of greater than 100:1 (Radovsky, unpublished). Arrhenotoky has been demonstrated for several nest-inhabiting species of *Androlaelaps*, including *A. centrocarpus* and *A. longipes* (Furman 1966). Neither thelytoky nor amphoterotoky has been confirmed for any dermanyssoids, and reference to amphoterotoky in *A. fahrenholzi* (Reytblat 1965) must be questioned. However, facultative parthenogenesis fits with the great preponderance of female *A. fahrenholzi* on hosts, and this may be an important adaptation with dependence on the discontinuous environment of the nest.

A. fahrenholzi (and other nest mites with similar biology) represents an interesting stage in the evolution of dermanyssoid parasitism. It has a highly adapted relationship with the host and the nest. Its characterization by various authors as a commensal, a facultative parasite, or a euryphagous schizophage are all valid to a degree, but none adequately describes the host–parasite relationship. The degree of dependence on the host has generally been missed, understandably in view of the dentate chelae, the readiness with which they feed on arthropods, and the crude and uncertain way in which they penetrate the skin of the host. *A. fahrenholzi* depends on blood or tissue fluids of vertebrates to reproduce; the source may be scabs, parasite feces, or even fed lice; it also penetrates the skin of young hosts, reopens wounds, and so forth.

The ability of some *Androlaelaps* to produce either eggs or larvae (*A. fahrenholzi*) or to produce eggs, larvae, or protonymphs (*A. casalis*) may also be an adaptation to the level of nutrients in a variable and quasi-parasitic mode of existence. Although Men (1959) was unable to relate the production of larvae versus protonymphs or the occasional egg to the nutritional level or any other factor, more study of qualitative and quantitative nutrition and its effect on reproduction is needed.

Androlaelaps species and other species that do not have mouthparts specifically adapted for blood feeding differ from the advanced hematophages in taking only moderate-sized meals; females of *A. casalis* increased in body weight with feeding by only 40–60% (Men 1959).

Haemogamasus

This taxon arose separately from *Androlaelaps*, but with its origins in the Hypoaspidinae and very likely also in the *Hypoaspis* complex. There are many parallels in the evolution of *Haemogamasus* and *Androlaelaps*, but there are interesting differences as well. In *Haemogamasus* a range of host and feeding relationships is encompassed from facultative, nonparasitic, and polyphagous nidicoles to true obligatory hematophages. This genus is both the beginning and the culmination of the haemogamasine series. Haemogamasinae did not lead to any other higher groups, and the few other genera in the subfamily besides *Haemogamasus* are small groups that either could have branched from *Haemogamasus* (*Brevisterna* and *Ischyropoda*) or are possibly separate in origin (*Eulaelaps*).

In *Haemogamasus* three principal levels of host association can be recognized: one for *H. pontiger* (Berlese) only, one for the remainder of the *H. reidi* Ewing group (Williams et al. 1978), and one for the *H. liponyssoides* Ewing group. Although it has all of the defining characteristics of the genus *Haemogamasus*, *H. pontiger* (= *H. oudemansi* Hirst) is morphologically conservative in several respects: the female chelicerae have the fully predatory form; the male chelicerae, although showing some tendency toward *reidi* group modifications, are chelate dentate with a distally free and relatively simple spermadactyl; hypertrichy of the idiosoma is moderate in comparison to other *Haemogamasus*; and the size is moderate in comparison to the generally rather large members of the genus. This is the only species of *Haemogamasus* that is typically found in free-living habitats. Large numbers may occur in the detritus on warehouse floors (Hughes 1961), and recorded free-living habitats include barley, sugar, wheat and rice straw, rice hulls, and flax tow (Evans and Till 1966). Under specimens examined from North America, besides four records from rodent nests or hosts, the following are recorded (Williamson et al. 1978, apparently from quarantine interception for the most part): grass packing, hay packing, beet seed, sod in ballast, old burlap bagging, rye straw, and grain spill. Nest records are more frequent than host records. In laboratory tests, Furman (1959a) recorded 63% of 38 *H. pontiger* feeding on heparinized mouse blood (compared to 100% of 500 *H. reidi*) and none of 20 feeding on dried blood (compared to all of 7 *H. reidi*). *H. pontiger* "can complete its development if only wheat germ is supplied as food" (Hughes 1961). *H. pontiger* appears to be a facultative nidicole with a very loose nest–host association. Although it has been inadequately studied, it appears to be a predator and saprophage without any parasitic tendencies; in this respect, it may be compared to some of the *Hypoaspis* found in nests, but perhaps with even less specialization for phoresy than such species as *Hypoaspis sardoa*.

Williams et al. (1978) established the *H. reidi* group for all those *Haemogamasus* that lack adaptation of the chelicerae for skin penetration as found in the *H. liponyssoides* group; although they dealt only with the North American fauna, the same classification can easily be extended to the Palaearctic and other regions where *Haemogamasus* occurs. Most biological studies of *H. reidi* in the United States were reported as *H. ambulans* (Thorell) prior to the clarification of the taxonomy of these species by Redington (1970).

Most *Haemogamasus* species belong in the *reidi* group, and, except for *H. pontiger*, they are all obligatory nidicoles. The female chelicerae in this group remain chelate-dentate (Fig. 9.2B), although some species have modifications not typical of free-living mites, such as the enlargement and modification of the pilus dentilis and the dorsal seta or reduction of the teeth, which presumably are related to dietary changes. The biology and feeding behavior in the *reidi* group were studied for *H. reidi* (*H. ambulans* of authors) (Furman 1959a, 1959b, 1968), *H. mandschuricus* Vitzthum (Goncharova and Buyakova 1960), and *H. citelli* Bregetova and Nelzina and *H. nidi* Michael (Nelzina and Danilova 1956). All of these species are associated primarily with nest-making rodents or the rodent-like pika (*Ochotona*), and all are primarily nest mites, found more frequently and in much greater numbers in the nest than on the host. *H. reidi* can reproduce, complete its development, and continue through repeated generations on a diet composed entirely of blood (heparinized or dried) or arthropods. This species will feed readily on flea feces, a variety of parasitic and free-living arthropods, and bird or reptile blood as well as that of mammals. In experiments with mites on live hosts, *H. reidi* fed from abraded skin and from scabs (Furman 1959b). In further tests, involving extended free access to suckling mice lacking umbilical or other scabs and with the anal-genital area taped over, a small percentage of mites appeared to have fed and there was a very low level of reproduction. Penetration of intact skin seems to be exceptional in *H. reidi*.

H. citelli, H. nidi, and *H. mandschuricus* also have a broad range of acceptable foods, comparable in extent to that of *H. reidi*. *H. mandschuricus* has a limited tendency to feed from the living host and a preference for young animals. However, it is not clear that this species penetrates unbroken skin (Goncharova and Buyakova 1960). *H. citelli* females maintained a constant rate of reproduction on free blood for 3–4 weeks and on tyroglyphoid mites for 2–3 weeks; in the former case, blood feeding was gradually reduced, and feeding and reproductive rates were restored by changing the available food to mites; in the latter case, the reproductive rate was restored by changing the diet to blood (Nelzina and Danilova 1956). Females on an all-blood diet showed an increasing tendency toward cannibalism, eventually destroying all young present (eggs, larvae, and nymphs). *H. nidi*, in contrast, was able to sustain reproduction on blood for 5 months but could not reproduce on a solely arthropod diet; however, it had the highest rate of reproduction on a mixed diet of blood and mites. During a 24-hour period of free access, 16% of female *H. citelli* engorged on newborn gophers and only 1–2% on adult gophers. Nelzina and Danilova (1956) recognized the possibility of feeding from preexisting breaks in the skin. Apparently they observed feeding on a host, because they noted that the depth of the puncture was slight, feeding took 15–19 minutes (compared to engorgement in 1–1.5 minutes on free blood), and feeding was not to satiation. *H. nidi* regularly became blood fed when given access to voles, but Nelzina and Danilova considered this to be from attacking scarified surfaces and withheld judgment on the ability of this species to puncture intact skin. They also came to the intriguing conclusion that bloodsucking by puncture of intact skin in *H. citelli* is seasonal and that it is most intensive in the spring when newborn gophers are in the nest.

The *H. reidi* group closely parallels the general trend in *Androlaelaps* with respect to feeding behavior and general biology. Men (1959b) interspersed species of these two taxa in presenting a graph of parasitic tendency based on decreasing entomophagy and increasing hematophagy, as follows: *A. longipes, H. citelli, A. casalis, H. nidi, A. fahrenholzi*. Although some of these relative placements may be questioned, the group comparison is significant. Also, *H. nidi* and *A. fahrenholzi* present an interesting parallel development of polyphagy combined with obligate hematophagy (in the sense that blood must be included for reproduction).

The *reidi* group parallels *Androlaelaps* in another way: the modification of male chelicerae for increasing their functional efficiency as organs of insemination and decreasing their trophic capability. *H. pontiger* has male chelicerae very much like those of free-living hypoaspidine predators. The *reidi*-group males exhibit a continuum of modifications, with the extreme form being chelae as specialized for sperm transfer, and hence as poorly suited to feeding, as those of any *Androlaelaps*. *H. reidi* itself is one of the least modified, but its chelae are essentially edentate on their free surfaces, and its movable chela is reduced. *H. thomomysi* Williams, *H. hirsutisimilis* Willmann, and *H. nidi* are some of the more extreme forms, with chelae that are quite variable but no longer can be considered chelate in a functional sense (see *H. ambulans*, Fig. 9.2*H*).

The *liponyssoides* group of *Haemogamasus* represents a stage in adaptation to hematophagy through cheliceral modification that surpasses any of the *Androlaelaps* as well as other Laelapinae. The development of slender edentate chelae that can operate together as the tip of chelicerae that puncture the skin, rather than tearing it, has evolved repeatedly in the dermanyssoids (Fig. 9.3*A–E*). It is observable in one of its crudest forms in the *liponyssoides* group. The group includes five species in North America and several in the Old World, and some morphological features suggest that it

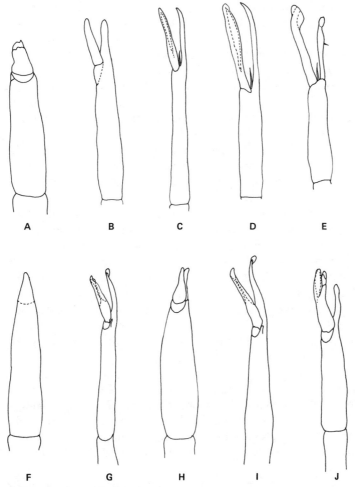

Figure 9.3 Development of chelicerae. (*a–e*) *Haemogamasus liponyssoides* Ewing; (*f–j*) *Steatonyssus antrozoi* Radovsky and Furman. (*a, f*) larvae; (*b, g*) protonymph; (*c, h*) deutonymph; (*d, i*) female; (*e, j*) male.

may be polyphyletic, that is, that even within this intrageneric grouping the slender chelae adapted for puncturing may have evolved more than once.

H. liponyssoides hesperus Radovsky was studied in the laboratory (Radovsky 1960). This subspecies is known from California and the eastern Palaearctic, whereas *H. l. liponyssoides* occurs in the eastern United States and parts of Mexico. *H. l. hesperus* is an obligatory hematophage. It will feed on heparinized free blood but only poorly. (Both adults and deutonymphs can be "force-fed" to full engorgement by immersion in a drop of blood; mites were maintained in this way by rinsing in distilled water after each blood immersion.) *H. l. hesperus* consistently feeds on both adult and suckling rodents and regularly penetrates the unbroken skin of sucklings. However, its feeding method is quite different from those highly specialized hematophagous mites that have been studied (e.g., Macronyssidae and Dermanyssidae). *H. l. hesperus* in colonies of hundreds generally produces only one or two wounds on a suckling mouse over a 24-hour period. The flow of blood from a small puncture is utilized by many mites to engorge fully. Males and deutonymphs kept separately can pierce the skin of sucklings; however, they are less capable of doing so than the females, and the isolated males and nymphs often feed poorly. The overall picture is of mites feeding communally where one mite (most likely a female in a mixed colony) will penetrate and feed, usually at a point with delicate skin (e.g., head, feet, or inguinal region); other mites will keep the puncture open and feed on blood flowing from the wound. This contrasts strikingly with macronyssids and dermanyssids, in which each individual normally penetrates unbroken skin regardless of the number feeding.

The trauma of the bite also supports the primitive nature of obligatory bloodsucking by *H. l. hesperus*. Usually the puncture in suckling mice is surrounded by an extensive purplish zone indicating subcutaneous hemorrhaging, even when feeding is by a single mite. This suggests that a spreading factor, as well as an anticoagulant, is present in the saliva. Obvious trauma of this sort is unusual in hematophages and is generally not found in the more highly adapted groups.

The comments on *H. liponyssoides* by Preobrazhenskaya and Preobrazhensky (1955) support some of the preceding observations on the feeding behavior of this species. They observed that *H. liponyssoides* is able to penetrate the skin of adult and newborn rodents to feed on blood and that this mite will not feed on arthropods even when they are crushed or cut up to expose fresh hemolymph. Men (1959b) noted that *H. liponyssoides* refuses free blood.

A typical attribute of blood-feeding nest parasites is the ability to take in much more than their own weight in blood at a single feeding. *Haemogamasus* species appear to present a graded series nicely coordinated with feeding behavior. This is my impression from observing these mites, but, since precise weighings of unfed versus engorged mites have not been made for any of the species, this can only be supported indirectly. Female dorsal plate size increases from *H. pontiger* through the rest of the *reidi* group to the *liponyssoides* group, which is presumably related to increased engorgement potential. The same sequence of relative enlargement with feeding is evident when these mites are maintained in the laboratory; *H. reidi* engorges more than *H. pontiger*, and *H. liponyssoides* engorges much more than *H. reidi*. Data on the idiosomal length and width of unfed versus engorged female *H. liponyssoides* (Radovsky 1959) can be extrapolated, assuming the same proportionate increase in body thickness as in width, to indicate at least a fourfold increase in volume with engorgement. *An-*

drolaelaps casalis takes in blood up to 40–60% of unfed body weight (Men 1959a) and, although *H. reidi* exceeds this, it appears to be closer to *A. casalis* than to *H. liponyssoides* in engorgement capacity.

Parasitic Adaptations

Haemogamasinae

This phylogenetic line from the hypoaspidines has had a limited adaptive radiation. At most, five genera are recognized. Besides *Haemogamasus*, previously treated, there are sufficient data to justify comment here only for *Eulaelaps* and *Brevisterna*.

Brevisterna utahensis (Ewing), in a small genus associated with Nearctic rodents, has been studied in the laboratory (Allred 1957; Furman 1959a). This mite has modified chelicerae, but not clearly of a puncturing type. It tends toward hematophagy, with an evident penchant for feeding on free liquid blood when provided. Although it is not a predator, *B. utahensis* may cannibalize its own eggs (and in one instance larvae), a behavior also sometimes associated with highly specialized gamasine hematophages. It will feed on dried blood, and an ability to penetrate the intact skin of a host has not been demonstrated. The protonymph in *Brevisterna* is apparently nonfeeding.

Two species of *Eulaelaps* have been studied in the laboratory: *E. stabularis* (Koch), an extremely widespread species both geographically and in host association (Kozlova 1959), and *E. cricetuli* Vitzthum, primarily associated with hamsters in the eastern USSR (Goncharova and Buyakova 1962). These are large, well-sclerotized mites that feed on a wide variety of materials. They are primarily predators and saprophages, but besides taking blood opportunistically both species were observed feeding on blood or lymph, apparently by penetrating the intact skin of suckling rodents, although this is probably an infrequent and secondary mode of feeding. These *Eulaelaps* appear to be comparable to the mites in the *Haemogamasus reidi* group that can use a wide range of foods. *E. stabularis* is often found in straw or poultry litter as well as in nests. *Eulaelaps* is like the hypoaspidines in having a feeding protonymph. This contrasts with *Haemogamasus* in which the protonymphal stage is uniformly brief and quiescent. Along with some morphological features, this raises a question about the placement of *Eulaelaps* in the Haemogamasinae.

Laelapinae

Laelapinae is cosmopolitan and highly successful, and it is the group from which a number of other parasitic subfamilies of Laelapidae and families of Dermanyssoidea may be derived. Because of the size, variety, and importance of this taxon, it is discussed extensively herein.

The genera accepted in the Laelapinae as restricted by Tipton (1960) are a reasonably compact and interrelated group, primarily associated with "rodents, particularly of the myomorph group." With respect to Tipton's (1960) revision, several genera that I discuss as laelapines, particularly associates of mammals other than rodents, were specifically excluded, omitted from discussion, or subsequently described; however, the concept of the subfamily and the principal diagnostic characters recognized are the same.

Laelapines are generally well-sclerotized mites, often with some expansion of body plates relative to free-living hypoaspidines, usually with chelate-dentate chelicerae in the female, and nearly always with male chelicerae modified as described before for *Androlaelaps* (fixed chela much reduced, movable chela fused with the greatly elongated spermadactyl) (Fig. 9.2*I*).

Laelaps, an extensive genus even when most narrowly interpreted, is now generally understood to include *Echinolaelaps*, another large complex of species, as well as *Eubrachylaelaps* and some other species groupings previously recognized as distinct genera. *Laelaps* species are parasites of myomorph rodents, particularly Muridae and Cricetidae (Tipton 1960; Furman 1972; Herrin and Tipton 1976). *Laelaps* could easily have derived from *Androlaelaps*, the least specialized laelapines, by increase in sclerotization; development of spiniform setae, particularly on the coxae; and generally a somewhat more intimate host association.

Relatively few laelapines other than *Androlaelaps* have been subjected to laboratory observations, and none extensively. Those that have been studied are in the genus *Laelaps*, and *L. (Echinolaelaps) echidninus* Berlese is perhaps the best known. This species has been used conveniently for studies on water balance, detailed morphology and fine structure, and other experimentation, partly because of the ease with which it can be collected from commensal rats and maintained on laboratory rats. *L. echidninus* feeds on defibrinated blood, is ovoviviparous, and has nonfeeding larvae and feeding protonymphs and deutonymphs (Owen 1956). Its requirement for food in both nymphal stages is characteristic of laelapines to the extent that it has been studied. *L. echidninus* will feed through the abraded skin of rodents and humans, on dried blood (as well as fresh or defibrinated blood), and on small arthropods, specifically Anoplura. Furman also noted that mites of this species will emerge from the fur and feed on "lachrymal fluids and exudates around the eyes" (Furman 1959a).

This picture of the host–parasite relationship can be extended from my observations of *L. echidninus* in laboratory colonies on white rats (Radovsky 1969 and unpublished data). Contrary to statements in the literature, *L. echidninus* can initiate wounds on the host. White rats in culture boxes heavily infested with this mite become scabby, particularly on the posterior dorsum where female mites tend to concentrate in the hair. Since there are no signs of similar scabs or even abrasions in these areas in uninfested rats, it appears that *L. echidninus* females can initiate wounds by tearing at the skin in the manner of *Androlaelaps fahrenholzi*. The feeding at the eyes is particularly interesting because one or several females leave their protected position in the hair of the posterior body and run directly to the eyes; after a period of feeding on secretions, they retrace their course into the pelage. The "purposeful" manner in which this is done indicates a specific complex of behavioral adaptations to utilize the secretions associated with the eyes, rather than a casual partaking of an encountered source of moisture or nutrients.

L. echidninus has progressed in its parasitic relationship beyond the majority of *Androlaelaps* species that have been studied. It will feed on arthropods, but this may be a relatively infrequent food source; Owen (1956) was unable to get *L. echidninus* to feed on a variety of small arthropods (except for cannibalism of its own larvae), and the arthropods offered and eaten in Furman's (1959a) tests were lice, a prey that may be encountered on the host and that, perhaps not incidentally, may contain blood. The association of females with the rodent host over extended periods distinguishes *L. echidninus* from most *Androlaelaps* species. This association goes beyond phoresy or visiting the host for the duration of a feeding period; *L. echidninus* is similar in the use of the host for harborage as well as food to some "host" fleas, as opposed to "nest" fleas.

Some species of Laelaps s. str. feed on free blood offered to them in the laboratory; for example, *L. nuttalli* Hirst adults and nymphs fed on whole blood and various fractions (red cells, plasma, serum) from both chicks and rodents, as did *A. fahrenholzi* and *L. echidninus* (Wharton and Cross 1957). Probably most *Laelaps* and other rodent-associated laelapines will feed on blood, at least opportunistically.

L. myonyssognathus Grochovskaya and Nguyen, a parasite of murine rodents, primarily *Bandicota* and *Rattus*, was studied in the laboratory in India (Mitchell 1968). This mite is intermediate between *Androlaelaps* and typical *Laelaps* in its diagnostic features, but it is currently assigned to *Laelaps* (*Laelaps*). An atypical feature of *L. myonyssognathus* is its cheliceral form: the second segment is somewhat elongate and attenuate, with slender chelae having reduced teeth. In fact, there is a wide range of cheliceral form in *Laelaps*, from short stout chelicerae with strong, heavily dentate chelae in some *Echinolaelaps* to somewhat elongate, slender chelicerae with slighter chelae and a reduction of teeth in some species of *Laelaps* (*Laelaps*), such as *L. agilis* Koch and *L. nuttalli*. Nevertheless, the partial convergence of *L. myonyssognathus* chelicerae with those of macronyssids and hirstionyssines is striking among *Laelaps* species. This form is indicative of adaptation for skin penetration and regular feeding on blood and/or lymph.

Mitchell (1968) fed *L. myonyssognathus* on defibrinated blood while studying its life cycle. Females engorged readily on abraded skin (mouse and human) but they did not feed when the open end of a tube containing them was placed on intact skin. *L. myonyssognathus* is generally larviparous, although it sometimes produces eggs, has feeding protonymphal and deutonymphal stages, and exhibits arrhenotokous parthenogenicity. *L. myonyssognathus* is attracted to blood and is stimulated and attracted by the carbon dioxide in human breath (Mitchell 1968). The CO_2 response was earlier reported for the macronyssid *Ornithonyssus bacoti* (Hirst) (Sasa and Wakasugi 1957) through similar tests in which CO_2-depleted breath was used as a control. However, response to breath and to CO_2 has been observed for a wide variety of mites, including *Haemogamasus reidi, H. liponyssoides,* and *Hirstionyssus* species (D. P. Furman, F. J. Radovsky, unpublished data).

Besides the limited laboratory studies, examination and analysis of collection data provide some additional information on laelapine host–parasite relationships. We generally find a preponderance of females of dermanyssoid parasites on or in the host; this may be associated with the limited need for male survival once copulation has taken place and with a generally short maturation period relative to the female reproductive life. Furthermore, the laelapines as a group, as noted for *Androlaelaps*, tend to be almost entirely represented by females on the host. I analyzed the collection records reported by Furman (1972) from an extensive survey in Venezuela (Handley 1976), in which nearly 25,000 hosts were examined, total recovery of ectoparasites was attempted, and relatively complete data on sex and stage of laelapines were provided. Summarizing the data (Furman 1972), I find the following (sex ratio given as the number of females per one male):

> *Androlaelaps* (8 species): sex ratio 16; immatures 5.3%
> *Laelaps* (15 species, omitting *L. dearmasi*): sex ratio 57; immatures < 1%
> *Gigantolaelaps* (13 species): sex ratio 118; immatures 0.6%
> *Tur* (5 species, omitting *T. apicalis*): sex ratio 11; immatures 0.4%
> *Mysolaelaps* (3 species): sex ratio 783; immatures 0.1%

These ratios and percentages are based on just over 1000 to nearly 7000 specimens for each genus. *Hymenolaelaps* with only 31 individuals of one species was represented entirely by females. The exceptions to this trend for females are as interesting as the trend itself. *Laelaps dearmasi* Furman and Tipton was the most frequently encountered laelapine, with nearly 19,000 specimens identified; 23% of the adults, nearly 1 in 4, were males, and immatures (mostly nymphs, some larvae) represented 11.2% of the mites of this species. Similarly, *Tur apicalis* Furman and Tipton (nearly 1800 specimens), had males represented as 26% of the adults; nymphs were 3.2% of the total. *Steptolaelaps heteromys* (Fox) (1470 specimens), the only species collected for its genus, had 40% males (sex ratio 1.5) and 5.4% nymphs. The tendency for a significant proportion of males (and to some extent immatures) of *L. dearmasi, T. apicalis,* and *S. heteromys* to be represented on the host is supported by other collections, so it is not a result of season, population buildup, or some other artifact. Apparently, although most laelapines are nest parasites in which only the female normally remains on the host when it forages outside the nest, there are a few species in which both sexes and the immatures associate with the host other than for feeding.

Among the Laelapinae there are several genera that have adaptations exceptional for this group. *Hymenolaelaps princeps* Furman, known only from the female parasitic primarily on murine rodents (Furman 1972), is exceptional in the degree to which the chelae have become slender and apparently modified for working together as the end of a puncturing chelicera, or possibly for taking up secretions. *Chrysochlorolaelaps benoiti* Evans and Till, parasitic on the African golden mole, is in another monotypic genus in which the chelicerae are quite far from the typical chelate-dentate type (Evans and Till 1965). We know very little about the feeding habits of most laelapines, and we can only surmise that a variety of host relationships are involved, including feeding on blood or lymph at points where the skin has been previously injured, feeding on various secretions, and combinations of these, but rarely with consistent penetration of intact skin.

Aetholaelaps and *Liponysella*, parasites of the lemurs of Madagascar, are intriguing and difficult to place. *Aetholaelaps*, with two species, fits nicely into the Laelapinae, despite its somewhat modified chelae; this classification is based not only on adult morphology but also on the structural evidence for an active and feeding deutonymphal stage. *Liponysella*, which is monotypic, is clearly related to *Aetholaelaps* in such features as the unusual form of the cheliceral processes. However, it is very similar to macronyssids in a number of characteristics of the female, including the diagnostic feature of a longitudinal ventral ridge on the palpal trochanter. The deutonymph of this species has been described (Zumpt 1950). Previously, I placed *Aetholaelaps* in the subfamily Laelapinae and *Liponysella* in the family Macronyssidae (Radovsky 1967, 1969), even though this implied a diphyletic origin for the latter family. Reexamining Zumpt's description of the deutonymph, I believe that it is probably of the laelapine type (see Macronyssidae below), and that *Liponysella* should be placed in the Laelapinae. This signifies that the *Aetholaelaps–Liponysella* complex is one more of the many examples of independent evolution of advanced piercing chelicerae, and that the adult morphology of *Liponysella* is a truly extraordinary example of parallel evolution with the Macronyssidae.

Notolaelaps and *Neolaelaps* also seem to be related to the Macronyssidae. These two taxa are the only members of the subfamily Laelapinae having a true rather than accidental association with bats. They are found on Pteropodidae, the family of Old World fruit bats that constitutes the suborder Megachiroptera. The monotypic *Notolaelaps*, known only from the female, is typically laelapine in essentially all respects. The chelae lack teeth, other than the dentoid curved tips, but they are similar to a number of other laelapines, including some *Laelaps*, in this regard. Nothing is known of its habits, but *Notolaelaps* appears to be related to *Neolaelaps* based on several features (Radovsky 1967). *Neolaelaps* is known from three species associated with a variety of pteropodids. This is a highly specialized genus with some extreme developments in setal spines and in which the spiracular atrium is enormously enlarged and associated with a complex peritreme. These mites appear to be blood feeders, including both nymphal stages, and their chelicerae are slender with edentate chelae. The male chelae are generally laelapine in form, but not approaching the extremes of spermadactyl modification found in some members of the subfamily. *Neolaelaps* has often been found on pupiparous flies parasitic on bats and apparently utilizes these Diptera phoretically—a very unusual adaptation among the Mesostigmata associated with mammals. *Notolaelaps* and *Neolaelaps* appear to be very near the stock from which the Macronyssidae evolved, and they are included in the discussion of that family.

Mesolaelapinae

Mesolaelaps, currently with some 10 species, was taken from its former placement in the Laelapinae s. str. to constitute a separate subfamily of the Laelapidae (Tenorio and Radovsky 1974). They look superficially like typical laelapines, except for the rather setose idiosoma. However, such characters as the hypoaspidine-like male chelicerae and the fimbriated tectum of the mouthparts indicate that they do not share a common ancestry with *Androlaelaps*, *Laelaps*, and their relatives. *Mesolaelaps* may represent a third line from the *Hypoaspis* complex. With more study, a number of other suprageneric taxa may have to be separated out of both the Laelapinae and the Hypoaspidinae to clarify systematic relationships. This would also add to our understanding of the origins and evolution of various parasitic associations.

   Mesolaelaps is found principally in New Guinea and Australia (the additional records are in the islands near New Zealand), and its hosts are marsupial bandicoots (Peramelidae) and, less frequently, murid rodents. There is no other significant biological information on the genus.

Alphalaelapinae

Alphalaelapinae is a monotypic subfamily notable for the extraordinary development of the male chelicerae (Tipton 1960). The female of *Alphalaelaps aplodontiae* (Jellison) has rather stout chelicerae with relatively unmodified dentate chelae, as in many of the laelapines, whereas the male has extremely elongate, slender, styletlike chelicerae (Fig. 9.2K), with elongation of the second segment (Fig. 9.2K) as in the females and nymphs of the bloodsucking Dermanyssidae. Dermanyssoid insemination takes place through sperm induction pores in the area of female coxae III–IV, which lead into an elongate tubulus annulatus passing to other organs connecting with a spermatheca. The specialization of the male chelicerae seen in *A. apolodontiae* seems a logical step in the evolution of nest parasites. Chelicerae of this kind presumably carry the sperm through the tubulus annulatus into some internal portion of the female reproductive system.

Why have other dermanyssoid parasites not developed similar chelicerae? Apparently the system that delivers the sperm packets to the atrium of the sperm induction pore is adequate so that selective pressure for a more complex system is absent or weak. In at least some groups, there may also be advantages in keeping the male chelicerae from interfering (through extreme specialization) with opportunistic feeding.

The specific host of *A. aplodontiae* is *Aplodontia rufa*, a species placed in a monotypic sciuromorph superfamily that is considered to be the least derived among extant rodents. *A. rufa* has a large faunule of parasites and nidicoles peculiar to it and in several instances highly distinctive morphologically; this suggests ecological isolation of the host as well as a long period of coevolution of the parasites with a taxonomically distinctive host.

Myonyssinae

Myonyssus, the single genus in the subfamily Myonyssinae, has an essentially Holarctic distribution and is associated with rodents, insectivores, and pikas (Lagomorpha). Myonyssines have slender chelicerae (the second segment is abruptly tapered near the base) and very slender, piercing chelae. These are rather large-bodied mites with extensive plates. They are not known to contain blood (Hughes 1959). The chelicerae appear to be adapted for penetrating skin, and *Myonyssus* species may feed shallowly on tissue fluids. However, no laboratory or incidental observations on living mites are available.

Based on collection records and the morphology of the nymphs, of the immatures, only the deutonymphal stage is active and feeding. The male chelicerae are very similar to those of the female, and the grooved spermadactyl is fused with the movable chela throughout its length.

Myonyssus has been classed with various other groups having "shearlike" chelae on piercing chelicerae. Its life cycle as well as its adult morphology contradicts the earlier placement of this taxon in or close to the Macronyssidae. The description of one species with coxal spurs (Strandtmann and Garrett 1970) gave support to the provisional placement of the hirstionyssine genera in the Myonyssinae (Radovsky 1966, 1967), but these spurs do not appear to be homologous with those in hirstionyssines. Myonyssinae continues to be treated as a monogeneric subfamily (Strandtmann and Wharton 1958; Evans and Till 1966; Radovsky 1969; Tenorio and Radovsky 1979), and it is one more example of the parallel development of similar piercing chelicerae in different groups of Dermanyssoidea.

Hirstionyssinae

The genera placed in the Hirstionyssinae were classed with the Macronyssidae until fairly recently (Evans and Till 1966; Radovsky 1966) based on the similar piercing-type chelicerae; basic differences in the developmental cycles of the two groups, as well as more detailed morphological analysis, make it quite clear that they are distinct and that the cheliceral similarities are convergent. Five genera are recognized: *Echinonyssus* (Fig. 9.4), *Trichosurolaelaps, Thadeua* (= *Australolaelaps*), *Patrinyssus*, and *Ancoranyssus*. *Echinonyssus*, the largest genus, which is essentially cosmopolitan, is a senior subjective synonym for *Hirstionyssus* (Domrow 1963; Tenorio and Radovsky 1979).

The hirstionyssine genus closest to the Laelapinae is *Trichosurolaelaps*, which is parasitic on syndactylous marsupials in Australia, and this genus has a considerable range of morphological variation. Some *Trichosurolaelaps* have sclerotization comparable to laelapines, a holotrichous setation on the dorsal plate, and mouthparts including chelicerae that are not significantly divergent from the more typical laelapines. However, the continuity with other species in the genus that clearly trend toward the specialized hirstionyssines supports the treatment of this and the other hirstionyssine genera as a separate subfamily. *Thadeua* also found on syndactylous marsupials in Australasia, *Patrinyssus* on the primitive rodent *Aplodontia* in western North America, and *Ancoranyssus* on a Southeast Asian porcupine are each a restricted genus with some particularly specialized features. *Echinonyssus*, despite its many species and wide distribution, is a rather compact genus, although there are unusually hypertrophied holdfast structures in some species. These mites are found primarily on a variety of rodents and insectivores, but some species have small carnivores and other small mammals as normal hosts.

All hirstionyssines are characterized by the presence of spurs (that arise directly from the cuticle, as opposed to spines that are developed from setae) on at least some of the coxae. Coxal spurs are found on a few species in several other groups, including Laelapinae, Myonyssinae, and Macronyssidae. However, it is only in the Hirstionyssinae that coxal spurs appear as an apparently monophyletic development characteristic of all species in the group. Spurs, like most strongly hypertrophied spines, which are also found in many hirstionyssines, apparently function as holding structures to catch on irregularities of the host dermis or on hairs and act to prevent dislodgement.

The distribution of Hirstionyssinae both geographically and among hosts suggests a great age for the group. Relevant distribution includes both the widespread geographic range of *Echinonyssus* and the location of several relictual genera associated with Australasian marsupials, North American *Aplodontia*, and others.

Hirstionyssines have several notable specializations for parasitism.

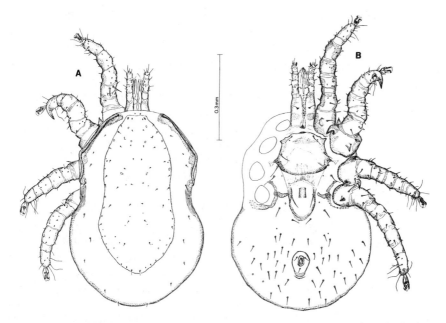

FIgure 9.4 *Echinonyssus liberiensis* (Hirst) female (Hirstionyssinae, Laelapidae). A. dorsum; B. venter. (Reproduced by permission of J. M. Tenorio and F. J. Radovsky. 1979. *J. Med. Entomol.* **16**:370–412.)

Their spurs and spines have been modified to an extreme degree in some forms. This specialization for attachment suggests that they are more closely host associated than most laelapines. *Echinonyssus*, in particular, has decreased in size. The size range (idiosomal length) for the genus is 400–950 μm (Herrin 1970), but most species are closer to the lower figure. Small size facilitates remaining on the host, both by reducing the chance for dislodgement as a result of normal host movements and as an aid to escaping direct host attack in grooming.

The life cycle of *Echinonyssus meridianus* (Zemskaya) has been studied in the greatest detail (Senotrusova 1962). This Palaearctic parasite of gerbils (Cricetidae) produces eggs; has brief, nonfeeding larval and protonymphal stages; and has both deutonymphs and adults that require blood meals—the former to molt and the latter, in the female, to produce eggs. The same results (Radovsky 1967, 1969; Radovsky and Budwiser, unpublished) were obtained with *E. breviseta* (Strandtmann and Morlan) from cricetid wood-rats in California. The morphology, especially the mouthparts, of immatures in other *Echinonyssus* and in other hirstionyssine genera indicates that the pattern of nonfeeding through the protonymphal stage and feeding in the deutonymphal stage is characteristic of the subfamily as a whole. The appearance of the gut contents in collected specimens indicates that hematophagy is a general subfamilial character.

Hirstionyssine mites may remain on the host for extended periods, beyond the requirements of dispersal and the time required for feeding. This varies widely within the subfamily. *Echinonyssus nasutus* Hirst females may have a close host relationship but, as in most Laelapinae, the adult female is the principal host-associated stage and sex. In fact, the male is not known for this species and the deutonymph has very rarely been encountered, although females are frequently taken on tree shrews (Tupaiidae) and *Rattus*. On the other hand, *Trichosurolaelaps* species as well as many *Echinonyssus* are often represented on the host by a significant proportion of males and deutonymphs as well as females.

Echinonyssus meridianus, which was taken throughout the year in a cold temperate area, had more females in the nest than on the host: host—females 37.5%, males 24%, deutonymphs 38.4%; nest —females 58.9%, males 25.3%, deutonymphs 15.9% (Senotrusova 1963). Since hirstionyssines as far as we know, and specifically *E. meridianus*, are rapid feeders (as opposed to slow feeders in the manner of macronyssid protonymphs), the reason for this nearly doubled percentage of deutonymphs on the host versus the nest is obscure. Actually, the raw figures suggest an even more intensive host association since the total mite collections were 4019 mites recovered from rodents and 801 mites found in the collected nests; although these figures are not directly comparable, nests were extensively examined, and the magnitude of the difference is impressive. Larvae and protonymphs were not found on the hosts in this study, and it is plausible that in this species newly molted deutonymphs seek out the host and remain on it, and that females leave the host primarily to oviposit.

The female chelicerae are variable in *Trichosurolaelaps*, ranging from those with relatively stout and bidentate chelae to slender, elongate, and edentate chelae. *Ancoranyssus trichys* Evans and Fain is a highly derived species in which the chelicerae are particularly characteristic: the first segment is elongate, more than twice the length of the second; the chelae are relatively short, broad, and laterally flattened; the movable chela is especially complex, with two of its projections fringed. (The chelicerae of *Ancoranyssus*, combined with its occurrence on Oriental porcupines, the questionable homology of its coxal spurs, and certain other features, indicate that this mite may not belong in the Hirstionyssinae, and the same features suggest a common phylogenetic origin for *Ancoranyssus* and the Hystrichonyssidae, which is discussed under that family.) However, the major trend in the subfamily toward piercing chelicerae with "shearlike" chelae is expressed in the numerous species of *Echinonyssus*, as well as *Thadeua*, *Patrinyssus*, and some *Trichosurolaelaps*.

The spurs on the coxae of hirstionyssines are of a characteristic type that suggests homologous origins for these structures within the group. These are relatively small, pointed projections on coxae II–III or II–IV. However, several types of additional spurs and setal modifications producing holdfast spines are found within the group. In *Echinonyssus*, the most prominent trend is the development of the anterior seta of coxa II into a recurved spinose process. This reaches its greatest expression in *E. nasutus*, which also has the unique feature of the anterior tip of the dorsal plate developed into a pointed and downward curving process extending anteriorly to the length of the palps. Among other Old World *Echinonyssus* there is a full series of intergrades from coxa II anterior seta being normally setose to its expression as a very large hook, with anteriorly directed shaft and recurved tip (Tenorio and Radovsky 1979). Other holdfast developments in *Echinonyssus* include a large posterior spur on coxa II [*E. creightoni* (Hirst)], and the two subterminal ventral setae on tarsus II, which are somewhat enlarged and peglike in many *Echinonyssus*, developed into large, pointed, posteriorly directed, spurlike structures [*E. liberiensis* (Hirst) (Fig. 9.4)]. Similar holdfast developments are found in other hirstionyssine genera. For example, in some *Trichosurolaelaps* and *Thadeua*, the anterior seta of coxa II has become a large spurlike process and *Ancoranyssus trichys* has developed a large double hook on genu I and retrograde spurs associated with the sternal plate.

Macronyssidae

The obligate parasites of the Macronyssidae (Figs. 9.3F–J; 9.5) are generally moderate-sized dermanyssoids, but range from quite small (just over 300 μm in some *Parichoronyssus*) to unusually large [up to 2500 μm in *Megistonyssus africanus* (Zumpt and Till)]. They are mostly nest or roost inhabitants. The general cheliceral structures (Figs. 9.2C; 9.3I) and the hematophagous (or tissue-fluid) diet of the females are closely parallel to those of some laelapids, particularly the hirstionyssines. Some macronyssid species show specific structural similarities to those of certain laelapines, and these similarities, together with their related host associations, suggest a macronyssid derivation from the laelapines. However, the family Macronyssidae represents one of the more significant evolutionary developments of the host–parasite relationship in the Dermanyssoidea by the modification of their life cycle and the structural and physiological consequences of this change.

The Macronyssidae are uniformly characterized by (1) a protonymph that attaches to the host for extended periods and, at least in some cases, feeds in a similar way to ixodid ticks, and (2) a deutonymph that is non-feeding and structurally "larviform." Among the structures that are influenced by this unusual life cycle, the chelicerae are most immediately illustrative (Fig. 9.3F–J). The feeding behavior of the protonymph has been studied in the greatest detail for *Chiroptonyssus robustipes*. Radovsky (1967) deduced the nature of feeding from periodic examinations of mites on the host and the condition of mites recovered from the host. Lavoipierre and Beck (1967) directly observed feeding by using a compound microscope and transilluminating the wing of the host. They injected Evans's blue into

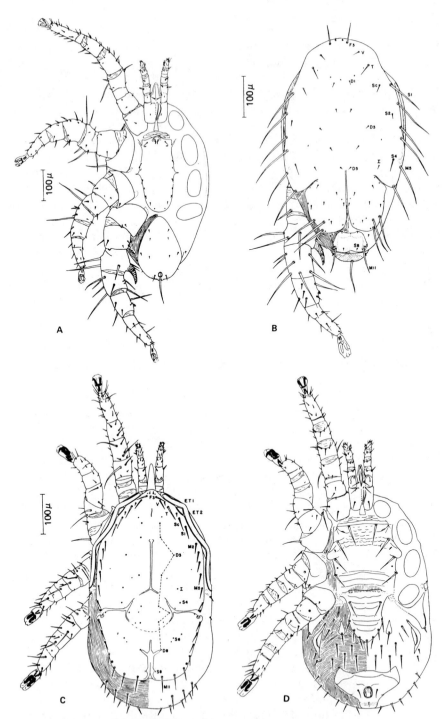

Figure 9.5 *Synasponyssus wenzeli* Radovsky and Furman (Macronyssidae). (*a, b*) Male; (*c, d*) female. (*a, d*) Ventral view; (*b, c*) dorsal view. (Reproduced by permission of F. J. Radovsky and D. P. Furman. 1969. *J. Med. Entomol.* **6**:385–393.)

bats on which mites were feeding, and sectioned the host wing in areas of mite feeding. The specific host of this mite is *Tadarida brasiliensis*, a molossid bat.

C. *robustipes* protonymphs feed for an extended period (3–6 days) on tissue fluids from vessels in which they induce increased permeability at the site of attachment. At the end of that time, they puncture a vessel and engorge rapidly (about 15–20 minutes) before leaving the host (Lavoipierre and Beck 1967). This feeding largely on acellular elements followed by rapid engorgement on blood is parallel to the feeding mode of ixodid ticks, first demonstrated for *Ixodes ricinus* (Linn.) by Lees (1952). During the long period of feeding prior to engorgement on whole blood, ixodids greatly increase (approximately double in *I. ricinus*) the cuticular thickness of their idiosomal parts not covered by fixed plates; when they rapidly engorge on whole blood, the cuticle is stretched to a point where it returns to its original thickness. The phenomenon of intrastadial modification of this magnitude resulting from cuticular growth has been termed *neosomy* (Audy et al. 1972). A neosomatic fixed flea of the genus *Tunga* has also been shown to feed on tissue fluids and white cells during the period of its massive cuticular development, and then to switch over to whole blood during its ovigerous period (Lavoipierre et al. 1979). It remains to be conclusively demonstrated that the same sort of cuticular thickening, in association with the long period of feeding on material other than whole blood, takes place in protonymphs of *C. robustipes*, but it is a logical inference (Radovsky 1967). Demonstration of such intrastadial cuticular growth would justify the application of the term neosomy to the protonymph of *C. robustipes* and most probably to a number of other macronyssids, if not to protonymphs in the family as a whole.

The Macronyssidae have been divided into two subfamilies, Macronyssinae and Ornithonyssinae (see Radovsky 1967, 1969), of which the latter is the more derived and specialized group and includes nearly all of the macronyssids on nonbat hosts as well as many bat parasites. Laboratory studies that confirm the extended feeding period of the protonymph on the host, as in *C. robustipes*, have been carried out for several ornithonyssines: *Ornithonyssus bacoti* (Bertram et al. 1946; Skaliy and Hayes 1949); *O. sylviarum* (Canestrini and Fanzago) and *O. bursa* (Berlese) (Sikes and Chamberlain 1954); *Ophionyssus natricis* (Gervais) (Camin 1953); *Pellonyssus passeri* Clark and Yunker (Clark and Yunker 1956); and *Steatonyssus antrozoi* Radovsky and Furman (Radovsky 1967) (Fig. 9.3F–J). The protonymphs of at least three *Radfordiella* species (Macronyssinae) have repeatedly been found embedded in obviously fixed feeding sites in the oral tissues of bats (Radovsky et al. 1971). *Chelanyssus* species and other *Chiroptonyssus* species besides *C. robustipes* have been found in fixed positions on the wings of bats. The frequency with which protonymphs of other macronyssids are found on the host and the morphology of the protonymph and the following deutonymphal stage indicate that slow-feeding protonymphs are characteristic of the entire family.

The protonymphs usually leave the host prior to molting, so that the macronyssid deutonymph is found in the nest or roost. The deutonymph is larviform in the sense of having generally weak sclerotization, poorly developed plates and setation, and ill-defined, obviously nonfunctional mouthparts. In most cases the ambulacral apparatus of the tarsi is similarly rudimentary. Adult macronyssids, which emerge after a relatively brief deutonymphal stage, are rapid feeders. Females of *C. robustipes* probe for a vessel, tear it, and then rapidly engorge (sometimes in 10–15 minutes) on blood seeping out of the hole in the vessel wall (Lavoipierre and Beck 1967). Most species lay eggs, but some are larviparous. The larvae do not feed and may molt to protonymphs in a day or so. Arrhenotoky is known for some species, such as *O. bacoti*, but *C. robustipes* apparently reproduces only after mating.

The prolonged feeding by macronyssid protonymphs allows for sufficient engorgement to carry the mite through a nonfeeding deutonymphal stage; consequently, preimaginal feeding is concentrated in a single period of contact with the host. A major advantage of parasitism is the abundance of food provided by the host, and many examples can be given of the concentration of trophic activity during the life cycle that abundance allows. Some dermanyssoids, for example, *Haemogamasus* and related genera, have feeding restricted to the deutonymph among the immatures. This requires an increased transfer of nutrients through the egg. However, the macronyssid life cycle, restricting feeding to the protonymph among the immatures, does not tend to increase the nutrient load per egg and may allow the female to maintain a relatively high reproductive rate.

In fact, because of the large quantity of nutrients that can be taken in with fixed feeding by the macronyssid protonymph, we should expect a higher reproductive rate among macronyssids than in Laelapinae, where both protonymph and deutonymph feed intermittently. Although data are limited, those available support these conclusions. Haemogamasines have a relatively large egg and a low reproductive rate; for example, a maximum of two eggs per day for individually held females of *H. liponyssoides* (Radovsky 1960) and a mean of one larva every two days for ovoviviparous *H. ambulans* (Furman 1959b). *L. myonyssognathus* produced larvae at a mean rate of less than one every three days (data from Mitchell 1968). The highest average among four species of *Androlaelaps* was one progeny every two days. By contrast, *C. robustipes* had an average reproductive rate of more than 3 eggs per day, as many as 5 eggs in one day, and 14 eggs after one blood meal (oviposition completed within 4 days) (Radovsky 1967). *Steatonyssus antrozoi* produced up to 6 eggs in 2.5 days following a single feeding.

That the protonymphal feeding mode allows macronyssids to maintain a higher reproductive rate is also indicated by the general retention of oviparity among macronyssids, especially in the most successful and polyxenic Ornithonyssinae. Those few macronyssines that have developed ovoviparity seem to have done so in relation to maintaining a more permanent ectoparasitic relationship because of the habits of their particular hosts.

The 26 genera of macronyssids are about equally divided between the two subfamilies; however, about 60% of the known species are in the more derived Ornithonyssinae. Macronyssinae is basically a bat-parasitizing group, and only the ditypic *Acanthonyssus* and monotypic *Argitis* are associated with rodents in the Neotropics. Numerous concurrences in parasite morphology and host association indicate a high level of phylogenetic coevolution and a great antiquity for the Macronyssidae.

As I have reviewed elsewhere (Radovsky 1967), bats may have existed in the Cretaceous period and are known from Paleocene fossils; apparently they had already separated into the two contemporary suborders (Megachiroptera and Microchiroptera) by that time. The oldest representatives of some modern families are known from the Upper Eocene or Lower Oligocene levels, and some more questionable identifications are even earlier. The occurrence of the laelapine genera *Notolaelaps* and *Neolaelaps* on Megachiroptera, which lack macronyssid parasites, and of the conservative macronyssine genera *Bewsiella* and *Ichoronyssus* on Microchiroptera, including some of the more primitive bats in that suborder, is evidence for a macronyssid origin on bats in the Paleocene epoch, if not earlier. It also appears that, as a result of the breaking up of Pangea, the three bat-parasitizing genera *Parichoronyssus*, *Radfordiella*, and *Macronyssoides* were isolated in South America where they evolved with the Phyllostomatoidea, while *Bewsiella* and *Ichoronyssus* along with the large genus *Macronyssus* and monotypic *Megistonyssus* evolved with the Vespertilionidae and the Rhinolophoidea in the Old World. Of these, *Macronyssus* eventually came into and spread through the New World with its vespertilionid hosts (Radovsky 1979). Secondary radiation, or isolation on particular hosts, subsequently gave rise to some other macronyssine genera in the New World, including *Nycteronyssus*, *Chiroecetes*, and *Synasponyssus* on bats and *Acanthonyssus* and *Argitis* on rodents.

While macronyssine radiation was still taking place, especially in the New World, the ornithonyssines arose, probably in the Old World and possibly with a *Cryptonyssus*-like ancestor. This must have been quite early, both because of the geographically widespread condition of the group and the secondary radiations they have undergone, as well as the fairly long time that must have passed for the development of the derivative family Rhinonyssidae (see below). Again, there apparently was a transfer of the Ornithonyssinae, including *Cryptonyssus* at some point, to the New World by the northern route. *Parasteatonyssus* and *Chelanyssus* specialized as molossid parasites in the Old World; *Chiroptonyssus* did so independently in the New World. *Steatonyssus* evolved as the most successful ornithonyssine genus on bats and became essentially cosmopolitan. *Ophionyssus*, which is found on reptiles (snakes and lizards), probably developed somewhere in the Holarctic region, and *Pellonyssus* on birds apparently developed in Africa; these genera also became cosmopolitan. *Ornithonyssus* evolved in the New World, probably in still-isolated South America; only with the transfer of hosts in post-Columbian times did several polyxenic species (*Ornithonyssus bacoti, O. sylviarum*, and *O. bursa*) transfer to become widespread in the Old World as well as the New World (Furman and Radovsky 1963). (It is unfortunate for nomenclatorial stability and for the understanding of ornithonyssine relationships that *Ornithonyssus* continues to be used as a dumping ground for hard-to-place species, since the genus is well defined by a clear set of characters.)

It has been proposed that the Macronyssidae arose as parasites of reptiles early in the Mesozoic era (Zemskaya 1969). However, since the macronyssids of reptiles are all highly derived ornithonyssines (*Ophionyssus* and *Draconyssus*), this theory appears to be incorrect.

Synasponyssus wenzeli Radovsky and Furman (Fig. 9.5), for which a monotypic genus was created, illustrates the coevolution of macronyssids and bats (Radovsky and Furman 1969). This Neotropical species is parasitic on a tree-roosting and primarily solitary relictual bat, *Thyroptera discifera*, which is one of two species in its genus and family; the family Thyropteridae is one of several small branches from the vespertilionoid stock in the Neotropics. *S. wenzeli* is highly conservative in certain respects, particularly in its retention on the dorsal plate of an unusually high number of primary setae (33, either as developed setae or trichoporal remnants). Furthermore, it appears to have an additional primary sensory seta not found in any other macronyssid in the main sensory field of tarsus I. *S. wenzeli* appears to be closest to such conservative macronyssine genera as *Bewsiella* and *Ichoronyssus*. At the same time, this species is perhaps the most highly specialized of any macronyssid in terms of deviation from the typical morphology of this and related families.

The specialized features in *S. wenzeli* relate to its permanent parasitism on a solitary host. The female is adapted both for the development and birth of a relatively large protonymph, which occupies a surprisingly large portion of the space within the idiosoma, and for clinging to the host. The male, of course, has adapted for the latter purpose only. Although both sexes are highly modified, *S. wenzeli* is the most sexually dimorphic of the macronyssids. The protonymph is less modified and closer to the general macronyssid type, except that its idiosomal plates are weakly sclerotized and reduced in area, presumably to assist in the birth of a relatively large nymph. Numbers of females, males, and protonymphs were taken in the single collection of this species, but no deutonymphs. Macronyssid deutonymphs are infrequently collected from the host, because in most cases the engorged protonymph leaves the host, and the deutonymphal stage is passed in the roost or nest. In *S. wenzeli*, the deutonymphs may be adapted for clinging to the host, where their development is brief and so they are unlikely to be collected. This situation is reported for *Macronyssus flavus* (Kolenati) (Zemskaya 1969) and may also be the case for several species in *Macronyssus* and related genera that have become permanent or nearly permanent parasites, with stronger claws and ovoviparity, but without the more extreme modifications that are seen in *Synasponyssus*.

The host association of *S. wenzeli* is further evidence of the antiquity of macronyssids and their coevolution with bats. Dr. R. Wenzel (in letter, cited by Radovsky and Furman 1969) noted that *Thyroptera* carries the most primitive genus of nycteribiid flies. The roosting behavior of *Thyroptera*, which is related to their distinctive morphology in having wing and foot suckers for clinging to leaves of banana, *Heliconia*, and so forth, results in a high level of isolation from other bats and a distinctive faunule of parasites.

Although the Rhinonyssidae are strictly parasites of the respiratory passages of birds and do not occur in mammals, their relationship to the Macronyssidae and the coevolutionary significance of that relationship is relevant here. The evidence that their life cycle is like that of the macronyssids strongly supports the conclusion that the avian nasal mites derived from the Macronyssidae (Radovsky 1964, 1966, 1967, 1969). Since none of the Rhinonyssidae have been reared and studied in the laboratory, this evidence is circumstantial. However, the morphological indications that the rhinonyssid protonymph feeds and that the deutonymph does not are convincing (Strandtmann 1961; Mitchell 1963; Domrow 1969), and it appears that the developmental sequence is the same throughout the family. Bregetova (1964, 1969) was among the first to stress the importance of life cycles in understanding the relationships of parasitic Mesostigmata, and she recognized the similar patterns and relationships of some Rhinonyssidae with the Macronyssidae. However, she placed these rhinonyssid species in a separate family on the basis of her interpretation of the life cycle of *Rhinonyssus* and some related genera as including functionally similar protonymphal and deutonymphal stages, a conclusion for which I have not been able to find any supporting evidence (Radovsky 1967, 1969; see also Domrow 1969).

Besides their similar life cycles, the hypothesis that Rhinonyssidae derived from Macronyssidae is supported by the morphological similarity of such rhinonyssid species as *Mesonyssoides ixoreus* Strandtmann and Clifford [original combination by Strandtmann and Clifford (1962), currently generic placement unsettled] to certain of the more derived ornithonyssines, particularly *Pellonyssus*, a widely distributed group of bird ectoparasites. The morphological changes necessary to derive *M. ixoreus* from *Pellonyssus* are relatively slight.

Bregetova (1964) noted the lack of correspondence between rhinonyssid genera and the higher categories of the birds they parasitize. Many of the mite genera are found in a wide range of host families and often cut across disjunct ordinal lines as well. This would support the assumption that the Rhinonyssidae derived from an advanced ornithonyssine, which suggests a relatively late origin in geologic time, but not too late, considering the wide distribution of rhinonyssids both geographically and in their range of avian hosts. If the rhinonyssids originated well into the Tertiary period, the avian orders would already have been established, but there still would be sufficient time to explain the wide distribution of the Rhinonyssidae. Having no competition for the niche they occupied and a generally uniform internal location, they must have had minimal barriers to transfer to and to establish themselves in new host groups.

Kethley and Johnston (1975) pointed out that ectoparasites may track "a resource that is a subunit of their host, and these subunits may be host-independently variable characters." Corollary to this, they concluded that "non-congruent host–parasite relationships" occur more commonly than had been thought. They use Rhinonyssidae as an example of noncongruence and note that while rhinonyssids (on birds) are derived from macronyssids (originating on bats), "It is unreasonable to attempt to derive the birds from the bats." This last quote typifies the misconceptions that may arise if resource tracking is considered without also giving weight to such factors as the time in the evolutionary history of a host group when it acquired a parasite group, or the effect of interspecific competition (or the lack of it) on parasites. Coevolution in the sense that "parasite interrelationships are congruent with host interrelationships" (Kethley and Johnston 1975) obviously does occur. It is, for example, quite evident in the relationship between macronyssids and the Chiroptera with which they primarily evolved, whereas there is little congruence of macronyssid interrelationships and those within the other host groups (reptiles, birds, rodents) to which they secondarily transferred after the major evolution of these host groups as they exist today. Similarly, the lack of congruence between rhinonyssid and avian relationships can be explained by the origin of this group of parasites after the principal radiation of birds, and their dissemination among hosts without competitive interference from established species of the same or an ecologically equivalent group. Of course, bats are not directly relevant to the evolution of Rhinonyssidae or their resource tracking, since the macronyssid ancestors of rhinonyssids had already transferred to birds before entering the nasal parasitope.

Dermanyssidae

The family Dermanyssidae is comprised principally of bird parasites. Of the two genera currently recognized, *Dermanyssus* includes some 16 species all found on birds, whereas *Liponyssoides* [= *Allodermanyssus*; see Sheals (1962)] includes seven valid species of which three are bird parasites and four are on rodents (Moss 1967, 1978; Domrow 1979); an undescribed *Liponyssoides* species is known from rodents in western North America.

Dermanyssids are most notably specialized for parasitism by the extremely long and slender chelicerae of the female and both nymphal stages (Fig. 9.2E). The lengthening is in the second segment, and the chelae are retained as minute structures at the tip of this segment (Fig. 9.2E). The chelicerae fit together and apparently function as a single styletiform tube in piercing the host tissues and sucking blood (Radovsky 1968, 1969).

Host associations in *Dermanyssus* range from the typical nest mites [including *D. gallinae* (De Geer) and *D. prognephilus* (Ewing)] that visit the host only for feeding, which have long legs and a high engorgement capacity, to species (such as *D. grochovskae* Zemskaya and *D. quintus* Vitzthum) that approach permanent parasitism, which have shorter, stouter legs, less engorgement capacity, attach their eggs to feathers, and are found more often on the host than in nests (Moss 1978). *Liponyssoides* of mammals are of the former type, that is, long-legged nest parasites. Few rodent-associated dermanyssids are known. There appears to be a greater tendency among Dermanyssoidea to develop permanent parasitism on birds and bats than on rodents and other small nonvolant mammals. It is notable that one of the two bird *Ornithonyssus* (*O. sylviarum*) is well advanced in the direction of host-restricted parasitism, while the mammalian parasites in the same genus are all nest forms (as far as we know).

Liponyssoides appears to be a less derived group than *Dermanyssus*. Moss (1978; also see Sheals 1962; Evans and Till 1964) points out differences between the two genera, and in each case the differing character state in *Liponyssoides* is plesiomorphic. The bird-parasitizing *Liponyssoides intermedius* (Evans and Till) was first described in *Dermanyssus* and has features indicating that it is transitional between rodent *Liponyssoides* and *Dermanyssus*. Thus the family Dermanyssidae appears to have evolved first on rodents and then transferred to birds.

The morphological evidence that dermanyssids are derived from the laelapine stock agrees with their apparent origin as rodent parasites. The dermanyssids do not appear to share any derived characters that suggest a common origin with other advanced dermanyssoid groups, and the structure of the male chelicerae is very much like that seen in such genera as *Laelaps* and *Androlaelaps*. This view is further supported by the life cycle: most advanced dermanyssoid parasites deviated in one way or another from having both nymphal stages active and feeding, but the latter situation is found in the Dermanyssidae as it is in most Laelapinae.

The impact on the host and host response has been examined only for some bird parasites among the dermanyssids. *Dermanyssus prognephilus* may significantly affect nestling weight gains and presumably survival after fledging, and parasitism may be an important component of the selective pressures determining clutch size in birds (Moss and Camin 1970). The influence of nest parasitism on the number of offspring produced at a time should also be examined for small mammals.

Halarachnidae

This family comprises two clearly separable groups of permanent cavity parasites: Raillietiinae and Halarachninae. The taxonomic relationship and placement of these two groups in one separate family suggested by

Radovsky (1969) must still be considered conjectural; it has been followed by some authors but not by others. However, as we discuss, subsequent information has tended to support that conclusion. Halarachnines are found in mammalian respiratory passages, while raillietiines are in mammalian ear canals. Considerable morphological disjunctions exist between Raillietiinae (five species) and Halarachninae (35 species) (Furman 1979). The joining of the two groups in one higher taxon was based to a large extent on their sharing a life cycle of an unusual kind: the larva is an active stage, frequently recovered and apparently functioning in interhost transfer, whereas both nymphal stages are largely suppressed, that is, nonfeeding, ephemeral, and at least sometimes with the deutonymph retaining the protonymphal skin so that both nymphal skins are cast together. Domrow (1981) states that larval structures in *Raillietia* support placement of the two subfamilies together.

Because of their separate parasitopes as well as the questions that remain about their relationships, the two subfamilies are discussed separately here.

Raillietiinae.   Of the four described species, only *Raillietia auris* (Leidy), commonly and widely found in domestic cattle, has been studied parasitologically beyond basic collection data. This and other species in the genus are parasites of the external ear, occurring consistently in the auditory meatus external to the tympanum. [Olsen and Bracken (1950) reported an instance of mites being found only in the tympanic bulla of a seriously affected heifer; in view of other reports, this must be at least tentatively considered as an abnormal infestation or an observational error.] There is an increase in discharge of cerumenal exudate in the ear and frequently a purulent discharge that may contain blood is mixed with the secretion. Many early authors noted behavioral signs that indicated considerable discomfort in infested cattle, and later studies have confirmed this effect.

The most informative study concerning the biology and host–parasite relationship of *R. auris* is that of Tsymbal and Litvishko (1955). They recovered 1138 mites from 45 cattle (57 ears) out of a total of 55 cattle (72 ears) autopsied, or 82% infestation rate by host and 79% by ear. Mites were present at all times of the year. Of the recovered mites, 82% were females, 8% males, and 10% larvae. No nymphal stages have been reported either in this or other studies of *R. auris* or in the few collections representing other species of *Raillietia*. Larvae as well as adults of both sexes have been found repeatedly in collections of *R. auris*. *R. auris* is larviparous, based on the many females observed to contain developed larvae. Since each nymphal instar is always represented at least by separate cuticle formation (see Halarachninae), we may anticipate that nymphs will be found when *R. auris* or other *Raillietia* species are sufficiently studied.

Feeding by *R. auris* is restricted to the adults, since the larva has mouthparts incapable of feeding, as in all other Gamasina. To whatever extent the male feeds, it obviously has no effect on the transfer of growth potential and energy, all of which must come through nutrients taken by the female and passed to the eggs.

The larva of *R. auris* has well-developed setation and appendages (including the elongate ambulacra), quite unlike the typical inactive dermanyssoid larvae. Since the larva is adapted beyond the needs of a transitional stage, it presumably functions as the transfer stage between hosts, as in at least some Halarachninae. Although it is obvious that the mobility of adult halarachnines is restricted by endoparasitic adaptations, some descriptions of *R. auris* (see Tsymbal and Litvishko 1955) indicate that the female is more highly modified than would be suggested by the usual dorsal-ventral depictions. The body is highly arched dorsally so that the height may be three-fourths or more the body length, and this unusual shape would interfere with locomotion outside of the ear.

The pathogenicity of *R. auris* (Tsymbal and Litvishko observed some degree of inflammation in 80% of the ears that they examined) suggests that this species infested cattle (i.e., *Bos taurus*) relatively recently, and that we are observing a poorly adapted host–parasite relationship.

Of the four other *Raillietia* species, *R. caprae* Quintero, Bassels, and Acevedo is found in goats, *R. hopkinsi* Radford and *R. whartoni* Potter and Johnston are found in African antelope of the genus *Kobus*, and these species are quite close to *R. auris* morphologically. However, *R. australis* Domrow, a parasite of the common wombat, is not only distinct from the other species in its host association but also in its morphology. This is most notable in the dorsal plate setation, *R. australis* having 32–33 pairs whereas the four species parasitizing bovids have 12–17 pairs. However, the reduction in relative size of the dorsal plate, restricting it largely or entirely to the podosomal region, agrees in all four species, and the general suite of characters morphologically defining the genus, in combination with the parasitope, leave little doubt that *R. australis* belongs to the same phyletic line with the other species (Potter and Johnston 1978).

The more conservative dorsal setation of *R. australis* and its presence in an Australian marsupial suggests that the raillietiine ear mites may have a truly ancient history of coevolution with mammals. Raillietiines are plesiomorphic in relation to halarachnines, and the apparent antiquity of the former group tends to support the concept that it gave rise to the halarachnines.

Halarachninae. Halarachnines are parasites of the respiratory tract in a wide range of mammalian groups (Furman 1979). In contrast to the raillietiines, the halarachnines have no tritosternum, the hypotrichy of their appendages is carried to the extent that some segments retain the larval condition in the adults, the epigynial plate is absent or a remnant, the peritreme is absent or vestigial, and the idiosomal form is frequently elongate and modified to an unusual degree. The separately derived Rhinonyssidae share a number of these traits, all of which may be understood in terms of an endoparasitic way of life.

Many of the adaptations found in internal acarine parasites such as the halarachnines are regressive in the sense that they involve loss or reduction of structures regularly found in free-living and most ectoparasitic relatives. It has been suggested that immunological responses by the host may play a role by selecting for reduction of some structures (plates, setae) that tend to increase the immunogenicity of the parasite (Fain 1977, 1979). Although such immunological pressures may be involved, I believe that the regressive adaptations have a more direct explanation, for the most part. Plates and elongate and numerous body setae no longer have a protective function against predators in endoparasites, and much of the need for closely distributed setae as tactile sensors is lost. Plates continue to retain a function in muscle attachment, and some tactile ability is needed even in the cavities of the host body, so that plates and some generally reduced setae are partially retained. Selection for regression is further increased by the need to reform the body mass to accommodate it to the spaces being occupied (Radovsky 1969). To the extent that host responses are involved, it seems more likely that simple irritation of the delicate mucosal surfaces by hard plates and long setae would result in host responses that would select for reduction of structures than that a specific immunological phenomenon, that is, an antigen-antibody reaction, occurs. Many groups of external parasites exhibit a degree of regression in plate and setal development that may be directly related to the reduced function of these structures, and which is certainly not likely to be related to irritation of the host. Acarine cavity parasites sometimes have little more intimacy with host tissues than ectoparasites that move freely about the surface of the host's body. It remains to be demonstrated that most cavity parasites stimulate antibody production by anything other than the structures that actually penetrate the tissues in feeding and the saliva introduced through these structures.

The host range of halarachnines is great, suggesting that the group is a very old one. It also includes a number of anomalies for which Furman (1979) suggested that, in some instances, separate genera may need to be established. Despite these anomalous associations, the host associations, as well as the parasitopes, can be seen as representing several major trends. *Pneumonyssus* with 14 species and *Rhinophaga* with six species are related genera primarily associated with Ethiopian and Oriental monkeys (Cercopithecidae) and apes (Pongidae). Most halarachnines, including *Rhinophaga*, are nasal mites, located in the nasal passages including the turbinates and sinuses and associated parts of the pharynx. *Pneumonyssus* is the only large group that is found principally in the lungs (all species except *P. duttoni*, which is in the trachea and bronchi). The major exceptions to the association of these two genera with higher Primates are the occurrence of two species of *Pneumonyssus* in African hyraxes (Procaviidae) and of two species of *Rhinophaga* in African porcupines (Hystricidae). One other species of *Pneumonyssus* is known only from captive phalangers (Phalangeridae) in New Guinea. The marsupial phalangers and Asiatic macaques (the genus of monkeys in which a *Pneumonyssus* species is found in Southeast Asia) overlap in their distributions on the islands of Timor and the Celebes, and the possibility of transfer between these host groups is geographically and ecologically reasonable. Domrow (1974), who described *P. capricornii* Domrow from a phalanger, also noted that the species is closest morphologically to *P. simicola* Banks from Asiatic macaques. The overall distribution of the genus suggests that a transfer from monkeys to marsupials was more likely to have occurred than its converse (Furman 1979), if indeed the phalanger infestation is established as being other than accidental.

The second major host trend in the halarachnines involves *Halarachne* with four species in hair seals (Phocidae) and *Orthohalarachne* with two species in fur seals (Otariidae) and walruses (Odobenidae). These are nasal mites except for *Orthohalarachne* adults, which occur primarily in the lungs. Six instances of light infestation by *Halarachne* were reported in wild sea otters (Mustelidae), approximately 3% of the animals that were examined for parasites, and one instance of a captive sea otter with an extremely heavy infestation that was "probably contributory to death" (Kenyon et al. 1965). The mites in all of these cases were considered indistinguishable from *H. miroungae* Ferris, at least in the broad sense, a parasite of the harbor seal, which often occurs in the same habitat with sea otters. Furman (1979) noted that the present view of marine carnivore evolution, in which the Otariidae/Odobenidae group is derived from the ursids and the Phocidae from the mustelids, suggests that *Orthohalarachne* and *Halarachne* may have evolved separately with their respective host groups to occupy the marine habitat. The occurrence of a *Halarachne* species in the mustelid sea otter "may represent a long-standing host–parasite association antedating the separation of the pre-mustelid, pre-phocid ancestral stem." However, Furman also notes that cross-infestation from seals is a plausible explanation of the sea otter infestations.

A third host trend involves *Zumptiella* with five species, primarily in rodents. Three species are in Holarctic tree squirrels (Sciuridae), one is in the Ethiopian spring hare (Pedetidae), and one is in an Ethiopian mongoose (Viverridae). The viverrid association may be the result of prey to predator transfer (Furman 1979), and it is not yet confirmed as a natural association.

The last genus of the subfamily is *Pneumonyssoides*, with four species that have a most peculiar assortment of hosts. Two species are in Ethiopian Suidae, the bush pig and the wart hog; one species is in Canidae, the domestic dog; and one species is in Neotropical monkeys (Cebidae). On the basis of the peculiar host and geographic distributions, Furman (1979) postulates that *Pneumonyssoides* is an unnatural grouping in which convergent evolution has led to morphological similarity. It is difficult to rationalize the occurrence in dogs as a transfer from wild suids, although dog-suid contact is quite plausible, because the question then arises as to why a suid-originated parasite widely distributed in domestic dogs has not transferred to domestic pigs.

The halarachnines have received attention because of the pathogenicity they can produce in various hosts. From the applied standpoint, this is of most concern in monkeys. Asian macaques, including essentially all imported rhesus monkeys, are regularly found to be infested with *Pneumonyssus simicola* (Hull 1970; Yunker 1973; Kim 1977). Also, African monkeys that may be used in laboratory research, particularly baboons, are often infested with other *Pneumonyssus* species of lung mites. *P. simicola* has been more intensively studied than any of the other species. These mites stimulate tuberclelike lesions in the lungs, and a number of female mites are typically found in each cyst. Bronchiolar and peribronchiolar inflammation is most evident histopathologically. Clinical signs are not apparent in most cases. However, it generally is believed that *P. simicola* can affect the health of its host severely, sometimes causing death in the case of massive infestation (Stone and Hughes 1969).

After the monkey lung mites, the most extensive pathological studies have concerned the two species of *Orthohalarachne* in the northern fur seal (*Callorhinus ursinus*). *O. attenuata* (Banks) and *O. diminuata* (Doetschman), the only species in the genus, have a wide host range and geographic distribution, but they often occur together in the same host species and the same individuals. Larvae of *O. attenuata* significantly outnumber adults in every age/sex category of the host, whereas larvae are even more predominant in collections of *O. diminuata* (e.g., an average of 1215 larvae and 8 adults in 66 subadult males examined) (Kim et al. 1980). Basically, the larvae of both species occur in the nasal turbinates, while the adults of *O. attenuata* are attached to the nasopharyngeal mucosa and the adults of *O. diminuata* are in the bronchi and bronchioles of the lungs.

There has been some difference of opinion on the extent to which *Orthohalarachne* species affect the health of their otariid hosts. Kim et al. (1980) concluded that heavy infestation by mites in the nasal turbinates of fur seals results in expiratory dyspnea, as a result of "chronic inflammation of the mucous membranes with erosion, hyperplasia and reactive mononuclear and eosinophilic cellular infiltration." They also observed marked signs in the lungs of heavily infested seals, including focal edema, congestion, fibrosis, and pneumonitis, along with bronchiolar involvement.

Pneumonyssoides caninum (Chandler and Ruhe) has received considerable attention because of its host. However, while various signs have been associated with the presence of this mite in dogs, it is a relatively benign parasite whose effects are generally limited to "excessive mucous production and hyperemia of the nasal mucosa" (Yunker 1973). This is further evidence of a long-standing association of this mite with canines, as opposed to a relatively recent transfer from a suid host.

Halarachnine–host interaction is viewed as a generally well-adapted relationship, despite the extreme pathology and even death attributable to some species of these mites. In the case of monkey lung mites, we do not know much at all about the course of infestations in nature; however, if it is anything like the observed severity in laboratory animals, we would expect this to have been observed more frequently in newly captured animals. The severe symptoms in lab animals are more the exception than the rule, and these are animals undergoing intermittent or continuous stress as a result of laboratory conditions (even the better range) and experimental regimes. The seal mites do appear to produce clinical symptoms in heavily infested wild hosts. In this case, it is perhaps pertinent that the host range of the two *Orthohalarachne* species, for which most observations on pathogenesis have been made, is quite wide. Are there host species with which there is a more long-standing association, and are the mites better adapted and less likely to produce symptoms in these? Or has the parasite sacrificed closer host adaptation in maintaining a broad host association over long geological periods?

Spinturnicidae

The mites of the family Spinturnicidae are morphologically among the most highly derived of the Dermanyssoidea. Most of their structures are specialized for ectoparasitism on bats, generally on the naked patagium (wing membrane) and uropatagium (tail membrane), and intermediate forms with more typical gamasines are not known. Rudnick (1960) revised the family and reviewed the literature up to that time. He pointed out the specificity of the various genera of Spinturnicidae for particular families of bats, and he suggested that this resulted from coevolution and, in relation to the host phylogeny, can form a basis for studying the phylogeny of the mites. Since that time, the spinturnicid–host associations have been used by a number of authors to support conclusions on the phylogenetic classification of bats. Laboratory data on the biology of spinturnicids are practically nonexistent. With a few exceptions, field observations have been cursory and incidental to collecting activities. The first in-depth study of the biology of spinturnicids based on field observations is by Deunff and Beaucournu (1981).

Spinturnicids are mites with large legs (thick and relatively long compared to the size of the idiosoma), usually bearing disproportionately large claws, and often with a characteristically diamond-shaped body. The diamond shape is associated with a poorly developed opisthosoma; the coxae of legs IV come almost to the posterior end of the body and coxae IV are close together (so that the coxae on each side form a markedly convex arc). The diamond-shaped small body and the large legs with strong claws provide a superb adaptation for holding and for rapid movement on flexible and inconstant surfaces like the bat wing and tail membranes. The laterally set claws can remain fixed as the body and legs flex to bring them closer together or further apart in corresponding to the folding or unfolding of the membrane.

The diamond-shaped body remains the rule for nymphs and males and newly molted females essentially throughout the family. However, females of some species and in certain genera develop a much-enlarged opisthosoma as they feed and become gravid. Domrow (1972) emphasized this characteristic in *Ancystropus*, in which the females have become fixed (see below), and it also appears to be typical of *Meristaspis, Periglischrus*, and some others. Furthermore, spinturnicids have also specialized for permanent parasitism on bat flying membranes by becoming nymphiparous and feeding on blood in all free stages (Rudnick 1960). Thus the normal stages are present, but egg and larva are intrauterine, females give birth to protonymphs, and the two nymphal stages and adults of both sexes feed on blood.

Recent systematic accounts of Spinturnicidae have resulted in the creation and the synonymizing of a number of genera. For the purposes of this chapter, the following genera are recognized: the Old World *Ancystropus* (=*Oncoscelus*), *Meristaspis*, *Eyndhovenia*, and *Paraperiglischrus*; the New World *Periglischrus* (=*Mesoperiglischrus*), *Cameronieta*, and *Paraspinturnix*; and the cosmopolitan *Spinturnix*. [Although the treatment of *Paraspinturnix* as a junior synonym of *Spinturnix* may be valid on morphological grounds (Domrow 1972), I have maintained *Paraspinturnix* because of its distinctive parasitope and the continued use by a number of authors.]

Most spinturnicids move about freely on the wing and/or tail membranes, but there are several exceptions involving the genera *Paraspinturnix*, *Ancystropus*, and *Meristaspis*. *Paraspinturnix globosus* Rudnick females are found in an extraordinary parasitope: within the anal orifice of bats of the genus *Myotis* in North America (Rudnick 1960). In studies of *M. velifer* from winter colonies, a single *P. globosus* was found in most hosts, while five of the infested bats each had two females (Reisen et al. 1976). The gravid females are essentially globular in form, with an idiosomal width approximating or even exceeding the length. This is much more pronounced in the fed and gravid females, which are normally observed, than in the occasionally seen newly molted females, which are more typically *Spinturnix*-like in form.

Most of the basic facts on the host–parasite relationships of *Ancystropus* and *Meristaspis*, including the parasitope, must be discovered or clarified. These genera are exclusively parasites of Megachiroptera, whereas all other Spinturnicidae are found on Microchiroptera. Surprisingly few systematic papers, even recent ones, give the parasitope for these genera. Rudnick (1960) cited Kolenati (1859, 1860) and Vitzthum (1941–1943) as sources for the concept that *Meristaspis* occurs on the eyelids and that *Ancystropus* is exclusively attached to the eyelids and canthi (eye angles). However, Rudnick also carefully pointed out that these are the only authorities to support these findings, and that he recovered *Meristaspis* from the host patagium and never found any around the eyelids. It also is not clear to what extent (perhaps entirely) Vitzthum was following Kolenati rather than any observations of his own. Kolenati changed his identifications for certain mites between papers and is not a particularly reliable source for parasitological data.

An *Ancystropus* species has been recovered from "the arm-pits of Malayan bats" (Domrow 1972), and there are extensive unpublished data on *Ancystropus* affixed with legs I deeply embedded in the tissues of the wing of its host (M. Lavoipierre, personal communication). Consequently, most evidence indicates that both *Meristaspis* and *Ancystropus* are largely, and perhaps entirely, associated with the host wing, although their feeding relationship may differ significantly from that of most spinturnicids. The association of one or both genera, or some other spinturnicid, with the eylids and canthi of pteropodids remains an open question, a possibility that cannot be rejected but that is certainly not yet established.

Kolenati (1860) first indicated that *Ancystropus* uses the forelegs to fix itself in the host tissues (see Rudnick 1960). Domrow (1972) recorded his *Ancystropus* species from the armpits of bats as "tightly embedded in the skin." He concluded that *Ancystropus kanheri* Hiregaudar and Bal also embeds legs I in the skin because of the frequency with which these legs are snapped off from recovered specimens. They are probably embedded at the level of the transverse cuticular ridge posterior to coxae II. The female of *A. kanheri* has a truly extraordinary anterior armature, with legs I much larger than other legs, large retrorse hooklike spurs on tarsus I and coxa I, anterior and posterior spurs on coxa II, a well-developed and anteriorly projecting dorsal plate and a well-sclerotized sternal plate—both of these plates shifted anteriad and ending on a line with the posterior margin of coxae II. Although this is an extreme, all females of *Ancystropus* and *Meristaspis* have thickened legs I with some specialized structures. This, with the observations of embedding cited, suggests that the females of each species in these two genera are fixed parasites, fastening to the host tissues by embedding the fore parts up to the base of legs I or beyond.

A remarkably analogous situation in some Rhinonyssidae provides further insight into the significance of the adaptations of the legs and the function of embedding them in host skin in *Ancystropus* and *Meristaspis*. The chelae in many rhinonyssids are "leaflike" and so poorly sclerotized that their ability to penetrate the mucosa of the nasal passages is questionable, as remarked by A. Fain (see discussion in Radovsky 1969). Intrigued by this question, Feider and Mironescu (1972) performed histological sections of three species of rhinonyssids in situ, selecting forms with poorly sclerotized chelicerae: *Mesonyssus gerschi* Feider and Mironescu, *M. hirsutus* Feider, and *Rhinonyssus colymbicola* Fain. In females of all of these species, legs I are extended more or less in parallel in front of the body and can penetrate with their modified claws into the mucosal tissues of the host. Feider and Mironescu (1972) concluded that the claws of leg I are preadapted to function in tissue penetration and that the delicate chelae may function in collection and conduction of mucus, epithelial cells, and blood to the oral opening of the mite.

One of the adaptive advantages of fixed parasitism is that a continuously feeding female has a higher reproductive potential than an intermittent feeder (Audy et al. 1972). However, fast-moving spinturnicids (e.g., most *Spinturnix*) may balance their lesser engorgement capability by being able to escape injury or destruction from preening by the host. Since preening can be very effective in bats, one may wonder how fixed parasites such as the female *Ancystropus* and *Meristaspis* are able to survive at all. Observations on an unrelated parasite may have a bearing on this question. The female of the flea *Tunga monositus* Barnes and Radovsky burrows into the pinna of a rodent ear, becomes pea sized and delicately saclike through neosomy, and is fully exposed to the claws of the host; yet, while the mouse scratches all around the parasite, the flea is nearly always left intact (Radovsky, unpublished; see also Barnes and Radovsky 1969; Lavoipierre et al. 1979). Rather than ignoring a nonirritating parasite, it appears that the host is avoiding injury to the parasite because this is presumably advantageous to the host. The advantage is perhaps to avoid the acute pathology resulting from destruction of the parasite within the tissues. This possibility should definitely be investigated.

Deunff and Beaucournu (1981), working in eastern France, not only carried out the first extensive studies of spinturnicid activities on the host but also, by studying several species on their respective hosts, were able to compare the parasite adaptations and to observe parameters that may be attributed to interactions between different spinturnicid species. The species studied were *Spinturnix myoti* (Kolenati) on *Myotis myotis*, *S. emarginatus* (Kolenati) on *Myotis emarginatus*, *S. mystacinus* (Kolenati) on *Myotis mystacinus*, and *Eyndhovenia euryalis* (Canestrini) and *Paraperiglischrus rhinolophinus* (Koch) on *Rhinolophus ferrumequinum*.

The three *Spinturnix* species are all located on the patagium, two exclusively, while *S. myoti* has some individuals on the uropatagium during the summer population peak. *E. euryalis* is also found on the patagium during the summer, but in winter this species remains in the body fur immediately adjacent to the wing membrane. Finally, in *P. rhinolophinus*, the females and nymphs are restricted to the uropatagium in the winter breeding season of this mite species; females occur principally on the uropatagium but also (a few) on the patagium in summer; and the males are found on the patagium.

Unlike the other four species studied by Deunff and Beaucournu (1981), all of which reproduce during the warm months and have a remnant population of overwintering adults, *P. rhinolophinus* reproduces on the hibernating host. The position of the females and nymphs on the uropatagium should be viewed relative to their specific location inside the highly vascularized and humid dorsal fold of the tail membrane, a protected site that may favor continued activity and feeding more than most of the host surfaces. The peak utilization of the same host is partitioned between this species and *E. euryalis* both spatially and temporally, with the latter species breeding on the patagium during the summer. Deunff and Beaucournu (1981) also suggested that *P. rhinolophinus* may mate in the summer, that is, in a different season from that of usual reproductive activity, based on the occasional finding in summer of females on the wings, the normal site for the males, and on a much higher male:female sex ratio in summer than in winter (0.8 versus 0.08).

The population buildups during the breeding season range from moderate (but very distinct) to extremely large. The most extreme in this respect of the mites studied is *S. myoti*. Winter lows of 40% incidence and less than four mites per infested host became summer highs of 100% incidence and an average of 19 mites per host. Furthermore, Deunff and Beaucournu (1981) confirmed that *S. myoti* may occur in large numbers on the substrate away from the host, an observation made for *Spinturnix* species in at least two previous papers (Holdenried et al. 1951; Tagiltsev 1971) but generally overlooked. They estimated that during part of July there were 6000 *S. myoti* on the bats of the colony and at least 25,000 on the guano below the roosting area. The dropping of the surplus mite population from the bats would protect the host, and hence the parasite, from the potentially fatal consequences of being overwhelmed by the mites. Deunff and Beaucournu (1981) suggested that the mites on guano are not necessarily a dead end for that part of the population, since young bats leaving to fly will often fall and crawl on the substrate and could easily acquire mites at that time. They also noted that the usual predominance of females is reversed with a 1.3:1 sex ratio on the guano, and they suggested that this may result from the greater movement of males in seeking out females for mating.

The phylogenetic coevolution of spinturnicids and their hosts has been discussed by several authors. Rudnick (1960) noted the apparent correspondence between the phylogenetic sequence in which spinturnicid genera can be placed and that of the families of bats that they principally parasitize. The mite sequence, with host families in parentheses, and proceeding from "oldest" to "youngest," is as follows: *Ancystropus* and *Meristaspis* (Pteropodidae); *Eyndhovenia* and *Paraperiglischrus* (Rhinolophidae); *Periglischrus* (Phyllostomatidae); *Spinturnix* (Vespertilionidae and Natalidae); and *Paraspinturnix* (Vespertilionidae). The indicated host associations remain basically valid. Dušbábek (1971) recognized two subfamilies with genera as follows: Spinturnicinae—*Ancystropus, Meristaspis* (with *Oncoscelus*), *Eyndhovenia, Spinturnix,* and *Paraspinturnix;* Periglischrinae—*Paraperiglischrus, Periglischrus* (with *Mesoperiglischrus*), and *Cameronieta.* He considered these two subfamilies to represent two separate lines of evolution characterized by different morphological and behavioral trends. Webb and Loomis (1977) suggested that phyllostomatoid spinturnicids (i.e., *Periglischrus* and *Cameronieta*) in the New World and rhinolophoid spinturnicids (i.e., *Eyndhovenia* and *Paraperiglischrus*) in the Old World each arose as offshoots from "a line common with that of *Spinturnix*."

Spinturnix is the one genus of the family that is found in both hemispheres. However, the New World species are relatively small in number (about seven compared to well over 20 in the Old World), are restricted to bats of the superfamily Vespertilionoidea, and do not fall into a clearly distinct New World grouping. By contrast, the *Periglischrus-Cameronieta* line is restricted to bats of the Phyllostomatoidea, a superfamily of Neotropical origin. *Spinturnix* apparently colonized the New World by way of the Behring route at a relatively late time, whereas the ancestors of the phyllostomatoid-associated spinturnicids were, as in the case of a similarly associated group of macronyssid genera, isolated with their hosts near the time of east-west continental separation. [Contrary to a statement in Radovsky (1979), I do not now concur with Webb and Loomis (1977) in the concept that "Dispersal of wing mites to the New World probably occurred on vespertilionids with subsequent infestation of phyllostomatids."]

Eyndhovenia and *Paraperiglischrus* are dissimilar genera occurring on rhinolophoid hosts. There is not a morphological basis for postulating a close common origin for them. *Paraperiglischrus* exhibits some similarities to *Periglischrus*, but these may be entirely convergent, although it is conceivable that they involve a common stem going back to the very early Cenozoic or late Mesozoic era.

Most spinturnicids are monoxenic or stenoxenic, although there are some notable exceptions. Their narrow specificity has permitted the development of some interesting conclusions or supporting data on host relationships, especially for the Phyllostomatoidea. Machado-Allison (1967) first suggested that mormoopines be placed in a separate family from the Phyllostomatidae, in relation to the divergence suggested by the primary association of *Cameronieta* with mormoopines, whereas *Periglischrus* is primarily phyllostomatid associated. Smith (1972) established Mormoopidae for *Mormoops* and *Pteronotus*, basing his conclusions mostly on a reevaluation of the morphology of the bats, but also citing additional evidence from Machado-Allison's studies on spinturnicid–host relationships. Similarly, Machado-Allison (1967) believed that the *Periglischrus* associated with vampire bats, Desmodontidae, indicated that the bat hosts should be included within the family Phyllostomatidae, and mammalogists generally agree with that conclusion at present. Dušbábek (1968) considered the particular *Periglischrus* association with *Brachyphylla* to indicate that this genus of bats belongs in the subfamily Phyllonycterinae, rather than its prior placement in Sternoderminae; Silva Taboa and Pine (1969) cited Dušbábek's work in support of morphological and behavioral evidence for placing *Brachyphylla* in the Phyllonycterinae.

Other examples can be given to indicate the applicability of spinturnicid associations to understanding host relationships. In this family, "resource tracking" of subunits of the host that involve "host-independently variable characters" (Kethley and Johnston 1975) may occur, but the congruence of host and parasite evolution suggests that it is the exception. The host–dependent tracking that takes place in the *Periglischrus-Cameronieta* group, for example, may relate to these being moderately large mites that regularly compete for the patagium as a whole. Such mites may not partition their available habitat and thus may consistently be involved in interspecies competition that results in specificity. As noted previously (see section on Macronyssidae), the timing of association with the host is important in the kind of tracking patterns observed, and spinturnicids have apparently been associated with bats since a time near the origin of this host group.

Dasyponyssidae, Manitherionyssidae, and Hystrichonyssidae

Three families of mammal mites, each known from one or two species, belong in the Dermanyssoidea but have highly derived characters that tend to obscure their relationships with other families. These are Dasyponyssidae on Neotropical armadillos (Edentata), Manitherionyssidae on Ethiopian pangolins (Pholidota), and Hystrichonyssidae on Oriental porcupines (Rodentia).

Species of Dasyponyssidae and Manitherionyssidae have been treated as a single taxon (Dasyponyssidae) because of superficial similarities, particularly the enlarged first pair legs, and perhaps because of similarities between their hosts. A separate family was established for *Manitherionyssus* (Radovsky and Yunker 1971), because the two groups lack other characters suggestive of a relationship, and their similarly scaly hosts are unrelated and are geographically isolated. The similarity of the forelegs appears "to be a case of convergent evolution of parasites of convergent hosts."

Dasyponyssidae contains two genera (*Dasyponyssus* and *Xenarthronyssus*) each based on a single species (Fonseca 1940; Radovsky and Yunker 1971). The hosts are all armadillos (Dasypodidae) but represent two genera and three species, spread over a geographic range from Panama to Uruguay. The two genera share characters clearly indicating their relationship, but they also are divergent in many respects. Consequently, it appears that the family has coevolved with its armadillo hosts over a long period. The nymphal stage associated in collections with adult dasyponyssids has been tentatively identified as the protonymph. This suggests the possibility of relation to the Macronyssidae, with a feeding protonymph and suppression of the deutonymph. There are no clear morphological indications of such a relationship, and some features, such as the very unusual and complex cheliceral structures, including long and highly developed arthrodial processes at the base of the movable chela, argue against it. However, it is conceivable that the dasyponyssid stock branched off of a very early macronyssid stock near the time when the Neotropics were separating from the southern continental land mass.

The life cycle and host–parasite relationship of dasyponyssids can only be inferred since there have been no direct observations. Nymphs have been seen in females, apparently protonymphs, and they were the same as the free nymphs studied. It appears that these mites develop *in utero* to a feeding nymphal stage, and most likely these are host-restricted parasites. The extreme development of legs I, and their general similarities to these legs in certain spinturnicids and rhinonyssids, suggests that they may be used in penetrating the tissues; certainly legs I are important holdfast structures. The other legs are weakly developed and have reduced claws. Presumably, dasyponyssids are relatively fixed parasites in any feeding stage, and they may attach to the softer tissues on the ventral side or between the scales of the host.

The family Manitherionyssidae includes a single species known only from the Cape Pangolin. This dermanyssoid is much more conservative than the dasyponyssids. However, its similarity with the latter family in having large legs I, with large sessile claws, and the relative slenderness and reduction of claws characterizing the other legs is striking. Furthermore, both families occur on mammalian hosts that separately have developed scales. A similar convergence between parasites of these two host groups occurs among intradermal fleas: a number of *Tunga* species, in which the female develops neosomatically within the skin, are found between the (supporting) abdominal scales of armadillos, whereas *Neotunga*, a quite unrelated flea also characterized by skin penetration and neosomy, is found only between the scales of African pangolins (Audy et al. 1972).

The family Hystrichonyssidae is also based on a single species, one with a highly modified body shape in the female, the only stage known (Keegan et al. 1960). (One collection of two specimens among the original three collections of this species was from a snake; it has generally been presumed that the rodents represented in the other two collections are the normal hosts.) The chelicerae are developed into styletlike structures, similar to those of the Dermanyssidae in their extreme length and slenderness and minute chelae. However, the morphological similarity stops there; whereas the major length of the shaft is derived from the second cheliceral segment in the Dermanyssidae, it results from elongation of the basal segment in *Hystrichonyssus* (Fig. 9.2F). In the latter, the basal segment is more than 90% of the cheliceral length, while the second segment is not unusually elongate. *Hystrichonyssus turneri* Keegan, Yunker, and Baker is quite a large mite, with an idiosomal length of about 1.5 mm or more; the chelicerae are about two thirds the length of the idiosoma. Further studies on the morphology and behavior of these mites would be particularly helpful in determining the degree to which the functional analogies of cheliceral adaptations may extend in relation to the Dermanyssidae.

Ancoranyssus was placed in the laelapid subfamily Hirstionyssinae when first described (Evans and Fain 1968). In the most recent revision of the subfamily (Tenorio and Radovsky 1979) this histionyssine placement is continued, but the chelicerae are described as separate and not proximally fused, contrary to the original description. *Ancoranyssus trichys* is similar to *H. turneri* in having the first segment of the chelicera unusually elongate relative to the second, although the proportion is only somewhat in excess of 2:1 in *A. trichys* compared to more than 10:1 in *H. turneri*. Nonetheless, the combination of the elongation of their first cheliceral segments and their occurrence on hystricid rodents in the Oriental region should cause us to look for a phylogenetic relationship between these two species, which also have reduced, hypotrichous dorsal plates and enlarged opisthosomas. On the other hand, *A. trichys* has an extraordinary complex of structures, involving the coxae, genua I, and the sternal plate region, for attachment to the host; none of these structures is similarly modified in *H. turneri*. I suggest that these two species are related but that they have taken quite different directions in their host relationships, resulting in major divergences in morphology. The female of *A. trichys* is adapted for fixed parasitism using the massive hooks on genua I, in particular, for fixation to the host skin. If *H. turneri* attaches to the host for extended periods, the mechanism for its doing so is not evident.

Spelaeorhynchidae

Spelaeorhynchidae does not appear to be derived from a dermanyssoid ancestor and it should not be placed in that superfamily. The type species was first considered to be a tick. Banks (1917) recognized the mesostigmate features of the family, but an ixodid relationship continued to be suggested by authors. In the most detailed analysis of *Spelaeorhynchus* to date (Fain et al. 1967), the group was considered to be much closer to the Mesostigmata, and its placement in that suborder has not been seriously questioned since. The same paper stated that "All of the characters except some highly specialized ones are typically laelaptoid [i.e., dermanyssoid]." It has also been suggested that there are more indications of a relationship of this family to the cohort Sejina than to the Gamasina (Radovsky 1969). It can at least be concluded that evidence is lacking to derive these highly specialized parasites from any of the parasitic lines within the Dermanyssoidea.

There are three named species recognized as valid in the family, all in the genus *Spelaeorhynchus*. They are closely similar morphologically, and they are known only from Neotropical bats of the families Phyllostomatidae and Mormoopidae (Webb and Loomis 1977). Only the female has been recovered from the host, normally attached to the pinna or tragus of the ear. Larvae dissected from gravid females have been described, but the male is unknown and only a single nymph has been collected and described. Fain et al. (1967) described the morphology of the females in detail, especially in relation to attachment and other adaptations for ectoparasitism. *Spelaeorhynchus* species all have females with an idiosoma that falls in the size range of 1440–1780 μm long by 1170–1560 μm wide.

However, "two young females" of *Spelaeorhynchus praecursor* Neumann were also described, apparently only recently attached to a host, and they were 1150 × 800 μm and 1140 × 870 μm. These relatively large mites have a very strongly sclerotized gnathosomal ring with its opening directed anteroventrad, within which two extremely short and stout chelicerae with strongly toothed chelae are contained (Fig. 9.2D). The chelicerae fasten to the host tissues, and the mite is additionally supported by an anterior dorsal plate and a sternal plate, the latter heavily sclerotized. The idiosoma projects from the host at an angle, with the ventral surface closer to the host surface. The slender legs project out from the body and are not attached; in fact, the weak ambulacra are known only from the "young females" that apparently have just acquired a host, and these structures are evidently modified to be easily shed by the established parasite. Another interesting adaptation is the posterior migration of the female genital opening, located well back on the opisthosoma and not far from the anus, apparently to assist in freeing young from the maternal parasite that is attached anteroventrally to the host.

The larvae dissected from females (Fain et al. 1967) have weakly sclerotized chelicerae and lack the gnathosomal specializations for host attachment found in the female; evidently the larva does not feed. The nymph of *S. praecursor* was described from a single specimen (Dušbábek 1970). This nymph is similar to the adult female in most features, including the loss of the ambulacra. Unfortunately, both chelicerae are missing. However, its morphology indicates that the nymph is parasitic.

In view of the development of the larva within the fixed female, the evidence for a parasitic nymphal stage, the apparent lack of adaptation of the legs for a free-living existence, and the lack of any collections of these mites from roosting areas, I suggest that *Spelaeorhynchus* species are permanent parasites. They have evolved on phyllostomatoid bats probably from a nondermanyssoid ancestor with particularly stout, heavily toothed chelicerae. The latter feature might be a necessary preadaptation for them to have developed their type of fixation method on the host, and chelicerae of this sort are characteristic of a number of mesostigmate groups, including the Sejina but excluding the Dermanyssoidea.

DISCUSSION AND CONCLUSIONS

The Mesostigmata are unusual in the degree to which they have become ecologically diversified while retaining a relatively uniform habitus. In this respect, they may have surpassed every other arthropod group, including the other acarine suborders. This is emphasized by comparing with their free-living relatives such parasitic families as the Platypsyllidae (Coleoptera), Nycteribiidae (Diptera), and Listrophoridae (Acariformes: Astigmata or Acaridida). Some of the more obvious features that appear to contribute to the ecological malleability combined with morphological conservatism in Mesostigmata are small size, flattened saclike body, lack of external segmentation joined with a simple system of body plates, cursorial legs, and multipurpose mouthparts.

The particular success of the Dermanyssoidea, in ecological breadth and especially in vertebrate relationships, makes them the principal subject of this chapter. The combination of two features in the *Hypoaspis* complex, from which all the parasitic groups in the superfamily can be derived, may contribute to this success: the ability to utilize a wide range of nutrients, and chelicerae that are more generalized than those in most other mesostigmate groups. Hypoaspidine nutrition has not been studied in detail, but the polyphagous nidicoles in *Androlaelaps*, *Eulaelaps*, and *Haemogamasus* can subsist on many different foods. The chelicerae seem to form the basis for ecological radiation in these mites in a somewhat analogous fashion to the modification of avian bill structures. Certain insular groups of birds, notably the drepanidid honeycreepers in Hawaii and the geospizine finches in the Galapagos, have occupied a wide spectrum of avian niches essentially through change in bill shape. The importance of the chelicerae in feeding mode and the variety of cheliceral types in the dermanyssoids are suggestive of this sort of adaptive radiation.

The apparent antiquity of several groups of dermanyssoid parasites of vertebrates suggests that preemption of general niches may also have been a factor in the special success of this group. That is, occupation of a number of nidicolous and parasitic niches by dermanyssoids may have been a deterrent to such specialization by other Mesostigmata over a very long geological time span.

There are intergrading forms in this group between free-living mites and nidicoles, between nidicoles and obligate parasites, between typical laelapines and typical hirstionyssines (e.g., *Trichosurolaelaps*), and so on. In the absence of any fossil record, one can only wonder how much change has occurred in these intergrades. In examining the Dermanyssoidea, in which much of the evolutionary history appears to be represented by extant species, it is tempting to suppose that a large array of "living fossils" is available to us for study. Although it is not known at what rate or to what extent intergradient dermanyssoids have changed, the existence of these ecological types does provide insights on the origins and evolution of parasitism that are not available in other groups. The sections on the *Hypoaspis* complex, *Androlaelaps*, and *Haemogamasus* are especially relevant to this point.

In addition to intergradient forms, modifications of the life cycle in dermanyssoids present some striking evidence of their evolutionary relationships. The suppression of either or both nymphal stages, so that they are no longer active and feeding, has occurred in a number of parasitic groups. Suppression of the protonymph alone has happened independently in the Haemogamasinae and the Hirstionyssinae, and probably also independently in the Myonyssinae. Suppression of the deutonymph while the protonymph remains active (and develops slow feeding, perhaps in relation to neosomy) is an unusual adaptation characteristic of the Macronyssidae, and also of the macronyssid-derived avian respiratory parasites in the Rhinonyssidae. In the endoparasitic Halarachnidae, both nymphal stages are suppressed, and the larva (as in at least some of the endoparasitic rhinonyssids) has taken on the function of an interhost transfer stage. These and other modifications (and conservatism) of the life cycle not only provide a strong basis for higher classification and evolutionary interpretation of the dermanyssoids, but also they are among the more fascinating and profitable areas for study of parasitic adaptations.

Congruence of host and parasite phylogenies is most evident in the Spinturnicidae, parasitic entirely on bats, and those Macronyssidae associated bats. These relationships have clearly proven fruitful in providing evidence on host evolution as well as on that of the parasites. They also are particularly important in that they have established the first dermanyssoid radiation on vertebrates as extremely ancient, starting in the very early Cenozoic or possibly the Mesozoic era.

A pattern of dermanyssoid evolution has emerged, and I have attempted to interpret it here. However, forms that are significant in understanding this pattern continue to be discovered. Our inventory of these is at best sketchy; the gaps in knowledge of the bionomics and basic host–parasite relationships are especially glaring. Following a period of considerable interest in the 1950s and the early 1960s, there has been a decline in biological studies of the vertebrate-associated Mesostigmata. This important area of research should be revitalized.

SUMMARY

The evolution of mammalian associates in the Mesostigmata has resulted in very diverse relationships, as facultative and obligatory parasites, nest-dwellers and host-dwellers, ectoparasites and endoparasites (respiratory system, external ear, anus), rapid feeders and fixed, long-term feeders. However, with a few minor exceptions, all of these close associates of mammals originated from a single group of related genera of basically predatory mites: the *Hypoaspis* complex (Laelapidae, Hypoaspidinae). The Laelapidae and all the derived families that include mammalian associates are in the Dermanyssoidea. Furthermore, the great majority of mesostigmate associates of birds and reptiles are ultimately derived from the *Hypoaspis* complex, which includes and also gave rise to several arthropod-associated groups.

At least three lines leading to vertebrate associations independently developed from the *Hypoaspis* complex. Mammals were the original vertebrate host group for each of these lines. One line leading to the Laelapinae appears to have radiated further to produce most of the extant groups associated with vertebrates.

The mammalian nest was central to the early evolution of parasitism in the Dermanyssoidea. This phenomenon is relatively easy to analyze because of a range of species that represents stages in the transition from predatory nidicoles to obligatory parasites. It is postulated that the nest not only provides a rich food source and the opportunity for phoresy (using the hosts for transfer between nests) but also facilitates contact between the sexes, thereby reducing trophic pressures on the male and allowing the male chelicerae to specialize more effectively for sperm transfer. Analysis of relationships between mesostigmate mites of some parasite families and subfamilies, and of the associated adaptive pathways, is also aided by the existence of intermediate or graded forms.

Families and, where appropriate, subfamilies of parasitic mesostigmate mites have been discussed separately herein with regard to relationships, adaptive evolution, and coevolution with hosts. Rodents and bats are especially important as hosts, beyond the relative species richness of these two orders. Coevolution with hosts appears to have been the rule for some groups, such as those Macronyssidae that are associated with bats, and the Spinturnicidae, which are restricted to bats. There is evidence that resource tracking has figured more prominently than taxonomic host tracking in several instances in which the host groups have apparently gone through their primary radiation prior to association with the parasite group; the Macronyssidae on nonchiropteran hosts, including rodents and other mammals, and the Rhinonyssidae, which are parasites in the respiratory tract of birds and are derived from advanced macronyssids, are examples.

The Dermanyssoidea clearly present unusual opportunities for the study of parasite evolution, and this potential is not currently being given adequate attention.

ACKNOWLEDGMENTS

Dr. W. W. Moss and Dr. JoAnn M. Tenorio commented on a draft of this chapter. Mr. Robert Domrow, Dr. Deane P. Furman, and Dr. Vernon J. Tipton made helpful comments on the tabulation of mammal-associated Mesostigmata. I am very grateful for the suggestions of all of these colleagues, some of which I followed. Obviously I must take responsibility for all opinions and certainly any errors that may appear.

REFERENCES

Allred, D. M. 1957. Notes on the life history and bionomics of the wood rat mite, *Brevisterna utahensis* (Acarina). *Trans. Am. Microsc. Soc.* **76**:72–78.

Audy, J. R., F. J. Radovsky, and P. H. Vercammen-Grandjean. 1972. Neosomy: Radical intrastadial metamorphosis associated with arthropod symbioses. *J. Med. Entomol.* **9**:487–494.

Baker, E. W., T. M. Evans, D. J. Gould, W. B. Hull, and H. L. Keegan. 1956. *A Manual of Parasitic Mites of Medical or Economic Importance.* Tech. Publ. National Pest Contr. Assoc. 170 pp.

Banks, N. 1917. New mites, mostly economic. *Entomol. News* **28**:193–199.

Barnes, A. M. and F. J. Radovsky. 1969. A new *Tunga* (Siphonaptera) from the Nearctic region with description of all stages. *J. Med. Entomol.* **6**:19–36.

Bertram, D. S., K. Unsworth, and R. M. Gordon. 1946. The biology and maintenance of *Liponyssus bacoti* Hirst, 1913, and an investigation into its role as vector of *Litomosoides carinii* to cotton rats and white rats, together with some observations on the infection in the white rats. *Ann. Trop. Med. Parasitol.* **40**:228–254.

Bregetova, N. G. 1956. *Gamasoid Mites (Gamasoidea).* Fauna SSR, Akad. Nauk SSR, Vol. 61. 247 pp.

Bregetova, N. G. 1964. *Some Problems of Evolution of the Rhinonyssoid Mites.* (Report presented at the First International Congress of Parasitology.) Nauka, Leningrad. 7 pp.

Bregetova, N. G. 1969. Ontogeny of gamasid mites as a basis of their natural system. Pages 289–295 in *Proc. 2nd Int. Congr. Acarol.*

Camin, J. H. 1953. Observations on the life history and sensory behavior of the snake mite, *Ophionyssus natricis* (Gervais) (Acarina: Macronyssidae). *Chic. Acad. Sci., Spec. Publ. No.* **10**. 75 pp.

Clark, G. M. and C. E. Yunker. 1956. A new genus and species of Dermanyssidae (Acarina: Mesostigmata) from the English sparrow with observations on its life cycle. *Proc. Helminthol. Soc. Wash.* **23**:93–101.

Denuff, J. 1977. Observations sur les Spinturnicidae de la région paléarctique occidentale (Acarina, Mesostigmata). Specificité, répartition et morphologie. *Acarologia* (Paris) **18**:602–617.

Deunff, J. and J. C. Beaucournu. 1981. Phenologie et variations du dermecos chez quelques espèces de Spinturnicidae (Acarina: Mesostigmata). *Ann. Parasitol. Hum. Comp.* **56**:203–224.

Domrow, R. 1963. New records and species of Austromalayan laelapid mites. *Proc. Linn. Soc. N.S.W.* **88**:199–220.

Domrow, R. 1969. The nasal mites of Queensland birds (Acari: Dermanyssidae, Ereynetidae, and Epidermoptidae). *Proc. Linn. Soc. N.S.W.* **93**:297–426 and 2 plates.

Domrow, R. 1972. Acari Spinturnicidae from Australia and New Guinea. *Acarologia* (Paris) **13**:552–584.

Domrow, R. 1973. New records and species of *Laelaps* and allied genera from Australia (Acari: Dermanyssidae). *Proc. Linn. Soc. N.S.W.* **98**:62–86.

Domrow, R. 1974. Notes on halarachnine larval morphology and new species of *Pneumonyssus* Banks (Acari: Dermanyssidae). *J. Aust. Entomol. Soc.* **13**:17–26.

Domrow, R. 1979. Dermanyssine mites from Australian birds. *Rec. West. Aust. Mus.* **7**:403–413.

Domrow, R. 1981. The genus *Raillietia* Trouessart in Australia (Acari: Dermanyssidae). *Proc. Linn. Soc. N.S.W.* **104**:183–193.

Dušbábek, F. 1968. Los acaros Cubanos de la familia Spinturnicidae (Acarina), con notas sobre su especificidad de hospederos. *Poeyana Inst. Biol.* (La Habana) Ser. A, **57**:1–31.

Dušbábek, F. 1970. On the Cuban species of the genus *Spelaeorhynchus. Acarologia* (Paris) **12**:258–261.

Dušbábek, F. 1971. Phylogeny of mites of the family Spinturnicidae Oudms. (Acarina). Pages 241–242 in *Proc. 13th Int. Congr. Entomol.* (1968).

Evans, G. O. 1955. A review of the laelaptid paraphages of the Myriapoda with descriptions of three new species (Acarina: Laelaptidae). *Parasitology* **45**:352–368.

Evans, G. O. 1957. An introduction to the British Mesostigmata (Acarina) with keys to families and genera. *Linn. Soc. J. Zool.* **43**:203–259.

Evans, G. O. and A. Fain. 1968. A new hirstionyssine mite from *Trichys lipura* Gunther. *Acarologia* (Paris) **10**:419–425.

Evans, G. O. and W. M. Till. 1962. The genus *Dermanyssus* de Geer (Acari: Mesostigmata). *Ann. Mag. Nat. Hist.*, Ser. 13, **5**:273–293.

Evans, G. O. and W. M. Till. 1964. A new species of *Dermanyssus* and a redescription of *Steatonyssus superans* Zemskaya (Acari: Mesostigmata). *Acarologia* (Paris) **6**:624–631.

Evans, G. O. and W. M. Till. 1965. A new laelapine mite from the golden mole *Chrysochloris stuhlmanni* Matschie. *Ann. Mag. Nat. Hist.*, Ser. 13, **8**:629–634.

Evans, G. O. and W. M. Till. 1966. Studies on the British Dermanyssidae (Acari: Mesostigmata). Part II. Classification. *Bull. Br. Mus (Nat. Hist.) Zool.* **14**(5):109–370.

Fain, A. 1977. Observations sur la spécificité des Acariens de la famille Myobiidae. Correlation entre l'evolution des parasites et de leurs hôtes. *Ann. Parasitol. Hum. Comp.* **52**:339–351.

Fain, A. 1979. Specificity, adaptation and parallel host–parasite evolution in acarines, especially Myobiidae, with a tentative explanation for the regressive evolution caused by the immunological reactions of the host. Pages 321–328 in J. G. Rodriguez (Ed.). *Recent Advances in Acarology*, Vol. 2. Academic Press, New York.

Fain, A., G. Anastos, J. H. Camin, and D. E. Johnston. 1967. Notes on the genus *Spelaeorhynchus*. Description of *S. praecursor* Neumann and of two new species. *Acarologia* (Paris) **9**:535–556.

Feider, Z. and I. Mironescu. 1972. Une modalité particulière de perfores la muqueuse nasale, utilisée par les acariens de la famille Rhinonyssidae Trouessart, 1895. *Acarologia* (Paris) **14**:21–31.

Fonseca, F. da. 1940. Notas de Acareologia. XXIX. *Dasyponyssus neivai*, gen. n., sp. n., acariano parasita de *Euphractus sexcinctus* (L.) (Acari, Dasyponyssidae, fam. n.). *Rev. Entomol. (Rio de J.)* **11**(1–2):104–119.

Furman, D. P. 1959a. Feeding habits of symbiotic mesostigmatid mites of mammals in relation to pathogen-vector potentials. *Am. J. Trop. Med. Hyg.* **8**:5–12.

Furman, D. P. 1959b. Observations on the biology and morphology of *Haemogamasus ambulans* (Thorell) (Acarina: Haemogamasidae). *J. Parasitol.* **45**:274–280.

Furman, D. P. 1966a. The spinturnicid mites of Panama. Pages 125–166 in R. L. Wenzel and V. J. Tipton (Eds.), *Ectoparasites of Panama*, Field Museum of Natural History, Chicago, IL.

Furman, D. P. 1966b. Biological studies on *Haemolaelaps centrocarpus* Berlese (Acarina: Laelapidae) with observations on its classification. *J. Med. Entomol.* **2**:331–335.

Furman, D. P. 1968. Effects of the microclimate on parasitic nest mites of the dusky footed wood rat, *Neotoma fuscipes* Baird. *J. Med. Entomol.* **5**:160–168.

Furman, D. P. 1972. Laelapid mites (Laelapidae: Laelapinae) of Venezuela. *Brigham Young Univ. Sci. Bull., Biol. Ser.* **17**(3):1–58.

Furman, D. P. 1979. Specificity, adaptation and parallel evolution in the endoparasitic Mesostigmata of mammals. Pages 329–337 in J. G. Rodriguez (Ed.). *Recent Advances in Acarology*, Vol. 2. Academic Press, New York.

Furman, D. P. and F. J. Radovsky. 1963. A new species of *Ornithonyssus* from the white-tailed antelope squirrel, with a rediagnosis of the genus *Ornithonyssus* (Acarina: Dermanyssidae). *Pan-Pac. Entomol.* **39**:75–79.

Goncharova, A. A. and T. G. Buyakova. 1960. Biology of *Haemogamasus mandschuricus* Vtzth. in Transbaikalia (far eastern area). (In Russian.) *Parazitol. Sb.* **19**:155–163.

Goncharova, A. A. and T. G. Buyakova. 1962. Biology of the gamasid mite *Eulaelaps cricetuli* Vitzthum in Transbaikalia. (In Russian.) *Zool. Zh.* **41**:139–143.

Hammen, L. van der. 1972. A revised classification of the mites (Arachnoidea, Acarida) with diagnoses, a key, and notes on phylogeny. *Zool. Meded.* (Leiden) **47**:273–292.

Handley, C. O., Jr. 1976. Mammals of the Smithsonian Venezuela Project. *Brigham Young Univ. Sci. Bull., Biol. Ser.* **20**:1–91.

Herrin, C. S. 1970. A systematic revision of the genus *Hirstionyssus* (Acari: Mesostigmata) of the Nearctic Region. *J. Med. Entomol.* **7**:391–437.

Herrin, C. S. and V. J. Tipton. 1975. Spinturnicid mites of Venezuela (Acarina: Spinturnicidae). *Brigham Young Univ. Sci. Bull., Biol. Ser.* **20**,1(2):1–72.

Herrin, C. S. and V. J. Tipton. 1976. A systematic revision of the genus *Laelaps* s. str. of the Ethiopian Region. *Great Basin Nat.* **36**(2):113–205.

Herrin, C. S. and C. E. Yunker. 1975. Systematics of the Neotropical *Hirstionyssus* mites with special emphasis on Venezuela (Acarina: Mesostigmata). *Brigham Young Univ. Sci. Bull., Biol. Ser.* **17**, 3(2):93–127.

Holdenried, R., F. C. Evans, and D. S. Longanecker. 1951. Host–parasite-disease relationships in a mammalian community in the central Coast Range of California. *Ecol. Monogr.* **21**:1–18.

Hughes, A. M. 1961. *The Mites of Stored Food.* Min. Agr. Fish., London, Tech. Bull. No. 9, vi + 287 pages.

Hughes, T. E. 1959. *Mites, or the Acari.* Athlone Press, London. 225 pp.

Hull, W. B. 1970. Respiratory mite parasites in nonhuman primates. *Lab. Anim. Care* **20**(2):402–406.

Karg, W. 1961. Ökologische Untersuchungen von edaphischen Gamasiden (Acarina, Parasitiformes). *Pedobiologia* **1**(1,2):53–74; 77–98.

Keegan, H. L. 1951. The mites of the subfamily Haemogamasinae (Acari: Laelaptidae) *Proc. U.S. Nat. Mus.* **101**:203–268.

Keegan, H. L., C. E. Yunker, and E. W. Baker. 1960. Malaysian Parasites. XLVI. *Hystrichonyssus turneri*, n. sp., n. g., representing a new subfamily of Dermanyssidae from a Malayan porcupine. *Inst. Med. Res. Fed. Malaya* **29**:205–208.

Kenyon, K. W., C. E. Yunker, and I. M. Newell. 1965. Nasal mites (Halarachnidae) in the sea otter. *J. Parasitol.* **51**:960.

Kethley, J. B. and D. E. Johnston. 1975. Resource tracking patterns in bird and mammal ectoparasites. *Misc. Publ. Entomol. Soc. Am.* **9**:231–236.

Kim, J. C. S. 1977. Pulmonary acariasis in Old World monkeys. *Vet. Bull.* **47**:249–255.

Kim, K. C., V. L. Haas, and M. C. Keyes. 1980. Populations, microhabitat preference, and effects of infestation of two species of *Orthohalarachne* (Halarachnidae: Acarina) in the northern fur seal. *J. Wildl. Dis.* **16**(1):45–51.

Kolenati, F. A. 1859. Beitrage zur Kentniss der Arachniden. *Sitzungsber. K. Akad. Wiss. Wien, Math.-Naturwiss. Cl.* **33**(1958):69–89.

Kolenati, F. A. 1860. Beitrage zur Kentniss der Arachniden. *Sitzungsber. K. Akad. Wiss. Wien, Math.-Naturwiss. Cl.* **40**:573–581.

Kozlova, R. G. 1959. Feeding habits of the mite *Eulaelaps stabularis* C. L. Koch, 1836 (Laelaptidae, Gamasides, Parasitiformes). (In Russian.) *Zool. Zh.* **38**:44–53.

Krantz, G. W. 1978. *A Manual of Acarology,* 2nd ed. Oregon State University Book Stores, Corvallis. 509 pp.

Lavoipierre, M. M. J. and A. J. Beck. 1967. Feeding mechanisms of *Chiroptonyssus robustipes*

(Acarina: Macronyssidae) as observed on the transilluminated bat wing. *Exp. Parasitol.* **20**:312–320.

Lavoipierre, M. M. J., F. J. Radovsky, and P. D. Budwiser. 1979. The feeding process of a tungid flea, *Tunga monositus* (Siphonaptera: Tungidae), and its relationship to the host inflammatory and repair response. *J. Med. Entomol.* **15**:187–217.

Lees, A. D. 1952. The role of cuticle growth in the feeding process of ticks. *Proc. Zool. Soc. Lond.* **121**:759–772.

Machado-Allison, C. E. 1967. The systematic position of the bats *Desmodus* and *Chilonycteris*, based on host–parasite relationships. *Proc. Biol. Soc. Wash.* **80**:223–226.

Men, Y. T. 1959a. Concerning the feeding of the mite *Haemolaelaps casalis* (Gamasoidea, Parasitiformes). (In Russian.) *Medskaya Parazitol.* **28**:477–481.

Men, Y. T. 1959b. Concerning the feeding of the mite *Haemolaelaps casalis* (Gamasoidea, Parasitiformes). (In Russian.) *Medskaya Parazitol.* **28**:603–609.

Micherdzinski, W. 1980. *Eine taxonomische Analyse der Familie Macronyssidae Oudemans, 1936. I. Subfamilie Ornithonyssinae Lange, 1958 (Acarina, Mesostigmata)*. Panstwowe Wydawnictwo Nankowe, Warsaw. 264 pp.

Mitchell, C. J. 1968. Biological studies on *Laelaps myonyssognathus* G. and N. (Acarina: Laelapidae). *J. Med. Entomol.* **5**:99–107.

Mitchell, R. W. 1963. Comparative morphology of the life stages of the nasal mite *Rhinonyssus rhinolethrum* (Mesostigmata: Rhinonyssidae). *J. Parasitol.* **49**:506–515.

Moss, W. W. 1967. Some new analytic and graphic approaches to numerical taxonomy, with an example from the Dermanyssidae (Acari). *Syst. Zool.* **16**:177–207.

Moss, W. W. 1968. An illustrated key to the species of the acarine genus *Dermanyssus* (Mesostigmata: Laelapoidea: Dermanyssidae). *J. Med. Entomol.* **5**:67–84.

Moss, W. W. 1978. The mite genus *Dermanyssus*: a survey, with description of *Dermanyssus trochilinis*, n. sp., and a revised key to the species (Acari: Mesostigmata: Dermanyssidae). *J. Med. Entomol.* **14**:627–640.

Moss, W. W. and J. H. Camin. 1970. Nest parasitism, productivity and clutch size in purple martins. *Science* **168**:1000–1003.

Nelzina, E. N. and G. M. Danilova. 1956. Materials on the biology of mites of the family Haemogamasidae (Gamasoidea, Parasitiformes). 1. Nutrition of *Haemogamasus citelli* Breg. et Nelz. and *H. nidi* Mich. (In Russian.) *Medskaya Parazitol.* **25**:352–358.

Nelzina, E. N., G. M. Danilova, and Z. I. Klinova. 1967. Gamasoid mites (Gamasoidea, Parasitiformes)—one of the basic components of burrow biocenoses. (In Russian.) *Parazitologiya* (Leningr.) **1**:412–421.

Newell, I. M. 1947. Studies on the morphology and systematics of the family Halarachnidae (Oudemans, 1906) (Acari: Parasitoidea). *Bull. Bingham Oceanogr. Collect. Yale Univ.* **10**(4):235–266.

Olsen, O. and F. K. Bracken. 1950. Occurrence of the ear mite, *Raillietia auris* (Leidy, 1872), of cattle in Colorado. *Vet. Med.* **45**:320–321.

Owen, B. L. 1956. Life history of the spiny rat mite under artificial conditions. *J. Econ. Entomol.* **49**:702–703.

Phillips, C. J., J. K. Jones, Jr., and F. J. Radovsky. 1969. Macronyssid mites in the oral mucosa of long-nosed bats: occurrence and associated pathology. *Science* **165**:1368–1369.

Potter, D. A. and D. E. Johnston. 1978. *Raillietia whartoni* sp. n. (Acari–Mesostigmata) from the Uganda kob. *J. Parasitol.* **64**:139–142.

Preobrazhenskaya, N. K. and A. A. Preobrazhensky. 1955. Experience in laboratory culturing of certain species of gamasoid mites—ectoparasites of rodents. (In Russian.) *Zool. Zh.* **34**:300–303.

Radovsky, F. J. 1959. M.S. thesis. University of California, Berkeley.

Radovsky, F. J. 1960. Biological studies on *Haemogamasus liponyssoides* Ewing (Acarina: Haemogamasidae). *J. Parasitol.* **46**:410–417.

Radovsky, F. J. 1964. *The Macronyssidae and Laelapidae (Acarina: Mesostigmata) Parasitic on Bats.* Ph.D. Dissertation, University of California, Berkeley. Pages 1–600.

Radovsky, F. J. 1966. Revision of the macronyssid and laelapid mites of bats: outline of classification with descriptions of new genera and new type species. *J. Med. Entomol.* **3**:93–99.

Radovsky, F. J. 1967. The Macronyssidae and Laelapidae (Acarina: Mesostigmata) parasitic on bats. *Univ. Calif. Publ. Entomol.* **46**:1–288.

Radovsky, F. J. 1968. Evolution and adaptive radiation of Gamasina parasitic on vertebrates (Acarina: Mesostigmata). (In Russian.) *Parazitologiya* (Leningr.) **2**:124–136.

Radovsky, F. J. 1969. Adaptive radiation in the parasitic Mesostigmata. *Acarologia* (Paris) **11**:450–483.

Radovsky, F. J. 1979. Specificity and parallel evolution of Mesostigmata parasitic on bats. Pages 347–353 in J. G. Rodriguez (Ed.), *Recent Advances in Acarology*, Vol. 2. Academic Press, New York.

Radovsky, F. J. and D. P. Furman. 1969. An unusual new genus and species of Macronyssidae (Acarina) parasitic on a disc-winged bat. *J. Med. Entomol.* **6**:385–393.

Radovsky, F. J., J. K. Jones, and C. J. Phillips. 1971. Three new species of *Radfordiella* (Acarina: Macronyssidae) parasitic in the mouth of phyllostomid bats. *J. Med. Entomol.* **8**:737–746.

Radovsky, F. J. and J. M. Tenorio. 1981. Ectoparasites of rodents. Pages 110–117 in D. M. Mueller-Dombois, K. W. Bridges, and H. L. Carson (Eds.). *Island Ecosystems: Biological Organization in Selected Hawaiian Communities. US/IBP Synthesis Series, No. 15.* Hutchinson Ross, Stroudsburg, Pennsylvania. 583 pp.

Radovsky, F. J., J. M. Tenorio, P. Q. Tomich, and J. D. Jacobi. 1979. Acari on murine rodents along an altitudinal transect on Mauna Loa, Hawaii. Pages 327–333 in *Proc. 4th Int. Congr. Acarol.* Saalfelden, Austria, 1974.

Radovsky, F. J. and C. E. Yunker. 1971. *Xenarthronyssus furmani* n. g. n. sp. (Acarina: Dasyponyssidae), parasites of armadillos, with two subspecies. *J. Med. Entomol.* **8**:135–42.

Redington, B. C. 1970. Studies on the morphology and taxonomy of *Haemogamasus reidi* Ewing, 1925. *Acarologia* (Paris) **12**:643–67.

Reisen, W. K., M. L. Kennedy, and N. T. Reisen. 1976. Winter ecology of ectoparasites collected from hibernating *Myotis velifer* (Allen) in Southwestern Oklahoma (Chiroptera: Vespertilionidae). *J. Parasitol.* **62**:628–635.

Reytblat, A. G. 1965. Biology of the gamasid mite *Haemolaelaps semidesertus* Breg. (Gamasoidea, Parasitiformes). (In Russian.) *Zool. Zh.* **44**:863–870.

Rudnick, A. 1960. A revision of the mites of the family Spinturnicidae (Acarina). *Univ. Calif. Publ. Entomol.* **17**(2):157–284.

Sasa, M. and M. Wakasugi. 1957. Studies on the effect of carbon dioxide as the stimulant on the tropical rat mite, *Bdellonyssus bacoti* (Hirst, 1913). *Jap. J. Exp. Med.* **27**:207–215.

Saunders, R. C. 1975. Venezuela Macronyssidae (Acarina: Mesostigmata). *Brigham Young Univ. Sci. Bull., Biol. Ser. 20,* **2**(2):75–90.

Senotrusova, V. N. 1962. Materials on the biology of the gamasoid mite *Hirstionyssus meridianus* Zem., 1951. (In Russian.) *Tr. Inst. Zool. Akad. Nauk Kaz. SSR* **16**:192–199.

Senotrusova, V. N. 1963. Ecology of the gamasid mite *Hirstionyssus meridianus* Zemsk. (Parasitiformes, Gamasoidea). (In Russian.) *Tr. Inst. Zool. Akad. Nauk Kaz. SSR* **19**:191–197.

Senotrusova, V. N. 1968. Biological peculiarities of the gamasoid mite *Ichoronyssus flavus* (Kolenati, 1956). (In Russian.) *Parazitologiya* (Leningr.) 2:339–341.

Sheals, J. G. 1962. The status of the genera *Dermanyssus, Allodermanyssus* and *Liponyssoides. Proc. 11th Int. Congr. Entomol., Vienna, 1960* 2:473–476.

Sikes, R. K. and R. W. Chamberlain. 1954. Laboratory observations on three species of bird mites. *J. Parasitol.* 40:691–697.

Silva Taboa, G. and R. H. Pine. 1969. Morphological and behavioral evidence for the relationship between the bat genus *Brachyphylla* and the Phyllonycterinae. *Biotropica* 1:10–19.

Skaliy, P. and W. J. Hayes. 1949. The biology of *Liponyssus bacoti* (Hirst, 1913) (Acarina: Liponyssidae). *Am. J. Trop. Med.* 29:759–772.

Smith, J. D. 1972. Systematics of the chiropteran family Mormoopidae. *Univ. Kans. Mus. Nat. Hist. Misc. Publ.* 56:1–132.

Stone, W. B. and J. A. Hughes. 1969. Massive pulmonary acariasis in the pig-tailed macaque. *Bull. Wildl. Dis. Assoc.* 5:20–22.

Strandtmann, R. W. 1961. The immature stages of the *Ptilonyssus* complex (Acari: Mesostigmata: Rhinonyssidae). *Proc. 11th Int. Congr. Entomol., Vienna, 1960* 1:283–286.

Strandtmann, R. W. and C. M. Clifford. 1962. A new genus and species of nasal mite from the varied thrush *Ixoreus neveus* (Gmelin) (Acarina: Rhinonyssidae). *J. Parasitol.* 48:723–725.

Strandtmann, R. W. and E. Garrett. 1970. The genus *Myonyssus* with description of a new species from Nepal (Mesostigmata: Laelapidae). *J. Med. Entomol.* 7:261–266.

Strandtmann, R. W. and G. W. Wharton. 1958. *A Manual of Mesostigmatid Mites Parasitic on Vertebrates.* The Inst. of Acarology, Contrib. No. 4., University of Maryland. 330 pp.

Tagiltzev, A. A. 1971. Arthropods collected from *Myotis mystacinus* Kohl. and *M. oxygnathus* Mont. in Zaysan Hollow. (In Russian.) *Parazitologiya* (Leningr.) 5:434–436.

Tenorio, J. M. 1982. Personal communication.

Tenorio, J. M. and F. J. Radovsky. 1973. Two new species of *Trichosurolaelaps* (Acarina: Laelapidae: Hirstionyssinae) from New Guinea. *J. Med. Entomol.* 10:147–157.

Tenorio, J. M. and F. J. Radovsky. 1974. The genus *Mesolaelaps* (Laelapidae: Mesolaelapinae, n. subfam.) with descriptions of two new species from New Guinea. *J. Med. Entomol.* 11:211–222.

Tenorio, J. M. and F. J. Radovsky. 1979. Review of the subfamily Hirstionyssinae, synonymy of *Echinonyssus* Hirst and *Hirstionyssus* Fonseca, and descriptions of four new species of *Echinonyssus* (Acari: Laelapidae). *J. Med. Entomol.* 16:370–412.

Till, W. M. 1963. Ethiopian mites of the genus *Androlaelaps* Berlese s. lat. *Bull. Brit. Mus. (Nat. Hist.) Zool.* 10(1):1–104.

Tipton, V. J. 1960. The genus *Laelaps* with a review of the Laelaptinae and a new subfamily Alphalaelaptinae. *Univ. Calif. Publ. Entomol.* 16(6):233–356.

Tsymbal, T. G. and N. T. Litvishko. 1955. Acariasis of the ear in cattle. (In Russian.) *Zool. Zh.* 34:1229–1241.

Vitzthum, H. G. 1940–1943. Acarina. Pages 1–1011 in H. G. Braun (Ed.). *Klassen und Ordnungen des Tierreichs,* 5(4), Book 5.

Waage, J. K. 1979. The evolution of insect/vertebrate associations. *Biol. J. Linn. Soc.* 12:187–224.

Webb, J. P., Jr. and R. B. Loomis. 1977. Ectoparasites. Pages 57–119 in *Biology of Bats of the New World Family Phyllostomatidae.* Part II. *Spec. Publ. Mus., Tex. Tech. Univ.* 13:1–364.

Wernz, J. G. and G. W. Krantz. 1976. Studies on the function of the tritosternum in selected Gamasida (Acari). *Can. J. Zool.* 54:202–213.

Wharton, G. W. and H. F. Cross. 1957. Studies on the feeding habits of three species of laelaptid mites. *J. Parasitol.* 43:45–50.

Williams, G. L., R. L. Smiley, and B. C. Redington. 1978. A taxonomic study of the genus *Haemogamasus* in North America, with descriptions of two new species (Acari: Mesostigmata, Laelaptidae). *Int. J. Acarol.* **4**:235–273.

Yunker, C. E. 1973. Mites. Pages 425–492 in R. J. Flynn (Ed.). *Parasites of Laboratory Animals.* Iowa State University Press, Ames.

Zemskaya, A. A. 1969. Types of parasitism in gamasoid mites. (In Russian.) *Med. Parazitol. Parazit. Bolezni* (1969) **4**:393–401.

Zumpt, F. 1950. Records of some parasitic Acarina from Madagascar, with description of a new *Chiroptonyssus* species. *Mem. Inst. Sci. Madagascar, Ser. A,* **4**:165–173.

Chapter 10

Ornithodoros (Ornithodoros savignyi (Audouin)

Tick and Mammal Coevolution, with Emphasis on *Haemaphysalis*

Harry Hoogstraal and Ke Chung Kim

From Research Project 3M161102BS10.AD.424, Naval Medical Research and Development Command, National Naval Medical Center, Bethesda, Maryland. The opinions and assertions contained herein are the private ones of the authors and are not to be construed as official or as reflecting the views of the Department of the Navy or of the naval service at large. (Request reprints from Medical Zoology Department, NAMRU-3, FPO, New York 09527.)

INTRODUCTION

The generally large acarines constituting the tick superfamily Ixodoidea apparently evolved as obligate parasites of Reptilia in the late Paleozoic or early Mesozoic era. During subsequent coevolution with birds and mammals, adaptations of most tick species have been conservative. Structural, developmental, physiological, ethological, and reproductive properties and processes have changed, but chiefly within narrow parameters. More radical adaptations characterize a small proportion of the world's mammal-parasitizing species. Most species that have burst their conservative evolutionary shackles are ixodids and parasitize livestock, some also feed on man and dogs. These species have been more intensely investigated than others; however, they are not entirely typical of their families. The other species that have adapted radically, argasid parasites of New World bats, are virtually unknown biologically.

About 800 tick species are divided into three families, 10 subfamilies, and 19 genera (Fig. 10.1). Most Argasidae ("soft ticks") retain basic biological patterns developed during their early history as reptile parasites. However, few argasids now parasitize reptiles (Hoogstraal and Aeschlimann 1982). Argasid biological adaptations involve chiefly resistance to dessication, host and microhabitat specificity, diapause, and longevity. Argasid life cycle adaptations occur in a few species of *Ornithodoros* and in the 11 species of the specialized subfamilies Otobinae, Antricolinae, and Nothoaspinae (Hoogstraal 1985).

In Ixodidae, the genus *Ixodes* (Prostriata, Ixodinae, about 217 species, worldwide) represents typical "hard ticks" with secondary biological and structural specializations, some of which probably developed during the Tertiary period. All other ixodid genera are in the Metastriata line. The tropical-subtropical *Aponomma* and *Amblyomma* (Amblyomminae, about 126 species) retain primitive structural characters. All but two of the 24 *Aponomma* species and 37 of the 102 *Amblyomma* species are reptile parasites (Hoogstraal and Aeschlimann 1982). There appears to be a close affinity between the *Amblyomma* faunas of Australia and South America. Only 17 species of the tropical-temperate *Haemaphysalis* (Haemaphysalinae, 156 species) retain "primitive" structural characters, and a few of these 17 species are known to be exceptional biologically. During the Tertiary period, numerous *Haemaphysalis* species coevolved with birds and mammals throughout much of the world (however, only five species occur in New

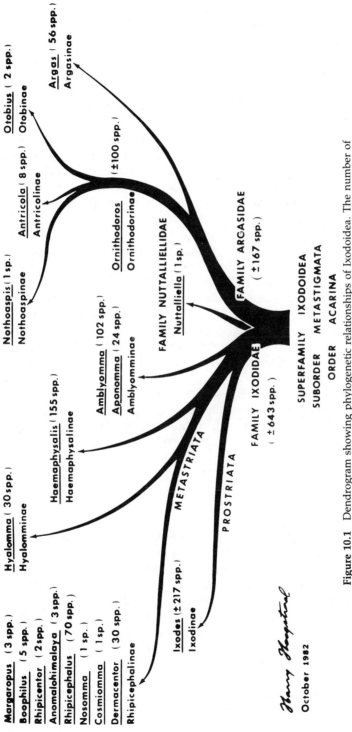

Figure 10.1 Dendrogram showing phylogenetic relationships of Ixodoidea. The number of species is given in parenthesis for each genus.

World). The distinctive host-related structural adaptations of *Haemaphysalis* ticks are discussed in detail hereinafter.

The Hyalomminae (genus *Hyalomma*, 30 species, in the Palearctic, Ethiopian, and Oriental regions) retain the primitive long palpi of the early reptile parasites and have a limited variety of hair-hooking spurs. One species, *H. (Hyalommasta) aegyptium* (Linn.), is entirely dependent on the tortoise, *Testudo*, for population survival, but immatures also parasitize birds and small mammals. Immatures of the subgenus *Hyalomma* often parasitize reptiles and birds but smaller-sized mammals are the chief hosts of most species in this subgenus. Adults parasitize chiefly Artiodactyla. The subgenus *Hyalomma* is highly adapted to arid and semiarid biotypes and to steppes and savannas with long dry seasons. Immatures of the subgenus *Hyalommina* infest chiefly rodents; adults parasitize artiodactyls; one is specific for the Indian porcupine.

The Rhipicephalinae (114 species, eight genera, all in the Palearctic, Oriental, and Ethiopian regions, except a few *Dermacentor* species in the Nearctic and Neotropical) virtually never (in terms of population survival) feed on reptiles or birds. Rhipicephalinae have evolved the most recently of all tick groups and are parasitic on mammals and tropical in distribution (except for several Holarctic *Dermacentor* species). Life cycle adaptations for parasitizing wandering, large mammals are notable in two-host *Rhipicephalus* species and in *Boophilus* and *Margaropus*; all species in the two last-named genera have a one-host life cycle.

In this chapter we present data for the host relationships of immatures and adults of *Haemaphysalis* species, the tick group most indicative of coevolution between ticks and mammals, together with analyses of body and appendage structures, and of life-cycle, biological, and distribution patterns, which furnish clues to the coevolution of these parasites and their hosts.

PREMAMMALIAN EVOLUTION OF TICKS

Ixodoidea are postulated to have evolved as obligate parasites of Reptilia in the warm, humid climate of the late Paleozoic or early Mesozoic era (Hoogstraal 1978). Large, glabrous reptiles living near each other were easily available hosts for ancestral ticks throughout the year. The ticks had four developmental stages, as they still do: egg, larva, nymph, and adult. Each postembryonic instar imbibed an enormous amount of blood or tissue material, emitted excess water with coxal fluids (Argasidae), or salivary fluids (Ixodidae), and digested the remainder by relatively sluggish physiological processes. There was no differentiated eye on the body surface. Questing for reptile hosts in early environments was uncomplicated; movement toward a suitable feeding site on the glabrous host was unimpeded by hairs or feathers.

Paleozoic-Mesozoic Reptilia radiated into numerous bizarre forms filling a variety of aquatic and terrestrial niches. Their more conservative tick parasites evolved along only two main lines (Fig. 10.1). The argasid line was represented by the genera *Argas* and *Ornithodoros*, partially as we know them today, and undoubtedly also by other genera that are now extinct. Other contemporary genera of the argasid line probably did not evolve until the Tertiary period. (No truly fossil ticks have been discovered.) The ixodid line was represented by primitive members of the branches that we now know as the genera *Ixodes, Aponomma, Amblyomma,* and *Haemaphysalis*. *Hyalomma* may have appeared later, close to the Cretaceous period of Mesozoic environmental stresses, when the Mesozoic reptile diversity was reduced to the Rhynchosauria and ancestors of modern birds and mammals. The Rhipicephalinae (*Dermacentor, Rhipicephalus, Boophilus,* and related genera) (Fig. 10.1) did not appear until the Tertiary period, when mammals and birds replaced reptiles as the dominant vertebrates. The few ixodid species (all in the subfamilies Hyalomminae or Rhipicephalinae) utilizing only one or two hosts in their life cycles did not evolve until after mammals appeared.

Early ixodid ticks were probably as large as the largest extant *Amblyomma* (females 10–12 mm long, males about a millimeter shorter, larvae about 1 mm long). A single bloodmeal for each postembryonic stage furnished food for ixodid larvae to change into nymphs, for nymphs to become adults, and for adults to mature and reproduce. Excessive quantities of host blood or tissue were required to meet these energy demands. Females took especially large meals in producing a single egg batch before dying and for eggs to develop and hatch into larvae capable of seeking their own food. Male ixodids needed less food to support reproductive processes and maintain life even while mating with one or more females. The early ixodid egg batch may have contained few eggs, each relatively large. The clue for this conjecture is furnished by the structurally primitive relict *Haemaphysalis (Alloceraea) inermis* Birula, which produces a batch of only 200 large eggs.

Argasids associated with Paleozoic-Mesozoic reptiles were probably 30–50% longer and broader than ixodids; their body volume distinctly exceeded that of ixodids. Three bloodmeals were insufficient to meet the energy requirements of these large ticks. Thus argasid nymphs underwent two or more instars, each with a separate bloodmeal on a separate host, rather than only one instar and one meal as in smaller ixodid nymphs. The large adult argasids took several bloodmeals, and the females oviposited after each full meal, thus differing distinctly from adult ixodids.

Primitive argasid and ixodid larvae could indulge in a leisurely feeding period. Their small size and concealed feeding sites in reptile skin folds sheltered them from being scraped off the active host. Early nymphs and adults probably also fed slowly for several days, but they fared badly. Few large nymphs and even fewer larger females could escape dislodgement while hanging from the host for several days like gradually enlarging beans. Accordingly, natural selection at an early stage of the tick evolutionary history resulted in unique adjustments in the length of their feeding periods. Small argasid larvae mostly continued to feed for several days, but larger nymphs and adults survived by feeding rapidly, in 30–60 minutes. Ixodid larvae, nymphs, and adults (females), on the other hand, fed slowly and gradually for several days. They reached their final large balloon shape only during the last 6–12 hours before disengaging from the host. Male ixodids also fed slowly (if at all) but took a smaller meal.

EVOLUTION OF MAMMALIAN TICKS

In the early Tertiary or the late Cretaceous period, some 70 million years ago, primitive bird and mammal lines exploded into numerous specialized orders replacing reptiles as the dominant terrestrial vertebrates. The new vertebrates filled more ecological niches and developed a greater variety of life-styles than did early reptiles. Most mammals and birds were much smaller than the majority of the early reptiles they replaced. Many Mesozoic ticks were probably unable to adapt to the new hosts and perished. Adaptive radiation in surviving tick lines paralleled that of the new vertebrates, but at a slower and more conservative tempo and rate. Distinct preferences for certain types of hosts among the biologically and ecologically disparate vertebrates developed in existing tick lines (genera and species groups). Existing lines diversified and new generic groups (Rhipicephalinae) evolved. As tick body size decreased, certain structures, biological properties, and behavior patterns were modified. There were few modifications in rates of physiological processes; some were more speedy, but most were even more sluggish than before. Different feeding patterns evolved in the various groups of Argasidae and Ixodidae. After the Pleistocene epoch, when man introduced domestic animal herds into the environment, these few species and species groups, in both families, were to achieve great veterinary and medical importance.

Most argasid species have remained sheltered in burrows or niches close to colonies, nests, roosts, dens, or caves frequently or seasonally revisited by birds or mammals. Thus, protected by microhabitats as well as by limited exposure time during feeding, and assured periodically of ample food from resting immature or adult birds or mammals, some large argasids have survived in association with large hosts such as porcupines, warthogs, wild pigs, and hyenas. Smaller argasids evolved together with small vertebrates such as martins, pigeons, tenrecs, rodents, and bats. Notably, the larvae, nymphs, and adults of each argasid species inhabiting a sheltered microhabitat feed on the same kind of host—the only one generally available in these situations. Hungry argasids may feed on exceptional hosts venturing into their habitat; some argasids survive on the atypical meal, but afterward most develop poorly if at all.

The Ixodidae have better adapted biologically and ecologically than the Argasidae to hazards associated with complete dependence on highly mobile birds and mammals in diverse habitats. For many reasons, size reduction was paramount to success. This reduction, in some cases, appears to be clearly related to the small size of the preferred host group. In other cases, the size reduction apparently results from a dominating evolutionary trend regardless of host size, as exemplified by the small *Haemaphysalis*, *Boophilus*, and *Margaropus* on characteristic hosts such as deer, antelopes, wild cattle, giraffes, and zebras.

The early *Aponomma* or *Aponomma*-like reptile-parasitizing ixodid was about 10 mm long with narrowly elongate, four-segmented palpi (Figs. 10.2, 10.3). The palpal segment 4 was separated from 3 by a suture and was situated apically, as in argasids. In contemporary *Aponomma*, the segment 4 (with its cluster of apical sensory setae) remains distinct and apical. However, the *Aponomma* species parasitizing modern reptiles, which are much smaller and less diverse than Mesozoic reptiles, have become small- or medium-sized ticks. An early step in ixodid body size reduction associated with birds and mammals occurred when the palpal segment 4 became a diminutive, ventrally directed appendage relocated in a protective pit of segment 3 (but retaining a cluster of sensory setae). This pit is apical in *H. (Alloceraea) vietnamensis* Hoogstraal and Wilson, the largest, most primitive species of the 155-member genus *Haemaphysalis* (Hoogstraal and Wilson 1966) (Fig. 10.4), but in more than 600 other ixodid species, this pit is ventral and subapical.

Today, adults of only *Aponomma*, *Amblyomma*, *Hyalomma*, some *Ixodes*, and a very few primitive *Haemaphysalis* have narrowly elongate palpi (Fig. 10.5). Elongate palpi persist in certain immature *Ixodes* and *Haemaphysalis* whose adult palpi have become short and compact. Evolutionary changes in structure have been slower and more conservative in immature than in adult ticks. Elongate palpi of immatures furnish clues to the early history of Ixodidae and to relationships between Ixodidae and Argasidae.

As new generic taxa, such as *Dermacentor*, *Rhipicephalus*, *Anomalohimalaya*, *Boophilus*, and other genera of Rhipicephalinae evolved together with mammals, adult palpi became short and compact (Fig. 10.4). Immature-stage palpi of *Dermacentor* and of some other rhipicephaline species remained elongate.

The typical ixodid three-host cycle (Fig. 10.6) became modified to a two-host or a one-host cycle in certain species of *Hyalomma* and Rhipicephalinae that inhabited environments where wandering mammal numbers were few, home ranges were extensive, or dry seasons were long and hot. The one-host cycle developed in response to the movements of widely wandering medium- or large-sized, forest- or steppe-inhabiting mammals or to

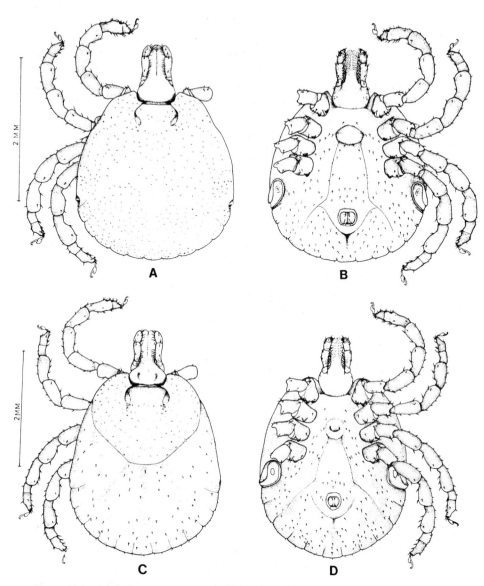

Figure 10.2 Adult *Aponomma varanensis* (Taiwan), a "Mesozoic-type" reptile parasite. (*A, B*) male; (*C, D*) female; dorsal and ventral views.

Figure 10.3 Adult *Aponomma varenensis*. (*A–H*) Male: (*A, B*) capitulum, dorsal and ventral views; (*C*) hypostome, ventral view; (*D*) genital area; (*E*) spiracular plate (A, anterior; D, dorsal); (*F*) coxae and trochanters I to IV; (*G*) femur IV, internal view; (*H*) tarsi I to IV, external view. (*I–P*) Female: (*I, J*) capitulum, dorsal and ventral views; (*K*) hypostome; (*L*) genital area; (*M*) spiracular plates; (*N*) coxae and trochanters I to IV; (*O*) femur IV; (*P*) tarsi I to IV.

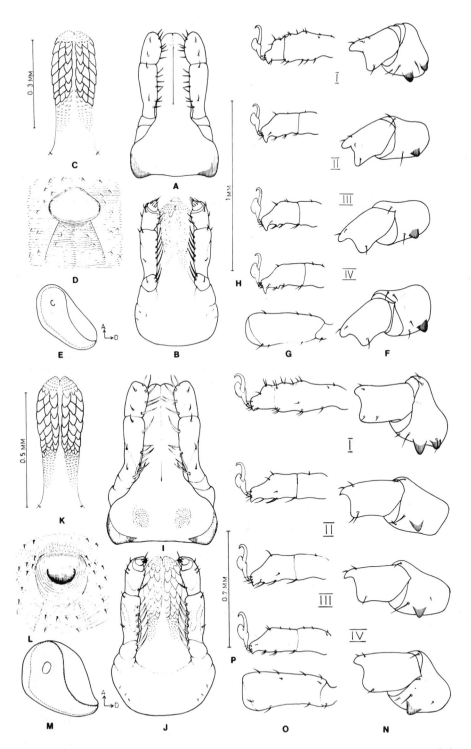

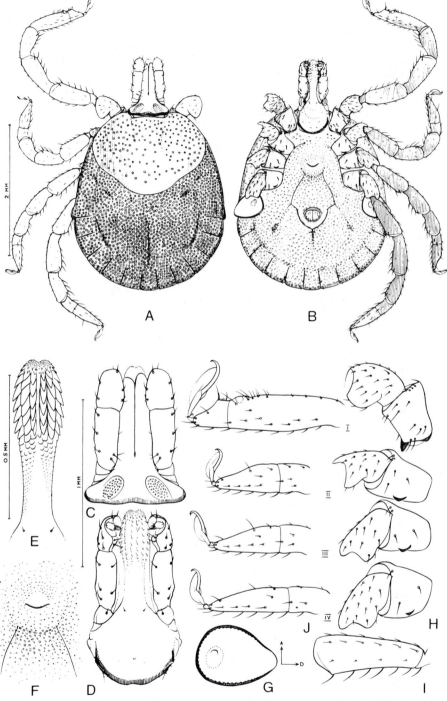

Figure 10.4 Female *Haemaphysalis* (*Alloceraea*) *vietnamensis,* structurally probably the most primitive member of the genus. (*A, B*) Dorsal and ventral views; (*C, D*) capitulum, dorsal, and ventral views; (*E*) hypostome, ventral view; (*F*) genital area; (*G*) spiracular plate (A, anterior; D, dorsal); (*H*) coxae and trochanters I to IV; (*I*) femur IV, internal view; (*J*) tarsi I to IV, external view.

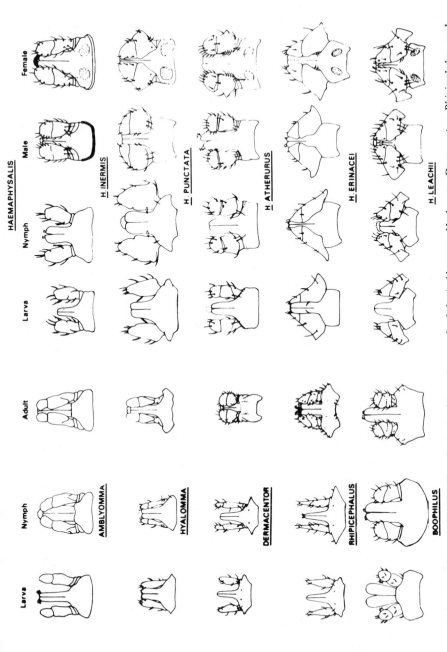

Figure 10.5 The capitulum (dorsal view) of immature and adult *Amblyomma*, *Hyalomma*, *Dermacentor*, *Rhipicephalus*, and *Boophilus* ticks (there is little sexual dimorphism of the capitulum in these genera), and of immature and adult *Haemaphysalis* of structurally primitive (SP) groups [*H. (Allocraea) inermis, H. (Aboimisalis) punctata*] and structurally advanced (SA) groups [*H. (Aborphysalis) atherurus, H. (Rhipistoma) erinacei, H. (R.) leachii*].

515

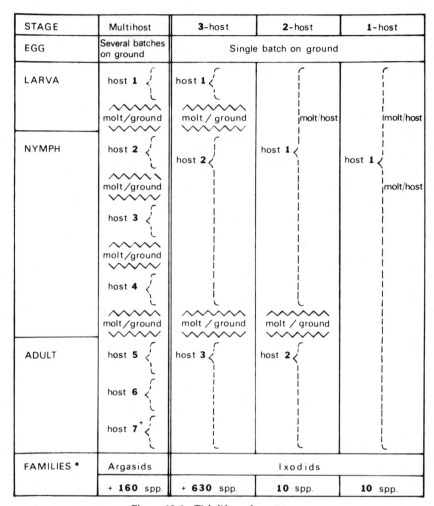

Figure 10.6 Tick life-cycle patterns.

feeding during winter, when small mammals moving under snow are unavailable to immature ticks of species that do not inhabit burrows (Hoogstraal 1978). In effect, winter activity is a substitute for diapause, which is an exceptionally common phenomenon in ticks.

HOST SPECIFICITY

The Ixodoidea show a high degree of strict host specificity, which is one of the important biological factors determining the ecological and geographical distribution and population densities of most ticks. However, the pattern of limited host specificity may be altered when physiologically accept-able domestic or feral mammals intrude into the primary host–parasite associations. This secondary host–parasite relationship often becomes so prevalent that it frequently obscures the pattern of host specificity and leads to incorrect conclusions. Hoogstraal and Aeschlimann (1982) re-viewed the patterns of tick–host associations and presented an analysis of host specificity among ticks of vertebrates. This analysis is the basis for the following discussion.

The host specificity of ticks is expressed by the degree of limited or strict association with the host: (1) strict-total specificity (ST); (2) moderate-total (MT); (3) strict-stage-stage (SSS); (4) strict/moderate-stage-stage (SMSS); (5) moderate-stage-stage (MSS); and (6) nonparticular (NP). Any particular type of host specificity is limited to a specific tick taxon.

The ST host specificity refers to the ticks whose adults and immatures both are strictly specific for the same limited host group, as in all genera of Argasidae and *Aponomma, Boophilus, Margarapus,* and a few species in other major genera of Ixodidae. The MT specificity indicates that both adults and immatures are strictly host specific but the host group is somewhat less limited as in *Ornithodoros, Ixodes, Amblyomma,* and *Rhipicephalus*. The SSS specificity represents the ticks whose adults and immatures are each strictly specific for different limited host groups; this occurs chiefly in Ixodidae (*Ixodes, Amblyomma, Hyalomma, Dermacentor, Nosomma,* and *Rhipicephalus*). In the SMSS, when adults are strictly specific to a host group, immatures are moderately specific to a different group, and con-versely, when immatures are strictly host specific, adults are moderately specific to a different host group, as in Ixodidae. The MSS ticks are those whose host specificity is moderately limited for both adults and imma-tures, again as in Ixodidae. The NP ticks readily accept nonspecific hosts as found in some hard ticks.

Monotremata. Four species are ST specific to monotremes: *Ixodes or-nithorhynchi* Lucas on platypus; three species on echidna—*Ornithodoros (Pavlovskyella)* sp., *Amblyomma (Adenopleura) echidnae* Roberts, and *Apo-nomma concolor* Neumann (most of *Aponomma* parasitic only on reptiles). Four species are MT: *Amblyomma* (three species), *Haemaphysalis* (one species); and three are NP.

Marsupialia. Twelve species are ST ticks representing *Ornithodoros (Pav-lovskyella), Ixodes (Entoplpiger), I. (Exopalpiger), I. (Sternalixodes), Haemaphy-salis (Ornithophysalis), Amblyomma,* and *Aponomma,* on wallabies, kan-garoos, dasyurids, bandicoots, and wombats. Eight species are MT (*Haemaphysalis* and *Ixodes*), and three are NT.

Insectivora. Madagascar tenrecs have eight ST species: *Argas* (one species), *Ixodes* (one species), *Haemaphysalis* (six species). About an equal number of *Ixodes* and *Haemaphysalis* are ST parasites of other insectivores.

Chiroptera. The 24 ST species parasitize the Old World bats, and 31 ST ticks are parasitic on the New World bats. They include *Argas* (*Carios*) (all six species), *A.* (*Chiropterargas*) (all four species), *Ornithodoros* (*Reticulinasus*) (all 11 species), *O.* (*Alectorobius*) (20 species), *O.* (*Subparmatus*) (all three species), *Antricola* (all seven species), and *Nothaspis* (monotypic) (Argasidae); *Ixodes* (*Eschatocephalus*) (two species), *I.* (*Lepidixodes*) (one species).

Primates. Two *Ixodes* species are ST parasites of Ethiopian *Colobus* and *Cercopithecus* monkeys. Malagasy lemurs have one *Ixodes* and one *Haemaphysalis* ST species.

Edentata and Pholidota. The Neotropical Edentata have two ST species and three SS (adult) species of *Amblyomma* (*Amblyomma*). In the same subgenus there are two other species that are MT parasites. Two *Amblyomma* (*Adenopleura*) species are ST parasites of pangolines or scaly anteaters. Other adenopleuran *Amblyomma* are specific for reptiles.

Lagomorpha. The Ochotonidae are host to two ST species of *Ixodes* and one MT species of *Ixodes* (*Pholeoixodes*). They also are parasitized by immatures of other *Ixodes* and *Dermacentor* and both adults and immatures of several *Haemaphysalis* and *Rhipicephalus* species. Rabbits and hares (Leporidae) are hosts of numerous immatures and some adults of ST, SMS, and NP species; ST—*Haemaphysalis* (two species), *Otobius* (one species), *Dermacentor* (one species), *Ixodes* (two species); SMS—*Haemaphysalis* (one species), *Amblyomma* (one species), *Ixodes* (two species), *Rhipicephalus* (five species).

Rodentia. Only limited data are available for host specificity in rodent ticks. There are 23 ST species of rodents in the Neotropical and 18 in the Nearctic regions. They represent *Ornithodoros*, *Ixodes*, and *Amblyomma*. In the Palearctic, 12 ST species comprise eight *Ixodes*, one *Argas*, one *Rhipicephalus*, and three *Anomalohimalaya*. The Malagasy Nesomyidae have five ST *Ornithodoros*, *Ixodes*, and *Haemaphysalis* species. Rodents are important hosts of many MT *Ornithodoros* and some ixodid species and for immatures of some 300 of the approximately 600 ixodid species with a three-host life-cycle pattern.

Carnivora. Adults of certain *Haemaphysalis* and *Rhipicephalus* species are primary parasites of Carnivora. *R.* (*R.*) *sanguineus* is an universal ST species for domestic dogs and *Ixodes* (two species) and *Haemaphysalis* (three species) are ST species. SS and MS (adult) species of Carnivora are *Ixodes* (seven species), *Amblyomma* (one species), *Dermacentor* (one species),

Haemaphysalis (17 species), *Rhipicephalus* (about 15 species), and *Rhipicentor* (two species).

Tubulidentata, Proboscidea, and Hyracoidea.   Little host specificity is found in ticks of Tubulidentata and Proboscidea. No host-specific tick is known for the aardvark, and adults of one *Amblyomma* and one *Dermacentor* species are SS parasites of the Ethiopian elephants. Six ST parasites are found on rock hyraxes (*Procavia* and *Heterohyrax*), which include four species of the *Haemaphysalis* (*Rhipistoma*) *orientalis* group, one *Ixodes*, and one *Rhipicephalus*. A single argasid ST species [*Ornithodoros* (*O*.) *procaviae* Theodor] is found on *Procavia* in the Negev Desert (Palearctic).

Perissodactyla.   Numerous SS or MS (adult) species of *Amblyomma*, *Hyalomma*, *Rhipicephalus*, *Dermacentor*, and *Boophilus* are parasitic on Equidae and Rhinocerotidae, with one ST species, *Margaropus winthemi* Karsch, on the zebra and the domestic horse. The Tapiridae have four *Amblyomma*, one *Ixodes*, and one *Dermacentor* SS species.

Artiodactyla.   At least 190 ixodid species (none are *Aponomma*, *Anomalohimalaya*, or *Rhipicentor*, and relatively few are *Ixodes*) and only six argasid ticks (five *Ornithodoros* and one *Otobius*) are parasitic on Artiodactyla. Most of these ticks are SS (adult) species. The ST species are *Ornithodoros* (*O*.) *porcinus* Walton on the Ethiopian suids; *Haemaphysalis traguli* Oudemans on the Oriental Tragulidae; *Dermacentor* (*Anocenter*) *nitens* Neumann on the Neotropical cervids; *Ornithodoros* (*Ornamentum*) *coriaceus* Koch on the Nearctic cervids; *O*. (*O*.) *indica* Rau and *Haemaphysalis* (*H*.) *birmaniae* Supino on the Oriental cervids; two *Margaropus* on the Ethiopian giraffids, *M. reidi* Hoogstraal and *M. wileyi* Walker; *Otobius megnini* Dugés on the Nearctic Antilocapridae; and *O*. (*Alveonasus*) *lahorensis* Neumann, *O*. (*Pavlovskyella*) *tholozani* (Laboulbène and Megnin), *Boophilus annulatus* (Say), and *B. kohlsi* Hoogstraal and Kaiser, on the Palearctic Bovidae, and *Boophilus decoloratus* (Koch) and *B. geigyi* Aeschlimann on the Ethiopian bovids.

HAEMAPHYSALINE TICKS

The 155 species of the genus *Haemaphysalis* (Fig. 10.1), parasitic on birds and mammals, constitute the most useful assemblage in the superfamily Ixodoidea for displaying numerous interrelated structural-biological clues to affinities between tick species and groups and to historical and contemporary host associations.

Only the genus *Ixodes* is larger (±217 species) than *Haemaphysalis*, but we know much less about *Ixodes* biology and immature stages. A number of structurally primitive species, each with specialized biological properties, provide indicators of the early history and contemporary survival and adaptation of *Haemaphysalis* species. No other tick genus shows this range of lucid clues or this variety of data for species of biological and evolutionary significance. An unusual proportion of this data derives from studies of species restricted to "remote" areas of the world.

Haemaphysalis sexual dimorphism is expressed by the usual ixodid characters: presence of porose areas on the female basis capituli, differences in the male and female scutums and external genital areas, and larger female body size (Figs. 10.11, 10.12, 10.15, 10.16, 10.20). However, *Haemaphysalis* ticks differ from those of other genera in that male capitular spurs and spurlike angles, and also coxal spurs, are almost invariably more luxurient than those of females. Comparing these special haemaphysaline characters in the same sex of different species, and between males and females of a single species, provides numerous valuable indicators of haemaphysaline and host associations and to the evolutionary dynamics of these relationships. Structural differences between larval, nymphal, and adult stages are equally valuable indicators of phylogenetic consociations and contribute basic criteria for differentiating haemaphysaline subgenera.

All *Haemaphysalis* species, so far as known, have a three-host type of life cycle. Hosts of immatures and adults of many species differ significantly: as do the hedgehog and fox, shrew and yak, mouse and lion, rat and boar, lizard and ibex, or bird and bison.

Male and female *Haemaphysalis* each furnish a minimum of 20 clearly definable, diverse structural characters for comparison. Each nymph provides at least 15 characters and each larva at least six. Different males and females in about 150 taxa (very few species are known by only one sex) provide about 6000 character states (units) for comparison. We also know of 125 different nymphs (1875 units) and 115 different larvae (690 units). With this total of 8565 units for morphological analysis, it has been possible to develop a subgeneric classification that is meaningful morphologically as well as biologically and reflects *Haemaphysalis*–host coevolution. Of the 155 known *Haemaphysalis* species, 150 easily fall into 16 subgenera (a few of which remain to be described).

Other ixodid genera (except *Ixodes*) contain fewer species with fewer structural indicators. After more than 200 *Ixodes* species have been better studied, a comparative phylogenetic and host analysis of *Ixodes* and *Haemaphysalis* should considerably enhance the understanding of tick and host coevolution.

STRUCTURALLY PRIMITIVE (SP) HAEMAPHYSALINES

The 17 structurally primitive (SP) haemaphysaline species in four subgenera (*Alloceraea, Allophysalis, Aboimisalis, Sharifiella*) link *Haemaphysalis* and other nonrhipicephaline ixodids and differ distinctly from 138 other haemaphysaline species in 12 other subgenera (Hoogstraal 1965, 1978). The most basic criterion of the 17 SP species is the presence, in each stage, or only in larvae and nymphs, of a lateral convexity of the basis capituli or of a projection from each side of the basis capituli. The next most basic criterion is the palpal structure. SP palpi are elongate and compact, but not basolaterally salient. SP capitular and leg spur development is, with few notable exceptions, exceedingly slight. In all of the other 138 haemaphysaline species, the basis capituli is rectangular, usually with posterior cornua; lateral projections are absent.

In the 17 species constituting SP subgenera, a stepwise progression from atypical to typical haemaphysaline characters is exhibited from species to species and from larva to nymph to adult, as well as between males and females of individual species (Fig. 10.7). This progression provides a clear picture of the structural evolution through history. Comparative criteria of SP and structurally advanced (SA) subgenera, closely associated with host and geographical relationships and biological properties, are the basic indicators of evolutionary processes within this large group of ixodid ticks.

Together with SP haemaphysalines, the single structurally intermediate (SI) subgenus *Herpetobia* (seven species) and the 11 SA subgenera exhibit a gradual but strong phylogenetic trend away from Amblyomminae-like characters of SP subgenera and set the genus *Haemaphysalis* apart structurally (but not biologically) from all other ixodid genera.

The dominating phylogenetic trend throughout the genus *Haemaphysalis* is away from laterally salient basis capituli and *Aponomma-Amblyomma* type elongate palpi toward (1) a rectangular basis capituli with posterior cornua, (2) compact to broadly basosalient palpi, and (3) various combinations of coxal and trochantal spurs and capitular spurs, angles, or emarginations. These hair-hooking devices (Figs. 10.8–10.10) assist the small tick in penetrating a maze of stiff hairs and spines to reach a feeding site on the host integument. A trend toward smaller bodies and capitula is also strong. In small-sized ticks, posterior cornua on a rectangular basis capituli and compact or short, broad palpi function more effectively to force a passage though feathers or fur than do a broad basis capituli and elongate palpi. Large primitive ticks parasitizing glabrous reptiles were not faced with this problem.

Subgenus *Alloceraea* Schultz

The *Alloceraea* species are *H. (A.) vietnamensis* Hoogstraal and Wilson of the Vietnam highlands; *H. (A.) kitaokai* Hoogstraal of Japan, Taiwan, and Hunan (China); *H. (A.) aponommoides* Warburton (Hoogstraal and Mitchell 1971) of highlands in the central and eastern Himalayan range and southern China; and *H. (A.) inermis* Birula of the southwestern USSR, northern Iran, Turkey, and eastern and southern Europe (in France probably introduced with East European deer for parks and hunting reserves).

H. (A.) vietnamensis SP characters are the most pronounced (Fig. 10.4). In

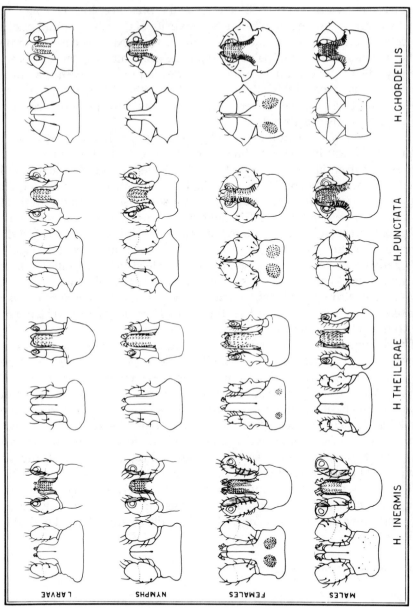

Figure 10.7 The capitulum (dorsal and ventral views) of the larva, nymph, male and female of selected species in structurally primitive *Haemaphysalis* subgenera.

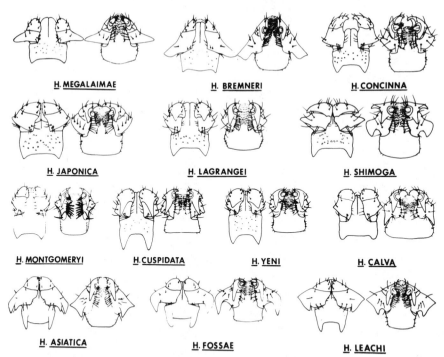

Figure 10.8 The male capitulum (dorsal and ventral views) of selected structurally advanced (SA) *Haemaphysalis* species, showing the broad palpi with hair-hooking spurs obsolete or minute in the *Ornithophysalis* species parasitizing birds (*H. megalaima*) or birds and marsupials (*H. bremneri*), and a unique palpal modification for grasping artiodactyl hairs (*H. concinna*), or a variety of hair-hooking spurs or slits for hooking around hairs of Artiodactyla (*H. japonica, lagrangei, shimoga, montgomeryi, cuspidata, yeni, calva*), or Carnivora (*H. asiatica, fossae, leachi*).

this large haemaphysaline (3.6 mm long), the basis capituli is short, laterally convex, and lacks cornua; the palpi are narrowly elongate (clavate), segment 3 lacks a ventral spur, and segment 4 is in an apical pit; the dental formula is 3/3; coxal spurs are small or obsolete; tarsi are narrowly elongate, and tarsus I has a huge claw; the body integument is leathery and the genital aperture is slitlike (Hoogstraal and Wilson 1966). Other female *Alloceraea* are structurally rather similar but distinctly smaller (2.5–2.9 mm long). The sambar deer or another large ungulate may be the host of adult *H.* (*A.*) *vietnamensis*, but it is now known only from vegetation.

Alloceraea males, nymphs, and larvae each also have a laterally convex or otherwise laterally projecting basis capituli lacking posterior cornua and elongate palpi lacking a ventral spur, but their dental formulae are 2/2. These are the only males with a 2/2 dental formula in the entire genus *Haemaphysalis*. The palpal segment 3 pit containing segment 4 is somewhat more posteriorly displaced in other *Alloceraea* species than it is in *H.* (*A.*) *vietnamensis*. The various spurs and spurlike angles of the body appendages (capitulum and legs) that function as hair-hooking devices in so many

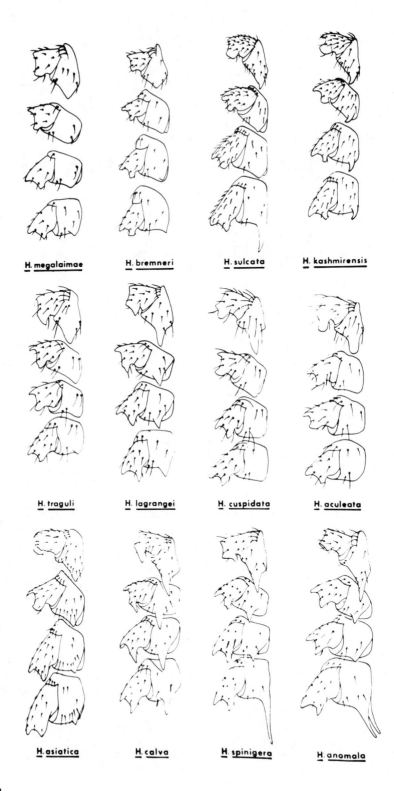

H. megalaimae   H. bremneri   H. sulcata   H. kashmirensis

H. traguli   H. lagrangei   H. cuspidata   H. aculeata

H. asiatica   H. calva   H. spinigera   H. anomala

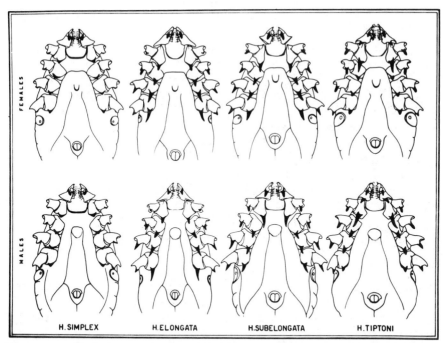

Figure 10.10 Ventral views of male and female *Haemaphysalis* parasites of tenrecs showing virtual absence of spines and hooking devices in *H.* (*Ornithophysalis*) *simplex* (which feed in the host ears) and the variety of palpal, coxal, and trochantal spurs and hooking devices in *H.* (*Elongiphysalis*) species which feed among the body spines and harsh hairs.

other haemaphysalines are obsolete or only very slightly developed in each *Alloceraea* stage [but moderately large in immature *H.* (*A.*) *kitaokai*].

Alloceraea structure apparently represents an *Aponomma*-like primitive prototype haemaphysaline that evolved in the late Paleozoic or early Mesozoic era (see "Premammalian Evolution of Ticks").

Ecologically and zoogeographically, contemporary *Alloceraea* are notably absent in humid tropical forests. Only the highland-inhabiting *H.* (*A.*) *vietnamensis* is found in the Oriental region east of the Indian subcontinent, where the genus *Haemaphysalis* probably originated. Many contemporary SA haemaphysalines are common in tropical Asian forests where two, three, or more species can be found infesting a single mammal. Competition from rapidly evolving SA species may have resulted in the extinction

<hr />

Figure 10.9 Male coxae and trochanters of selected structurally advanced *Haemaphysalis* species, showing absence or virtual absence of hair-hooking spurs in *H.* (*Ornithophysalis*) species (*megalaimae* and *bremneri*) parasitizing birds and marsupials, the slight development of these spurs in *H. traguli*, a parasite of the small-sized Mouse Deer (*Tragulus*), and the variety of hair-hooking spurs in species adapted to Artiodactyls.

of the SP structural type in tropical forests (except the subgenus *Sharifiella* in Madagascar).

The marginal tropical-temperate forest habitat (1450 m altitude) of *H. (A.) vietnamensis* apparently more closely approximates the early haemaphysaline habitat-climatic type than that of the other *Alloceraea* species. *H. (A.) kitaokai* is confined to temperate Japan and mountains of Hunan (China) and Taiwan, *H. (A.) aponommoides* to Himalayan and Southern Chinese highlands (2000–4900 m altitude), and *H. (A.) inermis* to temperate European lowlands. This distribution pattern appears to represent a general ecological shift (Darlington 1957) away from the ancestral ecological-environmental type of the genus. The presence of *H. (A.) vietnamensis* in a tropical-temperate forest margin appears to satisfy the Horton's postulate (1973) that "if environments similar to the ancestral environment still occur in the original center of dispersal, then primitive species are still likely to occur there."

Lizards are important hosts of immature *H. (A.) inermis* in Palearctic lowlands. However, reptiles are rare or absent in the highland and more northern habitats where other *Alloceraea* species have survived. Immatures of each contemporary *Alloceraea* species commonly feed on shrews and rodents; bird hosts have also been recorded. Adults chiefly parasitize Artiodactyla such as the yak, a variety of deer, wild and domestic sheep and goats, and occasionally carnivores (bear, wild cat, etc.). Adults also infest domestic cattle, buffaloes, and horses. Few other *Haemaphysalis* species are associated with domestic herbivores; the fact that all SP adults (except *Sharifiella*) feed on domestic animals is apparently significant in the survival of this relict group.

Notable biological peculiarities also characterize the *Alloceraea* species that have been studied in this respect. Female *H. (A.) inermis* (see Brumpt in Nuttall and Warburton 1915) and *H. (A.) kitaokai* deposit fewer than 1000 eggs (Kitaoka and Morii 1967), or less than 25% of the total egg production of most other *Haemaphysalis* species. Each egg and the resulting larva is unusually large for this genus. Immatures feed fully in 90 minutes to 6 hours [see also Nosek 1973, for *H. (A.) inermis*]; immatures of other *Haemaphysalis* species (and of most other ixodid species) require two to seven days to feed. In the cold, high Himalayas, adult *H. (A.) aponommoides* is active through much of the year (Hoogstraal and Mitchell 1971). However, adults of related species inhabiting warmer lower altitudes are active chiefly in winter. "Uncountable numbers" of adult *H. (A.) kitaokai* parasitize Japanese deer when the temperature is near or below 0°C (Kitaoka and Fujisaki 1972). Adult *H. (A.) inermis* quest for hosts in early and late winter and in spring, and are easily observed moving about on snow and on twigs and grass above the snow surface (Macicka 1958). This cold-season adult activity pattern, an adaptation to prevent desiccation during activity on hot, dry summer days, suggests that the water balance properties of SP species may not differ greatly from those of SA species in humid tropical forests. The unique leathery *Alloceraea* integument is probably also specially adapted for water conservation.

Among SP haemaphysalines, the chromosomes of only *H.* (*A.*) *kitaokai* have been studied (Oliver, Tanaka, and Sawada 1974). This species possesses fewer chromosomes ($♀ = 18 + XX$, $♂ = 18 + X$) than is typical of several studied species in the SA subgenera ($♀ = 20 + XX$, $♂ = 20 + X$). The current working hypothesis of these authors is that the "primitive" chromosome condition in the genus *Haemaphysalis*, and probably in most genera of the family Ixodidae, consists of a $2n$ number of 20 autosomes plus two sex chromosomes in females and 20 autosomes plus one sex chromosome in males. The absence of a pair of chromosomes in *H.* (*A.*) *kitaokai* might have reduced the genetic variability of this species and confined it structurally, and in some respect biologically, to a primitive type of existence.

Subgenus *Allophysalis* Hoogstraal

The eight species in the subgenus *Allophysalis* display a combination of SP structural conformity and of stepwise or abrupt advances in SP ranks. Certain *Allophysalis* nymphs and/or adults approach a SA pattern in part, but no larvae do so. These eight relict-type ticks avoid competition from other ticks in rocky biotopes in the 1600–3800 meter altitude range of Asian mountains. The *Allophysalis* species are *H.* (*A.*) *tibetensis* Hoogstraal of Tibet; *H.* (*A.*) *pospelovashtromae* Hoogstraal of southern USSR and Mongolia (Figs. 10.11–10.14); *H.* (*A.*) *danieli* Cerny and Hoogstraal of northern Pakistan and Afghanistan; *H.* (*A.*) *demidovae* Emel'yanova of Mongolia; *H.* (*A.*) *garhwalensis* Dhanda and Bhat of northern India and Nepal; *H.* (*A.*) *xinjiangensis* Teng of western China; *H.* (*A.*) *warburtoni* Nuttall of southern China and Nepal; and *H.* (*A.*) *kopetdaghica* Kerbabayev of Turkmen SSR and northern Iran.

Each immature *Allophysalis* (except the nymph of *warburtoni*) has a short, broadly angular basis capituli (Fig. 10.14). This probably illustrates one of many designs that appeared and disappeared during evolution from the short, rounded *Alloceraea* basis capituli pattern to the rectangular, cornua-bearing pattern of SA species.

Immature *Allophysalis* basis capituli patterns range from short, remarkably laterally expanded types (*pospelovashtromae, demidovae, danieli, garhwalensis*) to posterolaterally convex and cornua-bearing (*warburtoni* nymph: the first cornua seen in this genus). Immature palpi remain elongate (clavate), as in *Alloceraea*, but have become compact in *kopetdaghica*. The first ventral spur on immature palpi is seen in the *warburtoni* nymph. Some *Allophysalis* nymphs and larvae bear the first large coxal spurs seen in this genus [except for immature *H.* (*A.*) *kitaokai*].

The *Allophysalis* adult basis capituli is variously shaped and armed with unusually variable cornua. In each species, the female basis capituli con-

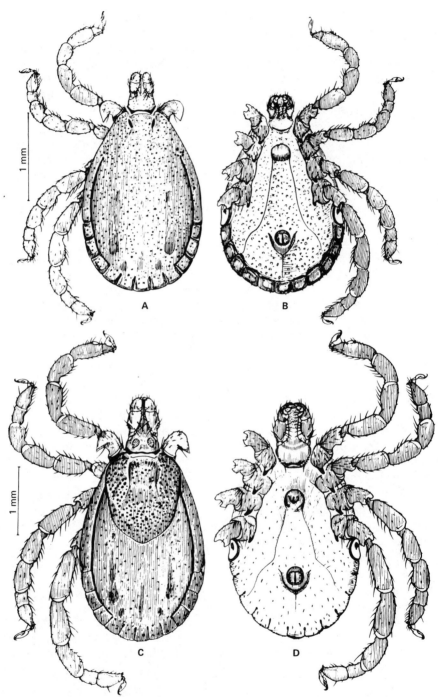

Figure 10.11 *Haemaphysalis* (*Allophysalis*) *pospelovashtromae*. (*A, B*) Male, dorsal and ventral views; (*C, D*) female, dorsal and ventral views.

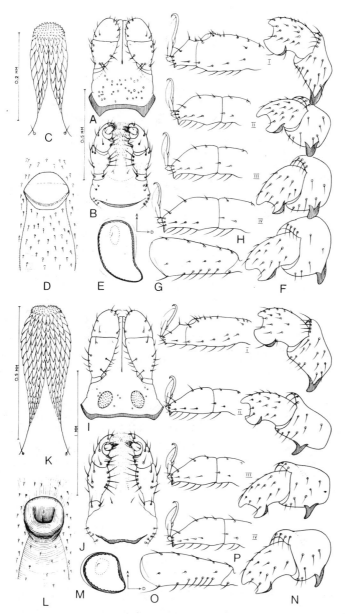

Figure 10.12 *Haemaphysalis* (*Allophysalis*) *pospelovashtromae*. (*A–H*) Male: (*A, B*) capitulum, dorsal and ventral views; (*C*) hypostome; (*D*) genital area; (*E*) spiracular plate; (*F*) coxae and trochanters I to IV; (*G*) femur IV, internal view; (*H*) tarsi I to IV. (*I–P*) Female: (*I, J*) capitulum, dorsal and ventral views; (*K*) hypostome; (*L*) genital area; (*M*) spiracular plate; (*N*) coxae and trodanter I to IV; (*O*) femur IV; (*P*) tarsi I to IV.

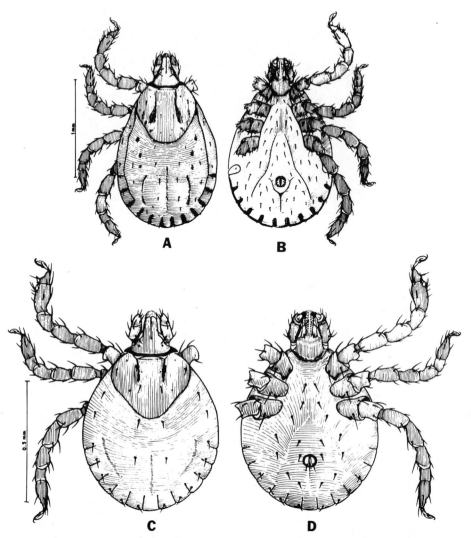

Figure 10.13 Immature *Haemaphysalis* (*Allphysalis*) *pospelovashtromae*. (A, B) Nymph, dorsal and ventral views; (C, D) larva, dorsal and ventral views.

forms more closely to the primitive (*Alloceraea*) type than does that of the male. Female *Allophysalis* palpi remain elongate, as in *Alloceraea*, but are only moderately elongate in *kopetdaghica*. Male palpi are shorter or compact, though not broadened. The adult palpus for the first time bears a small ventral spur which is, however, medially rather than posteriorly directed. This medially directed spur type is the primitive progenitor of that characterizing bird-parasitizing and primitive mammal-parasitizing species of the SA subgenus *Ornithophysalis*. Adult dental formulae are now 4/4, as in almost all other haemaphysalines, but 5/5 or 6/6 in some *warbur-*

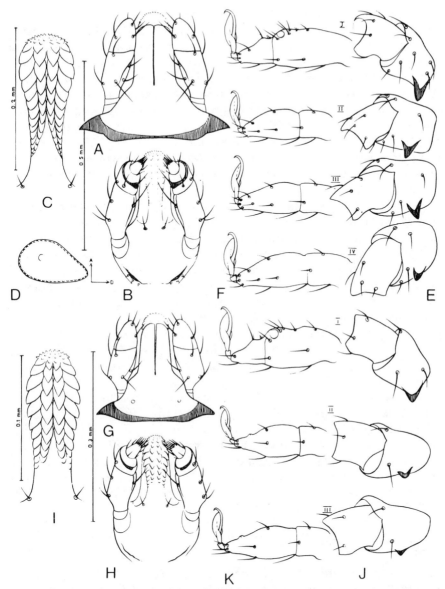

Figure 10.14 Immature *Haemaphysalis* (*Allphysalis*) *pospelovashtromae*. (*A–F*) Nymph: (*A, B*) capitulum, dorsal and ventral views; (*C*) hypostome; (*D*) spiracular plate; (*E*) Coxae and trochanters I to IV; (*F*) tarsi I to IV. (*G–K*) Larva: (*G, H*) capitulum, dorsal and ventral views; (*I*) hypostome; (*J*) coxae and trochanters II, III; (*K*) tarsi II, III.

toni. The 3/3 adult formula, earlier observed in *Alloceraea*, will be found again in adults of only a few SA species specially adapted to parasitizing tenrecs, rodents, and rabbits. Adult coxal spurs are now moderately sized or fairly large and exhibit various triangular, spatulate, lanceolate, or hook-like forms characteristic of haemaphysalines specialized for parasitizing Artiodactyla. All tarsi are short and bear an unusually large apicoventral hook. Female external genital structures differ distinctly from each other and also from those of all other haemaphysalines. Individual unfed male *H. (A.) warburtoni* may be small (2.2 mm long) but average 2.8 mm long, as in all other *Allophysalis* species except *H. (A.) tibetensis*, which is quite large (3.3 mm long). Unfed females are at least as large (3.4–3.8 mm long) as those of *H. (A.) vietnamensis*.

Each *Allophysalis* adult parasitizes members of the rich Asian-mountain artiodactyl fauna and less often, domestic artiodactyls, marmots, and humans. Adult activity is recorded chiefly during spring and fall, and that of immatures from late spring to fall. Adults are recorded from domestic goats, sheep, cattle, yaks, and a dog (one specimen), and from the wild goral, serow, thar, musk deer, and ibex. Immatures parasitize rodents, chiefly *Alticola, Cricetulus*, and *Marmota*, and also hares and the rock-dwelling pika *Ochotona* (Lagomorpha). Both adult and immature *H. (A.) kopetdaghica* were taken from an immature wild goat, *Capra hircus aegagrus* Erxleben, which was probably sick (Hoogstraal and Wassef 1979). Large ground-feeding birds such as the monal pheasant are also important hosts of immature *H. (A.) warburtoni* (Hoogstraal 1971).

The biology and ecology of *H. (A.) pospelovashtromae* in the USSR have been reviewed by Grebenyuk (1966) from literature and personal observations in Kirghiz SSR. Immatures infest a variety of alpine rodents, pikas, and hares, occasionally artiodactyls and carnivores, and are common on ground-feeding birds in the Caucasus. Its life cycle extends over three years (Sartbaev 1955; Ogandzhanian and Martirosian 1965). Notably, immatures feed for about seven days, thus differing from those of *Alloceraea*. Each female deposits at least 1000 eggs.

Subgenus *Aboimisalis* Santos Dias

Aboimisalis immatures retain the primitive SP basis capituli form. Adults are structurally more advanced and show a slight step in the generic trend to broad palpi. The species include: *H. (A.) cornupunctata* Hoogstraal and Varma of Nepal, northern India, Pakistan, and Afghanistan; *H. (A.) punctata* Canestrini and Fanzago of southwestern Asia and much of Europe (Figs. 10.15–10.18); *H. (A.) chordeilis* (Packard) of Canada and the United States; and *H. (A.) cinnabarina* Koch of Brazil. The Eurasian stem of this subgenus is represented by *H. (A.) cornupunctata*, confined to altitudes between 1600 and 3200 m, and by *H. (A.) punctata* of lowland semideserts, steppes, and open forests. The American species are structurally much like

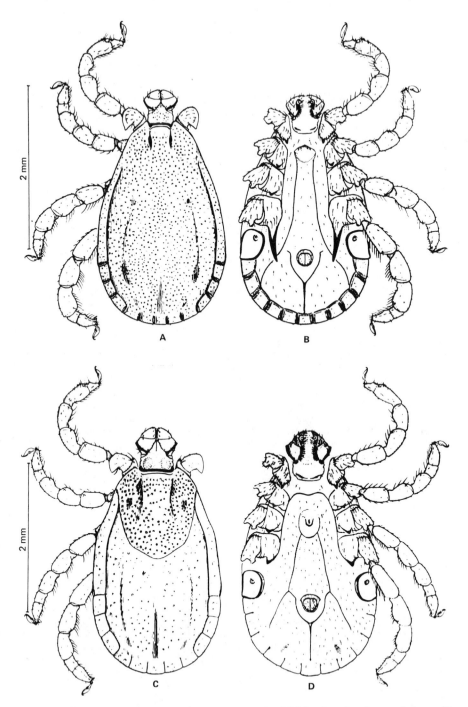

Figure 10.15 *Haemaphysalis (Aboimisalis) punctata.* (*A, B*) Male, dorsal and ventral views; (*C, D*) female, dorsal and ventral views.

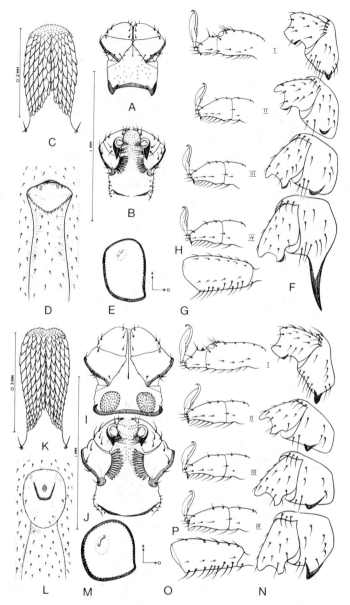

Figure 10.16 *Haemaphysalis (Aboimisalis) punctata.* (*A–H*) Male: (*A, B*) capitulum, dorsal and ventral views; (*C*) hypostome; (*D*) genital area; (*E*) spiracular plate; (*F*) coxae and trochanters I to IV; (*G*) femur IV, inner view; (*H*) tarsi I to IV. (*I–P*) Female: (*I, J*) capitulum, dorsal and ventral views; (*K*) hypostome; (*L*) genital area; (*M*) spiracular plate; (*N*) coxae and trochanters I to IV; (*O*) femur IV; (*P*) tarsi I to IV.

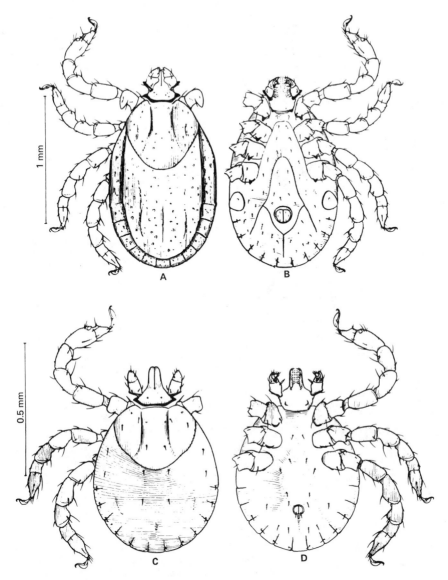

Figure 10.17 Immature *Haemaphysalis* (*Aboimisalis*) *punctata*. (*A, B*) Nymph; (*C, D*) larvae, dorsa and ventral views.

their Euroasian counterparts. *H.* (*A.*) *cinnabarina* is known only from two females collected before 1844 (Hoogstraal 1973). Notably, only two other haemaphysaline species (subgenus *Gonixodes*) occur in the Americas.

The lateral expansion of the immature *Aboimisalis* basis capituli is the most extreme of all SP forms, especially in nymphs (Fig. 10.5). Immature palpi have become compact; they lack salience in *H.* (*A.*) *cornupunctata* but

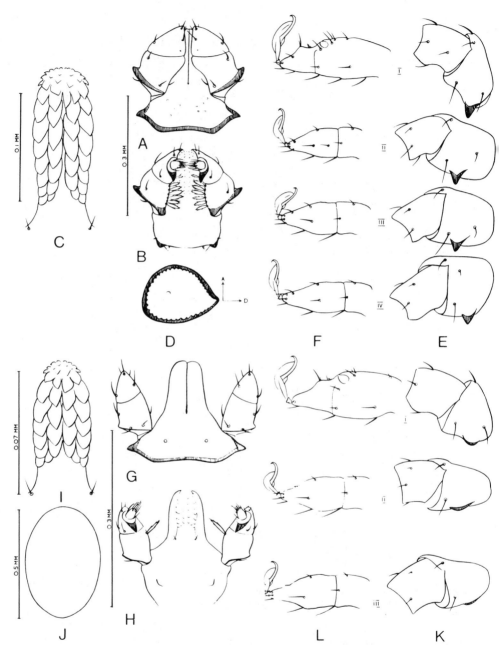

Figure 10.18 Immature *Haemaphysalis* (*Aboimisalis*) *punctata*. (*A–F*) Nymph: (*A, B*) capitulum, dorsal and ventral views; (*C*) hypostome; (*D*) spiracular plate; (*E*) coxae and trochanters I to IV; (*F*) tarsi I to IV. (*G–L*) Larva: (*G, H*) capitulum, dorsal and ventral views; (*I*) hypostome; (*J*) spiracular plate; (*K*) coxae and trochanters I to III; (*L*) tarsi I to III.

are slightly salient in *H. (A.) punctata* and *H. (A.) chordeilis*, and the ventral spur is slight or obsolete. Dental formulae are 2/2. Coxal spurs are obsolete to moderate sized.

Adult *Aboimisalis* differ from their SP-type immatures in having a rectangular basis capituli characteristic of SA species. Cornua extend from this base in each male but in females only in *H. (A.) cornupunctata*. Adult palpi are quite compact, slightly broader than in the few males of earlier subgenera in which the palpi tended to be more compact. Dental formulae are 4/4 to 6/6. Coxal spurs are moderate sized in females, but male coxae I to III spurs are much reduced and coxa IV spur is very long and lanceolate. This lanceolate spur type reappears in several SA species associated with Artiodactyla. The short tarsi are ventrally hooked as in *Allophysalis* and *Sharifiella*. The adult remains large in Eurasian species (length about 3.0 mm in males, about 3.4 mm in females) but is moderate-sized in the American species.

Biologically and ecologically, each *Aboimisalis* species differs distinctly. All our records (including unpublished) and those of Dhanda and Kulkarni (1969) for both immature and adult *H. (A.) cornupunctata* are from wild or domestic artiodactyls. Immatures and adults are sometimes taken from one artiodactyl host, but immatures are not found on insectivores or rodents examined in the same collecting localities in Nepal, India, Pakistan, and Afghanistan. Parasitism of the same kind of host by immatures and adults, seen among SP subgenera only in *Sharifiella*, occurs in few species of SA subgenera. All *H. (A.) cornupunctata* records are from the spring and fall seasons. Ecologically, *H. (A.) cornupunctata* is similar to most other SP ticks but quite different from the three other *Aboimisalis* species.

Adult *H. (A.) punctata* parasitize wild and domestic artiodactyls, rarely other vertebrates. Immatures feed chiefly on small hosts (Sartbaev 1961; Grebenyuk 1966; Tovornik 1970; Nosek 1971). Numerous ground-feeding birds are important hosts and carry nymphs and larvae when migrating (Hoogstraal et al. 1963, 1964). Immatures also parasitize lizards, insectivores, rodents, and hares, less often artiodactyls and carnivores. The ecological adaptability and geographic range of *H. (A.) punctata* are greater than in most other haemaphysalines. However, this species does not reach alpine levels or penetrate into the interior of humid temperate forests. Its life cycle usually extends over a 2-year period. In most areas adult *H. (A.) punctata* feed during fall and spring but in mild latitudes also in winter and/or summer. Immatures are active from spring to fall. *H. (A.) punctata* egg numbers 3000–5000 per female, and immature and adult feeding periods are similar to those generally reported for SA species. Larvae feed for three to five days, nymphs for four to seven days, and females for six or more days.

The North American *H. (A.) chordeilis*, which is especially common in Canada (Gregson 1956; Kohls 1960), infests grouse and other game birds but rarely mammals. There are no definitive biological or ecological studies of H. (A.) *chordeilis*. The long coxa IV spur of the male [similar to those of H. (A.) *punctata* and H. (A.) *cornupunctata*] sets this species apart from practically all other bird-infesting haemaphysalines, which are characterized by conservative spur development. The H. (A.) *chordeilis* spur probably reflects a dominant feature of *Aboimisalis* structure rather than a useful adaptation for parasitizing birds.

In review, among the 17 SP species, 15 have coevolved with Artiodactyla, 1 with birds, and 1 (see below) with tenrecs.

Subgenus *Sharifiella* Santos Dias

The single species of *Sharifiella*, H. (S.) *theilerae* Hoogstraal, conforms to SP group criteria but also displays unusual characters which are probably functionally adaptive to parasitizing the spiny tenrecs of Madagascar (see following section: *Haemaphysalis* and Tenrecs). The short, rounded basis capituli of each stage lacks cornua. The elongate palpi are either imperceptibly (larva) or slightly (nymph and adult) broadened posteriorly and lack a discrete ventral spur. Dental formulae are 2/2 in immatures and 3/3 in adults. A small spur near the external margin of each coxa, as well as a second spur near the internal margin of some coxae, are both unique in the entire genus. The very short, ventrally hooked tarsi bear large claws but extraordinarily small pulvilli. As is common among tropical haemaphysalines, this species is small (male 2.0 to 2.3 mm long). This is one of the few SP species from outside Eurasia and which parasitizes only small hosts. The SP subgenus *Sharifiella* is compared with SA tenrec-infesting haemaphysalines in the following section. These structurally and phylogenetically disparate haemaphysaline parasites of a single mammalian family provide rich clues to tick and mammal coevolution.

HAEMAPHYSALIS AND TENRECS (MALAGASY INSECTIVORES)

Six *Haemaphysalis* species are specific parasites of Malagasy tenrecs (Insectivora: Tenrecidae) (Hoogstraal et al. 1974; Uilenberg et al. 1980). Their hosts are the four coarse-haired, spiny tenrecs, *Setifer setosus* (Schreber), *Echinops telfairi* Martin, *Tenrec ecaudatus* (Schreber), and *Hemicentetes semispinosus* (Cuvier), and the soft-furred *Microgale* (*Nesogale*) *talazaci* (Major) (Eisenberg and Gould 1970) (Table 10.1).

Three of the six species parasitizing coarse-haired, spiny tenrecs—H. *elongata* Neumann, H. *subelongata* Hoogstraal, and H. *tiptoni* Hoogstraal—constitute the distinctive SA subgenus *Elongiphysalis* Hoogstraal, Wassef, and Uilenberg (Fig. 10.10). These are exclusive parasites of *S. setosus*, *T. ecaudatus*, and H. *semispinosus*. H. (E.) *subelongata* is a primary parasite of *T. ecaudatus*, the most widely distributed tenrec, which is found in a variety of habitats. H. (E.) *tiptoni*, apparently confined to rain forests, feeds chiefly on

Table 10.1   Host Associations (%) of *Haemaphysalis* Species on Tenrecs[a]

Tenrec species	H. (O.)[b] simplex	H. (E.)[b] elongata	H. (E.)[b] subelongata	H. (E.)[b] tiptoni	H. (S.)[b] theilerae
	79[c]	58[c]	24[c]	8[c]	10[c]
Setifer setosus	51	26	0	0	10
Tenrec ecaudatus	1	31	92	38	90
Hemicentetes semispinosus[d]	1	36	4	62	0
Echinops telfairi	38	2	0	0	0
Microgale talazaci[e]	6	3	0	0	0
"Tenrec" (unident.)	3	2	4	0	0

Source: After Hoogstraal et al. 1974.
[a] Percent association based on 179 lots in Hoogstraal Tick Collection.
[b] *H. (O.)* = *H. (Ornithophysalis)*; *H. (S.)* = *H. (Sharifiella)*; *H. (E.)* = *H. (Elongiphysalis)*.
[c] Number of lots.
[d] Previously listed as *H. semispinosus* and *H. nigriceps*.
[e] All ticks are immature.

H. semispinosus, but also on *T. ecaudatus*. *H. (E.) elongata* commonly infests each coarse-haired species except *E. telfairi* of the arid southwest. Only immatures of *H. (O.) simplex* and *H. (E.) elongata* are known from the soft-furred *Microgale talazaci*.

Three other haemaphysalines specific to coarse-haired tenrecs are *H. (Ornithophysalis) simplex* Neumann, *H. (O.) simplicima* Hoogstraal and Wassef (both SA species), and *H. (Sharifiella) theilerae* Hoogstraal (a SP species). *H. (O.) simplex* parasitizes *S. setosus* in different ecological zones in northwestern lowland and mountain forests (excluding rain forests) and cultivated areas. *H. (O.) simplex* also parasitizes *E. telfairi* in arid southwestern plains, as does *H. (O.) simplicima*. *H. (S.) theilerae* infests the common *T. ecaudatus*, and infrequently *S. setosus*, in eastern mountains and lowlands from northern to southern Madagascar.

H. (Elongiphysalis) species have unique multiple lanceolate coxal spurs for grasping coarse hairs and spines (Fig. 10.10). Certain trochantal spurs are also extraordinarily developed for this purpose. *H. (E.) elongata*, which is common on different tenrec species, displays a wide range of structural variation of the capitulum, and of the coxal and trochantal spurs; this polytypic species perhaps represents a diversification process.

H. (Ornithophysalis) simplex, which has minute spurs similar to those of most bird-infesting haemaphysalines, successfully coexists on *S. setosus* with the strikingly spurred *H. (E.) elongata*. *H. (O.) simplex* feeds in the practically hairless ears and does not invade the spiny-coarse pelage. *H. (E.) elongata*, on the other hand, feeds chiefly among the dorsal spines and coarse hairs (Hoogstraal 1953). *H.* (*O.*) *simplicima* appears to be a rare (perhaps unsuccessful) species; its spurs are all but obsolete, but we do not know its feeding site on the tenrec host.

Among the 20 species of *Ornithophysalis*, 11 parasitize birds or birds and mammals in the Oriental, Australian, Malagasy, Palearctic, and Ethiopian regions, two are confined to Australian marsupials, two to Australian-New Guinea rodents, two to Malagasy tenrecs, and four to Oriental rodents. In Madagascar, *H.* (*O.*) *madagascariensis* Colas-Belcour and Millot is a parasite of a large ground-feeding bird, the coucal. In northeastern Australia and southeastern New Guinea, *H.* (*O.*) *doenitzi* Warburton and Nuttall parasitizes ground-feeding birds. These lines of bird-infesting ticks apparently adapted to primitive mammals (marsupials and tenrecs in the Australian and Malagasy regions) and to certain rodents (in the Australian region) at an early period in mammalian evolution.

H. (*Sharifiella*) *theilerae* is a SP species and apparently the least successful of the six haemaphysaline parasites of tenrecs. The primitive, slightly specialized *H.* (*S.*) *theilerae* competes with highly specialized *H.* (*Elongiphysalis*) species which thrive, often in dense clusters, on the same hosts.

If ticks of the SP subgenera *Allophysalis*, *Alloceraea*, *Aboimisalis*, and *Sharifiella* had become extinct before our time, it would have been much more difficult to determine the steps in phylogeny and structural adaptations of *Haemaphysalis* ticks and in their coevolution with mammals.

STRUCTURALLY INTERMEDIATE (SI) HAEMAPHYSALINES

Subgenus *Herpetobia* Canestrini

The subgenus *Herpetobia* Canestrini (Hoogstraal and McCarthy 1965), a relict, pivotal branch in *Haemaphysalis* phylogeny, is structurally intermediate between SP and SA groups. All *Herpetobia* immatures and adults have a SA-pattern rectangular basis capituli. Their palpi, however, which are compact, but are slightly salient in certain nymphs, represent the forerunner of the broad palpi characterizing SA haemaphysalines. Aside from these critical phylogenetic characters, the four *Herpetobia* species differ quite considerably from each other structurally and biologically.

The type species of the subgenus, *H.* (*H.*) *sulcata* Canestrini and Fanzago, ranges from Kashmir, southern USSR, and southwestern Asia to Yemen and Sinai, and is also found in southern Europe (Grebenyuk 1966; Aeschlimann et al. 1968; Sacca et al. 1969; Tovornik and Brelih 1973; Rageau 1973; Hoogstraal and Valdez 1980). The other *Herpetobia* species occur in and near the western and central Himalayas. These are *H.* (*H.*) *kashmirensis* Hoogstraal and Varma of northern India, Pakistan, and Afghanistan (Hoogstraal and McCarthy 1965) (Figs. 10.19–10.22); *H.* (*H.*) *nepalensis* Hoogstraal of northern India, Nepal, and Tibet; and *H.* (*H.*) *sundrai* Sharif

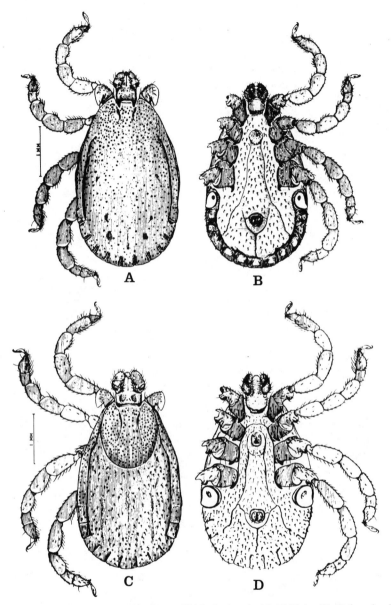

Figure 10.19 Adult *Haemaphysalis* (*Herpetobia*) *kashmirensis.* (*A, B*) Male; (*C, D*) female, dorsal and ventral views.

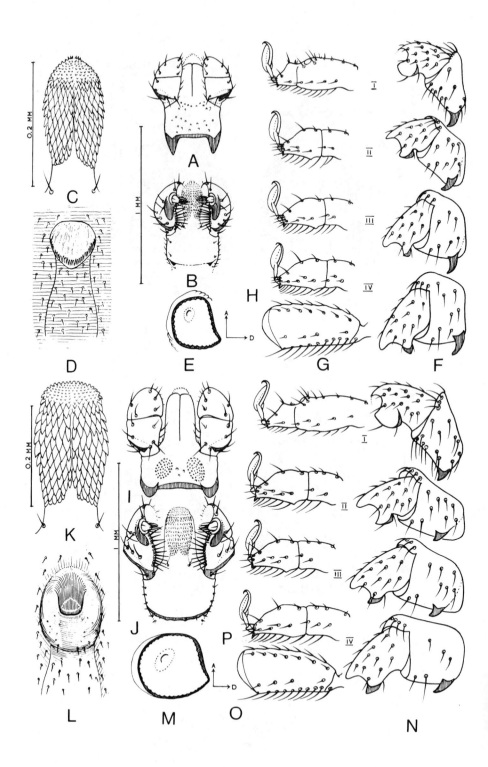

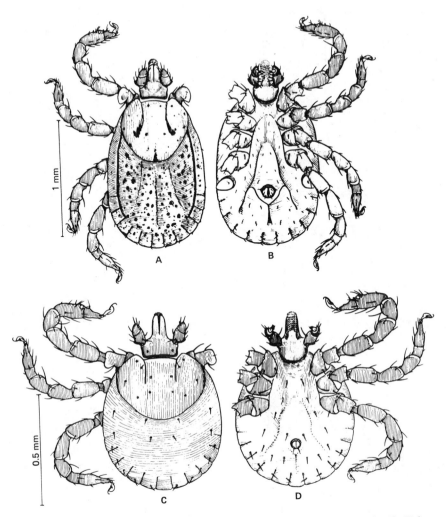

Figure 10.21 Immature *Haemaphysalis* (*Herpetobia*) *kashmirensis*. (*A, B*) Nymph; (*C, D*) larva, dorsal and ventral views.

Figure 10.20 Adult *Haemaphysalis* (*Herpetobia*) *kashmirensis*. (*A–H*) Male: (*A, B*) capitulum, dorsal and ventral views; (*C*) hypostome; (*D*) genital area; (*E*) spiracular plate; (*F*) coxae and trochanters I to IV; (*G*) femur IV; (*H*) tarsi I to IV. (*I–P*) Female: (*I, J*) capitulum, dorsal and ventral views; (*K*) hypostome; (*L*) genital area; (*M*) spiracular plate; (*N*) coxae and trochanters I to IV; (*O*) femur IV; (*P*) tarsi I to IV. This structurally intermediate species has a structurally advanced basis capituli but adult and immature palpi are essentially compact or only very slightly broadened.

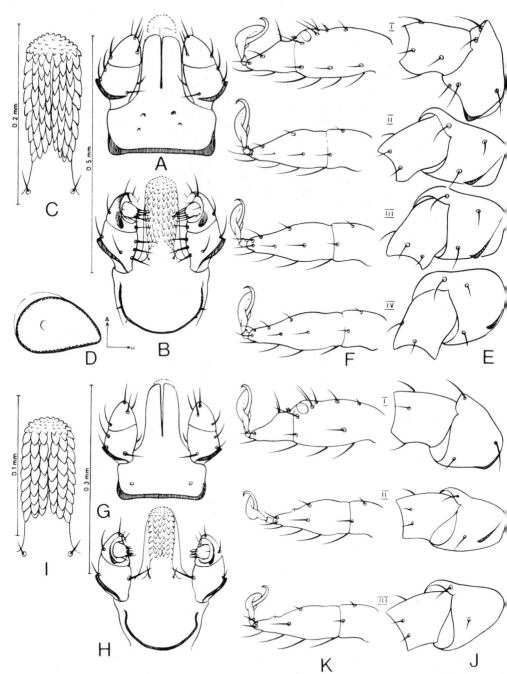

Figure 10.22 Immature *Haemaphysalis* (*Herpetobia*) *kashmirensis*. (*A–F*) Nymph: (*A, B*) capitulum, dorsal and ventral views; (*C*) hypostome; (*D*) spiracular plate; (*E*) coxae and trochanters I to IV; (*F*) tarsi I to IV. (*G–K*) Larva: (*G, H*) capitulum, dorsal and ventral views; (*I*) hypostome; (*J*) coxae and trochanters I to III; (*K*) tarsi I to III, showing the slight palpal broadening, the unusual 3/3 and 4/4 dental formula of the larva and nymph, respectively, and absence of spurs.

of northern India and Nepal (Dhanda and Bhat 1971) [*H. (H.)*] *himalaya* Hoogstraal is probably a junior synonym of *H. (H.) sundrai*]. This geographical-ecological pattern is much like those of the SP subgenera *Alloceraea* and *Allophysalis*; the western lowland dispersal of *H. (H.) sulcata* recalls that of *H. (A.) inermis* in the SP category.

The many structural variations in samples of each *Herpetobia* taxon [except *H. (H.) nepalensis*] indicate the need for laboratory and field studies to settle interrelated biological-taxonomic questions. As in SP subgenera, numerous *Herpetobia* species have probably become extinct. The four contemporary *Herpetobia* species form a truncated branch illustrating host-related structural developments which subsequently disappear or reappear in a few SA groups. The SA subgenus *Haemaphysalis* apparently evolved from *Herpetobia* and spread fanwise eastward through tropical forests of the Oriental region and into the temperate eastern Palearctic region (especially the USSR, China, Korea, and Japan).

Larval *Herpetobia* lack cornua and discrete coxal spurs, retaining the features of their reptile-parasitizing progenitors. The first larval palpal ventral spur is seen in *H. (H.) sundrai*. Immature dental formulae are conventionally 2/2, but are 3/3 and 4/4 in the *H. (H.) kashmirensis* larva and nymph and 3/3 in the *H. (H.) sundrai* nymph. *Herpetobia* nymphs differ in their degree of cornua development, palpal salience, chaetotaxy, and other ways, and are sensitive indicators of interrelated biological and mammalian host factors as well as taxonomic relationships.

Adult *Herpetobia* are moderate to large in size (male length 2.8–4.0 mm), though individual *H. (H.) sulcata* may be smaller. Their cornua are moderately large to large in males and obsolete or small in females. The compact adult palpi are usually ridged posterodorsally in *H. (H.) sulcata* and *H. (H.) kashmirensis* (an evolutionary experiment that failed to survive). Each species except *H. (H.) nepalensis* has a prominent but quite variable ventral spur on palpal segment 3. This ventral spur is exceptionally variable in different population samples of *H. (H.) sulcata*, but short and broad or internally directed in adults of other *Herpetobia* species. The 4/4 dental formulas of adult *H. (H.) sulcata* and *H. (H.) nepalensis* increase to 5/5 to 7/7 in the other species. *H. (H.) sulcata* male coxal spurs (Fig. 10.9) are lanceolate on IV, a moderately large hook on I, and conventionally triangular but variable in size on II and III. These characters are common among male haemaphysaline parasites of Artiodactyla. Female coxal spurs of *H. (H.) sulcata* are obsolete to moderate in size. *H. (H.) kashmirensis* male coxal spurs are hooklike (Fig. 10.9), and female spurs are either hooklike or triangular; those of *H. (H.) sundrai* are a conventional elongate triangle on I and a broad triangle on II, III, and IV. In *H. (H.) nepalensis*, unusual rounded ridges are the forerunners of the spurs of other species. *Herpetobia* female external genitalia are uniquely indented in *H. (H.) sulcata*, unusually subcircular in *H. (H.) nepalensis*, but conventionally rectangular in the other species. These and other unmentioned unusual and variable features among *Herpetobia* adults, and differences between immature and adult stage structural patterns, suggest a wide area for investigation of interrelated phylogenetic and host-adaptation factors.

Ecologically, *Herpetobia* is a subgenus of temperate open forests, steppes, semideserts, and rocky mountainsides. Adult *Herpetobia* parasitize wild artiodactyls (mouflon, sambar deer, chital, muntjac, thar, blackbuck, yak, cow-yak hybrid, etc.) as well as all domestic artiodactyls. Other hosts are uncommon.

Immature *H.* (*H.*) *nepalensis* and *H.* (*H.*) *sundrai* infest the same hosts as adults. Hosts of immature *H.* (*H.*) *kashmirensis* and *H.* (*H.*) *sulcata* differ greatly from those of adults. Larval and nymphal *H.* (*H.*) *kashmirensis* chiefly infest the lizard *Agama tuberculata*. We have a record of both immature and adult *H.* (*H.*) *kashmirensis* infesting a single *A. tuberculata* in northern India. Hosts of immature *H.* (*H.*) *sulcata* include numerous lizards, snakes, and tortoises, birds nesting in, on, or near the ground (roller, stonechat, nuthatch, lark, sparrow, etc.), and occasional small- or medium-sized mammals.

There are records (unpublished) of the immature and adult activity of *H.* (*H.*) *nepalensis* of the Himalayas from each month of the year. Adult *H.* (*H.*) *sulcata* are generally most active in spring and fall, and immatures from spring to early fall.

STRUCTURALLY ADVANCED (SA) HAEMAPHYSALINES

The 11 SA haemaphysaline subgenera contain a total of 132 species. The basis capituli of both immatures and adults in this group is rectangular (never expanded laterally) and usually bears posterior cornua. The palpi of immatures and adults mostly show some basal broadening of segment 2; some retain a compact (but not elongate) form. The salience begins with a slight extension of the posterior breadth and reaches the broadly triangular outline characterizing most species of this genus. The dental formula is 2/2 in immatures and 4/4 in adults, rarely 3/3 or 4/4 in the former, and occasionally 3/3 or 5/5 to 7/7 in the latter. Adults vary in length between 2.2 and 3.5 mm; a few are even smaller; but very few exceed 3.5 mm. All of these 132 species, except the two in the American subgenus *Gonixodes*, occur in the Old World. No SA populations occur above 1500 m altitude in the Himalayas or elsewhere, and few reach even this altitude. Most species occur in humid, wooded zones. The few species that inhabit semiarid environments are found chiefly in littoral zones, floodplains or riparian forests, vegetated valleys or foothills, or oases. Of the 132 SA species, six parasitize only birds and seven parasitize birds and small- or medium-sized mammals (see subgenus *Ornithophysalis*). Three species of the subgenus *Haemaphysalis* chiefly infest mammals but also feed on birds. (Immatures of one *Gonixodes* species frequently feed on birds as well as on leporids.)

The 115 remaining SA haemaphysaline species have coevolved with mammals (chiefly Artiodactyla and Carnivora, also Insectivora, Hyracoidea, Primates, Lagomorpha, and Rodentia) and never or practically never parasitize birds or reptiles.

The structural adaptations and radiation of SA haemaphysaline adults, and their coevolution with mammals should be studied in conjunction with Hoogstraal and Aeschlimann (1982), who discuss the ticks associated with each mammalian order and list most of the ticks specific to each order or family.

Haemaphysaline host-adapted structures are more pronounced in males than in females, presumably because females are somewhat larger, and possibly stronger, and do not wander on the host in search of several mating partners, as do males. These adaptations are less pronounced in nymphs, and even less so in larvae, presumably because these tiny ticks can slip between the host hairs and do not have to "fight their way" through the pelage to reach a feeding site.

Subgenus *Ornithophysalis* Hoogstraal and Wassef

The *Ornithophysalis*, an early SA assemblage of 20 Old World species, illustrates (1) phylogenetically abrupt palpal basal broadening but absence or mild development of spurs on the capitulum (Figs. 10.7, 10.8), coxae, and trochanters (Fig. 10.9) (characterizing bird-infesting haemaphysalines); (2) the generally slight development, if any, of these spurs and other hairhooking devices in species parasitizing the phylogenetically ancient Marsupialia and two species parasitizing the tenrec (Insectivora); and (3) the survival of *Ornithophysalis* characters, with only slight modifications, in Oriental species adapted to Rodentia (Hoogstraal and Wassef 1973).

Notably, haemaphysalines that have adapted to Rodentia in the Ethiopian region are members of the subgenus *Rhipistoma*, probably the most recent of haemaphysaline subgenera, in contrast to *Ornithophysalis*, one of the earliest types of SA haemaphysaline, which has no members confined to mammals in the Ethiopian region.

The host associations of the 20 *Ornithophysalis* species are as follows: Birds—*H. (O.) doenitzi* Warburton and Nuttall; *H. (O.) phasiana* Saito, Hoogstraal and Wassef; *H. (O.) madagascariensis* Colas-Belcour and Millot; *H. (O.) hoodi* Warburton and Nuttall; *H. (O.) megalaimae* Rajagopalan (Figs. 10.8, 10.9); *H. (O.) minuta* Kohls. Birds and marsupials—*H. (O.) bremneri* Roberts (Figs. 10.8, 10.9). Marsupials—*H. (O.) petrogalis* Roberts; *H. (O.) lagostrophi* Roberts. Tenrecs (Insectivora)—*H. (O.) simplex* Neumann (Fig. 10.10); *H. (O.) simplicima* Hoogstraal and Wassef. Birds and small-sized mammals (chiefly rodents)—*H. (O.) howletti* Warburton; *H. (O.) humerosa*

Warburton and Nuttall; *H.* (*O.*) *ornithophila* Hoogstraal and Kohls; *H.* (*O.*) *tauffliebi* Morel; *H.* (*O.*) *pavlovskyi* Pospelova-Shtrom. Rodents—*H.* (*O.*) *bandicota* Hoogstraal; *H.* (*O.*) *sciuri* Kohls; *H.* (*O.*) *kadarsani* Hoogstraal; and *H.* (*O.*) *ratti* Kohls.

H. (*O.*) *minuta*, known only from birds, is one of the smallest ixodid species. It has a typical *Ornithophysalis* facies but exceptional (for this subgenus) spurs on the palpi, coxae, and trochanters. This exception supports a well-founded generalization: the smaller the size of the adult tick (especially those of medium-sized and large-sized hosts) the more luxurient are certain of its spurs (and often the more atypical is its life cycle).

H. (*O.*) *humerosa* is extraordinarily modified for slipping between bird feathers or mammal hairs; it has an extremely narrowly elongate (louselike) body (see Roberts 1970; Fig. 39). Hair-hooking devices are virtually absent in *H.* (*O.*) *humerosa*; the cornua are small, all other spurs are minute or obsolete. A tendency toward narrow and elongate bodies also occurs in the tenrec parasite *H.* (*E.*) *elongata* (but here the spurs are extreme) and in the African carnivore parasite *H.* (*Rhipistoma*) *leachi* and related species.

Ornithophysalis, with its broad palpi, appears to have evolved abruptly from SP subgenera with compact palpi [see *H.* (*Aboimisalis*) *punctata*] when birds and mammals replaced reptiles as the world's dominant vertebrates. Significantly, of the 20 contemporary *Ornithophysalis* species, six parasitize only birds, five parasitize both birds and various mammals, three parasitize birds and marsupials or only marsupials, two parasitize only tenrecs, and four parasitize Oriental or Australian rodents. The only haemaphysalines which specifically parasitize the phylogenetically ancient tenrecs (Insectivora) are the SP species *H.* (*Sharifiella*) *theilerae* and the three SA species constituting the subgenus *Elongiphysalis*, which represents a highly specialized branch from *Ornithophysalis*. Except for the few species of the subgenus *Haemaphysalis* that parasitize birds as well as mammals, all other SA haemaphysaline adults are specific to mammals and never feed on birds, although immatures of a few SA species may infest birds. The four contemporary *Ornithophysalis* species specific to Oriental and Australian rodents suggest an early association between these ticks and mammals in the Oriental and Australian regions and, together with other evidence, with the origin of the genus *Haemaphysalis* in the Oriental region. The Australian and Oriental marsupial- and rodent-infesting species probably evolved from species like the contemporary bird-parasitizing *H.* (*O.*) *doenitzi* of the Oriental region and eastern New Guinea and Australia. The Malagasy tenrec-parasitizing species probably evolved from species like the contemporary bird (coucal)-infesting *H.* (*O.*) *madagascariensis*. *H.* (*O.*) *tauffliebi*, which parasitizes both birds and small- or medium-sized mammals in the Ethiopian region, probably evolved from a species similar to the common bird-parasitizing *H.* (*O.*) *hoodi* of this region. It is probably significant phylogenetically that the "basic" Oriental-Australian, Malagasy, and Ethiopian bird-parasitizing *Ornithophysalis* (*doenitzi, madagas-*

cariensis, hoodi) are difficult to distinguish from each other but that no such problem is presented by the closely related, more geographically restricted, species that parasitize certain birds or both birds and small mammals.

Subgenus *Haemaphysalis* Koch

The subgenus *Haemaphysalis* apparently evolved from *Herpetobia* (or a related, now extinct ancestor) in the Indian subregion of the Oriental region and fanned southeastward into the Oriental region and northeastward into the Palearctic region. The type species of the subgenus, *H. (H.) concinna* Koch, also ranges westward in wooded zones from Japan to northern Iran, Italy, and France.

The subgenus *Haemaphysalis* contains 16 contemporary species [including *H. (H.) japonica* Warburton (Fig. 10.8) with two subspecies] in six species groups. Four species in three groups (*flava, campanulata*, and *birmaniae* groups) remain in the Indian subregion [one of these four species, *H. (H.) campanulata* Warburton, is also found in the eastern Palearctic region]; there are two related species elsewhere in the Oriental region and six related species in the eastern Palearctic region. A few of the Palearctic species also occur in northern mountains of the Oriental region. Two groups (*concinna* and *japonica* groups) with three species are Palearctic. The sixth group, with a single species, *H. (H.) silacea* Robinson, is a relict in southeast Africa (Ethiopian region) (Hoogstraal 1963).

Wild and domestic Artiodactyla (various deer, wild pigs, the Serow, Goral, Blackbuck, etc.) are the chief hosts of most adults (11 of the 16 species) of the subgenus *Haemaphysalis*. Carnivora are secondary hosts of some of these 11 species.

Adults of the three members of the *campanulata* group are not restricted to Artiodactyla. *H. (H.) campanulata* parasitizes carnivores and also artiodactyls in Japan, Korea, northeastern China, the USSR, and southern India. *H. (H.) pentalagi* Pospelova-Shtrom parasitizes the Japanese Black Rabbit (*Pentalagus*) in the Amami Group, Japan (Hoogstraal and Yamaguti 1970). *H. (H.) verticalis* Itagaki, Noda and Yamaguchi parasitizes the Suslik (*Citellus*) and Jird (*Meriones*) in northeastern China (Emel'yanova and Hoogstraal 1973).

Host data for the *flava* group are limited. *H. (H.) indoflava* Dhanda and Bhat (Himalayan region and Madras of India) are recorded from dogs, and others are from the jackal, fox, wild pig, cattle, and man. The immature stages are unknown. Adult *H. (H.) flava* Neumann infest wild and domestic carnivores as well as Artiodactyla (wild pigs, deer, cattle) in Japan, Korea, the eastern USSR, and China. The third member, *H. (H.) megaspinosa* Saito, parasitizes Artiodactyla (wild pigs, deer, Serow) in Japan.

Immatures of the artiodactyl- and carnivore-parasitizing species generally feed on rodents and also on hedgehogs. Immatures of the widely distributed *H. (H.) concinna* (Japan to France) also frequently infest ground-feeding birds and are the only ones of this subgenus known to do so. Immatures and adults of the rabbit and rodent parasites *H. (H.) pentalagi* and *H. (H.) verticalis* feed on the same hosts.

The basis capituli of each stage, except some larvae, bears moderate to fairly large cornua. Small but distinctive *ventral* cornua are present in immatures of *H. (H.) indoflava* and *H. (H.) birmaniae,* and also appear in a less developed form in the immatures of a few other species in this subgenus. Ventral cornua are rare in the genus *Haemaphysalis* but reappear in immatures of both species of *Gonixodes* and in adults of *H. (G.) leporispalustris.* Immature *H. (H.) birmaniae* are also unusual in that they, like adults, parasitize Artiodactyla (rather than Rodentia). The functional adaptation of ventral cornua in relation to the hosts of adults and/or immatures in which they appear should be investigated. Unfortunately, immature *H. (H.) indoflava* are unknown.

Palpi of *Haemaphysalis* immatures and adults are only slightly advanced from the compact form characterizing *Herpetobia* [and the SP group *H. (A.) punctata;* see *H. (H.) japonica* in Fig. 10.8]. These palpi are slightly more elongate, either broader posteriorly (as in some *Herpetobia* nymphs) or campanulate with a moderate posterior flange, but are not broadly expanded (as in *Ornithophysalis, Elongiphysalis,* and *Rhipistoma*) except in *H. (H.) verticalis,* the Japanese Black Rabbit parasite. Ventral spurs of palpal segment 3 are mostly moderately large, but are small in *H. (H.) megaspinosa,* in which the exceptionally large coxal IV spur apparently compensates for the small palpal spur.

Except for the ventral spur of segment 3 and some posterior broadening, the palpi of the subgenus *Haemaphysalis* are not specialized for mammal-hair hooking—with a single unique exception. The *H. (H.) concinna* palpal segment 3, especially that of the male, is remarkable (Fig. 10.8). The internal margin of each segment 3 is uniquely concave so that the two palpi can surround a hair and function as a hair-grasping device. This host-adaptive feature, as well as the ability of *H. (H.) concinna* immatures to parasitize birds, may contribute to the fact that this species is "the most successful" (most widely distributed and prevalent) in this subgenus, which consists of mostly uncommon and geographically restricted species.

Adult coxal spurs are reduced in the rabbit- and rodent-parasitizing species of the subgenus *Haemaphysalis.* The coxa I spur of the artiodactyl and carnivore parasites is usually quite large, and the II, III, and IV spurs are variable (small to large). However, in both sexes of *H. (H.) megaspinosa,* the IV spurs are very large. In male *H. (H.) flava,* the IV spur is lanceolate as in a few species of *Aboimisalis* and *Herpetobia* and others which we discuss later. The female *H. (H.) flava* IV spur is larger than usual.

The trochanters do not bear ventral spurs in many *Haemaphysalis* species, but do so on leg I in *H. (H.) concinna* (♂), *H. (H.) flava* (♂, ♀), and one or both sexes of the *birmaniae* group: *birmaniae; darjeeling* Hoogstraal and Dhanda; *goral* Hoogstraal; *traubi* Kohls; *roubaudi* Toumanoff; *filippovae* Bolotin.

Possibly owing to the limited range of its palpal and other functional adaptations for coevolution with artiodactyls and carnivores, the subgenus *Haemaphysalis* in itself [except *H.* (*H.*) *concinna*] does not include an outstandingly successful assemblage of *Haemaphysalis* ticks.

H. (*H.*) *silacea* is treated here following the overview of the Oriental and Palearctic members of the subgenus *Haemaphysalis* because of its relict status in the fauna of southeast Africa (Zululand and eastern Cape Province). Structurally and biologically this species does not stand apart from the ordinary members of the subgenus *Haemaphysalis* (Hoogstraal 1963) except that its adults infest a greater variety of hosts than do the Oriental and Palearctic species. Hosts of the *silacea* adults are antelopes, livestock, mongooses, hares, and birds. *H.* (*H.*) *silacea* immature stages are known only from laboratory-reared specimens. Together with the bird-parasitizing *H.* (*Ornithophysalis*) *hoodi* and the bird- and mammal-parasitizing *H.* (*O.*) *tauffliebi*, this is probably the most ancient form of SA haemaphysaline in the Ethiopian region and represents a Gondawanian relict.

Subgenus *Gonixodes* Dugès

The two members of the subgenus *Gonixodes—H.* (*G.*) *leporispalustris* (Packard) (Alaska to Argentina) and *H.* (*G.*) *juxtakochi* Cooley (Mexico to Argentina)—are the only SA haemaphysalines of the Nearctic and Neotropical regions. [Sketches of both species appear in Cooley (1946); the *H. kochi* Arãgao in the Cooley report is a preoccupied name and synonymous with *H. juxtakochi*.] *Gonixodes* apparently evolved from the subgenus *Haemaphysalis* (or a related extinct branch). Both *Gonixodes* species have a typical SA basis capituli. Adult and immature *H.* (*G.*) *leporispalustris* and immature *H.* (*G.*) *juxtakochi* also have unusual ventral cornua, which we first observed in some species of the subgenus *Haemaphysalis*. The campanulate palpi are rather broadly expanded posteriorly in *H.* (*G.*) *leporispalustris* but rather elongate and only slightly expanded in *H.* (*G.*) *juxtakochi*. The palpal segment 3 ventral spur is short and internally directed (bird-parasitizing type) in *H.* (*G.*) *leporispalustris*, whose immatures parasitize birds, leporids, and less often other small mammals, whereas adults are restricted to leporids. These palpal spurs are exceptionally strong in adult *H.* (*G.*) *juxtakochi*, and strong in immatures. Most coxal spurs in this subgenus are rather well developed or quite large [coxa I spur of male and nymphal *H.* (*G.*) *juxtakochi*]. Adult *H.* (*G.*) *juxtakochi* chiefly parasitize Neotropical deer (*Mazama*); immatures primarily parasitize these deer and the agouti (*Dasyprocta*: Rodentia). The great differences in spur development of these two species are obviously associated with the bird feathers and leporid pelage invaded by *H. (G.) leporispalustris* and with the coarser pelage of the deer hosts by *H. (G.) juxtakochi.*

Subgenus *Kaiseriana* Santos Dias

Kaiseriana consists of 33 species [subspecies attributed to *H. (K.) cornigera* are tentatively considered as full species]. Adults are structurally specialized, some extremely, for parasitizing Artiodactyla; most also feed on Carnivora, but other hosts are exceptional. Immatures parasitize small mammals, especially Rodentia, and those of a few species feed on Artiodactyla and Carnivora as well as Rodentia. Only immatures of the biologically distinctive *H. (K.) longicornis* Neumann occasionally parasitize birds. Most *Kaiseriana* species have been illustrated and described or redescribed by Hoogstraal in the *Journal of Parasitology* since 1963; thus to save space, we omit the names of taxon authors and give this subgenus rather less attention than it deserves in relation to coevolution with mammals.

Twenty-seven of the 33 *Kaiseriana* species are Oriental in distribution [three of these (*hystricis, mageshimaensis, yeni*) extend into the eastern Palearctic and one (*bancrofti*) extends into Australia and New Guinea]; two are eastern Palearctic (*ias, longicornis*); one (*novaeguineae*) is Australian-Papuan; and three (*aciculifer, parmata, rugosa*) are Ethiopian.

The three *Kaiseriana* species in Australia-New Guinea require special consideration owing to the absence of native Artiodactyla and Carnivora in this region. *H. (K.) longicornis* (*bispinosa* group) was introduced into Australia with cattle from northeastern Asia within the last 100 years (Hoogstraal et al. 1968). *H. (K.) novaeguineae* (*cornigera* group) of New Guinea and Australia is unknown elsewhere; we suggest that Oriental populations either remain to be discovered (probably in the Indonesian archipelago) or became extinct after *novaeguineae* was introduced with domestic pigs or deer into New Guinea and/or Australia. *H. (K.) bancrofti* (*hylobatis* group) is known only from a single authenticated collection (unpublished) from the Oriental region (vegetation, Java) but is common on livestock, kangaroos, and other marsupials, and birds in certain coastal areas of Australia and New Guinea. Livestock or birds probably introduced *bancrofti* into the Australian region. *H. (K.) bancrofti* is closely related to only one other species, *hylobatis*, a seldom-seen Indonesian-Malaysian species with host data remarkably atypical for the rather strictly host-specific subgenus *Kaiseriana*. Our records of adult *hylobatis* (unpublished) are from wild pigs (*Sus*) (two collections), a Langur monkey (*Presbytis*) (one), the Banded civet (*Arctictis*) (one), a rodent (*Rattus*) (one), a gymnure (*Hylomys*: Insectivora: Erinaceidae) (one), a domestic dog (one), humans (two), a ground-frequenting bird (*Centropus*) (one), and vegetation (eight). Our three collections of immatures are from human (one), *Rattus* (one), and vegetation (one). An atypical host-related biological pattern, like this one of *hylobatis*, may be the reason why the related *bancrofti* is not known outside the

Australian region, except for our single collection from forest vegetation on Java (Oriental region).

The hallmark of *Kaiseriana* is a hair-hooking spur extending from the posterodorsal margin of the adult palpal segment 3 (Hoogstraal et al. 1965) (Fig. 10.8: *lagrangei, shimoga, cuspidata, yeni*). In *cornigera* group males this spur is supplemented or replaced by a gap in the external surface of the palpus (Fig. 10.8: *shimoga*); this efficient hair-grasping gap is formed by the broad posterior expansion and apical recurvature of the movable segment 3 and the anterior narrowing of segment 2. The generally small, often frail *cornigera* group adults have moderately to very large cornua on the basis capituli, an extraordinary variety of hair-hooking devices on the greatly broadened palpi (Fig. 10.8), and pronounced spurs on most or all coxae. The coxa IV spur is lanceolate in *spinigera* (Fig. 10.9), *novaeguineae, aciculifer,* and *rugosa*; double (scissorlike) in *anomala* (Fig. 10.9), *cornigera, shimoga taiwana,* and *ias*; and a combined short and large (lanceolate) spur in *psalistos.* Notably, in the West African *rugosa*, the species most distant from the Oriental origin of the *cornigera* group, the palpi are campanulate, and the posterodorsal margin of palpal segment 3 is uniquely recurved internally rather than medially spurred (Hoogstraal and El Kammah 1972).

Asian adults of the *cornigera* group are chiefly associated with deer (Sambar, Sika, Chital, Timor Deer, Muntjac, etc.), and also with wild bovines (Guar, Anoa, Banteng, Serow, etc.) and wild pigs (*Sus* species). A few adults parasitize wild carnivores and the domestic dog and man. The African members (*aciculifer* and *rugosa*) parasitize numerous artiodactyls (Bushbuck, Sitatunga, Reedbuck, Waterbuck, Kob, Impala, Hartebeest, Oribi, Gazelle, Duiker, Dik-dik, Buffalo, etc.) and infrequently carnivores. Asian and African domestic cattle and other livestock are infested when feeding in the vicinity of wild hosts of *cornigera* group ticks. Thus cattle introduced into previously unspoiled biotopes account for greater population densities of *anomala* in Uttar Pradesh and Himachel Pradesh (Hoogstraal et al. 1967, 1972); of *spinigera*, the chief vector of Kyasanur Forest disease virus (Togaviridae: *Flavivirus*), in Karnataka, India (Hoogstraal 1981); and of *aciculifer* in Africa (unpublished).

Kaiseriana species (*nadchatrami* group) with only moderately broadened campanulate adult palpi (as in the subgenus *Haemaphysalis*), but with a small dorsal spur on segment 3 (signifying the subgenus *Kaiseriana*), appear to have coevolved with wild pigs (*Sus* species), which are still abundant in many Oriental forests. In the *nadchatrami* group (*nadchatrami, semermis, kinneari, papuana, susphilippensis*), cornua and coxal spurs are unenlarged. Adults of this group infest other artiodactyls and carnivores but apparently prosper best where there are populations of pigs. The virtual lack of hair-hooking devices in pig-parasitizing haemaphysalines is correlated with the hosts' sparse hairs, which offer no obstacles to tick movement toward a feeding and mating site.

Other *Kaiseriana* species (*bispinosa* group) have campanulate adult palpi like those of the pig-associated species (*nadchatrami* group), but they have larger dorsal and ventral spurs on segment 3, a distinctively elongate spur on coxa I, fairly large spurs often present on the other coxae, and a distinct ventral spur on trochanter I (Figs. 10.8, 10.9: *lagrangei, cuspidata, yeni*). The *bispinosa* group evolved together with Oriental and eastern Palearctic deer, antelopes, and wild cattle; its species are *bispinosa, ramachandrai, renschi, lagrangei, davisi, luzonensis, longicornis, mageshimaensis, yeni, aculeata, borneata,* and *cuspidata.* Adults of the *bispinosa* group may infest carnivores and other medium to large mammals, but most are recorded from deer. The last three species listed (*aculeata* subgroup) show extreme development of *bispinosa* group characters. Notably, some *bispinosa* group adult spur and capitulum characters are extraordinarily well expressed in certain immatures, especially nymphs. However, these stages lack a dorsal spur on palpal segment 3.

The *bispinosa* group of Palearctic Japan and continental eastern Asia (*yeni, longicornis, mageshimaensis*) is especially interesting biologically. *H. (K.) yeni* extends into southern Japan from Vietnam (Oriental region), the other two are eastern Palearctic (eastern China, Korea, Soviet Far East, and Japan). *H. (K.) longicornis* has normal bisexual populations in the warmer southern part of its range and unique parthenogenetic populations in the colder northern part of its range (the dividing line is roughly at the latitude of Tokyo). The parthenogenetic reproductive pattern persists in *longicornis* that have been introduced into South Pacific islands, eastern Australia, and New Zealand (Hoogstraal et al. 1968). This species is the reservoir and vector of several pathogens of domestic animals and humans in eastern Asia and Japan and has become an important pest of livestock in certain areas. Parthenogenetic strains of *mageshimaensis* have been reared in the laboratory but are not recognized from field data. No other haemaphysalines are known to be parthenogenetic.

The *hystricis* group consists of two species, *hystricis* (Assam to southern Japan and highlands of Malaya and Sumatra) and *celebensis* (Indonesia: Sulawesi). Adults have moderately broad campanulate palpi and moderately large cornua, dorsal and ventral spurs on palpal segment 3 and most coxae and on trochanter I ventrally. The palpi are exceptional in that the posteroventral margin of palpal segment 2 forms a sharp angle near its insertion; the space between this spurlike angle and the palpal base can serve as a hair-hooking device. Hosts are the Sambar, Muntjac, wild pig, various carnivores, and occasionally livestock, dogs, and man, and also (for *celebensis*) the Anoa and Timor Deer.

The *parmata* group consists of a single African forest-dwelling species, *parmata*, which appears to have evolved from a *hystricis*-like progenitor but which lacks coxal spurs. As in *hystricis* group immatures, the *parmata* larval and nymphal coxa I spur is quite large. Adult *parmata* infest various antelopes, the Cape buffalo, and livestock in highlands from the southern Sudan and Ethiopia to Tanzania and into West Africa (Cameroun, Ghana,

Nigeria). Immatures are recorded from rodents, elephant shrews, small carnivores, hares, and birds. The unique absence of coxal spurs in an adult artiodactyl-parasitizing haemaphysaline may be a genetic anomaly associated with the great distance of the Ethiopian *parmata* from the Oriental origin of its progenitors.

Subgenus *Garnhamphysalis* Hoogstraal and Wassef

Garnhamphysalis of humid eastern Oriental forests, apparently a recent branch from prototype stock of the subgenus *Haemaphysalis*, is highly specialized for parasitizing the Sambar, *Cervus unicolor* subspp., other deer (*C. philippinus*), and less often wild pigs and carnivores (bear, civet). Most likely, the poorly known immatures chiefly parasitize rodents. *Garnhamphysalis* species are *calva* Nuttall and Warburton (Thailand, Malay Peninsula, Indonesia: Sumatra, and Borneo), *mjoebergi* Warburton (Borneo, Sumatra), and *rusae* Kohls (Philippines: Mindanao, Luzon) (Hoogstraal and Wassef 1981). In these generally medium-sized haemaphysalines with campanulate palpi, the dental formula is the usual 4/4 in females but 5/5 to 7/7 in males, and the capitular, coxal, and trochantal spur development, especially in males, is diverse and distinctive. For these characters, the large posteroventral spur of the *calva* palpal segment 2 and the large ventral spur of segment 3 are illustrated (Fig. 10.8), as are the elongate spurs of all coxae and of trochanters I to III (Fig. 10.9). The large coxal spurs of average-sized and larger male *mjoebergi* are much reduced in smaller specimens, and the 5/5 (or 6/6) dental formula typical of male *mjoebergi* is reduced to the female formula 4/4 in the smallest specimens (Hoogstraal and Wassef 1982). Small males with reduced spurs and denticle files may be less successful in reaching females attached to the host than average-sized and large males and may contribute less frequently to the species gene pool. The male *calva* and *rusae* resemble each other more closely than do the females of these species, a unique and inexplicable phenomenon in the genus *Haemaphysalis*.

Subgenus *Aborphysalis* Hoogstraal, Dhanda, and El Kammah

Aborphysalis species are *aborensis* Warburton (northeastern India, southern Nepal, Burma, Thailand, Laos, Vietnam); *kyasanurensis* Trapido, Hoogstraal, and Rajagopalan (southern India, Sri Lanka); *formosensis* Neumann (Taiwan, Philippines, southern Japan); *atherurus* Hoogstraal, Trapido, and Kohls (Thailand, Vietnam, Malaya); and *capricornis* Hoogstraal (Thailand) (Hoogstraal, Dhanda, and El Kammah 1971).

Deer and wild pigs are not infrequently recorded hosts of adult *Aborphysalis*, but the more numerous records from porcupines (Rodentia: Hystricidae: *Hystrix, Atherurus, Thecurus*), which are seldom examined for ectoparasites, suggest that these ticks coevolved with porcupines. This postulation is supported by the unusual differences between immature and adult *Aborphysalis* and by certain adult characters. The single collection of *capricornis* with authetic host data is from the serow, *Capricornis sumatraensis swettenhami* (Butler) (Bovidae). Immature *Aborphysalis* are recorded from porcupines, small carnivores, hares, rodents, and birds.

Aborphysalis palpi differ from those of most haemaphysalines except *Segalia* in being broadly expanded in either larvae and nymphs or only in nymphs, and compact or campanulate in adults. In other assemblages with broad immature-stage palpi, immature and adult palpi are equally broad (*Ornithophysalis*, *Elongiphysalis*, *Rhipistoma*). Adult *Aborphysalis* either have very small capitular, coxal, and trochantal spurs as in *capricornis*, or moderate-sized coxal spurs as in the male *atherurus*, or moderately large coxal spurs in both sexes as in *aborensis*, *formosensis*, *kyasanurensis*; and the males also have trochantal spurs ventrally as in *formosensis*. All capitular spurs (cornua and palpal segment 3 ventral spur) are much reduced, and no other hair-hooking devices except the leg spurs are present on these ticks. The capitulum does not function for hair-hooking, but the leg segments may do so (except in *capricornis*) with their sundry spurs, none of which are large. This unusual combination is likely to be associated with the variety of quills, bristles, and soft and hard hairs in different areas of the bodies of the three host genera. With the virtual absence of capitular hair-hooking devices, broad palpi would impede the straightforward progress of the adult haemaphysaline through the harsh pelage of these specialized hosts. The reversal of the usual immature-adult palpal form in *Aborphysalis* is postulated to be a functional adaptation for coexistence with spiny and quilled porcupines.

Subgenus *Segalia* Santos Dias

In a brief key to ixodid subgenera, Santos Dias (1963) proposed *Fonsecaia* for a single species, *H. montgomeryi* Nuttall. Later, Santos Dias (1968) considered *Fonsecaia* to be a *nomen bis lectum* and replaced it with *Segalia*. *H. montgomeryi* and *H. parva* (Neumann 1897), (not Neumann 1908), comprise the *parva* group of the subgenus *Segalia* (Hoogstraal 1983).

As in *Aborphysalis*, immature *Segalia* palpi are broader than those of adults. Immature *parva* and *montgomeryi* palpal segments 2 form large, unique posteroventral spurs and nymphal *montgomeryi* coxal spurs are unusually large for this developmental stage. The adult palpi are campanulate (male) and broadly campanulate (female) in *parva* and elongately campanulate in *montgomeryi*. The adult *parva* palpal segment 3 ventral spur and coxal spurs are only moderately large. However, the adult *montgomeryi* palpal segment 2 posteroventral margin forms an acute spur, and the segment 3 ventral spur is unusually elongate (Fig. 10.8), even more so in the female (unillustrated) than in the male. The male *montgomeryi* coxal and trochantal spurs are all extraordinarily large, and the female coxal spurs are moderately large. The adult dental formula is 6/6 or 7/7.

The close phylogenetic relationship of these two species and yet their highly disparate adults is revealed by the similarity of the distinctive female external genitalia, the immature stages, the campanulate adult palpi, and the comparative palpal forms of immatures and adults. Despite this close relationship, host-adaptive structures are mild or reduced in adult *parva* and extremely developed in adult *montgomeryi*. Notably, such structures are exceptionally well developed in immatures of both species.

The conservatively specialized adult *parva* chiefly parasitizes sheep and goats, but also parasitizes cattle, camels, donkeys, dogs, man, wild hares, and carnivores (badger, bear, fox, jungle cat, etc.). Immatures infest various burrowing rodents, as well as the hedgehog, hare, ground-feeding birds, lizards, and snakes, but rarely attack domestic sheep, goats, or cattle. This species inhabits steppe, semidesert, and eastern Mediterranean-type biotopes from northwestern Iran and southern USSR to Iraq, Syria, Lebanon, Jordan, Israel, Turkey, Bulgaria, Romania, Greece, Italy, and northern Libya. Notably, adults are most active during late fall, winter, and early spring.

The extravagantly specialized adult *montgomeryi* also chiefly parasitizes sheep and goats. There are few records from cattle, buffalo, dogs, or man. Data from two wild Bovidae, the Serow, *Capricornis sumatraensis thar*, and the Goral, *Naemorhedus goral*, are probably representative of the original hosts of adults in the subgenus *Segalia*. Immature *montgomeryi* infest various rodents, shrews, hedgehogs, ground-feeding birds, and small-sized carnivores. This highland species (1750–3800 m altitude) is common in northtern Pakistan and India, Nepal, Sikkim, and China (Tibet).

In retrospect, *parva* group immatures are structurally similar to the most recently evolved *Rhipistoma* immatures. Adults are similar to those of the subgenus *Haemaphysalis*, which evolved early in haemaphysaline SA developmental history, but are exceptionally unspecialized in *parva* and exceptionally specialized in *montgomeryi*. An intensive investigation into the functional anatomy, behavior, and host preferences of this group should be made. Comparative functional and behavioral studies of *Segalia* and *Aborphysalis* would increase our insight into haemaphysaline biology and phylogeny.

The Oriental *obesa* group of the subgenus *Segalia* parallels, in certain respects, the *parva* group. The *obesa* group includes *H. obesa* Larousse (northeastern India to Thailand, Vietnam, Cambodia, and northern Malaysia); *H. hirsuta* Hoogstraal, Trapido, and Kohls (Java, Sumatra, Sumbawa); and *H. sumatraensis* Hoogstraal, El Kammah, Kadarsan, and Anastos (Java, Sumatra). Immatures have broadly campanulate palpi with a moderately large ventral spur on segment 3 (but lacking the segment 2 ventral spur characterizing the *parva* group) and a moderately large spur on coxa I. Adult palpi are compact or compact-elongate and have a moderately large ventral spur on segment 3. Adult cornua are small or moderate sized; however, immature cornua are relatively large (except larval *hirsuta*). (Female *H. sumatraensis* external genitalia are not typical of this group; this species possibly should be categorized elsewhere.)

The *obesa* group adults were taken from large mammals, small mammals were virtually never examined for ticks in forests. Since adults and immatures were collected by sweeping from vegetation, we are uncertain of the host associations of immatures of this group.

Our 21 collections of adult and immature *H. obesa* are mostly from dense vegetation in humid forests where game animals, including the elephant, abound. Adult *H. obesa* were taken from the wild pig, tapir, Muntjac, Asiatic Black Bear, and Capped Monkey (*Presbytis*). Immatures were from the Ferret-badger (*Melogale personata*), Red Dog (*Cuon alpinus*), and two humans.

Our numerous adult and immature *H. hirsuta* are also mostly from forest vegetations (91 collections). Six other collections of adults and nine of immatures are from humans. Other adults have been taken from a domestic dog (one collection), pig (one), buffalo (two), and sheep (two), and from the wild pig (five), and Banting (*Bos javanicus*) (two). Immatures are from the Sambar Deer (*Cervus unicolor equinus*) (one) and Palm Civet (*Paradoxurus hermaphroditus*) (one).

Adult *H. sumatraensis* collections are from the Sambar Deer (ten collections), the Timor Deer (*C. timorensis*) (one), "*Cervus sp.*" (two), the tiger (two), the wild pig (two), the domestic dog (one), and forest vegetation (two). Immatures are recorded from the Palm Civet (three) and vegetation (one). Most hosts were slaughtered and examined for ticks during an extensive forest felling program in Lampung District, southern Sumatra.

Subgenus *Dermaphysalis* Hoogstraal, Uilenberg, and Klein

The monotypic subgenus *Dermaphysalis* is represented by a single male *H. (D.) nesomys* taken from an endemic rodent, *Nesomys rufus* Peters, in an eastern Madagascar mountain forest. Possibly branched from *Rhipistoma*, this minute (1.8 mm long, 1.0 mm broad) leathery, essentially spurless parasite with campanulate palpi and hypostome denticles reduced to knobs, is an extreme example of the reduction and modification of haemaphysaline (and other ixodid) characters in association with a small, soft-furred rodent. The historical isolation of this host and parasite in Madagascar and the role of this isolation in the developmental processes of the structurally singular *H. (D.) nesomys* should be considered.

Subgenus *Rhipistoma* Koch

Rhipistoma, probably the most recent haemaphysaline subgeneric assemblage, shows signs of contemporary speciating associated with small- and medium-sized Carnivora in the Oriental, Ethiopian, and Malagasy regions. Adults of 13 of the 27 *Rhipistoma* species are exclusive parasites of carnivores. Adults of the other 14 species are specialized for parasitizing hedgehogs (one species), lemurs (one species), hyraxes (four species), flying squirrels (one species), springhares (one species), ground squirrels (three species; one of these also carnivores), leporids (two species; one also carnivores), or hedgehogs and carnivores (one species with three or four subspecies).

All *Rhipistoma* immature-stage and adult palpi are broadly expanded posteriorly; many are modified by spurs or grooves as hair-hooking devices. The basis capituli has small to large cornua. The capitulum (basis capituli and palpi), not leg spurs, assists *Rhipistoma* movement among the host pelage. Coxal and trochantal spurs are unspecialized, generally small, or even obsolete.

The subgenus *Rhipistoma* was defined by Hoogstraal et al. (1963) and is illustrated herein (Fig. 10.5: *H. erinacei, H. leachi*; Fig. 10.8: *H. asiatica, H. fossae, H. leachi*). Five groups of species constitute the subgenus.

The *erinacei* group of the subgenus *Rhipistoma* consists of a single southern Palearctic species with three or four subspecies: *erinacei*, northern Africa, southeastern Italy to Bulgaria; *taurica*, Israel to the USSR and Afghanistan; *turanica*, Arabia to Pakistan and southern USSR; *ornata*, Israel— badly described, poorly known, status uncertain. Despite their structural simplicity (except for broadly triangular palpi), *erinacei* group ticks thrive in numerous semidesert and eastern Mediterranean-Turanian environments where hedgehogs and carnivores exist.

All *erinacei* group palpi are broadly expanded but otherwise simple; none have marginal spurs or grooves. Adult palpi have a strong ventral spur from segment 3; this spur is obsolete (or virtually so) on immature-stage palpi. As is generally true of carnivore- and hedgehog-parasitizing haemaphysalines, the body is narrowly elongate. Immatures parasitize rodents or hedgehogs. Adults are frequently found on hedgehogs and also on foxes and other carnivores, whose ears are their favored feeding sites. Hedgehogs may be parasitized by adults in the absence of carnivores, or may in fact be the principal hosts for all or some "strains" of *H. erinacei* subspp. This hypothesis should be investigated experimentally.

The structurally and biologically complex *asiatica* group of the subgenus *Rhipistoma* (Hoogstraal and Morel 1970) consists of 14 species in five subgroups which parasitize a variety of mammals in four regions of the Old World. Each of these 14 species deserves more intensive study.

The *H. (R.) asiatica* group probably evolved from a common ancestor in the Oriental region or from closely related ancestors in the Oriental and Palearctic regions. If it evolved from more than a single ancestral type, one probably gave rise to the *asiatica* subgroup of viverrid parasites in the Oriental region and also spread to the Malagasy region. The other, morphologically on the order of the *caucasica* subgroup, arose in the southwestern Palearctic (*caucasica* subgroup—leporid hosts), and spread to the Ethiopian region (*orientalis* subgroup—rock hyrax hosts; *calcarata* subgroup—ground squirrel hosts).

In the *asiatica* subgroup of the subgenus *Rhipistoma*, five species parasitize mostly viverrid carnivores (also felids and canids) in Madagascar (*fossae, eupleres, obtusa*), southern China and Burma to Borneo and Sumatra (*asiatica*), and the Near East including Lebanon, Israel, Iraq, and Oman (*adleri*). Immatures, where known, feed on rodents or insectivores. The broad palpi of adults bear luxurient spur or groove devices for hair grasping (Fig. 10.8). Immature-stage palpi of some species are also strongly spurred but in other species are merely broadened. As is typical of *Rhipistoma*, all leg spurs are small or obsolete.

The *lemuris* subgroup of *Rhipistoma* is represented by H. (R.) *lemuris*; adults and immatures of this group parasitize at least nine species of Malagasy lemurs. This broad, heavy-bodied haemaphysaline has uniquely tilted palpi forming a proximal hair-grasping groove which is supplemented by a large ventral spur on segment 3. Cornua are obsolete on the adult basis capituli but present on those of nymphs (which lack the palpal hair-grasping groove of adults). The unusually broad spur on coxa I points to a significant difference in the functional anatomy between H. (R.) *lemuris* and carnivore-parasitizing *Rhipistoma* species.

In the *caucasica* subgroup, H. (R.) *caucasica* adults infest *Lepus* spp. in southwestern USSR and northern Iran; H. (R.) *hispanica* adults infest *Oryctolagus cuniculus* in Spain and France (Camargue). Both species occasionally parasitize ground-feeding birds such as partridges; and *caucasica* is also recorded from the jackal. Immatures are poorly known, they feed either on the same host as adults or on rodents or hedgehogs. The capitulum pattern of H. (R.) *caucasica* is more conservative than that of H. (R.) *lemuris*, including moderately large to large cornua, and adult palpal properties of each species show certain somewhat mild modifications similar to those of H. (R.) *lemuris*. The coxal spurs are generally small.

The *orientalis* subgroup of *Rhipistoma* consists of four species parasitizing the Rock Hyrax, *Procavia capensis*, of eastern Africa: *bequaerti*, Sudan to Kenya; *orientalis*, Malawi to Mozambique; *hyracophila* and *cooleyi*, South Africa (Hoogstraal and Wassef 1981). The greatly broadened nymphal palpi are notable for an exceptionally strong spurlike angle on the posteroventral margin of segment 2, but are otherwise quite simple. The adult structures are notable because they are reduced or unusually modified hair-hooking spurs and angles on the capitulum and coxae. These properties reflect the close, "easy" association between immatures and adults of the *orientalis* subgroup and their den-inhabiting, colonial, slowly moving, soft-furred hosts. Host specificity is further demonstrated by the relatively numerous incidental records of *orientalis* subgroup specimens from the tree hyrax, *Dendrohyrax* spp., which has a much different life-style.

The absence of *orientalis* subgroup records from rock hyrax populations of West Africa, where different endemic ticks parasitize these animals, is difficult to explain. The northeastern African-Near East habitats of rock hyraxes are probably too dry for *Haemaphysalis* survival.

The 13 species constituting the *canestrinii* group (five described species) of the Oriental region and the closely related *leachi* group (eight described species) of the Ethiopian region are carnivore parasites, except for one species from Indonesia and two "rodent and carnivore" species from southern Africa. The *leachi* group is postulated to have been derived from a *canestrinii*-like progenitor (Hoogstraal 1971).

Most adults in both groups have an elongate body, strong cornua in males and moderate in females, strong posterior spurs from one or both posterior margins of the broad palpal segment 2, a strong ventral spur from palpal segment 3, small to moderate-sized coxal spurs, and no ventral spurs on the trochanters. The capitular spurs are more or less reduced in one carnivore-parasitizing species and in the three species that have reverted to parasitizing specialized burrow- or treehole-dwelling Rodentia.

The *canestrinii* subgroup (*canestrinii* group) consists only of *H. (R.) canestrinii* from eastern Pakistan, India (Madhya Pradesh, Assam), Nepal, Burma, northern Thailand, Vietnam, and Taiwan. Adults are recorded chiefly from Viverridae (civets, mongooses), but have also been recorded from Mustelidae (hog badger, ferret badger), Canidae (jackal, fox), and Felidae (jungle cat, fishing cat, tiger, leopard), with single records from a hare and a jungle fowl. Immatures are recorded from murid and sciurid rodents, civets, and a tree shrew. Our samples of adults from these various carnivore and other hosts from Pakistan to Taiwan are remarkably uniform structurally (Hoogstraal 1971).

The *indica* subgroup (*canestrinii* group) consists of four species, one of which is polytypic and may actually be two or more species (or subspecies) classified as a single taxon. *H. (R.) indica* ranges from Nepal, India, and Sri Lanka (Oriental region) westward into Pakistan, Iran, and Oman (Palearctic region). Hosts of adults are chiefly mongooses (Viverridae); other hosts are the genet, civet, jackal, fox, ratel, leopard, wild cat, and domestic dog. There are host records of immature stages from the wild cat, wild pig, and hare, and rodents, shrews, and hedgehogs. The Oriental-Palearctic distribution pattern of *H. (R.) indica* is unusual if not unique.

H. (R.) koningsbergeri of the *indica* subgroup is found in Southeast Asia (southern Thailand, Vietnam, Malaya), Borneo, and Indonesia (Java and Sumatra), where adults parasitize the kinds of hosts listed for *H. (R.) indica* and occasionally others (mouse deer, wild pig, tapir, etc.). The capitular spurs of adult *H. (R.) koningsbergeri* are mild in comparison to those of most other carnivore-parasitizing members of the subgenus *Rhipistoma*. Nevertheless, this parasite is quite successful in a wide area of Southeast Asia. The coxal spurs are in the lower range of moderate sized.

H. (R.) bartelsi of the *indica* subgroup parasitizes Javan treehole-dwelling flying squirrels, *Petaurista* spp. Adults appear to be miniatures of *H. (R.)*

koningsbergeri, but the larvae are quite dissimilar. As might be expected, the larval palpal segment 2 posterior margins are unspurred in *H. (R.) bartelsi* but strongly spurred (ventrally) in *H. (R.) koningsbergeri*. Spurs of adult *H. (R.) bartelsi* are even weaker than those of *H. (R.) koningsbergeri* and resemble those of *H. (R.) pedetes* and *H. (R.) zumpti* (*leachi* group).

H. (R.) heinrichi (*indica* subgroup) is a polytypic species, or possibly two or more species or subspecies—one in India, the other from Burma, southern China and Thailand to Laos and Vietnam. (More extensive field samples and laboratory investigations are required to solve this problem.) Our records from India include adults and immatures from the fox, *Vulpes gengalensis*, and immatures from *Rattus*. Elsewhere, most collections of adults and immatures are from the ferret-badger, *Melogale personata* (and *M. moschata* from China) (Mustelidae), with some adults from the hog badger, *Arctonyx collaris*, and the domestic dog, and immatures from different species of *Rattus*, civets, mongooses, and the tree shrew. *H. (R.) heinrichi* is a rather small haemaphysaline (males about 2 mm long) with strong capitular spurs and strong (for the subgenus *Rhipistoma*) spurs on coxae I and II. *H. (R.) heinrichi* represents a unique problem in tick speciation and coevolution with carnivores, but there is little or no chance in the near future to obtain materials and data to answer the numerous questions relating to this taxon.

The Ethiopian *leachi* group has long been exceptionally difficult to understand biologically and taxonomically. However, since clarifying the status of *H. (R.) leachi* (Hoogstraal 1958), it has been possible to recognize related taxa and to describe six other species in this group (at least three others not mentioned herein are in preparation for publication). On completing this series of a dozen reports from a 25-year study, the way is clear for a profound investigation of *Haemaphysalis*-carnivore coevolution in Africa.

The *leachi* subgroup (*leachi* group) consists of *H. (R.) leachi* (Egypt, Nile Valley to South Africa); *H. (R.) punctaleachi* (eastern Uganda to Liberia); *H. (R.) moreli* (Tanzania and Ethiopia to Senegal); and *H. (R.) paraleachi* (West Africa); followed by an undescribed species from the Sudan to South Africa. The hosts of adult *H. (R.) leachi* are any wild and domestic carnivores, and as far as we know they have no marked preference for any species, genus, or family (however, this question should be investigated more precisely). Immature *H. (R.) leachi* infest the common burrowing rodents of the African savannas and occasionally hedgehogs or small carnivores that share either a rodent burrow or have preyed on an infested rodent. The broad palpi of adult and immature *H. (R.) leachi* and related African species are much like those of the Oriental *H. (R.) canestrinii*; structural differences between species are relatively slight but easily recognizable. Immatures have bold spurs from palpal segment 3, either both dorsally and ventrally or only ventrally.

The host patterns of adults and immatures of *H. (R.) paraleachi* and of the undescribed African species are similar to those of *H. (R.) leachi*. However, our data for the West African *H. (R.) punctaleachi* are chiefly from forest-dwelling civets (Viverridae), as well as the genet, leopard, jackal, domestic dog, and (one record) antelope. The West and East African *H. (R.) moreli* also appears to be chiefly associated with civets and genets only in savanna environments, and it also feeds on other wild and domestic carnivores. Immature *H. (R.) punctaleachi* are unknown; those of *H. (R.) moreli* are from burrowing rodents.

The *spinulosa* subgroup of the *leachi* group remains under study. *H. (R.) spinulosa* (and closely related undescribed species) feed on most carnivore populations of Africa; the immatures of some of these ticks parasitize the same hosts as adults do. *H. (R.) norvali* is a hedgehog parasite of Zimbabwe.

The *pedetes* subgroup (*leachi* group) consists of two very mildly spurred species associated with specialized colonial rodents in southern Africa and is frequently found on small-sized carnivores coexisting with or preying on these rodents. *H. (R.) zumpti* (eastern South Africa, Basutoland, Zambia, Malawi) inhabits the extensive burrow systems excavated by ground squirrels (*Xerus*), and also inhabited by the yellow mongoose, *Cynictis penicillata*, and at least seven other species of genets, mongooses, and other small-sized carnivores. The closely related *H. (R.) pedetes* parasitizes the springhare, *Pedetes capensis* (Rodentia: Pedetidae), in Transvaal. A few records of adult *H. (R.) pedetes* are from carnivores of the genera *Felis*, *Ictonyx*, and *Herpestes*. Immatures of this species are unknown.

SUMMARY

The ticks of the superfamily Ixodoidea are postulated to have evolved as obligate parasites of Reptilia in the warm, humid climate of the late Paleozoic or early Mesozoic era. During this period, their basic physiological patterns were established, as was the multihost developmental pattern of the Argasidae and the three-host developmental pattern of the Ixodidae.

In the family Argasidae, structural clues to coevolution with post-Paleozoic hosts are obscured by the dominating structural and physiological adaptations of these ticks for preserving a favorable water balance in inclement environments and for survival despite long intervals between bloodmeals. Nevertheless, numerous species of *Argas*, *Ornithodoros*, *Antricola*, and *Nothoaspis* have coevolved with cave-dwelling bats. Other *Ornithodoros* species are associated with certain groups of burrowing or den-inhabiting mammals, fewer with marine birds, tree-hole nesting birds, and reptiles. *Argas* species that do not parasitize bats are specific parasites of birds, a few cave- or den-inhabiting mammals, or the giant tortoise.

In the family Ixodidae, many species of the subfamily Amblyomminae (i.e., 22 of the 24 species of *Aponomma*; 37 of the 102 species of *Amblyomma*) parasitize contemporary reptiles. When Aves and Mammalia replaced Reptilia as dominant vertebrates in the Tertiary period, these 59 species retained most of the properties of earlier reptile parasites. However, most of these species are probably smaller than Mesozoic Amblyomminae and the *Amblyomma* may have somewhat more distinct eyes. The two contemporary *Aponomma* species not associated with reptiles have adapted to the echidna (Monotremata) and the wombat (Marsupilia); 64 of the 102 *Amblyomma* species have adapted to mammals or (immatures) to birds and/or mammals. One *Amblyomma* species is chiefly recorded from marine birds, but its structure recalls that of *Amblyomma* species of marine and land iguanas and the giant tortoise.

The subfamilies Ixodinae (*Ixodes*, ± 217 species), Haemaphysalinae (*Haemaphysalis*, 155 species), and Hyalomminae (*Hyalomma*, 30 species) were well established before the Tertiary period. Almost all recent species of these three subfamilies have adapted to mammals, birds, or birds and mammals. Adults and/or immatures of a few species also parasitize reptiles. Only single species in the genera *Haemaphysalis* and *Hyalomma* appear to depend for survival chiefly on lizards and tortoises, respectively.

The Rhipicephalinae evolved much later, probably in the Cretaceous period (when reptile lines were reduced to the Rhynchosauria and ancestors of modern birds and mammals appeared) or in the early Tertiary. (Hyalomminae probably also evolved within this general time frame.) The approximately 115 contemporary Rhipicephalinae species are confined to the Palearctic, Oriental, and Ethiopian regions. There are no indigenous Rhipicephalinae in Madagascar, Australia or New Guinea; and only a few *Dermacentor* spp. in the New World. Rhipicephalinae genera are *Dermacentor* (30 species), *Rhipicentor* (two species), *Boophilus* (five species, and *Margaropus* (three species). Species in these genera are virtually never associated with reptiles or birds.

Most adult Rhipicephalinae are only or chiefly associated with Artiodactyla, and fewer are also or chiefly with Carnivora and/or Perissodactyla. Exceptions are *Anomalohimalaya*, which have reverted to parasitizing arvicolid rodents (*Alticola* spp.) at high altitudes in the Himalaya and outlying mountain ranges, and a few of the 70 species of *Rhipicephalus* that parasitize hyraxes or certain rodents or leporids. Immature Rhipicephalinae chiefly parasitize rodents; some also infest insectivores and/ or leporids, or carnivores. In the genus *Rhipicephalus*, immatures of a few highly specialized species feed together with adults on Artiodactyla or Perissodactyla.

The genus *Haemaphysalis* (155 species) is the most useful assemblage in the superfamily Ixodoidea for displaying interrelated structural-biological clues to affinities between species and species groups and to historical and contemporary host associations. In a group of 17 structurally primitive species, some having *Aponomma*-like "Mesozoic-type" characters, there is a stepwise progression, from structurally primitive (SP) to structurally intermediate (SI) to structurally advanced (SA) species, that reflects stages in *Haemaphysalis* coevolution with birds and various groups of mammals. This stepwise progression, from atypical (SP) to typical (SA) haemaphysaline structure, is exhibited from species to species, as well as from larva to nymph to adult, and between males and females of individual species. Among the 17 SP species, adults of Mesozoic-type relicts chiefly parasitize Artiodactyla, mostly in Himalayan and outlying Palearctic and Oriental alpine and subalpine zones. These relict species either lack pronounced coxal spurs (as did Mesozoic reptile parasites) or have bizarre, sometimes exceptionally large coxal spurs (hair-hooking devices) to assist movement through the dense heavy pelage of the artiodactyl host. [On the capitulum, specialized hair-hooking devices developed only after the narrowly elongate SP palpi changed to compact or posteriorly broadened (SA type).] The primeval SP species are the largest in the genus; the dominating trend toward size reduction operates simultaneously with palpal broadening.

Mammal-parasitizing haemaphysalines have a variety of hair-hooking devices—spurs on the coxal and trochanter spurs, basis capituli spurs (cornua), and palpal spurs and marginal indentations or slits—to assist the small tick in reaching a feeding site or a mate on the hairy host. These devices are especially luxuriant in smaller-sized (physically weaker) species and in males, which are smaller than females and are promiscuous mate-seekers, moving about on the host more than females do. In bird-parasitizing haemaphysalines there is little or no development of these devices except (atypically) in minute-sized species. Tenrec-parasitizing species that feed among the spines and harsh hairs of the body have luxuriant capitular, coxal, and trochanter spurs, but those species that feed in the glabrous ears are virtually spurless (like bird-infesting species).

Patterns of SP and SA development and of hair-hooking devices generally furnish valuable clues to haemaphysaline–mammal coevolution. In other tick genera which have fewer species, no contemporary SP representatives, and a larger size (reducing the need for hooking-devices), clues to coevolution are chiefly derived from biological data and seldom if ever from structural data.

REFERENCES

Aeschlimann, A., P. A. Diehl, G. Eichenberger, R. Immler, and N. Weiss. 1968. Les tiques (Ixodoidea) des animaux domestiques au Tessin. *Rev. Suisse Zool.* **75**:1039–1050.

Cooley, R. A. 1946. The genera *Boophilus, Rhipicephalus*, and *Haemaphysalis* (Ixodidae) of the New World. *Bull. Natl. Inst. Health* **187**:1–54.

Darlington, P. J., Jr. 1957. *Zoogeography: The geographical distribution of animals.* John Wiley, New York. 675 pp.

Dhanda, V. and H. R. Bhat. 1971. *Haemaphysalis (Herpetobia) sundrai* Sharif (Ixodoidea: Ixodidae), a tick parasitizing sheep in the western Himalayas, redescription of female, description of male, and ecological observations. *J. Parasitol.* **57**:646–650.

Dhanda, V. and S. M. Kulkarni. 1969. Immature stages of *Haemaphysalis cornupunctata* Hoogstraal and Varma, 1962 (Acarina: Ixodidae) with new host and locality records, and notes on its ecology. *Orient. Insects* **3**:15–21.

Doss, M. A. and G. Anastos. 1977. *Index—Catalogue of Medical and Veterinary Zoology.* Spec. Publ. No. 3, Ticks and tickborne diseases. III. Checklist of families, genera, species, and subspecies of ticks. U. Md/USDA, Washington, D.C.

Doss, M. A., M. M. Farr, K. F. Roach, and G. Anastos. 1974. *Index—Catalogue of Medical and Veterinary Zoology.* Spec. Publ. No. 3. Ticks and Tickborne diseases. I. Genera and Species of Ticks. Part 1–3. U. Md/USDA, Washington, D.C.

Eisenberg, J. F. and E. Gould. 1970. *The Tenrecs: A Study in Mammalian Behavior and Evolution.* Smithson. Contr. Zool. **27**:1–138, Washington, D.C.

Emel'yanova, N. D. and H. Hoogstraal. 1973. *Haemaphysalis verticalis* Itagaki, Noda, and Yamaguchi: Rediscovery in China, adult and immature identity, rodent hosts, distribution and medical relationships (Ixodoidea: Ixodidae). *J. Parasitol.,* **59**:724–733.

Grebenyuk, R. V. 1966. *Ixodid ticks (Parasitiformes, Ixodidae) of Kirgizia.* (In Russian.) *Akad. Nauk Kirgiz. SSR,* Inst. Biol., Frunze. 328 pp.

Gregson, J. D. 1956. *The Ixodoidea of Canada.* Publ. Dep. Agric. Canada, (930). 92 pp.

Hoogstraal, H. 1953. Ticks (Ixodoidea) of the Malagasy faunal region (excepting the Seychelles). Their origins and host-relationships; with descriptions of five new *Haemaphysalis* species. *Bull. Mus. Comp. Zool. Harv.,* **111**:37–113.

Hoogstraal, H. 1958. Notes on African *Haemaphysalis* ticks. IV. Description of Egyptian populations of the yellow dog-tick, *H. leachii leachii* (Audouin 1827), (Ixodoidea, Ixodidae). *J. Parasitol.* **44**:548–558.

Hoogstraal, H. 1963. Notes on African *Haemaphysalis* ticks. V. Redescription and relationships of *H. silacea* Robinson, 1912, from South Africa (Ixodoidea, Ixodidae). *J. Parasitol.* **49**: 830–837.

Hoogstraal, H. 1965. Phylogeny of *Haemaphysalis* ticks. *Proc. 12th Int. Congr. Entomol.* (London, July 1964), pp. 760–761.

Hoogstraal, H. 1971. *Haemaphysalis (Allophysalis) warburtoni* Nuttall: Description of immature stages, adult structural variation, and hosts and ecology in Nepal, with a redefinition of the subgenus *Allophysalis* Hoogstraal (Ixodoidea: Ixodidae). *J. Parasitol.* **57**:1083–1095.

Hoogstraal, H. 1973. Redescription of the type material of *Haemaphysalis (Aboimisalis) cinnabarina* (revalidated) and its junior synonym *H. (A.) sanguinolenta* described by Koch in 1844 from Brazil (Ixodoidea: Ixodidae). *J. Parasitol.* **59**:379–383.

Hoogstraal, H. 1978. Biology of ticks. Pages 3–14 in J. K. H. Wilde (Eds.). *Tick-borne Diseases and Their Vectors.* Proc. Int. Conf. (Edinburgh, September–October 1976).

Hoogstraal, H. 1981. Changing patterns of tickborne diseases in modern society. *Annu. Rev. Entomol.* **26**:75–99.

Hoogstraal, H. 1985. Argasid and Nuttalliellid ticks as parasites and vectors. *Adv. Parasitol.* **24**: (in press).

Hoogstraal, H. and A. Aeschlimann. 1982. Tick–host specificity. *Bull. Soc. Entomol. Suisse* **55**:5–32.

Hoogstraal, H. and K. M. El Kammah. 1972. Notes on African *Haemaphysalis* ticks. X. *H. (Kaiseriana) aciculifer* Warburton and *H. (K.) rugosa* Santos Dias, the African representatives of the *spinigera* subgroup (Ixodoidea: Ixodidae). *J. Parasitol.* **58**:960–978.

Hoogstraal, H. and V. C. McCarthy. 1965. Hosts and distribution of *Haemaphysalis kashmirensis* with descriptions of immature stages and definition of the subgenus *Herpetobia* Canestrini (resurrected). *J. Parasitol.* **51**:674–679.

Hoogstraal, H. and R. M. Mitchell. 1971. *Haemaphysalis (Alloceraea) aponommoides* Warburton (Ixodoidea: Ixodidae), description of immature stages, hosts, distribution, and ecology in India, Nepal, Sikkim, and China. *J. Parasitol.* **57**:635–645.

Hoogstraal, H. and P. C. Morel. 1970. *Haemaphysalis (Rhipistoma) hispanica* Gil Collado, a parasite of the European rabbit, redescription of adults, and description of immature stages (Ixodoidea: Ixodidae). *J. Parasitol.* **56**:813–822.

Hoogstraal, H. and R. Valdez. 1980. Ticks (Ixodoidea) from wild sheep and goats in Iran and medical and veterinary implications. *Fieldiana: Zool.*, n.s., **6**:1–16.

Hoogstraal, H. and H. Y. Wassef. 1973. The *Haemaphysalis* ticks (Ixodoidea: Ixodidae) of birds. 3. *H. (Ornithophysalis)* subgen. n.: Definition, species, hosts, and distribution in the Oriental, Palearctic, Malagasy, and Ethiopian faunal regions. *J. Parasitol.* **59**:1099–1117.

Hoogstraal, H. and H. Y. Wassef. 1979. *Haemaphysalis (Allophysalis) kopedaghica*: Identity and discovery of each feeding stage on the wild goat in northern Iran (Ixodoidea: Ixodidae). *J. Parasitol.* **65**:783–790.

Hoogstraal, H. and H. Y. Wassef. 1981. *Haemaphysalis (Garnhamphysalis)* subgen. nov. (Acarina: Ixodidae): Candidate tick vectors of hematozoa in the Oriental region. *Parasitol. Trop. Soc. Protozool.* (Spec. Publ. 1): 117–124.

Hoogstraal, H. and H. Y. Wassef. 1982. *Haemaphysalis (Garnhamphysalis) mjoebergi*: Identity, structural variation and biosystematic implications, deer hosts, and distribution in Borneo and Sumatra (Ixodoidea: Ixodidae). *J. Parasitol.* **68**:138–144.

Hoogstraal, H. and N. Wilson. 1966. Studies on Southeast Asian *Haemaphysalis* ticks (Ixodoidea, Ixodidae). *H. (Alloceraea) vietnamensis* sp. n., the first structurally primitive haemaphysalid recorded from southern Asia. *J. Parasitol.* **52**:614–617.

Hoogstraal, H. and N. Yamaguti. 1970. *Haemaphysalis (H.) pentalagi* Pospelova-Shtrom, a parasite of the Japanese black rabbit: Redescription of the male and description of the female, nymph, and larva (Ixodoidea: Ixodidae). *J. Parasitol.* **56**:367–374.

Hoogstraal, H., V. Dhanda, and H. R. Bhat. 1972. *Haemaphysalis (Kaiseriana) anomala* Warburton (Ixodoidea: Ixodidae) from India: Description of immature stages and biological observations. *J. Parasitol.* **58**:605–610.

Hoogstraal, H. and V. Dhanda, and K. M. El Kammah. 1971. *Aborphysalis*, a new subgenus of Asian *Haemaphysalis* ticks; and identity, distribution, and hosts of *H. aborensis* Warburton (resurrected) (Ixodoidea: Ixodidae). *J. Parasitol.* **57**:748–760.

Hoogstraal, H., G. M. Kohls, and H. Trapido. 1967. Studies on Southeast Asian *Haemaphysalis* ticks (Ixodoidea, Ixodidae). *H. (Kaiseriana) anomala* Warburton: Redescription, hosts and distribution. *J. Parasitol.* **53**:196–201.

Hoogstraal, H., H. Trapido, and G. M. Kohls. 1965. Southeast Asian *Haemaphysalis* ticks (Ixodoidea, Ixodidae). *H. (Kaiseriana) papuana nadchatrami* spp. n. and redescription of *H. (K.) semermis* Neumann. *J. Parasitol.* **51**:433–451.

Hoogstraal, H., H. Y. Wassef, and G. Uilenberg. 1974. *Haemaphysalis (Elongiphysalis) elongata* Neumann subgen. n. (Ixodoidea: Ixodidae): Structural variation, hosts, and distribution in Madagascar. *J. Parasitol.* **60**:480–498.

Hoogstraal, H., F. H. S. Roberts, G. M. Kohls, and V. J. Tipton. 1968. Review of *Haemaphysalis (Kaiseriana) longicornis* Neumann (resurrected) of Australia, New Zealand, New Caledonia, Fiji, Japan, Korea, and northeastern China and USSR, and its parthenogenetic and bisexual populations (Ixodoidea, Ixodidae). *J. Parasitol.* **54**:1197–1213.

Hoogstraal, H., M. N. Kaiser, M. A. Traylor, E. Guindy, and S. Gaber. 1963. Ticks (Ixodidae) on birds migrating from Europe and Asia to Africa, 1959–1961. *Bull. W.H.O.* **28**:235–262.

Hoogstraal, H., M. A. Traylor, S. Gaber, G. Malakatis, E. Guindy, and I. Helmy. 1964. Ticks (Ixodidae) on migrating birds in Egypt, spring and fall 1962. *Bull. W.H.O.* **30**:355–367.

Horton, D. R. 1973. Primitive species. *Syst. Zool.*, **22**:330–333.

Kitaoka, S. and K. Fujisaki. 1972. On the winter-active ticks, *Haemaphysalis kitaokai* Hoogstraal, 1969 and *H. megaspinosa* Saito, 1969. (In Japanese.) *Jap. J. Vet. Sci.* **34** (Supp.):173–174. (In English: NAMRU3-T635.)

Kitaoka, S. and T. Morii. 1967. The biology of *Haemaphysalis (Alloceraea) ambigua* Neumann,

1901 with description of immature stages (Ixodoidea, Ixodidae). *Natl. Inst. Anim. Health Q.*, Tokyo, **7**:145–152.

Kohls, G. M. 1960. Records and new sysonymy of New World *Haemaphysalis* ticks, with descriptions of the nympha and larva of *H. juxtakochi* Cooley. *J. Parasitol.*, **46**:355–361.

Macicka, O. 1958. On the bionomics of *Haemaphysalis inermis* in our home country. (In Slovak.) *Cslka Parasit.* **5**:121–124. (In English: NAMRU3-T54.)

Nosek, J. 1971. The ecology, bionomics, and behavior of *Haemaphysalis* (*Aboimisalis*) *punctata* tick in Central Europe. *Z. Parasitenk.*, **37**:198–210.

Nosek, J. 1973. Some characteristic features of the life history, ecology and behaviour of the ticks *Haemaphysalis inermis*, *H. concinna*, and *H. punctata*. *Proc. 3rd Int. Congr. Acarol.* (Prague, August 31–September 6, 1971), pp. 479–482.

Nuttall, G. H. F. and C. Warburton. 1915. *Ticks. A monograph of the Ixodoidea*. Part III. The genus *Haemaphysalis*. Cambridge University Press, London, pp. 349–550.

Ogandzhanian, A. M. and B. A. Martirosian. 1965. Occurrence of the tick *Haemaphysalis warburtoni* Nutt. (Acarina: Ixodidae) in Armenia. (In Russian; Armenian summary.) *Izv. Akad. Nauk Armyan. SSR, s. Biol. Nauk*, **18**:69–72. (In English: NAMRU3-T206).

Oliver, J. H., Jr., K. Tankaka, and M. Sawada. 1974. Cytogenetics of ticks (Acari: Ixodoidea). 14. Chromosomes of nine species of Asian haemaphysalines. *Chromosoma* **45**:445–456.

Rageau, J. 1973. Repartition geographique et role pathogene des tiques (Acariens: Argasidae et Ixodidae) en France. (Mater. II. Symp. Med. Vet. Acaroent., Gdansk, October 21–23, 1971). (Polish summary.) *Wiad. Parazyt.* **18**:707–718; disc. p. 719 (1972).

Roberts, F. H. S. 1970. *Australian ticks.* Commonwealth Scientific and Industrial Research Organization, Melbourne, Australia. 267 pp.

Sacca, G., M. L. Mastrilli, M. Balducci, P. Verani, and M. C. Lopes. 1969. Studies on the vectors of arthropod-borne viruses in central Italy: Investigations on ticks. *Ann. 1st. Sup. Sanita* **5**:21–28, plate I.

Santos Dias, J. A. T. 1963. Sobre a validade do subgenero *Delphyiella* Travassos Dias, 1955 (Acarina-Ixodoidea). *Mem. Estud. Mus. Zool. Univ. Coimbra* (286). 18 pp.

Santos Dias, J. A. T. 1968. Um novo nome para o subgenero *Fonseccaia* Travassos Dias, 1963 (Acarina-Ixodoidea), nomen bis lectum. *Rev. Cienc. Vet.*, **1**, s.A.:173–175.

Sartbaev, S. K. 1955. Materials on the biology of *Haemaphysalis warburtoni* Nutt., 1912 under Kirgiz conditions. (Preliminary information.) (In Russian.) *Tr. Inst. Zool. Parazit. Akad. Nauk Kirgiz. SSR*, **4**:121–127. (In English: NAMRU3-T12.)

Sartbaev, S. K. 1961. Ticks of the genus *Haemaphysalis* in Kirgizia. (In Russian.) *Tr. 4th Konf. Prirod. Ochag. Bolez. Vop. Parazit. Kazakh. Respub. Sred. Azii* (September 15–20, 1959), **3**:484–488. (English translation by Plous, F. K., Jr., edited by N. D. Levine, 1968. University of Illinois Press, Urbana, Chicago and London, pp. 341–343). (In English: NAMRU3-T120.)

Teng, K.-F. 1980. Two new species of *Haemaphysalis* from China (Acarina: Ixodidae). (In Chinese; English summary.) *Acta Zootax. Sin.* **5**:144–149.

Tovornik, D. 1970. Ecological system of arboviral infections in Slovenia and certain other regions of Yugoslavia. (In Slovenian; German summary.) *Slov. Akad. Znan. Umet.*, Ljubljana. Razprave **13**(1):1–81.

Tovornik, D. and S. Brelih. 1973. Angaben über einige Zeckenarten aus Jugoslawischen gebieten des Adriabereiches. (Mater. II. Symp. Med. Vet. Acaroent., Gdansk, October 21–23, 1971.) (Polish summary.) *Wiad. Parazyt* **18**(4–6):731–734 (1972).

Uilenberg, G. H., Hoogstraal, and J. M. Klein. 1980. Les tiques (*Ixodoidea*) de Madagascar et leur role vecteur. *Arch. Inst. Pasteur, Madagascar*, (spec. no., 1979). 153 pp.

Demodex molossi Desch, Lukoschus, and Nutting

Chapter 11

Prostigmata–Mammalia

Validation of Coevolutionary Phylogenies

William B. Nutting

INTRODUCTION

Several thousand species of Prostigmata (Acinetida) maintain a symbiotic relationship with species of the Mammalia that suggests some degree of evolutionary bonding, possibly extending back to mid-Jurassic times. *Demodex marsupiali* Nutting, Lukoschus, and Desch, from the South American opossum *Didelphis marsupialis*, which is host-species specific, differs markedly from another undescribed host-specific demodicid, *Demodex* sp., from the North American *Didelphis virginiana* (Nutting et al. 1980). Lombert (1979) in reviewing the Myobiidae, notes that several species have been recovered from Australian marsupials, but within this family group the degree of symbiotic specificity is in dispute.

Such disagreements concerning symbiotic organisms from many metazoan phyla (Nutting 1968) and prostigmate mites from eutherian or metatherian mammals indicate that studies of poorly known groups of parasites are either ill-considered or incomplete. Cardinal requisites for coevolutionary bonding, which should be observed, are suggested by Nutting (1975). The practice of reporting host specificity for a single or even statistically valid differential recovery of symbiote from one species of host must be abandoned, unless it is shored up by careful consideration of such major biological interrelationships as discussed in this chapter.

The following account addresses the general problem of coevolution and symbiosis, some precautionary concerns in coevolutionary assessment, possible coevolutionary linkage between families of Prostigmata and orders of Mammalia, and my evaluation, based on mite–host environmental relationships, of those prostigmate species that appear to be adjunct coevolved indicators of mammalian evolution (Fig. 11.1). Likewise, mammalian host species should be valid indicators of prostigmate evolution, thus indicating coevolved mite–mammal phylogenies.

Each major section in this chapter is written as a self-sustaining unit, despite mandatory redundancies, in an attempt to provide models useful for any studies of arthropod–mammalian coevolution to document or deny symbiophylogeny. The classifications of Krantz (1978) for Prostigmata and Eisenberg (1981) for mammals are followed here with some modifications.

COEVOLUTION AND SYMBIOSIS: DISCUSSION AND DEFINITIONS

Coevolved Specificities and Phylogenies

The term coevolution in studies of organism associations has been very loosely applied by taxonomically oriented entomologists and acarologists (*mea culpa*) to document phylogenetic symbiotic parallelisms or congruences (Nutting 1974; Kethley and Johnston 1975) in terms of host

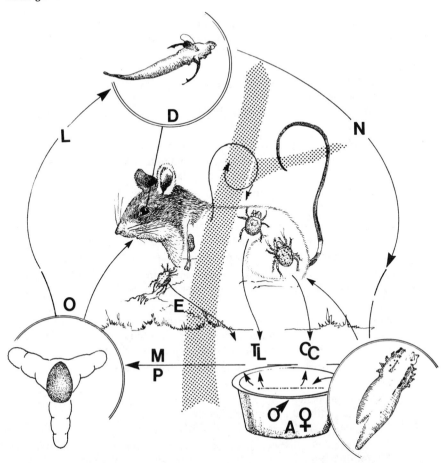

Figure 11.1 Presentational Construct—An artistic conception of the Prostigmata–Mammalian coevolution. The outer sequence shows basic mite life cycle: ova (O), larva (L), nymph (N), adults (A), of symbiotic prostigmate demodicidae mites (D), magnified and modified in keeping with the life cycle and speciation of mammals. Other capitals (C, L, E, M, P, C, T) are designates of current mite families of major concern in relation to mammalian phylogeny. The circular insets are magnifications. The shaded area represents speciation. (Original drawing by Margaret Nutting.)

specificity. There is (1) little confirmation of taxonomic assessments for the coevolved partners (both taxa), (2) superficiality of studies, (3) paucity of laboratory or field studies of the interlocked biologies of the symbiotic participants, and (4) lack of validation of the symbiotic species specifications (Nutting 1968).

In my view, *organic coevolution* should be defined as *symbiotic bonding between only two discrete species of organisms that is necessary, in at least one life cycle stage, for the survival of one or both species before and after speciation of at least one species of the partnership.* This may be caused by macromutational

Table 11.1 Terms and Types of Symbiosis Used in Conjunction with a Discussion of Coevolution and Symbiophylogenies in Prostigmata-Mammalia[a]

I. *General Terms*

A. Coevolution: Symbiotic bonding of two species for survival (one[+]), and during speciation (one[+]).

B. Symbiosis: Intimate persistent bonding of two distinct species (symbiote and symbiont).

1. Symbiote: Small, less active, dependent species.

   a. Synhospitaly: two or more congeneric symbiote species with one symbiont species (diads, triads, etc.).

   b. Polyhospitaly: sharing one symbiont species by several different symbiote genera.

2. Symbiont: Large, more active, independent species.

   a. Symbioty: one symbiote with two or more (diad, triad, etc.) congeneric symbiont species (possible symbiospecificity).

   b. Polybioty: one symbiote species shared by different symbiont genera or higher symbiont categories (possible asymbiospecificity).

3. Stadial symbiosis: Either symbiote (infective) or symbiont (infested) shares an instar or stage. Monostadial to holostadial with prefix denotations, as larvistadial or pubistadial, etc.

C. Symbiotic specificity: Persistent intertaxa symbioses of phylogenetically related symbiotes with phylogenetically related symbionts.

1. Taxa specificity: Suprageneric specificities (as class, subclass, order, family) without assurance of either asymbiospecificity or symbiospecificity or cases of *incerta sedis* (Nutting 1968).

2. Asymbiospecificity (ecologic specificity): Intergeneric to higher taxa multiple synhospitaly to polysynhospitaly or synbioty to polysynbioty as lack of "host species specificity" or asymbiospecificity. (Host specificity used in single-species symbioses should be qualified—suspect, plausible, likely). Asymbiospecificity suggests "reticulate" coevolution, higher taxa specificity, but unlikely symbiophylogeny. Microgeographic synhospitalic coevolution unlikely.

3. Symbiospecificity (genetic specificity): Multiple synhospitaly and/or synbioty at symbiote and/or symbiont generic or species level. Host-species specificity for symbiote and symbiote-species specificity for symbiont. Symbiospecificity suggests "riverine" coevolution, higher taxa symbiospecificity, and potential symbiophylogeny. Microgeographic synhospitalic coevolution likely.

D. Symbiophylogeny: Symbiote–symbiont coevolutionary symbiospecificity to at least the multiple species level of the highest symbiotic taxons under consideration.

572

Table 11.1 *(Continued)*

II. *Specific Symbioses: Prostigmata–Mammalia*
  A. Asymbiotic:
    1. Amensal, etc.: Fortuitous, transient or casual relationship without special advantage or disadvantage to either organism.
  B. Symbiotic: Intimate (necessary) relationship between two different species: the symbiote (mite) and symbiont (mammal).

	Symbiote	Symbiont
1. Phoretic	Transport benefit	No harm to host (casual biting)
2. Mutualistic	Mites may feed on other parasitic arthropods or waste of host	Benefit host by reducing parasites or by debris
3. Commensalistic	May gain nutrients and housing protection	No harm to host (casual abrasion or biting)
4. Inquilistic	Housing benefit	No harm to host (casual abrasion)
5. Parasitic	Nutrients gained, substrate and housing	Cellular destruction, toxins or pathology
6. (A)Symbiospecific	Qualifiers of 1–5 above either (asymbiospecific) by **reticulate** or (Symbiospecific) by **riverine** coevolution, the latter documenting symbiophylogenies.	

[a]See Tables 11.2 and 11.3 for specific examples and text for more precise definitions.

(punctuational) sympatric speciation, dichotomous allopatric Neo-Darwinian speciation, species formation by introgressive hybridization, and/or linear (genetic drift) speciation of both or either symbiotic species. Such a definition of coevolution admits microgeographic evolution (as plausible in synhospitaly), but neither rules out ecologic specificity (i.e., species sharing by one partner with other taxa) nor rules in interphylogenetic congruences for one or both partners in coevolution. An unbroken chain of speciation must be documented and confirmed for the coevolutionary partners at least to the family level, without ecologic specificity of either partner, before we can be assured of symbiotic coevolution and validate mutual phylogenetic congruences (Table 11.1).

In arthropod–mammalian coevolution, the absence of clear fossil records of symbiotics at the pre- and post-speciation levels, losses by extinction of nearest neighbor symbiotic partners, and the fact that the structural, functional (physiological), and behavioral characteristics may be little changed for either symbiotic partner as speciation occurs lends further difficulty to resolution for coevolution at the species or even generic specificity level. The specific and generic specificity should be documented at first for genetic specificity (Nutting 1968) in multiple nearest neighbor groups before we can use any symbiotic partnerships to validate phylogenies of symbiotes and symbionts (Nutting and Desch 1979). At any taxonomic level we must beware of convergent evolution, microgeographic (topologic or visceral) evolution, and of symbiotic relict species for both ecto- and endosymbiotic relationships.

Records of arthropod synhospitaly, which indicates microgeographic (topologic or visceral) speciation of the symbiotes (Nutting 1968), suggest that the rates of species formation may differ markedly between symbiotic partners. Thus we would anticipate species sharing of one or both participants at least to the generic or intergeneric level, even in instances of very specialized, coevolved bondings. For metazoan symbiotes, especially those allied with terrestrial organisms, the major hurdle in survival seems to be the adaptive measures for transference between symbionts and topologic or visceral habitat colonization rather than the other (often homogeneous) biological attributes of the symbionts. This would favor ecologic specificity for the symbiotes, unless adapted by natural selection to a very restrictive habitat on the symbiont or very subtle mechanisms of species finding (host or symbiont selectivity).

To be most useful in phylogenetic studies, the documentation of symbiotic coevolution must be extended to include all branches, including those currently extinct, of the symbiotic partners with the degree of genetic specificity of the symbiotes. Thus any current studies of mammalian arthropods are obviously doomed to a high degree of speculation.

Taxa specificity at or above the family level may have been developed recently in evolution and is more likely to be ecologic rather than genetic in nature and related to recent introductions of either symbiotic organism. Ideally, we should eventually find a primitive arthropod symbiote on present-day primitive reptiles closely related to primitive mammals which shows a solid pattern of biological adaptations matching the biological characteristics of monotreme, metatherian, and eutherian mammalian symbiotes. These can then be followed by comparative studies of all facets of the interlocked biologies for multiple genetically isolated and matching specificities, and eventually establish reciprocal symbiotic (coevolved) phylogenies (symbiophylogenetic coevolution—Table 11.1). Unfortunately, common ground in terminology for studies of such symbioses has not been attained by the scientific community.

(A)Symbiospecificities and Stadial Specificities

Symbiosis, directly translated as "living together" is used here in much the original sense of DeBary (1879), except that we include "a consistent and intimate relationship between two distinct species with survival benefits occurring to one or both species." This does not distinguish between ecologic or genetic specificity for either symbiotic partner.

Read (1970) preferred the term *symbiote* for either symbiotic partner, especially the dependent organism, and alternatively *host* for the independent partner. On the other hand, Dindal (1975) suggested *symbiont* for either, with appended A or B depending on energy utilization. He noted that oscillations may occur between the two symbionts so that on some occasions symbiont A may be an energy user reducing symbiont B's supply whereas at other times B may utilize A's energy.

Because neither Read's nor Dindal's designations would clarify symbiotic commensalism, I prefer to restrict *symbiote* to the smaller, more dependent, and/or less active partner, and *symbiont* (host) for the larger, more independent, and/or more active partner. For congeneric sharing (one symbiont species symbiotic with two or more congeneric symbiotes), Eichler's (1966) term *synhospitaly* is useful, and *symbioty* (one symbiote shared by two or more congeneric symbionts) is suggested here (Table 11.1). These terms may be especially useful in Mammalia–Arthropoda coevolution to focus our attention on comparative topologic or visceral (microgeographic) evolution of the symbiotes (Nutting and Desch 1979) and their degree of structural, physiological, and behavioral changes which match the patterns of specificity, synhospitaly, or symbioty. One must consider also the physical, chemical, and biotic factors of the symbiont's environment which may have some adverse effect on the symbiote.

The intermesh of symbiote–symbiont life cycles also presents problems in symbiosis. The stages in the arthropod life cycle (even the eggs with, as in ticks, hypersymbiotes) may run the gamut of symbiosis with mammals. In turn the structural, physiological, and behavioral attributes of the mammal change from egg through immature to bisexual adults. Many arthropods select certain topologic sites on mammals; the mammalian immune response and reproductive physiology is not well established in the immatures; and the intimacy of nursing probably may make retransfer of certain symbiotes impossible in immatures and male mammals.

For such as the symbiotic arthropods, we propose mono-, poly- or holostadial, with, if need be, some specification of the stages involved and type of symbiosis (as larvistadial symbiotes for trombiculid mite parasites). For the symbionts, mammals, the terms monostadial, polystadial, and holostadial also can be used with more specific qualifiers for immatures and type of symbiosis (as embryostadial or pubistadial symbionts).

Whether mono- or holostadial, true symbiospecificity must be validated, at least at the generic level of one or the other, as all or none! Statistical recoveries of out-symbiotic bondings, if reproduction (preferably to the F_2 generation) occurs in the symbiote for both extrageneric symbionts, should thus be asymbiotic. Obviously monogeneric symbionts would be most easily tested.

Genetic specificity (Nutting 1968) can now be equated with symbio-specificity both as indicative of a coevolved and assured mutual, specific symbiotic bonding replacing the symbiote-oriented host species specificity. Ecologic specificity can be replaced by asymbiospecificity for the many instances of symbionts with shared, overlapping, ecologies sharing sym-biotes (symbioty, etc.) with similar habitat requirements. Mayr (1957) speculated that as parasites (symbiotes) became better known, most would be found capable of surviving on a wide range of hosts. Within asymbio-specificity a symbiote may show symbiont (host) selectivity, or may even fail to reproduce through inadequacy of the habitat provided by the sym-biont, including cross-reaction or response to priorly acquired immune responses. Such truly or falsely asymbiotic organisms would be useless unless laboratory experiments were performed, even though basically coevolved with a single higher taxon of the Mammalia, because it could not validate the pattern of symbiotic phylogenies within that taxon. Laboratory attempts at symbiote transference with naive symbionts could aid in set-tling this issue.

Again, the value of such terms and neologisms lies in drawing research attention to focal points of use in coevolutionary analysis which have been neglected in most studies of symbiotic phylogenies (Table 11.1).

Symbioses: Prostigmata–Mammalia

The following types of symbioses have been discovered to date for Prostig-mata–Mammalia (Table 11.1) and are suspected to be significant in the problem of coevolution and for reciprocal validation of phylogenies:

1. *Phorecism:* Several prostigmate mites, such as *Pediculochelus* sp., are provided with holdfast structures and have been captured on mammals. Although they occasionally produce a nonspecific mild dermatitis, these prostigmate mites are so far asymbiospecific, relying on a wide range of organisms for transport—thus useless for coevolution studies. Survival in such phoretic mites is merely a matter of relocation to potentially greener fields.

2. *Mutualism:* This subdivision of symbiosis is defined as an intimacy between the symbiote and symbiont at the level of advantages for both with no harm to either. When a symbiotic relationship involves organisms such as *Demodex* sp., especially associated with man, many workers con-sider it as a mutualistic attachment to the host. They suggest that these mites operate as automatic chimney sweeps clearing the hair follicles of debris and ingesting excess sebum! If this were true, both organisms would benefit and no harm would accrue to either. Similarly, for ereynetids such as *Ricardoella limacum* (Schrank) their propensity for subsisting on snail slime has become adapted in their postgenitor species for clearing phlegm from the mouth cavity of vertebrates. However, in either of the above cases, the minute amounts of sebum or mucus ingested by the mites could hardly help the host. Baker (1970) has settled the ereynetid issue by showing that *R. limacum* constructs a stylostome, through which they suck up digested host tissue and blood elements: thus a case of parasitism. For demodicids, it has been confirmed that all destroy cells of the host: again parasitism.

Vercammen-Grandjean and Rak (1968) present a plausible instance that could be interpreted as mutualism for the Cheyletiellidae. Here the dog provides substrate for *Cheyletiella yasguri* Smiley, and in turn these mites consume the parasitic louse-flies of this host: thus benefiting both symbiote and symbiont (dog). In so doing, however, this acarine is now known to damage the host and thus seems capable of adventitious parasitism.

The parasitic *Demodex canis* Leydig, in a like manner, was reported to be decimated by *Cheyletiella parasitivorax* (Mégnin): another case of possible mutualism (Kuscher 1940). In most of these cases, however, the prostigmate–mammal relationship is decidedly not symbiotically host species specific (*C. yasguri* is also found on cats and man); the mites are not markedly modified, are surficial wanderers, and furthermore accept a wide range of arthropods as food supply. All such mites could well qualify as free-living predators, merely using the mammal as a handy decoy for other organisms which also may respond asymbiotically or symbiotically to the mammalian skin conditions. This does not rule out the possibility of the super-symbiotic condition known as hyperparasitism which could be operative in evolution in such instances as in the synergistic relationships among dogs (*Canis familiaris*), bacteria (*Staphylococcus* spp.), and the prostigmatic mites, *D. canis* (see Nutting 1975).

Should any instance of mutualism between prostigmates and mammals be proven as host species specific, this would indicate an evolutionary set of adaptations that could provide a key to coevolution. So far no mites seem to have adapted beyond the speculative level (Table 11.2).

3. *Commensalism:* This is classically defined as an intimate symbiotic association in which the symbiote attains benefits without damage to the symbiont (host). *Chelacaropsis moorei* Baker (Cheyletiellidae) illustrates this condition as a predatory symbiote on any microarthropod located in the fur of mammals (Summers and Price 1970). This species is relatively unmodified and similar to other free-living predatory cheyletids of the genus *Cheyletus* (Fig. 11.2), although they seem able to penetrate the skin of a mammal either defensively or to obtain nourishment as facultative parasites. In fact, with slight modifications, cheyletids could become obligate mammalian parasites, as for *Cheyletiella* spp., which cause mange or eczyma in dogs and man (von Bronswijk and de Kreek 1976; Olsen and Roth 1947). Accordingly, they are noted as showing "host specificity." Such cases, however, should better be considered host selective with likely asymbiospecificity.

The recovery of living mites from house dust, animal shelters, mammal

Table 11.2 Records of Prostigmatic Mites Associated with Mammals[a]

(Cohort) Family	(A)Symbiosis[b]	Representative	Authority	
(Pachygnathina)				
Pediculochelidae	Phorecism	*Pediculochelus* sp.	Krantz	(1978)
(Eupodina)				
Ereynetidae	Parasitism	*Speleognathus bovis*	Fain	(1963)
(Speleognathinae)	(Inquilism?)			
Tydeidae	Asymbiotic	*Tydeus* sp.	Thor	(1933)
(Eleutherengonina)				
(Subcohort: Heterostigmae)				
Pyemotidae	Asymbiotic	*Pyemotes* sp.	Treat	(1975)
Pygmephoridae	Asymbiotic	*Pygmephorus* sp.	Gurney and Hussey	(1967)
Tarsonemoidea	Asymbiotic	*Tarsonemus* sp.	Hewitt et al.	(1973)
(Subcohort: Raphignathae)				
(Superfamily: Cheyletoidea)				
Cheyletidae	Commensalism	*Chelacaropsis moorei*	Summers and Price	(1970)
Cheyletiellidae	Mutualism by	*Cheyletiella yasguri*	Vercammon-Grandjean	(1968)
	predation?		and Rak	
	Parasitism	*Cheyletiella yasguri*	Smiley	(1965)

Myobiidae	*Myobia musculina*	Parasitism[c]	Fain	(1975)
Psorergatidae	*Psorergates* sp.	Parasitism[c]	Fain et al.	(1966)
Demodicidae	*Demodex folliculorum*	Parasitism[c]	Desch and Nutting	(1972)
(Superfamily: Tetranychoidea)				
Tetranychidae	*Bryobia* sp.	Asymbiotic	Southcott	(1973)
(Subcohort: Parasitengonae)				
Erythraeidae	*Balaustium* sp.	Asymbiotic	Newell	(1963)
Trombiculidae	*Eutrombicula alfred-dugesi*	Parasitism (stadial)	Williams	(1946)
Leeuwenhoekiidae	*Apolonia tigipioensis*	Parasitism (stadial)	Krantz	(1978)
Thermacaridae	*Thermacarus* sp.	Phorecism	Krantz	(1978)
(Subcohort: Anystae)				
Pterygosomatidae	*Pimeliaphilus* sp.	Asymbiotic	Baker et al.	(1956)
Anystidae	*Anystis* sp.	Asymbiotic	Southcott	(1973)

Source: Taxonomic array following Krantz 1978.

[a](A)Symbiotic designations determined by W. B. Nutting based on reports by the respective authorities.

[b]All recorded as biting or lesion producing; thus contact, defensive or facultative parasitism, even in phoretic, asymbiotic or commensal species.

[c]Holostadial.

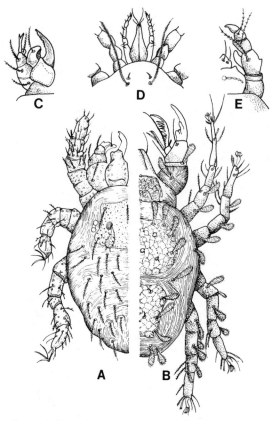

Figure 11.2   General morphology showing basic body form (all much as *A* plus *B*) and some significant features of (*A*) Trombiculidae (larva, dorsal half); (*B*) Cheyletidae (adult, dorsal half); (*C*) Leeuwenhoekiidae (palp and chelicera, dorsal); (*D*) Ereynetidae (anterodorsal); and (*E*) Cheyletiellidae (dorsal palp). (Redrawn and modified after W. B. Nutting in *Mammalian Diseases and Arachnids*, Vol. 1, 1984, with permission of CRC Press, Inc., Boca Raton, Florida.)

nests, and bedding (see Baker and Wharton 1952), and from sputum (Carter et al. 1944) and urine (Mekie 1926) offers no proof of commensalism, especially if their morphology shows little modification from the free-living form. Because such species as *Pyemotes ventricosus* (Newport) which can wander freely across the skin surface, show little morphological modifications for close symbiotic relationships with mammalian hosts, they must be held suspect as little more than transient or casual associates.

4.   *Inquilism:* The Ereynetidae show an evolutionary sequence that involves a relatively free-living species that passes through a "regressive" evolutionary stage, even to reduction or neoteny (Fain 1965), to a very specialized species with extreme regressive characters that survives in the respiratory tract of mammals. If this species subsists only on mucoid products (produced energy source by symbiont) without damaging the host, it could be considered an inquilistic relationship. On the other hand, if an ereynetid intermediate in this evolutionary scheme, living in the mouth cavity, subsists solely on food ingested by the mammalian host, then this could be considered commensalism. In either hypothetical case, it is difficult to see how symbiospecificity, of major concern to coevolution, could be assured. A further extension of this sequence, however, would recognize parasitism should the symbiotes be discovered consuming cells of its habitat. An examination of the mouthparts and feeding mechanisms of the Ereynetidae shows that any of these options is plausible, and only concentrated attention to the biology of any symbiote, as accomplished by Baker (1973), can settle the status of such species.

5. *Parasitism:* Many symbiotic Prostigmata are markedly modified to take direct advantage of the mammal's energy resources and in so doing cause damage to these symbionts: thus they are parasitic. Two examples show this condition, one the clearly ectoparasitic myobiids, *Myobia* spp., which move about the skin surface puncturing the epidermis for cell cytoplasm and lymph and even rupturing capillaries to utilize serum and blood corpuscles (Wharton 1960). They have slight morphological changes from the free-living condition; the forelegs are modified for clasping hair, and the hind legs are modified with claws for maintaining host contact. The endoparasitic *Demodex brevis* Akbulatova is found deep in the Meibomian (tarsal) glands where it feeds on gland cells (English and Nutting 1981). It has remarkable modifications for invasion of the ducts of the Meibomian complex.

Grosshans et al. (1974) confirmed the endoparasitic role of *D. brevis* in man. These mites not only destroy sebaceous cells, but also penetrate the dermis producing an inflammatory response and granuloma (Nutting 1976a). Recently a similar *Demodex* species has been found as a minor pathogen deep in the subdermal fascia of the white-tailed deer (Nutting 1975). All of these demodicids are markedly modified: for example, most are cigar shaped to take advantage of the cavelike habitat, the pilosebaceous complex of the mammalian symbiont. Besides these usual habitats, *D. canis* has been found in the lymph nodes (Canepa and daGrana 1941; French 1964). Furthermore, an undescribed demodicid was found in the lining of the esophagus in the grasshopper mouse (Nutting et al. 1973).

Parasitism is the best matching type of symbiosis currently available for studies of coevolution, especially to explore the possibility of symbiospecificity. Representatives of several families of Prostigmata show some type of parasitism (Table 11.2), but so far the Ereynetidae, Myobiidae, Psorergatidae, Demodicidae, and Trombiculidae (*sensu latu*) are apparently obligate parasites to some degree with, thus, marked potentials for coevolutionary phylogenetic studies.

Many prostigmatic symbiotes are often wrongly called parasitic if they are simply found in association with a host organism. In the Ereynetidae, if they feed solely on mucus and do no damage in the feeding process, they are not true parasites. Thus parasitism is restricted to those symbiotes that damage the symbiont. In lieu of firm biological (especially histological) studies we have weak assurances about the kind and degree of symbiosis for many prostigmatic mites. Only by histological examination have the demodicids been found positively parasitic in man (Desch and Nutting 1972). The cause of discovered lesions has been determined for only a few parasitic species of Prostigmata, as in the "nesting" of psorergatids (Luko-schus 1967a). In most cases concentrated studies of a multiple series of histological preparations of prostigmatic mites *in situ* are required to confirm their parasitic roles. We must be sure that other organisms, skin diseases, or even cases of neurogenic dermatitis, as in symbiophobia, do not provide false readings (Nutting and Beerman 1983).

Even with histological data it is difficult to separate facultative from obligate parasitism, especially in relatively free-living species. Many pro-stigmatic mites, such as the adults of *Balaustium* spp., may bite mammals and possibly sustain life processes using mammalian tissue fluids: such mites may possibly be facultative monostadial (adult only) parasites. At the other end of the spectrum, and most pertinent to coevolution, are those mite species such as the demodicids whose entire life cycle (holostadial) clearly shows a long, intimate, and necessary parasitic bond to the host mammals at the histological level. On the other hand, only the larvae of the Trombiculidae are (mono- or larvistadial) obligate parasites producing sol-itary larval lesions in a wide range of vertebrates. The assurance of the species is difficult even in whole-mounts; several species may be parasitiz-ing the same host animal, and thus connection with the lesion and species or even transmitted disease is also difficult.

The time span of a symbiotic relationship can only be accurately as-sessed by fossil records which are difficult to come by for the mammalian skin-dwelling Prostigmata. Our best estimates, however, can be obtained, as suggested previously, by careful studies of the most primitive, least specialized parasites. From these an assessment can be made of those holostadial parasitic symbiotes that are most specialized, markedly adapted to their symbiont habitat, and shared with like species of closely related more modern symbionts. In this light, species of the Myobiidae and Demodicidae are both found as holostadial obligate parasites on primitive marsupials (as Didelphidae), with markedly similar representative species on advanced eutherians such as Primates. Furthermore, they are both only known from the Mammalia. Demodicids, however, are much more special-ized in all biological features, matching a more specialized and restrictive mammalian skin habitat, and, in records so far established, show host species specificity (symbiospecificity). Myobiids, thus, seem asymbio-specific parasites and demodicids symbiospecific.

The possibility that a degree of pathogenesis may be useful as an index of the time span of such parasitic associations should also be explored (see the section on habitat and pathogenesis) because it is well documented in protozoan symbiotes that those most recently established on the host tend to produce the most damage. In this view, parasitic Prostigmata that produce the most virulent diseases would only be recently adapted as symbiotes and thus relatively poor markers of mammalian evolution. They, as the astigmatic *Sarcoptes* spp., may also accept a wide range of hosts—thus developing asymbiospecificity rather than symbiospecificity. However, each mammal responds somewhat differently to each symbiote.

6. *(A)Symbiospecifism*: These terms are suggested for subdivisional categories of symbiosis applicable, also, as qualifiers for paragraphs 1 through 5 (as, for example, symbiospecific mutualism). Other symbiotic categories focus on the symbiote, whereas these two terms include, also, the symbiont, thus accentuating more securely the degree of symbiotic interdependence in evolution.

Asymbiospecifism is exemplified by the independent recovery of one species of myobiid on six different species of *Sorex* and one on at least four different mammalian genera (see Lombert 1979). Symbiospecifism is confirmed at the histological level by several investigators (Desch and Nutting 1972; Grosshans et al. 1974, etc.) for *Demodex folliculorum* (Simon) and *D. brevis*, and only from *Homo sapiens*.

As with other symbiotic relationships, single confirmed cases of prostigmate–mammal symbiospecificity, although instructive in terms of biological adaptations between symbiont and symbiote, are relatively useless in inferring or validating the phylogenies of either partner group. What we need are multiple, confirmed, and comparatively detailed symbiospecificities to document symbiophylogenetic coevolution.

In the absence of clear cases of symbiospecificity for phorecism, mutualism, and commensalism in the Prostigmata–Mammalia, we must look to the parasitic cases of symbiosis that have evolved genetic patterns of mutual bonding interactions between symbiote and symbiont at the most specific level relatively free of distortions of the extrahost environment. Even these multiple instances of the adaptations that indicate symbiospecificity for both symbiotic partners must be carefully assessed to be of value in establishing coevolutionary phylogenies.

This series of plausible kinds and degrees of symbiosis points to the fact, as Caullery (1952) noted, that the extremes of symbiotic relationships can be identified with ease but that we should expect overlapping or not clearly distinguishable cases in many instances. This will certainly apply to prostigmatic symbiotes attempting to survive on their mammalian symbionts.

Adjuncts for Coevolutionary Assessment

The following critical topics for the coevolved Prostigmata–Mammalia are pertinent to evaluation of the foregoing six categories of symbiotic relations that aid in elucidation of coevolution at the symbiophylogenetic level. These topics will clarify some of the terms and concepts used in the foregoing discussion.

Specialized versus Generalized Coevolution

As a taxon the Mammalia have very generalized body surface (skin) characteristics: they are homoiotherm vertebrates with hair (as in hair follicles) and skin glands (mammae, etc.) in epidermal layers overlying a mesodermally derived, layered dermis. The dermis is composed of connective tissues through which the blood vessels, nerves, and blood elements ramify. Smooth muscles, fascia, and alveolar tissues are often found beneath the dermis. The epidermal and dermal components are reduced, with several structures wanting, in the large head, esophageal, or protodeal cavities.

Against this backdrop of the generalized mammalian skin—the "first line of defense" against symbiotic arthropods (mites)—symbiotic myobiids, psorergatids, and demodicids, all of which have similar basic structures, are differentially adapted for coevolution. One would surmise that the most generalized arthropod symbiotes capable of holdfast feeding on the thin-skinned surface areas of the mammalian symbiont would be asymbiospecific and found on a wide range of mammals (the case with separate species of myobiids) whenever the mammalian habitats (as dens) or niches overlap. Specialized symbiotes which require a specialized habitat, as the hair follicle, could be symbiospecific.

Both the mammalian skin and prostigmate mites are subject to natural selection by the extrasymbiont environment, which may be especially severe for prostigmates during transference, colonization, and translocation.

Consideration of structures, functions, and behaviors indicates that, of the prostigmate symbiotes, the Myobiidae are the most generalized and *Demodex* the most specialized, whereas, even though maintaining generalized skin components, the mammals have become specialized for skin component change and distribution, and for behavior, from the more generalized primitive, small rodents to such as the huge pinnipeds. Thus both mammals and these prostigmatic mites show a gradation from generalized to specialized, with a series of intermediate symbiotes sequentially specialized for transfer, translocation, and colonization.

Transference, Translocation, and Colonization

The more specialized and isolated the mammal, the less likely that a generalized asymbiospecific mite such as a myobiid will be found to infest it. On the other hand, the coevolved specialized symbiospecific mite should be located on its host regardless of host species isolation.

Prostigmate symbiotes are subjected to severe selection pressure during transfer between hosts and therefore must develop special adaptations for survival. A generalized symbiote would be translocated most easily on a generalized and topologically homogeneous mammalian skin with little selection pressure because of its generalized adaptive mechanisms for transfer.

Intersymbiont species colonization could be anticipated even between specialized symbionts for such as the asymbiospecific myobiids. Markedly specialized symbiotes such as the symbiospecific *Demodex* spp. would need multiple transfer mechanisms to overcome the selection pressure during transference, translocation, and colonization.

Intergenerational transfer of symbiotes, often matrilinear, as postulated for the specialized *Demodex* spp., can be fatal for symbiospecific arthropods if the symbiont parents are extirpated. This need not be true for an asymbiospecific arthropod, such as for a generalized symbiote as *Myobia* spp., which could readily transfer from even the dead symbiont to any other host species with similar ecology. *D. canis* dies with its host (Nutting 1950), and adult myobiids move to extrahost areas as the carcass cools.

Microgeographic Evolution

It is expected that in all holostadial parasites (as Myobiidae, Psorergatidae, Demodicidae) of mammals that show high levels of synhospitaly, polyhospitaly would prove to be microgeographically (microtopologically) speciated. Comparative studies of synhospitaly and polyhospitaly show, however, that the taxonomically generalized myobiids, which are able to transfer whenever mammals overlap ecologically, are both synhospitalic and polyhospitalic, whereas within the genera of the closely related *Soricidex*, *Demodex*, and *Pterodex* there are "monad" to "quatrad" synhospitalies (Nutting 1974) and a few of polyhospitaly. Such comparative studies should be followed up by concentrating on the relationship of diagnostic symbiote characteristics as adapted to the topologically distributed differences in the host skin, on the assurance that mite reproduction for each species occurs, and on the assurance (or not) of intergeneration of transfer and colonization. This would aid in separating the asymbiophyletic from the symbiophyletic kinds of relationships for our Prostigmata–Mammalia.

Physiological Compatibility

Although little is known about the symbiote–symbiont physiological compatibility, Fain (1979) has suggested that elements of this will prove most important, especially in the Myobiidae, for host specificity and host–parasite phylogeny. Other than the T-cell immunodeficiency postulated for *D. canis* in dogs (Scott et al. 1974), we can find no secure evidence to date of a physiological natural selective mechanism for Prostigmata–Mammalia cases of symbiosis. Furthermore, this cellular immunodeficiency has only been reported for massive generalized cases of demodectic mange in a stress-prone domesticated breed of mammal. We should bear Fain's suggestion well in mind, but we must have more extensive testing of this hypothesis, in both laboratory and field, before we are able to separate arthropod–mammalian coevolved phylogenetic patterns of symbioses from assured coevolved symbiophyletic symbiotic patterns.

These adjunct aids are used in the assessments of the Prostigmata–Mammal coevolution that follows with special attention to the Myobiidae, Psorergatidae, and Demodicidae.

TAXA OF CONCERN

Even for the taxonomically best-known prostigmate associates of mammals, the degree of absence of symbiospecificity is difficult to assess because so little attention has been paid to comparative mite or mammal biologies. Taxa of concern for Prostigmata–Mammalia thus primarily represent the mites and their associated mammals with reported records of recovery and with some suggestions as to need for studies of coevolution.

Prostigmata

As of 1983, we can locate only 18 families of symbiotic Prostigmata (Table 11.2) out of the approximately 120 prostigmate families mentioned in Krantz (1978). In the table, mite taxa are presented in the taxonomic sequence of Krantz with a judgment for each family of the kind, presence, or absence of symbiosis.

A single pediculochelid, *Pediculochelus* sp., recovered from a rodent, noted in Krantz (1978), is closely allied to several other species with proven phoresy. This representative of the Pachygnathina may be closely related to the fossil *Protacarus crani* Hirst (1923) recovered from Devonian strata. Other related pachygnathine mites are not only phoretic but also feed on vegetable debris, and are commonly found in litter, soils, or grasslands. These facts indicate their remarkably long and premammalian history of free-living detritus feeding. No pathogenesis in mammals has been reported for any member of this group. In fact, Price (1973) notes that the chelicerae of one member of *Pediculochelus* phoretic on bees are weakly chelate and so hardly fitted for vertebrate parasitism. One might expect, however, that the implied long history of such phoretic species of debris-feeding mites could produce species capable of biting and even feeding on mammalian tissue fluids especially of thin-skinned ground dwellers such as rodents. This accentuates the possibility that many of the asymbiotic prostigmate species associated with mammals may have not only structural but also physiological barriers to utilizing tissue fluids of animals. This may also indicate that those which bite mammals may be doing so

Table 11.3   Selected Instances of Asymbiospecificity (Ecologic Specificity) in the Ereynetidae[a]

Symbiont (Host) Order Species	Symbiote (Parasite) Species	Locale
Avian: Galliformes— ground-dwelling birds		
Gallus gallus	*Speleognathopsis galli*	Ruanda-Urundi
Numida meleagris	*Speleognathopsis galli*	Ruanda-Urundi
Mammalian: Artiodactyla— ground-grazing bovid		
Bos taurus(?)	*Speleognathus bovis*(?)	
Mammalian: Chiroptera— roost niche-sharing(?) bats		
Eidolon helvum	*Neospeleognathopsis chiropteri*	Ruanda-Urundi
Epomophorus labiatus minor	*Neospeleognathopsis chiropteri*	Ruanda-Urundi
Rousettus aegyptiacus	*Neospeleognathopsis chiropteri*	Egypt
Mammalian: Rodentia— ground-dwelling rodents		
Gerbilliscus bohmi	*Paraspeleognathopsis galliardi*	Ruanda-Urundi
Dasymys incomtus	*Paraspeleognathopsis galliardi*	Ruanda-Urundi
Pelomys fallax	*Paraspeleognathopsis galliardi*	Ruanda-Urundi
Arvicanthis abyssinicus	*Paraspeleognathopsis galliardi*	Ruanda-Urundi

[a]Revised and assembled from Fain 1963.

defensively, as do many spiders, rather than to utilize host nutrients. The fact remains, however, that even such phoretic species seem admirably preadapted for parasitism in or on mammals.

The eupodine Ereynetidae are parasitic on both invertebrates (Baker 1973) and vertebrates. Baker described histologically the formation of a stylostome, such as is well known in trombiculid larvae, which certainly use it for tissue penetration and feeding. The lawrencarinate ereynetids are hematophagous and commonly found in the respiratory tract of amphibians and reptiles (Lawrence 1952). Fain (1962) records that *Lawrencarus eweri* Fain parasitizes six species of the amphibian *Bufo*: hardly host-species specific. Speleognathines are of interest here because the members parasitize birds as well as mammals (Table 11.3). Fain (1963) records that closely related taxa are found in ground-feeding avian species and in rodents. Furthermore, a few speleognathines are recorded as "libre" which suggests that they could invade many ground-dwelling vertebrates. In bats, *Neospeleognathopsis chiropteri* (Fain) is found in three different genera (*Eidolon*, *Epomophorus*, and *Rousettus*) from Africa (Fain 1963) that could be in close contact on roosting. Thus with the ecological overlap of hosts, the ability to survive free of the host, the recovery of one species from several host genera, and the homogeneity and absence of isolating mechanisms in the vertebrate respiratory tract, ereynetids show a high level of ecological specificity: thus they are not symbiospecific in mammals.

The other eupodine group, the Tydeidae, seems even more weakly tied to mammals in terms of specificity: the record of *Tydeus molestus* (Moniez) biting man, by Thor (1933), appears a casual fortuitous (asymbiotic) event.

Within the Eleutherengonina, three families of the subcohort Heterostigmae, one of the superfamily Tetranychoidea, and two of the subcohort Anystae seem similarly asymbiotic with mammals (Table 11.2). In the subcohort Parasitengonae, Newell (1963) indicates that *Balaustium* sp. is a fortuitous mammalian associate whose life-style is dependent on plant feeding. I have also identified a member of this genus which bit a student as he sat in a meadow filled with forbs and grasses.

The trombiculids, as adults, are predators on arthropods, but only as larvae parasitize vertebrates: thus they are monostadial parasites. Most are surficial skin feeders but a few are found intranasally, especially in bats and rodents (Nadchatram 1970). Several species have also been noted as having invaded hair follicles as larvae or as having penetrated the dermis in rodents and shrews (Sweatman 1971). Trombiculids seem capable of accepting a wide range of terrestrial vertebrates. *Neotrombicula autumnalis* (Shaw) for example, will accept man, rodents, and birds (Krantz 1978), and so is certainly not symbiospecific with mammals. Traub and Wisseman (1974) suggest that bird and reptile chiggers cause more severe tissue reactions in man. This may indicate some nonspecific physiological adaptation of recent vintage in mammals. It does, however, seem much too broad an observation to be used to document coevolution at the present time in view of the often-related differential response to chigger bites from the identical species of trombiculids in man.

The remarkable response of trombiculids to host carbon dioxide and amino acids, plus the formation of a stylostome from products of both host and parasites indicates a long association with mammals. However, Nadchatram (1970) was able to demonstrate a solid relationship between host ecology and a nidicolous habit in several trombiculids. All of this indicates a high level of ecological specificity or asymbiospecificity with little if any potential for establishing coevolutionary parallelism.

The leeuwenhoekiids are phylogenetically closely related to the Trombiculidae. Their larvae attack a wide range of vertebrates, including bats, birds, reptiles, and even an invertebrate (Krantz 1978). They have remarkable mouth parts, called "chelostyles" by Vercammon-Grandjean (1966), long legs, and are large. There are reports of multiple host acceptance by one species (*Apolonia tigipioensis* Torres and Braga for man and birds). These data indicate an ambicolous life-style rather than host symbiospecificity. The chelostyles, in immatures and adults, are more developed than those in trombiculids, and thus more suited for parasitism or predation. The wide range of hosts indicates asymbiospecificity.

Several families of the superfamily Cheyletoidea are apparently well adapted symbiotically to vertebrates and are potentially the most likely Prostigmata as adjunct markers of vertebrate phylogeny. However, cheyletids and cheyletiellids are not useful as phylogenetic indicators. The cheyletids are notorious predators of arthropods, and even *Cheyletiella parasitivorax* (Mégnin) has been found preying on other arthropods (Kuscher 1940). Although von Bronswijk and de Kreek (1976) strongly suggest a plausible host specificity, they admit the difficulty of species determination because of the variability of diagnostic characters. In overview, both taxa are little modified from free-living forms in morphology and life-style. Current recoveries from mammals (e.g., *Cheyletiella yasguri* on man, dogs, cats, etc.) indicate that these two familial groups are fortuitous associates rather than genetically specific markers of mammalian evolution.

The three cheyletoidean families, Myobiidae (Lombert 1979), Psorergatidae (Giesen 1982), and Demodicidae (Nutting and Desch 1979), have recently been reevaluated in terms of their host specificity with mammals. In the discussion of mammalian taxa that follows, and the remainder of this chapter, these three families are used exclusively in our consideration of coevolution in the Prostigmata–Mammalia.

Mammalia

All 4000+ members of the Class Mammalia are eligible for symbiosis with any member of the 18 families of Prostigmata noted before and in Table 11.2. All of these symbiotic mites either cling to hairs (e.g., myobiids) or surface epidermis (e.g., pediculochelids), or penetrate epidermis (e.g., trombiculids), occasionally into the dermis (e.g., psorergatids), or as semiendoparasites (e.g., demodicids) in the pilosebaceous-glandular systems. The mammalian skin complex is homogeneous for the mites structurally and physiologically, and thus ecologically, although it is topologically modified by the evolved life-style of each species. An indication of this homogeneity in the Mammalia is that the trombiculid larvae can locate, penetrate, feed, and survive on most mammals, from large, thick-skinned artiodactyls to small, thin-skinned rodents, and demodicids are suspect for surviving on all mammals.

Although anticipating additions to the families of Prostigmata symbiotic with mammals, I elect to consider only those mammalian taxa from which the Myobiidae, Psorergatidae, and Demodicidae have been recovered to date (Table 11.4). These three prostigmates are the only known taxa whose members are all holostadial parasites restricted to mammals and coevolved at either the asymbiospecific or symbiospecific level, thus potentially symbiophylogenetic. In this light, mammalian "taxa of concern" merely represent those so parasitized, whereas all others are of major research concern

Table 11.4 Mammalian Taxa, Class to Suborder, from Which Symbiotic Species of Myobiids, Psorergatids, and Demodicids Have Been Recovered[a]

	Class Mammalia
Subclass Monotremata	None, despite a few careful examinations both gross and histological (W.B.N.).
Subclass Metatheria	
Order Marsupialia	*Australomyobia necopina* Domrow
	Demodex marsupiali Nutting, Lukoschus and Desch
Subclass Eutheria	
Order Insectivora	*Protomyobia nodosa* Jameson
	Psorergates sorici Lukoschus
	Demodex talpae Hirst
Order Chiroptera	
Suborder Microchiroptera	*Neomyobia myoti* (Dusbabek)
	Psorergatoides molossi Lukoschus, Rosmalen, and Fain
	Demodex aelleni Fain
Suborder Megachiroptera	*Ewingana molossi* Dusbabek
	Psorergatoides kerivoulae Fain
	Demodex molossi Desch, Lukoschus and Nutting
Order Rodentia	*Myobia apodemi* Uchikawa
	Psorobia hystrici Till
	Demodex aurati Nutting
Order Primates	*Psorobia cercopitheci* Zumpt and Till
	Demodex folliculorum (Simon)
Order Carnivora	
Suborder Pinnipedia	*Demodex zalophi* Dailey and Nutting
Suborder Fissipedia	*Psorobia mustelae* Lukoschus
	Demodex cati Megnin
Order Artiodactyla	*Psorobia ovis* Womersley
	Demodex bovis Stiles
Order Lagomorpha	*Radfordia cricetulus* Fain and Hyland
	Psorobia sp. Giesen, 1982
	Demodex cuniculus Pfeiffer
Order Perissodactyla	*Demodex caballi* Desch and Nutting
Order Hyracoidea	*Demodex* sp.—in hand (W.B.N.)
Order Edentata	*Demodex* sp.—in hand (W.B.N.)

[a]Mite representatives from Lombert, 1979 (myobiids), Giesen, 1982 (psorergatids), list in Appendix I or the literature. Mammalian taxa simplified and modified from Eisenberg 1979 and Honacki, Kinman, and Koeppi 1982.

because they may well provide critical, new information for interpretation of coevolution.

Of the three subclasses of mammals—Monotremata, Metatheria, Eutheria—only the monotremes are free, so far, of these cheyletoidean mites. The metatherians (Order Marsupialia) and eutherians are parasitized by myobiids and demodicids, and the latter by all three Prostigmata. Absence of all three from the monotremes and psorergatids from the Metatheria undoubtedly is a function of lack of investigation.

The Orders from which representative symbiotes of all three families have been recovered include the Chiroptera, Insectivora, Rodentia, and Lagomorpha. Psorergatid and demodicid species also share the Orders Primates, Carnivora, and Artiodactyla. The Demodicidae are found as symbiotes with species in the Orders Perissodactyla, Edentata, and Hyracoidea. The following summarizes and indicate symbioses in known mammalian taxa:

Myobiidae—Some 300 + species have been recovered from 2 subclasses, 5 orders, (approximately) 35 families, (approximately) 46 genera, and (approximately) 160 species of mammals (Lombert 1979).

Psorergatidae—About 100 species have been recovered from the Subclass Eutheria, 7 orders, 27 families, (approximately) 70 genera, and (approximately) 115 species of mammals (Giesen 1982).

Demodicidae—Approximately 150 species have been recovered from 2 subclasses, 11 orders, 35 + families, (approximately) 60 genera, and (approximately) 75 species of mammals (see Summary and Tables 11.1 and 11.2).

It is apparent from these figures that synhospitaly must be common in at least the Myobiidae and Demodicidae. If we allow for this, some 3900 species of the 4000 + mammals may have new species of each of these mite families in need of discovery, description, and clarification of the interlocked mite-mammalian biologies. Some precautionary notes are obviously in order for both present and future attempts to resolve the problems of Prostigmata–Mammalia coevolution and symbiophylogeny.

LIMITS AND CAUTIONS

Limits to our understanding of acarine host specificity with vertebrate symbionts and some procedures useful for future studies of this topic were discussed by Nutting (1968). Since then, as evidenced by this book on coevolution, the number of publications alleging host specificity has increased markedly, especially for the Prostigmata–Mammalia (see dates in the references). Despite this new information, we are nearly as limited in our interpretation of true symbiospecificity with mammals as we were in 1968 because of the small size and cryptic habits that make prostigmate mites difficult to recover.

Multiple recoveries of a species of mite from a mammalian species, even those adduced by associated signs of pathogenesis to be symbiotic, are not adequate to assure either symbiosis or symbiospecificity. For most Prostigmata, especially in suspect cases of synhospitaly, histological studies of the mammalian skin and demonstration of mite life cycle stages *in situ* (as Nutting and Desch 1972) are needed to clearly discriminate the symbiote species and lend credence to the level of symbiosis. Many prostigmate mites are not only capable of utilizing the same habitat (e.g., hair follicle) but also of sharing the habitat with other organisms such as fungi, bacteria, and the astigmatid mites such as *Audicoptes* sp. (Lavoipierre 1964).

In resolving the problems of coevolution for each symbiont mammal and symbiote, prostigmate mites must be described precisely at the species level. Unfortunately, the precision of both symbiote and symbiont descriptions is woefully weak, and both mammalogists and acarologists are incapable of judging each other's taxonomic decisions. Even within each discipline, specialists disagree or often use esoteric criteria (such as karyotypes and single to few setae) with little regard to providing descriptions based on adequate samples of multiple mensurable characteristics as free as possible from environmental distortion. Many mammals and mites are often described on the basis of a very few (>10) specimens of often only one stage (as in trombiculids) or of an individual of only one sex. Taxonomic accounts of symbiotic mites should include the number of host individuals by sex and approximate age or stage, as well as locus and altitude of capture. They should also include the number of mites recovered from one or each host individual, lest mite–host ecologies distort these characteristics.

The difficulties in mammalian taxonomy are illustrated by the current disagreement among mammalian taxonomists as to whether *Didelphis marsupialis* in South America is distinct from *D. virginiana* in North America (Gardner 1973): most of their morphological features are identical, including dentition, and the markedly different karotypes of both seem to share transitional karyotypes with specimens in Central America. Meanwhile, the resident demodicids on each also share this taxonomic indecision! However, such taxonomic indecision could be resolved or at least be provisionally stabilized by accurate assessment of the symbiotic arthropods. For the *Didelphis* problem a new demodicid has been found from *D. virginiana*, distinctly different from that reported from *D. marsupialis*. Distinction between two symbionts has been suggested by using demodicids from sympatric species of *Peromyscus* (Nutting and Desch 1979).

Taxonomic discrimination for symbiotic Prostigmata is usually difficult because of unconfirmed species distinctions, inadequate samples and descriptions for life cycle stages, and a lack of genetically stable characters.

In the trombiculids, as "stadial symbiotes" of mammals, many new species have been described, based *only* on small numbers of larvae (see Krantz 1978) and primarily on the basis of chaetotaxy. A careful study by Goksu et al. (1960) shows that in *"Trombicula akamushi"* statistical analysis was needed for chaetotaxy used in descriptions and identification of this species. They aver (p. 208) that " . . . the taxonomist . . . should beware of relying upon characters capable of variation." Nonetheless, most prostigmatids are still described and keyed out on the basis of variable and unreliable characters using small samples.

In the Psorergatidae measurements of body length and width and setal length in males and females of seven species show remarkable overlap (Kok et al. 1971). In the largest sample ($n = 49$) for females of *Psorergates mycromydis* Lukoschus, Fain, and Beaujean, the range of variation in length was 35 μm on an animal of approximately 150 μm.

Castro and Hobart (1980) demonstrate very clearly that the measurable dimensions of such soft-bodied organisms are markedly fluctuant depending on the mounting media used and the distortion produced by handling or coverslip pressure. They examined *Demodex bovis* Stiles and found a significant difference in the total lengths of female mites using polyvinyl alcohol in lactic phenol, 255.5 ± 8.8 μm; 85% saline solution, 244.2 ± 14.9 μm; and in Hoyer's medium, 229.1 ± 18.9 μm ($n = 20, 20, 23$, respectively). In our work on demodicids, we have always randomly discarded obviously deranged specimens, although these were useful for obtaining stadial ratios. I do, however, recommend that specimens be obtained from several individual hosts, several regions of the host body, and/or several lesions to avoid sampling bias established by transference weeding, habitat weeding, and the likelihood that mites in an individual lesion, such as those produced by *D. caprae* Railliet, may all be descendents of one selected parthenogenetic female.

Numerical taxonomy has not been used for symbiotic prostigmates and may not be useful with all parasitic forms because in some species (i.e., certain Demodicidae) numbers of measurable characters are drastically reduced. Furthermore, this technique has an added disadvantage: if numbers of characters are low, environmentally vagile characters, such as those deranged in preparation and those with highly mutation-prone structures, may need to be lumped with the few genetically stable characters. These last can only be discovered by statistical analysis, wanting in most species descriptions.

More attention must be paid by acarologists to systematic studies of mites of hybrid host mammals, demes or subspecies of the same host species, and broadly dispersed host species in which antipodal specimens show noteworthy physiological or structural differences. To the first, Santa Gertruda cattle in Australia share *D. bovis* with *Bos taurus* (Nutting, unpublished), and domesticated cats, thought to be hybrids, have so far in our studies provided only *D. cati* Mégnin. Both examples, however, are inconclusive because the mites of these hybrids and of their parent species have not been studied extensively.

Adequate samples of *D. brevis* and *D. folliculorum* Simon of *Homo sapiens* have now been recovered and well studied from "pure blood" Australian aborigines (Nutting and Green 1976), Eskimos (Nutting, unpublished data), Japanese (Hakugawa 1978), and Europeans (several authors), but weakly studied in all other racial groups. Despite ancient and unknown miscegenations, it appears that members of widely dispersed demes and antipodal skin-different hosts share identical demodicids.

It seems plausible here that if *Psorergates* spp. are only ecologically specific, the intimate association between domesticated mammals, such as cattle and sheep, could provide occasions for interhost species transference. For host-tissue-feeding organisms, the change in diet, temperature, and so forth, could then effect characteristics of low genetic stability, thus leading to marked intraspecific variation. In the description of *P. bos*, Johnston (1964) used only the characteristics of size and ventral setal morphology based on too few specimens to distinguish it from *P. ovis*. The tabular listing by Lukoschus (1969) for these two species notes identical measurements for five characteristics with a size difference of only 11–32 μm. He did not indicate the specimen number for these values, although a range is presented for total length.

All of these instances lead us to the conclusion that we currently should use a statistically valid sample for the analysis and description of both mammals and mites. We should also avoid subspecies in our accounts until more precise studies have been carried out on sufficient samples. These studies may include in vitro culture of mites with special attention to the F_2 generation of parents (P) selected from either extreme for many measurements and suspect for genetic stability. If held under controlled conditions, such studies could also reveal unsuspected changes in gross morphology.

Accordingly, the lacunae in the study of prostigmatic symbiotic mites include the following:

1. *Status as symbiote:* Whether the species is a true parasite, commensal, mutual, or of another type; and within these categories whether it is facultative or obligate; also, degree of host specificity, and especially symbiospecificity (host species specificity).

2. *Synhospitaly* (Eichler 1966): Two or more species of congeneric mites are often found on one host species in the parasitic Myobiidae, Psorergatidae, Demodicidae, and Trombiculidae. Which (or both, or all) are host-species specific? Here, each symbiote stage must be located in each prior stage to insure life cycle sequences for each synhospitalic species.

3. *Sexual dimorphism and parthenogenesis:* A definitive case of arrhenotoky in the Demodicidae was discovered by extensive reas-

sessment of sex ratios in *D. caprae*. Thelytoky is also a possibility in other members of this genus. Such parthenogenesis not only aids in symbiote transference but also speeds up the rate of coevolution. Table 11.5 provides illustrative sex ratios.

4. *Life cycle:* Even within a genus, as *Demodex*, stage elision has been noted (Nutting et al. 1971), and neoteny is present in two different

Table 11.5 Records of Recovery Listed by Stage for Selected Myobiids, Psorergatids, and Demodicids[a,b]

Family Species (Citation)	Ova	Larvae	Proto-nymph	Nymph	Male/Female (ratio)
Myobiidae					
Myobia sp. (this report, Desch pers. comm.)	7	7	14	25	1 3 (1:3)
Pteropimyobia pahangensis (Fain and Lukoschus, 1979)		21 ova and immatures			12 10 (1:1)
Psorergatidae					
Psorergates cinereus	1	3	3	3	7 11 (1:1.5)
P. canadensis	1	3	0	1	6 11 (1:1.8)
P. maniculatus (Kok et al. 1971)	1	1	0	7	10 11 (1:1)
Demodicidae					
Ophthalmodex sp. (this report)	21	20	—	11	21 73 (1:3.4)
Demodex odocoilei (Desch and Nutting, 1974)	347	71	73	156	90 551 (1:6.1)
D. brevis (Desch and Nutting, 1972)	170	52	20	64	157 537[c] (1:3.4)
D. sp. (rat) (this report, Desch pers. comm.)	167	—	102 —	85	186 136[d] (1.4:1)
D. gapperi (Nutting et al. 1971)	10	6	—	10	16 31 (1:1.8)
D. cafferi (Nutting and Guilfoy, 1979)	287	141	130	250	13 413 (1:32)

[a]For hosts see authority.
[b]Sex ratios indicative of reproductive mechanisms for successful transference in coevolution.
[c]Plausible arrhenotoky.
[d]Potential thelytoky.

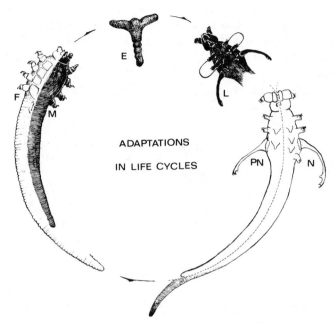

Figure 11.3 Drawings of the stages in the life cycle of *Demodex molossi*. The adult male (M) and female (F) are typical in morphology for the genus, but the egg (E) has large anterior holdfast prominences, also present in the larva (L). The huge clawlike legs III in the larva, the ventral shelflike scutes, and prominent palps are found also as holdfasts in the protonymphal (PN) and nymphal (N) stages. The latter has stumplike nonclawed legs IV. (Redrawn and modified from Desch et al., 1972.)

cases of Ereynetidae (Fain 1974) in which apparent viviparity and behavioral differences exist (Baker 1973). Demodicids, which reside in tarsal glands, commonly have marked modifications in immatures which match, as in *D. marsupiali*, the morphology of their habitat. Adults of the undescribed symbiote from *Didelphis virginiana* differ only moderately from *D. marsupiali*, but immatures differ markedly. This accentuates the need for exploring the full life cycle of any symbiote.

5. *Holdfast mechanisms:* Immature instars and even ova, as again in demodicids, may have some remarkably modified holdfast structures. Why do they need to be so redundant? *Demodex molossi* Desch, Lukoschus, and Nutting has three specialized sets on the larval stage (Fig. 11.3) (Desch et al.1972). If hair follicle characteristics are identical among symbionts of different orders, why should holdfast mechanisms be different?

6. *Behavioral traits:* Transference either on warmed skin in vitro or between laboratory animals of the same skin characteristics may determine the likelihood of host infestation and coevolutionary bonding.

Cross-breeding studies could also be attempted in the ectoparasitic myobiids.

7. *Ratios of life stages:* An adequate sample of skin-invasive parasites is difficult to recover to determine sex or stage ratios. In *Demodex gapperi* Nutting, Emejuaiwe, and Tisdel we were able to dissolve entire eyelids of the symbiont red-backed vole in KOH 5% at 87°C, and thus expose the entire population. In studies of large papules, *D. caprae* revealed immatures in the center and adults along the periphery. We exhausted the entire contents, mixed carefully, and took a substantial aliquot and spread this well in Hoyer's medium. As suggested by Lebel and Nutting (1973b), instars were randomly tallied until the least abundant scores 20. This has only been accomplished for a few demodicids.

8. *Culture attempts:* This is mandatory for such organisms as *Demodex* spp. which have marked modifications that exactly match the host habitat and so, possibly respond to environmental factors. Nutting (unpublished) found no changes in morphology of instars of *D. caprae* maintained for 100 days in micropore filter tubes on embryonic Golden hamster skin placed beneath the skin of chickens. All stages and hamster skin cells were viable at sacrifice and, therefore, development assumed to have taken place.

For the mammalian mite habitat, the epidermal–epithelial and pilosebaceous complex, we briefly consider the following:

1. *Variability and localization:* Often in cerumen and similar glands, there is a clinal shift of anatomy from normal hair follicles with sebaceous glands apically in the ear, to markedly changed multiple and massive cerumen glands with hair and follicular epithelial reduction deep in the concha. Where is the optimum habitat for the remarkably specialized *Demodex marsupiali*? The ova in this species is bomb shaped, the larvae have large holdfast supracoxal spines, but the adults are very similar to those of primates. So far, the mammalian tarsal glands in bats and rodents are the loci of the most bizarre prostigmate parasites in all stages of the mite life cycle. Thus eyelids should be studied extensively in all mammals for evidences of symbiotic adaptations. These coupled with the eyebrow area skin possess the greatest variety of habitats for ectoparasitic mites.

2. *Sex, age, and condition of host:* The pilosebaceous complex and epithelia are markedly different physically and biochemically in the young mammal as compared to the adult. In all life stages the morphology and chemistry of the habitat may also shift under seasonal, hormonal, or behavioral cycling. Unthrifty, diseased, stressed, or crowded mammals may also influence the level of mite infestation

by either increasing or decreasing mite reproduction and rate of development.

3. *Differential distribution of habitat:* In mammals, differential distribution of the epidermal habitats available to prostigmatic mites has not been clearly elucidated for most mammalian species. Many of these mites have been missed, such as *Ophthalmodex artibei* Lukoschus and Nutting, because of the failure to examine the large epithelial-lined head cavities.

4. *Patterns of maintenance:* The incarceration or domestication of mammals, as well as close association of mixed species (as man and pets) may suggest ecologic specificity of mites if transfers are effected.

5. *Foster nursing or hand rearing:* Many mammals are amenable to foster nursing and/or gnotobiotic rearing. Either of these, or their combination, would help settle the problem of host acceptance of a mite for transference. These should also be used to settle the problem of symbiote colonization. Similar studies of parent species of domesticated mammalian hybrids should be useful in assessing symbiospecificity.

6. *Symbiont behavior:* Little is available in published literature with regard to host behavior, especially in association with other species, such as during predation, nursing, copulation, and so on which may permit transference of symbiotic mites.

7. *Symbiont ecology:* Mammalian migration patterns (especially bats), population expansions, and the fortuitous or willful introduction of exotic species by man should all be cataloged at the local scene of coevolutionary studies. Lack of maintenance of niche overlap for any mammal may clarify the degree or absence of coevolutionary bonding.

For both mammals and mites, misidentification, mislabeling, and compositing samples from multiple species have all distorted the accuracy of coevolutionary assessment. In coevolution studies, especially for the minute Prostigmata, collaboration of mammalogist and entomologist (acarologist) must take place, preferably from symbiont capture and symbiote identification to publication of symbiophylogenetic analyses.

STRUCTURAL CUES AND FUNCTIONAL MECHANISMS

Although more than 550 myobiids, psorergatids, and demodicids have been identified to be symbiotic with more than 250 species of mammals, attention for the mites has been nearly exclusively on morphology. From these studies by extrapolation, and the few observations of living and fossil prostigmates, we can, however, attempt to match their coadapted structures, functions, and behaviors with the physical, chemical, and biotic interactive features of mammals. Comparative studies of these mites, ex-free-living fossils, and extant modern prostigmates may determine which are most precisely symbiospecifically adapted and thus most useful in validating mammalian evolution.

The pachygnathine fossil *Protacarus crani* Hirst has structures similar to extant free-living prostigmate mites. They provide the basis to construct a generalized, free-living type with preadapted structures and mechanisms from which these mammalian parasites may have evolved.

Such free-living generalized forms probably evolved prior to the origin of mammals and must have possessed the following preadaptive prostigmate attributes (see Fig. 11.4*A*):

1. Generalized structures, with small body (about 1 mm in length), long, multisegmented legs (about six segments), thick exoskeleton (2 μm±), tracheal respiratory tract, setose body, palps, sexual dimorphism, incomplete cleavage, and chelicerae.
2. Arachnidan functions, with tearing or puncturing mouthparts, chelate or styletiform; salivary secretions for predigestion; pumping mechanisms for liquid ingestion; extracellular digestion; complete digestive tract; waste as guanine expelled through an anus; and respiratory system obtaining oxygen and eliminating carbon dioxide and ammonia.
3. Crevice-seeking and thigmotactic, phototactic, chemotactic, and stereotactic behaviors, with appropriate sensoria.
4. Life cycles with oval, larval, protonymphal, deutonymphal, and tritonymphal stages; fertilization internal and ova deposited on or in the habitat.

Further evidence on origins of these mites should be obtained by studies of cheyletoidian families of such as the Cloacaridae, parasitic on reptiles. Of the three families, only the myobiids possess nearly all of the features, except eyes, of the primitive free-living Prostigmata; both psorergatids and demodicids are markedly modified for all features, with *Demodex* spp. most modified in all stages of the life cycle. The following features of these taxa are clues to coevolution.

Structural Features

Shortened Life Cycle and Morphology

Within the genus *Demodex* we find the omission of the protonymph in the Meibomian gland-dwelling *D. gapperi* (Nutting et al. 1971) and a foreshortening from the ova to the nymphal stage of the life cycle in *D. caviae*

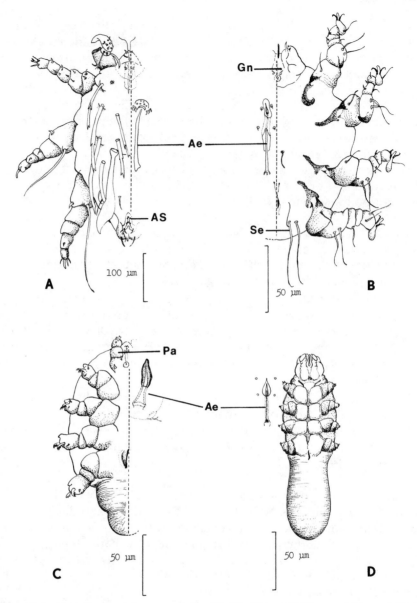

Figure 11.4 Ventral views of representative (*A*) Myobiidae; (*B*) Psorergatidae; and (*C, D*) Demodicidae (*C, Ophthalmodex; D, Demodex*). Ae, male genitalia dorsal; Gn, gnathosoma; Ch, chelicera; Pa, palp; As, female genitalia; Se, setae; I–IV legs. (Redrawn and modified from W. B. Nutting, in *Mammalian Diseases and Arachnids*, Vol. 1, 1984, with permission of CRC Press Inc., Boca Raton, Florida.)

(Bacigalupo and Roveda 1954). Protonymphal stages are also omitted in *Ophthalmodex artibei* Lukoschus and Nutting (1979). Myobiids (adults and immatures) and psorergatid adults are heavily setose, at least on legs, but demodicids are naked.

Most ova are elongate-oval, except there is a marked elongation in some *Demodex*. Immatures are rounded in the psorergatids and spindle shaped in the demodicids, whereas adults are rounded in the psorergatids, elongate spindle shaped or tortoise shaped in the demodicids, and rectangular in the myobiids. Legs are long and six jointed (5 + segments) in myobiid, psorergatid, and a few demodicid adults, and short (three segments) in all other demodicid adults. The front legs or hind legs III and IV of myobiids are often modified and equipped with processes (primitive) or pincers (advanced) as holdfast structures. These are intricately formed and positioned, and in hair-clasping types apparently adapted to a specific hair size (Fig. 11.4).

Ova

Ova of some species from each of these families are deposited in or on the habitat. Narrowly ovate eggs of some myobiids are secured to hairs by glandular secretions. Most psorergatid and a few demodicid ova are ovate; they are laid in parent-cut epidermal pits, sebaceous glands, or in crypts of the large head-cavity epithelia. Demodicid ova in hair follicles are often spindle shaped, whereas those of psorergatids are always oval. Certain demodicid ova located in the Meibomian ducts have bizarre projections (i.e., anterior prongs), unusual shapes (i.e., long snakelike), and are operculate, with larvae bearing epistomal egg teeth (Desch et al. 1971).

Immatures

Myobiid and psorergatid immatures mirror the morphology of adults except their legs are less prominent; in fact, they are mere stubs in some psorergatids. The psorergatid larvae, as well as adults, often invade the hair follicle—the myobiids only rarely and as larvae. Demodicid immatures resident in hair follicles, epidermal pits, and sebaceous glands are spindle shaped with either acuminate or rounded opisthosomal termini. Again, the demodicids in the Meibomian ducts are singularly modified, often with multiple holdfast processes, pronglike legs, and long, laterally flexible terminally spined palps. These last carry over from larvae through nymphs to adults.

Body outlines of demodicid immatures are often unusual. Lukoschus (personal communication) has recently described a nymphal stage that is not only "kangaroolike" in morphology but also has a supernumerary opisthosoma.

Digestive Tract

The myobiid digestive tract is probably complete; at least an anus is evident. Such is not the case in demodicids and psorergatids: no known species of either has an anus. In *Demodex folliculorum* the esophagus dead-ends against large endodermal midgut cells, which apparently obtain nutrients pinocytotically, process them intracellularly, and then deposit birefringent waste crystals of phosphorylated guanine in concretions (Desch and Nutting 1977). Sequestration of wastes and obviation of an anus are undoubtedly important to demodicid survival so that persistent hair follicle fouling does not occur. Cytophagy is the rule in all three families and hematophagy occurs only in myobiids. Tissue fluids and cytoplasm probably are preprocessed by salivary products in psorergatids and demodicids (Fig. 11.5). Myobiids are structurally most approximate to the free-living condition for obtaining food, processing food, and eliminating waste.

Specialized Structures

Specialized sensillae (thigmotactic?) are present in all three taxa; eyes are absent; and although there are no proven chemosensoria, the subgnathosomal pits in psorergatids and demodicids may subserve this function. Male demodicids and psorergatids have four dorsal podosomal tubercles which are certainly tactile sensitive.

Myobiid females possess a pair of aedeagal claspers related to mating on the skin surface: these are absent in the other families. Primitive myobiids, like *Archaeomyobia*, are clasperless (Lombert 1979).

In the hair-follicle and epidermal-pit-dwelling demodicids, a peculiar opisthosomal organ (proctodeum of Desch et al. 1970), along with other morphological features, is a useful taxonomic discriminant in separating and identifying the species. These structures are quite unusual in shape (flower- to saclike; multiple stamenlike to snakelike), have sexual dimorphism, and are absent in demodicids resident primarily in sebaceous glands, in most chiropterans, and also often in male mites (Desch 1973). Both ordinal and familial specificity are reported on the basis of these structures (Nutting and Desch 1979).

Supracoxal spines vary in size and shape in demodicids, are often absent in immatures, and may also be sensory, but are modified as holdfast structures in the immature stages of *Demodex marsupiali* (Nutting et al. 1980) and *Soricidex dimorphus* Bukva (1982).

Mechanisms, Including Behavior

Several mechanisms help these prostigmate mites survive a parasitic lifestyle, including modifications from the free-living (ancestral) condition: mobile structures, toky, and transference.

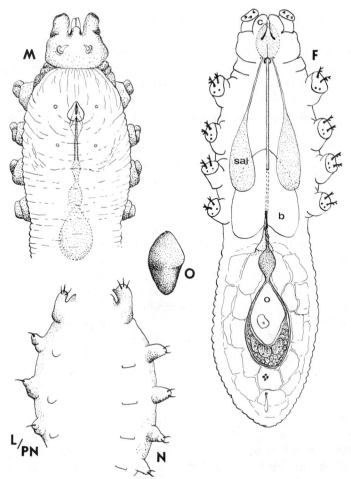

Figure 11.5 Structural details of stages in the life cycle of a typical *Demodex* sp.. O, ovum; M, anterodorsal male, showing position of genitalia and reproductive system. F, ventral aspect of female showing c, chelicerae; sal, salivary system; b, brain; O, ovary with reproductive system; guanine crystals (four dots) are directly anterior to fingerlike opisthosomal organ. Larval (L) and protonymphal (PN) idiosomal ventral views, left half, and nymphal (N) ventral view, right half, show stumplike legs and abbreviated ventral scutes. (Redrawn and modified from W. B. Nutting drawing submitted to the *Scientific American* based on originals by Margaret Nutting.)

Mobile Structures

Claws, hooks, spines, and even enlarged setae (as in myobiids) are probably tactile sensory as well as holdfast in nature. Palpal spines in demodicids are used to rip through the eggshell. We speculate that the opistomal egg teeth operate like wedges against the opercular groove in *Demodex longissimus* Desch, Nutting, and Lukoschus (Desch et al. 1972).

Large hooklike motile legs III in the larvae of *Demodex molossi* Desch, Nutting, and Lukoschus penetrate the tarsal duct epithelia securing the mites against the flushing action of secretions. Also useful are alar processes, ventral scutes, and large laterally flexible palps with spines to maintain mouthparts in the epithelium against the forces evoked by the motions of the eyelid while feeding.

In demodicid mites, extension of all body segments, appendages, and even egg laying are dependent on muscular contractions in the mite body that pull the exoskeleton against haemocoelic fluids (as hydroskeleton). Flexion in appendages is produced by contraction of minute, intersegmental muscles. This is probably also true in myobiids and psorergatids.

Toky

The peculiar sex ratios of demodicids puzzled us for several years. Lebel and Nutting (1973) surmised that thelytoky may be responsible in *Demodex caprae*, and Desch (1973) discovered that males were haploid (2 chromosomes = 2N); thus unfertilized ova yield males parthenogenetically. Recently, an undescribed species of demodicid in a rat showed all the evidences of thelytoky; for example, males larger and more numerous. From the published records and by extrapolation (Table 11.5) it appears that neither myobiids nor psorergatids are parthenogenetic. Parthenogenesis may be a mechanism for insuring transfer between hosts (see the section, Coevolution and Symbiosis).

Transference

The various measures of transference between hosts, the selection of appropriate loci on the hosts, and the sensory mechanisms used in these relocations are all critical to coevolutionary survival.

Longer legs, thick exoskeleton, rectangular shape, holdfast structures, atoky, and surface-wandering habits in myobiid adults demark them as basically ectoparasitic, not well fitted for moving into the pilosebaceous complex or burrowing. They are capable of transferring rapidly between hosts during loose contact, perhaps only of hair or thin membranes, as in chiropteran wings touching. They may have acquired, as have trombiculids, a response to CO_2 gradients or possibly use temperature gradients as primary sensory cues. Ova on cast-off hairs may also seed the young in the mammalian niche: a mechanism of transference not shared by psorergatids or demodicids.

Demodicids and psorergatids are quite different from myobiids, especially in body shape (spindle to oval shaped). They have a very thin exoskeleton, and burrowing or pilosebaceous-dwelling propensities. Thin exoskeletons make it impossible for demodicids to survive extrahost for more than a few hours, whereas myobiids, with thick exoskeletons, are able to survive for days in a moist environment. The opisthosomal organ in demodicids may subserve water balance. These are only in species that reside in the hair follicles or epidermal pits.

Other Host–Mite Relations

Host–mite relations may also serve as clues to the degree of coevolution validity of using either symbiote or symbiont as markers of the other's phylogeny. The following relationships exist for myobiids, psorergatids, demodicids, and their mammalian symbionts.

Host–Mite Ecologies

One test of symbiospecificity is the failure of parasite transfer between the hosts of the same habitat. Such tests can be observed under domestication (farms, zoos), including pets, which are notorious for transfer of organisms to man, as in *Sarcoptes* sp.. In the wild, nidicolous mites (many mesostigmatic dermanyssids) are noteworthy as interhost parasites of nest-sharing or communal-roosting species. We have found four species of bats of three genera "back to back" in a cave, but their demodicids are discrete. Myobiids of one of these, *Myotis* sp., are shared by several other genera of bats (Table 11.6). The possibilities for interspecies transfer have increased in bats and rats because of their broad migration patterns and man-made transfers, respectively.

Predator–prey relations in mammals are another niche-sharing intimacy which should be examined to determine ecological specificity in mites. Jameson (1970) has reported that *Protomyobia claparedei* (Poppe) is present on *Blarina brevicauda* and also on *Sorex cinereus*, and that *B. brevicauda* preys on *S. cinereus*. In our extensive exploration of demodicids of felines, some of which are known mousers, we have recently found rodentine demodicids, but these are surficial epidermal feeders. Whether or not these are synhospitalic species that were passed on during hybridization in cats or acquired from rodents remains for further research explication.

Representative records of multisymbiont species (symbioty to polybioty) and multisymbiote species infestations (synhospitaly or polyhospitaly) of prostigmate–mammal relations are listed in Table 11.6, and 11.7. Reports of synhospitaly of demodicids by Hirst (1919) for rodents or shrews, Fain's (1960) records for bats, and intermittently, even today (1983), by medico-veterinary practitioners for mutual domestic animal–man infestations are neither taxonomically confirmed nor examined histologically for their locus in the mammalian skin complex.

Records to date of symbioty for myobiids, psorergatids, and demodicids strongly suggest asymbiospecificity in the first two and symbiospecificity in demodicids, at least the more markedly adapted, that reside in the hair follicle.

Table 11.6   Records of Multiple Host Utilization (Polysymbioty) by Selected Species of Prostigmata of Coevolutionary Concern with Mammalia

Family	Prostigmata	Mammalia	Authority
Myobiidae	*Pteracarus chalinolobus*	*Myotis daubentoni*	Womersley 1941
		Chalinolobus gouldii	Womersley 1941
		Nycticeius greyi	Womersley 1941
—One sp. on 3 genera of Chiroptera; 1 sp. on 6 sp. of *Sorex*—Lombert 1979			
Psorergatidae	*Psorergates muricola*	*Lophyuromys aquilus*	Fain 1961
		Otomys irrogatus	Fain 1961
		Hybomys univettatus	Fain et al. 1966
		Apodemus sylvaticus	Fain et al. 1966
		Mus musculus	Hirst 1919
—One sp. on 5 genera of Rodentia; 1 sp. on 2 sp. of *Myotis*—Fain 1959			
Demodicidae			
	Stomatodex corneti	*Barbastella barbastellus*	Fain 1960
		Myotis myotis	Fain 1960
		Myotis dasycnemus	Fain 1960
—One sp. (or subspp.) on 2 genera; 1 subspp. on 2 species—Fain 1960			
	Demodex nanus[a]	*Rattus rattus*	Hirst 1919
		Rattus norvegicus	Hirst 1919
—One sp. on 2–4 spp. of genera as *Rattus*—paper in progress (Desch, pers. comm.)			

[a]Not confirmed in our lab. (Desch, pers. comm.); no assured multiple hosts to date for any species of the *Demodex*—we do anticipate, however, that conservative hair-follicle-dwelling species may carry through in speciation to share multiple hosts, but rarely if ever multiple genera.

Habitat and Adaptation

The myobiids are generally very similar to free-living mites, loosely adapted to the host habitat. The major adaptations are holdfast organs, including hair-adherent ova; processes, claspers, and setae to hold to hairs; aedeagae, claws, or hooks on tarsi for surface purchase; and long chelicerae for puncturing in both adults and immatures. The smaller size of immatures does make follicular invasion feasible. Also, less epithelial keratin is therein as a barrier to feeding with their less developed chelicerae.

Psorergatids have retained long legs as adults but have remained small enough to burrow epidermal pits of mite body size and invade hair follicles. Ova and nearly legless immatures are protected in these habitats, and the burrowing habit makes "nesting" possible with little if any screening effect against the oval body shape. Similar sequences of habitat-related events and adaptations seem to exist in the tortoise-shaped demodicids, as

Table 11.7 Selected Examples of Mammalian Species Synhospitaly (Congeneric Sharing) and Polyhospitaly (Intergeneric Sharing) of Symbiotic Prostigmata[a]

Prostigmata Family	Mammalia	Prostigmata	Authority
Myobiidae	*Myotis myotis*	*Myotimyobia myotis*	Lombert 1979
		M. klapaleki	
		Neomyobia myoti	
		N. klapaleki	
		Pteracarus chalinobolus	
		P. minutus	
		P. submedianus	
Psorergatidae	*Microtus agrestis*	*Psorergates microti*	Fain et al. 1966
		P. musculinus	Michael 1889
		P. agrestis	Lukoschus et al. 1967
Demodicidae[b]	*Homo sapiens*	*Demodex brevis*[c]	Desch and Nutting 1972
		D. folliculorum[c]	
	Mesocricetus auratus	*D. aurati*[c]	Nutting 1961
		D. criceti[c]	
	Carollia perspicillata	*D. carolliae*	Desch et al. 1971
		D. longissimus	Desch et al. 1972

[a]This implies either ecologic asymbiospecificity or genetic symbiospecificity with microgeographic evolution evolutionary convergence.
[b]Up to 4 spp. of *Demodex* on 1 host sp. by F. S. Lukoschus (pers. comm.).
[c]Confirmed by other authors.

Ophthalmodex. In both of these groups, waste is intracellularly stored reducing the chances of habitat fouling and host cellular derangements, although the latter is a surface inhabitant of the large head cavities.

In *Demodex* spp. all stages are modified in structure and function to fit the confines and physiology of the host. As in *Demodex criceti* Nutting and Rauch, small, short chelicerae permit the mite to make burrows—but only in thin nonkeratinized epithelia or in ducts, distended follicles, and skin glands of the pilosebaceous system. The species dwelling in epidermal pits are short and "squat"; the elongate gland dwellers are intermediate; and the duct and follicle dwellers are markedly elongate; all are spindle shaped in contour, even the ova in many instances (Fig. 11.6).

The peculiar opisthosomal organs found in this group seem to be linked with host skin habitat and physiology because only pit and follicle dwellers possess them. Furthermore, they all store waste crystals intracellularly,

MAMMALIAN SKIN HABITATS:

Topologic Sites of
Prostigmatic Symbiotes

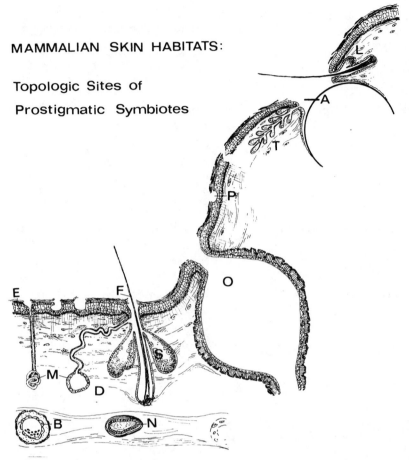

Figure 11.6 Mammalian skin habitats of certain prostigmatic mites showing topologically different sites of infestation. O, oral cavity; A, ocular or aural cavity; P, epidermal pits; F, hair in follicle; E, epidermis; D, dermis; B, blood vessels; N, lymph node; S, sebaceous holocrine gland; M, mixed apocrine and/or eccrine glands; L, eyelid; T, tarsal gland. Ectoparasitic myobiids are located broadly on E or in upper hair follicle; psorergatids are located in E and P, follicle, S, and E of A; *Ophthalmodex* spp. or *Stomatodex* spp. and *Demodex* spp. are widely dispersed in follicles, S or P, or all other sites.

have few setae or sensillae, and even occasionally conform to host skin habitat by reduction in life-cycle stages. In these species the diameter of the pilosebaceous entryway seems to be an important selective factor (especially in regard to the diameter of adult transfer stages) in evolution (Table 11.8). It appears that these parameters and the high mutation rate are of prime importance to the establishment of synhospitaly in this group by way of microgeographic evolution.

These facts suggest that these mites are specifically adapted to particular

Table 11.8 Characteristics of Stages in the Life Cycle of Three Demodicids (*D. gapperi, D. longissimus, D. molossi*) Restricted to the Ducts of the Tarsal Glands[a]

Order Host Demodicid	Rodentia *Clethrionomys gapperi* *D. gapperi*	Chiroptera *Carollia perspicillata* *D. longissimus*	Chiroptera *Molossus molossus* *D. molossi*
Duct diameter (cross-section)	50 μm	120 μm	220 μm
Stages (length × width)			
Ovum	295.8 ± 23.6 x 29.9 ± 4.7	805.1 ± 84.1 x 48.4 ± 4.3	238.1 ± 9.5 x 66.7 ± 3.8
Larva	284.9 ± 36.1 x 34.5 ± 7.7	908.4 ± 116.4 x 49.7 ± 5.6	580.3 ± 93.0 x 64.8 ± 8.7
Protonymph	—	896.1 ± 131.4 x 57.9 ± 6.5	625.7 ± 99.5[b] x 74.4 ± 8.7
Nymph	346.9 ± 39.8 x 41.0 ± 8.8	839.0 ± 96.4 x 67.5 ± 9.2	667.3 ± 95.0 x 83.2 ± 9.5
Adult Male	262.3 ± 12.9 x 46.9 ± 7.1	722.7 ± 61.8 x 73.8 ± 4.0	451.8 ± 37.2 x 82.5 ± 5.6
Female	360.8 ± 28.5 x 43.7 ± 3.5	790.0 ± 149.7 x 80.1 ± 4.1	582.8 ± 126.4 x 91.0 ± 5.4

Source: Data rearranged from Desch et al. 1972 and Nutting et al. 1971.

[a]All measurements represent means and standard deviations in micrometers for 20 specimens of each.

[b]Only 17 specimens.

microhabitats in specific hosts: most relevant to symbiospecificity and studies of phylogeny.

Impact on Host Habitat

Members of all three taxa, the myobiids in mice, the psorergatids in sheep, and the demodicids in feral and laboratory animals, have been incriminated as producing marked pathology in host mammals. The few feral animals with psorergatid mange seem to have been in poor physical condition prior to the mite-caused lesions, as Hirst (1919) also noted in demodicid infestations of domesticated animals. Otherwise, in wild mammals psorergatid populations are usually small, and the mite-caused lesions are few and localized.

It seems apparent that the range of pathology must be examined from the coevolved impact of an individual mite to major multimite pathogenesis expressed by host humoral or cellular responses with, also, any impact on the mites at both gross and histological levels. Theoretically the least damage to either symbiote or symbiont would indicate the longest span of coevolutionary association.

HABITAT AND PATHOGENESIS

Myobiids, psorergatids, and demodicids are all holostadial parasites limited to mammals. Both adults and immatures obtain their food by puncturing the skin with needlelike chelicerae and sucking up cell cytoplasm and tissue fluids which have been preorally changed by salivary products. The habitats of all three groups are basically restricted to tissues of ectodermal origin.

These symbiotic mites are so minute, in all stages, that only very minor pathology of the epidermis and pilosebaceous complex is evident in histological sections at the feeding site of the individual mite. Nutting and Sweatman (1970) noted the lack of cellular reactions on penetration into the dermis of several dozen *Demodex antechini* Nutting and Sweatman. Minor cell damage by individuals or few mites is rapidly repaired by mitoses of the epidermal stratum germinativum. When mites of these three families, especially demodicids and psorergatids, are trapped within the pilosebaceous complex they reproduce, and resultant large numbers produce lesions such as the large benign tumors (up to 3 cm) for *D. antechini* or large papules (up to 5 cm) in *D. caprae* (Nutting 1975).

The skin habitats of these three families of parasitic mites, the pathology produced by each mite species, and the cellular and biochemical response of the mammalian species parasitized can be used to test the hypothesis that the mite species least pathologic to its mammalian symbiont species is the most coevolved for its role in symbiospecificity. Conversely, the more pathogenesis and host reaction the more likelihood of asymbiospecificity.

The major habitats of the mammalian skin utilized by mite species of these three taxa are represented by Figure 11.6. These include (1) the external epidermis, (2) large-cavity epithelia, (3) epithelia of the hair follicle, (4) holocrine (sebaceous) glands and duct epithelia, (5) apocrine (sudoriferous) glands and duct epithelia, (6) eccrine (sweat) glands and duct epithelia, and (7) mixed apocrine–eccrine glands and duct epithelia. The myobiids may puncture the dermis and blood vessels, whereas psorergatids and demodicids have been found in the dermis after penetration of the stratum germinativum. *D. canis* in heavy infestations in the dog can be located in the circulatory system including the lymph nodes (Canepa and daGrana 1941).

In psorergatids and demodicids follicular crowding may lead to hyperplasia which continues as a renewing resource of food, investing the mites in a papular lesion. This host cellular response to the included mites is a very unusual example of a symbiote–symbiont "cooperation" which may be of coevolutionary symbiophyletic significance.

The pathologic condition known as "demodectic mange" by *D. canis* and *Staphylococcus* spp. is an often fatal disease and reputedly associated with an immune response through a T-cell deficiency (Scott et al. 1974). This pathogenic condition, we believe, is atypical of coevolved feral animal symbioses, because dogs are domesticated species long reared under changed genetic and environmental conditions affecting the structure, function, behavior, and physiology of the skin. This relationship, therefore, should have little significance to studies of coevolution.

Although structures of all skin habitats are basically similar among monotremes, metatherians, and eutherians, they differ in size, type, physiology, and topologic distribution, especially for glands, ducts, and the presence or absence of inert hairs of keratin. The hairs are also of different types, either continuously formed, as are human scalp hairs, or possessing a pattern of origin (anagen stage), growth (catagen), and resting (telogen) stages prior to molt. In juvenile mammals initial pilosebaceous components often differ markedly in structure and physiology from the adults. Glandular structures also produce secretions that vary to some degree in chemistry, especially with the cycles of reproduction. The skin components may vary drastically in thickness and keratinization, both topologically and among different species of mammals (Montagna and Parakkal 1974).

As ectoparasitic, primarily skin-surface-dwelling organisms, the adult myobiids and some immatures merely cruise the outer epidermis, puncturing to the dermis, often at the hair–epidermal interface. Cell cytoplasm, dermal tissue fluids, and the contents of capillaries are pumped into the gut with wastes which are presumably lost through the anus. Feces from immature stages may be pathologic in hair follicles, particularly in large in situ populations. Ova are secured to hairs, and larvae either move into the upper reaches of the hair follicle or cling to the skin where they feed much as adults. Feeding, except in massive infestations, produces little host cellular response except for some minute keratinization of the local epidermis. Huge populations, however, produce loss of telogen stage hairs and probably in some areas of multiple invasion of the follicle, an inflammatory response (dermatitis in mice; Baker et al. 1956). The fact that mice groom and scratch areas of the body suggests that they itch and that they also may develop localized immune reactions to salivary products of the mites.

Normally, myobiid populations are small with no gross sign of lesions. All but ova actively transfer from host to host, whereas adults and the ova could be lost on hair molting. This is borne out by our recent study that all but four adults and many ova departed from the cold carcass of *Peromyscus* sp. on which large numbers of immatures existed (Table 11.5). In no case are large lesions reported for members of this family under feral conditions. Again, however, such as laboratory mice are prone to develop myobiid mange.

Psorergatid populations are also usually small. Based on morphological considerations, we believe that adults are the only transfer stage and move across the mammalian surface or between mammals. They seem to select non- or weakly keratinized areas of the skin and either cut body-sized pits, as in some rodents, or move into the hair follicles to puncture and ingest epithelial cell contents. In the hair follicles they may burrow nests or produce keratin-filled lesions with pitted hyperplastic epithelial lesions (see Lukoschus 1967). Psorergatids deposit huge ova, nearly two-thirds of adult size in pit or nest luminae. They hatch and larvae appear. The larvae molt to become protonymphs and then nymphs, all with stubby legs, and then develop to long-legged adults. Aside from epidermal pit or nest hyperplasia and keratin produced by adults, low population density causes no gross damage to the host (Kok et al. 1971). Adults burrow into the stratum granulosum and here apparently produce progeny which in turn produce the large "nests" whose centers are either fluid filled if there are few mites or keratin filled if there are many adults. Should these be mechanically ruptured or should excess mites bore to the dermis, host inflammatory responses are produced, followed by granulomata with giant cells which ingest the mites phagocytically (Flynn and Jaraslow 1956). Also, in case of multiple mite death deep in situ in the epithelial pockets, we would expect an immune response established in the host. Related to this, the fact that these adults have very minute chelicerae (about 20 μm) suggests that salivary products may contain potent enzymes (Sweatman 1971) that aid in burrowing and could also establish immune reactions.

The turtle-shaped demodicids, as *Ophthalmodex* spp., inhabit the large epithelial-lined head cavities of the host. *Ophthalmodex artibei* occasionally invades and distorts the lachrymal gland ducts (Lukoschus and Nutting

1979). *Ophthalmodex* adults puncture and feed on epidermal cells and abrade the conjunctiva with claws (Lukoschus et al. 1980). Ova were found free in the rippled epithelium beneath the eyelids. Larvae and nymphs with stubby bilobed large-clawed legs are cytophagous on epithelial cells beneath the lids. The epithelial lining of lachrymal glands, when invaded by adults, showed moderate cell destruction and flattening with leucocyte infiltration near their greatest distension. Considering that the mites were alive, and have small chelicerae and no anus, we believe all changes in the duct were due to capillary stenosis and perivascular infiltration by pressure on the epithelium rather than a true inflammatory or immune response. Only adults probably transfer to other hosts through direct contact.

For *Demodex* spp., the patterns of habitat use and pathology vary with all known stages, especially among synhospitalics with those semiendoparasites most closely matching the limitations of their habitat. Mites have now been found in all seven habitats described earlier and also (1) free in the dermis and in subdermal fascia, (2) in blood vessels in various internal organs, and (3) in the reticuloendothelia of spleen and lymph nodes (Nutting 1975). Rapid dehydration and death easily occur because of their small, elongate body, short (three-segment) legs, and thin exoskeleton (1.5 hours at 20°C in RH 50%) (Nutting 1976a, b). This indicates that rapid transfer and habitat matching or cutting pits in the epidermis must take place rapidly. As for transfer behavior, *D. canis* does not in any stage leave a cold carcass but dies in situ (Nutting 1950).

Concerning pathogenesis, demodicids distort the host–mite habitats or extrahabitat areas much more severely than do psorergatids. Although they are basically cytophagous (Nutting and Rauch 1963), in heavy infestations they decimate sudoriferous gland cells and sebaceous gland cells and plug Meibomian (tarsal) gland ducts. They penetrate dermis and capillaries and thus produce an inflammatory response followed by granuloma. They also produce large benign tumors, nodules, and papules, and reduce eccrine gland epithelia. In association with bacteria and mite death, they undoubtedly produce a short-term immune response (Scott et al. 1974). Under normal conditions they are topologically distributed in their species-specific habitat (see Distribution).

Massive infestations of demodicids produce pathologies much beyond those occurring even in the heaviest infestations of myobiids, psorergatids, or "ophthalmodecids." This could at first glance suggest a greater degree of maladaptation to the mammalian habitats. Upon review of our records, however, we find that, in all but chiropterans, the most markedly diseased animals are domesticated: either farm or laboratory animals or mutual pets (as man and dog). In comparing mite habitat derangements in chiropterans and domesticated animals, we find only limited and localized macrolesions in bats (one to three per bat), whereas such hosts as cattle may have several thousand nodules with 16 to several thousand mites each. Furthermore, in rodents and man there is an adaptation related to host hormones: males have larger populations of demodicids than do females; and a relationship exists between stress and hydrocorticosteroids in dog (G. Muller, personal communication) and in man (Hakugawa 1978).

The following conclusions are drawn from the discussion:

1. Myobiids are true ectoparasites with superficial habitats and very little pathology evident. Thus they are little species specific to habitat or host damage of mammal species. Microanalysis, histology of lesions, and further studies of their biology may show a broad-range immune response, suggesting that they recently or weakly coevolved with hosts, much as in trombiculids. Certainly at the individual mite level they are much more pathogenic than individual demodicids.

2. Psorergatids are semiendoparasitic, except during transfer. They commonly, even with small populations, produce marked pathology. They seem more recently coevolved with hosts than do the demodicids; however, this is far from certain because there is so little evidence for or against the establishment of immunity in either group.

3. The demodicids, such as *Ophthamodex*, *Rhinodex*, and *Stomatodex* (Fain 1959b, c), are limited to a moist, undifferentiated habitat of mammalian head cavities with an unlimited food supply of renewing, nonkeratinized cells. No large populations have been found to assess the degree of pathogenesis. Tentatively, they are considered somewhat more highly coevolved than psorergatids.

4. The demodicids are semiendoparasites, and under feral condition their infestations are low with only very minor pathology, at the expense of renewable resources. (Domesticated mammal pathology is, we believe, aberrant in coevolution.) Their adaptations for habitat use, barring changes in host physiology, are very host specific and precise, thus indicating a long, unbroken chain of coevolution with mammals.

Based on this analysis of host–mite adaptations for interactions, including pathologies in their shared habitats, all three mite families seem coevolved at the ordinal level (ordinal specificity). Under marked genetic or environmental changes, as in laboratory, domesticated, or emaciated feral animals, the most marked pathologies occur with the spindle-shaped demodicids (e.g., tumors in mammals and mite destruction by host giant cells). Myobiid–mouse relations under laboratory conditions produce only erythematous mange.

In healthy feral mammals, however, with "normal" small populations of mites, pathologic interactions occur, in order of most to least symbiont damage, from the Myobiidae, Psorergatidae plus tortoise-shaped Demodicidae, and the spindle-shaped semiendoparasitic Demodicidae. Thus the hypothesis of "least damage longest domiciled" would seem to indicate that these spindle-shaped demodicids are the most intimately coadapted with mammals, symbiospecific, and potential indicators of symbiospecificity, whereas the others are asymbiospecific.

DISTRIBUTION

All known extant myobiids, psorergatids, and demodicids, as holostadial parasitic symbiotes restricted to mammals and thus at the symbiont mammalian subclass level, have coevolved either by way of symbiospecificity or asymbiospecificity. Their current mutual geographic distributions, especially of endemic coevolved species, lend some corroborative evidence for distinguishing between these coevolved patterns of symbiosis in our attempt to determine coevolved phylogenies. The few "records of the rocks" (geologic distribution) and even introduced nonendemics are also aids in this analysis.

The record of *Protacarus crani* from the Devonian sandstone in Europe (Hirst 1923) shows that a free-living, possibly pachygnathid, prostigmate mite (Krantz 1978) had evolved well before the origin of terrestrial vertebrates. It undoubtedly shared all of the preadaptive features for parasitism with mammals discussed elsewhere in this chapter. Records from the Chiapas amber of the Miocene epoch in Mexico of a mesostigmatic mite, *Dendrolaelaps* sp., congeneric with extant Mesostigmata (Hirschmann 1971) and eight astigmatic mites (Woolley 1971) differing little from modern forms, indicate a worldwide geographic distribution of mites with a latent, even nidicole, potential for symbiosis with the Mammalia.

It appears that there was an adequate time span for the mites preadapted for parasitism to become symbiotic with premammalian reptiles as parasitic modern Prostigmata such as the cheyletoidean Cloacaridae. Either these last or the cheyletoidean Ophioptidae are prime candidates as reptilian symbiotes that may have given rise even in Jurassic times to either or all of the three symbiotic prostigmate cheyletoidean families under consideration. By at least the Miocene epoch potentially nidicolous or asymbiospecific mites at generic levels were already established at an essentially modern level. It seems plausible that preadapted precursors of prostigmatic mites have a long evolutionary history stemming from Pangean times to today and have possibly provided asymbiospecific and symbiospecific species from reptilian ancestral symbiotes as the mammals evolved and spread to all parts of the world.

In reviewing the geographic distribution, we elect to concentrate upon the somewhat isolated land mass Australia, with a limited number of endemic mammalian orders [Marsupialia, Rodentia (Muridae), and Chiroptera], the recent invasion by man (Primates) and his companion dingo

(Carnivora), and the historic influx of man-transported mammals, especially Rodentia, Carnivora, and Artiodactyla.

Our speculations on geologic distribution would predict eventual recovery of at least one of the cheyletid-related symbiotes from the Australasian Monotremata: so far no such records are available. The Metatheria, however, and its sole order, Marsupialia, are well represented by the parasitic myobiids and demodicids (Table 11.9).

Fain (1979) has reiterated the view (Fain 1974; Lombert 1979), based primarily on characters of the holdfast legs I, that the Myobiidae have coevolved with three major (ordinal) groups: the (1) most primitive marsupialians, (2) insectivorans and chiropterans, and (3) the rodents. This seems borne out by specimens from Australia even though the Insectivora are endemically absent. One should note, however, that the Chiroptera are both endemic in and wide ranging between many continents, with habitats shared by rodents and marsupials, so that further studies may reveal cross-transference between symbiont groups. Furthermore, it should be noted that such species as *Hipposiderobia belli* Fain and Lukoschus from Australian bats show rodent-type clasping legs I (Fain and Lukoschus 1979a, b), perhaps resulting from either convergence or interspecies transference. Chiropetra also share all three families of parasitic mites.

Several researchers have recovered species of myobiids, possibly specific, to species, genera, or families of bats and rodents from Australia, Europe, or Japan (Jameson and Dusbabek 1971; Dusbabek 1969; and Uchikawa, personal communication). These reports all indicate phylogenetic linkages with host geographic distribution. Numerous records exist for myobiid species that share the hosts of the same genera and families (see Lombert 1979), but they are difficult to document as having either generic or familial specificity because of the many instances of poly-hospitaly and polybioty (Tables 11.6 and 11.7). The cosmopolitan *Myotis myotis* shares roosting sites with other bats and also several species and genera of myobiids.

It seems plausible that a myobiid host ecologically isolated for a long time or a sympatric host species with shared roosting isolated from other mammals could coevolve either symbiospecificity or asymbiospecificity. Although no clear evidence exists, as yet, the recovery of such mites as *Radfordia* sp., usually a rodent myobiid, from a lagomorph suggests that such specificity would soon disappear under ecological overlap of unusual host combinations. This may well be happening with the introduction of nonmurid rodents and domesticated mammals to Australia.

The psorergatid, *Psorergates ovis* Womersley, are well known from introduced Australian sheep (Womersley 1941). It is plausible, also, that transfer of psorergatids occurred between a predatory shrew and an insectivorous shrew. The Psorergatidae are known from the predatory Carnivora (Lukoschus 1967). Thus it is reasonable to expect dingos to share species with sheep. Furthermore, the morphologically similar demodicids, as *Ophthalmodex* spp., in different parts of the world are also found in Rodentia, Chiroptera, and Primates. For both taxa, the evidence from geographic distribution, although suggesting coevolution with mammals, suggests that no host specificity is present.

In Australia demodicids such as *D. brevis* are host species specific for most domesticated introduced mammals, including man. The dasyurid demodex, *D. antechini*, is related to *D. marsupiali* of South America (Nutting et al. 1908b), but several bats (as *D. macroglossi*, Desch 1982) and a wallaby (Oppong, personal communication) have their own discrete species. In fact, the records of two host species, geographically associated and sharing the same demodicid (Table 11.6), have not been confirmed by modern criteria for species distinctions.

We would expect that well-isolated mammalian demes, sibling species, species recently sympatric, or even hybrids may harbor identical demodicids. However, two sympatric rodents, *Peromyscus leucopus* and *P. maniculatus*, possess different demodicids (Nutting and Desch 1979), whereas the initial study shows that demodicids of sympatric *Rattus* spp. apparently differ only in size (Desch, unpublished data).

The few records of topologic distribution for the free-wandering myobiids seem to indicate that there are few on-host isolating barriers to interhost transfer. Hairless and heavily keratinized areas of infested mammals, which could create a barrier to myobiids, are on the appendages including tails, thus sustaining nondisjunct distribution. In *Peromyscus leucopus*, myobiids are found in any haired area of the body, whereas in several species of bats the immatures and adults of mites, such as *Pteracarus* spp., are limited to the wing membranes, which are sparsely haired (Fain and Lukoschus 1979a, b). Considering the moderate range of holdfast and feeding modifications and records of polyhospitaly in these long-legged essentially free-living forms, there are few selective barriers to either topologic distribution or to interhost transfer. Even competitive exclusion is unlikely in their coevolution with mammals because resources are always overabundant and readily available on the symbiont surface.

Similarly, the thick keratinization of epidermis may be the only barrier to psorergatids. They show, as do the demodicids, a higher recovery rate for the head region of mammals (Lukoschus 1967a, b). For these taxa, this topologic selectivity may be related to interhost transference which, unlike myobiids, may require closer contact for a long period, such as during nursing.

Members of the genus *Demodex* have a very intimate tie between the various and often disjunct skin habitats (see the section, "Habitat and Pathogenesis") and specific species. Certain *Demodex* species are now known to be restricted to the Meibomian glands, such as *D. gapperi* (Nutting et al. 1971); to the sebaceous glands or their derivatives, such as *D. brevis* (Desch and Nutting 1972); or to the epidermis, such as *D. criceti* (Nutting 1975). Except for these last, which cut body-sized pits as housing,

Table 11.9 Representative Prostigmatic Parasites, with Authority, Showing Recovery for Mammalian Order[a]

Mite Family (Subfamily)	Mammalian Order	Mite Species	Authority
I. Myobiidae see Lombert 1979	Metatheria		
	Marsupialia	*Archaemyobia inexpecta*	Jameson 1955
	Eutheria		
	Insectivora	*Eadiea desmanae*	Lukoschus 1969
	Chiroptera	*Ewingana molossi*	Dusbabek 1968
	Rodentia	*Idiurobia idiuri*	Fain 1973
	Lagomorpha	*Radfordia cricetulus*	Fain and Hyland 1970
II. Psorergatidae see Giesen 1982	Eutheria		
	Insectivora	*Psorergates sorici*	Lukoschus 1968
	Chiroptera	*Psorergates nycteris*	Fain 1959
	Rodentia	*Psorergates simplex*	Tyrell 1883
	Primates	*Psorergates cercopitheci*	Zumpt and Till 1955
	Carnivora	*Psorergates mustelae*	Lukoschus 1969
	Artiodactyla	*Psorergates ovis*	Womersley 1941
	Lagomorpha	*Psorobia* sp.	Giesen 1982

III. Demodicidae
see list this report

Eutheria		
Chiroptera	*Ophthalmodex artibei*	Lukoschus and Nutting 1979
Primates	*Rhinodex baeri*	Fain 1959
(Rodentia)	undescribed	Lukoschus, pers. comm.
Metatheria		
Marsupialia	*Demodex marsupiali*	Nutting et al. 1980
Eutheria		
Insectivora	*Demodex talpae*	Hirst 1921
Chiroptera	*Demodex aelleni*	Fain 1960
Primates	*Demodex saimiri*	Lebel and Nutting 1973
Rodentia	*Demodex glareoli*	Hirst 1919
Carnivora	*Demodex canis*	Leydig 1859
Lagomorpha	*Demodex cuniculi*	Pfeiffer 1903
Perissodactyla	*Demodex caballi*	Desch and Nutting 1978
Artiodactyla	*Demodex bovis*	Stiles 1892
Edentata	*Demodex sp.* }	in hand, U. Mass. Lab. W.B.N.
Hyracoidea	*Demodex sp.* }	

[a]See Figure 11.8.

619

topologic distribution and survival of mites seem related by the correlation between mite diameter and that of the free-space entryway to the habitat (Table 11.8). The many cases of synhospitaly now under study at the histologic level (Desch and Nutting 1972) for *Demodex* spp. show species-specific topologic distribution suggesting microgeographic evolution (Nutting and Desch 1979). This topic is explored more fully later in this chapter.

Few recoveries or analyses have been made in attempts to link the distribution of symbiotic prostigmate mites with the influence of host ecologies in coevolution. Representatives of all three families have been found on at least one terrestrial host of short (about 2-year) to long (about 10-year) life span, and from tropical, temperate, and arctic zones. Symbiote extinction due to symbiont, environmental ecologies may be responsible for the absence of prostigmate parasites on such as the monotreme Duck-billed platypus.

Myobiids, ectoparasites with thick exoskeletons, seem able to withstand desert conditions, as they are reported from the Kangaroo rat, *Dipodomys* sp. (Howell and Elzinga 1962). Those restricted to the wings of night-flying bats, such as *Pteracarus* sp., evidently withstand remarkable wing velocities and changes in temperature (Fain and Lukoschus 1979a, b). Burrowing moles and rodents which hibernate are also infested (Lombert 1979). There are no records of recovery from marine mammals or any mammal in conjunction with potentially selective seasonal or diurnal changes, or are there any records of conditions under which interhost transfer occur. Myobiids are reported to date from only five orders of mammals, all of which are nearly totally haired.

Psorergatids are known from muskrats and beavers, both freshwater inhabitants, as well as from the burrowing moles and rodents (Lukoschus 1967a, b). Records have also been obtained from mammals of varied food habits, from herbivores to insectivores to carnivores. Transfer between predator and prey seems "ecologically" likely. Psorergatids, although not known from the Order Marsupialia, are present on six mammalian orders including the thick-skinned Artiodactyla.

For demodicids, we now have records from a marine mammal (*Demodex zalophi* Dailey and Nutting 1979), several fresh water species, fruit and insectivorous bats, desert and rain forest mammals, and so on. Several carnivores studied in our laboratory so far show no species sharing with their prey. Demodicids are known from 11 orders including a wide variety of skin and hair structures.

It is plausible that, except during transfer, myobiids are most affected by the extrahost or skin surface environmental conditions. During transfer very short-term contact at any orientation would seem adequate for even host interspecies transference. The other two taxa, as semiendoparasites, are protected from the extrahost environmental changes except during transfer, which can only transpire given adequate long-term contact such as during nursing, grooming, or copulating. For all of these small-bodied (average 1 mm) mites, we anticipate several thousand new species will be described in the next few years! To settle some of these problems of distribution and coevolution, research attention to measures of transference should be focused on the endemic mammals before these are either exterminated, expatriated, or contacted by introduced mammals.

Structural features of the mite habitats, such as hair follicle diameters (previously discussed), which provide the physical selection factors of the ecology of these symbiotic mites are noted extensively elsewhere in this chapter. So far only demodicids have been maintained extrahost to assess any physiological influences on mite characteristics, survival, and distribution. The environmental lipids versus mite lipids have only been studied for *Demodex caprae* (Stromberg 1968). These mites survived 100 days as avian skin subplants, and their taxonomic features remained stable throughout the full life cycle when compared to freshly obtained specimens (Nutting, data unpublished). Similar subplants of several demodicids on like species-infested mammals failed to show any lethal immune reactions. The Stromberg study suggests that host lipids may inhibit mite development and reproduction, not survival (Nutting 1975). To verify this we need extensive experimental work to explain the extrahost, symbiont and symbiote, biochemical and behavioral factors for their effect on topologic or intersymbiont transfer distribution.

The few following observations are very important and needed in future studies of mite survival and distribution with regard to coevolution:

1. Study the effect of sex hormones and hydrocorticosteroids on inhibition or increase of demodicid reproduction (Gray 1968; Hakugawa 1978), and the importance in this of seasonal timing of peak populations to distribution between hosts.

2. Search for intimate multiple species aggregation on bats in roosts, in zoos, or even in homes, which house multiple mammalian pets and man. Both interspecies and topologic studies are needed, especially at the level of skin histology.

3. Foster nursing or hybrid crossing of infested mammals; examine natural hybrids with respect to mechanisms of transference and locus of mite colonization.

4. Culture mites in subskin cylinders or tissue culture to settle taxonomic distinctions and to use for a variety of studies such as immunology and transfer.

5. Examine for the symbiont species from isolated or unusual environments, or from mammalian orders from which myobiids, psorergatids, or demodicids have not been recovered.

As suggested by Nutting (1968) investigators of these minute parasites need to combine taxonomic studies with delineation of the distribution and characteristics of the habitat, especially with experimental work such as that just suggested and in the section, "Limits and Cautions." It seems clear that only experimental rather than observational work can provide proof for patterns of coevolutionary phylogeny in terms of current mite–mammal species distribution.

Distribution patterns seem to indicate that myobiids, psorergatids, and the free-moving, large-cavity-dwelling demodicids (*Ophthalmodex, Stomatodex, Rhinodex*) are asymbiospecific. Only members of the genera, *Demodex, Pterodex*, and the new (Bukva 1982) *Soricidex*, are likely to be symbiospecific with mammals. Special research attention should be paid to the hair-follicle-dwelling members of *Demodex* because the entryway to the pilosebaceous complex appears to be a physical, natural selection barrier in part responsible for both maintaining symbiospecificity and permitting microgeographic speciation of the symbiotes.

COEVOLUTIONS, SYMBIOSPECIFICITIES, AND SYMBIOPHYLOGENIES

Previous sections in this chapter concentrate most attention on the myobiid, psorergatid, and demodicid symbiotes despite the fact that the mammalian symbionts provide all of their life support systems and have coevolved with them, possibly even to shared phylogenies.

This section discusses, reviews, and expands on some coevolved relationships that seem critical to solving prostigmate–mammalian problems like the degree of coevolution, of symbiospecificity, and of phylogenetic parallelisms.

Coevolved Ecologies

It seems clear from the fact that certain members of all three mite families have been obtained as totally dependent parasites from species of the primitive Order Insectivora that coevolution is assured for each at least at the level of order. Ordinal specificity for myobiids is suggested by Fain (1979); ordinal and familial specificity for demodicids is noted as plausible by Nutting and Desch (1979), and host–species specificities by many authors for representatives of all three families (Nutting 1968; Lombert 1979; Giesen 1982).

Whether or not these specificities are "beyond reasonable scientific doubt" these mites must, at one time in evolution, have been adapted to the adapted components of the mammalian skin and to mechanisms for insuring transference among or between host individuals with like skin characteristics, with like niches, and like habitats. Thus the coevolutionary pathways were open in two directions early in mammalian evolution for either phyletic or pseudophyletic parallelisms dependent on the environment shared by both symbiote (during transfer and colonization) and mammalian symbiont (during mite transfer or niche utilization).

Ecologic Adaptations

Although remarkable adaptive radiations in many mammalian taxa led to marked differences in internal organs, behavior, and longevity, the physiology and structure of the skin have remained relatively unchanged. Histological studies made in our laboratory of skins from several haired and hairless body surface areas of representatives of seven orders reveal relatively minor differences in:

1. Hair (or none) size, types, distribution, and ontogeny.
2. Epidermal and dermal thickness in ontogeny.
3. Gland size, types, distribution, and ontogeny.
4. Structures aiding physiological homeostasis in ontogeny.

See Montagna and Parakkal (1974) and Marples (1965) for structure and functions of the skin. These adaptations are, however, topologically modified and differentially distributed.

The locus of the symbiote, the ecological surroundings or the physical, chemical, and biotic surface conditions provided by the symbiont, even in homoiotherms, will determine the survival of the symbiote. In this sense semiendoparasites such as demodicids will have a less fluctuant environment than the ectoparasitic myobiids if skin features are equal. In either case, the most critical ecological period would be that during transfer between hosts. Geographic distribution of mammals changed adventitiously or by man have produced shared ecological niches and thus opened the way to asymbiotic mite transfer.

Demodicids, thin-exoskeletoned, and so subject to desiccation, would be less likely to survive during transfer than thick-exoskeletoned myobiids. Recovery of the myobiid *Radfordia* sp. from the desert-dwelling *Dipodomys* spp. (Howell and Elzinga 1962) may be related to such environmental resistance, whereas the recent report of *Demodex zalophi* found in the marine California sea lion definitely shows survival during transfer under a wide range of physical and chemical host-environmental conditions both terrestrial and aquatic. Myobiids move rapidly from a dead host during field collecting whereas *D. canis* dies in situ in dead dogs (Nutting 1950).

In all respects the generalized myobiids seem better adapted to host transfers and ecologies in terms of coevolution than do the other three groups, which seems related to their generalized morphology and adaptation to a surficial skin habitat (Fig. 11.7).

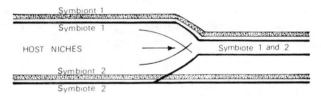

Figure 11.7 Diagrammatic representation of ecologic specificity (asymbiospecificity) as suggested for most prostigmatic mites—even the holostadial myobiids, psorergatids, and the demodicids of the genera *Ophthalmodex*, *Rhinodex*, and *Stomatodex*. When host niches overlap transfer may be effected by the mites. (Modified and redrawn from Nutting, 1968.)

Mite Habitats

The conservative nature of the mammalian skin and its adnexae would at first seem a poor isolating mechanism barring marked coadaptation to the host environments, especially for ectoparasites. Features matching myobiids to habitat are little modified from the free-living preadapted mites except for holdfast prominences or pincers on legs, elongation of chelicerae, and some exaggeration of tarsal claws. Demodicids, on the other hand, have evolved a spindle-shaped morphology (especially width) which shows a remarkable correlation in adults with the entryway to their habitat. They, especially in immature and egg stages, are often strikingly modified to match the confines of their habitat, and even as adults they have highly specialized measures of survival, both external (holdfasts) and internal (opisthosomal organs, parthenogenesis, waste removal) mechanisms.

The elongate demodicids seem to have evolved chemosensory selection, resource partitioning for synhospitalic species, arrhenotoky as an aid to transference, and adaptive biochemical controls to population dynamics: all related to habitat conditions.

In ecological terms, ectoparasitic myobiids are more exposed and unprotected environmentally and more reproductively selected (*r*-select) than the markedly protected endoparasitic, spindle-shaped demodicids (*K*-select). They both are pathogenic, but the former feed deep in the dermis often puncturing capillaries and thus probably produce minor cell reactions and, if numerous, possibly a T-cell and local immune response. However, the demodicids, in small numbers, consume only renewing epithelial cells and are normally buffered by layered epidermis from the host internal milieu. When, however, large numbers of these mites break through into the dermis or rupture capillaries, an inflammatory response occurs with "predation" by host giant cells (Nutting 1976a, b). Small (about 5-mm diameter) lesions in goats caused by *D. caprae* present an unusual, mainly lipid-filled habitat with a boundary of renewable resources. These usually rupture on enlargement after which a granuloma is formed with the phagocytic giant cells. The host cellular, immune, and foreign body re-

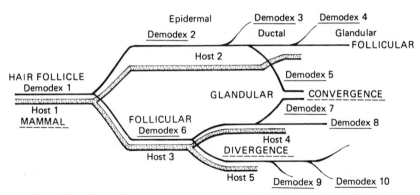

Figure 11.8 Plausible pathways of coevolutionary bonding (symbiospecificity) in mite–mammal associations (symbioses) and microgeographic evolution (as in *Demodex* spp.). (Modified and redrawn from Nutting, 1968.)

sponses to dermal invasion seem also conservative, being essentially identical in all mammals we have examined. Furthermore, in demodicids, possibly owing to genetically oriented breeding and environmental management, major cases of pathogenesis are known only from domesticated or laboratory-maintained mammals.

Microgeographic Evolution

Multiple cases of synhospitaly are found in each of the three mite families (Table 11.7) with at least congeneric "triads" in demodicids (Nutting 1974), and these last each in a different microhabitat on one individual animal. This phenomenon suggests microgeographic evolution (Nutting and Desch 1979) but not necessarily by way of symbiospecific coevolution (Fig. 11.8).

Because of the variety and often disjunct distribution of habitats on *one* mammal, we can look on the mammal as a wandering Galapagos archipelago with each island (e.g., an eyelid) having several differing habitats. If the initial inhabitants have a broad acceptance of ecological factors and the ability to move between islands (such as myobiids on any haired epidermal surface) little microgeographic evolution should occur. On the other hand, initial inhabitants, such as demodicids (which penetrate hair follicles), could become isolated and speciate either on one island or by successive intertransfers among the various islands, taking up other niches. Islands with no vacant niches may cause species extinctions or further species changes by competitive exclusions or hybridizations. Before the archipelagic mass disintegrated altogether, they would have to find another such area or all would be lost. Furthermore, the multiple species on the original archipelago would be only moderately screened in myobiids but drastically screened for spindle-shaped demodicids. Natural selection and successive species formations could occur, therefore, on one or successive hosts of a single species.

In comparative evolutionary terms the average mammalian mutation rate is 1×10^{-5} (Mange and Mange 1980), whereas Lebel and Desch (1974) reported 1×10^{-6} for *Demodex caprae* from specimens in a single mite-entrapped lesion. We would anticipate that a thorough sequential study of demodicid populations would show, with arrhenotoky and habitat entry-way screening, a much higher mutation and selection rate and at least twice the mammalian rate of skin-habitat change. In mammals, as man, with extended longevity, microgeographic speciation from demodicids could conceivably occur on one host individual.

Obviously, if mites overpopulate and destroy their environment (pathology) they perish, or likewise if competitively pressed, they may become extinct or move to new habitats.

Recovery records, especially under pathological conditions, show that myobiids have a wide distribution on the host body, synhospitaly, with little resource partitioning, and therefore little likelihood of microgeographic evolution. For demodicids, 50% of 134 mite species share congeners on one symbiont mammal species (Nutting 1974). This suggests that here a microgeographic pattern of coevolution occurs. If this be true, we must be cautious in our selection of species for documenting or predicting phylogenies (Table 11.9).

Symbiospecificities

The circumstantial evidence, admittedly still anemic, points at present to a loose gradient of asymbiospecificity (ecological specificity) beginning with the myobiids, leading to the marked and synhospitalically coadapted demodicids of the genus *Demodex*, with each of its known members symbiospecific (genetic specificity—host species specificity; compare Figs. 11.7 and 11.8).

Myobiids have the biological features that permit deep penetration to rupture of capillaries, tissue degradation by salivary products, and possible fecal contamination producing erythema. They indicate a symbiosis somewhat recent and not yet well adapted to low incidence of infestation, synhospitaly without marked resource partitioning, and multiple hosts sharing the same mite species or genera. All indicate a near-nidicole and non-host-species specificity (asymbiospecificity).

In well-isolated mammals, myobiids undoubtedly evolved species diagnostic of generic or higher coevolutionary specificity. If, however, any other mammal species came to share a myobiid-infested niche (of such or other mammal species), the generalized, weakly adapted features and mechanisms would not prevent transfer and/or viable coinfestations. Further modification on the new host, if in turn isolated, could produce a new species of myobiid, but again the degree of adaptive change needed would be small. Thus there is no barrier to out-taxa transfer. The quantity of antigen injected by a single or few myobiids would seem to indicate that even the host physiology or mechanisms such as an immune response would be inadequate to prevent intertaxa infestations.

The records of myobiid polyhospitaly indicate that although myobiids have become structurally, functionally, physiologically, and coevolutionarily adapted to metatherian and eutherian mammals, they, because of host niche overlap and host species swapping, have become asymbiospecific.

Similarly, but with a lesser degree of assurance, we find the psorergatids and the tortoise-shaped demodicids (as *Ophthalmodex* spp.) asymbiospecific. Although the latter are the most poorly known of all prostigmate symbiotes, they are somewhat generalized as free-moving parasites of the large, epithelial-lined head cavities of (only) mammals. In several respects, especially the long legs in adults, they seem more closely allied to the psorergatids than to the more specialized, elongate demodicids. The better-known psorergatids, as adults, also have long legs useful in transference. The records of symbiotes in psorergatids, plus comparative morphology of the adults which indicates intermediate (between myobiids and demodicids, *sensu restrictu*) reproductive rate selection, suggest asymbiospecificity for these two groups.

Throughout this chapter, evidences for host-species specificity (symbiospecificity) for the elongate species of the Demodicidae has been presented (*ad nauseum?*). This includes, especially in *Demodex* spp., the large number (11) of ordinal representative symbiont species parasitized by distinct species of this one genus, the remarkable number (50%) of synhospitalic associations with discrete species of symbiotes, and the remarkably modified structures and mechanisms coevolved with a specific mammal and distinctive for each species of *Demodex*. They all indicate symbiospecificity (coevolved species specificity).

Symbiophylogenies

Obviously at the field, laboratory, and intellectual levels we need immensely more physical and mental attention to the problem of Prostigmata–Mammalia coevolution, especially to establish secure evidence of symbiospecificity. However, the following observations sum up our current stand on coevolved symbiophylogenies (see Fig. 11.8, Table 11.9).

Myobiids are the most generalized and *r*-select of the three groups of mites. Although the largest number of species are described, they represent symbiotic bonding (to date) for the fewest (five) ordinal groups in two subclasses of Mammalia. Within these admittedly coevolved (*sensu latu*) symbioses the lack of symbiospecificity strongly suggests that they cannot be used as phylogenetic indexes of mammalian phylogeny.

Psorergatids and possibly such tortoise-shaped demodicids as *Ophthalmodex* spp. are the next largest group (120+) available for comparative studies, although their members (including the demodicids noted) to date are known from seven orders of the Eutheria. They are all less specialized (rounded and no anus yet discovered, but long-legged adults) than the elongate demodicids (as *Demodex* spp.). These also seem midway between the demodicids and myobiids for reproductive rate selection. Evidence of asymbiospecificity for the psorergatids establish them as currently unlikely as valid indicators for mammalian phylogeny (Fig. 11.7).

The coevolved (only with mammals), elongate demodicids (especially *Demodex* spp.) are the most modified, most *K*-select, and most symbiospecifically synhospitalic of all Prostigmata now known. Furthermore, this one genus shares discrete and often synhospitalic species with discrete matching or shared species of eleven orders of mammals. Such symbiospecificity indicates symbiophyletic specificities (Fig. 11.8).

As currently known (1983), all evidence, especially that for symbiospecificity, supports the view that the genus *Demodex* can be used to validate mammalian phylogeny, at least at the subclass level. They are, then, symbiophylogenetic with metatherian and eutherian mammals, with both, at least theoretically, predictive of each other's phylogenies (Fig. 11.9).

CONCLUSIONS

Of the 18 known families of symbiotic Prostigmata, only the Myobiidae, Psorergatidae, and Demodicidae are specifically coevolved with the Mammalia [subclass specificity, and order (ordinal) specificity]; species of each are structurally, functionally, and behaviorally adapted for survival as holostadial parasites of the mammalian skin. This last has apparently evolved a low-grade humoral or cellular immunity response and foreign body reaction only to major generalized mite pathogenesis. In turn, the physiology of the host, at least sex hormones for demodicids, has some inhibitory or stimulatory effect on mite reproduction. No evidence of physiological incompatibility in natural feral symbiosis was found in this analysis. The physical structures of the topologically differentiated mammalian skin complex, especially the diameter and distribution of the pilosebaceous apparatus, plus mechanisms of transference and translocation, are critical in natural selection to symbiotic coevolution and microgeographic evolution found only (to date) in the genus *Demodex* (Demodicidae).

Assessment in this chapter of the coevolved symbiotic mammalian representatives in relation to representatives of these three mite families leads to the following conclusions:

1. All three have symbiotic representatives that show subclass specificity and ordinal specificity—this either asymbiospecifically by way of reticulate evolution or symbiospecificity (or riverine and/or microgeographic evolution) derived in coevolution.

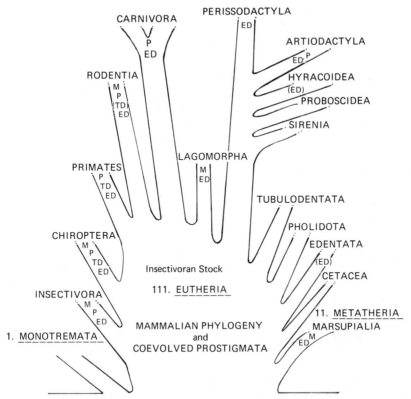

Figure 11.9 Diagrammatic representation of a "bush" of mammalian phylogeny showing subclasses (I Monotremata, II Metatheria, and III Eutheria), and ordinal taxa. Records of recovery of Myobiidae (M), Psorergatidae (P), genera *Ophthalmodex*, *Stomatodex*, or *Rhinodex* (TD), and *Demodex* (ED), are superimposed on the ordinal branches. Undescribed species in parentheses. (Recast and redrawn from Nutting, 1974.)

2. Currently, representatives of the genus *Demodex* (Demodicidae) only are symbiont species specific, and family specific with certain mammals. These, therefore, demonstrate coevolved family specificity and symbiospecificity (host species specificity or genetic specificity) with the Mammalia.

3. Currently, the known parasitic Myobiidae, Psorergatidae, and tortoise-shaped demodicids (as *Ophthalmodex* spp.) are asymbiospecific (ecologic specificity) or not host-species specific with their cosymbiotic mammals thus showing reticulate evolution and not useful as symbiophyletic markers [(a)symbiophylogenic].

4. Certain representatives of the genus *Demodex* show symbiogenic topologic speciation as well as symbiospecificity and thus serve as valid markers of mammalian phylogeny.

5. Conversely, mammalian symbionts can be used as valid markers of members of the genus *Demodex* and possibly microgeographic speciation with the family Demodicidae. Thus, symbiophylogenetically, mammals are potential markers of demodicid phylogeny.

6. Currently, based on information in this chapter, certain prostigmate members of the genus *Demodex* symbiotic with certain members of the subclasses Metatheria and Eutheria share coevolved symbiophylogenies (phylogenetic congruences).

SUMMARY

Analysis of "coevolution and symbiosis" for Arthropoda–Mammalia suggests the utility of redefining and replacing terms in current use. Current terms and/or their replacements include: *symbiote*, the arthropod species; *symbiont*, the mammalian species; and *asymbiospecific* for ecologic specificity and *symbiospecificity* for host-species specificity or genetic specificity. Multiple comparative and coevolved partners proven symbiospecific are needed to be predictive of phylogenetic congruences (both partners), here termed *symbiophylogeny*.

In examination of reports for 18 families of Prostigmata with special symbiotic species symbiotic with species of Mammalia, only the holostadially parasitic, restricted to mammalian subclass, Myobiidae, Psorergatidae, and Demodicidae, are securely coevolved with (only) mammals.

Considerations of mutual adaptations and interactions of structure, function (physiology), behavior, and for survival suggest that only the spindle-shaped *Demodex*-like demodicids have coevolved in phyletic congruence with the mammalian species of Metatheria and Eutheria: all thus are symbiophylogenic and either likely valid markers or predictors of the other's phylogeny.

The following list summarizes the Prostigmata–Mammalia associations for the Demodicidae (see Lombert 1979 for Myobiidae and Giesen 1982 for Psorergatidae). Undescribed species await adequate stage samples prior to description. Some pre-1960 inadequate descriptions are deleted. Documented corrections or additions are welcomed by the author.

Demodicidae	Mammalia	References
	METATHERIA	
	Marsupialia	
Demodex antechini	*Antechinus stuartii*	Nutting and Sweatman (1970)
D. marsupiali	*Didelphis marsupialis*	Nutting et al. (1980)
—at least 4 undescribed spp. on hand.		

Demodicidae	Mammalia	References

EUTHERIA
Insectivora

Demodex talpae	*Talpa europaea*	Hirst (1921)
D. soricinus	*(Plecotus auritus)*	Hirst (1918)
Soricidex dimorphus	*Sorex araneus*	Bukva (1982)

—at least 4 undescribed spp. on hand.

Chiroptera

Stomatodex rousetti	*Rousettus leachi*	Fain (1960)
S. corneti	*Barbastella barbastellus*	Fain (1960)
S. corneti subsp.	*Myotis myotis*	Fain (1960)
myotis?	*Myotis dasycneme*	Fain (1960)
Ophthalmodex carolliae	*Carollia perspicillata*	Lukoschus et al. (1980)
O. molossi	*Molossus molossus*	Lukoschus et al. (1980)
O. artibei	*Artibeus lituratus*	Lukoschus and Nutting (1979)
Demodex macroglossi	*Macroglossus minimus*	Desch (1981)
D. bicaudatus	*Macroglossus minimus*	Kniest and Lukoschus (1981)
D. aelleni	*Myotis daubentoni*	Fain (1960)
D. chiropteralis	*Plecotus auritus*	Hirst (1921)
D. phyllostomatis	*Phyllostomus hastatus*	Leydig (1859)
D. lacrimalis	*(Apodemus sylvaticus)*	Lukoschus and Jongman (1974)
D. melanopteri	*Eptesicus melanopterus*	Lukoschus et al. (1972)
D. longissimus	*Carollia perspicillata*	Desch et al. (1972)
D. molossi	*Molossus molossus*	Desch et al. (1972)
D. carolliae	*Carollia perspicillata*	Desch et al. (1971)
Pterodex carolliae	*Carollia perspicillata*	Lukoschus et al. (1980)

(only species known to date)
—at least 44 undescribed spp. on hand.

Primates

Rhinodex baeri	*Galago senegalensis*	Fain (1959)
Stomatodex galagoensis	*Galago senegalensis*	Fain (1959)
Demodex folliculorum	*Homo sapiens*	Nutting and Desch (1972)
D. brevis	*Homo sapiens*	Nutting and Desch (1972)
D. saimiri	*Saimiri sciureus*	Lebel and Nutting (1973)

—at least 16 undescribed spp. on hand.

Demodicidae	Mammalia	References
Rodentia		
Demodex caviae	*Cavia cobaya*	Bacigalupo and Roveda (1954)
D. aurati	*Mesocricetus auratus*	Nutting (1961)
D. criceti	*Mesocricetus auratus*	Nutting and Rauch (1958)
D. sciurinus	*Sciurus vulgaris*	Hirst (1923)
D. gliricolens	*Arvicola amphibius*	Hirst (1921)
D. glareoli	*Evotomys glareolus*	Hirst (1919)
D. longior	*Apodemus sylvaticus*	Hirst (1918)
D. apodemi	*Apodemus sylvaticus*	Hirst (1918)
D. lacrimalis	*Apodemus sylvaticus*	Lukoschus and Jongman (1974)
D. muscardini	*Muscardinus avellanarius*	Hirst (1919)
D. nanus	*Rattus rattus* (*Rattus norvegicus?*)	Hirst (1918)
D. ratti	*Rattus norvegicus*	Hirst (1919)
D. huttereri	*Apodemus agrarius*	Mertens et al. (1984)
D. arvicolae var. *musculi* (?)	*Mus musculus*	Mertens et al. (1984)
D. arvicolae	*Microtus* (*Arvicola*) *agrestis*	Zschokke (1888)
D. erinacei	(*Erinaceus europaeus*)	Hirst (1921)
D. melanopteri	(*Eptesicus melanopterus*)	Lukoschus et al. (1984)
D. gapperi	*Clethrionomys gapperi*	Nutting et al. (1971)
D. peromysci	*Peromyscus leucopus*	Lombert et al. (1983)

—at least 55 undescribed spp. on hand.

	Carnivora	
Demodex melesinus	*Taxidia taxus*	Hirst (1921)
D. ermineae	*Mustela erminea*	Hirst (1919)
D. cati	*Felis domestica*	Megnin (1877)
D. canis	*Canis familiaris*	Leydig (1859)
D. zalophi	*Zalophus californicus*	Dailey and Nutting (1981)

—at least 6 undescribed spp. on hand.

	Lagomorpha	
D. cuniculi	*Oryctolagus cuniculus*	Pfeiffer (1903)

—at least 6 undescribed spp. on hand.

Demodicidae	Mammalia	References
	Perrisodactyla	
Demodex caballi	*Equus caballus*	Desch and Nutting (1978)
D. equi	*Equus caballus*	Railliet (1895)
	Artiodactyla	
Demodex pseudaxisi	*Cervus hortulorum*	Shpringol'ts-shmidt (1937)
D. caprae	*Capra hircus*	Railliet (1895)
D. odocoilei	*Odocoileus virginianus*	Desch and Nutting (1974)
D. ghanensis	*Bos taurus*	Oppong et al. (1975)
D. ovis	*Ovis aries*	Railliet (1895)
D. bovis	*Bos taurus*	Stiles (1892)
D. phylloides	*Sus scrofa*	Czoker (1879)
D. cafferi	*Syncerus caffer*	Nutting and Guilfoy (1979)

—at least 5 undescribed spp. on hand.
  Unreported (but in hand):
    Edentata: one undescribed sp.
    Hyracoidea: one undescribed sp.

ACKNOWLEDGMENTS

Major effort for this chapter was expended by my wife, Margaret Nutting. She produced the fine art work, some of the technical work, and the typing of many drafts. Thanks are inadequate but here profusely extended!

This chapter has relied heavily on labors of and past discussions with Dr. Clifford Desch, Dr. Fritz Lukoschus, and Dr. Alex Fain, all well known to acarologists for their distinguished research in acarology. Thanks are due also to several other friends and colleagues, Drs. Muller (California), Uchikawa (Japan), and Oppong (Ghana) listed as personal communications in the text, and students—names often as coauthors—who have provided friendly stimulus or aid in past years.

The editor, Dr. K. C. Kim, and his editorial assistant, Kelly Morris, have also helped by picking up errors in my initial manuscript which allowed me time to rethink the problems discussed in this revision.

A value judgment has been made in omitting the lists of Myobiidae and Psorergatidae by Lombert (1979) and Giesen (1982). These are "in-house" publications loaned to the author by Dr. Fritz Lukoschus—for which my gratitude—and will soon be revised and published. This ethical judgment is coupled also with the judgment expressed in this chapter that myobiids and psorergatids are asymbiospecific, thus currently not useful as valid phylogenetic markers with the mammalia.

Any errors of fact or aberrations of speculation in this chapter are the sole responsibility of the author and should not be attributed to any of the aforementioned individuals.

REFERENCES

Bacigalupo, J. and R. J. Roveda. 1954. *Demodex caviae* n. sp. (English summary). *Rev. Med. Vet.*, Buenos Aires **36**:149–153.

Baker, E. W., T. M. Evans, D. J. Gould, W. B. Hull, and H. L. Keegan. 1956. *A Manual of Parasitic Mites of Medical or Economic Importance*. Natl. Pest Control Assoc. Tech. Publ.

Baker, E. W. and G. W. Wharton. 1952. *An Introduction to Acarology*. Macmillan, New York.

Baker, R. A. 1970. The food of *Riccardoella limacum* (Schrank) (Acari: Trombidiformes). *J. Nat. Hist.* **4**:521–530.

Baker, R. A. 1973. Notes on the internal anatomy, the food requirements and development in the family Ereynetidae (Trombidiformes). *Acarologia* **15**:43–52.

Bronswijk, J. E. M. H. von and E. J. de Kreek. 1976. *Cheyletiella* (Acari: Cheyletiellidae) of dog, cat and domesticated rabbit, a review. *J. Med. Entomol.* **13**:315–327.

Bukva, V. 1982. *Soricidex dimorphus* G. N., Sp. N. (Acari: Demodicidae) from the common shrew, *Sorex araneus*. *Folia Parasitol.* (Praha) **29**:343–351.

Canepa, E. and A. da Grana. 1941. La presencia del *Demodex folliculorum* Owen en los ganglios linfaticos de perros demodecticos. *Rev. Fac. Agron. Vet.* (Buenos Aires) **9**:109–114.

Carter, H. F., G. Wedd, and V. St. E. D'Abrera. 1944. The occurrence of mites (Acarina) in human sputum and their possible significance. *Indian Med. Gaz.* **79**:163–168.

Castro, J. J. de and J. Hobart. 1980. Comments upon the identification of Demodicid mites from cattle with special reference to *Demodex bovis*. In de Castro, J. J. Studies on mites of the genus *Demodex* (Acari: Demodicidae) on cattle in North Wales. Masters Thesis, University of Wales. Appendix 2.

Caullery, M. 1952. *Parasitism and Symbiosis*. Sedgwick and Jackson, London.

Cheng, T. C. 1971. *Aspects of the Biology of Symbiosis*. University Park Press, Baltimore.

Czokor, J. 1879. Die Haarsackmilben des Schweines, *Demodex phylloides*, eine neue Varietat. *Oesterr. Vierteljahrssch. Wiss. Vet.* **51**:133–185.

Dailey, M. and W. Nutting. 1979. *Demodex zalophi* sp. nov. (Acari: Demodicidae) from *Zalophus californianus*, the California sea lion. *Acarologia* **21**:421–428.

DeBary, D. L. 1879. *Die Erscheinung der Symbiose*. K. J. Trubner, Strassburg.

Desch, C. E. 1973. The biology of demodicids of man. Ph.D. Thesis, University of Massachusetts (Amherst).

Desch, C. E. 1981. Parasites of Western Australia: *Demodex macroglossi* spec. nov. (Acari: Prostigmata: Demodicidae) of *Macroglossus lagochilus* (Chiroptera: Pteropodidae). *Rec. West. Aust. Mus.* **9**:41–47.

Desch, C. E. 1982. Personal communication. University of Connecticut, Hartford.

Desch, C. E. and W. B. Nutting. 1971. Demodicids (Trombidiformes: Demodicidae) of medical and veterinary importance. *Proc. 3rd Intl. Congr. Acarol.*, Prague, pp 499–505.

Desch, C. and W. B. Nutting. 1972. *Demodex folliculorum* (Simon) and *D. brevis* Akbulatova of man: redescription and reevaluation. *J. Parasitol.* **58**:169–177.

Desch, C. and W. B. Nutting. 1974. *Demodex odocoilei* spec. nov. from the white-tailed deer, *Odocoileus virginianus*. *Can. J. Zool.* **2**:219–222.

Desch, C. E. and W. B. Nutting. 1977. Morphology and functional anatomy of *Demodex folliculorum* (Simon) of man. *Acarologia* **19**:422–462.

Desch, C. E., and W. B. Nutting. 1978. Redescription of *Demodex caballi* (= *D. folliculorum* var. *equi* Railliet, 1895) from the horse, *Equus caballus*. *Acarologia* **20**:235–240.

Desch, C. E., J. O'Dea, and W. B. Nutting. 1970. The proctodeum—a new key character for demodicids (Demodicidae). *Acarologia* **12**:522–526.

Desch, C., R. R. Lebel, W. B. Nutting, and F. Lukoschus. 1971. Parasitic mites of Surinam I. *Demodex carolliae* sp. nov. (Acari: Demodicidae) from the fruit bat (*Carollia perspicillata*). *Parasitology* **62**:303–308.

Desch, C., W. B. Nutting, and F. S. Lukoschus. 1972. Parasitic mites of Surinam VII. *Demodex longissimus* n. sp., from *Carollia perspicillata*, and *D. molossi* n. sp., from *Molossus molossus* (Demodicidae: Trombidiformes): Meibomian complex inhabitants of neotropical bats (Chiroptera). *Acarologia* **15**:35–53.

Dindal, D. L. 1975. Symbiosis: Nomenclature and Proposed Classification. *Biologist* **57**:129–142.

Dusbabek, F. 1968. Two new species of the Genus Ewingana (Acarina: Myobiidae) from Cuba. *Folia Parasitol.* (Praha) **15**:67–74.

Dusbabek, F. 1969. To the phylogeny of genera of the family Myobiidae (Acarina). *Acarologia* **11**:537–584.

Eichler, W. 1966. Two new evolutionary terms for speciation in parasitic animals. *Syst. Zool.* **15**:216–218.

Eisenberg, J. F. 1981. *The Mammalian Radiations*, University of Chicago Press, Chicago, IL.

English, F. and W. B. Nutting. 1981. Demodicosis of ophthalmic concern. *Am. J. Ophthalmol.* **91**:362–372.

Fain, A. 1959a. Les acariens psoriques des chauves-souris. IX. Nouvelles observations sur le genre Psorergates Tyrell. *Bull. Ann. Soc. R. Entomol. Belg.* **95**:232–248.

Fain, A. 1959b. Deux nouveaux genres d'acariens vivant dans l'epaisseur des muqueuses nasale et buccale chez un lemurien (Trombidiformes: Demodicidae). *Bull. Ann. Soc. R. Entom. Belg.* **95**:263–273.

Fain, A. 1959c. Les acariens psoriques des chauves-souris. III. Le Genre Psorergates Tyrell, (Trombidiformes-Psorergatidae). *Bull. Ann. Soc. R. Entom. Belg.* **95**:54–69.

Fain, A. 1960. Les Acariens psoriques parasites des chauves-souris. XII. La famille Demodicidae Nicolet. *Acarologia* **2**:80–87.

Fain, A. 1961. Notes sur le genre *Psorergates* Tyrell. Description de *Psorergates ovis* Womersley et d'une espece nouvelle. *Acarologia* **3**:60–71.

Fain, A. 1962. Les Acariens Parasites Nasicoles des Batraciens Revision des Lawrencarinae Fain, 1957 (Ereynetidae: Trombidiformes). *Inst. R. Sci. Nat. Belg. Bull.* **38**:1–69 (# 25).

Fain, A. 1963. Chaetotaxie et classification des Speleognathinae. *Inst. R. Sci. Nat. Belg. Bull.* **39**:1–80.

Fain, A. 1965. Quelques aspects de l'endoparasitisme par les Acariens. *Ann. Parasitol.* (Paris) **40**:317–327.

Fain, A. 1973a. Nouveaux taxa dans la famille Myobiidae (Acarina: Trombidiformes). *Rev. Zool. Bot. Afr.* **87**:614–621.

Fain, A. 1973b. Notes sur quelques nouveaux Acariens parasites de mammiferes (Myobiidae: Trombidiformes). *Bull. Ann. Soc. R. Entomol. Belg.* **109**:216–218.

Fain, A. 1974. Observations sur les Myobiidae parasites des rongeurs evolution parallele hotes–parasites (Acariens: Trombidiformes). *Acarologia* **16**:441–475.

Fain, A. 1979. Specificity, adaptation and parallel host–parasite evolution in acarines, especially Myobiidae, with a tentative explanation for the regressive evolution caused by the immunological reactions of the host. In J. G. Rodriguez (Ed.). *Recent Advances in Acarology*, Vol. 2. Academic, New York.

Fain, A. and K. Hyland. 1970. Notes on the Myocoptidae of North America with description of a new species on the Eastern Chipmunk. *Tamias striatus* Linnaeus. *J. N. Y. Entomol. Soc.* **78**:80–87.

Fain, A. and F. S. Lukoschus. 1979a. Parasites of Western Australia VI. Myobiidae parasitic on bats (Acarina: Prostigmata). *Rec. West. Aust. Mus.* **7**:61–107.

Fain, A. and F. Lukoschus. 1979b. Parasites of Western Australia. VIII. Myobiidae parasitic on bats (Acarina: Prostigmata). *Rec. West. Aust. Mus.* **7**:287–299.

Fain, A., F. Lukoschus, and P. Hallmann. 1966. Le genre Psorergates chez les murides. Description de trois especes nouvelles (Psorergatidae: Trombidiformes). *Acarologia* **8**:251–274.

Flynn, R. J. and B. N. Jaroslow. 1956. Nidification of a mite (*Psorergates simplex* Tyrrell, 1883: Myobiidae) in the skin of mice. *J. Parasitol.* **42**:49–52.

French, F., Jr. 1964. *Demodex canis* in canine tissues. *Cornell Vet.* **54**:271–290.

Gardner, A. L. 1973. The systematics of the genus *Didelphis* (Marsupialia, Didelphidae) in North and Middle America. *Spec. Publ., Mus. Tex. Tech. Univ.* **4**:81.

Giesen, K. M. T. 1982. A Review of Psorergatidae with description of a new species. *Scriptie* No. 30, Lab. voor Aquatische Oecologie, Katholieke Universiteit, Nijmegen. 86 pp.

Goksu, K., G. W. Wharton, and C. E. Yunker. 1960. Variation in populations of laboratory-reared *Trombicula* (*Leptotrombidium*) *akamushi* (Acarina: Trombiculidae). *Acarologia* **2**:199–209.

Gray, J. 1968. Effect of host sex and hormone treatment on *Demodex caprae* and *D. aurati*. Masters Thesis, University of Massachusetts (Amherst).

Grosshans, E. M., M. Kremer, and J. Maleville. 1974. *Demodex folliculorum* and histogenesis of granulomatous Rosacea. *Hautarzt* **25**:166–177.

Gurney, B. and N. W. Hussey. 1967. *Pygmephorus* species (Acarina: Pyemotidae) associated with cultivated mushrooms. *Acarologia* **9**:353–358.

Hakugawa, S. 1978. *Demodex folliculorum* Infection on the Face. *W. Jap. Dermatol.* **40**:275–284 (supplement).

Hewitt, R., G. I. Barrow, D. C. Miller, F. Turk, and S. Turk. 1973. Mites in the personal environment and their role in skin disorders. *Br. J. Dermatol.* **89**:401–409.

Hirschmann, W. 1971. A fossil mite of the genus *Dendrolaelaps* (Acarina, Mesostigmata, Digamasellidae) found in amber from Chiapas, Mexico. *Univ. Cal. Publ. Entomol.* **63**:69–70.

Hirst, S. 1918a. On four new species of the genus *Demodex*, Owen. *Ann. Mag. Nat. Hist.* **2**:145–146.

Hirst, S. 1918b. On the origin and affinities of the Acari of the family Demodecidae, with brief remarks on the morphology of the group. *Ann. Mag. Nat. Hist.* **1**:400.

Hirst, S. 1919. Studies on Acari, No. 1. The genus *Demodex* Owen. *London, Brit. Mus.* (*Nat. Hist.*) pp. 1–44.

Hirst, S. 1921. On some new or little known acari, mostly parasitic in habit. *Proc. Zool. Soc. London* **2**:357–378.

Hirst, S. 1923. On some arachnid remains from the Old Red Sandstone (Rhynie Chert Bed, Aberdeenshire). *Ann. Mag. Nat. Hist.* **12**(9):455.

Honacki, J. H., K. E. Kinman, and J. W. Koeppi. 1982. *Mammal Species of the World*. Allen Press and Assoc. Syst. Collections, Lawrence, KA.

Howell, J. P. and R. J. Elzinga. 1962. A new *Radfordia* (Acarina: Myobiidae) from the Kangaroo Rat and a key to the known species. *Ann. Ent. Soc. Am.* **55**:547–555.

Jameson, E. W. 1955. A summary of the genera of Myobiidae (Acarina). *J. Parasitol.* **41**:407–416.

Jameson, E. W. 1970. Notes on some myobiid mites (Acarina: Myobiidae) from old world insectivores (Mammalia: Soricidae and Talpidae). *J. Med. Entomol.* **7**:79–84.

Jameson, E. W. and F. Dusbabek. 1971. Comments on the myobiid mite genus Protomyobia. *J. Med. Entomol.* **8**:33–36.

Johnston, D. E. 1964. *Psorergates bos*, a new mite parasite of domestic cattle (Acari–Psorergatidae). *Res. Circ. 129* (Ohio Agric. Exper. Sta., Wooster, Ohio). 1–7 pp.

Kethley, J. B. 1970. A revision of the family Syringophilidae (Prostigmata: Acarina). *Contrib. Am. Entomol. Inst.* **5**:1–76.

Kethley, J. B. and D. E. Johnston. 1975. Resource tracking patterns in bird and mammal ectoparasites. *Misc. Publ. Entomol. Soc. Am.* **9**:231–236.

Kok, N. J. J., F. S. Lukoschus, and F. V. Clulow. 1971. Three new itch mites from Canadian small mammals (Acarina: Psorergatidae). *Can. J. Zool.* **49**:1239–1248.

Krantz, G. W. 1978. *A Manual of Acarology*, 2nd ed. Oregon State University Book Store, Corvallis.

Kuscher, A. 1940. Raubmilben beim Hund (*Cheyletiella parasitivorax* in Dogs). *W. Tieraerztl. Monatsschr.* **27**:10–16.

Lavoipierre, M. M. J. 1964. A new family of acarines belonging to the suborder Sarcoptiformes parasitic in the hair follicles of primates. *Ann. Natal Mus.* **16**:191–198.

Lawrence, R. F. 1952. A new parasitic mite from the nasal cavities of the South African toad *Bufo regularis* Reuss. *Proc. Zool. Soc.* (London) **121**:747–752.

Lebel, R. R. and C. E. Desch. 1974. Karyotype and anomalous development in *Demodex caprae*. *Proc. 4th Intl. Congress Acarology*. Akademiai Kiado, Budapest, pp. 525–529.

Lebel, R. R. and W. B. Nutting. 1973a. Demodectic mites of subhuman primates I. *Demodex saimiri* sp. n. (Acari: Demodicidae) from the squirrel monkey, *Saimiri sciureus*. *J. Parasitol.* **59**:719–722.

Lebel, R. R. and W. B. Nutting. 1973b. Population dynamics in a parasitic mite, *Demodex caprae* (Trombidiformes: Demodicidae). *Proc. 3rd Intl. Congress Acarology, Prague*. W. Junk L. V., The Hague, pp. 517–521.

Leydig, F. 1859. Ueber Haarsackmilben und Kratzmilben. *Arch. Naturg*, Berlin **1**:339–354.

Lombert, H. A. P. M. 1979. Morfologische ontwikkelingsreeksen, gastheerspecificiteit en parallele ontwikkeling bij de Myobiidae (Acarina: Trombidiformes). Ph.D. Thesis, Zoology, Catholic University, Nijmegen.

Lukoschus, F. 1982. Personal communication. Department of Zoology, Catholic University, Nijmegen.

Lukoschus, F. S. 1967a. Kratzmilben an Spanischen Kleinsaugern (Psorergatidae: Trombidiformes). *Rev. Iber. Parasitol.* **27**:203–224.

Lukoschus, F. 1967b. *Psorergates* (Psorobia) *mustelae* spec. nov. Eine neue Kratzmilbe von *Mustela nivalis* L. (Acarina: Psorergatidae). *Sonderdr. Zool. Anz.* **183**:111–118.

Lukoschus, F. 1968. Neue Kratzmilben von einheimischen Insektivoren (Psorergatidae: Trombidiformes). *Tijdsch. Entomol.* **111**:75–88.

Lukoschus, F. S. 1969. *Eadiea desmanae* spec. nov. (Acarina: Myobiidae) von *Galemys pyrenaicus*. *Acarologia* **11**:575–584.

Lukoschus, F. S. and R. H. G. Jongman. 1974. *Demodex lacrimalis* spec. nov. (Demodicidae: Trombidiformes) from the Meibomian glands of the European wood mouse *Apodemus sylvaticus*. *Acarologia* **16**:27–281.

Lukoschus, F. S. and W. B. Nutting. 1979. Parasitic mites of Surinam XIII *Ophthalmodex artibei* gen. nov., spec. nov. (Prostigmata: Demodicidae) from *Artibeus lituratus* with notes on pathogenesis. *Intl. J. Acarol.* **5**:299–304.

Lukoschus, F. S., R. H. G. Jongman, and W. B. Nutting. 1972. Parasitic mites of Surinam XII. *Demodex melanopteri* sp. n. (Demodicidae: Trombidiformes) from the Meibomian glands of the neotropical bat *Eptesicus melanopterus*. *Acarologia* 14:54–58.

Lukoschus, F. S., A. G. Woeltjes, C. E. Desch, and W. B. Nutting. 1980a. Parasitic mites of Surinam XX: *Pterodex carolliae* gen. nov., spec. nov. (Demodicidae) from the fruit bat, *Carollia perspicillata*. *Intl. J. Acarol.* 6:9–14.

Lukoschus, F. S., A. G. Woeltjes, C. E. Desch, and W. B. Nutting. 1980b. Parasitic mites of Surinam XXXV: Two new *Ophthalmodex* spp. (*O. carolliae, O. molossi*: Demodicidae) from the bats *Carollia perspicillata* and *Molossus molossus*. *Intl. J. Acarol.* 6:45–50.

Mange, A. and E. Mange. 1980. *Genetics: Human Aspects*. Saunders, New York.

Marples, M. J. 1965. *The Ecology of the Human Skin*. Charles C. Thomas, Springfield, IL.

Mayr, E. 1957. Evolutionary aspects of host specificity among parasites of vertebrates. In *Premier Symposium sur la spécificité parasitaire des parasites de Vertébrés*. Université de Neuchatel, Paul Attinger S.A.

Megnin, J. P. 1877. Monographie de la tribu des sarcoptides psoriques qui comprend tous les acariens de la gale de l'homme et des animaux. *Rev. Mag. Zool.* 5:46–213.

Mekie, E. D. 1926. Parasitic infection of the urinary tract. *Edinb. Med. J.* 33:708–719.

Michael, A. D. 1889. On some unrecorded parasitic Acari found in Great Britain. *J. Linn. Soc.* (*Zool.*) *London* 20:400–406.

Montagna, W. and P. Parakkal. 1974. *The Structure and Function of Skin*. Academic, New York.

Muller, G. 1982. Personal communication. Walnut Creek, California.

Nadchatram, M. 1970. Correlation of habitat, environment and color of chiggers and their potential significance in the epidemiology of scrub typhus in Malaya (Prostigmata: Trombiculidae). *J. Med. Entomol.* 7:131–144.

Newell, I. M. 1963. Feeding habits in the genus *Balaustium* (Acarina: Erythraeidae), with special reference to attacks on man. *J. Parasitol.* 49:498–502.

Nutting, W. B. 1950. Studies on the genus *Demodex* Owen (Acari, Demodicoidea Demodicidae). Ph.D. Thesis, Cornell University, Ithaca, New York.

Nutting, W. B. 1961. *Demodex aurati* sp. nov. and *D. criceti*, ectoparasites of the golden hamster (*Mesocricetus auratus*). *Parasitology* 51:515–522.

Nutting, W. B. 1968. Host specificity in parasitic acarines. *Acarologia* 10:165–180.

Nutting, W. B. 1974. Synhospitaly and speciation in the Demodicidae (Trombidiformes). *Proc. 4th Intl. Acarology Congress*. Akademiai Kiado, Budapest, pp. 267–272.

Nutting, W. B. 1975. Pathogenesis associated with hair follicle mites (Acari: Demodicidae). *Acarologia* 17:493–507.

Nutting, W. B. 1976a. Hair follicle mites (Acari: Demodicidae) of man. *Intl. J. Dermatol.* 15:79–98.

Nutting, W. B. 1976b. Hair follicle mites (*Demodex* spp) of medical and veterinary concern. *Cornell Vet.* 66:214–231.

Nutting, W. 1984. Biology of Prostigmata. In W. B. Nutting (Ed.), *Mammalian Diseases and Arachnids*, Vol. 1. CRC Press, Boca Raton, FL.

Nutting, W. B. and H. Beerman. 1983. Demodicosis and Symbiophobia: Status, terminology and treatments. *Intl. J. Dermatol.* 22:13–17.

Nutting, W. B. and M. Dailey. 1980. Demodicosis in the California sea lion, *Zalophus californianus*. *J. Med. Entomol.* 17:344–347.

Nutting, W. B. and C. E. Desch. 1979. Relationships between mammalian and demodicid phylogeny. *Rec. Adv. Acarol.* 2:339–345.

Nutting, W. B. and A. Green. 1976. Pathogenesis associated with hair follicle mites (*Demodex* spp.) in Australian Aborigines. *Br. J. Dermatol.* 94:307–312.

Nutting, W. B. and F. M. Guilfoy. 1979. *Demodex cafferi* n. sp. from the African buffalo, *Syncerus caffer. Intl. J. Acarol.* **5**:9–14.

Nutting, W. B. and H. Rauch. 1958. *Demodex criceti* n. sp. (Acarina: Demodicidae) with notes on its biology. *J. Parasitol.* **44**:328–333.

Nutting, W. B. and H. Rauch. 1963. Distribution of *Demodex aurati* in the host (*Mesocricetus auratus*) skin complex. *J. Parasitol.* **49**:323–329.

Nutting, W. B. and G. K. Sweatman. 1970. *Demodex antechini* sp. nov. (Acari: Demodicidae) parasitic on *Antechinus stuartii* (Marsupialia). *Parasitology* **60**:425–429.

Nutting, W. B., S. O. Emejuaiwe, and M. O. Tisdel. 1971. *Demodex gapperi* sp. n. (Acari: Demodicidae) from the red-backed vole, *Clethrionomys gapperi. J. Parasitol.* **57**:660–665.

Nutting, W. B., L. C. Satterfield, and G. E. Cosgrove. 1973. *Demodex* sp. infesting tongue, esophagus, and oral cavity of *Onychomys leucogaster*, the grasshopper mouse. *J. Parasitol.* **59**:893–896.

Nutting, W., J. Andrews, and C. Desch. 1980a. Studies in Symbiosis: Hair follicle mites of mammals and man. *J. Biol. Ed.* (England) **13**:315–321.

Nutting, W., F. Lukoschus, and C. Desch. 1980b. *Demodex marsupiali* sp. nov. from *Didelphis marsupialis. Zool. Med.* **56**:83–90.

Olsen, S. J. and H. Roth. 1947. En mide *Cheyletiella parasitivorax* hos kat, foraarsagende hududslet hos mennesker. *Medlemsbl. Dansk. Dyrlaegeforen.* **11**:207–217.

Oppong, E. N. W. 1973. Personal communication. University of Ghana, Legon, Accra, Ghana.

Oppong, E. N. W., R. P. Lee, and S. A. Yasin. 1975. *Demodex ghanensis* sp. nov. (Acari, Demodicidae) parasitic on West Africa cattle. *Ghana J. Sci.* **15**:39–43.

Pfeiffer, F. 1903. Acarus folliculorum cuniculi. *Berl. Tieraerztbl. Wchnschr.* **19**:155–156.

Price, D. W. 1973. Genus *Pediculochelus* (Acarina: Pediculochelidae), with notes on *P. raulti* and descriptions of two new species. *Ann. Entomol. Soc. Am.* **66**:302–307.

Prietsch, J. 1886. *Demodex folliculorum* bei einem Samburhirsch. *Berl. Vet. K. Sachs.* **30**:89.

Railliet, A. 1895. *Traite de Zoologie Medicale et Agricole.* 2nd ed. Paris.

Read, C. P. 1970. *Parasitism and Symbiology.* Ronald Press, New York.

Scott, D. W., B. R. H. Farrow, and R. D. Schultz. 1974. Studies on the therapeutic and immunologic aspects of generalized demodectic mange in the dog. *J. Am. Anim. Hosp. Assoc.* **10**:233–244.

Shpringol'ts-Shmidt, A. I. 1937. (Ectoparasites of some species of far eastern deer.) (Russian text.) *Vest.* (26) *Dal'nevostoch. Fil. Akad. Nauk SSSR.* **26**:133–140.

Smiley, R. L. 1965. Two new species of the genus Cheyletiella. *Proc. Entomol. Soc. Wash.* **67**:75–79.

Southcott, R. V. 1973. *Survey of Injuries to Man by Australian Terrestrial Arthropods.* R. V. Southcott, Mitcham, S. Australia (Private publication.).

Stiles, C. 1892. On *Demodex folliculorum* var. *bovis* in American cattle. *Can. Entomol.* **24**:286–290.

Stromberg, B. 1968. *Demodex caprae*: An investigation of lipids and exoskeleton. Masters Thesis, University of Massachusetts, Amherst.

Summers, F. M. and D. W. Price. 1970. Review of the mite family Cheyletidae. *Univ. Calif. Publ. Entomol.* **61**:1–153.

Sweatman, G. K. 1971. Mites and Pentastomes. Pages 3–64 in J. W. Davis and R. C. Anderson (Eds). *Parasitic Diseases of Wild Mammals.* Iowa State University Press, Ames.

Thor, S. 1933. Acarina. Tydeidae, Ereynetidae. *Tierreich* **60**:1–82.

Traub, R. and C. L. Wisseman. 1974. The ecology of chigger-borne rickettsiosis (scrub typhus). *J. Med. Entomol.* **11**:237–303.

Treat, A. E. 1975. *Mites of Moths and Butterflies.* Cornell University Press, Ithaca, pp. 239–270.

Tyrell, J. B. 1883. On the occurrence in Canada of two species of parasitic mites. *Proc. Can. Inst. Toronto* **1**:342–343.

Uchikawa, K. 1981. Personal communication. Shinshu University, Matsumoto City, Japan.

Vercammen-Grandjean, P. H. 1966. *Whartonia pachywhartoni* n. sp., an extraordinary parasite of a Brazilian bat (Leeuwenhoekiidae—Acarina). *Acarologia* **8**:282–284.

Vercammen-Grandjean, P. H. and H. Rak. 1968. *Cheyletiella yasguri* Smiley 1965, un parasite de canides aux Etats-Unis et hyperparasite d'Hippobascide en Iran (Acarina: Cheyletidae). *Ann. Parasitol.* **43**:405–412.

Wharton, G. W. 1960. Host–parasite relationships between *Myobia musculi* (Schrank, 1781) and *Mus musculus* Linnaeus, 1758. Libro Homenaje Caballero y Caballero, pp. 571–575. (Issued Aug. 26, Beltsville Parasitol. Lab.)

Williams, R. W. 1946. A contribution to our knowledge of the bionomics of the common North American chigger, *Eutrombicula alfreddugesi* (Oudemans) with a description of a rapid collecting method. *Am. J. Trop. Med.* **26**:243–250.

Womersley, H. 1941. Notes on the Cheyletidae of Australia and New Zealand with description of new species. *Rec. S. Austral. Mus.* **7**:51–64.

Woolley, T. A. 1971. Fossil oribatid mites in amber from Chiapas, Mexico (Acarina: Oribatei—Cryptostigmata). Pages 91–99 in *Studies of Fossiliferous Amber Arthropods of Chiapas, Mexico*, Part 2. University of California Press, Berkeley.

Zschokke, E. 1888. Die Haarbalmilben bei der Feldmaus. *Schweiz. Arch. Tierheilkd* **20**:69–82.

Zumpt, F. and W. Till. 1955. The mange-causing mites of the genus *Psorergates* (Acarina: Myobiidae) with description of a new species from the South African monkey. *Parasitology* **45**:269–274.

Chapter **12**

Ursicoptes americanus Fain and Johnston

Evolution of Astigmatid Mites on Mammals

Alex Fain and Kerwin E. Hyland, Jr.

INTRODUCTION

During the Fifth International Congress of Acarology a symposium was held which dealt with the specificity, adaptation, and parallel evolution of hosts and their parasitic acarines. It was shown that in some families of mites, such as the Myobiidae, specificity and parallel evolution are well marked and can be used to evaluate the degree of primitiveness of the hosts as well as the relationships existing between certain hosts or groups of hosts (Fain 1979b).

The study of evolution of both host and parasite has revealed that some groups of parasitic mites are almost as old as their hosts (Fain 1976a, 1977a). Specificity is more marked in permanent parasites than in temporary ones. The pilicolous specialization has produced a particularly strong specificity not only in mites (e.g., Myobiidae and Listrophoroidea) but also in some insects such as the lice.

It seems probable that many parasitic mites have been derived from species living in the nests of birds or mammals. This appears particularly true for some parasitic Astigmata such as the Psoroptoidea which live on mammals and the Analgoidea on birds, both of whose ancestors were probably the nidicolous Pyroglyphidae living in contact with these animals. As a matter of fact, members of this family of mites are morphologically closer to these parasitic groups of acarines than to the free-living Astigmata of the family Acaridae. In the Pyroglyphidae the regressive evolution seems to have preceded the infestation of the host. We have called this phenomenon *preadaptation* (Fain 1979b).

PARASITISM IN THE ASTIGMATA

Morphological Adaptation to Parasitism

In the adaptation to parasitism two different kinds of phenomena are involved which are independent of each other: one is constructive, the other is regressive (Fain 1969). The constructive adaptations involve the hypertrophy of existing organs and are adaptations toward particular conditions of the parasitic way of life. They occur more frequently in ectoparasites than in endoparasites and are especially important in mites that attach themselves to the hair or the skin. These phenomena are secondary adaptations or specializations and not directly related to the phylogeny of the animal.

In contrast, regressive phenomena are characterized by the progressive disappearance of external structures such as tarsal claws, sclerotized plates, and setae. The regression is particularly well marked in the endoparasites, and as a rule it is more important in parasites living on more highly evolved hosts than in those parasitizing more primitive animals. This kind of evolution is directly related to the phylogeny of the parasite.

Regressive Evolution in the Parasitic Mites

It is well known that the general trend in the evolution of animals is toward complication or complexity of structures. In the parasitic mites, however, this rule is not respected, and their evolution is fundamentally of the regressive type. Generally the importance of the regression in the parasite is correlated with the degree of evolution of the host. Host and parasite have a parallel evolution, but they go in opposite directions.

Fain (1979b) surmised that regressive evolution is related to the immunological reactions of that part of the host that tends to kill and reject the parasite. To escape from this rejection the mite always tends to select the less antigenic and therefore the most regressed phenotype.

Biological Adaptations of Mites to Parasitism

Parasitic life usually produces an acceleration of the postembryonic development. Generally the free-living and the ectoparasitic mites are oviparous. In the true endoparasites ovoviviparity or viviparity is the rule. The remainder of the developmental cycle is also accelerated. Some or all of the nymphal stages may disappear in the endoparasitic forms. It is to be noted, however, that acceleration of the cycle may also be observed in some highly adapted ectoparasites such as the larval Trombiculidae or the nidicolous Hypoderidae, whose deutonymphs are tissue parasites of birds or mammals.

The ectoparasitic Astigmata are generally oviparous. Their life cycles may be summarized as follows: egg, prelarva, protonymph, tritonymph, and adult. The prelarva is only represented by an ecdysial organ composed of two small sclerotized conical structures situated on a transparent membrane (Fain 1977b). This organ is also present in the free-living Astigmata: Acaridae, Glycyphagidae, and Pyroglyphidae (Fain and Herin 1979). In the endoparasitic Astigmata the prelarva is absent and viviparity is the rule. This is the case in the Gastronyssidae living in the respiratory tract or in the stomach of bats. In *Sarcoptes scabiei* (Linnaeus), which is not a true endoparasite because it lives in the external noncellular layer of the epidermis, the prelarva is absent.

PARASITIC ASTIGMATIC MITES

We wish to consider here the more important families of Astigmata parasitic on mammals.

Family Listrophoridae

Listrophorids are fur mites that are permanently attached to the bases of hairs during all stages of their development. The organ of attachment consists of two striated chitinous membranes situated in the anterior region of the sternum (Fig. 12.1). These mites probably feed on fatty substances produced by the hair follicles. Their life cycle consists of egg, protonymph, tritonymph, and adults (Fain 1971, 1973; Fain and Hyland 1974).

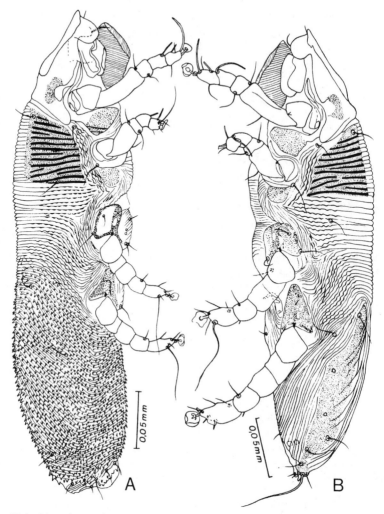

Figure 12.1 *Listrophorus phenacomys* Fain and Hyland (lateral view). (*A*) Female, (*B*) male. (From Fain and Hyland 1974. *Bull. Inst. R. Sci. Nat. Belg.* **50**:1–69. By permission of the Royal Institute of Natural Sciences of Belgium.)

The host specificity at the species level is generally strict. The same genus may be present in two or three families of the same order of hosts or more rarely in two different orders of hosts.

In the genus *Afrolistrophorus* the dorsum bears a large postscapular median shield which indicates that this genus is primitive. It consists of twelve species on Afrotropical rodents, eight species on Oriental rodents, one species on a European rodent, one species on *Mus musculus*, one species on a rodent of Patagonia, and one species on the marsupial, *Lestoros inca*.

The genus *Geomylichus* is also primitive but more specialized than *Af-*

rolistrophorus. It contains 13 species all confined to the New World of which eight live on Geomyoidea, four on Cricetidae (Hesperomyinae), and one on the rabbit, but this was probably accidental (Fain et al. 1978).

The genus *Prolistrophorus* has a postscapular shield which has partially disappeared. It is therefore more regressed and thus more evolved than the preceding genera. It contains 12 species living on American Cricetidae or on Echymyidae.

In the genus *Listrophorus* the postscapular shield is divided into two dorsolateral shields, which indicates that it is more regressed and more evolved than *Prolistrophorus*. The genus *Listrophorus* contains 13 species in North America which live mostly on Microtidae. In Europe it is represented by nine species four of which are also represented in North America. All these species live on Microtidae. This taxon is absent in the Afrotropical and Oriental regions as well as in Madagascar and Australia. Six other genera are found in the rodent families Aplodontidae, Cricetidae, Spalacidae, and Sciuridae.

The Insectivora are parasitized by 11 species belonging to four genera (*Asiochirus*, *Olistrophorus*, *Dubininetta*, and *Echinosorella*). All of these species are endemic on Insectivora except one which lives on a rodent. In the Lagomorpha (family Leporidae) there is only one endemic genus, *Leporacarus*, which has three species. In the Carnivora there are four genera represented by 10 species. In three of these genera (*Lynxacarus*, *Lutracarus*, and *Hemigalichus*) the postscapular shield is well developed, which suggests a primitive condition.

Lynxacarus contains five species living on Carnivora and four species living on other mammals (one on a rodent, and three on *Tupaia* spp.). *Lutracarus* is monotypic and is specialized for *Lutra canadensis* (Schreber). It is more evolved than *Lynxacarus*. *Hemigalichus* is also monotypic and lives on an Asiatic carnivore of the genus *Hemigalus*.

The fourth genus, *Carnilistrophorus*, is completely devoid of a postscapular shield and is therefore the most evolved of the entire family. It contains five species, of which three live on Afrotropical Carnivora, one on Macroscelididae, and one on the genus *Myospalax*.

The family Sciuridae is host to four genera (*Sciurochirus*, *Metalistrophorus*, *Pteromychirus*, and *Aeromychirus*) made up of 10 species. All of these species are highly specialized and are restricted to these rodents.

Family Chirodiscidae

The family Chirodiscidae is divided into four subfamilies of which the most important is the subfamily Labidocarpinae. It contains 19 genera and 147 species. Sixteen of these genera (with 128 species) are found on bats; the others are specialized for other mammals [e.g., *Soricilichus*, *Schizocarpus*, and *Lutrilichus* (see list of genera in Appendix A)].

All of these species are compressed laterally and attach to the hair of their hosts by means of legs I and II which are strongly modified and bear large striated chitinous membranes.

The postembryonic development in the Labidocarpinae from bats is very peculiar. There are two morphologically distinct lines: one for the female and one for the male. The larva, protonymph, and tritonymph of the female line are modified in that they bear copulatory lobes. In the male line all the stages are normal and lack the copulatory lobes (Fain 1971).

Family Myocoptidae

The family Myocoptidae consists of 50 species grouped in six genera, of which four are endemic for rodents, one for insectivores, and one for marsupials (Fain 1970; Fain et al. 1970). These mites attach to the hair of their hosts by means of their posterior legs (legs III and IV in female and leg III in male) which are strongly modified into clasper organs. Development is the same as for the Listrophoridae. This family is generally cosmopolitan, but so far it has not been recorded from Madagascar or Australia.

Family Atopomelidae

The Atopomelidae is a large family made up of 45 genera and 351 species. The great majority of these species are found only in tropical or subtropical regions. Most of the genera are endemic to a particular family of hosts. Among them, 24 are endemic to marsupials (six genera on American and 18 on Australian marsupials). The Australian marsupials contain genera and species different from those found on American marsupials. However, the genus *Didelphoecius* is widespread in American Didelphidae and is very close to *Dasyurochirus* living in Australian Dasyuridae, suggesting that the hosts are also close to each other.

These mites also infest certain primitive Neotropical mammals of the families Caviidae, Echimyidae, Capromyidae, and Solenodontidae. They are also found on Old World mammals such as Talpidae, Soricidae, Erinaceidae, and Tenrecidae; as well as on rodents such as Nesomyidae and Bathyergidae, or on primates like Lemuridae (Fain 1972, 1976).

The postembryonic development has been described for *Chirodiscoides caviae* Hirst. It resembles that of the Labidocarpinae, with two morphologically different lines of immatures, a male and a female line (Fain 1979a). Host specificity is generally well marked, but less so than in the Myobiidae.

Family Psoroptidae

The family Psoroptidae is economically very important because it contains several species producing mange in various domestic animals, especially in cattle. It is divided into nine subfamilies (see Appendix A).

The subfamily Psoroptinae contains all of the species parasitic on domestic animals. The life cycles of both *Psoroptes* and *Chorioptes* have been

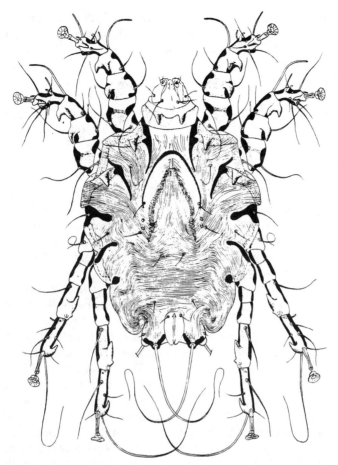

Figure 12.2 *Gaudalges caparti* Fain; female (ventral view). (From Fain 1963. *Bull. Inst. R. Sci. Nat. Belg.* **39**:1–125. By permission of the Royal Institute of Natural Sciences of Belgium.)

elucidated and found to be similar to those of the Labidocarpinae and the Atopomelidae (Fain 1975). There are two different lines of development, one female, the other male. The protonymph and tritonymph of the female line have copulatory lobes, whereas the nymphs in the male line are devoid of these lobes (Sweatman 1958; Fain 1964b).

The Psoroptidae are skin mites. They are characterized by the presence of an apical recurved, conical process on the anterior legs and sometimes on the posterior legs as well (Fig. 12.2). Their bodies are flattened dorsoventrally, and in the most primitive genera the idiosoma and the legs bear retrose processes serving to attach them to the host skin. These processes for attachment are lacking in the Psoroptinae. In most of the genera, legs IV and sometimes also legs III are reduced, sometimes very strongly so. The Psoroptinae live on the surface of the skin, generally at the base of thick crusts produced by the host. They parasitize four orders of hosts. All the other subfamilies are peculiar to one order of host except for the Psoralginae which are found on Edentata and on Australian Marsupialia (Vombatidae) (Fain 1963, 1965b) (see list of genera in Appendix A).

Family Sarcoptidae

The family Sarcoptidae contains several genera carrying important agents of scabies which affect both man and animals. The family has been divided into four subfamilies, the most important being the subfamily Sarcoptinae which contains six genera: *Sarcoptes, Prosarcoptes, Cosarcoptes, Pithesarcoptes, Trixacarus,* and *Tychosarcoptes.*

It has been suggested that the genus *Sarcoptes* contains only one species, *S. scabiei,* and that all of the 30 other species described in this genus are synonymous with the latter. All of these "species" are based on variable morphological characters without any taxonomic value. *Sarcoptes scabiei* is a very variable species, and it possesses a mixture of both stable and variable morphological characters (Fig. 12.3).

The great variability seen in *S. scabiei* (Linnaeus) suggests that this species is not completely adapted to any of the present hosts but remains in a continuous process of adaptation in different hosts. The cause of this instability is probably related to the large number and variety of hosts that the mite infests. This mite produces scabies in more than 40 different hosts belonging to seven orders of mammals. The variability of this species is probably the result of the continuous interbreeding of the strains living on man and animals (Fain 1968, 1978). The development of *S. scabiei* comprises the following stages: egg, larva, protonymph, tritonymph, and adults.

The family Sarcoptidae is distinguished from the Psorptidae by the absence of an apical process on the tarsi. This process is replaced by several short spines (Fig. 12.4) which allow the mite to form a tunnel in the corneous layers of the skin of the hosts. They produce deeper lesions than the Psoroptidae and are often more difficult to find.

Other genera of Sarcoptidae, such as *Notoedres* and *Trixacarus,* contain species that cause mange in domestic animals.

Family Teinocoptidae

This family contains two genera and 17 species, all parasitic on bats. They are regressed, and in the females leg IV is either strongly reduced (*Teinocoptes*) or absent (*Chirobia*) (Fain 1959).

Family Lemurnyssidae

The family consists of two genera. The monotypic genus *Lemurnyssus* lives in the nasal cavity of *Galago senegalensis* (Galagidae) in Africa; the genus *Mortelmansia,* which has three species, lives in the same habitat in South

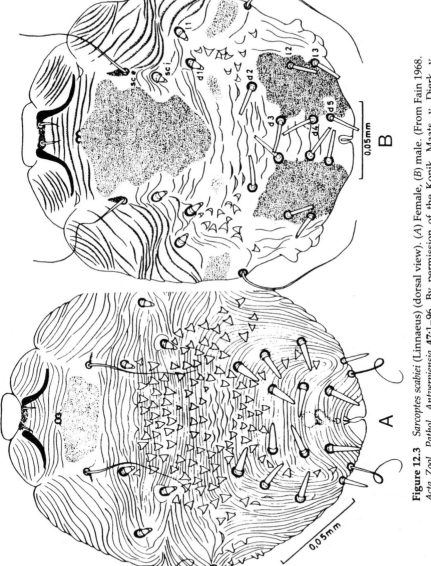

Figure 12.3 *Sarcoptes scabiei* (Linnaeus) (dorsal view). (A) Female, (B) male. (From Fain 1968. *Acta Zool. Pathol. Antverpiensia* **47**:1–96. By permission of the Konik. Maats. v. Dierk. v. Antwerp.)

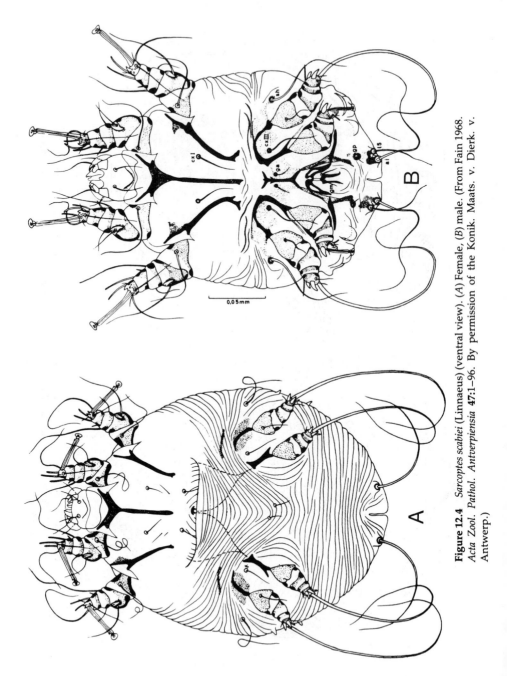

Figure 12.4 *Sarcoptes scabiei* (Linnaeus) (ventral view). (*A*) Female, (*B*) male. (From Fain 1968. *Acta Zool. Pathol. Antverpiensia* **47**:1–96. By permission of the Konik. Maats. v. Dierk. v. Antwerp.)

American monkeys. The presence of very closely related mites in both Afrotropical galagids and Neotropical monkeys suggests the existence of some relationship between these primates (Fain 1964a).

Family Rhyncoptidae

The family Rhyncoptidae strongly modified and specialized. They are fixed in the hair follicles of their hosts by the anterior half of the body while the posterior part is free and outside the follicle. This family comprises one genus, *Rhyncoptes*, containing four species of which one lives on a rodent of the family Hystricidae in South Africa, two other species on Neotropical monkeys, and one species on an Afrotropical monkey of the genus *Cercopithecus* (Fain 1965a).

Family Audycoptidae

Mites of the family Audycoptidae are morphologically closely related to the Rhyncoptidae, but their anterior legs are considerably less well developed, and the mites are completely embedded in the hair follicle. Three genera, *Audycoptes*, *Uriscoptes*, and *Saimirioptes*, have been described so far with five species distributed among the host families Cebidae, Ursidae, and Procyonidae.

Family Gastronyssidae

The gastronyssids are divided into two subfamilies: Gastronyssinae, endemic to bats, and Yunkeracarinae, endemic to rodents (Fain 1964c, 1967a). The Gastronyssinae contains five genera (*Gastronyssus*, *Rodhainyssus*, *Opsonyssus*, *Eidolonyssus*, and *Mycteronyssus*) and 20 species. *Gastronyssus bakeri* Fain lives attached to the gastric mucosa of a fruit bat in Africa. The species of the other genera live either attached to the cornea of the eye or in the nasal cavities of bats.

The subfamily Yunkeracarinae with two genera and three species parasitizes the nasal cavities of rodents.

Families Acaridae and Glycyphagidae

The Acaridae and Glycyphagidae are not true parasites, however, some species, mainly those of the family Acaridae, may cause contact dermatitis in man. Some species of the family Glycyphagidae produce hypopi which may attach to the hair or invade the hair follicles of their hosts, primarily rodents. Their presence may result in a pathological condition.

Most of these species feed on fungi that develop in decaying organic material. They infest the nest of insects, birds, or rodents. They are also found in stored food (grain, cheese, dried fruit, etc.), especially when this food has been stored under poor conditions (high level of moisture) and has become moldy (Hughes 1976).

Persons who have repeated contact with foodstuff infested with these mites may develop a skin allergy known as contact dermatitis. The most important species are *Tyrophagus putrescentiae* (Shrank), which causes copra itch; *Acarus siro* Linnaeus, the agent of baker's itch and of cheese mite dermatitis; *Carpoglyphus lactis* (Linnaeus), which produces dried fruit mite dermatitis, and *Glycyphagus domesticus* (De Geer), which causes grocer's itch (Baker et al. 1956).

Some Glycyphagidae produce heteromorphic deutonymphs or hypopi which attach to the hair of mammals, mainly rodents. These hypopi are phoretic nymphs which serve for both the dispersal of the species and for its survival under adverse conditions. In some species such as *Aplodontopus sciuricola* Hyland and Fain these hypopi invade the hair follicle or the subcutaneous tissues of their hosts (mainly rodents) and may cause skin lesions (Tadkowski and Hyland 1979).

Family Pyroglyphidae

Most of the Pyroglyphidae are free-living. They infest birds' nests and are devoid of pathological action. A few species live on the skin of birds and are true parasites. There is also a series of species that lives in house dust and is responsible for respiratory allergies, mainly bronchial asthma (Voorhorst et al. 1964; Spieksma 1967; Oshima 1967). These house-dust mites constitute a serious health problem in all countries of the world (Wharton 1976).

Thus far the Pyroglyphidae comprises 14 genera and 35 species. Most of these species live in birds' nests, one species has been found in the nest of a rodent, and 10 species are found mostly or exclusively in house dust.

The most important species in relation to house-dust asthma is *Dermatophagoides pteronyssinus* Trouessart (Figs. 12.5–12.7). It is present throughout the world, and its true habitat is the house (Fain 1966, 1967b). This mite is present in dust on the floor especially in living rooms and bedrooms. Recent studies have demonstrated that mattresses are the primary breeding grounds for this mite. It is especially abundant in wet maritime climates such as the coastal areas of England, Belgium, and the Netherlands.

Another species, frequently associated with *D. pteronyssinus* is *D. farinae* Hughes. In certain drier climates (central parts of the United States and Europe) *D. farinae* may become more abundant than *D. pteronyssinus*. Two other species also found frequently in house dust are *Euroglyphus maynei* (Cooreman) and *Hirstia domicola* Fain, Oshima, and Van Bronswÿk, both of which have a more local distribution than the two preceding ones.

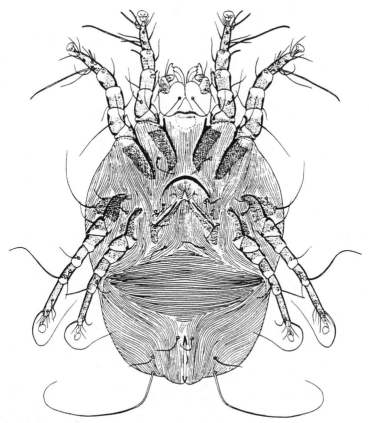

Figure 12.5 *Dermatophagoides pteronyssinus* (Trouessart); female (ventral view). (From Fain 1966. *Acarologia* **8**:302–327. By permission of the Centre National de la Recherche Scientifique.)

CONCLUSIONS AND SUMMARY

The parallel evolution of host and parasite is well marked in many parasitic mites, especially in permanent parasites that spend their entire lives associated with the host. Specificity is closely related to the degree of permanency of the parasitism and is particularly strict in the pilicolous mites (e.g., Myobiidae and Listrophoroidea). Specificity and parallel evolution can be used to evaluate the degree of evolution of the host and the relationships existing among certain hosts.

Some groups of parasitic mites are almost as old as their hosts. It appears that many groups of acarines, especially the Astigmata, originated from species living in the nests of mammals or birds. Free-living nidicolous species of the family Pyroglyphidae are probably the ancestors of most of the parasitic Astigmata living on both mammals and birds.

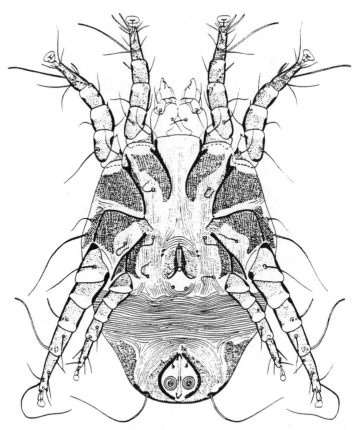

Figure 12.6 *Dermatophagoides pteronyssinus* (Trouessart); male (ventral view). (From Fain 1966. *Acarologia* **8**:302–327. By permission of the Centre National de la Recherche Scientifique.)

The adaptation of a free-living mite to a parasitic life is both morphological and biological. Two different kinds of phenomena, independent of each other, are involved in morphological adaptation: one is constructive, the other regressive. Constructive adaptation involves the hypertrophy and modification of existing organs favoring particular conditions of the parasitic life, especially the attachment to the hair or the skin of the host. This kind of adaptation represents secondary phenomena or specialization and is not directly related to the phylogeny of the parasite. Regressive adaptation is characterized by the progressive disappearance of external structures (e.g., claws, hairs, shields). Regression is particularly well marked in endoparasitic mites and as a rule is more important in parasites living on highly evolved hosts than in those living on more primitive animals. This kind of evolution is directly related to the phylogeny of the host.

The most prominent biological adaptation to parasitism is an accelera-

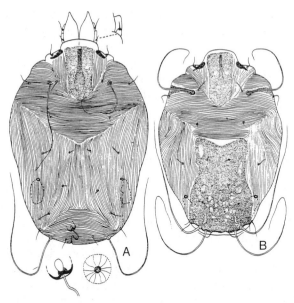

Figure 12.7 *Dermatophagoides pteronyssinus* (Trouessart) (dorsal view). (*A*) Female, (*B*) male. (From Fain 1966. *Acarologia* **8**:302–327. By permission of the Centre National de la Recherche Scientifique.)

tion of postembryonic development. In certain parasitic mites some or all of the nymphal stages may disappear completely. A rudimentary prelarva is present in the free-living and in most of the ectoparasitic Astigmata, but it is absent in the endoparasitic Astigmata.

Host–parasite parallel evolution is analyzed in the most important families of Astigmata parasitic on mammals. These mites belong to 15 families, 162 genera, and 849 species (through 1979).

In the Listrophoridae there is generally a good correlation between the degree of regression of the mite and the degree of evolution of the host. In the genus *Afrolistrophorus* the dorsum bears a large postscapular shield which indicates that the taxon is primitive. It contains 21 species, most of which are from Afrotropical rodents. The genus *Prolistrophorus* has a postscapular shield slightly eroded in its median part, which means it is more regressed, and hence more evolved, than *Afrolistrophorus*. It contains 12 species living on American Cricetidae or Echimyidae. In the genus *Listrophorus* the median part of the postscapular shield has completely disappeared, and there are only two lateral, but completely separated, postscapular shields. This genus, which contains 22 species living on Microtidae, is more regressed than *Prolistrophorus*. It appears from these observations that the most regressed, and hence the most evolved, genus of mites lives on the most evolved hosts (Microtidae).

Another family, Chirodiscidae, contains 147 species grouped into 19 genera. Most of these species live on bats. They are highly specialized fur mites which have a laterally compressed body. Their postembryonic development is very unusual in that there are two morphologically different lines—one female and one male.

The largest group of fur mites (351 species and 45 genera) is the family Atopomelidae. This group is almost entirely tropical or subtropical. Most of the species infest rodents, but the family also infests marsupials, both Australian and American. All the genera found associated with marsupials are restricted to marsupials, however the genus *Didelphoecius*, which is widespread in American Didelphidae, is very close to the genus *Dasyurochirus* which lives on Australian Dasyuridae. This suggests that the hosts are also close to each other.

The Psoroptidae and Sarcoptidae are skin mites. The family Psoroptidae is large, divided into nine subfamilies each confined to one order of hosts, except the Psoralginae which is represented on American Edentata and on Australian Marsupialia. All of these mites live attached to the surface of the skin. The family Sarcoptidae contains numerous species producing mange in man and animals. It has been suggested that the genus *Sarcoptes* contains only one variable species, *S. scabiei*, which is able to parasitize man, numerous domestic mammals, and even wild mammals living in captivity.

In addition to these groups of parasitic Astigmata, there is a family of free-living mites, the Acaridae, which contains species causing contact dermatitis in man. Moreover, the deutonymphs of some Glycyphagidae may produce mange in rats. Another mite family, the Pyroglyphidae, lives in house dust and in the nests of birds and mammals. A few species are parasitic on birds, but most of them are free-living. *Dermatophagoides pteronyssinus*, a cosmopolitan species living in house dust, is the most important species of this family because it is frequently associated with allergic bronchial asthma in man.

REFERENCES

Baker, E. W., T. M. Evans, D. J. Gould, W. B. Hull, and H. L. Keegan. 1956. *A Manual of Parasitic Mites of Medical or Economic Importance.* Nat. Pest Control Assoc., New York.

Fain, A. 1959. Les Acariens psoriques des chauves-souris. IV. Le genre *Teincoptes* Rhodain. Creation d'une nouvelle famille Teincoptidae (Sarcoptiformes). *Rev. Zool. Bot. Afr.* 59:118–136.

Fain, A. 1963. Les Acariens producteurs de gale chez les Lemuriens et les singes avec une étude des Psoroptidae (Sarcoptiformes). *Bull. Inst. R. Sci. Nat. Belg.* 39:1–125.

Fain, A. 1964a. Les Lemurnyssidae parasites nasicoles de Lorisidae africains et des Cebidae sud-américains. Description d'une espèce nouvelle. (Acarina: Sarcoptiformes). *Ann. Soc. Belg. Med. Trop.* 44:453–458.

Fain, A. 1964b. Le développement postembryonnaire chez les Acaridiae parasites cutanés des Mammifères et des Oiseaux (Acarina: Sarcoptiformes). *Acad. R. Belg. Cl. Sci. Bull.,* 5E Série, 50:19–34.

Fain, A. 1964c. Chaetotaxie et classification des Gastronyssidae avec description d'un nouveau genre parasite nasicole d'un Ecureuil sudafricain (Acarina: Sarcoptiformes). *Rev. Zool. Bot. Afr.* **70**:40–52.

Fain, A. 1965a. A review of the family Rhynciptidae Lawrence parasitic on Porcupines and Monkeys (Acarina: Sarcoptiformes). *Adv. Acarol.* **2**:135–159.

Fain, A. 1965b. Les acariens producteurs de gale chez les Edentés et les Marsupiaux (Psoroptidae et Lobalgidae Sarcoptiformes). *Inst. R. Sci. Nat. Belg. Mem.* **41**:1–41.

Fain, A. 1966. Allergies respiratoires produites par un Acarien (*Dermatophagoides pteronyssinus*) vivant dans les poussières des habitations. *Bull. Acad. R. Med. Belg.* **6**(6–7):479–499.

Fain, A. 1967a. Observations sur les Rodhainyssinae. Acariens parasites des voies respiratoires des Chauves-Souris (Gastronyssidae: Sarcoptiformes). *Acta Zool. Pathol. Antverpiensia* **44**:3–35.

Fain, A. 1967b. Le genre *Dermatophagoides* Bogdanov, 1864—Son importance dans les allergies respiratoires et cutaneés chez l'homme (Psoroptidae: Sarcoptiformes). *Acarologia* **9**:179–225.

Fain, A. 1968. Etude de la variabilité de Sarcoptes scabiei avec une révision des Sarcoptidae. *Acta Zool. Pathol. Antverpiensia* **47**:1–196.

Fain, A. 1969. Adaptation to Parasitism in Mites. 2nd International Congress of Acarology in Sutton Bonington (England), 19–25 July 1967. *Acarologia* **11**:429–449.

Fain, A. 1970. Les Myocoptidae en Afrique au Sud du Sahara (Acarina: Sarcoptiformes). *Ann. Mus. R. Afr. Cent. Ser. Quarto Zool.* **179**(8):1–67.

Fain, A. 1971. Les Listrophoridés en Afrique au Sud du Sahara (Acarina: Sarcoptiformes). II. Familles Listrophoridae et Chirodiscidae. *Acta Zool. Pathol. Antverpiensia* **54**:1–231.

Fain, A. 1972. Les Listrophoridés en Afrique au Sud du Sahara (Acarina: Sarcoptiformes). III. Famille Atopomelidae. *Ann. Mus. R. Afr. Cent. Ser. Quarto Zool.* **197**(8):1–200.

Fain, A. 1973. Les Listrophoridés d'Amérique Neotropicale (Acarina: Sarcoptiformes). I. Familles Listrophoridae et Chirodiscidae. *Bull. Inst. R. Sci. Nat. Belg.* **49**:1–149.

Fain, A. 1975. Nouveaux taxa dans les Psoroptinae. Hypothèse sur l'origine de ce groupe (Acarina, Sarcoptiformes, Psoroptidae). *Acta Zool. Pathol. Antverpiensia* **61**:91–118.

Fain, A. 1976a. Ancienneté et Spécificité des acariens parasites. Evolution parallèle. Hôtes–Parasites. *Acarologia* **17**:396–374.

Fain, A. 1976b. Faune de Madagascar. Arachnides, Acariens, Astigmata, Listrophoroidea. *Off. Rech. Sci. Tech. Outre-Mer* (Paris) **42**:1–131.

Fain, A. 1977a. Observations sur la spécificité des Acariens de la famille Myobiidae. Corrélation entre l'évolution des parasites et de leurs hôtes. *Ann. Parasitol. Hum. Comp.* **52**:339–351.

Fain, A. 1977b. The prelarva in the Pyroglyphidae (Acarina: Astigmata). *Int. J. Acarol.* **3**:115–116.

Fain, A. 1978. Epidemiological problems of scabies. *Int. J. Dermatol.* **17**:20–31.

Fain, A. 1979a. Les Listrophorides d'Amérique Neotropical (Acarina: Astigmates). II. Famille Atopomelidae. *Bull. Inst. R. Sci. Nat. Belg.* **51**:1–158.

Fain, A. 1979b. Specificity, adaptation and parallel host–parasite evolution in acarines, especially Myobiidae, with a tentative explanation for the regressive evolution caused by the immunological reactions of the hosts. *Proc. 5th Int. Congr. Acarol.*, U.S.A. 1978; *Rec. Adv. Acarol.* **2**:321–328.

Fain, A. and A. Herin. 1979. La prélarve chez les Astigmates. *Acarologia* **20**:566–571.

Fain, A. and K. E. Hyland. 1974. The Listrophoroid Mites in North America. II. The Family Listrophoridae. *Bull. Inst. R. Sci. Nat. Belg.* **50**:1–69.

Fain, A., A. J. Munting, and F. S. Lukoschus. 1970. Les Myocoptidae parasites des rongeurs en Hollande et en Belgique (Acarina: Sarcoptiformes). *Acta Zool. Pathol. Antverpiensia* **50:**67–172.

Fain, A., J. O. Whitaker, T. G. Schwan, Jr., and F. S. Lukoschus. 1978. Notes on the genus *Geomylichus* Fain, 1970 (Astigmata: Listrophoridae) and descriptions of six new species. *Int. J. Acarol.* **4:**115–124.

Hughes, A. M. 1976. *The Mites of Stored Food and Houses.* Ministry of Agriculture, Fisheries, and Food. Tech. Bull. 9. London.

Oshima, S. 1967. Studies on the genus *Dermatophagoides* as flour mites. *Jap. J. Sanit. Zool.* **18:**213–215.

Spieksma, F. Th. M. 1967. The house-dust mite *Dermatophagoides pteronyssinus* (Trouessar 1897), producer of the house-dust allergen (Acari: Psoroptidae). Thesis. Leiden.

Sweatman, G. K. 1958. On the life history and validity of the species in *Psoroptes*, a genus of mange mites. *Can. J. Zool.* **36:**905–929.

Tadkowski, T. M. and K. E. Hyland. 1979. The developmental stages of *Aplodontopus sciuricola* (Astigmata) from *Tamias striatus* L. (Sciuridae) in North America. *Proc. 4th Int. Congr. Acarol.* **1974:**321–326.

Voorhorst, R., M. I. A. Spieksma-Boezeman, and F. Th. M. Spieksma. 1964. Is a mite (*Dermatophagoides* sp.) the producer of the house-dust allergen? *Allerg. Asthma* **10:**329–34.

Wharton, G. W. 1976. House dust mites. Review article. *J. Med. Entomol.* **12:**557–621.

PART FOUR

OVERVIEW

Chapter 13

Parasitism and Coevolution

Epilogue

Ke Chung Kim

INTRODUCTION

The chapters of this book raise the following types of questions concerning the associations of parasitic arthropods and mammals: Are parasites specifically associated with one, a few, or many host species? Are there any

A contribution from the Frost Entomological Museum, Department of Entomology, The Pennsylvania State University, University Park, PA (AES Proj. No. 2594).

consistent patterns in the associations between lineages of parasites and hosts? Have specific associations persisted for long periods of evolutionary or geologic time? With what other parasites do particular species co-occur in a community? Are specific associations limited to particular geographic regions? How are parasite species specially adapted to their hosts? Are there reciprocal adaptations between parasites and their hosts? Are the specific associations coevolutionary?

Many of these questions are answered for most of the obligate mammalian parasites within the Insecta and Acari (Chapters 4–12). The associations of parasitic arthropods and mammals are highly evolved ecological systems with dynamic ecological and genetic relationships among the associating species. These relationships can be observed in various evolutionary patterns of adaptation, host specificity, geographical distribution, and phylogeny of associations between parasites and their hosts. Although little direct evidence is presented for coevolution or reciprocal coadaptation between parasitic arthropods and mammals, there are many examples that can be inferred to be coevolutionary. Many of the relationships discussed must be the result of coevolutionary processes, considering their intimate associations and adaptations (Chapters 1, 3, 4, 8, 11).

In this epilogue, I describe a scenario for the evolution of parasitic arthropod and mammal associations based on observed evolutionary patterns and coevolutionary perspectives of parasitic arthropods.

EVOLUTIONARY PATTERNS OF PARASITIC ARTHROPODS

Parasitic arthropods that have obligatory relationships with mammalian hosts include members of the insect orders Hemiptera, Coleoptera, Diptera, Anoplura, Mallophaga, and Siphonaptera, and the acarine orders Parasitiformes (Mesostigmata and Metastigmata) and Acariformes (Prostigmata and Astigmata). The Anoplura, Mallophaga, and many Mesostigmata are permanent parasites, whereas the others are temporary parasites (see Chapter 1). Parasitic arthropods are specific to particular mammalian taxa. Host specificity may occur at various host taxonomic levels; some parasites are monoxenous, whereas others are specific to congeneric species or host taxa at the familial or even ordinal level (Nutting 1968; Marshall 1981a; Kim, Chapter 1).

Parasitic arthropods of mammals have many morphological, behavioral, and physiological adaptations for promoting their physical association with the host and their capacity to feed on it. These involve reproduction, life cycle, host finding, microhabitat selection, and feeding. Many of these parasites are haematophagous; others are keratinophagous or mucophagous. Overdispersion, which helps to maintain a stable parasite–host system, is a common pattern of distribution for parasitic arthropods. Population structure usually does not correspond to a pyramid of numbers of individuals by life stages because of overlapping generations and temporal changes in demographic and environmental forces (Kim, Chapter 1). Sex ratio is usually unbalanced, with females predominating (Marshall 1981b; Kim, Chapter 1).

Host Associations and Specificity

Most parasitic insects are monoxenous, but some are oligoxenous. The Polyctenidae (Hemiptera), Arixenidae (Dermaptera), Nycteribiidae, and Streblidae (Diptera) are exclusive parasites of bats (Chiropetra), whereas the Siphonaptera, Cuterebridae (Diptera), Anoplura, parasitic Coleoptera, and all hemimerid Dermaptera are parasitic on the Rodentia. The dipteran Hippoboscidae, Gasterophilidae, Hypodermatidae, Oestridae, Calliphoridae, and Sarcophagidae are primarily parasitic on large herbivores, particularly the Artiodactyla (Kim and Adler, Chapter 4).

Most associations between Anoplura and mammals appear to have begun before major host cladogenesis. The sucking lice have coevolved with their hosts and speciated following the adaptive radiation of the hosts, but their associations and distribution have been modified by emigration and extinction. Although the cladogenesis of Anoplura closely parallels mammalian phylogeny, evolutionary rates differ between sucking lice and their hosts and among parasite taxa within the same group. Sucking lice parasitic on the monotyphlan Insectivora represent original infestations, whereas infestations of the lipotyphlans by Hoplopleuridae and Polyplacidae are secondary. The polyplacid associations with prosimian Primates are secondary, but associations of anthropoid Primates with *Pediculus*, *Pthirus*, and *Pedicinus* are primary. The Rodentia harbor 70% of the known Anoplura including most Enderleinellidae, Hoplopleuridae, and Polyplacidae. The primary hosts of these lice are concentrated in the three families Sciuridae, Cricetidae, and Muridae. The Cervidae and Bovidae are primary hosts of *Solenopotes* and *Linognathus*, respectively. Infestations of *Linognathus* on the fissiped Carnivora are secondary, whereas the Echinophthiriidae are endemic to the Carnivora and are primary parasites of aquatic carnivores. The associations of Pecaroecidae with Tayassuidae, Microthoraciidae with Camelidae, and Ratemiidae with Equidae are remnants of a much larger diversity (Kim and Ludwig 1978a; Kim, Chapter 5).

The Mallophaga are predominantly parasitic on birds (85% of the total diversity), but some are found on limited groups of mammals. Of the mammalian Mallophaga, the Amblycera are primarily parasitic on Australian and Neotropical marsupials and Neotropical rodents, the Ischnocera are permanent parasites of various mammals, and the Rhyncophthirina are parasitic on African and Oriental elephants and African wart hogs (Emerson and Price, Chapter 6; Kim, Chapter 7).

A close congruence exists between the diversity patterns of the parasitic Psocodea and their Carnivora hosts. Their diversity patterns suggest that the trichodectids (Mallophaga) radiated closely following the evolution of the Fissipedia and that the echinophthiriids (Anoplura) evolved parallel with the Pinnipedia radiation (Kim et al. 1975; Kim, Chapter 7).

The fleas (order Siphonaptera) are primarily parasites of mammals, having about 100 species on birds. The majority of fleas are restricted to a single host species or host genus (or a set of related species). The stephanocircids, doratopsyllids, and certain pygiopsyllids of marsupials date back to at least the Cretaceous period. Some pulicids evolved in the Oligocene epoch along with their sciurid hosts (Traub 1972, 1980). Fleas are parasites of terrestrial mammals only as adults, but their larvae are associated with the host's dwellings. Thus, aquatic mammals such as whales, manatees, and seals, and those terrestrial mammals which have no dens and vast home ranges lack indigenous fleas. Flea distribution and host associations are quite distinct, and the level of their evolutionary development is well correlated with that of the host (Traub, Chapter 8).

The mesostigmatid mites have diverse relationships with mammals as facultative and obligatory parasites, nest dwellers, and host dwellers, or ectoparasites and endoparasites. Almost all of these mammalian associates are in the Dermanyssoidea derived from the *Hypoaspis* complex. Certain mesostigmate mites are exclusively parasitic on particular mammalian hosts: Spelaeorhynchidae and Spinturnicidae on bats, Hystrichonyssidae on rodents, Halarachnidae on Pinnipedia and Primates as respiratory mites, and Dasyponissidae on edentates. The Macronyssidae, Dermanyssidae, and Laelapidae, however, are parasitic on numerous hosts, having various life styles (Radovsky, Chapter 9).

The Ixodoidea are among the oldest obligate ectoparasites of vertebrates. Ticks were already parasitic on reptiles in the late Paleozoic or early Mesozoic eras (Hoogstraal 1978). Host associations and structural adaptations provide an insight into their evolution (Hoogstraal and Kim, Chapter 10). Numerous argasids are associated with cave-dwelling bats, but other *Argas* and *Ornithodoros* species have much wider associations, such as with burrowing or den-inhabiting mammals, marine birds, tree-hole-nesting birds or reptiles.

In the Ixodidae, although many amblyommine ticks (*Aponemma* and *Amblyomma*) are still parasitic on contemporary reptiles, two contemporary *Aponema* species have adapted to the echidna (Monotremata) and the wombat (Marsupialia), and 64 of the 102 *Amblyomma* species are parasitic on birds and mammals. Almost all recent species of Ixodinae, Haemaphysalinae, and Hyalomminae are are parasitic on birds and mammals, and only a few species, perhaps primitive, parasitize reptiles. The ixodid Rhipicephalinae evolved much later than other ticks, perhaps in the Cretaceous or early Tertiary period, along mammalian phyletic lines. Contemporary rhipicephaline ticks are virtually never associated with birds or reptiles. Most adults are associated primarily with Artiodactyla and fewer with Carnivora or Perissodactyla, whereas rhipicephaline immatures chiefly parasitize rodents, although some infest insectivores, leporids, or carnivores (Hoogstraal and Kim, Chapter 10).

The symbiotic Prostigmata include 18 families, of which the Myobiidae, Psorergatidae, and Demodicidae are obligatory and mostly permanent parasites. They are structurally, functionally, and behaviorally adapted to the mammalian skin. All three taxa are specific to mammalian hosts at the ordinal level; the demodicid *Demodex* is the sole taxon host specific at the species level (Nutting, Chapter 11).

The parasitic Astigmata originated from the free-living nidicolous species of Pyroglyphidae. Five taxa, Bakerocoptidae, Chirorhynchobiidae, Gastronyssidae, Rosensteinidae, and Teinocoptidae, are exclusively parasitic on bats, whereas the Myocoptidae, Pneumocoptidae, Pyroglyphidae, and Yunkeracaridae are specific to the Rodentia. The Lemurnyssidae and Audicoptidae are respiratory parasites. Other parasitic prostigmatid mites infest a wide range of mammals: for example, the Listophoridae are parasites of Insectivora, Scandentia, Lagomorpha, Carnivora, Rodentia, and Marsupialia; the Chirodiscidae are parasitic on the Insectivora, Chiroptera, Carnivora, Rodentia, and Primates; and the Psoroptidae are parasitic on the Insectivora, Carnivora, Perissodactyla, Artiodactyla, Primates, Marsupialia, and Edentata (Fain and Hyland, Chapter 12).

Parasitic Adaptations

Parasitic arthropods must remain in or on the mammalian host to sustain specific parasite–host relationships. They have acquired various morphological, behavioral, and physiological adaptations primarily geared for promoting physical association with the host and feeding capability on or in it. Some adaptations are general responses to parasitism, whereas others are specific to particular traits and reactions of the host. These adaptations include microhabitat selection. For example, the Audycoptidae (*Andycoptes* and *Saimirioptes*) are found in the lip tissue of the South American squirrel monkey (Lavoipierre 1964; Fain 1968), where they feed on sebaceous materials around sinus hair and other hair follicles. *Demodex folliculorum* Owen is commonly found in the hair follicles of the forehead and eyebrows of man (Nutting, Chapter 11), and *Orthohalarachne attenuata* (Banks) adults are found in the nasopharynx of seals (Kim et al. 1980).

There are two different trends in morphological adaptations: progressive and regressive. Progressive adaptation involves hypertrophy and modification of organs for particular conditions of parasitic life, especially for attachment to the host hair or skin, whereas regressive adaptation shows the gradual reduction and disappearance of anatomical structures such as claws, hairs, or shields (Fain 1969; Fain and Hyland, Chapter 12). Some adaptations are common to most parasitic arthropods although unrelated phylogenetically. These include both progressive and regressive adaptations in size and shape of the body, head, eyes, setae, sensoria, and other traits (Marshall 1980, 1981a; Kim and Adler, Chapter 4).

Most parasitic insects and acarines are dorsoventrally flattened, with the head and other structures such as antennae, mouthparts, and the thorax modified. Others like fleas and some nycteribiid and streblid flies are laterally compressed to allow them to move about readily in dense hair. In some taxa, such as fleas, polyctenids, platypsyllids, and bat flies, setal adaptation is striking, the setae being modified into combs or ctenidia (Marshall 1980, 1981a; Kim and Adler, Chapter 4). The mouthparts of parasitic arthropods are also highly modified. In haematophagous insects different parts of the mouthparts are modified into stylets to aid in sucking blood. For example, in Anoplura the maxillae form the piercing fascicle (three stylets) making the food channel, and the hypopharynx and the labium are attached posteriorly to the walls of the enclosing sac forming the salivary channel (Kim, Chapter 1). In parasitic Mesostigmata the chelicerae are the major component of the feeding organs with the terminal chelae modified into slender shafts (Radovsky, Chapter 9).

Another major morphological adaptation of parasitic arthropods are the attachment or holdfast organs (Fain 1969; Radovsky 1969; Marshall 1981a; Kim, Chapter 1). Development of attachment organs varies greatly among different taxa. In Anoplura and Mallophaga the tibia, tarsus, and claws are modified as the holdfast organs (Kim, Chapter 1). In the Ixodidae and some Mesostigmata, on the other hand, attachment involves the hypostome, corniculi, and chelicerae, and in other mites, such as chiggers and Spelaeorhynchidae, the highly modified chelicerae, tarsal suckers, and episthosomal sucker or claspers are used (Fain 1969; Dubinina 1969; Radovsky, Chapter 9; Fain and Hyland, Chapter 12).

In parasitic Astigmata there are significant correlations between the degree of regressive adaptations in mites and the degree of evolution of the hosts. For example, of Listrophoridae parasitic on rodents, the primitive *Afrolistrophorus* bears a large postscapular shield on the dorsum, whereas *Prolistrophorus* and *Listrophorus*, which are found on the most evolved microtine hosts, have the postscapular shield reduced and its median part absent. In other words, parasitic mites found on the most evolved hosts are the most regressed in many taxonomic characters (Fain 1969; Fain and Hyland, Chapter 12).

Among the prominent biological adaptations to parasitism are those involving reproduction and life cycle. Ovovivipary and vivipary frequently occur in the Polyctenidae, parasitic Dermaptera, pupiparous Diptera, endoparasitic mites like Halarachnidae, Rhinonyssidae, and Entonyssidae, and some ectoparasitic mites (Fain 1969; Marshall 1981a; Kim, Chapter 1; Fain and Hyland, 12). In rabbit fleas, reproductive cycles are closely synchronized with the reproductive cycles of the hosts, and host hormones directly influence the maturation of flea eggs and oviposition (Rothschild and Ford 1972; Marshall 1981a). In certain parasitic mites some or all of the nymphal stages may disappear (Fain 1969; Furman and Smith 1973; Kim et al. 1980; Kim, Chapter 1), and furthermore the annual life cycles of seal lice (Echinophthiriidae) are closely synchronized with the annual migration and breeding patterns of their hosts (Kim 1975, and Chapter 1).

Geographical Distribution

The contemporary Camelidae, consisting of *Camelus* (two species) and *Lama* (two species), are infested with a monotypic family of sucking lice, the Microthoraciidae (*Microthoracius*). Of the four known species of *Microthoracius*, one species is found on the African/Asian *Camelus* and three species on the South American *Lama*, which are limited to the Andean highlands and are also infested with the trichodectid mallophagan *Bovicola* (Kim 1982 and Chapter 5). The present distribution of camelid/microthoraciid associations poses intriguing questions. How is it that monotypic parasites are found on such closely related mammalian hosts in such widely separated areas? When did the *Bovicola/Lama* association begin in relation to *Microthoracius*?

The camelids are traceable to late Eocene ancestors in North America where camelids were abundant throughout the Tertiary period with 21 genera (eight genera in Camelidae). The Oligocene *Poebrotherium* goes back more than 35 million years. *Camelus* and *Lama* have been taxonomically distinct for several million years (Gauthier-Pilters and Dagg 1981). Migration of camelids to Asia occurred in the late Pliocene epoch, and further dispersal to Europe and North Africa and independently from North America to South America took place during the Pleistocene (Simpson 1945). Considering the taxonomy and distribution of lice and the historical biogeography of Camelidae, the following conclusions can be made: (1) camelids were infested with *Microthoracius* long before their Pliocene and Pleistocene expansion from North America to Eurasia, Africa, and South America; (2) the diversity of *Microthoracius* was much larger during the late Tertiary period; (3) many species of microthoraciids became extinct along with the North American camelids during the Pleistocene and Pliocene (Kim and Ludwig 1978a; Gauthier-Pilters and Dagg 1981; Kim 1981, and Chapter 5; (4) *Bovicola* became established on the llamas long after the camelids migrated to South America (Kim, Chapter 7).

As shown by the camelid/microthoraciid associations, present patterns of geographical distribution are the historical manifestations of climatic, geomorphological, and environmental changes that establish new physical barriers. Furthermore, the present geographical distribution of parasitic arthropods has been shaped by (1) successive host availability, (2) host dispersal and migration, (3) host transfer, (4) host speciation, and (5) host extinction (Hopkins 1957; Pielou 1979).

To explore such historical manifestations, the associations of parasitic arthropods and mammals must be analyzed critically on the basis of sound taxonomy of the parasites and their hosts, the geological and geographical distributions of the hosts, and the historical patterns of land-mass movements and climatic changes.

The parasitic hemipteran Polyctenidae, consisting of five genera and 32 species, occur on five of the 17 families of Microchiroptera of which the largest family, Vespertilionidae, conspicuously lacks them. They are geographically limited at the species or species-group level (Kim and Adler, Chapter 4). Similarly, the other two bat parasites, Nycteribiidae and Streblidae, show distinct distribution patterns at lower taxonomic levels (Kim and Adler, Chapter 4); for example, there is no streblid taxon common between the Old and New Worlds (Wenzel et al. 1966; Wenzel 1976). The monotypic louse flies *Allobosca* and *Proparabosca* are ectoparasites of Ethiopian Primates, whereas *Austrolfersia* and *Ortholfersia* are parasitic on Australian marsupials. Although *Lipoptena* is found worldwide on Artiodactyla, the monotypic *Neolipoptena* is confined to the Western North American Cervidae (Maa 1963).

Infestations of Malagasy Primates with the polyplacids *Lemurpediculus* and *Phthirpediculus*, and the inclusion of *Polyplax brachyuromyis* Kim and Emerson in the Ethiopian *Polyplax jonesi* group suggest the African connection of endemic Malagasian Anoplura (Kim and Emerson 1974; Kim 1982, and Chapter 5). The data support the invasion of Madagascar by African mammals by chance dispersal over side stretches of sea in the post-early Miocene epoch (Coryndon and Savage 1973) and also suggest that the ancestral Malagasy mammals already had polyplacids at the time of invasion (Traub 1980).

The distribution of Anoplura and Mallophaga on Carnivora shows distinct patterns (Kim, Chapter 7). The Anoplura are centered in the Holarctic and Oceanic regions, whereas the mammalian Mallophaga are concentrated in the Paleotropical and Neotropical realms. The Amblycera have an austral origin where the Protomallophaga existed in the early Cretaceous period (Kim and Ludwig 1978b, 1982; Traub 1980; Kim, Chapter 5). The distribution of the ischnoceran *Felicola* group and *Suricatoecus*, which are primarily parasitic on feloids, is centered in the Ethiopian region, whereas the distribution of *Trichodectes*, which are primarily canoid parasites, is centered in the Neotropical region with "spill-overs" to the Nearctic or other regions. *Neotrichodectes* occurs on Procyonidae and Mustelidae in the Nearctic region (Kim, Chapter 7). The amblyceran Trimenoponidae are found on Neotropical land mammals. The four species of *Cummingsia* are confined to South American marsupials (three species) and a rodent (one species). Again, the amblyceran family Gyropidae is found only on Neotropical mammals (Emerson and Price, Chapter 6).

The fleas are of ancient lineage and coevolved closely with their hosts. Marsupials had stephanocircid and doratopsylline fleas when they migrated to Australia from South America by way of Antarctica (Traub 1980).

The distribution of spilopsylline (Pulicidae) and leptopsyllid rabbit fleas suggests dispersal by way of the North Atlantic prior to the closing of the route in the Mid-Eocene epoch (Traub, Chapter 8).

Of the hard tick Amblyomminae, most *Aponomma* and more than 30% of *Amblyomma* parasitize contemporary reptiles, with the others parasitic on birds and mammals. The two contemporary *Aponomma* species not associated with reptiles have adapted to the monotreme echiidna and the marsupial wombat in the Australian region. Most other ixodids (Ixodinae, Haemaphysalinae, and Hyalomminae) are parasitic on birds and mammals worldwide. The Rhipicephalinae are primarily parasitic on mammals; about 115 contemporary species are confined to the Palearctic, Oriental, and Ethiopian regions, and no indigenous rhipicephaline ticks occur in the New World, Madagascar, Australia, or New Guinea. These data suggest that the Rhipicephalinae evolved in the early Tertiary period much later than other ticks and long after mammals were well established in the Palearctic, Oriental, and Ethiopian regions.

THE COEVOLUTIONARY PERSPECTIVE OF PARASITISM

The evolution of parasite–host associations is a multispecies process in which two principal associates, a parasite species and a host species, undergo evolutionary interaction, while interacting also with other associating species in a parasite community on the host. As a parasite lives and obtains its food from the host, it has acquired morphological and biological adaptations to promote its parasitic relationship with the host. At the same time, a parasite must evolve ecological strategies to promote coexistence with other associating species in the community (Kim, Chapter 1). Thus a parasite–host association involves the evolution of interspecific interactions and also population and community dynamics, and the coevolutionary perspective must include both types of interspecific processes, a parasite versus the host and a parasite versus other associates.

Coevolution

The term *coevolution*, first used by Ehrlich and Raven (1964), affords a range of conceptual variations as long as the use of the term is clearly defined. Coevolution is defined narrowly by Janzen (1980) as "any evolutionary change in a trait of the individuals in one population in response to a trait of the individuals of a second population, followed by evolutionary response by the second population to the change in the first." This definition emphasizes the specificity and reciprocity in evolution of specific traits linked between two interacting species, and is widely accepted by evolutionary biologists (e.g., Thompson 1982; Futuyma and Statkin 1983). On the other hand, the term coevolution used by parasitologists refers to phylogenetic processes (e.g., Brooks 1979, 1981; Chapters 8, 10, and 11). Brooks (1979) defined it as "a combination of two processes: *co-accommodation* between host and with no implication of host or parasite speciation and *co-speciation*, indicating concomitant host and parasite speciation." However, coevolution is considered here as a process and a manifestation of the evolution of interacting species and is defined as reciprocal evolutionary change in interacting species in a parasite community involving both the parasite species versus the host and the parasite species versus other parasites.

Inherent in the coevolutionary concept of parasitism is the gene-for-gene relationships between parasites and hosts (Person 1959; Rapport and Person 1980; Thompson 1982). The gene-for-gene concept introduced by Flor (1942, 1955) with the flax/rust system, and also called the matching gene theory by Gallun and Khush (1980), was defined by Person, Sambroski, and Rohringer (1962): "A gene-for-gene relationship exists when the presence of a gene in one population is contingent on the continued presence of a gene in another population, and where the interaction between the two genes leads to a single phenotypic expression by which the presence and absence of the relevant gene in either organism may be reorganized."

Several examples have been documented for plants and pathogens (Person 1959, 1967; Flor 1971; Day 1974; Burnett 1975; Gallun et al. 1975; Vanderplank 1982; Barrett 1983). Recently, gene-for-gene interaction was also observed between wheat and the Hessian fly *Myetiola desructor* (Gallun 1977; Gallun and Rhush 1980). However, no direct observation has been made for parasitic arthropods and vertebrates as yet.

Coadapted Tolerance

Most relationships between parasites and their hosts are stable ecological systems where the parasite load is small and tolerable for the host and where host responses are sustainable for parasites. Such a state of parasite–host relationships is reached only after a period of active interactions molded by selection and adaptation (Sprent 1962; Chapter 1).

Through interactive processes hosts develop behavioral and immune responses to prevent invasion and subsequent establishment of parasites. In the face of this adverse selection pressure, parasites acquire counterstrategies including various immune evasions against some or all host responses (Dineen 1963a, b; Ogilvie and Wilson 1976). These stepwise interactions ultimately lead to a dual modification of evolutionary strategies in both parasites and hosts, operating through natural selection. In this process, parasite antigens are stabilized and modified toward conformity with related substances, and corresponding antigen-combining sites in the host are selectively obliterated. The outcome of this process is the development of stable ecological systems where the host is rendered tolerant to the parasites. This phenomenon is called "adaptation tolerance" by Sprent (1962, 1969) and "molecular mimicry" by Damian (1979). Vane-Wright (1976) considers it antergic aggressive mimicry where the host is both the operator and model and the parasite is the mimic.

Parasite–host association is usually a multipartite system in which many parasite species and the host interact. The interaction molds the structure of a parasite community. One parasite species may exclude other species, limit their abundance, or even interfere with their reproductive potential. The hypothesis proposed by Schad (1966) suggests that cross-immunity promotes the adaptation of the parasite to its environment and limits the abundance of a competing species indirectly through the immune responses of the host. For example, in the tapeworm/white mouse system, the abundance of one species in an individual host is limited by a previous infection with another species (Heyneman 1962).

Host Specificity

Parasitic arthropods such as Anoplura and Mallophaga in highly evolved parasite–host relationships are characterized by a high degree of host specificity. A particular parasite species is specifically parasitic on a certain host species or on phylogenetically related, usually congeneric, host species (Kim, Chapter 1). This intimate relationship is accompanied by behavioral, physiological, and morphological adaptations.

Close parasite–host associations likely involve immunological specificity (Dineen 1963a, b; Damian 1964; Schad 1966). In immunological cross-reactions, a particular antigen for the parasite evokes the production of a specific antibody that will react to that antigen or to closely related antigens that occur in phylogenetically related species (Roitt 1980). Furthermore, host specificity in such close associations also involves gene-for-gene interactions between parasites and hosts (Clarke 1976; Rapport and Person 1980). Thus host specificity represents a complex of genetic, immunological, physiological, behavioral, ecological, and morphological interactions between parasites and their hosts, and coevolutionary relationships may involve one or more of these interactions.

Development of Coevolutionary Relationships

Analyses of host specificity, population patterns, and adaptations in parasitic arthropods provide the following assumptions about their parasite–host relationships (Kim, Chapter 1):

1. Genetic polymorphism exists in the ability of parasites to attack hosts and also in the ability of hosts to resist parasites (Clarke 1976; Damian 1979).

2. Parasitic arthropod populations are host limited and depend on the numbers of susceptible hosts (Clarke 1976).

3. The distribution of parasitic arthropods among hosts is overdispersed (Crofton 1971a, b).

4. Morphological and biological adaptations involved in host specificity are genetically fixed.

As the parasitic arthropod/mammal association goes through a series of transitional states from free-living to an obligate, permanent parasitic state, host specificity evolves and becomes more intimate by acquiring morphological and biological adaptations to promote physical association with the host and the capacity of feeding on it (Kim, Chapter 1). When founder populations of an opportunistic species, which may be parasites, predators, or scavengers, first invade the "vacant" host, the outcome is either colonization (establishment of an association) or extinction (Kim, Chapter 1). Naturally, colonizers are biological generalists with a high intrinsic rate of increase, short generation time (MacArthur 1960, 1972), and preadaptations suited for the parasitic mode of life (Pickett 1976). Even for the colonizing species, founder populations could face extinction because of genetic variation in the ability of founders to attack hosts and also that of hosts to resist parasites. As colonization takes root and the parasite community becomes stable, the parasite–host association becomes obligatory by coadaptive evolutionary processes through which the ecotopes of associating species are narrowed and their biological adaptations are molded by selection. This process also involves immunologic cross-reactions between hosts and parasites (Sprent 1962; Dineen 1963a, b; Damian 1962, 1964) which may become genetically fixed to become gene-for-gene interactions. These interactions are coadaptive among interacting associates (Clarke 1976; Vanderplank 1982).

Approaches to the Study of Coevolution

Many intimate associations between parasitic arthropods and mammals are undoubtedly the result of evolutionary interactions and perhaps coevolution between parasites and their hosts (Mitter and Brooks 1983; Holmes 1983), but they also involve coevolution of interacting parasite species within a parasite community (Connell 1980). It is relatively easy to demonstrate that a parasite species has adapted to particular characters of the host (Kellogg 1913; Metcalf 1929; Hopkins 1949; Fain 1969; Kim and Ludwig 1978b, 1982; Traub 1980; and Chapters 1, 4, 5, 6, 7, 8, 9, 10, 11, and 12 of this book). However, the demonstration of other evolutionary interactions, such as host adaptation to parasite infestation, coevolution of parasite and host, and pairwise coevolution among associating parasites, is more difficult because it requires evidence that the host and other associating species evolved in response to the parasite adaptation. For example, the long-tailed field mouse, *Apodemus sylvaticus*, is infested with nine species of Siphonaptera, two species of Anoplura, 12 species of Acari, and one species each of Coleoptera and Ixodidae (Elton et al. 1931), any combination of which may actually be associated with a particular host individual. It is relatively easy to document that the tibio-tarsal complex of the sucking louse *Hoploplura affinis* (Burmeister) evolved in response to the host's hair characteristics, and its microhabitat restriction is a function of the host's grooming behavior. It is, however, difficult to demonstrate that *A. sylvaticus* evolved the hair characteristics, grooming behavior, and particular immune reaction in response to the louse infestation, or that *H. affinis* evolved a certain behavior pattern of microhabitat restriction in response to the ecological characteristics of associating ectoparasites such as the tick *Ixodes tenuirostris* (Neumann). It is even more difficult to demonstrate the counteradaptations or reciprocal evolutionary change in the parasite (e.g., *H. affinis*) for coevolution (Kim, Chapter 1).

Several lines of research in evolutionary biology provide evidence of coevolution (Gilbert and Raven 1975; Slatkin and Maynard Smith 1979; Thompson 1982; Futuyma and Slatkin 1983; Levin 1983). Coevolution could be studied by direct natural history observations and experimentation within natural communities (e.g., Janzen 1966, 1967), by observation of the direct responses to interactions or genetic changes (e.g., Fenner and Ratcliffe 1965; Barrett 1983), by phylogenetic analysis of parallel cladogeneses (e.g., Kim et al. 1975; Brooks 1979, 1981; Mitter and Brooks 1983; Kim, Chapter 7), by analysis of taxonomic characters involved in coadaptation (e.g., Fain 1979; Hopkins 1949; Clay 1949; Kim and Ludwig 1978b, 1982; Holmes 1983; and Chapters 8–12 of this book), and by studying host specificity and geographical distribution (e.g., Hopkins 1949, 1957a, b; Clay 1947, 1957; Traub 1980; and Chapters 4, 5, 6, 7, and 8).

Theoretical modeling of coevolution in the host–parasite system has also provided an important approach in population genetics and evolutionary ecology (e.g., Mode 1958; Gilbert 1979; Roughgarden 1979; Slatkin and Maynard Smith 1979; Anderson and May 1982; Levin 1983).

ORIGIN AND PHYLOGENY OF PARASITE–HOST ASSOCIATIONS

The evolution of parasitic arthropods is tied closely to the biology and evolution of their hosts and may occur by many different routes. Thus parasite–host associations observed in different parasitic insects and acarines represent various stages of evolutionary succession. Some associations such as those observed in the trichodectid *Geomydoecus* (Timm 1983) and the sucking lice *Pedicinus* (Kuhn and Ludwig 1967) are highly evolved with specialized parasites closely adapted to their hosts. Others are still loosely established and fluid such as those found in many parasitic mites

(Radovsky, Chapter 9). Intimacy of host associations indicates that they are either very old in the evolutionary sense or relatively recent but with rigorous mutual selection and stepwise coadaptation.

The phylogeny of certain parasitic insects closely parallels the phylogeny of their mammalian hosts (e.g., Kim and Ludwig 1978b, 1982; Timm 1983), whereas many parasitic mites show noncongruence of parasite and host phylogenies (e.g., Kethley and Johnston 1975). Congruence between parasite and host phylogenies has impressed parasitologists for a long time (Kellogg 1913; Metcalf 1929; Eichler 1948), leading to the formulation of a number of so-called "Rules" (see Inglis 1971; Brooks 1979, 1981), such as Fahrenholz's Rule (see Stammer 1957), Szidat's Rule (see Szidat 1956), Manter's Rule (see Manter 1955, 1966), Eichler's Rule (see Eichler 1948); they may be collectively called the phylogenetic tracking model. On the other hand, for associations with noncongruence of parasite and host phylogenies, Kethley and Johnston (1975) proposed the resource tracking hypothesis. Although they appear conflicting, the phylogenetic and resource tracking models are not opposing hypotheses. They are based on different assumptions of parasite dispersal and instead represent opposite ends of a continuum for parasite–host associations (Timm 1983).

The degree of intimacy in parasite–host associations is a relative measure which can be determined by a combination of qualitative and quantitative adaptations, host specificity (%), and dispersal (route and rate) (Kim, Chapter 1). Morphological and biological adaptations and host specificity of parasitic arthropods are well documented in the chapters of this book (Chapters 4–12). However, little is known about the route and rate of dispersal for different parasitic arthropods, which requires ecological studies (Kethley and Johnston 1975; Timm 1983; Kim 1975 and Chapter 1).

Preadaptations

Many different parasitic arthropods such as ticks, mites, and lice were already present before the great radiation of the Cenozoic mammals during the Paleocene and Eocene epochs (Petrunkevitch 1955; Hennig 1969; Kim and Ludwig 1982). Some of these arthropods were already parasitic on premammalian hosts such as reptiles (Hoogstraal and Kim, Chapter 10), and others were scavengers, saprophages, or even chance feeders. Many of these were associated with nests, roosts, dung, or other habitats of their associated host animals (Balashov 1984; Kim, Chapter 1).

Analyses of parasitic arthropod–mammal associations suggest that many were established at the time of the Cenozoic mammal radiation and that the parasites have closely evolved with their hosts. Other associations were established relatively recently, long after the major Eocene radiation. Regardless of when their association was established, the parasitic arthropods (or their progenitors) that successfully colonized the host are likely to have been opportunistic species (Kim, Chapter 1).

Opportunistic species are usually ecological and physiological generalists with a high intrinsic rate of increase and short generation time (MacArthur 1960, 1972). They have preadaptations for physical association with the host and the ability to feed on it by which they can adapt themselves to the parasitic mode of life. They can rapidly increase in numbers when environmental factors are favorable and can shift their realized niches with morphological and physiological preadaptations to avoid competition and to make use of new adaptive opportunities (Pickett 1976).

As the founder populations of opportunistic species invade new habitats (hosts), they are subjected to rigorous differential selection pressures. This process may lead to selective extinction because the populations have varying abilities to counteract new host defenses (Foin et al. 1975). Furthermore, the success of their colonization also depends on the structure of the parasite community on the host that the founder populations invaded and the genetic capacity of the hosts to counter the parasite invasion thus stabilizing the relationship. Otherwise, extreme success of the parasites may cause extinction of the host.

Resource Tracking and Cophylogeny

Parasitic arthropods took a number of evolutionary routes before their associations with mammalian hosts became obligatory (Fig. 13.1). Insects such as Anoplura established associations with their hosts long before the Cenozoic mammal radiation and evolved along with the host cladogenesis (Kim and Ludwig 1982), whereas many parasitic mites colonized their hosts only during the late Tertiary or even the Pleistocene epoch although they were present long before (Radovsky, Chapter 9; Nutting, Chapter 11).

Several models for the origin and evolution of parasite–host associations are depicted in the phylogenetic trees of Figure 13.1 (Ludwig and Kim 1981). The host lineage (capital letters) is superimposed by that of the parasites (lower case). Here, the host stock $A-U$ is infested with the parasite stock having two primary lineages, 1 and 2, where the host line $A-C$ had never been colonized by parasites (*primary absence*) (Fig. 13.1, 4), but host line $R-U$ is free of the parasites because of the extinction of j^* before R cladogenesis (*secondary absence*) (Fig. 13.1, 5).

Resource Tracking Model

This model includes both *original invasion* (or *infestation*) (Fig. 13.1, 1) and *primary invasion* (or *infestation*) (Fig. 13.1, 2 and 3) by major host changes. These host changes of parasitic arthropods occur between hosts that share similar topographic and anatomical features. In these processes, the para-

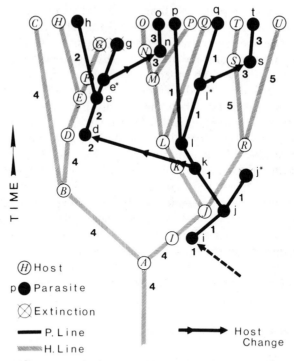

Figure 13.1 Origin and phylogeny of host–parasite associations. 1. Original invasion or infestation. 2. Primary invasion or infestation. 3. Secondary invasion or infestation. 4. Primary absence. 5. Secondary absence. Capital letters (*A–U*) in open circles represent host species; solid circles with lower case letters (d–t) represent parasites; lower-case letters with asterisks represent anagenetic species involved in extinction or host change.

sites track resources without regard to host relatedness (Kethley and Johnston 1975). The underlying assumption of the model is that host transfer occurs between two host taxa of distant or no phylogenetic relationship (Ludwig and Kim 1981). However, host transfers could occur between taxonomically related hosts contributing to congruent host–parasite relationships (Kethley and Johnston 1975). The term *original invasion* (Fig. 13.1, 1) refers to the infestation of a host lineage by a parasite stock that was never associated with the host stock, whereas *primary invasion* (Fig. 13.1, 2 and 3) is defined as the first infestation of a host line (or taxon) by a parasite taxon that previously infested other lineages of the same host stock. A host lineage (e.g., the host line *R, S, T, U*) that is secondarily free of parasites by extinction may be infested by another parasite taxon (the parasite line l*–s, t) later on in its evolution; this phenomenon may be called *secondary invasion* (or *infestation*) (Fig. 13.1, 3 from the parasite line l–l*). The parasite line l–p follows the host line *L–P*, and the parasite p occurs on both the hosts *O* and *P*, whereas the parasite line n–o originates by host change from the parasite e*, whose progeny g became extinct along with the host G, and the parasite o co-occurs on the host O with the parasite p. Similarly, the parasite line s–t originates from the parasite l* by host change and establishes on the host line S–T whose ancestors I and J were infested with progenitors of the parasite line that later became extinct (Fig. 13.1, j*).

Phylogenetic Tracking Model

This model includes both parasite anagenesis (Fig. 13.1, d–e, i–j, j–l, l–q) and cladogenesis (Fig. 13.1, e–h and g, j–k and j*, l–p and q) in parallel with host evolution. When these obligatory relationships become coevolutionary, the evolution of parasite–host associations encompasses two processes (Brooks 1979): *co-accommodation* defined as the "mutual adaptation of a given parasite species and its host(s) through time" (Fig. 13.1, the host–parasite lines D/d–G/g, J/j–L/l, and L/l–Q/q) and *co-speciation* referring to concomitant parasite and host speciation (Fig. 13.1, the host–parasite lines E/e–H/h and G/g, J–j–J–j* and K/k, L/l to M–P/p and Q/q). It results in patterns that support Fahrenholz's Rule which states that "parasite phylogeny mirrors host phylogeny" (Brooks 1979), Szidat's Rule that the more primitive host harbors the more primitive parasites (Szidat 1956, 1960), and Eichler's Rule that the higher the taxonomic category of the host taxon the more parasite genera it harbors (Eichler 1948; Inglis 1971).

Now, let us consider the Anoplura–Mammalia association as an example for the evolution of parasite–host associations following the phylogenetic trees in Figure 13.1. Primary absence (4) is found in Marsupialia, Edentata, Insectivora, and Chiroptera (Kim and Ludwig 1978a), and secondary absence (5) occurs in Sirenia, Carnivora, and Cetacea (Kim and Ludwig 1978a; Kim, Chapter 5). The Protanoplura successfully colonized early mammals during the early Paleocene or perhaps the late Cretaceous period (*original invasion*; Fig. 13.1, 1) (Kim and Ludwig 1982). As the Cenozoic radiation of recent mammals opened up numerous adaptive zones, different taxa of Anoplura colonized various host lineages (*primary invasion*; Fig. 13.1, 1, 2) such as Pedicinidae (*Pedicinus*) on the primate Cercopithecidae, Hybophthiriidae (*Hybophthirius*) on Tubulidentata, Neolinognathidae (*Neolinognathus*) on Macroscelididae, Enderleinellidae on Sciuridae, and Hoplopleuridae and Polyplacidae on Rodentia. Mammalian hosts that had not harbored previously or lost sucking lice through extinction were invaded later by anopluran taxa parasitic on the other host lineages (*secondary invasion* or *infestation*; Fig. 13.1, 3). Secondary invasion is found in the associations of *Linognathus* with Canidae, the polyplacid *Sathrax* with the menotyphlous insectivore Tupaiidae, and *Polyplax* species with insectivores. Multispecies infestation (Fig. 13.1, parasites o and p on the host line N–O) also occurs commonly in Anoplura; for example, *Neohaematopinus* and *Enderleinellus* on Sciuridae, and *Polyplax* and *Hoplopleura* on the rodent Muridae (Kim and Ludwig 1978a; Kim 1982).

ACKNOWLEDGMENTS

I am indebted to Peter H. Adler and Allen L. Norrbom for their input. My discussion with them has helped to improve this chapter. My thanks are also due to Allen L. Norrbom and Charles W. Pitts, Jr. for reading the manuscript and also to Thelma Brodzina for her excellent typing effort.

REFERENCES

Anderson, R. M. and R. M. May. 1982. Coevolution of hosts and parasites. *Parasitology* **85**:411–426.

Balashov, Yu. A. 1984. Interaction between bloodsucking arthropods and their hosts, and its influence on vector potential. *Annu. Rev. Entomol.* **29**:137–156.

Barrett, J. A. 1983. 7. Plant–fungus symbioses. Pages 137–160 in D. J. Futuyma and M. Slatkin (Eds.), *Coevolution*. Sinauer Associates, Sunderland, MA.

Brooks, D. R. 1979. Testing the context and extent of host–parasite coevolution. *Syst. Zool.* **28**:299–307.

Brooks, D. R. 1981. Hennig's parasitological methods: a proposed solution. *Syst. Zool.* **30**:229–249.

Burnett, J. H. 1975. *Mycogenetics*. John Wiley, New York.

Clarke, B. 1976. The ecological genetics of host–parasite relationships. Pages 87–103 in A. E. R. Taylor and R. Muller (Eds.), *Genetic Aspects of Host–Parasite Relationships*, Symp. Brit. Soc. Parasitol. Vol. 14.

Clay, T. 1949. Some problems in the evolution of a group of ectoparasites. *Evolution* **3**:279–299.

Clay, T. 1957. The Mallophaga of birds. Pages 120–157 in J. T. Baer (Ed.), *First Symposium on Host Specificity Amongst Parasites of Vertebrates*, Inst. Zool., University of Neuchatel, Neuchatel.

Connell, J. H. 1980. Diversity and the coevolution of competitors, or the ghost of competition past. *Oikos* **35**:131–138.

Coryndon, J. C. and R. J. G. Savage. 1973. The origin and affinities of African mammal fauna. Pages 121–135 in N. F. Hughes (ed.), *Organisms and Continents Through Time*. Spec. Pap. Paleontol. No. 12, The Paleontological Association, London.

Crofton, H. D. 1971a. A quantitative approach to parasitism. *Parasitology* **62**:179–193.

Crofton, H. D. 1971b. A model of parasite–host relationships. *Parasitology* **63**:343–364.

Damian, R. T. 1962. A theory of immunoselection for eclipsed antigens of parasites and its implications for the problem of antigenic polymorphism in man. *J. Parasitol.* **48**:16.

Damian, R. T. 1964. Molecular mimicry: Antigen sharing by parasite and host and its consequences. *Am. Nat.* **98**:129–149.

Damian, R. T. 1979. Molecular mimicry in biological adaptation. Pages 103–126 in B. B. Nickol (Ed.), *Host–Parasite Interfaces*, Academic Press, New York.

Day, P. R. 1974. *Genetics of Host–Parasite Interactions*. W. H. Freeman, San Francisco, CA.

Dineen, J. K. 1963a. Immunological aspects of parasitism. *Nature* **197**:268–269.

Dineen, J. K. 1963b. Antigenic relationships between host and parasite. *Nature* **197**:471–472.

Dubinina, Kh. V. 1969. Certain adaptations of listrophorid mites (Fam. Listrophoridae) to parasitism in the hair cover of hosts-rodents. *Proc. Ind. Int. Congr. Acarol.* **1967**:299–300.

Ehrlich, P. R. and P. H. Raven. 1964. Butterflies and plants: a study in coevolution. *Evolution* **18**:586–608.

Eichler, W. 1948. Some rules in ectoparasitism. *Ann. Mag. Nat. Hist.* **12**:588–598.

Elton, C., E. G. Ford, and J. R. Baker. 1931. 36. The health and parasites of a wild mouse population. *Proc. Zool. Soc. London* 1931 (3)657–721.

Fain, A. 1968. Notes sur trois acariens remarquables (Sarcoptiformes). *Acarologia* **10**:276–291.

Fain, A. 1969. Adaptation to parasitism in mites. *Acarologia* **11**:429–449.

Fenner, F. and F. N. Ratcliffe. 1965. *Myxomatosis.* Cambridge University Press, Cambridge.

Foin, T. C., J. W. Valentine, and F. J. Ayala. 1975. Extinction of taxa and Van Valen's law. *Nature* (London) **257**:514–515.

Furman, D. P. and A. W. Smith. 1973. In vitro development of two species of *Orthohalarachne* (Acarina: Holarachnidae) and adaptations of the life cycles for endoparasitism in mammals. *J. Med. Entomol.* **10**:414–416.

Futuyma, D. J. and M. Slatkin (Eds.). 1983. *Coevolution.* Sinauer Associates, Sunderland, MA.

Flor, H. H. 1942. Inheritance of pathogenicity *Melampsora lini. Phytopathology* **32**:653–669.

Flor, H. H. 1955. Host–parasite interaction in flax rust—its genetics and other implication. *Phytopathology* **45**:680–685.

Flor, H. H. 1971. Current status of the gene-for-gene concept. *Annu. Rev. Phytopathol.* **9**:275–296.

Gallun, R. L. 1977. The genetic basis of Hessian fly epidemics. *Ann. N.Y. Acad. Sci.* **287**: 223–229.

Gallun, R. L. and G. S. Khush. 1980. Genetic factors affecting expression and stability of resistance. Pages 64–85 in F. G. Maxwell and P. R. Jennings (Eds.), *Breeding Plants Resistant Insects*, Wiley-Interscience, New York.

Gallun, R. L., K. J. Starks, and W. D. Guthrie. 1975. Plant resistance to insects attacking cereals. *Annu. Rev. Entomol.* **20**:337–357.

Gauthier-Pilters, H. and A. I. Dagg. 1981. *The Camel. Its Evolution, Ecology, Behavior, and Relationship to Man.* The University of Chicago Press, Chicago, IL.

Gilbert, L. E. 1979. Development of theory in the analysis of insect–plant interactions. Pages 117–154 in D. J. Horn, R. D. Mitchell, and G. R. Stairs (Eds.), *Analysis of Ecological Systems*, Ohio State University Press, Columbus.

Gilbert, L. E. and P. H. Raven. 1975. *Coevolution of Animals and Plants.* University of Texas Press, Austin.

Hennig, W. 1969. *Die Stammesgeschichte der Insekten.* Verlag von Waldemar Kramer in Frankfurt an Main.

Heyneman, D. 1962. Studies on helminth immunity. II. Influence of *Hymenolepsis nana* (Cestoda: Hymenolepididae) in dual infections with *H. diminuata* in white mice and rats. *Exp. Parasitol.* **12**:7–18.

Holmes, J. C. 1983. 8. Evolutionary relationships between parasitic helminths and their hosts. Pages 161–185 in D. J. Futuyma and M. Slatkin (eds.), *Coevolution.* Sinauer Associates, Sunderland, MA.

Hoogstraal, H. 1978. Biology of Ticks. Pages 3–14 in J. K. H. Wilde (ed.), *Ticke-Borne Diseases and Their Vectors. Proc. Internat. Conf.* (Edinburgh, September–October 1976).

Hopkins, G. H. E. 1949. The host associations of the lice of mammals. *Proc. Zool. Soc. London* **119**:387–604.

Hopkins, G. H. E. 1957a. Host associations of Siphonaptera. Pages 64–87 in J. G. Baer (Ed.), *First Symposium on Host Specificity Amongst Parasites of Vertebrates*, Inst. Zool., University of Neuchatel, Neuchatel.

Hopkins, G. H. E. 1957b. The distribution of Phthiraptera on mammals. Pages 88–119 in J. G.

Baer (Ed.), *First Symposium on Host Specificity Amongst Parasites of Vertebrates*, Inst. Zool., Univ. Neuchatel, Neuchatel.

Inglis, W. G. 1971. Speciation in parasitic nematodes. *Advances in Parasitology* 9:185–224.

Janzen, D. H. 1966. Coevolution of mutualism between ants and acacias in Central America. *Evolution* 20:249–275.

Janzen, D. H. 1967. Interaction of the bull's-horn acacia (*Acacia cornigera* L.) with its ant inhabitant (*Pseudomyrmex ferruginea* F. Smith) in eastern Mexico. *Univ. Kansas Sci. Bull.* 47:315–558.

Janzen, D. H. 1980. When is it coevolution? *Evolution* 34:611–612.

Kellogg, V. L. 1913. Distribution and species-forming of ecto-parasites. *Am. Nat.* 47:129–158.

Kethley, J. B. and D. E. Johnston. 1975. Resource tracking patterns in bird and mammal ectoparasites. *Misc. Publ. Entomol. Soc. Am.* 9:231–236.

Kim, K. C. 1975. Ecology and morphological adaptation of the sucking lice (Anoplura: Echinophthiriidae) on the northern fur seal. *Rapp. Reun. Cons. Perm. Int. Expl. Mer* 169:504–515.

Kim, K. C. 1982. Host specificity and phylogeny of Anoplura. Deux. Symp. Spec. Parasitaire des Parasit. Vertebr. 13–17 Avril 1981, *Mem. Mus. Nat. Hist. Nat.*, N.S., Ser. A., Zool., 123:123–127.

Kim, K. C. and K. C. Emerson. 1974. A new *Polyplax* and records of sucking lice (Anoplura) from Madagascar. *J. Med. Entomol.* 11:107–111.

Kim, K. C. and H. W. Ludwig. 1978a. The family classification of the Anoplura. *Syst. Entomol.* 3:249–284.

Kim, K. C. and H. W. Ludwig. 1978b. Phylogenetic relationships of parasitic Psocodea and taxonomic position of the Anoplura. *Ann. Entomol. Soc. Am.* 71:910–922.

Kim, K. C. and H. W. Ludwig. 1982. Parallel evolution, cladistics, and classification of parasitic Psocodea. *Ann. Entomol. Soc. Am.* 75:537–548.

Kim, K. C., V. L. Haas, and M. C. Keyes. 1980. Populations, microhabitat preference and effects of infestation of two species of *Orthohalarachne* (Halarachnidae: Acarina) in the northern fur seal. *J. Wildl. Dis.* 16:45–52.

Kim, K. C., C. A. Repenning, and G. V. Morrejohn. 1975. Specific antiquity of the sucking lice and evolution of otariid seals. *Rapp. Reun. Cons. Perm. Int. Expl. Mer* 169:544–549.

Kuhn, H. J. and H. W. Ludwig. 1967. Die Affenläuse des Gattung *Pedicinus*. *Zeitschr. Zool. Syst. Evol.* 5:144–297.

Lavoipierre, M. M. J. 1964. A new family of Acarines belonging to the Suborder Sarcoptiformes parasitic in the hair follicles of Primates. *Ann. Natal Mus.* 16:191–208.

Levin, S. A. 1983. Some approaches to the modelling of coevolutionary interactions. Pages 21–65 in M. H. Nitecki (Ed.), *Coevolution*. University of Chicago Press, Chicago, IL.

Ludwig, H. W. and K. C. Kim. 1981. Proposal for definitions of host–parasite relationships (unpublished).

Maa, T. C. 1963. Genera and species of Hippoboscidae (Diptera): types, synonymy, habitats, and natural groupings. *Pac. Inst. Monogr.* 6:1–186.

MacArthur, R. H. 1960. On the relative abundance of species. *Am. Nat.* 94:25–36.

MacArthur, R. H. 1972. *Geographical Ecology. Patterns in the Distribution of Species*. Harper & Row, New York.

Manter, H. W. 1955. The zoogeography of trematodes of marine fishes. *Exp. Parasitol.* 4:62–86.

Manter, H. W. 1966. Parasites of fishes as biological indicators of recent and ancient conditions. Pages 59–71 in J. E. McCauley (Ed.), *Host–Parasite Relationships*. Oregon State University Press, Corvallis.

Marshall, A. G. 1980. The function of combs in ectoparasitic insects. Pages 79–87 in R. Traub and H. Starke (Eds.), *Fleas*, Proc. Int. Conf. Fleas, Ashton Wold/Peterborough/UK/21–25 June 1977. A. A. Balkema, Rotterdam.

Marshall, A. G. 1981a. *The Ecology of Ectoparasitic Insects*. Academic Press, London.

Marshall, A. G. 1981b. The sex ratio in ectoparasitic insects. *Ecol. Entomol.* **6**:155–174.

Metcalf, M. M. 1929. Parasites and the aid they give in problems of taxonomy, geographical distribution, and paleogeography. *Smithson. Misc. Coll.* **81**(8):1–36.

Mitter, C. and D. R. Brooks. 1983. 4. Phylogenetic aspects of coevolution. Pages 65–98 in D. J. Futuyma and M. Slatkin (Eds.), *Coevolution*. Sinauer Associates, Sunderland, MA.

Mode, C. J. 1958. A mathematical model for the co-evolution of obligate parasites and their hosts. *Evolution* **12**:158–165.

Nutting, W. B. 1968. Host specificity in parasitic acarines. *Acarologia* **10**:165–180.

Ogilvie, B. M. and R. J. M. Wilson. 1976. Evasion of the immune response by parasites. *Br. Med. Bull.* **32**:177–181.

Person, C. 1959. Gene-for-gene relationships in host:parasite systems. *Can. J. Bot.* **37**:1101–1130.

Person, C. 1967. Genetic aspect of parasitism. *Can. J. Bot.* **45**:1193–1203.

Person, C. D., J. Samborski, and R. Rohringer. 1962. The gene-for-gene concept. *Nature* **194**:561–562.

Petrunkevitch, A. 1955. Arachnida. Chelicerata with sections on Pycnogonida and *Palaeoispus*. In R. C. Moore (Ed.), *Treatise on Invertebrate Paleontology. Part P, Arthropods* **2**:42–102.

Pielou, E. C. 1979. *Biogeography*. John Wiley, New York.

Pickett, S. T. A. 1976. Succession: an evolutionary interpretation. *Am. Nat.* **110**:107–119.

Radovsky, F. J. 1969. Adaptive radiation in the parasitic Mesostigmata. *Acarologia* **11**:450–478.

Rapport, D. J. and C. O. Person. 1980. Games that genes play: host–parasite interactions in a game-theoretic context. *Evol. Theor.* **4**:275–287.

Roitt, I. M. 1980. *Essential Immunology*. Blackwell Scientific Publ., Oxford.

Roughgarden, J. 1979. *Theory of Population Genetics and Evolutionary Ecology: An Introduction*. Macmillan, New York.

Rothschild, M. and R. Ford. 1972. Breeding cycle of the flea *Cediopsylla simplex* is controlled by breeding cycle of host. *Science* **178**:625–626.

Schad, G. A. 1966. Immunity, competition, and natural regulation of helminth populations. *Am. Nat.* **100**:359–364.

Simpson, G. G. 1945. The principles of classification and a classification of mammals. *Bull. Am. Mus. Nat. Hist.* **85**:1–350.

Slatkin, M. and J. Maynard Smith. 1979. Models of Coevolution. *Q. Rev. Biol.* **54**:233–263.

Sprent, J. F. A. 1962. Parasitism, immunity and evolution. Pages 149–165 in G. W. Leeper (Ed.), *The Evolution of Living Organisms*, Symp. R. Soc. Victoria, Melbourne University Press, Melbourne.

Sprent, J. F. A. 1969. Evolutionary aspects of immunity in zooparasitic infections. In G. J. Jackson, D. Herman, and I. Singer (Eds.), *Immunity to Parasitic Animals* **1**:3–62.

Stammer, H. J. 1957. Gedanken zu den parasito-phyletischen Regeln und zur Evolution der parasiten. *Zool. Anz.* **159**:255–267.

Szidat, L. 1956. Der Marine Charakter der Parasitenfauna der Susswasserfische des Stromsystems des Rio de la Plata und ihre Deutung als Reliktfauna des Tertiaren Tethys-Meeres. *Proc. 14th Int. Congr. Zool.* **1953**:128–138.

Thompson, J. N. 1982. *Interaction and Coevolution*. John Wiley, New York., N.Y.

Timm, R. M. 1983. Fahrenholz's rule and resource tracking: a study of host–parasite coevolution. Pages 225–265 in M. H. Nitecki (Ed.), *Coevolution*. University of Chicago Press, Chicago, IL.

Traub, R. 1980. The zoogeography and evolution of some fleas, lice and mammals. Pages 95–172 in R. Traub and H. Starcke (Eds.), *Fleas*. A. A. Balkema, Rotterdam.

Vanderplank, J. E. 1982. *Host–Pathogen Interactions in Plant Disease*. Academic Press, New York.

Vane-Wright, R. I. 1976. A unified classification of mimetic resemblances. *Biol. J. Linn. Soc.* **8**:25–56.

Vanzolini, P. E. and L. Guimaraes. 1955. Lice and the history of South American land mammals. *Rev. Brasil. Entomol.* (Sao Paulo) **3**:13–46.

Wenzel, R. L. 1976. The streblid batflies of Venezuela (Diptera: Streblidae). *Brigham Young Univ. Sci. Bull., Biol. Ser.* **20**:1–177.

Wenzel, R. L., V. L. Tipton, and A. Kiewlicz. 1966. The streblid batflies of Panama (Diptera: Calypterae: Streblidae). Pages 405–675 in R. L. Wenzel and V. J. Tipton (Eds.), *Ectoparasites of Panama*. Field Mus. Nat. Hist., Chicago, IL.

Appendix A

List of Parasitic Arthropods Associated with Mammals

Appendix A is a composite of parasite–host lists prepared for various taxa by several authors:

Hemiptera, Dermaptera, Coleoptera, Diptera (Chapter 4): K. C. Kim and P. H. Adler
Anoplura (Chapter 5): K. C. Kim
Mallophaga (Chapter 6): K. C. Kim
Siphonaptera (Chapter 8): R. Traub
Mesostigmata (Chapter 9): F. J. Radovsky
Ixodoidea (Chapter 10): K. C. Kim
Prostigmata (Chapter 11): K. C. Kim
Astigmata (Chapter 12): A. Fain and K. E. Hyland.

This list is by no means complete, although our best efforts have been made. Some published records may represent stragglers or contamination and need further verification. Nevertheless, they are included for some of the parasitic groups in the list.

Families and genera of parasitic arthropods are listed alphabetically and the classification system and names of Honacki et al. (*Mammal Species of the World: A Taxonomic and Geographic Reference*, Allen Press and Association of Systematics Collections, 1982) are followed here for mammal hosts. The number of species is indicated, where possible, in parentheses next to the name of the taxon and this refers only to those associated with mammals; for example, Hippoboscidae is given as having 48 species, although 149 additional species are known from birds. Mammalian hosts are listed in their approximate order of importance as hosts for each of the listed parasite taxa except Ixodides, where they are alphabetically listed. When the host association involves more than one order for each genus of parasitic arthropods, mammalian taxa of the same order are separated by commas but those of different orders are separated by semicolons. Major references used to compile the list for each higher taxon are given in parentheses next to the ordinal or subordinal name, and the bibliographic data appear at the end of Appendix A(1). Those hosts with asterisks indicate very rare association or stragglers. The literature review for the lists of most taxa was completed in December 1981, but the literature survey for

Anoplura and Mallophaga was completed up to the end of December 1982. Thanks are due to K. C. Emerson, James E. Keirans, and Robert L. Smiley for their kind review of the lists for chewing lice, ticks, and the Prostigmata, respectively.

Parasitic Arthropods	Mammals (Hosts)
Class INSECTA	
Order HEMIPTERA (Maa 1964; Marshall 1981[1])	
Fam. POLYCTENIDAE (32)	CHIROPTERA.
Subfam. Polycteninae (8)	
Eoctenes (7)	Emballonuridae, Megadermatidae, Nycteridae.
Polyctenes (1)	Megadermatidae.
Subfam. Hesperocteninae (24)	
Androctenes (3)	Rhinolophidae.
Hesperoctenes (16)	Molossidae.
Hypoctenes (5)	Molossidae.
Order DERMAPTERA (Marshall 1981)	
Fam. ARIXENIIDAE (5)	CHIROPTERA.
Arixenia (2)	Molossidae.
Xenaria (3)	Molossidae.
Fam. HEMIMERIDAE (11)	RODENTIA.
Araeomerus (2)	Cricetidae (*Beamys*).
Hemimerus (9)	Cricetidae (*Cricetomys*).
Order ANOPLURA (Kim and Ludwig 1978)	
Fam. ECHINOPHTHIRIIDAE (12)	PINNIPEDIA, CARNIVORA.
Antarctophthirus (6)	Otariidae, Odobenidae, Phocidae.
Echinophthirius (1)	Phocidae.
Latagophthirus (1)	Mustelidae (*Lutra*).
Lepidophthirus (2)	Phocidae.
Proechinophthirus (2)	Otariidae.
Fam. ENDERLEINELLIDAE (49)	Sciuridae.
Enderleinellus (43)	Sciuridae.
Atopophthirus (1)	Petauristinae (*Petaurista*).
Microphthirus (1)	Petauristinae (*Glaucomys*).
Phthirunculus (1)	Petauristinae (*Petaurista*).
Werneckia (3)	Finambulini.
Fam. HAEMATOPINIDAE (22)	ARTIODACTYLA, PERISSODACTYLA.
Haematopinus (22)	Bovidae, Suidae, Cervidae; Equidae.
Fam. HAMOPHTHIRIIDAE (1)	DERMOPTERA.
Hamophthirius (1)	Cynocephalidae (*Cynocephalus*).
Fam. HOPLOPLEURIDAE (134)	RODENTIA, INSECTIVORA, LAGOMORPHA.
Subfam. Hoplopleurinae (122)	RODENTIA, LAGOMORPHA.
Hoplopleura (117)	Arvicolidae, Cricetidae, Muridae, Octodontidae, Sciuridae; Ochotonidae.

Parasitic Arthropods	Mammals (Hosts)
Pterophthirus (5)	Caviidae, Echimyidae.
Subfam. Haematopinoidinae (12)	INSECTIVORA, RODENTIA.
Ancistroplax (3)	Soricidae.
Haematopinoides (1)	Talpidae.
Schizophthirus (7)	Gliridae.
Typhlomyophthirus (1)	Cricetidae (*Typhlomys*).
Fam. HYBOPHTHIRIDAE (1)	TUBULIDENTATA.
Hybophthirus (1)	Orycteropodidae.
Fam. LINOGNATHIDAE (69)	ARTIODACTYLA, CARNIVORA, HYRACOIDEA.
Linognathus (51)	Bovidae, Giraffidae; Canidae.
Solenopotes (10)	Bovidae, Cervidae.
Prolinognathus (8)	Procaviidae.
Fam. MICROTHORACIIDAE (4)	ARTIODACTYLA.
Microthorcius (4)	Camelidae.
Fam. NEOLINOGNATHIDAE (2)	MACROSCELIDEA.
Neolinognathus (2)	Macroscelididae.
Fam. PECAROECIDAE (1)	ARTIODACTYLA.
Pecaroecus (1)	Tayassuidae.
Fam. PEDICINIDAE (16)	ANTHROPOID PRIMATES.
Pedicinus (16)	Cercopithecidae.
Fam. PEDICULIDAE (2)	ANTHROPOID PRIMATES.
Pediculus (2)	Hominidae, Cebidae, Hylabatidae, *Pongidae.
Fam. POLYPLACIDAE (177)	RODENTIA, INSECTIVORA, LAGOMORPHA, PROSIMIAN PRIMATES, SCANDENTIA.
Ctenophthirus (1)	Echimyidae.
Cuyana (1)	Chinchillidae.
Docophthirus (1)	Tupaiidae.
Eulinognathus (23)	Cricetidae, Dipodidae, Bathyergidae, Ctenomyidae, Pedetidae.
Fahrenholzia (13)	Heteromyidae.
Galeophthirus (1)	Caviidae.
Haemodipsus (6)	Leporidae.
Johnsonpthirus (4)	Sciuridae.
Lemurphthirus (3)	Lemuridae, Lorisidae.
Lemurpediculus (2)	Lemuridae.
Linognathoides (9)	Sciuridae.
Mirophthirus (1)	Cricetidae (*Typhlomys*).
Neohaematopinus (28)	Sciuridae, Cricetidae (*Neotoma*), Heteromyidae (*Dipodomys*).
Phthirpediculus (2)	Indriidae.
Polyplax (76)	Cricetidae, Muridae, Rhyzomyidae, Sciuridae.

Parasitic Arthropods	Mammals (Hosts)
Proenderleinellus (1)	Cricetidae (*Cricetomys*).
Sathrax (1)	Tupaiidae.
Scipio (4)	Petromyidae, Thryonomyidae.
Fam. PTHIRIDAE (2)	ANTHROPOID PRIMATES.
Pthirus (2)	Hominidae, Pongidae.
Fam. RATEMIIDAE (2)	PERISSODACTYLA.
Ratemia (2)	Equidae.

Order MALLOPHAGA (Emerson and Price 1981, 1985)
    Suborder AMBLYCERA

Fam. ABROCOMOPHAGIDAE (1)	RODENTIA.
Abrocomophaga (1)	Abrocomidae.
Fam. BOOPIIDAE (43)	MARSUPIALIA, CARNIVORA.
Boopia (16)	Dasyuridae, Peramelidae, Vombatidae.
Heterodoxus (14)	Canidae, Viverridae; Macropodidae.
Latumcephalum (3)	Macropodidae.
Macropophila (4)	Macropodidae.
Paraboopia (1)	Macropodidae.
Paraheterodoxus (3)	Macropodidae.
Phacogalia (2)	Dasyuridae.
Fam. GYROPIDAE (67)	RODENTIA, ARTIODACTYLA, PRIMATES.
Aotiella (1)	Cebidae.
Gliricola (31)	Capromyidae, Caviidae, Echimyidae.
Gyropus (19)	Abrocomidae, Caviidae, Cricetidae, Ctenomyidae, Echimyidae.
Macrogyropus (4)	Caviidae, Dasyproctidae; Tayassuidae.
Monothoracius (2)	Caviidae, Dasyproctidae.
Phtheiropoios (8)	Abrocomidae, Chinchillidae, Ctenomyidae.
Pitrufquenia (1)	Capromyidae, Myocastoridae.
Protogyropus (1)	Caviidae.
Fam. TRIMENOPONIDAE (11)	RODENTIA, MARSUPIALIA.
Chinchillophaga (1)	Caviidae.
Cummingsia (4)	Caenolestidae, Didelphidae; Cricetidae.
Harrisonia (1)	Echimyidae.
Hoplomyophilus (1)	Echimyidae.
Philandesia (3)	Chinchillidae.
Trimenopon (1)	Caviidae.

    Suborder ISCHNOCERA

Fam. TRICHODECTIDAE (334)	ARTIODACTYLA, RODENTIA, CARNIVORA, EDENTATA, HYRACOIDEA, PERISSODACTYLA, PRIMATES.

Parasitic Arthropods	Mammals (Hosts)
Bovicola (31)	Bovidae, Camelidae, Cervidae; Equidae.
Cebidicola (3)	Cebidae.
Damalinia (18)	Bovidae, Cervidae, Tragulidae.
Dasyonyx (15)	Procaviidae.
Eurytrichodectes (2)	Procaviidae.
Eutrichophilus (11)	Erethizontidae.
Felicola (27)	Felidae, Herpestidae, Hyaenidae, Protelidae, Viverridae.
Geomydoecus (99)	Geomyidae.
Lorisicola (1)	Lorisidae.
Lutridia (3)	Mustelidae.
Lymeon (2)	Bradypodidae.
Neofelicola (4)	Viverridae.
Neotrichodectes (10)	Mustelidae, Procyonidae.
Parafelicola (6)	Viverridae.
Procavicola (32)	Procaviidae.
Procaviphilus (7)	Procaviidae.
Stachiella (9)	Mustelidae.
Suricatoecus (15)	Canidae, Herpestidae, Viverridae.
Trichodectes (16)	Canidae, Mustelidae, Procyonidae, Ursidae, Viverridae.
Tricholipeurus (23)	Cervidae, Bovidae.
Fam. TRICHOPHILOPTERIDAE (2)	PRIMATES.
Trichophilopterus (2)	Indriidae, Lemuridae.
Suborder RHYNCHOPHTHIRINA	
Fam. HAEMATOMYZIDAE (2)	PROBOSCIDEA, ARTIODACTYLA.
Haematomyzus (2)	Elephantidae, Suidae.
Order COLEOPTERA (Marshall 1981)	
Fam. LEPTINIDAE (6)	RODENTIA.
Leptinillus (1)	Aplodontidae, Castoridae (*Castor*), Cricetidae.
Leptinus (5)	Muridae.
Fam. PLATYPSYLLIDAE (2)	RODENTIA, INSECTIVORA.
Platypsyllus (1)	Castoridae (*Castor*).
Silphopsyllus (1)	Talpidae (*Desmana*).
Fam. STAPHYLINIDAE (57) Amblyopinini	MARSUPIALIA, RODENTIA.
Amblyopinodes (15)	Caviidae, myomorph rodents (mostly Cricetidae).
Amblyopinus (34)	Marsupialia, Rodentia.
Edrabius (6)	Ctenomyidae.
Megamblyopinus (2)	Ctenomyidae.

Order DIPTERA (Maa 1963, 1966, 1969; Marshall 1981; Papavero 1977; Wenzel and Tipton 1966; Zumpt 1965)

Parasitic Arthropods	Mammals (Hosts)
Fam. CALLIPHORIDAE	ARTIODACTYLA, CARNIVORA, PERISSODACTYLA, PRIMATES, PROBOSCIDEA, RODENTIA, TUBULIDENTATA, AND OTHER MAMMALS.
Auchmeromyia	Hominidae.
Booponus	Bovidae, Cervidae.
Chrysomya	Bovidae; Equidae; Hominidae; Elephantidae; others.
Cochyliomyia	Bovidae; others.
Cordylobia	Bovidae; Canidae, Felidae, Viverridae; Cebidae; Cricetidae, Muridae; others.
Elephantoloemus	Elephantidae.
Pachychoeromyia	Suidae; Canidae, others; Orycteropodidae.
Fam. CUTEREBRIDAE	ARTIODACTYLA, LAGOMORPHA, PRIMATES, RODENTIA.
Alouattamyia	Primates.
Cuterebra	Leporidae; Cricetidae.
Dermatobia	Bovidae, Cervidae, Suidae; Canidae; Equidae; others.
Montemyia	Unknown.
Pseudogametes	Unknown.
Rogenhofera	Rodentia.
Fam. GASTEROPHILIDAE	PROBOSCIDEA, PERISSODACTYLA.
Cobboldia	Elephantidae.
Gasterophilus	Equidae.
Gyrostigma	Rhinocerotidae.
Neocuterebra	Elephantidae.
Platycobboldia	Elephantidae.
Rodhainomyia	Elephantidae.
Ruttenia	Elephantidae.
Fam. HIPPOBOSCIDAE (48)	ARTIODACTYLA, CARNIVORA, PERISSODACTYLA, PRIMATES, MARSUPIALIA.
Allobosca (1)	Indriidae, Lemuridae.
Austrolfersia (1)	Macropodidae.
Hippobosca (7)	Bovidae, Camelidae, Giraffidae; Canidae, Felidae, Hyaenidae, Viverridae; Equidae.
Lipoptena (30)	Bovidae, Cervidae, Tragulidae.
Melophagus (3)	Bovidae.
Neolipoptena (1)	Cervidae.
Ortholfersia (4)	Macropodidae.
Proparabosca (1)	Indriidae.

Parasitic Arthropods	Mammals (Hosts)
Fam. HYPODERMATIDAE	ARTIODACTYLA, LAGOMORPHA, PERISSODACTYLA, RODENTIA.
Hypoderma	Bovidae, Cervidae.
Oedemagna	Cervidae.
Oestroderma	Lagomorpha.
Oestromyia	Lagomorpha; Rodentia.
Pallasiomyia	Bovidae.
Pavlovskiata	Bovidae.
Portschinskia	Lagomorpha (?); Rodentia.
Przhevalskiana	Bovidae.
Strobiloestrus	Artiodactyla.
Fam. NYCTERIBIIDAE (256)	CHIROPTERA (Old World).
Archinycteribia (3)	Pteropodidae.
Basilia (103)	Phyllostomidae, Emballonuridae. Vespertilionidae.
Conotibia (1)	Vespertilionidae.
Cyclopodia (25)	Pteropodidae.
Dipseliopoda (3)	Pteropodidae.
Eucampsipoda (12)	Pteropodidae.
Hershkovitzia (3)	Thyropteridae.
Leptocyclopodia (22)	Pteropodidae.
Nycteribia (27)	Rhinolophidae, Vespertilionidae.
Penicillidia (24)	Rhinolophidae, Vespertilionidae.
Phthiridium (32)	Rhinolophidae.
Stereomyia (1)	Vespertilionidae.
Fam. OESTRIDAE	ARTIODACTYLA, MARSUPIALIA, PERISSODACTYLA, PROBOSCIDEA.
Acrocomyia	Cervidae.
Cephalopina	Camelidae.
Cephenemyia	Cervidae.
Gedoelstia	Cervidae.
Kirkioestrus	Artiodactyla.
Loewioestrus	Bovidae.
Oestroides	Artiodactyla.
Oestrus	Bovidae.
Pharyngobolus	Elephantidae.
Pharyngomyia	Bovidae, Cervidae.
Procephenemyia	Cervidae.
Rhinoestrus	Equidae.
Suinoestrus	Suidae.
Tracheomyia	Macropodidae.
Fam. SARCOPHAGIDAE	ARTIODACTYLA, CARNIVORA, PERISSODACTYLA, PRIMATES, AND OTHERS.
Wohlfahrtia	Bovidae; Canidae, Mustelidae; Hominidae; Equidae; others.

Parasitic Arthropods	Mammals (Hosts)
Fam. STREBLIDAE (221)	CHIROPTERA (New World).
Anatrichobius (1)	Vespertilionidae.
Anostrebla (5)	Phyllostomidae.
Ascodipteron (18)	Emballonuridae, Megadermatidae, Rhinolophidae, Rhinopomatidae, Vespertilionidae.
Aspidoptera (3)	Phyllostomidae.
Brachyotheca (2)	Megadermatidae.
Brachytarsina (25)	Emballonuridae, Rhinolophidae, Vespertilionidae.
Eldunnia (1)	Phyllostomidae.
Exastinion (3)	Phyllostomidae.
Joblingia (1)	Vespertilionidae.
Mastoptera (2)	Phyllostomidae.
Megastrebla (8)	Pteropodidae.
Megistopoda (3)	Phyllostomidae.
Metalasmus (1)	Phyllostomidae.
Neotrichobius (4)	Phyllostomidae.
Noctiliostrebia (5)	Noctilionidae.
Nycterophila (5)	Natalidae, Phyllostomidae.
Paradyschiria (5)	Noctilionidae.
Paraeuctenodes (2)	Phyllostomidae.
Parastrebla (1)	Phyllostomidae.
Paratrichobius (7)	Phyllostomidae.
Phalcophila (1)	Natalidae, Phyllostomidae.
Pseudostrebla (3)	Phyllostomidae.
Raymondia (18)	Megadermatidae, Nycteridae, Rhinolophidae, Vespertilionidae.
Raymondioides (1)	Rhinolophidae.
Speiseria (3)	Phyllostomidae.
Stizostrebla (1)	Phyllostomidae.
Strebla (24)	Emballonuridae, Phyllostomidae.
Synthesiostrebla (1)	Phyllostomidae.
Trichobioides (1)	Phyllostomidae.
Trichobius (65)	Furipteridae, Molossidae, Natalidae, Phyllostomidae, Vespertilionidae.
Xenotrichobius (1)	Noctilionidae.

Order SIPHONAPTERA (Traub, Chapter 8)
    CERATOPHYLLOIDEA

Fam. ANCISTROPSYLLIDAE (3)	ARTIODACTYLA.
Ancistropsylla (3)	Cervidae.
Fam. CERATOPHYLLIDAE (311)	RODENTIA, LAGOMORPHA, INSECTIVORA, CARNIVORA.
Aenigmopsylla (1)	Sciurinae.
Amphalius (3)	Ochotonidae.

Parasitic Arthropods	Mammals (Hosts)
Callopsylla (*Callopsylla*) (13)	Arvicolidae, Ochotonidae, Petauristinae (1).
C. (*Paracallopsylla*) (1)	Mustelidae.
C. (*Typhlocallopsylla*) (1)	Talpidae.
Ceratophyllus (*Rosickyiana*) (1)	Mustelidae.
Citellophilus (14)	Sciurinae.
Dactylopsylla (*Dactylopsylla*) (7)	Geomyidae.
D. (*Spicata*) (8)	Geomyidae.
Diamanus (2)	Sciurinae.
Foxella (4)	Geomyidae.
Hollandipsylla (1)	Petauristinae.
Jellisonia (9)	Neotomini.
Kohlsia (18)	Neotominae, Sciurinae (2), Oryzomyini (?)(1).
Libyastus (15)	Sciurinae.
Macrostylophora (19)	Sciurinae, Petauristinae (2).
Malaraeus (12)	Arvicolidae, Neotomini.
Megabothris (18)	Arvicolidae, Sciurinae (1); Mustelidae (?)(1).
Megathoracipsylla (1)	Sciurinae.
Miriampsylla (1)	Gliridae.
Monopsyllus (*Amonopsyllus*) (10)	Sciurinae, Gliridae (1).
M. (*Monopsyllus*) (9)	Sciurinae, Petauristinae, Neotomini (2), Gliridae (1); Ochotonidae (1).
M. (*Paramonopsyllus*) (2)	Ochotonidae.
Myoxopsylla (2)	Gliridae.
Nosopsyllus (*Gerbillophilus*) (16)	Gerbillinae.
N. (*Nosinius*) (1)	Gerbillinae.
N. (*Nosopsyllus*) (31)	Murinae, Sciurinae, Arvicolidae, Gerbillinae (1).
N. (*Penicus*) (1)	Murinae.
Opisocrostis (6)	Sciurinae.
Opisodasys (*Opisodasys*) (3)	Sciurinae, Neotomini.
O. (*Sciuropsylla*) (4)	Sciurinae.
Orchopeas (9)	Sciurinae, Neotomini.
Oropsylla (6)	Sciurinae.
Paraceras (7)	Sciurinae, Murinae; Mustelidae.
Pleochaetis (16)	Neotomini, Sigmodontini (3), Sciurinae (2).
Rostropsylla (1)	Sciurinae.
Rowleyella (1)	Sciurinae.
Smitipsylla (1)	Petauristinae.
Spuropsylla (1)	Sciurinae.
Syngenopsyllus (1)	Sciurinae.
Tarsopsylla (1)	Sciurinae.
Thrassis (11)	Sciurinae, Arvicolidae (1), Heteromyidae (1).

Parasitic Arthropods	Mammals (Hosts)
Traubella (1)	Neotomini.
Fam. CHIMAEROPSYLLIDAE (25)	RODENTIA, MACROSCELIDEA.
Chiastopsylla (14)	Murinae, Otomyinae.
Chimaeropsylla (2)	Macroscelidae.
Cryptopsylla (1)	Bathyergidae.
Demeillonia (2)	Macroscelidae.
Epiremia (1)	Otomyinae, Murinae.
Hypsophthalmus (3)	Otomyinae, Murinae.
Macroscelidopsylla (1)	Macroscelidae.
Praopsylla (1)	Murinae.
Fam. COPTOPSYLLIDAE (19)	RODENTIA.
Coptopsylla (19)	Gerbillinae (Cricetidae).
Fam. ISCHNOPSYLLIDAE (108)	CHIROPTERA.
Alectopsylla (1)	Vespertilioninae.
Allopsylla (2)	Molossidae.
Araeopsylla (9)	Molossidae, Embellonuridae.
Chiropteropsylla (2)	Embellonurid, Rhinopomatid, Rhinolophid and Megadermatid Bats.
Coorilla (2)	Molossidae.
Dampfia (1)	Vespertilioninae.
Hormopsylla (5)	Molossidae, Vespertilioninae.
Ischnopsyllus (*Hexactenopsylla*) (7)	Vespertilioninae.
I. (*Ischnopsyllus*) (15)	Vespertilioninae.
Lagaropsylla (16)	Molossidae, Rhinolophinae.
Mitchella (1)	Vespertilioninae.
Myodopsylla (12)	Vespertilionidae, Molossidae.
Nycteridopsylla (16)	Vespertilionidae.
Oxyparius (1)	Miniopterinae, Vespertilioninae, Nycteridae.
Porribius (4)	Molossidae, Vespertilioninae.
Ptilopsylla (1)	Molossidae.
Rhinolophopsylla (4)	Rhinolophinae, Vespertilionidae.
Rothschildopsylla (1)	Noctilionidae, Molossidae.
Serendipsylla (1)	Megadermatidae.
Sternopsylla (4)	Molossidae, Vespertilioninae.
Thaumapsylla (3)	Pteropodidae.
Fam. LEPTOPSYLLIDAE (210)	RODENTIA, LAGOMORPHA, INSECTIVORA, MACROSCELIDEA, CARNIVORA.
Aconothobius (3)	Ochotonidae.
Acropsylla (3)	Murinae.
Amphipsylla (45)	Cricetini, Arvicolidae, Myospalacinae.
Brachyctenonotus (1)	Myospalacinae.
Brevictenidia (1)	Ochotonidae.
Caenopsylla (3)	Ctenodactylidae; Canidae; Macroscelididae.

Parasitic Arthropods	Mammals (Hosts)
Calceopsylla (1)	Myospalacinae.
Chinghaipsylla (1)	Ochotonidae (?).
Conothobius (1)	Ochotonidae.
Cratynius (*Angustus*) (1)	Erinaceidae.
C. (*Cratynius*) (3)	Erinaceidae.
Ctenophyllus (4)	Ochotonidae.
Desertopsylla (1)	Dipodidae.
Dolichopsyllus (1)	Aplodontidae.
Frontopsyllus (*Frontopsylla*) (20)	Ochotonidae; Arvicolidae.
F. (*Mafrontia*) (1)	Dipodidae.
F. (*Profrontia*) (3)	Murinae.
Geusibia (8)	Ochotonidae.
Hopkinsipsylla (1)	Dipodidae.
Jordanopsylla (1)	Neotomini.
Leptosylla (*Leptosylla*) (9)	*Murinae.*
L. (*Pectinoctenus*) (6)	Murinae.
Mesopsylla (6)	Dipodidae.
Minyctenopsyllus (1)	Sciurinae.
Ochonothobius (3)	Ochotonidae.
Odontopsyllus (3)	Leporidae.
Ophthalmopsylla (*Cystipsylla*) (4)	Dipodidae.
O. (*Eremedosa*) (1)	Cricetini, Murinae.
O. (*Opthalmopsylla*) (4)	Dipodidae.
Paractenopsyllus (8)	Nesomyinae; Tenecidae.
Paradoxopsyllus (33)	Murinae, Arvicolidae, Cricetini, Gerbillinae.
Peromyscopsylla (18)	Arvicolidae, Neotomini, Murinae.
Phaenopsylla (6)	Cricetini.
Sigmactenus (4)	Murinae.
Tsaractenus (1)	Nesomyinae (?).
Typhlomyopsyllus	Cricetidae (*Typhlomys*).
Fam. VERMIPSYLLIDAE (36)	ARTIODACTYLA,CARNIVORA, LAGOMORPHA, PERISSODACTYLA.
Chaetopsylla (*Arctopsylla*) (2)	Ursidae, Hyaenidae.
C. (*Chaetopsylla*) (23)	Ursidae, Procyonidae, Mustelidae, and other Carnivora; Ochotonidae (1).
Dorcadia (2)	Bovidae, Cervidae.
Vermipsylla (8)	Bovidae; Equidae.
Fam. XIPHIOPSYLLIDAE (7)	RODENTIA.
Xiphiopsylla (7)	Murinae, Otomyinae, Rhizomyidae.
HYSTRICHOPSYLLOIDEA	
Fam. HYSTRICHOPSYLLIDAE (535)	RODENTIA, INSECTIVORA, LAGOMORPHA, MARSUPIALIA, CARNIVORA.
Acedestia (1)	Peramelidae.
Adoratopsylla (*Adoratopsylla*) (3)	Didelphidae.

Parasitic Arthropods	Mammals (Hosts)
A. (*Tritopsylla*) (2)	Didelphidae.
Agastopsylla (4)	Akodontini, Phyllotini.
Anomiopsyllus (12)	Neotomini. (Nests)
Atyphloceras (6)	Arvicolidae, Neotomini.
Callistopsyllus (1)	Neotomini.
Carteretta (2)	Heteromyidae.
Catallagia (14)	Arvicolidae, Neotomini.
Chiliopsylla (1)	Microbiotheriidae (?).
Conorhinopsylla (2)	Sciurinae, Neotomini. (Nests)
Corrodopsylla (4)	Soricidae.
Corypsylla (3)	Talpidae.
Ctenoparia (3)	Akodontini, Oryzomyini.
Ctenophthalmus (*Alloctenus*) (3)	Soricidae.
C. (*Ctenophthalmus*) (14)	Arvicolidae, Murinae; Insectivora.
C. (*Ducictenophthalmus*) (1)	Arvicolidae.
C. (*Ethioctenophthalmus*) (41)	Murinae.
C. (*Euctenophthalmus*) (37)	Arvicolidae, Murinae.
C. (*Geoctenophthalmus*) (4)	Rhyzomyidae (*Tachyoryctes*).
C. (*Idioctenophthalmus*) (3)	Soricidae.
C. (*Medioctenophthalmus*) (12)	Arvicolidae.
C. (*Metactenophthalmus*) (2)	Arvicolidae.
C. (*Nearctoctenophthalmus*) (3)	Arvicolidae.
C. (*Neoctenophthalmus*) (1)	Myospalacinae.
C. (*Palaeoctenophthalmus*) (8)	Spalacidae, Arvicolidae, Cricetini.
C. (*Paractenophthalmus*) (1)	Gerbillinae.
C. (*Sinoctenophthalmus*) (11)	Arvicolidae.
C. (*Spalacoctenophthalmus*) (7)	Spalacidae.
Delotelis (2)	Arvicolidae.
Dinopsyllus (*Cryptoctenopsyllus*) (1)	Bathyergidae.
D. (*Dinopsyllus*) (25)	Murinae; Soricidae.
Doratopsylla (5)	Soricidae.
Eopsylla (1)	Petauristinae (?).
Epitedia (7)	Neotomini, Sciurinae.
Genoneopsylla (6)	Ochotonidae.
Hystrichopsylla (17)	(Many small rodents and insectivores).
Idilla (1)	Dasyuridae.
Listropsylla (9)	Murinae, Gerbillinae.
Megarthroglossus (12)	Neotominae, Sciurinae. (Nests)
Meringis (18)	Heteromyidae.
Nearctopsylla (*Beringiopsylla*) (11)	Soricidae, Talpidae.
N. (*Chinopsylla*) (1)	Sciurinae (?).
N. (*Nearctopsylla*) (2)	Mustelidae.
N. (*Neochinopsylla*) (1)	Myospalacinae.
Neopsylla (50)	Sciurinae, Murinae, Arvicolidae, Cricetini.
Neotyphloceras (2)	(Very broad host range)
Palaeopsylla (44)	Soricidae, Talpidae; Arvicolidae (1).

Parasitic Arthropods	Mammals (Hosts)
Paraneopsylla (6)	Arvicolidae; Ochotonidae.
Paratyphloceras (1)	Aplodontidae.
Phalacropsylla (5)	Neotomini.
Rhadinopsylla (*Actenophthalmus*) (45)	Arvicolinae, Neotomini, Sciurinae, and others.
R. (*Micropsylla*) (2)	Neotomini.
R. (*Micropsylloides*) (1)	Talpidae; Myospalacinae (?).
R. (*Ralipsylla*) (2)	Sciurinae, Cricetini (?).
R. (*Rhadinopsylla*) (7)	Gerbillinae.
Rothschildiana (2)	Murinae.
Stenischia (5)	Talpidae, Soricidae; Murinae, Arvicolidae.
Stenistomera (3)	Neotomini.
Stenoponia (17)	(Many small rodents and insectivores).
Strepsylla (8)	Neotomini, Arvicolidae.
Tamiophila (1)	Sciurinae.
Trichopsylloides (1)	Aplodontidae.
Typhloceras (2)	Murinae.
Wagnerina (5)	Ochotonidae; Cricetini, Arvicolidae.
Wenzella (2)	Heteromyidae. (Nests)
Xenodaeria (1)	Soricidae.
Fam. MACROPSYLLIDAE (2)	RODENTIA, MARSUPIALIA.
Macropsylla (1)	Murinae; Dasyuridae.
Stephanopsylla (1)	Dasyuridae (?).
Fam. PYGIOPSYLLIDAE (163)	RODENTIA, MARSUPIALIA, INSECTIVORA, MONOTREMATA, SCANDENTIA.
Acanthopsylla (18)	Scansorial Murinae; Dasyuridae.
Afristivalius (14)	Murinae.
Astivalius (1)	Murinae.
Austropsylla (1)	Macropodidae (?).
Aviostivalius (1)	Murinae.
Bibikovana (11)	Murinae.
Bradiopsylla (1)	Tachyglossidae.
Choristopsylla (4)	Phalangeridae, Petauridae, Burramyidae.
Ctenidiosomus (4)	Caenolestidae; Cricetidae.
Ernestinia (1)	Murinae; Dasyuridae.
Farhangia (2)	Sciurinae.
Gryphopsylla (*Gryphopsylla*) (1)	Murinae (spinose).
G. (*Migrastivalius*) (1)	Murinae (scansorial).
Idiochaetis (1)	Murinae (giant rats).
Lentistivalius (*Destivalius*) (1)	Tupaiidae.
L. (*Lentistivalius*) (7)	Murinae, Sciurinae; Soricidae.
Lycopsylla (2)	Vombatidae.
Medwayella (14)	Callosciurini.
Metastivalius (9)	Murinae (8); Dasyuridae (1).

Parasitic Arthropods	Mammals (Hosts)
Muesebeckella (2)	Petauridae.
Nestivalius (1)	Murinae.
Obtusifrontia (3)	Murinae; Dasyuridae.
Orthopsylloides (2)	Murinae.
Papuapsylla (19)	Murinae.
Parastivalius (5)	Peramelidae.
Pygiopsylla (6)	Peramelidae, Dasyuridae, Petauridae; Murinae.
Rectidigitus (4)	Petamelidae, scansorial marsupials; murines.
Smitella (1)	Murinae (scansorial).
Stivalius (5)	Murinae.
Striopsylla (3)	Peramelidae.
Tiflovia (2)	Murinae.
Traubia (4)	Murinae; Dasyuridae, Peramelidae.
Uropsylla (1)	Dasyuridae.
Wurunjerria (1)	Petauridae.
Zyx (1)	Murinae.
Fam. STEPHANOCIRCIDAE (39)	MARSUPIALIA, RODENTIA.
Barreropsylla (1)	Akodontini.
Cleopsylla (2)	Caenolestidae, South American Hesperomyinae.
Coronapsylla (1)	Dasyuridae.
Craneopsylla (1)	South American Hesperomyinae; Didelphidae.
Nonnapsylla (1)	Caviidae.
Plocopsylla (17)	South American Hesperomyinae.
Sphinctopsylla (6)	South American Hesperomyinae; Caenolestidae.
Stephanocircus (7)	Peramelidae, Dasyuridae, Phalangeridae; Murinae.
Tiarapsylla (3)	Chinchillidae, Ctenomyidae, Caviidae.
MALACOPSYLLOIDEA	
Fam. MALACOPSYLLIDAE (2)	EDENTATA.
Malacopsylla (1)	Dasypodidae.
Phthiropsylla (1)	Dasypodidae.
Fam. RHOPALOPSYLLIDAE (103)	RODENTIA, EDENTATA.
Delostichus (6)	Octodontidae, Abrocomidae, Caviidae, and other Phiomorphs.
Ectinorus (*Dysmicus*) (11)	Caviidae, Octodontidae, Phyllotini, Akodontini.
E. (*Ectinorus*) (12)	Octodontidae, Ctenomyidae, Phyllotini.
Eritranis (1)	Caviidae.
Listronius (3)	Akodontini.
Panallius (1)	Caviidae.

Parasitic Arthropods	Mammals (Hosts)
Polygenis (46)	Sigmodontini, Oryzomyini, Thomasomyini.
Rhopalopsyllus (6)	Dasyproctidae; Dasypodidae; Tayassuidae.
Scolopsyllus (1)	(?).
Tetrapsyllus (11)	Phiomorph rodents and South American Hesperomyinae.
Tiamastus (5)	Caviidae, Ctenomyidae, Phyllotini.
PULICOIDEA	
Fam. PULICIDAE (155)	RODENTIA, ARTIODACTYLA, CARNIVORA, EDENTATA, HYRACOIDEA, INSECTIVORA, LAGOMORPHA, MARSUPIALIA.
Aphropsylla (2)	(?).
Archaeopsylla (2)	Erinaceidae.
Cediopsylla (4)	Leporidae.
Centetipsylla (1)	Tenrecidae.
Ctenocephalides (11)	Felidae, Canidae; Procaviidae (Some species are catholic).
Delopsylla (1)	Pedetidae.
Echidnophaga (21)	Many species have a very broad host range.
Euhoplopsyllus (4)	Leporidae.
Hoplopsyllus (2)	Leporidae; Rodentia.
Moeopsylla (1)	Suidae.
Nesolagobius (1)	Leporidae.
Parapulex (2)	Murinae.
Pariodontis (2)	Hystricidae.
Procaviopsylla (6)	Procaviidae.
Pulex (*Juxtapulex*) (2)	Dasypodidae; Suidae.
P. (*Pulex*) (4)	Suidae, some catholic; Geomyidae (1).
Pulicella (1)	(?).
Spilopsyllus (1)	Leporidae.
Synopsyllus (5)	Nesomyinae, Murinae; Tenrecidae.
Synosternus (7)	Erinaceidae; Gerbillinae, Sciuridae, (catholic—1).
Xenopsylla (75)	Murinae, Otomyinae, Gerbillinae, Sciuridae, Bathyergidae (1), (some catholic).
Fam. TUNGIDAE (22)	CARNIVORA, CHIROPTERA, EDENTATA, PRIMATES, RODENTIA.
Hectopsylla (10)	Caviidae, Chinchillidae; Mustelidae.
Neotunga (2)	Manidae.
Rhynchopsyllus (1)	Molossidae, Vespertilionidae.
Tunga (*Brevidigita*) (2)	Murinae.
T. (*Tunga*) (7)	Suidae; Murinae; and large mammals including humans.

Parasitic Arthropods	Mammals (Hosts)

Class ARACHNIDA
 Subclass ACARI
  Order PARASITIFORMES (Krantz 1970; Whittaker and Wilson 1974)

   Suborder MESOSTIGMATA$_2$
    Dermanyssoidea (833)

Fam. LAELAPIDAE (548)	VARIOUS MAMMALIA.
Subfam. Laelapinae (331)	RODENTIA, INSECTIVORA, MARSUPIALIA, CHIROPTERA, PRIMATES AND OTHER MAMMALS.
Aetholaelaps (2)	Lemuroidea.
Andreacarus (10)	Muridae; Marsupialia.
Androlaelaps (125)	Rodentia; other Mammalia.
Cavilaelaps (1)	Caviidae.
Chrysochlorolaelaps (1)	Chrysochloridae.
Domrownyssus (1)	Dasyuridae; Muridae (intranasal).
Gigantolaelaps (20)	Cricetidae, other Rodentia.
Gnatholaelaps (1)	Procaviidae; Sciuridae.
Hymenolaelaps (1)	Cricetidae.
Hyperlaelaps (2)	Muroidea.
Laelaps (125)	Muroidea, other Rodentia.
Liponysella (1)	Lemuroidea.
Longolaelaps (2)	Muridae.
Mysolaelaps (4)	Muroidea.
Nakhoda (1)	Erinaceidae.
Neolaelaps (3)	Pteropodidae.
Neoparalaelaps (1)	Caviidae.
Notolaelaps (1)	Pteropodidae.
Ondatralaelaps (2)	Cricetidae.
Oryctolaelaps (2)	Talpidae.
Peramelaelaps (1)	Peramelidae.
Radfordilaelaps (1)	Pedetidae.
Rhodacantha (2)	Dasyuridae (ear canals).
Rhyzolaelaps (3)	Rhizomyidae.
Sinolaelaps (1)	Cricetidae (Platycanthimyidae).
Steptolaelaps (2)	Heteromyidae, Muroidea.
Tricholaelaps (3)	Cricetidae (Platycanthimyidae), Muridae, Sciuridae.
Tropilaelaps (1)	Muroidea? ("field rats")
Tylolaelaps (1)	Rhizomyidae.
Tur (10)	Echimyidae, other Rodentia.
Subfam. Mesolaelapinae (10)	MARSUPIALIA, RODENTIA.
Mesolaelaps (10)	Peramelidae; Muridae.
Subfam. Haemogamasinae (59)	RODENTIA, INSECTIVORA, LAGOMORPHA, MARSUPIALIA.

Parasitic Arthropods	Mammals (Hosts)
Acanthochela (1)	Didelphidae.
Brevisterna (3)	Cricetidae, Sciuridae.
Eulaelaps (12)	Rodentia; Insectivora; Ochotonidae.
Haemogamasus (40)	Rodentia; Insectivora; Ochotonidae.
Ischyropoda (3)	Geomyoidea, Cricetidae, Sciuridae.
Subfam. Alphalaelapinae (1)	RODENTIA.
Alphalaelaps (1)	Aplodontidae.
Subfam. Myonyssinae	RODENTIA, LAGOMORPHA, INSECTIVORA.
Myonyssus (8)	Muroidea; Ochotonidae; Soricoidea.
Subfam. Hirstionyssinae (139)	RODENTIA, MARSUPIALIA, INSECTIVORA, CARNIVORA.
Ancoranyssus (1)	Hystricidae.
Echinonyssus (120)	Rodentia; Insectivora; Carnivora.
Patrinyssus (1)	Aplodontidae.
Thadeua (5)	Macropodidae.
Trichosurolaelaps (12)	Phalangeroidea, Perameloidea.
Fam. MACRONYSSIDAE (148)	CHIROPTERA, RODENTIA, AND OTHER MAMMALS.
Subfam. Macronyssinae (63)	CHIROPTERA, RODENTIA.
Acanthonyssus (2)	Cricetidae, Echimyidae.
Argitis (1)	Cricetidae.
Bewsiella (2)	Rhinolophoidea.
Chiroecetes (1)	Phyllostomidae.
Ichoronyssus (2)	Vespertilionidae, Rhinolophoidea.
Macronyssus (40)	Vespertilionidae, Rhinolophoidea.
Macronyssoides (2)	Phyllostomidae.
Megistonyssus (1)	Rhinolophidae.
Nycteronyssus (1)	Phyllostomidae.
Parichoronyssus (4)	Phyllostomidae, Emballonuridae.
Radfordiella (6)	Phyllostomoidea.
Synasponyssus (1)	Thyropteridae.
Subfam. Ornithonyssinae (85)	CHIROPTERA, RODENTIA AND OTHER MAMMALS.
Chelanyssus (3)	Molossidae.
Chiroptonyssus (3)	Molossidae.
Cryptonyssus (7)	Vespertilionidae; Gliridae.
Lepidodorsum (1)	Cricetidae.
Lepronyssoides (2)	Echimyidae, Caviidae.
Mitonyssus (2)	Noctilionidae, Molossidae.
Ornithonyssus (20)	Rodentia; Marsupialia; other Mammalia.
Parasteatonyssus (4)	Molossidae.
Steatonyssus (40)	Vespertilionoidea, Rhinolophoidea, Emballonuroidea.
Trichonyssus (3)	Vespertilionidae.

Parasitic Arthropods	Mammals (Hosts)
Fam. DERMANYSSIDAE (4)	RODENTIA.
Liponyssoides (4)	Muridae, other Rodentia.
Fam. HALARACHNIDAE (40)	PRIMATES, CARNIVORA, ARTIODACTYLA, RODENTIA, HYRACOIDEA, MARSUPIALIA.
Subfam. Raillietiinae (5)	ARTIODACTYLA, MARSUPIALIA.
Raillietia (5)	Bovidae; Vombatidae.
Subfam. Halarachninae (35)	PRIMATES, CARNIVORA, RODENTIA, HYRACOIDEA, ARTIODACTYLA.
Halarachne (4)	Phocidae, Mustelidae.
Orthohalarachne (2)	Otariidae, Odobenidae.
Pneumonyssoides (4)	Suidae; Canidae; Cebidae.
Pneumonyssus (14)	Cercopithecidae, Pongidae; Procaviidae.
Rhinophaga (6)	Cercopithecidae, Pongidae; Hystricidae.
Zumptiella (5)	Sciuridae; Pedetidae, Viverridae.
Fam. SPINTURNICIDAE (89)	CHIROPTERA.
Ancystropus (8)	Pteropodidae.
Cameronieta (6)	Mormoopidae.
Eyndhovenia (1)	Rhinolophidae.
Meristaspis (8)	Pteropodidae.
Paraperiglischrus (8)	Rhinolophoidea.
Paraspinturnix (1)	Vespertilionidae.
Periglischrus (22)	Phyllostomoidea.
Spinturnix (35)	Vespertilionoidea, Emballonuridae.
Fam. DASYPONYSSIDAE (2)	EDENTATA.
Dasyponyssus (1)	Dasypodidae.
Xenarthronyssus (1)	Dasypodidae.
Fam. MANITHERIONYSSIDAE (1)	PHOLIDOTA.
Manitherionyssus (1)	Manidae.
Fam. HYSTRICHONYSSIDAE (1)	RODENTIA.
Hystrichonyssus (1)	Hystricidae.
UNASSIGNED, PROBABLY NOT DERMANYSSOIDEA	
Fam. SPELAEORHYNCHIDAE (3)	CHIROPTERA.
Spelaeorhynchus (3)	Phyllostomoidea.
Fam. ASCIDAE (?) (3)	RODENTIA, INSECTIVORA.
Myonyssoides (3)	Bathyergidae; Chrysochloridae.

Suborder IXODIDES (= METASTIGMATA) (Hoogstraal and Aeschlimann 1982; Doss et al. 1974; Doss and Anastos 1977; Keirans 1984)

Fam. ARGASIDAE (108)

Subfam. Argasinae (12)	*ARTIODACTYLA, *CARNIVORA, CHIROPTERA, HYRACOIDEA, LAGOMORPHA, *PERISSODACTYLA, RODENTIA.

Parasitic Arthropods	Mammals (Hosts)
Argas (12)	Suidae; Canidae, Felidae; Emballonuridae, Molossidae, Noctilionidae, Nycteridae, Phyllostomidae, Pteropodidae, Rhinolophidae, Rhinopomatidae, Vespertilionidae; Procaviidae; Leporidae; Equidae; Cricetidae, Muridae.
Subfam. Antricolinae (8)	CHIROPTERA.
Antricola (8)	Molossidae, Mormoopidae, Phyllostomidae, Vespertilionidae.
Subfam. Nothaspinae (1)	CHIROPTERA.
Nothaspis (1)	Mormoopidae (probably).
Subfam. Ornithodorinae (85)	ARTIODACTYLA, CARNIVORA, CHIROPTERA, INSECTIVORA, LAGOMORPHA, MARSUPIALIA, PERISSODACTYLA, PRIMATES, RODENTIA.
Ornithodoros (85)	Bovidae, Camelidae, Cervidae, Suidae, Tayassuidae; Canidae, Felidae, Mustelidae; Emballonuridae, Molossidae, Mormoopidae, Natalidae, Noctilionidae, Nycteridae, Phyllostomidae, Pteropodidae, Rhinopomatidae, Vespertilionidae; Erinaceidae, Soricidae; Leporidae; Didelphidae; Equidae; Cebidae, Cercopithecidae; Arvicolidae, Caviidae, Chinchillidae, Cricetidae, Ctenodactylidae, Dipodidae, Echimyidae, Heteromyidae, Hystricidae, Muridae, Sciuridae.
Subfam. Otobiinae (2)	ARTIODACTYLA, *CARNIVORA, LAGOMORPHA, *RODENTIA.
Otobius (2)	Antilocaprinae, Bovidae, Cervidae; *Canidae, *Felidae; Leporidae, *Ochotonidae; *Cricetidae.
Fam. IXODIDAE (577)	
Subfam. Amblyomminae (64)	ARTIODACTYLA, CARNIVORA, CHIROPTERA, EDENTATA, INSECTIVORA, LAGOMORPHA, MACROSCELIDEA, MARSUPIALIA, MONOTREMATA, PERISSODACTYLA, PHOLIDOTA, PRIMATES, PROBOSCIDEA, RODENTIA, SCANDENTIA.
Amblyomma (62)	Bovidae, Camelidae, Cervidae, Giraffidae, Hippopotamidae, Suidae, Tayassuidae;

Parasitic Arthropods	Mammals (Hosts)
	Canidae, Felidae, Herpestidae, Hyaenidae, Mustelidae, Procyonidae, Protelidae, Ursidae, Viverridae; Mormoopidae, Noctilionidae, Nycteridae, Phyllostomidae, Vespertilionidae; Bradypodidae, Dasypodidae, Myrmecophagidae; Erinaceidae, Tenrecidae; Leporidae; Macroscelididae; Dasyuridae, Didelphidae, Macropodidae; Ornithorhynchidae, Tachyglossidae; Equidae, Rhinocerotidae, Tapiridae; Manidae; Callithricidae, Cebidae, Cercopithecidae, Galagidae, Hylobatidae; Elephantidae; Arvicolidae, Caviidae, Chinchillidae, Cricetidae, Dasyproctidae, Echimyidae, Erethizontidae, Geomyidae, Heteromyidae, Hydrochaeridae, Hystricidae, Muridae, Sciuridae, Thryonomyidae; Tupaiidae.
Aponomma (2)	*Bovidae, *Suidae; *Canidae, *Herpestidae; *Peramelidae, Vombatidae; Tachyglossidae; Anomaluridae, Hystricidae, Muridae.
Subfam. Haemaphysalinae (155)	ARTIODACTYLA, CARNIVORA, CHIROPTERA, EDENTATA, HYRACOIDEA, INSECTIVORA, LAGOMORPHA, MACROSCELIDEA, MARSUPIALIA, MONOTREMATA, PERISSODACTYLA, PHOLIDOTA, PRIMATES, RODENTIA, SCANDENTIA, TUBULIDENTATA.
Haemaphysalis (155)	Bovidae, Camelidae, Cervidae, Giraffidae, Suidae, Tragulidae; Canidae, Felidae, Hyaenidae, Herpestidae, Mustelidae, Protelidae, Ursidae, Viverridae; Vespertilionidae; Myrmecophagidae; Procaviidae; Erinaceidae, Soricidae, Tenrecidae; Leporidae, Ochotonidae; Macroscelididae; Dasyuridae, Macropodidae, Peramelidae, Petauridae; Tachyglossidae; Equidae, Tapiridae; Manidae; Cebidae, Cercopithecidae, Galagidae, Hominidae, Hylobatidae, Lemuridae, Lorisidae; Agoutidae, Anomaluridae, Arvicolidae, Bathyergidae, Chinchillidae, Cricetidae, Dasyproctidae,

Parasitic Arthropods	Mammals (Hosts)
	Dipodidae, Echimyidae, Erethizontidae, Geomyidae, Heteromyidae, Hystricidae, Muridae, Pedetidae, Rhizomyidae, Sciuridae, Spalacidae; Tupaiidae; Orycteropodidae.
Subfam. Hyalomminae (30)	ARTIODACTYLA, CARNIVORA, INSECTIVORA, LAGOMORPHA, PERISSODACTYLA, PRIMATES, RODENTIA.
Hyalomma (30)	Bovidae, Camelidae, Cervidae, Giraffidae, Hippopotamidae, Suidae; Canidae, Felidae, Herpestidae, Ursidae; Erinaceidae, Soricidae; Leporidae, Ochotonidae; Equidae, Rhinocerotidae; Cercopithecidae; Agoutidae, Arvicolidae, Cricetidae, Dipodidae, Hystricidae, Muridae, Sciuridae.
Subfam. IXODINAE (217)	ARTIODACTYLA, CARNIVORA, CHIROPTERA, DERMOPTERA, EDENTATA, HYRACOIDEA, INSECTIVORA, LAGOMORPHA, MACROSCELIDEA, MARSUPIALIA, MONOTREMATA, PERISSODACTYLA, PHOLIDOTA, PRIMATES, RODENTIA, SCANDENTIA.
Ixodes (217)	Bovidae, Camelidae, Cervidae, Giraffidae, Hippopotamidae, Suidae, Tragulidae; Canidae, Felidae, Herpestidae, Mustelidae, Procyonidae, Protelidae, Ursidae, Viverridae; Molossidae, Phyllostomidae, Rhinolophidae, Rhinopomiatidae, Vespertilionidae; Cynocephalidae; Dasypodidae; Procaviidae; Chrysochloridae, Erinaceidae; Soricidae, Talpidae, Tenrecidae; Leporidae, Ochotonidae; Macroscelididae; Dasyuridae, Didelphidae, Macropodidae, Myrmecobiidae, Peramelidae, Petauridae, Phalangeridae, Phascolarctidae, Vombatidae; Ornithorhynchidae, Tachyglossidae; Equidae, Rhinocerotidae, Tapiridae; Manidae; Cebidae, Cercopithecidae, Galagidae, Lemuridae, Lorisidae; Abrocomidae, Agoutidae, Apolodontidae, Arvicolidae,

Parasitic Arthropods	Mammals (Hosts)
	Capromyidae, Chinchillidae, Cricetidae, Ctenomyidae, Dasyproctidae, Dipodidae, Echimyidae, Erethizontidae, Geomyidae, Gliridae, Heteromyidae, Hystricidae, Muridae, Myocastoridae, Pedetidae, Rhizomyidae, Sciuridae, Spalacidae, Thryonomyidae, Zapodidae; Tupaiidae.
Subfam. Rhipicephalinae (111)	ARTIODACTYLA, CARNIVORA, CHIROPTERA, EDENTATA, HYRACOIDEA, INSECTIVORA, LAGOMORPHA, MACROSCELIDEA, MARSUPIALIA, PERISSODACTYLA, PHOLIDOTA, PRIMATES, PROBOSCIDEA, RODENTIA, SCANDENTIA, TUBULIDENTATA.
Anomalohimalaya (3)	Bovidae, Camelidae, Cervidae, Suidae; Canidae, Felidae; Soricidae; Leporidae; Dasyuridae, Didelphidae; Equidae; Arvicolidae, Erethizontidae, Muridae.
Boophilus (5)	Bovidae, Camelidae, Cervidae, Giraffidae, Suidae; Canidae, Felidae, Ursidae; Leporidae; Dasyuridae, Didelphidae, Macropodidae; Equidae, Tapiridae; Cercopithecidae; Erethizontidae, Hydrochaeridae, Hystricidae, Muridae.
Cosmiomma (1)	Bovidae, Hippopotamidae; Rhinocerotidae.
Dermacentor (31)	Bovidae, Camelidae, Cervidae, Hippopotamidae, Suidae, Tayassuidae; Canidae, Felidae, Herpestidae, Mustelidae, Procyonidae, Ursidae, Viverridae; Vespertilionidae; Soricidae, Talpidae; Leporidae, Ochotonidae; Macroscelididae; Dasyuridae, Didelphidae; Equidae, Rhinocerotidae, Tapiridae; Cercopithecidae; Elephantidae; Arvicolidae, Caviidae, Cricetidae, Dipodidae, Erethizontidae, Gliridae, Heteromyidae, Hydrochaeridae, Hystricidae, Muridae, Sciuridae, Spalacidae, Zapodidae; Tupaiidae.
Margaropus (3)	Bovidae, Cervidae, Giraffidae; *Canidae; Equidae.
Nosomma (1)	Bovidae, Suidae; Canidae, Ursidae.
Rhipicentor (2)	Bovidae; Canidae, Felidae, Viverridae; Erinaceidae; Hystricidae.

Parasitic Arthropods	Mammals (Hosts)
Rhipicephalus (65)	Bovidae, Camelidae, Cervidae, Giraffidae, Hippopotamidae, Suidae, Tragulidae; Canidae, Felidae, Herpestidae, Hyaenidae, Mustelidae, Protelidae, Ursidae, Viverridae; Rhinolophidae; Myrmecophagidae; Procaviidae; Erinaceidae, Soricidae; Leporidae, Ochotonidae; Macroscelididae; Macropodidae; Equidae, Rhinocerotidae; Manidae; Callithricidae, Cerocopithecidae, Galagidae, Pongidae; Elephantidae; Arvicolidae, Cricetidae, Ctenodactylidae, Dipodidae, Gliridae, Hydrochaeridae, Hystricidae, Muridae, Pedetidae, Sciuridae, Thryonomyidae, Zapodidae.

Order ACARIFORMES (Whitaker and Wilson 1974; Krantz 1978; Nutting Chapter 11; Fain and Hyland, Chapter 12; Smiley 1977, Smiley and Whitaker 1984, Smiley 1984)

Suborder PROSTIGMATA

Fam. CHEYLETIDAE	RODENTIA, CARNIVORA, CHIROPTERA, LAGOMORPHA, MARSUPIALIA.
Alliea	Muridae.
Cheyletiella	Canidae, Felidae, Mustelidae; Leporidae; Muridae.
Cheyletonella	Vespertilionidae.
Cheyletus	Cricetidae, Heteromyidae, Muridae, Sciuridae; Didelphidae.
Criokeron	Sciuridae; Tupaiidae.
Eucheyletia	Leporidae; Cricetidae.
Eucheyletiella	Ochotonidae.
Nihelia	Viverridae; Lemuridae.
Teinocheylus	Rodents.
Fam. DEMODICIDAE	ARTIODACTYLA, CARNIVORA, CHIROPTERA, INSECTIVORA, LAGOMORPHA, MARSUPIALIA, PERISSODACTYLA, PRIMATES, RODENTIA.
Demodex (48)	Bovidae, Cervidae, Suidae; Canidae, Felidae, Mustelidae, Otariidae; Molossidae, Phyllostomidae, Pteropodidae, Vespertilionidae; Talpidae; Leporidae; Dasyuridae, Didelphidae; Equidae; Cebidae, Hominidae; Arvicolidae, Caviidae, Cricetidae, Gliridae, Muridae, Sciuridae.

Parasitic Arthropods	Mammals (Hosts)
Ophthalmodex (3)	Molossidae, Phyllostomidae.
Pterodex (1)	Phyllostomidae.
Rhinodex (1)	Galagidae.
Soricidex (1)	Soricidae.
Stomatodex (3)	Pteropodidae, Vespertilionidae; Galagidae.
Fam. SPELEOGNATHIDAE	CHIROPTERA, ARTIODACTYLA, RODENTIA, PRIMATES, MARSUPIALIA.
Hipposideroptes (1)	Rhinolophidae.
Neospeleognathopsis (2)	Pteropodidae, Rhinolophidae, Vespertilionidae; Didelphidae.
Speleochir (5)	Nycteridae, Phyllostomidae.
Paraspeleognathopsis (3)	Lorisidae; Gliridae, Muridae.
Speleognathus (1)	Bovidae.
Speleorodens (5)	Cricetidae, Muridae, Sciuridae.
Speleomys (2)	Cricetidae (Gerbillinae), Muridae.
Fam. MYOBIIDAE	CHIROPTERA, RODENTIA, INSECTIVORA, MARSUPIALIA.
Acanthopthirius	Vespertilionidae.
Amorphacarus	Soricidae; Sciuridae.
Archemyobia	Didelphidae.
Blarinoba	Soricidae; Cricetidae, Muridae.
Eadiea	Talpidae.
Ewingana	Molossidae, Vespertilionidae.
Eutalpacarus	Talpidae.
Myobia	Cricetidae, Muridae.
Protomyobia	Soricidae; Zapodidae.
Pteracarus	Vespertilionidae.
Radfordia	Molossidae; Soricidae; Arvicolidae, Cricetidae, Heteromyidae, Muridae, Zapodidae.
Fam. PSORERGATIDAE (7)	ARTIODACTYLA, CHIROPTERA, PRIMATES, RODENTIA.
Psorergates (7)	Bovidae; Emballonuridae, Megadermatidae, Nycteridae, Phyllostomidae, Rhinolophidae, Vespertilionidae; Cercopithecidae; Hystricidae.
Fam. PYGMEPHORIDAE (52)	RODENTIA, INSECTIVORA, LAGOMORPHA.
Pygmephorus (52)	Soricidae, Talpidae; Leporidae; Arvicolidae, Cricetidae, Geomyidae, Hystricidae, Sciuridae.

Parasitic Arthropods	Mammals (Hosts)
Suborder ASTIGMATA	
Fam. ATOPOMELIDAE (351)	RODENTIA, INSECTIVORA, MARSUPIALIA, PRIMATES, CARNIVORA.
Atellana (1)	Phalangeridae.
Atopomelopsis (1)	Peramelidae.
Atopomelus (3)	Erinaceidae, Soricidae, Talpiidae.
Austrobius (1)	Dasyuridae (?), Myrmecobiidae (?).
Austrochirus (7)	Peramelidae, Phalangeridae.
Bathyergolichus (3)	Bathyergidae.
Campylochiropsis (3)	Petauridae.
Campylochirus (8)	Petauridae.
Capromylichus (1)	Capromyidae.
Capromysia (1)	Capromyidae.
Centetesia (2)	Tenrecidae.
Chirodiscoides (11)	Caviidae, Echimyidae.
Cubanochirus (1)	Solenodontidae.
Cystostethum (35)	Macropodidae.
Dasyurochirus (14)	Dasyuridae, Myrmecobiidae, Peramelidae.
Didelphilichus (2)	Didelphidae.
Didelphoecius(18)	Didelphidae.
Distroechurobia (1)	Burramyidae.
Domingoecius (1)	Capromyidae.
Dromiciolichus (1)	Microbiotheriidae.
Euryzygomysia (1)	Echimyidae.
Isothricola (5)	Echimyidae.
Koalachirus (1)	Phalangeridae.
Labidopygus (2)	Dasyuridae.
Lemuroptes (4)	Primates.
Listrocarpus (9)	Callithrichidae, Cebidae.
Listrophoroides (159)	Cricetidae (Nesomyidae); Viverridae; Tenrecidae; Lemuridae.
Metadidelphoecius (1)	Didelphidae.
Micropotamogalichus (1)	Tenrecidae.
Murichirus (23)	Macropodidae, Phalangeridae; Muridae.
Myocastorobia (1)	Capromyidae.
Neodasyurochirus (1)	Dasyuridae.
Notoryctobia (1)	Notoryctidae.
Oryzomysia (8)	Cricetidae (Hesperomyinae).
Petaurobia (3)	Petauridae.
Petrogalochirus (3)	Macropodidae, Phalangeridae.
Phalangerobia (1)	Phalangeridae.
Plagiodontochirus (1)	Capromyidae.
Procytostethum (1)	Macropodidae.
Prodidelphoecius (1)	Didelphidae.
Sclerochiroides (1)	Peramelidae, Petauridae.

Parasitic Arthropods	Mammals (Hosts)
Sclerochiropsis (1)	Thylacomyidae.
Scolonoticus (4)	Dasyuridae, Petauridae.
Teinochirus (1)	Muridae.
Tenrecobia (2)	Tenrecidae.
Fam. AUDYCOPTIDAE (5)	CARNIVORA, PRIMATES.
Audycoptes (2)	Cebidae.
Saimirioptes (1)	Cebidae.
Ursicoptes (2)	Procyonidae, Ursidae.
Fam. BAKEROCOPTIDAE	CHIROPTERA.
Bakerocoptes	Pteropodidae.
Fam. CHIRODISCIDAE (151)	CARNIVORA, CHIROPTERA, INSECTIVORA, PRIMATES, RODENTIA.
Subfam. Chirodiscinae (1)	
Chirodiscus (1)	(?) Bird.
Subfam. Schizocoptinae (2)	INSECTIVORA.
Schizocoptes (2)	Chrysochloridae.
Subfam. Lemuroeciinae (1)	PRIMATES.
Lemuroecius (1)	Cheirogaleidae.
Subfam. Labidocarpinae (147)	CHIROPTERA, PRIMATES, CARNIVORA, RODENTIA, INSECTIVORA.
Adentocarpus (1)	Vespertilionidae.
Afrolabidocarpus (5)	Emballonuridae, Rhinolophidae.
Alabidocarpus (31)	Emballonuridae, Megadermatidae, Molossidae, Noctilionidae, Nycteridae, Phyllostomidae, Pteropodidae, Rhinolophidae, Vespertilionidae.
Asiolabidocarpus (3)	Megadermatidae, Rhinolophidae.
Dentocarpus (15)	Emballonuridae, Molossidae, Phyllostomatidae, Rhinolophidae, Vespertilionidae.
Labidocarpellus (8)	Pteropodidae.
Labidocarpoides (5)	Rhinolophidae.
Labidocarpus (7)	Megadermatidae, Rhinolophidae.
Lawrenceocarpus (6)	Phyllostomidae.
Lutrilichus (5)	Mustelidae, Viverridae.
Olabidocarpus (20)	Emballonuridae, Megadermatidae, Molossidae, Nycteridae, Rhinolophidae, Vespertilionidae.
Parakosa (7)	Molossidae, Phyllostomatidae.
Paralabidocarpus (16)	Emballonuridae, Furipteridae, Megadermatidae, Nycteridae, Phyllostomidae, Rhinolophidae.
Pteropiella (1)	Pteropodidae.
Rynconyssus (1)	Galagidae.
Schizocarpus (12)	Castoridae.

Parasitic Arthropods	Mammals (Hosts)
Schizolabicarpus (1)	Rhinolophidae.
Soricilichus (2)	Soricidae.
Trilabidocarpus (1)	Molossidae.
Fam. CHIRORHYNCHOBIIDAE (2)	CHIROPTERA.
Chirorhynchobia (2)	Phyllostomidae.
Fam. GALAGALGIDAE	PRIMATES.
Galagalges	Galagidae.
Fam. GASTRONYSSIDAE (23)	CHIROPTERA, RODENTIA.
Subfam. Gastronyssinae (1)	CHIROPTERA.
Gastronyssus (1)	Pteropodidae.
Subfam. Rodhainyssinae (19)	CHIROPTERA.
Eidolonyssus (1)	Pteropodidae.
Mycteronyssus (1)	Pteropodidae.
Opsonyssus (7)	Pteropodidae, Rhinolophidae.
Rodhainyssus (10)	Megadermatidae, Molossidae, Nycteridae, Vespertilionidae.
Subfam. Yunkeracarinae (3)	RODENTIA.
Sciuracarus (1)	Sciuridae.
Yunkeracarus (2)	Muridae.
Fam. LEMURNYSSIDAE (4)	PRIMATES.
Lemurnyssus (1)	Galagidae.
Mortelmanssia (3)	Callithricidae, Cebidae.
Fam. LISTROPHORIDAE (110)	RODENTIA, CARNIVORA, INSECTIVORA, LAGOMORPHA, MARSUPIALIA, SCANDENTIA.
Aeromychirus (2)	Sciuridae.
Afrolistrophorus (24)	Cricetidae, Muridae; Caenolestidae.
Aplodontochirus (1)	Aplodontidae.
Asiochirus (3)	Soricidae.
Carnilistrophorus (5)	Mustelidae, Viverridae; Macroscelididae; Cricetidae.
Dubininetta (3)	Talpidae.
Echinosorella (1)	Erinaceidae.
Geomylichus (13)	Cricetidae, Geomyidae, Heteromyidae.
Hemigalichus (1)	Viverridae.
Leporacarus (3)	Leporidae.
Listrophorus (18)	Arvicolidae, Cricetidae.
Lutracarus (1)	Mustelidae.
Lynxacarus (9)	Felidae, Mustelidae; Muridae.
Metalistrophorus (4)	Sciuridae.
Olistrophorus (4)	Soricidae; Muridae.
Prolistrophorus (12)	Cricetidae, Echimyidae, Muridae.
Pteromychirus (1)	Sciuridae.
Quasilistrophorus (1)	Cricetidae.
Sciurochirus (3)	Sciuridae; Tupaiidae.
Sclerolistrophorus (3)	Cricetidae (Hesperomyinae).

Parasitic Arthropods	Mammals (Hosts)
Spalacarus (2)	Spalacidae.
Fam. LOBALGIDAE (5)	EDENTATA, MARSUPIALIA, RODENTIA.
Coendalges (1)	Erethizontidae.
Echimytricalges (3)	Didelphidae; Echimyidae.
Lobalges (1)	Bradypodidae.
Fam. MYOCOPTIDAE (50)	RODENTIA, MARSUPIALIA,
Criniscansor (5)	Cricetidae, Dipodidae, Muridae.
Dromiciocoptes (1)	Microbiotheriidae.
Gliricoptes (8)	Gliridae.
Myocoptes (18)	Arvicolidae, Cricetidae, Muridae.
Sciurocoptes (2)	Sciuridae.
Trichoecius (16)	Cricetidae, Muridae.
Fam. PNEUMOCOPTIDAE (4)	RODENTIA.
Pneumocoptes (4)	Rodentia.
Fam. PSOROPTIDAE (58)	
Subfam. Psoroptinae (23)	ARTIODACTYLA, CARNIVORA, INSECTIVORA, LAGOMORPHA, PERISSODACTYLA, RODENTIA.
Caparinia (5)	Hyaenidae, Mustelidae; Erinaceidae; Cricetidae.
Choriopsoroptes (2)	Bovidae.
Chorioptes (5)	Bovidae; Mustelidae, Ursidae; Leporidae; Equidae.
Choriotodectes (1)	Bovidae.
Echimyalges (1)	Echimyidae.
Myoproctalges (1)	Dasyproctidae.
Otodectes (1)	Canidae, Felidae.
Psorochorioptes (1)	Bovidae.
Psoroptes (5)	Bovidae; Leporidae; Equidae.
Trouessalges (1)	Tayassuidae.
Subfam. Nasalialginae (1)	PRIMATES.
Nasalialges (1)	Cercopithecidae.
Subfam. Listropsoralginae (14)	MARSUPIALIA, RODENTIA.
Listropsoralges (2)	Didelphidae.
Listropsoralgoides (1)	Echimyidae.
Petauralges	Petauridae.
Subfam. Makialginae (8)	PRIMATES.
Daubentonialges (1)	Daubentoniidae.
Gaudalges (3)	Indriidae, Lemuridae.
Lemuralges (1)	Indriidae, Lemuridae.
Makialges (3)	Lemuridae.
Subfam. Paracoroptinae (6)	PRIMATES.
Pangorillalges (2)	Pongidae.
Paracoroptes (4)	Cercopithecidae.
Subfam. Cebalginae (7)	PRIMATES.

Parasitic Arthropods	Mammals (Hosts)
Alouattalges (1)	Cebidae.
Cebalges (1)	Cebidae.
Cebalgoides (1)	Callithricidae, Cebidae.
Fonsecalges (2)	Callithricidae, Cebidae.
Procebalges (1)	Cebidae.
Schizopodalges (1)	Cebidae.
Subfam. Marsupialginae (1)	MARSUPIALIA.
Marsupialges (1)	Didelphidae.
Subfam. Cheirogalalginae (1)	PRIMATES.
Cheirogalalges (1)	Cheirogaleidae.
Subfam. Psoralginae (7)	EDENTATA, MARSUPIALIA.
Acaroptes (2)	Vombatidae.
Edentalges (3)	Bradypodidae, Myrmecophagidae.
Psoralges (2)	Bradypodidae, Myrmecophagidae.
Fam. RHYNCOPTIDAE (4)	PRIMATES, RODENTIA.
Rhyncoptes (4)	Callithricidae, Cebidae, Cercopithecidae; Hystricidae.
Fam. SARCOPTIDAE (63)	MARSUPIALIA, PRIMATES, RODENTIA, AND OTHERS.
Subfam. Sarcoptinae (7)	PRIMATES, RODENTIA, AND OTHERS.
Cosarcoptes (1)	Cercopithecidae.
Pithesarcoptes (1)	Cercopithecidae.
Prosarcoptes (1)	Cebidae (captivity), Cercopithecidae.
Sarcoptes (1)	Humans and all the domestic mammals.
Trixacarus (2)	Caviidae, Muridae.
Tychosarcoptes (1)	?
Subfam. Notoedrinae (52)	CARNIVORA, CHIROPTERA, INSECTIVORA, LAGOMORPHA, PRIMATES, RODENTIA.
Chirnyssoides (8)	Phyllostomidae.
Chirnyssus (2)	Emballonuridae, Vespertilionidae.
Chirophagoides (1)	Mystacinidae.
Notoedres (29)	Felidae, Procyonidae; Emballonuridae, Molossidae, Vespertilionidae, Leporidae; Muridae, Sciuridae.
Nycteridocoptes (12)	Pteropodidae, Rhinolophidae, Vespertilionidae.
Suncicoptes (1)	Soricidae.
Subfam. Caenolestocoptinae (1)	MARSUPIALIA.
Caenolestocoptes (1)	Caenolestidae.
Subfam. Diabolicoptinae (3)	MARSUPIALIA.
Diabolicoptes (1)	Dasyuridae.
Satanicoptes (2)	Dasyuridae.
Fam. TEINOCOPTIDAE (17)	CHIROPTERA.
Chirobia (3)	Pteropodidae.
Teinocoptes (14)	Pteropodidae.

1. MAJOR REFERENCES

Doss, M. A. and G. Anastos. 1977. Tick and tick-borne diseases. III. Checklist of families, genera, species, and subspecies of ticks. *Index-Catalogue of Medical and Veterinary Zoology,* USDA, Spec. Publ. No. 3, 97 pp.

Doss, M. A., M. A. Farr, K. F. Roach, and G. Anastos. 1974. Ticks and tick-borne diseases. I. Genera and species of ticks. Parts 1–3; II. Hosts, Parts 1–3. *Index-catalogue of Medical and Veterinary Zoology,* USDA, Spec. Publ. No. 3.

Emerson, K. C. and R. D. Price. 1981. A host–parasite list of the Mallophaga on mammals. *Misc. Publ. Entomol. Soc. Am.* **12**(1):1–72.

Hoogstraal, H. and A. Aeschlimann. 1982. Tick–host specificity. *Bull. Soc. Entomol. Suisse* **55**:5–32.

Kim, K. C. and H. W. Ludwig. 1978. The family classification of the Anoplura. *Syst. Entomol.* **3**:249–284.

Krantz, G. W. 1978. *A Manual of Acarology,* 2nd ed. Oregon State University Book Stores, Corvallis, OR.

Keirans, J. E. 1984. Personal communication.

Maa, T. C. 1963. Genera and species of Hippoboscidae (Diptera): types, synonymy, habitats, and natural groupings. *Pacif. Ins. Monogr.* **6**:1–186.

Maa, T. C. 1964. A review of the Old World Polyctenidae (Hemiptera: Cimicoidea). *Pacif. Ins.* **6**:494–516.

Maa, T. C. 1966. Studies in Hippoboscidae (Diptera). Part 1. *Pacif. Ins. Monogr.* **10**:1–148.

Maa, T. C. 1969. Studies in Hippoboscidae (Diptera). Part 2. *Pacif. Ins. Monogr.* **20**:1–312.

Marshall, A. G. 1981. *The Ecology of Ectoparasitic Insects.* Academic Press, London.

Papavero, N. 1977. In E. Schimitschek and K. A. Spencer (Eds.), *The World Oestridae (Diptera), Mammals and Continental Drift,* Vol. 14, Series Entomologica. W. Junk, The Hague.

Smiley, R. L. 1977. Further studies on the family Cheyletiellidae (Acarina). *Acarologia* **19**:225–241.

Smiley, R. L. 1984. Personal communication.

Smiley, R. L. and J. O. Whitaker, Jr. 1984. Key to New and Old World *Pygmephorus* species and descriptions of six new species (Acaari: Pygmephoridae). *Int. J. Acarol.* **10**(2):59–73.

Wenzel, R. L. and V. J. Tipton. 1966. *Ectoparasites of Panama.* Field Mus. Nat. Hist., Chicago, IL.

Whitaker, J. O. and N. Wilson. 1974. Host and distribution lists of mites (Acari), parasitic and phoretic, in the hair of wild mammals of North America, north of Mexico. *Amer. Mid. Nat.* **91**:1–67.

Zumpt, F. 1965. *Myiasis in Man and Animals in the Old World.* Butterworths, London.

2. MESOSTIGMATA

Strandtmann and Wharton (1958) last attempted in a single publication to encompass all of the genera and species of Mesostigmata associated with vertebrates. No such effort is made here, although some of the same questions about classification and what has been published have had to be answered in preparing this list. This has been a significant problem only for the Laelapinae, for which the world genera were last summarized by Tipton (1960). Some genera, synonymies, and/or species descriptions have likely been missed for the Laelapinae, in particular. The list should be viewed as a general guide to higher classification and a summary of host associations, not as a revisionary work.

The subfamily Hypoaspidinae is omitted, although some hypoaspidines are associated with vertebrates as nidicoles and sometimes also as phoronts, that are also found in other habitats at least occasionally. Other mesostigmates (e.g., some Parasitidae) that are commonly found in nests of vertebrates strictly as predators or saprophages are also excluded.

Many judgments about generic ranges and validity are implicit in the manner of listing. It is not practical to discuss or elaborate on these decisions, but it should be noted that *Laelaps* is treated in a broad sense to include *Echinolaelaps* and *Eubrachylaelaps*, and *Androlaelaps* includes *Haemolaelaps*.

The significant mammalian host records and those verified are listed for a mite taxon, but records considered accidental or incidental are excluded. For example, Dasyuridae (Marsupialia) are not listed as hosts for *Steatonyssus* (Macronyssidae), although one *Steatonyssus* species is known only from four females taken from a member of the family Dasyuridae. There are at least two other cases where macronyssids in bat-parasitizing genera involved type series recovered from non-chiropteran hosts that proved to be accidental associations.

Appendix B

List of Mammals and Their Parasitic Arthropods

Ke Chung Kim

The following list is compiled on the basis of the preceding *List of Parasitic Arthropods Associated with Mammals*. The mammal classification and names of Honacki et al. (*Mammal Species of the World: A Taxonomic and Geographic Reference*, Allen Press and Association of Systematics Collection, 1982) are followed here, and mammal orders and families are alphabetically listed.

The following abbreviations are used for the ordinal and subordinal names of parasitic arthropods:

Class INSECTA	Class ARACHNIDA, Subclass ACARI
HEM: Hemiptera	MES: Mesostigmata, Parasitiformes
DER: Dermaptera	IXO: Ixodides, Parasitiformes
ANO: Anoplura	PRO: Prostigmata, Acariformes
MAL: Mallophaga	AST: Astigmata, Acariformes
COL: Coleoptera	
DIP: Diptera	
SIP: Siphonaptera	

This list is prepared as a reference for parasitic arthropod–mammal associations. Although a best effort was made to include all the published records, this list is by no means complete. Many published records cited in the following list may represent stragglers or contamination and need further verification as to their validity or accuracy. Thanks are due to K. C. Emerson and Robert Traub for their review of this list.

Mammals (Hosts)	Parasitic Arthropods
Order ARTIODACTYLA	ANO: Haematopinidae, Linognathidae, Microthoraciidae, Pecaroecidae; MAL: Gyropidae; Trichodectidae; Haematomyzidae; DIP: Calliphoridae, Cuterebridae, Hippoboscidae, Hypodermatidae, Oestridae, Sarcophagidae; SIP: Ancistropsyllidae, Pulicidae, Rhopalopsyllidae, Vermipsyllidae; MES: Halarachnidae (Raillietiinae, Halarachninae); IXO: Argasidae, Ixodidae; PRO: Demodicidae, Psorergatidae, Speleognathidae; AST: Psoroptidae (Psoroptinae).
Fam. BOVIDAE (inc. Antilocaprinae)	HAEMATOPINIDAE: *Haematopinus;* LINOGNATHIDAE: *Linognathus, Solenopotes;* TRICHODECTIDAE: *Bovicola, Damalinia, Tricholipeurus;* HIPPOBOSCIDAE: *Hippobosca, Lipoptena, Melophagus;* CALLIPHORIDAE: *Booponus, Cochliomyia, Cordylobia, Chrysomya;* SARCOPHAGIDAE: *Wohlfahrtia;* CUTEREBRIDAE: *Dermatobia;* OESTRIDAE: *Loewioestrus, Oestrus, Pharyngomyia;* HYPODERMATIDAE: *Hypoderma, Pallasiomyia, Pavlovskiata, Przhevalskiana;* VERMIPSYLLIDAE: *Dorcadia, Vermipsylla;* RAILLIETIINAE: *Raillietia;* ARGASIDAE: *Ornithodoros, Otobius;* IXODIDAE: *Amblyomma, Aponomma; Haemaphysalis; Hyalomma; Ixodes; Anomalohimalaya, Boophilus, Dermacentor, Margaropus, Nosomma, Rhipicentor, Rhipicephalus;* DEMODICIDAE: *Demodex;* SPELEOGNATHIDAE: *Speleognathus;* PSORERGATIDAE: *Psorergates;* PSOROPTINAE: *Choriopsoroptes, Chorioptes, Choriotodectes, Psorochorioptes, Psoroptes.*
Fam. CAMELIDAE	MICROTHORACIIDAE: *Microthoracius;* TRICHODECTIDAE: *Bovicola;* HIPPOBOSCIDAE: *Hippobosca;*

Mammals (Hosts)	Parasitic Arthropods
	OESTRIDAE: *Cephalopina;* ARGASIDAE: *Ornithodoros;* IXODIDAE: *Amblyomma; Haemaphysalis;* *Hyalomma; Ixodes; Anomalohimalaya,* *Boophilus, Dermacentor, Rhipicephalus.*
Fam. CERVIDAE (inc. Moschinae)	HAEMATOPINIDAE: *Haematopinus;* LINOGNATHIDAE: *Solenopotes;* TRICHODECTIDAE: *Bovicola, Tricholipeurus;* HIPPOBOSCIDAE: *Lipoptena, Neolipoptena;* CALLIPHORIDAE: *Booponus;* CUTEREBRIDAE: *Dermatobia;* OESTRIDAE: *Acrocomyia, Cephenemyia,* *Gedoelstia, Pharyngomyia, Procephenemyia;* HYPODERMATIDAE: *Hypoderma,* *Oedemagna;* ANCISTROPSYLLIDAE: *Ancistropsylla;* VERMIPSYLLIDAE: *Dorcadia;* ARGASIDAE: *Ornithodoros, Otobius;* IXODIDAE: *Amblyomma; Haemaphysalis;* *Hyalomma; Ixodes; Anomalohimalaya,* *Boophilus, Dermacentor, Margaropus,* *Rhipicephalus;* DEMODICIDAE: *Demodex.*
Fam. GIRAFFIDAE	LINOGNATHIDAE: *Linognathus;* HIPPOBOSCIDAE: *Hippobosca;* IXODIDAE: *Amblyomma; Haemaphysalis;* *Hyalomma; Ixodes; Boophilus, Margaropus,* *Rhipicephalus.*
Fam. HIPPOPOTAMIDAE	IXODIDAE: *Amblyomma; Hyalomma; Ixodes;* *Dermacentor, Cosmiomma, Rhipicephalus.*
Fam. SUIDAE	HAEMATOPINIDAE: *Haematopinus;* HAEMATOMYZIDAE: *Haematomyzus;* CALLIPHORIDAE: *Pachychoeromyia;* OESTRIDAE: *Suinoestrus;* PULICIDAE: *Moeopsyllus, Pulex (Pulex),* *P. (Juxtapulex);* HALARACHNINAE: *Pneumonyssoides;* ARGASIDAE: *Argas, Ornithodoros;* IXODIDAE: *Amblyomma, Aponomma;* *Haemaphysalis; Hyalomma; Ixodes;* *Anomalohimalaya, Boophilus, Dermacentor,* *Nosomma, Rhipicephalus;* DEMODICIDAE: *Demodex.*
Fam. TAYASSUIDAE	PECAROECIDAE: *Pecaroecus;* GYROPIDAE: *Macrogyropus;* RHOPALOPSYLLIDAE: *Rhopalopsyllus;*

Mammals (Hosts)	Parasitic Arthropods
	ARGASIDAE: *Ornithodoros;*
	IXODIDAE: *Amblyomma, Dermacentor;*
	PSOROPTINAE: *Trouessalges.*
Fam. TRAGULIDAE	TRICHODECTIDAE: *Damalinia;*
	HIPPOBOSCIDAE: *Lipoptena;*
	IXODIDAE: *Haemaphysalis; Ixodes;*
	Rhipicephalus.
Order CARNIVORA	ANO: Echinophthiriidae, Linognathidae;
(inc. Pinnipedia)	MAL: Boopiidae; Trichodectidae;
	DIP: Calliphoridae, Cuterebridae,
	Hippoboscidae, Sarcophagidae;
	SIP: Ceratophyllidae, Chimaeropsyllidae,
	Hystrichopsyllidae, Leptopsyllidae,
	Pulicidae, Tungidae, Vermipsyllidae;
	MES: Halarachnidae (Halarachninae),
	Laelapidae (Hirstionyssinae:
	Echinonyssus);
	IXO: Argasidae, Ixodidae;
	PRO: Cheyletidae (incl. *Teinocheylus),*
	Demodicidae;
	AST: Atopomelidae, Audicoptidae,
	Chirodiscidae, Listrophoridae,
	Myocoptidae, Psoroptidae
	(Psoroptinae), Sarcoptidae
	(Notoedrinae).
Fam. CANIDAE	LINOGNATHIDAE: *Linognathus;*
	BOOPIIDAE: *Heterodoxus;*
	TRICHODECTIDAE: *Suricatoecus,*
	Trichodectes;
	HIPPOBOSCIDAE: *Hippobosca;*
	CALLIPHORIDAE: *Cordylobia,*
	Pachychoeromyia;
	SARCOPHAGIDAE: *Wohlfahrtia;*
	CUTEREBRIDAE: *Dermatobia;*
	LEPTOPSYLLIDAE: *Caenopsylla;*
	PULICIDAE: *Ctenocephalides;*
	HALARACHNINAE: *Pneumonyssoides;*
	ARGASIDAE: *Argas, Ornithodoros, Otobius;*
	IXODIDAE: *Amblyomma, Aponomma;*
	Haemaphysalis; Hyalomma; Ixodes;
	Anomalohimalaya, Boophilus, Dermacentor,
	Margaropus, Nosomma, Rhipicentor,
	Rhipicephalus;
	CHEYLETIDAE: *Cheyletiella;*

Mammals (Hosts)	Parasitic Arthropods
	DEMODICIDAE: *Demodex*; PSOROPTINAE: *Otodectes*.

Fam. FELIDAE	TRICHODECTIDAE: *Felicola*;
	HIPPOBOSCIDAE: *Hippobosca*;
	CALLIPHORIDAE: *Cordylobia*;
	PULICIDAE: *Ctenocephalides*;
	ARGASIDAE: *Argas, Ornithodoros, Otobius*;
	IXODIDAE: *Amblyomma; Haemaphysalis;*
	  Hyalomma; Ixodes; Anomalohimalaya,
	  Boophilus, Dermacentor, Rhipicentor,
	  Rhipicephalus;
	CHEYLETIDAE: *Cheyletiella*;
	DEMODICIDAE: *Demodex*;
	LISTROPHORIDAE: *Lynxacarus*;
	PSOROPTINAE: *Otodectes*;
	NOTOEDRINAE: *Notoedres*.

Fam. HERPESTIDAE	TRICHODECTIDAE: *Felicola, Suricatoecus*;
	IXODIDAE: *Amblyomma, Aponomma;*
	  Haemaphysalis; Hyalomma; Ixodes;
	  Dermacentor, Rhipicephalus.

Fam. HYAENIDAE	TRICHODECTIDAE: *Felicola*;
	HIPPOBOSCIDAE: *Hippobosca*;
	VERMIPSYLLIDAE: *Chaetopsylla*
	  (*Arctopsylla*);
	IXODIDAE: *Amblyomma; Haemaphysalis;*
	  Rhipicephalus;
	PSOROPTINAE: *Caparinia*.

Fam. MUSTELIDAE	ECHINOPHTHIRIIDAE: *Latagophthirus*;
	TRICHODECTIDAE: *Lutridia,*
	  Neotrichodectes, Stachiella, Trichodectes;
	SARCOPHAGIDAE: *Wohlfahrtia*;
	CERATOPHYLLIDAE: *Callopsylla*
	  (*Paracallophsylla*), *Ceratophyllus*
	  (*Rosickyiana*), *Megabothris, Paraceras*;
	VERMIPSYLLIDAE: *Chaetopsylla*
	  (*Chaetopsylla*);
	HYSTRICHOPSYLLIDAE: *Nearctopsylla*
	  (*Nearctopsylla*);
	TUNGIDAE: *Hectopsylla*;
	HALARACHNINAE: *Halarachne*;
	ARGASIDAE: *Ornithodoros*;
	IXODIDAE: *Amblyomma; Haemaphysalis;*
	  Ixodes; Dermacentor, Rhipicephalus;
	CHEYLETIDAE: *Cheyletiella*;
	DEMODICIDAE: *Demodex*;

Mammals (Hosts)	Parasitic Arthropods
	CHIRODISCIDAE: *Lutrilichus;* LISTROPHORIDAE: *Carnilistrophorus,* *Lynxacarus;* PSOROPTINAE: *Caparinia, Chorioptes.*
Fam. ODOBENIDAE	ECHINOPHTHIRIIDAE: *Antarctophthirus;* HALARACHNINAE: *Orthohalarachne.*
Fam. OTARIIDAE	ECHINOPHTHIRIIDAE: *Antarctophthirus,* *Proechinophthirus;* HALARACHNINAE: *Orthohalarachne;* DEMODICIDAE: *Demodex.*
Fam. PHOCIDAE	ECHINOPHTHIRIIDAE: *Antaractophthirus,* *Echinophthirius, Lepidophthirus;* HARACHNINAE: *Halarachne.*
Fam. PROCYONIDAE (inc. *Ailurus*)	TRICHODECTIDAE: *Neotrichodectes,* *Trichodectes;* VERMIPSYLLIDAE: *Chaetopsylla* (*Chaetopsylla*); IXODIDAE: *Amblyomma; Ixodes; Dermacentor;* AUDYCOPTIDAE: *Ursicoptes;* NOTOEDRINAE: *Notoedres.*
Fam. PROTELIDAE	TRICHODECTIDAE: *Felicola;* IXODIDAE: *Amblyomma; Haemaphysalis;* *Ixodes; Rhipicephalus.*
Fam. URSIDAE (inc. *Ailuropoda*)	TRICHODECTIDAE: *Trichodectes;* VERMIPSYLLIDAE: *Chaetopsylla* (*Chaetopsylla*), *C.* (*Arctopsylla*); IXODIDAE: *Ambylomma; Haemaphysalis;* *Hyalomma; Ixodes; Boophilus, Nosomma,* *Dermacentor, Rhipicephalus;* AUDYCOPTYIDAE: *Ursicoptes;* PSOROPTINAE: *Chorioptes.*
Fam. VIVERRIDAE	BOOPIIDAE: *Heterodoxus;* TRICHODECTIDAE: *Felicola, Neofelicola,* *Parafelicola, Suricatoecus, Trichodectes;* HIPPOBOSCIDAE: *Hippobosca;* CALLIPHORIDAE: *Cordylobia;* HALARACHNINAE: *Zumptiella;* IXODIDAE: *Amblyomma; Haemaphysalis;* *Ixodes; Dermacentor, Rhipicentor,* *Rhipicephalus;* CHEYLETIDAE: *Nihelia;* ATOPOMELIDAE: *Listrophoroides;* CHIRODISCIDAE: *Lutrilichus;* LISTROPHORIDAE: *Carnilistrophorus,* *Hemigalichus.*

Mammals (Hosts)	Parasitic Arthropods
Order CHIROPTERA	HEM: Polyctenidae; DER: Arixeniidae; DIP: Nycteribiidae, Streblidae; SIP: Ischnopsyllidae, Tungidae; MES: Laelapidae, Macronyssidae (Macronyssinae, Ornithonyssinae), Spelaeorhynchidae, Spinturnicidae; IXO: Argasidae, Ixodidae; PRO: Cheyletidae, Demodicidae, Myobiidae, Psorergatidae, Speleognathidae; AST: Bakerocoptidae, Chirodiscidae, Chirorhynchobiidae, Gastronyssidae (Gastronyssinae, Rodhainyssinae), Sarcoptidae (Notoedrinae), Teinocoptidae.
Fam. EMBALLONURIDAE	POLYCTENIDAE: *Eoctenes;* NYCTERIBIIDAE: *Basilia;* STREBLIDAE: *Ascodipteron, Brachytarsina, Strebla;* ISCHNOPSYLLIDAE: *Araeopsylla, Chiropteropsylla;* MACRONYSSINAE: *Parichoronyssus;* ORNITHONYSSINAE: *Steatonyssus;* SPINTURNICIDAE: *Spinturnix;* ARGASIDAE: *Argas, Ornithodoros;* PSORERGATIDAE: *Psorergates;* CHIRODISCIDAE: *Afrolabidocarpus, Alabidocarpus, Dentocarpus, Olabidocarpus, Paralabidocarpus;* NOTOEDRINAE: *Chirnyssus, Notoedres.*
Fam. FURIPTERIDAE	STREBLIDAE: *Trichobius;* CHIRODISCIDAE: *Paralabidocarpus.*
Fam. MEGADERMATIDAE	POLYCTENIDAE: *Eoctenes, Polyctenes;* STREBLIDAE: *Ascodipteron, Brachyotheca, Raymondia;* ISCHNOPSYLLIDAE: *Chiropteropsylla, Serendipsylla;* CHIRODISCIDAE: *Alabidocarpus, Asiolabidocarpus, Labidocarpus, Olabidocarpus, Paralabidocarpus;* RODHAINYSSINAE: *Rodhainyssus.*
Fam. MOLOSSIDAE	POLYCTENIDAE: *Hesperoctenes, Hypoctenes;* ARIXENIIDAE: *Arixenia, Xenaria;* STREBLIDAE: *Trichobius;*

Mammals (Hosts)	Parasitic Arthropods
	ISCHNOPSYLLIDAE: *Allopsylla, Araeopsylla, Coorilla, Hormopsylla, Legardopsylla, Myodopsylla, Porribius, Ptilopsylla, Rothschildopsylla, Sternopsylla;* TUNGIDAE: *Rhynchopsyllus;* ORNITHONYSSIDAE: *Chelanyssus, Chiroptonyssus, Mitonyssus, Parasteatonyssus;* ARGASIDAE: *Argas, Antricola, Ornithodoros;* IXODIDAE: *Ixodes;* DEMODICIDAE: *Demodex, Ophthalmodex;* MYOBIIDAE: *Ewingana, Radfordia;* CHIRODISCIDAE: *Alabidocarpus, Dentocarpus, Olabidocarpus, Parakosa, Trilabidocarpus;* RODHAINYSSINAE: *Rodhainyssus;* NOTOEDRINAE: *Notoedres.*
Fam. MORMOOPIDAE	SPINTURNICIDAE: *Cameronieta;* ARGASIDAE: *Antricola, Nothaspis* (?), *Ornithodoros;* IXODIDAE: *Amblyomma.*
Fam. MYSTACINIDAE	NOTOEDRINAE: *Chirophagoides.*
Fam. NATALIDAE	STREBLIDAE: *Nycterophila, Phalcophila, Trichobius;* ARGASIDAE: *Ornithodoros.*
Fam. NOCTILIONIDAE	STREBLIDAE: *Noctiliostrebia, Paradyschiria, Xenotrichobius;* ISCHNOPSYLLIDAE: *Rothschildopsylla;* ORNITHONYSSINAE: *Mitonyssus;* ARGASIDAE: *Argas, Ornithodoros;* IXODIDAE: *Amblyomma;* CHIRODISCIDAE: *Albidocarpus.*
Fam. NYCTERIDAE	POLYCTENIDAE: *Eoctenes;* STREBLIDAE: *Raymondia;* ISCHNOPSYLLIDAE: *Oxyparius;* ARGASIDAE: *Argas, Ornithodoros;* IXODIDAE: *Amblyomma;* SPELEOGNATHIDAE: *Speleochir;* PSORERGATIDAE: *Psorergates;* CHIRODISCIDAE: *Alabidocarpus, Olabidocarpus, Paralabidocarpus;* RODHAINYSSINAE: *Rodhainyssus.*
Fam. PHYLLOSTOMIDAE (inc. Desmodontinae)	NYCTERIBIIDAE: *Basilia;* STREBLIDAE: *Anostrebla, Aspidoptera, Eldunnia, Exastinion, Mastoptera, Megistopoda, Metalasmus, Neotrichobius,*

Mammals (Hosts)	Parasitic Arthropods
	Nycterophila, Paraeuctenodes, Parastrebla, Paratrichobius, Phalcophila, Pseudostrebla, Speiseria, Stizostrebla, Strebla, Synthesiostrebla, Trichobioides, Trichobius; MACRONYSSINAE: *Chirocetes, Macronyssoides, Nycteronyssus, Parichoronyssus, Radfordiella;* SPINTURNICIDAE: *Periglischrus;* SPELAEORHYNCHIDAE: *Spelaeorhynchus;* ARGASIDAE: *Argas, Antricola, Ornithodoros;* IXODIDAE: *Amblyomma; Ixodes;* DEMODICIDAE: *Demodex, Ophthalmodex, Pterodex;* SPELEOGNATHIDAE: *Speleochir;* PSORERGATIDAE: *Psorergates;* CHIRODISCIDAE: *Alabidocarpus, Dentocarpus, Lawrenceocarpus, Parakosa, Paralabidocarpus;* CHIROPHYNCHOBIIDAE: *Chirorhynchobia;* NOTOEDRINAE: *Chirnyssoides.*
Fam. PTEROPODIDAE	NYCTERIBIIDAE: *Archinycteribia, Cyclopodia, Dipseliopoda, Eucampsipoda, Leptocyclopodia;* STREBLIDAE: *Megastrebla;* ISCHNOPSYLLIDAE: *Thaumapsylla;* LAELAPIDAE: *Neolaelaps, Notalaelaps;* SPINTURNICIDAE: *Ancystrepus, Meristaspis;* ARGASIDAE: *Argas, Ornithodoros;* DEMODICIDAE: *Demodex, Stematodex;* SPELEOGNATHIDAE: *Neospeleognathopsis;* BAKEROCOPTIDAE: *Bakerocoptes;* CHIRODISCIDAE: *Alabidocarpus, Labidocarpellus, Pteropiella;* GASTRONYSSINAE: *Gastronyssus;* RODHAINYSSINAE: *Eidolonyssus, Mycteronyssus, Opsonyssus;* NOTOEDRINAE: *Nycteridocoptes;* TEINOCOPTIDAE: *Chirobia, Teinocoptes.*
Fam. RHINOLOPHIDAE (inc. Hipposiderinae)	POLYCTENIDAE: *Androctenes;* NYCTERIBIIDAE: *Nycteribia, Penicillidia, Phthiridium;* STREBLIDAE: *Ascodipteron, Brachytarsina, Raymondia, Raymondioides;* ISCHNOPSYLLIDAE: *Chiropteropsylla, Legaropsylla, Rhinolophopsylla;* MACRONYSSINAE: *Bewsiella, Ichoronyssus, Macronyssus, Megistonyssus;*

Mammals (Hosts)	Parasitic Arthropods
	ORNITHONYSSINAE: *Steatonyssus;* SPINTURNICIDAE: *Eyndhovenia,* *Paraperiglischrus;* ARGASIDAE: *Argas;* IXODIDAE: *Ixodes, Rhipicephalus;* SPELEOGNATHIDAE: *Hipposideroptes,* *Neospeleognathopsis;* PSORERGATIDAE: *Psorergates;* CHIRODISCIDAE: *Afrolabidocarpus,* *Alabidocarpus, Asiolabidocarpus,* *Dentocarpus, Labidocarpoides, Olabidocarpus,* *Paralabidocarpus, Schizolabicarpus;* RODHAINYSSINAE: *Opsonyssus;* NOTOEDRINAE: *Nycteridocoptes.*
Fam. RHINOPOMATIDAE	STREBLIDAE: *Ascodipteron;* ISCHNOPSYLLIDAE: *Chiropteropsylla;* ARGASIDAE: *Argas, Ornithodoros;* IXODIDAE: *Ixodes.*
Fam. THYROPTERIDAE	NYCTERIBIIDAE: *Hershkovitzia;* MACRONYSSINAE: *Synasponyssus.*
Fam. VESPERTILIONIDAE	NYCTERIBIIDAE: *Basilia, Conotibia,* *Nycteribia, Penicillidia, Stereomyia;* STREBLIDAE: *Anatrichobius, Ascodipteron,* *Brachytarsina, Joblingia, Raymondia,* *Trichobius;* ISCHNOPSYLLIDAE: *Alectopsylla, Dampfia,* *Hormopsylla, Ischnopsyllus (Ischnopsyllus), I.* *(Hexactenopsylla), Mitchella, Myodopsylla,* *Nycteridopsylla, Oxyparius, Porribius,* *Rhinolophopsylla, Sternopsylla;* TUNGIDAE: *Rhynchopsyllus;* MACRONYSSINAE: *Cryptonyssus,* *Ichoronyssus, Macronyssus, Steatonyssus,* *Trichonyssus;* SPINTURNICIDAE: *Paraspinturnix,* *Spinturnix;* ARGASIDAE: *Argas, Antricola, Ornithodoros;* IXODIDAE: *Amblyomma; Haemaphysalis;* *Ixodes; Dermacentor;* CHEYLETIDAE: *Cheyletonella;* DEMODICIDAE: *Demodex, Stomatodex;* SPELEOGNATHIDAE: *Neospeleognathopsis;* MYOBIIDAE: *Acanthopthirius, Ewingana,* *Pteracarus;* PSORERGATIDAE: *Psorergates;* CHIRODISCIDAE: *Alabidocarpus,* *Dentocarpus, Olabidocarpus;*

Mammals (Hosts)	Parasitic Arthropods

RODHAINYSSINAE: *Rodhainyssus;*
NOTOEDRINAE: *Chirnyssus, Notoedres,*
   Nycteridocoptes.

Order DERMOPTERA

ANO: Hamophthiriidae;
IXO: Ixodidae.

   Fam. CYNOCEPHALIDAE

HAMOPHTHIRIIDAE: *Hamophthirius;*
IXODIDAE: *Ixodes.*

Order EDENTATA

MAL: Trichodectidae;
SIP: Malacopsyllidae, Pulicidae,
       Rhopalopsyllidae, Tungidae;
MES: Dasyponyssidae;
IXO: Ixodidae;
AST: Lobalgidae, Psoroptidae
       (Psoralginae).

   Fam. BRADYPODIDAE

TRICHODECTIDAE: *Lymeon;*
IXODIDAE: *Amblyomma;*
LOBALGIDAE: *Lobalges;*
PSORALGINAE: *Edentalges, Psoralges.*

   Fam. DASYPODIDAE

MALACOPSYLLIDAE: *Malacopsylla,*
   Phthiropsylla;
PULICIDAE: *Pulex (Juxtapulex);*
RHOPALOPSYLLIDAE: *Rhopalopsylla;*
TUNGIDAE: *Tunga (Tunga);*
DASYPONYSSIDAE: *Dasyponyssus,*
   Xenarthronyssus;
IXODIDAE: *Amblyomma, Ixodes.*

   Fam. MYRMECOPHAGIDAE

TUNGIDAE: *Tunga (Tunga);*
IXODIDAE: *Amblyomma, Haemaphysalis,*
   Rhipicephalus;
PSORALGINAE: *Edentalges, Psoralges.*

Order HYRACOIDEA

ANO: Linognathidae;
MAL: Trichodectidae;
SIP: Pulicidae;
MES: Halarachnidae (Halarachninae),
       Laelapidae;
IXO: Argasidae, Ixodidae;

   Fam. PROCAVIIDAE

LINOGNATHIDAE: *Prolinognathus;*
TRICHODECTIDAE: *Dasyonyx,*
   Eurytrichodectes, Procavicola, Procaviphilus;
PULICIDAE: *Ctenocephalides, Procaviopsylla;*
LAELAPIDAE: *Gnatholaelaps;*
HALARACHNINAE: *Pneumonyssus;*
ARGASIDAE: *Argas;*
IXODIDAE: *Haemaphysalis; Ixodes;*
   Rhipicephalus.

Mammals (Hosts)	Parasitic Arthropods
Order INSECTIVORA	ANO: Hoplopleuridae, Polyplacidae; COL: Platypsyllidae; SIP: Ceratophyllidae, Leptopsyllidae, Hystrichopsyllidae, Pygiopsyllidae, Pulicidae; MES: Laelapidae (Laelapinae; Haemogamasinae: *Eulaelaps, Haemogamasus*; Hirstionyssinae: *Echinonyssus*, Myonyssinae), Ascidae (?); IXO: Argasidae, Ixodidae; PRO: Demodicidae, Myobiidae, Pygmephoridae; AST: Atopomelidae, Chirodiscidae, Listrophoridae, Psoroptidae (Psoroptinae), Sarcoptidae (Notoedrinae).
Fam. CHRYSOCHLORIDAE	LAELAPIDAE: *Chrysochloralaelaps*; ASCIDAE: *Myonyssoides* (?); IXODIDAE: *Ixodes*; CHIRODISCIDAE: *Schizocopotes*.
Fam. ERINACEIDAE	LEPTOPSYLLIDAE: *Cratynius* (*Cratynius*), *C.* (*Augustus*); PULICIDAE: *Archaeopsylla, Synosternus*; LAELAPIDAE: *Nakhoda*; ARGASIDAE: *Ornithodoros*; IXODIDAE: *Amblyomma; Haemaphysalis; Hyalomma; Ixodes; Rhipicentor, Rhipicephalus*; ATOPOMELIDAE: *Atopomelus*; LISTROPHORIDAE: *Echinosorella*; PSOROPTINAE: *Caparinia*.
Fam. SOLENODONTIDAE	ATOPOMELIDAE: *Cubanochirus*.
Fam. SORICIDAE	HOPLOPLEURIDAE: *Ancistroplax*; HYSTRICHOPSYLLIDAE: *Corrodopsylla, Ctenophthalmus* (*Alloctenus*), *C.* (*Idioctenophthalmus*), *Dinopsyllus* (*Dinopsyllus*), *Doratopsylla, Nearctopsylla* (*Beringiopsylla*), *Palaeopsylla, Stenischia, Xenodaeria*; PYGIOPSYLLIDAE: *Lentistivalius* (*Lentistivalius*); LAELAPIDAE: *Myonyssus*; ARGASIDAE: *Ornithodoros*; IXODIDAE: *Haemaphysalis; Hyalomma; Ixodes; Anomalohimalaya, Dermacentor, Rhipicephalus*;

Mammals (Hosts)	Parasitic Arthropods
	DEMODICIDAE: *Demodex*; MYOBIIDAE: *Amorphacarus, Blarinobia, Protomyobia, Radfordia*; PYGMEPHORIDAE: *Pygmephorus*; ATOPOMELIDAE: *Atopomelus*; CHIRODISCIDAE: *Soricilichus*; LISTROPHORIDAE: *Asiochirus, Olistrophorus*; NOTOEDRINAE: *Suncicoptes*.
Fam. TALPIDAE	HOPLOPLEURIDAE: *Haematopinoides*; PLATYPSYLLIDAE: *Silphopsyllus*; CERATOPHYLLIDAE: *Callopsylla (Typhlocallopsylla)*; HYSTRICHOPSYLLIDAE: *Corypsylla, Nearctopsylla (Beringiopsylla), Palaeopsylla, Rhadinopsylla (Miropsylloides), Stenischia*; LAELAPINAE: *Oryctolaelaps*; IXODIDAE: *Haemaphysalis*; *Ixodes*; *Dermacentor*; DEMODICIDAE: *Demodex*; MYOBIIDAE: *Eadiea, Eutalpacarus*; PYGMEPHORIDAE: *Pygmephorus*; ATOPOMELIDAE: *Atopomelus*; LISTROPHORIDAE: *Dubininetta*.
Fam. TENRECIDAE	LEPTOPSYLLIDAE: *Paractenopsyllus*; PULICIDAE: *Centetipsylla, Synopsyllus*; IXODIDAE: *Amblyomma*; *Ixodes*; ATOPOMELIDAE: *Centetesia, Listrophoroides, Micropotamogalichus, Tenrecobia*.
Order LAGOMORPHA	ANO: Hoplopleuridae, Polyplacidae; DIP: Cuterebridae, Hypodermatidae; SIP: Ceratophyllidae, Hystrichopsyllidae, Leptopsyllidae, Pulicidae, Vermipsyllidae; MES: Laelapidae (Haemogamasinae; Myonyssinae); IXO: Argasidae, Ixodidae; PRO: Cheyletidae, Demodicidae, Listrophoridae; AST: Psoroptidae (Psoroptinae), Sarcoptidae (Notoedrinae).
Fam. LEPORIDAE	POLYPLACIDAE: *Haemodipsus*; CUTEREBRIDAE: *Cuterebra*; LEPTOPSYLLIDAE: *Odontopsyllus*; PULICIDAE: *Cediopsylla, Euhoplopsyllus, Hoplopsyllus, Nesolagobius, Spilopsyllus*;

Mammals (Hosts)	Parasitic Arthropods

	ARGASIDAE: *Argas, Ornithodoros, Otobius;*
	 IXODIDAE: *Amblyomma; Haemaphysalis;*
	   Hyalomma; Ixodes; Anomalohimalaya,
	   Boophilus, Dermacentor, Rhipicephalus;
	 CHEYLETIDAE: *Cheyletiella;*
	 DEMODICIDAE: *Demodex;*
	 PYGMEPHORIDAE: *Pygmephorus;*
	 LISTROPHORIDAE: *Leporacarus;*
	 PSOROPTINAE: *Chorioptes, Psoroptes;*
	 NOTOEDRINAE: *Notoedres.*

Fam. OCHOTONIDAE	HOPLOPLEURIDAE: *Hoplopleura;*
	 CERATOPHYLLIDAE: *Amphalius, Callopsylla*
	   (Callopsylla), Monopsyllus (Monopsyllus), M.
	   (Paramonopsyllus);
	 LEPTOPSYLLIDAE: *Aconothobius,*
	   Brevictenidia, Chinghaipsylla, Conothobius,
	   Ctenophyllus, Frontopsylla (Frontopsylla),
	   Geusibia, Ochonothobius;
	 VERMIPSYLLIDAE: *Chaetopsylla*
	   (Chaetopsylla);
	 HYSTRICHOPSYLLIDAE: *Genoneopsylla,*
	   Paraneopsylla, Wagnerina;
	 HAEMOGAMASINAE: *Eulaelaps,*
	   Haemogamasus;
	 MYONYSSINAE: *Myonyssus;*
	 ARGASIDAE: *Otobius;*
	 IXODIDAE: *Haemaphysalis, Hyalomma,*
	   Ixodes, Dermacentor, Rhipicephalus;
	 CHEYLETIDAE: *Eucheyletiella.*

Order MACROSCELIDEA	ANO: Neolinognathidae; SIP: Chimaeropsyllidae, Leptopsyllidae; IXO: Ixodidae; AST: Listrophoridae.

Fam. MACROSCELIDIDAE	NEOLINOGNATHIDAE: *Neolinognathus;*
	 CHIMAEROPSYLLIDAE: *Chimaeropsylla,*
	   Demeillonia, Macroscelidopsylla;
	 LEPTOPSYLLIDAE: *Caenopsylla;*
	 IXODIDAE: *Amblyomma; Haemaphysalis;*
	   Ixodes; Dermacentor, Rhipicephalus;
	 LISTROPHORIDAE: *Carnilistrophorus.*

Order MARSUPIALIA	MAL: Boopiidae, Trimenoponidae; COL: Staphylinidae; DIP: Hippoboscidae, Oestridae; SIP: Hystrichopsyllidae, Macropsyllidae, Pygiopsyllidae, Stephanocircidae;

Mammals (Hosts)	Parasitic Arthropods
	MES: Halarachnidae (Raillietiinae); Laelapidae (Laelapinae: *Andreacarus*; Mesolaelapinae; Haemogamasinae; Hirstionyssinae); Macronyssidae (Ornithonyssinae; *Ornithonyssus*);
	IXO: Argasidae, Ixodidae;
	PRO: Cheyletidae, Demodicidae, Myobiidae, Speleognathidae;
	AST: Atopomelidae, Listrophoridae, Lobalgidae, Myocoptidae, Psoroptidae (Listropsoralginae, Marsupialginae, Psoralginae), Sarcoptidae (Sarcoptinae, Caenolestocoptinae, Diabolicoptinae).
Fam. BURRAMYIDAE	PYGIOPSYLLIDAE: *Choristopsylla*; ATOPOMELIDAE: *Distoechurobia*.
Fam. CAENOLESTIDAE	TRIMENOPONIDAE: *Cummingsia*; PYGIOPSYLLIDAE: *Ctenidiosomus*; STEPHANOCIRCIDAE: *Cleopsylla*, *Sphinctopsylla*; LISTROPHORIDAE: *Afrolistrophorus*; CAENOLESTOCOPTINAE: *Caenolestocoptes*.
Fam. DASYURIDAE	BOOPIIDAE: *Boopia*, *Phacogalia*; HYSTRICHOPSYLLIDAE: *Idilla*; MACROPSYLLIDAE: *Macropsylla*, *Stephanopsylla* (?); PYGIOPSYLLIDAE: *Acanthopsylla*, *Pygiopsylla*, *Traubia*, *Uropsylla*; STEPHANOCIRCIDAE: *Coronapsylla*, *Stephanocircus*; LAELAPINAE: *Domrownyssus*, *Rhodacantha*; IXODIDAE: *Amblyomma*; *Haemaphysalis*; *Ixodes*; *Anomalohimalaya*, *Boophilus*, *Dermacentor*; DEMODICIDAE: *Demodex*; ATOPOMELIDAE: *Austrobius*, *Dasyurochirus*, *Labidopygus*, *Neodasyurochirus*, *Scolonoticus*; DIABOLICOPTINAE: *Diabolicoptes*, *Santanicoptes*.
Fam. DIDELPHIDAE	TRIMENOPONIDAE: *Cummingsia*; HYSTRICHOPSYLLIDAE: *Adoratopsylla* (*Adoratopsylla*), *A.* (*Tritopsylla*); STEPHANOCIRCIDAE: *Craneopsylla*;

Mammals (Hosts)	Parasitic Arthropods
	HAEMOGAMASINAE: *Acanthochela;* ARGASIDAE: *Ornithodoros;* IXODIDAE: *Ambylomma; Ixodes;* *Anomalohimalaya, Boophilus, Dermacentor;* CHEYLETIDAE: *Cheyletus;* DEMODICIDAE: *Demodex;* SPELEOGNATHIDAE: *Neospeleognathopsis;* MYOBIIDAE: *Archemyobia;* ATOPOMELIDAE: *Didelphilichus,* *Didelphoecius, Metadidelphoecius,* *Prodidelphoecius;* LOBALGIDAE: *Echimytricalges;* LISTROPSORALGINAE: *Listropsoralges;* MARSUPIALGINAE: *Marsupialges.*
Fam. MACROPODIDAE	BOOPIIDAE: *Boopia, Heterodoxus,* *Latumcephalum, Macropophila, Parabooia,* *Paraheterodoxus;* HIPPOBOSCIDAE: *Austrolfersia, Ortholfersia;* OESTRIDAE: *Tracheomyia;* PYGIOPSYLLIDAE: *Austropsylla* (?); HIRSTIONYSSINAE: *Thadeua;* IXODIDAE: *Amblyomma; Haemaphysalis;* *Ixodes; Boophilus, Rhipicephalus;* ATOPOMELIDAE: *Cystostethum, Murichirus,* *Petrogalochirus, Procytostethum.*
Fam. MICROBIOTHERIIDAE	HYSTRICHOPSYLLIDAE: *Chiliopsylla* (?); ATOPOMELIDAE: *Dromiciolichus;* MYOCOPTIDAE: *Dromiciocoptes.*
Fam. MYRMECOBIIDAE	IXODIDAE: *Ixodes;* ATOPOMELIDAE: *Austrobius* (?), *Dasyurochirus.*
Fam. NOTORYCTIDAE	ATOPOMELIDAE: *Notoryctobia.*
Fam. PERAMELIDAE	BOOPIIDAE: *Boopia;* HYSTRICHOPSYLLIDAE: *Acedestia;* PYGIOPSYLLIDAE: *Parastivalius,* *Pygiopsylla, Rectidigitus, Striopsylla,* *Traubia;* STEPHANOCIRCIDAE: *Stephanocircus;* LAELAPINAE: *Peramelaelaps;* HIRSTIONYSSINAE: *Trichosurolaelaps;* MESOLAELAPINAE: *Mesolaelaps;* IXODIDAE: *Aponomma, Haemaphysalis,* *Ixodes;* ATOPOMELIDAE: *Atopomelopsis,* *Austrochirus, Dasyurochirus, Sclerochiroides.*

Mammals (Hosts)	Parasitic Arthropods
Fam. PETAURIDAE	PYGIOPSYLLIDAE: *Choristopsylla, Muesebeckella, Pygiopsylla, Wurunjerria;* IXODIDAE: *Haemaphysalis, Ixodes;* ATOPOMELIDAE: *Campylochiropsis, Campylochirus, Petaurobia, Sclerochiroides, Scolonoticus;* LISTROPSORALGINAE: *Petauralges.*
Fam. PHALANGERIDAE	PYGIOPSYLLIDAE: *Choristopsylla;* STEPHANOCIRCIDAE: *Stephanocircus;* HIRSTIONYSSINAE: *Trichosurolaelaps;* IXODIDAE: *Ixodes;* ATOPOMELIDAE: *Atellana, Austrochirus, Koalachirus, Murichirus, Petrogalachirus, Phalangerobia.*
Fam. PHASCOLARCTIDAE	IXODIDAE: *Ixodes.*
Fam. THYLACOMYIDAE	ATOPOMELIDAE: *Sclerochiropsis.*
Fam. VOMBATIDAE	BOOPIIDAE: *Boopia;* PYGIOPSYLLIDAE: *Lycopsylla;* RAILLIETIINAE: *Raillietia;* IXODIDAE: *Aponomma; Ixodes;* PSORALGINAE: *Acaroptes.*
Order MONOTREMATA	SIP: Pygiopsyllidae; IXO: Ixodidae.
Fam. ORNITHORHYNCHIDAE	IXODIDAE: *Amblyomma; Ixodes.*
Fam. TACHYGLOSSIDAE	PYGIOPSYLLIDAE: *Bradiopsylla;* IXODIDAE: *Amblyomma, Aponomma; Haemaphysalis; Ixodes.*
Order PERISSODACTYLA	ANO: Haematopinidae, Ratemiidae; MAL: Trichodectidae; DIP: Calliphoridae, Cuterebridae, Gasterophilidae, Hippoboscidae, Hypodermatidae, Oestridae, Sarcophagidae; SIP: Vermipsyllidae; IXO: Argasidae, Ixodidae; PRO: Demodicidae; AST: Psoroptidae (Psoroptinae).
Fam. EQUIDAE	HAEMATOPINIDAE: *Haematopinus;* RATEMIIDAE: *Ratemia;* TRICHODECTIDAE: *Bovicola;* HIPPOBOSCIDAE: *Hippobosca;* CALLIPHORIDAE: *Chrysomya;* SARCOPHAGIDAE: *Wohlfahrtia;* GASTEROPHILIDAE: *Gasterophilus;* CUTEREBRIDAE: *Dermatobia;*

Mammals (Hosts)	Parasitic Arthropods
	OESTRIDAE: *Rhinoestrus;* VERMIPSYLLIDAE: *Vermipsyllus;* ARGASIDAE: *Argas, Ornithodoros;* IXODIDAE: *Amblyomma; Haemaphysalis;* *Hyalomma; Ixodes; Anomalohimalaya,* *Boophilus, Dermacentor, Margaropus,* *Rhipicephalus;* DEMODICIDAE: *Demodex;* PSOROPTINAE: *Chorioptes, Psoroptes.*
Fam. RHINOCEROTIDAE	GASTEROPHILIDAE: *Gyrostigma;* IXODIDAE: *Amblyomma; Hyalomma; Ixodes;* *Cosmiomma, Dermacentor, Rhipicephalus.*
Fam. TAPIRIDAE	IXODIDAE: *Amblyomma; Haemaphysalis;* *Ixodes; Boophilus, Dermacentor.*
Order PHOLIDOTA	SIP: Tungidae; MES: Manitherionyssidae; IXO: Ixodidae.
Fam. MANIDAE	TUNGIDAE: *Neotunga;* MANITHERIONYSSIDAE: *Manitherionyssus;* IXODIDAE: *Amblyomma; Haemaphysalis;* *Ixodes; Rhipicephalus.*
Order PRIMATES	ANO: Pedicinidae, Pediculidae, Polyplacidae, Pthiridae; MAL: Gyropidae, Trichodectidae, Trichophilopteridae; DIP: Calliphoridae, Cuterebridae, Hippoboscidae, Sarcophagidae; SIP: Tungidae; MES: Halarachnidae (Halarachninae), Laelapidae; IXO: Argasidae, Ixodidae; PRO: Cheyletidae, Demodicidae, Speleognathidae, Psorergatidae; AST: Atopomelidae (*Lemuroptes*), Audycoptidae, Chirodiscidae, Galagalgidae, Lemurnyssidae, Rhyncoptidae, Psoroptidae, (Nasalialginae, Makialginae, Paracoroptinae, Cebalginae, Cheirogalalginae), Sarcoptidae (Sarcoptinae, Notoedrinae).
Fam. CALLITHRICIDAE	IXODIDAE: *Amblyomma; Rhipicephalus;* ATOPOMELIDAE: *Listrocarpus;* LEMURNYSSIDAE: *Mortelmanssia;* RHYNCOPTIDAE: *Rhyncoptes;* CEBALGINAE: *Cebalgoides, Fonsecalges.*

Mammals (Hosts)	Parasitic Arthropods
Fam. CEBIDAE	PEDICULIDAE: *Pediculus;* GYROPIDAE: *Aotiella;* TRICHODECTIDAE: *Cebidicola;* CALLIPHORIDAE: *Cordylobia;* HALARACHNINAE: *Pneumonyssoides;* ARGASIDAE: *Ornithodoros;* IXODIDAE: *Amblyomma; Haemaphysalis;* *Ixodes;* DEMODICIDAE: *Demodex;* ATOPOMELIDAE: *Listrocarpus;* AUDYCOPTIDAE: *Audicoptes, Saimirioptes;* LEMURNYSSIDAE: *Mortelmanyssus;* RHYNCOPTIDAE: *Rhyncoptes;* CEBALGINAE: *Alouttalges, Cebalges,* *Cebalgoides, Fonsecalges, Procebalges,* *Schizopodalges.* SARCOPTINAE: *Prosarcoptes.*
Fam. CERCOPITHECIDAE	PEDICINIDAE: *Pedicinus;* HALARACHNINAE: *Pneumonyssus,* *Rhinophaga;* ARGASIDAE: *Ornithodoros;* IXODIDAE: *Amblyomma; Haemaphysalis;* *Hyalomma; Ixodes; Boophilus, Dermacentor,* *Rhipicephalus;* PSORERGATIDAE: *Psorergates;* RHYNCOPTIDAE: *Rhyncoptes;* NASALIALGINAE: *Nasalialges;* PARACOROPTINAE: *Paracoroptes;* SARCOPTINAE: *Cosarcoptes, Pithesarcoptes,* *Prosarcoptes.*
Fam. CHEIROGALEIDAE	CHIRODISCIDAE: *Lemuroecius;* CHEIROGALALGINAE: *Cheirogalalges.*
Fam. DAUBENTONIIDAE	MAKIALGINAE: *Daubentonialges.*
Fam. GALAGIDAE	IXODIDAE: *Amblyomma; Haemaphysalis;* *Ixodes; Rhipicephalus.* DEMODICIDAE: *Rhinodex, Stomatodex;* CHIRODISCIDAE: *Rhynconyssus;* GALAGALGIDAE: *Galagalges;* LEMURNYSSIDAE: *Lemurnyssus.*
Fam. HOMINIDAE	PEDICULIDAE: *Pediculus;* PTHIRIDAE: *Pthirus;* CALLIPHORIDAE: *Auchmeromyia,* *Chrysomya;* SARCOPHAGIDAE: *Wohlfahrtia;* TUNGIDAE: *Tunga (Tunga);* IXODIDAE: *Haemaphysalis;*

Mammals (Hosts)	Parasitic Arthropods
	DEMODICIDAE: *Demodex*;
	SARCOPTINAE: *Sarcoptes*.
Fam. HYLOBATIDAE	PEDICULIDAE: *Pediculus*;
	IXODIDAE: *Amblyomma; Haemaphysalis*.
Fam. INDRIIDAE	POLYPLACIDAE: *Phthirpediculus*;
	TRICHOPHILOPTERIDAE: *Trichophilopterus*;
	HIPPOBOSCIDAE: *Allobosca, Proparabosca*;
	MAKIALGINAE: *Gaudalges, Lemuralges*.
Fam. LEMURIDAE	POLYPLACIDAE: *Lemurphthirus,*
	Lemurpediculus;
	TRICHOPHILOPTERIDAE: *Trichophilopterus*;
	LAELAPIDAE: *Aetholaelaps, Liponysella*;
	IXODIDAE: *Haemaphysalis; Ixodes*;
	CHEYLETIDAE: *Nihelia*;
	ATOPOMELIDAE: *Listrophoroides*;
	MAKIALGINAE: *Gaudalges, Lemuralges,*
	Makialges.
Fam. LORISIDAE	POLYPLACIDAE: *Lemurphthirus*;
	TRICHODEITIDAE: *Lorisicola*;
	IXODIDAE: *Haemaphysalis; Ixodes*;
	SPELEOGNATHIDAE: *Paraspeleognathopsis*.
Fam. PONGIDAE	PEDICULIDAE: *Pediculus*;
	PTHIRIDAE: *Pthirus*;
	HALARACHNINAE: *Pneumonyssus,*
	Rhinophaga;
	IXODIDAE: *Rhipicephalus*;
	PARACOROPTINAE: *Pangorillalges*.
Order PROBOSCIDEA	MAL: Haematomyzidae;
	DIP: Calliphoridae, Gasterophilidae,
	Oestridae;
	IXO: Ixodidae.
Fam. ELEPHANTIDAE	HAEMATOMYZIDAE: *Haematomyzus*;
	CALLIPHORIDAE: *Chrysomya,*
	Elephantoloemus;
	GASTEROPHILIDAE: *Cobboldia, Neocuterebra,*
	Platycobboldia, Rodhainomyia, Ruttenia;
	OESTRIDAE: *Pharyngobolus*;
	IXODIDAE: *Amblyomma; Dermacentor,*
	Rhipicephalus.
Order RODENTIA	DER: Hemimeridae;
	ANO: Enderleinellidae, Hoplopleuridae,
	Polyplacidae;
	MAL: Abrocomophagidae, Gyropidae,
	Trimenoponidae, Trichodectidae;

Mammals (Hosts)	Parasitic Arthropods
	COL: Leptinidae, Platypsyllidae, Staphylinidae;
	DIP: Calliphoridae, Cuterebridae, Hypodermatidae;
	SIP: Ceratophyllidae, Chimaeropsyllidae, Coptopsyllidae, Hystrichopsyllidae, Leptopsyllidae, Macropsyllidae, Pulicidae, Pygiopsyllidae, Rhopalopsyllidae, Stephanocircidae, Tungidae, Xiphiopsyllidae;
	MES: Dermanyssidae, Halarachnidae (Halarachninae), Hystrichonyssidae, Laelapidae (Alphalaelapinae; Laelapinae: *Androlaelaps*, *Gigantolaelaps*, *Laelaps*, *Tur*; Haemogamasinae: *Eulaelaps*, *Haemogamasus*; Hirstionyssinae: *Echinonyssus*; Mesolaelapinae), Macronyssidae (Macronyssinae; Ornithonyssinae: *Ornithonyssus*), Ascidae (?);
	IXO: Argasidae, Ixodidae;
	PRO: Cheyletidae, Demodicidae, Speleognathidae, Myobiidae, Pygmephoridae;
	AST: Atopomelidae, Chirodiscidae, Gastronyssidae (Yunkeracarinae), Listrophoridae, Lobalgidae, Myocoptidae, Pneumocoptidae (*Pneumocoptes*), Psoroptidae (Psoroptinae, Listropsoralginae), Rhyncoptidae, Sarcoptidae (Sarcoptinae, Notoedrinae).
Fam. ABROCOMIDAE	ABROCOMOPHAGIDAE: *Abrocomophaga*; GYROPIDAE: *Gyropus*, *Phtheiropoios*; RHOPALOPSYLLIDAE: *Delostichus*; IXODIDAE: *Ixodes*.
Fam. AGOUTIDAE	IXODIDAE: *Haemaphysalis*; *Hyalomma*; *Ixodes*.
Fam. ANOMALURIDAE	IXODIDAE: *Aponomma*; *Haemaphysalis*.
Fam. APLODONTIDAE	LEPTINIDAE: *Leptinillus*; LEPTOPSYLLIDAE: *Dolichopsyllus*; HYSTRICHOPSYLLIDAE: *Paratyphloceras*, *Trichopsylloides*; LAELAPIDAE: *Alphalaelaps*;

Mammals (Hosts)	Parasitic Arthropods
	HIRSTIONYSSINAE: *Patrinyssus;* IXODIDAE: *Ixodes;* LISTROPHORIDAE: *Aplodontochirus.*
Fam. ARVICOLIDAE	HOPLOPLEURIDAE: *Hoplopleura;* CERATOPHYLLIDAE: *Callopsylla (Callopsylla), Malaraeus, Megabothris, Nosopsyllus (Nosopsyllus), Thrassis;* LEPTOPSYLLIDAE: *Amphipsylla, Frontopsylla (Frontopsylla), Paradoxopsylla, Peromyscopsylla;* HYSTRICHOPSYLLIDAE: *Atyphloceras, Catallagia, Ctenophthalmus (Ctenophthalmus), C. (Ducictenophthalmus), C. (Euctenophthalmus), C. (Medioctenophthalmus), C. (Metactenophthalmus), C. (Nearctoctenophthalmus), C. (Palaeoctenophthalmus), C. (Sinoctenophthalmus), Deltotelis, Neopsylla, Palaeopsylla, Paraneopsylla, Rhadinopsyila (Actenophthalmus), Stenischia, Strepsylla, Wagnerina;* ARGASIDAE: *Ornithodoros;* IXODIDAE: *Amblyomma; Haemaphysalis; Hyalomma; Ixodes; Anomalohimalaya, Dermacentor, Rhipicephalus;* DEMODICIDAE: *Demodex;* MYOBIIDAE: *Radfordia;* PYGMEPHORIDAE: *Pygmephorus;* LISTROPHORIDAE: *Listrophorus, Mycoptes.*
Fam. BATHYERGIDAE	POLYPLACIDAE: *Eulinognathus;* CHIMAEROPSYLLIDAE: *Cryptopsylla;* HYSTRICHOPSYLLIDAE: *Dinopsyllus (Cryptoctenopsyllus);* PULICIDAE: *Xenopsylla;* ASCIDAE: *Myonyssoides* (?); IXODIDAE: *Haemaphysalis;* ATOPOMELIDAE: *Bathyergolichus.*
Fam. CAPROMYIDAE	GYROPIDAE: *Gliricola, Pitrufquenia;* IXODIDAE: *Ixodes;* ATOPOMELIDAE: *Capromylichus, Capromysia, Domingoecius, Myocastorobia, Plagiodontochirus.*
Fam. CASTORIDAE	LEPTINIDAE: *Leptinillus;* PLATYPSYLLIDAE: *Platypsyllus;* CHIRODISCIDAE: *Schizocarpus.*

Mammals (Hosts)	Parasitic Arthropods
Fam. CAVIIDAE	HOPLOPLEURIDAE: *Pterophthirus;* POLYPLACIDAE: *Galeophthirus;* GYROPIDAE: *Gliricola, Gyropus,* 　*Macrogyropus, Monothoracius, Protogyropus;* TRIMENOPONIDAE: *Chinchillophaga,* 　*Trimenopon;* STAPHYLINIDAE: *Amblyopinoides;* STEPHANOCIRCIDAE: *Nonnapsylla,* 　*Tiarapsylla;* RHOPALOPSYLLIDAE: *Delostichus,* 　*Ectinorus (Dysmicus), Eritranis, Panallius,* 　*Tiamastus;* TUNGIDAE: *Hectopsylla;* LAELAPIDAE: *Cavilaelaps, Neoparalaelaps;* ORNITHONYSSINAE: *Lepronyssoides;* ARGASIDAE: *Ornithodoros;* IXODIDAE: *Amblyomma; Dermacentor;* DEMODICIDAE: *Demodex;* ATOPOMELIDAE: *Chirodiscoides;* SARCOPTINAE: *Trixacarus.*
Fam. CHINCHILLIDAE	POLYPLACIDAE: *Cuyana;* GYROPIDAE: *Phtheiropoios;* TRIMENOPONIDAE: *Philandesia;* STEPHANOCIRCIDAE: *Tiarapsylla;* TUNGIDAE: *Hectopsylla;* ARGASIDAE: *Ornithodoros;* IXODIDAE: *Amblyomma; Haemaphysalis;* 　*Ixodes.*
Fam. CRICETIDAE	HEMIMERIDAE: *Araeomerus, Hemimerus;* HOPLOPLEURIDAE: *Hoplopleura,* 　*Typhlomyophthirus;* POLYPLACIDAE: *Eulinognathus,* 　*Mirophthirus, Neohaematopinus, Polyplax,* 　*Proenderleinellus;* GYROPIDAE: *Gyropus;* TRIMENOPONIDAE: *Cummingsia;* LEPTINIDAE: *Leptinillus;* STAPHYLINIDAE: *Amblyopinodes;* CALLIPHORIDAE: *Cordylobia;* CUTEREBRIDAE: *Cuterebra;* CERATOPHYLLIDAE: *Jellisonia, Kohlsia,* 　*Malaraeus, Monopsyllus (Monopsyllus),* 　*Opisodasys (Opisodasys), Orchopeas,* 　*Nosopsyllus (Nosopsyllus),* 　*N. (Gerbillophilus), N. (Nosinius), Pleochaetis,* 　*Traubella;* CHIMAEROPSYLLIDAE: *Chiastopsylla,* 　*Epiremia, Hypsophthalmus;*

Mammals (Hosts)	Parasitic Arthropods
	COPTOPSYLLIDAE: *Coptopsyllus,* *Coptopsylla;*

COPTOPSYLLIDAE: *Coptopsyllus,*
  Coptopsylla;
LEPTOPSYLLIDAE: *Amphipsylla,*
  Brachyctenonotus, Calceopsylla,
  Jordanopsylla, Ophthalmopsylla (Eremedosa),
  Paractenopsyllus, Paradoxopsyllus,
  Peromyscopsylla, Phaenopsylla, Tsaractenus,
  Typhlomyopsyllus;
XIPHIOPSYLLIDAE: *Xiphiopsylla;*
HYSTRICHOPSYLLIDAE: *Agastopsylla,*
  Anomiopsyllus, Atyphloceras, Callistopsyllus,
  Catallagia, Conorhinopsylla, Ctenoparia,
  Ctenophthalmus (Neoctenophthalmus),
  C. (Palaeoctenophthalmus),
  C. (Paractenophthalmus), Epitedia, Listropsylla,
  Megarthroglossus, Nearctopsylla
  (Neochinopsylla), Neopsylla, Phalacropsylla,
  Rhadinopsylla (Rhadinopsylla),
  R. (Actenophthalmus), R. (Micropsylla),
  R. (Micropsylloides) (?), R. (Ralipsylla) (?),
  Stenistomera, Strepsyall, Wagnerina;
STEPHANOCIRCIDAE: *Barreropsylla,*
  Cleopsylla, Craneopsylla, Plocopsylla,
  Sphinctopsylla;
RHOPALOPSYLLIDAE: *Ectinorus*
  (Ectinorus), E. (Dysmicus), Listronius,
  Polygenis, Tetrapsyllus, Tiamastus;
PULICIDAE: *Synopsyllus, Synosternus,*
  Xenopsylla;
TUNGIDAE: *Tunga (Tunga);*
LAELAPIDAE: *Gigantolaelaps, Hymenolaelaps,*
  Ondatralaelaps, Sinolaelaps, Tricholaelaps;
HAEMOGAMASINAE: *Brevisterna,*
  Ischyropoda;
MACRONYSSINAE: *Acanthonyssus, Argitis;*
ORNITHONYSSINAE: *Lepidodorsum;*
ARGASIDAE: *Argas, Ornithodoros, Otobius;*
IXODIDAE: *Amblyomma; Haemaphysalis;*
  Hyalomma; Ixodes; Dermacentor, Rhipicentor;
CHEYLETIDAE: *Cheyletus, Eucheyletia;*
DEMODICIDAE: *Demodex;*
SPELEOGNATHIDAE: *Speleomys,*
  Speleorodens;
MYOBIIDAE: *Blarinobia, Myobia, Radfordia;*
PYGMEPHORIDAE: *Pygmephorus;*
ATOPOMELIDAE: *Listrophoroides,*
  Oryzomysia;

Mammals (Hosts)	Parasitic Arthropods

	LISTROPHORIDAE: *Afrolistrophorus,*
	Carnilistrophorus, Criniscansor, Geomylichus,
	Listrophorus, Myocoptes, Prolistrophorus,
	Quasilistrophorus, Sclerolistrophorus,
	Trichoecius;
	PSOROPTINAE: *Caparinia.*
Fam. CTENODACTYLIDAE	LEPTOPSYLLIDAE: *Caenopsylla;*
	ARGASIDAE: *Ornithodoros;*
	IXODIDAE: *Rhipicephalus.*
Fam. CTENOMYIDAE	POLYPLACIDAE: *Eulinognathus;*
	GYROPIDAE: *Gyropus, Phtheiropoios;*
	STAPHYLINIDAE: *Edrabius,*
	Megamblyopinus;
	STEPHANOCIRCIDAE: *Tiarapsylla;*
	RHOPALOPSYLLIDAE: *Ectinorus*
	(Ectinorus), Tiamastus;
	IXODIDAE: *Ixodes.*
Fam. DASYPROCTIDAE	GYROPIDAE: *Macrogyropus, Monothoracius;*
	RHOPALOPSYLLIDAE: *Rhopalopsyllus;*
	IXODIDAE: *Amblyomma; Haemaphysalis;*
	Ixodes;
	PSOROPTINAE: *Myoproctalges.*
Fam. DIPODIDAE	POLYPLACIDAE: *Eulinognathus;*
	LEPTOPSYLLIDAE: *Desertopsylla,*
	Frontopsylla (Mafrontia), Hopkinsipsylla,
	Mesopsylla, Ophthalmopsylla
	(Ophthalmopsylla), O. (Cystipsylla);
	ARGASIDAE: *Ornithodoros;*
	IXODIDAE: *Haemaphysalis; Hyalomma; Ixodes;*
	Dermacentor, Rhipicephalus.
	MYOCOPTIDAE: *Criniscansor.*
Fam. ECHIMYIDAE	HOPLOPLEURIDAE: *Pterophthirus;*
	POLYPLACIDAE: *Ctenophthirus;*
	GYROPIDAE: *Gliricola, Gyropus;*
	TRIMENOPONIDAE: *Harrisonia,*
	Hoplomyophilus;
	LAELAPIDAE: *Tur;*
	MACRONYSSINAE: *Acanthonyssus;*
	ORNITHONYSSINAE: *Lepronyssoides;*
	ARGASIDAE: *Ornithodoros;*
	IXODIDAE: *Amblyomma; Haemaphysalis;*
	Ixodes;
	ATOPOMELIDAE: *Chirodiscoides,*
	Euryzygomysia, Isothricola;
	LISTROPHORIDAE: *Echimytricalges,*
	Prolistrophorus;

Mammals (Hosts)	Parasitic Arthropods
	PSOROPTINAE: *Echimyalges;*
	PSOROPTINAE: *Echimyalges;*
	LISTROPSORALGINAE: *Listropsoralgoides.*
Fam. ERETHIZONTIDAE	TRICHODECTIDAE: *Eutrichophilus;*
	IXODIDAE: *Amblyomma; Haemaphysalis; Ixodes; Anomalohimalaya, Boophilus, Dermacentor;*
	LOBALGIDAE: *Coendalges.*
Fam. GEOMYIDAE	TRICHODECTIDAE: *Geomydoecus;*
	CERATOPHYLLIDAE: *Dactylopsylla (Dactylopsylla), D. (Spicata), Foxella;*
	PULICIDAE: *Pulex (Pulex);*
	HAEMOGOMASINAE: *Ischyropoda;*
	IXODIDAE: *Amblyomma; Haemaphysalis; Ixodes;*
	PYGMEPHORIDAE: *Pygmephorus;*
	LISTROPHORIDAE: *Geomylichus.*
Fam. GLIRIDAE	HOPLOPLEURIDAE: *Schizophthirus;*
	CERATOPHYLLIDAE: *Miriampsylla, Monopsyllus (Monopsyllus), M. (Amonopsyllus), Myoxopsylla;*
	ORNITHONYSSINAE: *Cryptonyssus;*
	IXODIDAE: *Ixodes; Dermacentor, Rhipicephalus;*
	DEMODICIDAE: *Demodex;*
	SPELEOGNATHIDAE: *Paraspeleognathopsis;*
	MYOCOPTIDAE: *Gliricoptes.*
Fam. HETEROMYIDAE	POLYPLACIDAE: *Fahrenholzia, Neohaematopinus;*
	CERATOPHYLLIDAE: *Thrassis;*
	HYSTRICHOPSYLLIDAE: *Carteretta, Meringis, Wenzella;*
	LAELAPIDAE: *Steptolaelaps;*
	ARGASIDAE: *Ornithodoros;*
	IXODIDAE: *Amblyomma; Haemaphysalis; Ixodes; Dermacentor;*
	CHEYLETIDAE: *Cheyletus;*
	MYOBIIDAE: *Radfordia;*
	LISTROPHORIDAE: *Geomylichus.*
Fam. HYDROCHAERIDAE	IXODIDAE: *Amblyomma; Boophilus, Rhipicephalus.*
Fam. HYSTRICIDAE	PULICIDAE: *Pariodontis;*
	HIRSTIONYSSINAE: *Ancoranyssus;*
	HALARACHNINAE: *Rhinophaga;*

Mammals (Hosts)	Parasitic Arthropods

	HYSTRICHONYSSIDAE: *Hystrichonyssus*;
	ARGASIDAE: *Ornithodoros*;
	IXODIDAE: *Amblyomma, Aponomma*;
	   Haemaphysalis; Hyalomma; Ixodes; Boophilus,
	   Dermacentor, Rhipicentor, Rhipicephalus;
	PSORERGATIDAE: *Psorergates*;
	PYGMEPHORIDAE: *Pygmephorus*;
	RHYNCOPTIDAE: *Rhyncoptes.*

Fam. MURIDAE	HOPLOPLEURIDAE: *Hoplopleura*;
	POLYPLACIDAE: *Polyplax*;
	LEPTINIDAE: *Leptinus*;
	CALLIPHORIDAE: *Cordylobia*;
	CERATOPHYLLIDAE: *Nosopsyllus*
	   (*Nosopsyllus*), *N.* (*Penicus*), *Paraceras*;
	CHIMAEROPSYLLIDAE: *Chiastopsylla,*
	   Epiremia, Hypsophthalmus, Praopsylla;
	LEPTOPSYLLIDAE: *Acropsylla, Frontopsylla*
	   (*Profrontia*), *Leptopsylla* (*Leptopsylla*), *L.*
	   (*Pectinoctenus*), *Paradoxopsyllus,*
	   Peromyscopsylla, Sigmactenus;
	XIPHIOPSYLLIDAE: *Xiphiopsylla*;
	HYSTRICHOPSYLLIDAE: *Ctenophthalmus*
	   (*Ctenophthalmus*), *C.* (*Ethioctenophthalmus*),
	   C. (*Euctenophthalmus*), *Dinopsyllus*
	   (*Dinopsyllus*), *Listropsylla, Neopsylla,*
	   Rothschildiana, Stenischia, Typhloceras;
	MACROPSYLLIDAE: *Macropsylla*;
	PYGIOPSYLLIDAE: *Acanthopsylla,*
	   Afristivalius, Astivalius, Aviostivalius,
	   Bibikovana, Gryphopsylla (*Gryphopsylla*), *G.*
	   (*Migrastivalius*), *Idiochaetis, Lentistivalius*
	   (*Lentistivalius*), *Metastivalius, Nestivalius,*
	   Obtusifrontia, Orthopsylloides, Papuapsylla,
	   Pygiopsylla, Rectidigitus, Smitella, Stivalius,
	   Tiflavia, Traubia, Zyx;
	STEPHANOCIRCIDAE: *Stephanocircus*;
	PULICIDAE: *Parapulex, Synopsyllus,*
	   Xenopsylla;
	TUNGIDAE: *Tunga* (*Tunga*), *T.* (*Brevidigita*);
	LAELAPIDAE: *Andreacarus, Domrownyssus,*
	   Hyperlaelaps, Laelaps, Longolaelaps,
	   Mysolaelaps, Steptolaelaps, Tricholaelaps,
	   Tropilaelaps (?); *Mesolaelaps; Myonyssus*;
	DERMANYSSIDAE: *Liponyssoides*;
	ARGASIDAE: *Argas, Ornithodoros*;

Mammals (Hosts)	Parasitic Arthropods
	IXODIDAE: *Amblyomma, Aponomma; Haemaphysalis; Hyalomma; Ixodes; Anomalohimalaya, Boophilus, Dermacentor, Rhipicephalus;* CHEYLETIDAE: *Alliea, Cheyletus;* DEMODICIDAE: *Demodex;* SPELEOGNATHIDAE: *Paraspeleognathopsis, Speleomys,* Speleorodens; MYOBIIDAE: *Blarinobia, Myobia, Radfordia;* ATOPOMELIDAE: *Murichirus, Teinochirus;* YUNKERACARINAE: *Afralistrophorus, Lynxacarus, Olistrophorus, Prolistrophorus;* MYOCOPTIDAE: *Criniscansor, Myocoptes, Trichoecius;* SARCOPTINAE: *Trixacarus;* NOTOEDRINAE: *Notoedres.*
Fam. MYOCASTORIDAE	GYROPIDAE: *Pitrufquenia;* IXODIDAE: *Ixodes.*
Fam. OCTODONTIDAE	HOPLOPLEURIDAE: *Hoplopleura;* RHOPALOPSYLLIDAE: *Delostichus, Ectinorus (Ectinorus), E. (Dysmicus).*
Fam. PEDETIDAE	POLYPLACIDAE: *Eulinognathus;* PULICIDAE: *Delopsylla;* LAELAPIDAE: *Radfordilaelaps;* HALARACHNINAE: *Zumptiella;* IXODIDAE: *Haemaphysalis; Ixodes; Rhipicephalus.*
Fam. PETROMYIDAE	POLYPLACIDAE: *Scipio.*
Fam. RHIZOMYIDAE	POLYPLACIDAE: *Polyplax;* XIPHIOPSYLLIDAE: *Xiphiopsyllus* HYSTRICHOPSYLLIDAE: *Ctenophthalmus (Geoctenophthalmus);* LAELAPIDAE: *Rhyzolaelaps, Tylolaelaps;* IXODIDAE: *Haemaphysalis; Ixodes.*
Fam. SCIURIDAE	ENDERLEINELLIDAE: *Atopophthirus, Enderleinellus, Microphthirus, Phthirunculus, Werneckia;* HOPLOPLEURIDAE: *Hoplopleura;* POLYPLACIDAE: *Johnsonpthirus, Linognathoides, Neohaematopinus, Polyplax;* CERATOPHYLLIDAE: *Aenigmopsylla, Callopsylla (Callopsylla), Citellophilus, Diamanus, Hollandipsylla, Kohlsia, Libyastus, Macrostylophora, Megabothris, Megathoracipsylla, Monopsylla (Monopsylla),*

Mammals (Hosts)	Parasitic Arthropods
	M. (*Amonopsyllus*), *Nosopsyllus* (*Nosopsyllus*), *Opisorrostis, Opisodasys* (*Opisodasys*), *O.* (*Sciuropsylla*), *Orchopeas, Oropsylla, Paraceras, Pleochaetus, Rostropsylla, Rowleyella, Smitipsylla, Spuropsylla, Syngenopsyllus, Tarsopsylla, Thrassis;* LEPTOPSYLLIDAE: *Minyctenopsyllus;* HYSTRICHOPSYLLIDAE: *Conorhinopsylla, Eopsylla* (?), *Epitedia, Megarthroglossus, Nearctopsylla* (*Chinopsylla*) (?), *Neopsylla, Rhadinopsylla* (*Ralipsylla*), *Tamiophila;* PYGIOPSYLLIDAE: *Farhangia, Lentistivalius* (*Lentistivalius*), *Medwayella;* PULICIDAE: *Synosternus, Xenopsylla;* LAELAPIDAE: *Gnatholaelaps, Tricholaelaps;* HAEMOGAMASINAE: *Brevisterna, Ischyropoda;* HALARACHNINAE: *Zumptiella;* ARGASIDAE: *Ornithodoros;* IXODIDAE: *Amblyomma; Haemaphysalis; Hyalomma; Ixodes; Dermacentor, Rhipicephalus;* CHEYLETIDAE: *Cheyletus, Criokeron;* DEMODICIDAE: *Demodex;* SPELEOGNATHIDAE: *Speleorodens;* MYOBIIDAE: *Amorphacarus;* PYGMEPHORIDAE: *Pygmephorus;* YUNKERACARINAE: *Sciuracarus;* LISTROPHORIDAE: *Aeromychirus, Metalistrophorus, Pteromychirus, Sciurochirus;* NOTOEDRINAE: *Notoedres.*
Fam. SPALACIDAE	HYSTRICHOPSYLLIDAE: *Ctenophthalmus* (*Palaeoctenophthalmus*), *C.* (*Spalacoctenophthalmus*); IXODIDAE: *Haemaphysalis; Ixodes; Dermacentor;* LISTROPHORIDAE: *Spalacarus.*
Fam. THRYONOMYIDAE	POLYPLACIDAE: *Scipio;* IXODIDAE: *Amblyomma; Ixodes; Rhipicephalus.*
Fam. ZAPODIDAE	IXODIDAE: *Ixodes; Dermacentor, Rhipicephalus;* MYOBIIDAE: *Protomyobia, Radfordia.*

Mammals (Hosts)	Parasitic Arthropods
Order SCANDENTIA	ANO: Polyplacidae; SIP: Pygiopsyllidae; IXO: Ixodidae; AST: Listrophoridae.

Fam. TUPAIIDAE	POLYPLACIDAE: *Docophthirus, Sathrax;*
	PYGIOPSYLLIDAE: *Lentistivalius*
	   (*Destivalius*);
	IXODIDAE: *Amblyomma; Haemaphysalis;*
	   Ixodes; Dermacentor;
	CHEYLETIDAE: *Criokeron;*
	LISTROPHORIDAE: *Sciurochirus.*

Order TUBULIDENTATA	ANO: Hybophthiridae; DIP: Calliphoridae; IXO: Ixodidae.
Fam. ORYCTEROPODIDAE	HYBOPHTHIRIDAE: *Hybophthirus;* CALLIPHORIDAE: *Pachychoeromyia;* IXODIDAE: *Haemaphysalis.*

Index

Scientific names are in *italics*; lightface for arthropods and **boldface** for mammals and other hosts. Page numbers in **boldface** indicate pages on which primary discussion appears in text and *italic* numbers indicate pages with illustrations. Species names are indexed under each generic name and specific epithets are not cross-indexed. Only class, order, and family names are indexed for Appendix A and B.